Staphylococcus

Genetics and Physiology

Edited by

Greg A. Somerville

School of Veterinary Medicine and Biomedical Sciences
University of Nebraska-Lincoln
Lincoln, NE
USA

Caister Academic Press

Caister Academic Press
Norfolk, UK

www.caister.com

British Library Cataloguing-in-Publication Data
A catalogue record for this book is available from the British Library

ISBN: 978-1-910190-49-4 (paperback)
ISBN: 978-1-910190-50-0 (ebook)

Cover design adapted from photographs provided by Dr John 'Dustin' Loy.

Contents

Contributors

Markus Bischoff
Saarland University
Institute of Medical Microbiology and Hygiene
Homburg
Germany

markus.bischoff@uniklinikum-saarland.de

Susan Boyle-Vavra
Department of Pediatrics
Section of Infectious Disease
University of Chicago
Chicago, IL
USA

sboyle@peds.bsd.uchicago.edu

Aisling F. Brown
Host Pathogen Interactions Group
School of Biochemistry and Immunology
Trinity College Dublin
Ireland

aibrown@tcd.ie

Robert S. Daum
Department of Pediatrics
Section of Infectious Disease
University of Chicago
Chicago, IL
USA

rdaum@peds.bsd.uchicago.edu

Susanne Engelmann
TU Braunschweig
Institut für Mikrobiologie
Mikrobielle Proteomik
Braunschweig
Germany

susanne.engelmann@helmholtz-hzi.de

Ronald S. Flannagan
Department of Microbiology and Immunology
University of Western Ontario
London, ON
Canada

rflanna2@uwo.ca

Dorte Frees
Department of Veterinary Disease Biology
University of Copenhagen
Frederiksberg
Denmark

df@sund.ku.dk

Stephan Fuchs
Institut für Mikrobiologie
Universität Greifswald; and
Abteilung Nosokomiale Infektionen
Robert Koch-Institut
Wernigerode
Germany

fuchss@rki.de

Angelika Gründling
Section of Microbiology and MRC Centre for
Molecular Bacteriology and Infection
Imperial College London
London
UK

a.grundling@imperial.ac.uk

Mélissa Hannauer
Department of Microbiology and Immunology
University of Western Ontario
London, ON
Canada

mhannaue@uwo.ca

Michael Hecker
Institut für Mikrobiologie
Universität Greifswald
Greifswald
Germany

hecker@uni-greifswald.de

David E. Heinrichs
Department of Microbiology and Immunology
University of Western Ontario
London, ON
Canada

deh@uwo.ca

Mathias Herrmann
Institute of Medical Microbiology and Hygiene
Homburg
Germany

mathias.herrmann@uks.eu

Hanne Ingmer
Department of Veterinary Disease Biology
University of Copenhagen
Frederiksberg
Denmark

hi@sund.ku.dk

Jodi A. Lindsay
Institute of Infection and Immunity
St George's University of London
Cranmer Terrace
London
UK

jlindsay@sgul.ac.uk

John Dustin Loy
School of Veterinary Medicine and Biomedical Sciences
University of Nebraska-Lincoln
Lincoln, NE
USA

jdloy@unl.edu

Rachel M. McLoughlin
Host Pathogen Interactions Group
School of Biochemistry and Immunology
Trinity College Dublin
Ireland

rachel.mcloughlin@tcd.ie

Richard A. Proctor
University of Wisconsin School of Medicine and Public Health
Ann Arbor, MI
USA

rap@wisc.edu

Pascale Romby
Universite de Strasbourg, CNRS, IBMC
Architecture et Réactivité de l'ARN
Strasbourg
France

p.romby@ibmc-cnrs.unistra.fr

Patrick M. Schlievert
Department of Microbiology
Carver College of Medicine
University of Iowa
Iowa City, IA
USA

patrick-schlievert@uiowa.edu

Jessica R. Sheldon
Department of Microbiology and Immunology
University of Western Ontario
London, ON
Canada

jsheldo2@uwo.ca

Mark S. Smeltzer
Department of Microbiology and Immunology
University of Arkansas for Medical Sciences
Little Rock, AR
USA

smeltzermarks@uams.edu

Greg A. Somerville
School of Veterinary Medicine and Biomedical Sciences
University of Nebraska-Lincoln
Lincoln, NE
USA

gsomerville3@unl.edu

Preface

In a letter to Robert Hooke, Isaac Newton wrote that "If I have seen further than others, it is by standing upon the shoulders of giants." This quote highlights the fundamental nature of science; specifically, that it is an incremental process based on the cumulative discoveries of generations of scientists. It is with this in mind that the authors of this book have attempted to summarize the progress in *Staphylococcus aureus* research by acknowledging the contributions of scientists past (our mentors), present (our colleagues), and future (our students and post-docs). While we may stand on the shoulders of giants, it is with the understanding that past observations can only be viewed in the present context with an appreciation for how evolution has changed *S. aureus*. This book provides current perspectives on *S. aureus* clinical significance in humans and animals and how this significance was established through bacterial physiology, genetics, and evolution to influence the host–pathogen interaction through virulence factors, immune modulation, and antibiotic resistance.

As the editor of this book, I am honoured to have worked with such an outstanding group of scientists and I look forward to our future collaborations. Hopefully, at our next scientific gathering, we can all be together and enjoy a glass of wine or beer to celebrate our accomplishment.

Greg A. Somerville, PhD
Lincoln, NE, USA

History of *Staphylococcus aureus* 1

Richard A. Proctor

Abstract
The history of *Staphylococcus aureus* and the history of bacterial infectious diseases parallel one another. *S. aureus* was one of the first pathogens described, which is not surprising as it was and is still one of the most common and feared infections in humans. This chapter traces the history of *S. aureus* from 1880 to the present.

Introduction
The magnitude of reviewing the history of staphylococci became apparent when I began to collect and select what to include. It is by necessity selective, which means apologies are already due to colleagues' work that was not quoted, and focused almost exclusively to *S. aureus* rather than other species of staphylococci. To provide depth and to trace a longitudinal description, only one or two examples from each topic is included. When searching on each subtopic, the key words often elicited over 2000 articles each, which meant that great selectivity occurred even within a given subtopic. To provide a detailed perspective, the story begins with the earliest publications and end yesterday so as to present the scope of the scientific accomplishments. More in depth reviews on each of the topics and subtopics are referenced at the start of each section.

Discovery: identification and classification

Early work on staphylococci, including speciation
The first reports on staphylococci date from 1880, when Alexander Ogston reported upon the isolation of *S. aureus* from a surgical wound infection (Ogston, 1880). The organisms were able to produce an abscess when injected into guinea pigs and mice (Ogston, 1881). Also in 1880, Louis Pasteur produced abscesses in animals injected with pus from human staphylococcal infections (Pasteur, 1880). In the following year, Ogston coined the term *Staphylococcus* for the genus (Ogston, 1882). In 1884, Rosenbach divided the genus into the species *aureus* and *albus* (Cowan, 1954). These designations remained until 1939, when Cowan differentiated *S. epidermidis* as a separate species based upon coagulase testing (slide agglutination) (Cowan, 1939), and this was further defined with serological testing in 1964 (Cowan, 1964).

In 1965, the Baird-Parker scheme was introduced for the classification of staphylococci and micrococci (Baird-Parker, 1965), wherein 'the Gram-positive and catalase-positive cocci are best separated into, the genus *Staphylococcus* and the genus *Micrococcus* on the ability of members of *Staphylococcus* to grow and produce acid from glucose anaerobically.' This remained the standard scheme until 1975, when Kloos and Sheffler offered a simplified scheme for the identification of *Staphylococcus*, which became the standard for clinical microbiology laboratories (Kloos, 1975). An expanded testing for speciation was adopted in 1999 (Freney, 1999), that included defining a *Staphylococcus* as 'Gram-positive coccus that forms clusters, produces catalase, appropriate cell wall and G+C content of 30–40 mol%. Speciation was based on pigment, growth requirements, fermentative and oxidative activity on carbohydrates, novobiocin susceptibility, enzymatic activities (nitrate reductase, alkaline phosphatase, arginine dihydrolase, ornithine decarboxylase, urease, cytochrome oxidase, staphylocoagulase in rabbit plasma, heat-stable nuclease, amidases, oxidases, clumping factor, and haemolytic activity on sheep or bovine blood agar' (Freney, 1999).

Rapid identification and susceptibility testing

Over time, evidence-based selection of antibiotic therapy has been championed to improve outcomes from infections. Thermonuclease has proven to be the best test to identify staphylococci as *aureus* (Lachica, 1976), and this test was applied to the identification of *S. aureus* from blood cultures (Madison, 1983; Ratner, 1985). Other direct testing of blood cultures to identify *S. aureus* included the coagulase test, latex agglutination, and lysostaphin susceptibility (Essers *et al.*, 1980; Engels *et al.*, 1981; Knight *et al.*, 1983), but these rapid identification tests have now been surpassed by PCR-based methods (Brakstad *et al.*, 1995; Del Vecchio *et al.*, 1995).

In hopes of reducing mortality, rapid diagnostics for strains having the methicillin-resistant *S. aureus* (MRSA) phenotype have been sought. This has proven difficult, as expression of the MRSA phenotypic requires multiple factors. In 1989, FemA was described as an extra factor that was necessary for the phenotypic expression of methicillin resistance (Berger-Bächi *et al.*, 1989). Since 1989, more than 22 factors have now been described as needed for the MRSA phenotype, which makes developing a PCR diagnostic test very difficult (Berger-Bächi *et al.*, 1998; McCallum *et al.*, 2010; Proctor and Smith, 2014). While PCR is most rapid for identification, separation of MRSA from methicillin susceptible *S. aureus* (MSSA) is the Achilles heel of PCR testing because of the complexity of the MRSA phenotype. Thus, PCR-based testing has not been FDA approved for clinical susceptibility testing (Proctor and Smith, 2014). The lack of approved PCR diagnostic tests means that diagnosis of MRSA relies on phenotypic tests, with the most rapid phenotypic tests for methicillin resistance involving chromogenic substrates in the agar (Gaillot, 2000). These phenotypic tests cut the time for providing susceptibility results by 12 h on positive blood cultures; however, it has not translated into decreased mortality (Nicolsen *et al.*, 2013). An FDA-approved phage-based test provided susceptibility results 24 hours faster than traditional agar-based testing methods (Proctor and Smith, 2014), although no clinical studies have been performed to show that the more rapid results improve mortality.

Strain identification for epidemiology using phage typing

To facilitate epidemiological investigations, primarily in hospitals, various methods have been developed to identify strains of *S. aureus*. In 1961, standardized methods for phage typing were established (Blair *et al.*, 1961) and used in 1962 as a method for defining strains involved in an outbreak (Parker, 1962). Phage typing was a clear advance over phenotypic characterization of strains and provided valuable information about outbreaks, especially within hospitals.

Molecular biological methods for staphylococci: typing

Phage typing has been replaced by a number of molecular methods such as pulsed-field gel electrophoresis (PFGE), SpA-typing (Freney *et al.*, 1999), Agr-typing, multilocus sequence typing (MLST), staphylococcal cassette chromosome *mec* (SCC*mec*)-typing, and whole-genome sequencing (Mediavilla *et al.*, 2012). These methods have provided more detailed local to global surveillance and epidemiological information. This information is reviewed in other chapters in this book and will not be considered further in this chapter.

Mapping of the *S. aureus* genome

Mapping of the *S. aureus* genome has been a tremendous help to scientists across the world in studying this pathogen. Much of the early mapping was carried out by Peter Pattee who developed physical maps of the genome by studying frequency of transfer of markers such amino acid auxotrophies (Bannantine *et al.*, 1996; Smith *et al.*, 1967). This careful and tedious work continued for the next two decades with plasmids and auxotrophic sites being painstakingly mapped (Pattee, 1981; Löfdahl, 1983; Lee *et al.*, 1985; Projan, 1985; Jones, 1987).

As modern sequencing technology emerged, the structural genes on the plasmids and chromosome could be integrated into a complete genomic package. On 18 January 1996, Human Genome Sciences (HGS, Rockville, MD, USA) announced that it had determined the first genome sequence of an *S. aureus* strain. This was the first time that HGS had used its large scale sequencing technology on an organism of non-human origin. Much to the disappointment of the staphylococcal community, HGS did not share the sequence with the scientific community as it had planned to leverage its information with industrial partners, e.g. vaccine development with MedImmune, Inc. Consequently, several other groups then undertook sequencing with governmental support so that the data would be freely available. This resulted in sequences of *S. aureus* strains N315 and Mu50 being published from Japan (Kuroda *et al.*, 2001), *S. aureus* strain 8325 from the University of Oklahoma (Iandolo *et al.*, 2002), *S. aureus* strain COL from TIGR in Rockville, MD, USA (Gill *et al.*, 2005), an EMRSA-16 clone (MRSA252) and an invasive MSSA strain (MSSA476) from the Sanger Institute in the United Kingdom (Holden *et al.*, 2004), and a USA300 clone from the University of California, San Francisco General Hospital (Diep, 2006). In the end, knowledge of the complete sequence proved to have limited value commercially, but the freely available sequence helped to accelerate progress in all areas of staphylococcal research in the long run.

Pathogenesis

Understanding the complex relationships between the host and *S. aureus* as a harmless commensal in the nose yet a virulent pathogen is a challenge whose history dates from 1894, shortly after staphylococci were first identified. These complex interactions with the human host require a multitude of factors and space constraints allow only a very few to be considered in this chapter.

Colonization

In 1944, Miles *et al.* associated nasal colonization with wound infections (Miles *et al.*, 1944). Jumping ahead to 2001, von Eiff *et al.* established that nasal colonization relates to the incidence of *S. aureus* bacteraemia in hospitalized patients (2001). Defining the relationship between colonization and invasive infection has led to hospital infection control programmes that have reduced *S. aureus* bacteraemia rates by 63% from 2002 to 2013 (Mitchell *et al.*, 2014). Universal decolonization, especially in intensive care units, is now recommended in US hospitals as it has had a similar impact on critically ill patients (Calfee *et al.*, 2014).

ACME and colonization

Two striking features of the USA300 epidemic strain was its ability to cause longer-term colonization of the skin and its enhanced ability to transmit from person to person (reviewed in Miller *et al.*, 2008). The success of the highly virulent and epidemic USA300 clones is in part due to its arginine catabolic mobile element (ACME) that was found in the SCC*mec* genomic island (Diep *et al.*, 2006; Goering *et al.*, 2007). Interestingly, research in rodent skin infection and pneumonia models failed to show that ACME was important in these infection models (Montgomery *et al.*, 2009); however, later work established the ACME was important for persistence on the skin and in abscesses (Planet *et al.*, 2013; Thurlow *et al.*, 2013). Normally, the arginine deiminase pathway is active only anaerobically in *S. aureus*, but ACME encodes additional arginine deiminase pathway genes that function aerobically. This allows strains carrying ACME to neutralize acid on the skin and in abscesses, thereby allowing for greater survival (Thurlow *et al.*, 2013).

Virulence factors (toxins, adhesins, enzymes, etc.) in *S. aureus*

Because virulence factors are reviewed in a later chapter, they are only discussed here for their historical significance. The first bacterial two-component toxin described and the first staphylococcal toxin identified was the Panton–Valentine leucocidin (PVL) (Panton *et al.*, 1932). α-Haemolysin (Hla) was selected because of its very long history of study and because a specific receptor has been identified. Toxic shock syndrome toxin-1 (TSST-1) is given as an example of a superantigen toxin. Protein A (SpA) is reviewed as an example of anti-host defence factor (see Spaan *et al.*, 2013) for reviews on the many other factors produced by *S. aureus* to neutralize the host immune responses). Fibronectin binding protein (Fnb) is provided as an example of host tissue adhesin. Polysaccharide intercellular adhesin (PIA) is an example of a polysaccharide for accumulation in biofilm and anti-phagocytic factor, which will also allow a brief discussion of *S. epidermidis*.

Toxins

Panton–Valentine leucocidin (PVL)

A brief history is recounted here, and a more detailed account of PVL is available on line (Lina, 2005).

In 1894, van de Velde coined the term 'leucocidin' for products produced by *S. aureus* that damaged rabbit neutrophils in the pleural space (van de Velde, 1894). In 1895, his group published studies on anti-leucocidin serum that protected rabbits from death in a pleural infection model (Denys *et al.*, 1895). We jump forward to 1922, when the haemolytic and leucocytic properties of PVL were separated (Julanielle, 1922). In 1932, Panton and Valentine confirmed that PVL was not associated with haemolytic activity, but it was associated with severe infections in four patients (Panton *et al.*, 1932). The term 'PVL' was first used by Joyce Wright, who distinguished PVL from other leucotoxins (Wright *et al.*, 1936). In 1957, Gladstone and van Heyningen showed the selective action of PVL on granulocytes and macrophages, which could be distinguished from α-haemolysin and δ-haemolysin and that had both haemolytic and leucocytotoxic activity (Gladstone *et al.*, 1957; Hunt *et al.*, 1958). In 1959, Woodin separated fast- (F) from slow- (S) eluting PVL fractions using an ion exchange column, which gave rise the term LukS-PV and LukF-PV for the two components of PVL (Woodin, 1959). This work was one of the first reports to suggest that the toxin was made from two distinct components. Publications in 1991–1992 from Yoshiyuk Kamio's group identified the phage origin of the toxin components and provided the sequence of the PVL toxins (Rahman *et al.*, 1991, 1992). Clinical correlations with disease severity were again reported in 1995 by Yves Piémont's group wherein they found a strong association between PVL and the development of furuncles in children (Cribier *et al.*, 1992). Further association with disease severity was reported by Gillet *et al.*, when a necrotizing, haemorrhagic pneumonia in patients with a preceding viral infection was first reported in 2002 (Gillet *et al.*, 2002). Controversy arose concerning the importance of PVL, but publications of results that showed major and surprising differences between animal models and especially the human neutrophil susceptibility to the toxin helped to resolve much of the conflict (Löffler *et al.*, 2010, 2013). More clarity came when it was shown that anti-PVL antibodies did not prevent re-infection in children (Hermos *et al.*, 2010), but these antibodies do mitigate the severity of disease and mortality (Rasigade *et al.*, 2011).

α-Toxin or α-haemolysin (Hla)

In 1928, 21 children in Australia were given a diphtheria immunization that was contaminated by *S. aureus*. This caused the death of 12 of the children and abscesses at the injection site of the survivors. The response was linked by Sir McFarlane Burnet to a single, heat-labile antigenic substance that caused haemolysis (Burnet, 1930). In 1935, Glenny and Stevens coined the term α-toxin for the rabbit erythrocyte specific toxin (Glenny, 1935). We move ahead to 1962, when a long debate began about the value of anti-α-toxin antibodies and the role of α-toxin in pathogenesis of *S. aureus* infections (Koenig *et al.*, 1962). While testing the Smith diffuse strain, an atypical *S. aureus* strain that is heavily encapsulated, Koenig *et al.* found that whole organism immunization of mice protected in an intraperitoneal infection model, but anti-α-toxoid antibodies did not (Koenig *et al.*, 1962). Just a year later, anti-α-toxoid immunization provided protection from infection in a murine suture model of infection and α-toxin mutants produced only very minimal lesions (Taubler *et al.*, 1963). Naturally, the differences in the animal model played a role in the contradictory results. In 1983, the α-toxin gene was identified (Kehoe, 1983), paving the way for showing that a mutation in *hla* reduced the virulence of *S. aureus* in a murine mammary gland model (Bramley *et al.*, 1989). Also in 1983, high levels of anti-α-toxin antibodies failed to protect patients from recurrent furunculosis (Christensson, 1983), but as noted above, these antibodies reduced the severity of the infections (Rasigade, 2011). Interestingly, anti-α-toxin antibodies were also shown to protect cultured bovine endothelial cells from lysis (Vann *et al.*, 1988). Between 2007 and 2008 more data were provided for Hla having an important role in the pathogenesis of *S. aureus* pneumonia (Bubeck Wardenburg *et al.*, 2007, 2008), and the value of anti-α-toxin active and passive antibodies to reduce pneumonia severity (Ragle, 2009). Similarly, in a rabbit model of severe sepsis, Hla was found to be critical for mortality (Crémieux *et al.*, 2041). Consistent with the critical role of α-toxin in infections, monoclonal antibodies against Hla showed promise for the treatment of pneumonia and dermonecrosis in rabbit models (Hua, 2014); however, clinical trials are needed to extend the protective effects in animal models to amelioration of human disease.

As the debate over the role of Hla in the pathogenesis of *S. aureus* infection slowly resolved, progress continued over the mechanism of action. In 2006, studies from Bayley's lab provided the mechanisms for the assembly of α-toxin into the pore that causes lysis of red blood cell (Jayasinghe *et al.*, 2006; Raychaudhuri *et al.*, 2011). In addition, Hla was found to lyse human polymorphonuclear leucocytes (PMNs) (Pang *et al.*, 2010); although, Hla has proven to have more than just lytic activities. First, the binding site for Hla was defined as ADAM10 (a disintigren and metalloprotease 10) (Inoshima *et al.*, 2011); when Hla binds to ADAM10 in endothelial cells it activates the host metalloprotease activity, digesting the endothelium–cadherin linkages (Powers *et al.*, 2012). Moreover, α-toxin activates the intracellular inflammatory pathways by activating NOD2 (nucleotide-binding oligomerization domain containing 2) (Hruz, 2009) and the NLRP3-inflammasome (Craven *et al.*, 2009) that induce host cell necrosis, thereby increasing tissue damage directly and through the inflammatory response. Hla also stimulates the Th17 pathway, adding to the inflammatory response by recruiting neutrophils (Frank *et al.*, 2012).

Finally, the phenol-soluble modulin (PSM) locus, *psmα*, regulates production of α-toxin during infection (Berube *et al.*, 2014); hence, synergy between Hla and PSMs, which damage neutrophils, enhances the toxicity of Hla (Peschel, 2013). In total, Hla has proven to be much more than just a lytic toxin and its effects are coordinated with PSMs to produce disease.

Superantigens: TSST-1

Toxic shock syndrome (TSS) was described in children in 1978 (Todd, 1978). Seven children (aged 8–17 years) presented with a high fever, headache, confusion, conjunctival hyperaemia, a scarlatiniform rash, subcutaneous oedema, vomiting, watery diarrhoea, oliguria, and a propensity to acute renal failure, hepatic abnormalities, disseminated intravascular coagulation, and severe prolonged shock. One patient died, one had gangrene of the toes, and all had fine desquamation of affected skin and peeling of palms and soles during convalescence. Five patients were studied prospectively and *S. aureus* related to phage-group I was isolated from mucosal (nasopharyngeal, vaginal, tracheal), or sequestered (empyema, abscess) sites, but not from blood. This organism produced an exotoxin which causes a positive Nikolsky sign in the new-born mouse and which was biochemically, pathologically and immunologically distinct from phage group II staphylococcal exfoliatin. In retrospect, there were earlier reports of TSS in 1927 and 1962, wherein it was called staphylococcal scarlet fever (Stevens, 1927; Feldman, 1962). In 1980, TSS was first linked to tampon use (Davis, 1980). Surveillance for TSS in Wisconsin detected 38 cases with onsets from September 1975 through June 1980, with the rate of TSS among menstruating women younger than 30 years being 2.4–3.3 times the rate among menstruating women who were 30 or older. In 1981, TSS was found to result from enterotoxin F, which is the alternative name for toxic shock syndrome toxin-1 (TSST-1) (Bergdoll *et al.*, 1981).

In 1985, the gene responsible for TSS was discovered using phage transfer of TSST-1 production between *S. aureus* strains, which allowed the recipient strain to produce TSS in rabbits (Rasheed *et al.*, 1985). In the same year, the TSS gene (*tst*) was cloned into a plasmid and studied in a rabbit model (de Azavedo *et al.*, 1985). These studies demonstrated that *tst* was responsible for a TSS illness in rabbits.

The basis for the severity of TSS (Chesney *et al.*, 1984) became apparent as the story of staphylococcal superantigens unfolded. In 1989, the term 'superantigen' was introduced for the ability of *S. aureus* toxins to interact with and stimulate T cells without the protein entering into the antigen-binding pocket (Choi *et al.*, 1989). The ability of TSST-1 to stimulate a huge percentage of T-cells and macrophages explained the severe signs and symptoms because of the massive and unbridled release of cytokines. Antibodies against tumour necrosis factor (TNF) protected primates from the lethal effects of TSST-1, showing that this important cytokine was critical to the illness (Hinshaw *et al.*, 1992).

Other severe septic syndromes such as purpura fulminans and Waterhouse–Friderichsen syndrome were described in 2005 (Adem *et al.*, 2005; Kravitz *et al.*, 2005), and it is likely that staphylococcal superantigens have a role in these syndromes (Brosnahan *et al.*, 2011). Thus, superantigens help to explain why many *S. aureus* infections are so severe.

Protein A (SpA or Jensen's antigen): an anti-host defence factor

S. aureus contains many factors that help it avoid host defences, allowing it to become part of the normal flora as well as to produce invasive infections. The history of protein A (SpA) is selected as an example of the immune evasions factors.

In 1958, Klaus Jensen (Jensen, 1958) showed that a cell wall antigen from *S. aureus* caused precipitation in normal human serum. The reaction occurred with γ-globulin and was believed to be due to the existence of cross-reacting antibodies against a broad range of targets. In 1962–1963, Löfkvist and Sjöquist (reviewed in Sjöquist *et al.*, 1967), as well as Yoshida (1963), provided data that Jensen's surface antigen was a protein and that it bound to the Fc portion, rather than the usual antigen binding site, of IgG (reviewed in Sjöquist *et al.*, 1967; Uhlen, 1984). The name protein A was given in 1964 (Grov *et al.*, 1964).

In 1969, it was discovered that SpA interfered with more than just IgG, but it also resulted in complement depletion (Sjöquist *et al.*, 1969) and covered surface structures that would normally activate the alternative complement pathway (Spika *et al.*, 1981). In 1977, protein A was shown to cause heterogeneity of the opsonic requirements for neutrophil phagocytosis (Verhoef *et al.*, 1977). In 1979, it was found that the IgG–SpA interaction caused platelet injury, which would facilitate thromboembolic complications of *S. aureus* infections (Hawiger *et al.*, 1979). The role for SpA in the pathogenesis of *S. aureus* infections has continued to expand as a modulator of the host immune response. SpA binds the TNF receptor (TNFR1) (Gomez *et al.*, 2004), the EGF receptor (EGFR), and stimulates ADAM17 in the lung, stimulating pro-inflammatory and cytokine regulatory pathways (Parker *et al.*, 2012; Ahn *et al.*, 2014). SpA also causes shedding of the TNFR1 thereby subverting the host immune response (Giai *et al.*, 2013). Murein hydrolases release SpA, which cross-link B-cell receptors thereby modifying host adaptive immune response (Becker *et al.*, 2014). SpA blocks TLR2 activation, which reduces the innate immune response (Hilmi *et al.*, 2014) and, finally, SpA binds fibronectin (Zuo, 2014) and von Willebrand factor, thereby allowing *S. aureus* to colonize host tissues (Hartleib *et al.*, 2000).

One of the unique features of SpA was its variable size (Sjödahl, 1977). Identification of the SpA gene in 1983 (Löfdahl *et al.*, 1983; Uhlen *et al.*, 1984) provided an explanation for the multiple sizes of SpA due to ability to alter lengths of the repeated segments within the gene (Hartleib *et al.*, 2000). This information then allowed for SpA-typing (Pickenhahn *et al.*, 1987; Frenay *et al.*, 1996), which is a very important epidemiological tool.

Adhesins: fibronectin, an essential host protein

In 1978, Pentti Kuusela described fibronectin binding to *S. aureus* (Kuusela, 1978), and soon more details of *S. aureus*–fibronectin interactions followed (Mosher *et al.*, 1980; Proctor *et al.*, 1982; Speziale *et al.*, 1982). When the fibronectin–*S. aureus* adhesins were first described, questions arose because excess quantities of the ligand would not disrupt the attachment between the bacterial and the host proteins. This was later found to be due to the fact that there was a 'dock, lock, and latch' mechanism such that they could not be displaced (House-Pompeo *et al.*, 1996; Foster *et al.*, 2014).

While some early thoughts were that fibronectin was a non-specific opsonin when added to vials containing phagocytes and *S. aureus*, the reduction in colony counts mostly came from clumping of the organisms wherein a clump of bacteria would produce a single colony on a plate rather than a true decrease in the numbers of organisms (Proctor *et al.*, 1982). That being said, fibronectin was important for maintaining macrophage bactericidal activity (Proctor *et al.*, 1985; Proctor, Prendergast *et al.*, 1985), but there was sufficient fibronectin in the basement membrane and on host cells to promote this effect even when serum fibronectin levels were low. For a period of time, fresh-frozen plasma was given to patients in surgical intensive care units to boost macrophage function by supplying fibronectin, which depleted this product in blood banks; however, the practice stopped because animal models and clinical studies did not support its use as a stimulant to the phagocytes (Mosher *et al.*, 1983).

The role of the *S. aureus* fibronectin receptor in pathogenesis was suggested when the number of fibronectin receptors was directly correlated with the ability of *S. aureus* to produce invasive infection (Proctor *et al.*, 1984). Fibronectin also was shown to be important for allowing *S. aureus* to adhere to endothelial cells (Hamill *et al.*, 1986; Proctor, 1987). Moreover, mutants in the staphylococcal fibronectin receptor showed reduced adherence to traumatized heart valves (Kuypers *et al.*, 1989). Further evidence for the importance of fibronectin binding came when the expression of the fibronectin binding protein in *Lactococcus lactis* allowed adherence to heart valves (Que, 2001).

A plethora of host binding proteins called MSCRAMMs (microbial surface component-recognizing adhesive matrix molecules) (Foster *et al.*, 2014) and SERAMs (secretable expanded repertoire adhesive molecules) (Chavakis, 2005) are present on *S. aureus* strains. Like fibronectin, these adhesive factors allow *S. aureus* to bind to a wide variety of host tissues.

Global regulators: *agr* as an example of regulation of virulence factors

Early studies using chemical mutagenesis suggested that there was a common regulatory mechanism for multiple toxins and surface proteins because spontaneous revertants gained multiple phenotypic changes in exoprotein production (Forsgren *et al.*, 1971; Yoshikawa *et al.*, 1974). A transposon mutant that showed reduced Hla production (Brown *et al.*, 1980) proved to have a mutation in the regulatory locus, *agr* (accessory gene regulator) (Recsei *et al.*, 1986). The mutation in *agr* resulted in decreased Hla production. Because other genes had been identified, i.e. *hla* and *tst* (de Azavedo, 1985; Pattee, 1986), this allowed the separation of the *agr* locus from the structural genes for Hla and TSST-1. Confirmation of these results was obtained during studies of a Tn*551* insert into a single locus that decreased production of several exoproteins, yet it increased SpA (Morfeldt *et al.*, 1988). The basis for all of these phenotypic changes being attributable to the *agr* locus became clear as the locus contained a two-component, quorum-sensing system wherein an auto-inducing peptide was the extracellular quorum signal and an RNA was the intracellular regulator [reviewed in Novick (2003)].

A non-transcribed RNA called RNAIII was found to be the intracellular signalling molecule for the Agr locus. The first portion of RNAIII is translated to form δ-haemolysin (Janzon, 1989), but the rest of RNAIII is not transcribed and its role in *agr*-signalling was first suggested in 1991 (Vandenesch *et al.*, 1991). It was confirmed as the signal in the Agr system when RNAIII itself, rather than any protein synthesis, was able to restore both the positive and negative regulatory functions in an *agr* null strain (Novick *et al.*, 1993). The regulation of exoprotein expression was found to be more complex when a staphylococcal accessory regulator, SarA, was described that regulated *agr* (Cheung, 1992). Even greater complexity was revealed when a mutation in another regulator gene, repressor of toxin (*rot*), was reported to allow toxin production in an *agr*-deleted strain (McNamara, 2000).

RNAIII effected gene expression through RNA–RNA interactions. The 5′ end of RNAIII positively regulates the translation of *hla* by competing directly with an intramolecular RNA secondary structure that sequesters the *hla* ribosomal binding site (Novick, 1993; Morfeldt *et al.*, 1995). The RNAIII-dependent *spa* expression was not only repressed at the transcriptional level (Novick, 1993; Gao, 2004), but also at the post-transcriptional level via an antisense RNA mechanism. The interaction between RNAIII and the mRNA of *hla* was the first description of an antisense RNA that stimulates translation of its target mRNA (Morfeldt *et al.*, 1995). Inhibition of *rot* translation by RNAIII was found to be a key feature of *agr* function (Geisinger *et al.*, 2006). An endonuclease that associated with RNAIII was found to degrade *spa* mRNA by the double-strand-specific endoribonuclease III (RNaseIII) (Huntzinger *et al.*, 2005; Romilly *et al.*, 2012). Thus, a mechanism was proposed: at low cell density, *spa* mRNA is efficiently translated and protected from degradation; at high cell density, RNAIII synthesis is induced by a quorum-sensing peptide, allowing binding to *spa* mRNA and occlusion of the ribosome binding site. This inhibition of translation favours degradation of *spa* mRNA involving RNase III at the initial step. RNase III-induced cleavages in the formed duplex and in the long hairpin II of *spa* mRNA may render the mRNA accessible to other endo- or exo-ribonucleases for rapid degradation.

While the quorum sensing peptide of the *agr* system could explain some of the growth phase-dependent changes in RNAIII levels (Vandenesch *et al.*, 1991), metabolic signals have been noted for decades. The impact of glucose and pH on expression of *agr* was noted by Regassa and coworkers (Regassa and Betley, 1992; Regassa *et al.*, 1992). Other metabolic signals that altered Agr responses are still unknown and were designated as black boxes (Novick, 2003). Anaerobic growth had also been known to suppress expression of haemolysins and toxic shock toxin production (Somerville *et al.*, 2009). Electron transport-defective *S. aureus*, which grew as small colony variants, had virtually absent RNAIII (Proctor *et al.*, 2006). Hence, an association between anaerobic metabolism and Agr activity is clear, but the signals remain unknown. Recently, some signals have been reported wherein metabolic regulators like CodY and RpiRc define how central metabolism overrides the quorum-sensing-dependent regulation of RNAIII transcription and virulence determinant synthesis (Majerczyk *et al.*, 2008; Zhu *et al.*, 2011; Roux *et al.*, 2014). A detailed understanding of how metabolism influences Agr activity may allow for controlling the virulence of *S. aureus*.

Intracellular pathogen

Shortly after the discovery of *S. aureus*, the concept that it could be an intracellular pathogen was introduced in 1886 when Wyssokowitsch reported that endothelial cells contained microorganisms (Wyssokowitsch, 1886). Similarly, in 1924, Domagk described pulmonary endothelial cells packed with staphylococci (reviewed in Altschul, 1954). In 1937 and 1943, studies of experimental endocarditis also reported intracellular *S. aureus* (Nedzel, 1937; MacNeal *et al.*, 1943). Later, the uptake of *S. aureus* by cultured endothelial cells was demonstrated using transmission electron microscopy (Ogawa, 1985). The first quantitative study of *S. aureus* uptake by endothelial cells was reported in 1986, and the suggestion was made that this might be a mechanism for staphylococcal persistence in endocarditis (Hamill *et al.*, 1986).

A role for Hla in the lysis of endothelial cells by intracellular *S. aureus* was shown by using mutants in *hla* that failed to damage the cells and by anti-Hla antibodies that protected the cultured endothelial cells from destruction by the Hla-producing wild type strain (Vann *et al.*, 1988). This led to the hypothesis that strains defective in Hla production might be a cause of persistent infections wherein organisms would hide from immune defences within host cells. A search for *hla* mutants in clinical *S. aureus* strains from persistent infections did not reveal *hla* mutants, rather bacteria that were defective in electron transport were discovered (Proctor, 1995). These organisms grew slowly, hence, the name small colony variants (SCVs), and had low levels of Hla due to the lack of RNAIII production. When laboratory selected SCVs were cultured with bovine endothelial cells, they were phagocytosed, but they were minimally toxic to the cells (Balwit, 1994). Further studies showed that *S. aureus* formed at a surprisingly high frequency within the intracellular milieu (Vesga *et al.*, 1996), perhaps due to the resistance of SCVs to host cationic antimicrobial peptides (Koo *et al.*, 1996; Samuelsen *et al.*, 2005; Glaser *et al.*, 2014). Of note, SCVs are especially well adapted as an intracellular pathogen as they have increased host cell adhesins, especially fibronectin binding protein that enhances internalization of *S. aureus* via host $\alpha_5\beta_1$ integrin (Fraunholz *et al.*, 2012), and reduced toxin production, so as to avoid damaging the host cell (Sendi *et al.*, 2009; Grosz *et al.*, 2014).

Our understanding of SCV formation and intracellular persistence were greatly extended when it was observed that *S. aureus* cultured with mammalian cells underwent reversible phenotypic switching to metabolically dormant SCVs (Tuchscherr, 2011). The intracellular organisms had diminished Hla and AgrA levels, which fell as the length of the incubation continued, while FnbA increased. Importantly, inflammatory cytokines decreased as the organisms switched from rapid growth to SCVs (Tuchscherr, 2011). This transition to an SCV state resulted in tissue culture transcriptional profiles that were strikingly similar to those found in tissues where the same strains of *S. aureus* were used to challenge mice and to those from patients with chronic *S. aureus* infections. Assays of cytokines in the murine model also showed a pattern very similar to the tissue culture models, wherein the temporal decreases in inflammatory cytokine expression decreased as the number of SCVs increased. While electron transport defects were described in intracellular persisting SCVs (Proctor *et al.*, 2006), the majority of the strains recovered by Tuchscherr *et al.* were not in this group of SCVs (Tuchscherr *et al.*, 2011). The pathways used to form these SCVs is under investigation, but several mechanisms other than electron transport and thymidylate biosynthetic defects have been described (Proctor, 2014). Finally, SCVs escape the phagosome (Giese *et al.*, 2011) and evade host defences by failing to alert the immune system of their presence by reduced activation of hypoxia-inducible factor (HIF) (Werth, 2010) and TLR2 (Hilmi, 2014). Thus, SCVs are very well adapted for persistence within host cells (Löffler *et al.*, 2014).

In addition to non-professional phagocytes, *S. aureus* were also reported to be able to survive within professional phagocytes. In 1952, Rogers and Tompsett (Rogers *et al.*, 1952) observed that *S. aureus* survived within human neutrophils and postulated that this might be a mechanism for developing metastatic infections. These studies were extended by Gresham *et al.*, wherein neutrophils loaded with *S. aureus* were able to establish infections in naive animals (Gresham *et al.*, 2000). A role for *agr* was found where the confined space within the phagosome stimulated *S. aureus* to produce virulence factors, but the neutrophil also responded by oxidizing the Agr autoinducing peptide (Rothfork *et al.*, 2004; Carnes *et al.*, 2010). Thus, multiple mechanisms for *S. aureus* survival within host cells, e.g. neutralization of host defences factors, escape from the phagosome, subversion of autophagy, and induction of apoptosis, allow *S. aureus* to be an intracellular pathogen (Fraunholz, 2012).

Biofilm

As noted in the introduction, this historical review focuses primarily on *S. aureus*, but the story of biofilm formation is more of an *S. epidermidis* than *S. aureus* story. Because the topic is large and covered in another chapter in this book, only a few of the highlights of biofilm history will be covered. In 1981, dramatic scanning electron micrographs of *S. epidermidis* biofilms on catheters harvested from human infections were published (Peters *et al.*, 1981). From these photographs, it was intuitively clear why these massive collections of organisms were so resistant to clearance by phagocytes and antibiotics (reviewed in Heilmann, 2006). In 1987, the ability of *S. epidermidis* to undergo phase variation provided a mechanism for the tremendous accumulation of bacteria on biomaterials (Christensen *et al.*, 1987). In 1996, the structure of the exopolysaccharide matrix of the biofilm (i.e. polysaccharide intercellular adhesion) was first published (Mack *et al.*, 1996). While these studies focused upon the ability of *S. epidermidis* to accumulate, more recent work has examined the ability of the organisms to be released (Lauderdale *et al.*, 2009). Recent reviews cover the biomolecular mechanisms (Laverty *et al.*, 2013), genetic regulation (Cue *et al.*, 2012), evasion of host defences (Hanke *et al.*, 2012; Scherr *et al.*, 2014), maturation/remodelling (Dai *et al.*, 2012; Otto, 2013), and novel treatments (Kiedrowski, 2011; Gomes, 2014) of staphylococcal biofilms.

Urinary tract infections

The only other non-*aureus Staphylococcus* covered in this review is *S. saprophyticus*, which is the second most common cause of urinary tract infections (UTIs) (Hovelius *et al.*, 1984; Raz *et al.*, 2005). In 1962, coagulase-negative staphylococci were reported as an aetiological agent of acute UTIs (Torres Pereira, 1962). Separation of staphylococcal UTI species was accomplished in 1968 (Mitchell, 1968), and further acceptance of its role in UTIs was published in 1974 (Maskell, 1974). Rectal and vaginal colonization was associated with UTIs (Latham, 1983). By 1984, it had been established as the second most common pathogen causing urinary tract infections in young women (Hovelius *et al.*, 1984). Its pathogenic potential was emphasized further, when it was noted to be a cause of renal stones (Fowler, 1985). Virulence factors include adhesins, a haemolysin, and exopolysaccharide production (Gatermann, 1986). In 1989, novobiocin resistance was accepted as a screen for identification of *S. saprophyticus* in urine specimens (McTaggart *et al.*, 1989). Epidemiological studies suggested that it occurred most frequently in young, sexually active woman (Gupta *et al.*, 1999).

Vaccines and immunity

A dream to make a protective vaccine against *S. aureus* has become more pressing as antibiotic resistance and virulence has increased. A number of recent reviews are available concerning the history of the clinical trials and vaccine immunity (Projan *et al.*, 2006; Schaffer *et al.*, 2009; Jansen *et al.*, 2013; Fowler *et al.*, 2014); hence, only a brief overview is provided here.

Role of antibodies in *S. aureus* immunity

In 1975, an examination the antibody titres in patients with acute appendicitis who developed *S. aureus* wound infections showed no correlation between pre-existing antibody levels and development of infection (Akatov *et al.*, 1975). The next human study involved the use of a whole heat-killed *S. aureus* (Poole-Warren *et al.*, 1991). While immunization increased alpha-haemolysin antibodies, it did not protect from peritonitis or exit site infections in chronic ambulatory peritoneal dialysis patients. Multiple other clinical trials, based upon increasing antibody titres to improve opsonophagocytic activity, have failed (Fowler *et al.*, 2014). While an argument has been made that the antibodies were not shown to increase killing of *S. aureus*, rather just facilitate uptake (Jansen *et al.*, 2013), this argument can be questioned because patients with agammaglobulinaemia do not have an increased incidence of *S. aureus* infections (Fowler *et al.*, 2014).

Animal models for vaccine development

Questions have been raised about the adequacy of animal models used for vaccine development (Kim *et al.*, 2014; Salgado-Pabon, 2014). Naturally, the animal model that predicts a human response will clearly be the correct model, and this will only be confirmed with a successful human vaccine clinical trial. Short of that, human epidemiological data are available wherein the anti-toxin antibody levels to enterotoxins, Hla, TSST-1, and PVL show a direct correlation to reduced disease severity, but not prevention of infection (reviewed in Fowler, 2014). These human data show very good concordance with animal model data, especially in rabbit models, wherein titres of anti-toxin antibodies also show less severe disease. One note of caution comes with the use of murine models for studying PVL or enterotoxins because mice are intrinsically highly resistant to these toxins, thereby making them a poor choice of species to study the role of staphylococcal toxins in human infections (Löffler *et al.*, 2010; Salgado-Pabon *et al.*, 2014).

No established biomarker for staphylococcal immunity

Another note of caution comes from the fact that all successful bacterial vaccines have only been developed when a clear biomarker that predicts protective human immunity to the pathogen is known (Proctor, 2012). For *S. aureus*, we have only limited knowledge of human staphylococcal protective immunity, hence, limited knowledge of biomarkers that correlate to good outcomes. Except for antitoxins, there is not yet a marker for protective, i.e. disease preventing, immunity. Currently, there are human studies that relate cytokine (IL-1β, IL-10, TNF-α) production with outcome of *S. aureus* bacteraemia (van Dissel *et al.*, 1998; Pichereau *et al.*, 2012; Rose *et al.*, 2012). These studies suggest that the cytokine profile might provide a biomarker that would predict vaccine efficacy, but large prospective human clinical studies are needed to show that these and other potential biomarkers will predict outcome.

Antibiotic efficacy and resistance

A huge part of the history of *S. aureus* is the dramatic impact of antibiotic therapy upon reducing mortality and growing antibiotic resistance that threatens these medical advances. There are multiple reviews concerning all types of antibiotic resistance in *S. aureus* (Chambers, 2003, 2005; Gonzalez, 2003; DeLeo *et al.*, 2009; Howden *et al.*, 2010; Rybak *et al.*, 2013; Gardete *et al.*, 2014), hence, this historical review will only focus upon cell wall active antibiotics and upon a few historical milestones.

The impact of penicillin upon mortality in *S. aureus* bacteraemia was striking. In the pre-penicillin era (1936–1955), mortality of *S. aureus* bacteraemia was ~90% (Smith *et al.*, 1960). With the advent of penicillin use in 1944, the mortality dropped to 28%, but it rose to 54% by 1954 as penicillinase-producing strains became widespread (Rubin, 1999). The problem of penicillinase had been anticipated, because in 1940, Abraham and Chain (Abraham *et al.*, 1988) described an enzyme from bacteria was able to destroy penicillin. Fleming then reported upon *in vitro* resistance (Fleming, 1942) and isolation of penicillinase followed (Kirby, 1944; Bondi *et al.*, 1948).

The first report of a penicillin treatment failure due to penicillinase came in 1949 with a case of acute osteomyelitis (Jeffery *et al.*, 1949). Methicillin was introduced in 1959, which resulted in a marked drop in mortality to 30% (Rubin *et al.*, 1999). In 1961, the first clinical case of MRSA was reported (Jevons, 1961), which was followed by a report in 1963 of the first cluster of MRSA cases (Stewart *et al.*, 1963). The number of MRSA isolates in hospitals increased dramatically from 4% in 1984 to 42% in 1997 (NNIS 1997). The impact of resistance on attributable mortality due to bacteraemia was equally dramatic with MRSA showing 10-fold higher mortality than methicillin susceptible *S. aureus* (MSSA) (O'Kane *et al.*, 1998). With a greater awareness of MRSA and with the widespread development of intensive care units, which could provide extensive supportive care for septic patients, mortality slowly decreased to ~20% for *S. aureus* bacteraemia (Rubin *et al.*, 1999; Edmond *et al.*, 1999).

In 1998, community acquired-MRSA (CA-MRSA) was reported in children with no identified predisposing risk factors (Herold *et al.*, 1998). These *S. aureus* strains were unusual as they were clindamycin susceptible compared to hospital acquired-MRSA strains. Two pulse-field gel electrophoresis types, USA300 and USA400 became predominant amongst the CA-MRSA in the USA (McDougal *et al.*, 2003). After pressure from the US Senate, a study of the mortality rate of the new strains of *S. aureus* was performed by the CDC and mortality as high as 75% with bacteraemia was reported (Klevens *et al.*, 2007). From 2002–2005, a number of severe syndromes was reported by the apparently resurgent and more virulent *S. aureus* strains (Chambers, 2005): Necrotizing fasciitis (Miller *et al.*, 2005), haemorrhagic pneumonia (Gillet, 2002), Waterhouse–Friderichsen syndrome (Adem *et al.*, 2005), and purpura fulminans (Kravitz, 2005).

In 1996, the first vancomycin intermediate susceptibility *S. aureus* (VISA) strain was reported in Japan (Centres for Disease Control and Prevention, 1997; Hiramatsu *et al.*, 1997). Soon thereafter, VISA was found in USA (Smith *et al.*, 1999). This had also been anticipated as *S. aureus* with increased minimal inhibitory concentrations (MICs) were seen in a patient receiving teicoplanin therapy for endocarditis (Kaatz *et al.*, 1990), and strains could be selected *in vitro* for vancomycin resistance (Boyle-Vavra *et al.*, 2000; Pfeltz *et al.*, 2000). Fully vancomycin-resistant *S. aureus* (VRSA) with MIC $\geq 32\,\mu g/ml$ were first reported in 2002 from a patient that had received long- term vancomycin treatment. The VRSA strain had acquired *vanA* from an enterococcus species (Centres for Disease Control and Prevention, 2002; Chang *et al.*, 2003; Tenover *et al.*, 2004; Gardete *et al.*, 2014). A more disturbing VRSA case followed wherein no vancomycin exposure had occurred, suggesting that VRSA could develop without antibiotic pressure (Whitener *et al.*, 2004).

Currently, the mortality rates for MRSA and MSSA bacteraemia seem to have stabilized at 20–22% (Tom *et al.*, 2014). This reduction in mortality probably has to due to the awareness of MRSA, the emphasis on early therapy, new drugs, novel combinations of antibiotics, and further improvements in ICU technologies (Rose *et al.*, 2013). The desire to understand the complexity of the MRSA phenotypes has driven a far better understanding of cell wall biosynthesis, thereby, opening new opportunities for antibiotic development (Roemer and Boone, 2013; Roemer *et al.*, 2013). This has in fact already produced compounds that greatly enhance the activity of β-lactam antibiotics against MRSA (Tan *et al.*, 2012; Mann *et al.*, 2013). Also, defining the mechanisms of action and resistance for multiple antibiotics has allowed for new therapeutics based upon defensins (Wilmes *et al.*, 2014).

Finally, persistent infection, even when apparently effective antibiotic therapy is used, is another type of resistance. In many organisms, slow growth and resistance to antibiotics can be linked (Wood *et al.*, 2013; Proctor *et al.*, 2014), and this is also found in *S. aureus* (Lechner *et al.*, 2012) because of changes in metabolism. We have found that antibiotics are profoundly less effective in clearing SCVs from biomaterials (Chuard *et al.*, 1997) and from host tissues (Proctor *et al.*, 2006; Kahl, 2014). Specific studies of antibiotics against *S. aureus* SCVs show that several mechanisms are involved, e.g. intracellular location, reduced uptake, and reduced activity due to slow cell wall metabolism (Garcia *et al.*, 2013). Thus, antibiotic resistance is very likely to be as much of the future as it has been of the history of *S. aureus*.

Meetings

Part of the history of *S. aureus* also relates to the meetings where investigators exchange information. While investigators studying streptococci had the Lancefield Society for the exchange of information and collaboration, progress on staphylococcal research lagged due to a lack of a well-supported meeting for gathering staphylococcal researchers. The International Symposium on Staphylococci and Staphylococcal Infections (ISSSI) meeting was started by Professor Janusz Jeljaszewicz (Parnowska, 2006) and held in Poland partially fulfilled this need, but the limited funding did not allow significant participation from workers outside of Europe. To address this need, the Staphylococcal Diseases Gordon Research Conference was first held 1989 (Richard Proctor and Peter Pattee, Chairs). Since then, the ISSSI and Gordon Conferences have been held in alternating years around the globe, which has greatly stimulated the exchange of data, collaborations, and progress in addressing the challenges of staphylococcal diseases.

Conclusions

One conclusion of this review is that the history of staphylococci can no longer be told in a single chapter. Most subtopic headings of this chapter, when searched on PubMed, produced more than 2500 citations, with the total number of citations over 45,000. While it is clear that many of the citations were listed more than one time, the amount of work and the number of highly dedicated and creative investigators is truly impressive. This attests to seriousness of the infections caused by staphylococci and the progress that has been made since the first publications in 1880 by Ogston and Pasteur (Ogston, 1880; Pasteur, 1880). Although more than a century has passed since the first intracellular staphylococci were noted and toxins studied, and more than seventy years since penicillinase was suggested, we still have much to learn about the pathogenesis and treatment of staphylococci. Developing a vaccine to prevent staphylococcal infections would be a great accomplishment, but we may need to be satisfied at finding one that reduces virulence. In all likelihood, we will still be facing challenges from staphylococci for the next century.

References

Abraham, E.P., and Chain, E. (1988). An enzyme from bacteria able to destroy penicillin. 1940. Rev. Infect. Dis. *10*, 677–678.

Adem, P.V., Montgomery, C.P., Husain, A.N., Koogler, T.K., Arangelovich, V., Humilier, M., Boyle-Vavra, S., and Daum, R.S. (2005). *Staphylococcus aureus* sepsis and the Waterhouse–Friderichsen syndrome in children. N. Engl. J. Med. *353*, 1245–1251.

Ahn, D.S., Parker, D., Planet, P.J., Nieto, P.A., Bueno, S.M., and Prince, A. (2014). Secretion of IL-16 through TNFR1 and calpain-caspase signaling contributes to MRSA pneumonia. Mucosal. Immunol. *7*, 1366–1374.

Akatov, A.K., Nurmanbetova, U.E., and Khatenever, M.L. (1975). [Antistaphylococcal bacteriostatic activity of the sera from surgical patients.] Zh. Mikrobiol. Epidemiol. Immunobiol. *10*, 55–58.

Altschul, R. (1954). *Endothelium, its Development, Morphology, Function, and Pathology.* New York: MacMillan Company.

Baird-Parker, A.C. (1965). The classification of staphylococci and micrococci from world-wide sources. J. Gen. Microbiol. *38*, 363–387.

Balwit, J.M., van Langevelde, P., Vann, J.M., and Proctor, R.A. (1994). Gentamicin-resistant menadione and hemin auxotrophic *Staphylococcus aureus* persist within cultured endothelial cells. J. Infect. Dis. *170*, 1033–1037.

Bannantine, J.P., and Pattee, P.A. (1996). Construction of a chromosome map for the phage group II *Staphylococcus aureus* Ps55. J. Bacteriol. *178*, 6842–6848.

Becker, S., Frankel, M.B., Schneewind, O., and Missiakas, D. (2014). Release of protein A from the cell wall of *Staphylococcus aureus*. Proc. Natl. Acad. Sci. U.S.A. *111*, 1574–1579.

Bergdoll, M.S., Crass, B.A., Reiser, R.F., Robbins, R.N., and Davis, J.P. (1981). A new staphylococcal enterotoxin, enterotoxin F, associated with toxic-shock-syndrome *Staphylococcus aureus* isolates. Lancet *1*, 1017–1021.

Berger-Bächi, B., Barberis-Maino, L., Strässle, A., and Kayser, F.H. (1989). FemA, a host-mediated factor essential for methicillin resistance in *Staphylococcus aureus*: molecular cloning and characterization. Mol. Gen. Genet. *219*, 263–269.

Berger-Bächi, B., and Tschierske, M. (1998). Role of fem factors in methicillin resistance. Drug Resist. Updat. *1*, 325–335.

Berube, B.J., Sampedro, G.R., Otto, M., and Bubeck Wardenburg, J. (2014). The *psmα* locus regulates production of *Staphylococcus aureus* alpha-toxin during infection. Infect. Immun. *82*, 3350–3358.

Blair, J.E., and Williams, R.E. (1961). Phage typing of staphylococci. Bull. World Health Organ. *24*, 771–784.

Bondi, A., and Dietz, C.C. (1948). The Susceptibility of Penicillinase-producing Bacteria to Penicillin: I. Factors Influencing Susceptibility. J. Bacteriol. 55, 843–847.

Boyle-Vavra, S., Berke, S.K., Lee, J.C., and Daum, R.S. (2000). Reversion of the glycopeptide resistance phenotype in *Staphylococcus aureus* clinical isolates. Antimicrob. Agents Chemother. *44*, 272–277.

Brakstad, O.G., andMaeland, J.A. (1995). Direct identification of *Staphylococcus aureus* in blood cultures by detection of the gene encoding the thermostable nuclease or the gene product. APMIS*103*, 209–218.

Bramley, A.J., Patel, A.H., O'Reilly, M., Foster, R., and Foster, T.J. (1989). Roles of alpha-toxin and beta-toxin in virulence of *Staphylococcus aureus* for the mouse mammary gland. Infect. Immun. *57*, 2489–2494.

Brosnahan, A.J., and Schlievert, P.M. (2011). Gram-positive bacterial superantigen outside-in signaling causes toxic shock syndrome. FEBS J. *278*, 4649–4667.

Brown, D.R., andPattee, P.A. (1980). Identification of a chromosomal determinant of alpha-toxin production in *Staphylococcus aureus*. Infect. Immun. *30*, 36–42.

Bubeck Wardenburg, J., Patel, R.J., and Schneewind, O. (2007). Surface proteins and exotoxins are required for the pathogenesis of *Staphylococcus aureus* pneumonia. Infect. Immun. *75*, 1040–1044.

Bubeck Wardenburg, J., and Schneewind, O. (2008). Vaccine protection against *Staphylococcus aureus* pneumonia. J. Exp. Med. *205*, 287–294.

Burnet, F.M. (1930). The production of staphylococcal toxin. J. Patho.l Bacteriol. *33*, 1–16.

Calfee, D.P., Salgado, C.D., Milstone, A.M., Harris, A.D., Kuhar, D.T., Moody, J., Aureden, K., Huang, S.S., Maragakis, L.L., and Yokoe, D.S. (2014). Strategies to prevent methicillin-resistant *Staphylococcus aureus* transmission and infection in acute care hospitals: 2014 update. Infect. Control Hosp. Epidemiol. *35*, 772–796.

Carnes, E.C., Lopez, D.M., Donegan, N.P., Cheung, A., Gresham, H., Timmins, G.S., and Brinker, C.J. (2010). Confinement-induced quorum sensing of individual *Staphylococcus aureus* bacteria. Nat. Chem. Biol. *6*, 41–45.

Centers for Disease Control and Prevention(CDC). (1997). Reduced susceptibility of *Staphylococcus aureus* to vancomycin–Japan, 1996. MMWR. Morb. Mortal. Wkly Rep. *46*, 624–626.

Centers for Disease Control and Prevention(CDC). (2002). *Staphylococcus aureus* resistant to vancomycin–United States, 2002. MMWR. Morb. Mortal. Wkly Rep. *51*, 565–567.

Chambers, H.F. (2003). Solving staphylococcal resistance to beta-lactams. Trends Microbiol. *11*, 145–148. Chambers, H.F. (2005). Community-associated MRSA–resistance and virulence converge. N. Engl. J. Med. *352*, 1485–1487.

Chang, S., Sievert, D.M., Hageman, J.C., Boulton, M.L., Tenover, F.C., Downes, F.P., Shah, S., Rudrik, J.T., Pupp, G.R., Brown, W.J., *et al.* (2003). Infection with vancomycin-resistant *Staphylococcus aureus* containing the *vanA* resistance gene. N. Engl. J. Med. *348*, 1342–1347.

Chavakis, T., Wiechmann, K., Preissner. K.T., and Herrmann, M. (2005). *Staphylococcus aureus* interactions with the endothelium: the role ofbacterial "secretable expanded repertoire adhesive molecules" (SERAM) in disturbing host defense systems. Thromb Haemost. *94*, 278–285.

Chesney, P.J., Bergdoll, M.S., Davis, J.P., and Vergeront, J.M. (1984). The disease spectrum, epidemiology, and etiology of toxic-shock syndrome. Annu. Rev. Microbiol. *38*, 315–338.

Cheung, A.L., Koomey, J.M., Butler, C.A., Projan, S.J., and Fischetti, V.A. (1992). Regulation of exoprotein expression in *Staphylococcus aureus* by a locus (*sar*) distinct from *agr*. Proc. Natl. Acad. Sci. U.S.A. *89*, 6462–6466.

Choi, Y.W., Kotzin, B., Herron, L., Callahan, J., Marrack, P., and Kappler, J. (1989). Interaction of *Staphylococcus aureus* toxin "superantigens" with human T cells. Proc. Natl. Acad. Sci. U.S.A. *86*, 8941–8945.

Christensen, G.D., Baddour, L.M., and Simpson, W.A. (1987). Phenotypic variation of *Staphylococcus epidermidis* slime production *in vitro* and *in vivo*. Infect. Immun. *55*, 2870–2877.

Christensson, B., Hedström, S.A., and Kronvall, G. (1983). Antibody response to alpha- and betahemolysin from *Staphylococcus aureus* in patients with staphylococcal infections and in normals. Acta Pathol. Microbiol. Immunol. Scand. B*91*, 351–356.

Chuard, C., Vaudaux, P.E., Proctor, R.A., and Lew, D.P. (1997). Decreased susceptibility to antibiotic killing of a stable small colony variant of *Staphylococcus aureus* in fluid phase and on fibronectin-coated surfaces. J. Antimicrob. Chemother. *39*, 603–608.

Cowan, S.T. (1939). Classification of staphylococci by slide agglutination. The Journal of Pathology and Bacteriology. *48*, 169–173.

Cowan, S.T., Shaw, C., and Williams, R.E. (1954). Type strain for *Staphylococcus aureus* Rosenbach. J. Gen. Microbiol. *10*, 174–176.

Cowan, S.T., and Steel, K.L.(1964). Comparison of differentiating criteria for staphylococci and micrococci. J. Bacteriol. *88*, 804–805.

Craven, R.R., Gao, X., Allen, I.C., Gris, D., Bubeck Wardenburg, J., McElvania-Tekippe, E., Ting, J.P., and Duncan, J.A. (2009). *Staphylococcus aureus* alpha-hemolysin activates the NLRP3-inflammasome in human and mouse monocytic cells. PLoS One*4*, e7446.

Crémieux, A.C., Saleh-Mghir, A., Danel, C., Couzon, F., Dumitrescu, O., Lilin, T., Perronne, C., Etienne, J., Lina, G., and Vandenesch, F. (2014). α-Hemolysin, not Panton–Valentine leukocidin, impacts rabbit mortality from severe sepsis with methicillin-resistant *Staphylococcus aureus* osteomyelitis. J. Infect. Dis. *209*, 1773–1780.

Cribier, B., Prévost, G., Couppie, P., Finck-Barbançon, V., Grosshans, E., and Piémont, Y. (1992). *Staphylococcus aureus* leucocidin: a new virulence factor in cutaneous infections? An epidemiological and experimental study. Dermatology *185*, 175–180.

Cue, D., Lei, M.G., and Lee, C.Y. (2012). Genetic regulation of the intercellular adhesion locus in staphylococci. Front. Cell. Infect. Microbiol. *2*, 38.

Dai, L., Yang, L., Parsons, C., Findlay, V.J., Molin, S., and Qin, Z. (2012). Staphylococcus epidermidis recovered from indwelling catheters exhibit enhanced biofilm dispersal and "self-renewal" through downregulation of agr. BMC Microbiol. *12*, 102.

Davis, J.P., Chesney, P.J., Wand, P.J., LaVenture, M., and LaVenture, M. (1980). Toxic-shock syndrome: epidemiologic features, recurrence, risk factors, and prevention. N. Engl. J. Med. *303*, 1429–1435.

de Azavedo, J.C., Foster, T.J., Hartigan, P.J., Arbuthnott, J.P., O'Reilly, M., Kreiswirth, B.N., and Novick, R.P. (1985). Expression of the cloned toxic shock syndrome toxin 1 gene (*tst*) *in vivo* with a rabbit uterine model. Infect. Immun. *50*, 304–309.

DeLeo, F.R., and Chambers, H.F. (2009). Reemergence of antibiotic-resistant *Staphylococcus aureus* in the genomics era. J. Clin. Invest. *119*, 2464–2474.

Del Vecchio, V.G., Petroziello, J.M., Gress, M.J., McCleskey, F.K., Melcher, G.P., Crouch, H.K., and Lupski, J.R. (1995). Molecular genotyping of methicillin-resistant *Staphylococcus aureus* via fluorophore-enhanced repetitive-sequence PCR. J. Clin. Microbiol. 33, 2141–2144.

Denys, J., and van de Velde, H. (1895). *Sur la production d'une antileucocidine chez les lapin vaccinés contre le staphylocoque pyogène.* La Cellule. *11*, 359–372.

Diep, B.A., Gill, S.R., Chang, R.F., Phan, T.H., Chen, J.H., Davidson, M.G., Lin, F., Lin, J., Carleton, H.A., Mongodin, E.F., *et al.* (2006). Complete genome sequence of USA300, an epidemic clone of community-acquired meticillin-resistant *Staphylococcus aureus*. Lancet *367*, 731–739.

Edmond, M.B., Wallace, S.E., McClish, D.K., Pfaller, M.A., Jones, R.N., and Wenzel, R.P. (1999). Nosocomial bloodstream infections in United States hospitals: a three-year analysis. Clin. Infect. Dis. *29*, 239–244.

Engels, W., Kamps, M.A., and van Boven, C.P. (1981). Rapid and direct staphylocoagulase assay that uses a chromogenic substrate for identification of *Staphylococcus aureus*. J. Clin. Microbiol. *14*, 496–500.

Essers, L., and Radebold, K. (1980). Rapid and reliable identification of *Staphylococcus aureus* by a latex agglutination test. J. Clin. Microbiol. *12*, 641–643.

Feldman, C.A. (1962). Staphylococcal scarlet fever. N. Engl. J. Med. *267*, 877–888.

Fleming, A. (1942). *In vitro* tests of penicillin potency. The Lancet. **239**(6199), 732–733.

Forsgren, A., Nordström, K., Philipson, L., and Sjöquist, J. (1971). Protein A mutants of *Staphylococcus aureus*. J. Bacteriol. *107*, 245–250.

Foster, T.J., Geoghegan, J.A., Ganesh, V.K., and Höök, M. (2014). Adhesion, invasion and evasion: the many functions of the surface proteins of *Staphylococcus aureus*. Nat. Rev. Microbiol. *12*, 49–62.

Fowler, J.E. (1985). *Staphylococcus saprophyticus* as the cause of infected urinary calculus. Ann. Intern. Med. *102*, 342–343.

Fowler, V.G., and Proctor, R.A. (2014). Where does a *Staphylococcus aureus* vaccine stand? Clin. Microbiol. Infect. *20* (Suppl. 5), 66–75.

Frank, K.M., Zhou, T., Moreno-Vinasco, L., Hollett, B., Garcia, J.G., and Bubeck Wardenburg, J. (2012). Host response signature to *Staphylococcus aureus* alpha-hemolysin implicates pulmonary Th17 response. Infect. Immun. *80*, 3161–3169.

Fraunholz, M., and Sinha, B. (2012). Intracellular *Staphylococcus aureus*: live-in and let die. Front. Cell. Infect. Microbiol. *2*, 43.

Frénay, H.M., Bunschoten, A.E., Schouls, L.M., van Leeuwen, W.J., Vandenbroucke-Grauls, C.M., Verhoef, J., and Mooi, F.R. (1996). Molecular typing of methicillin-resistant *Staphylococcus aureus* on the basis of protein A gene polymorphism. Eur. J. Clin. Microbiol. Infect. Dis. *15*, 60–64.

Freney, J., Kloos, W.E., Hajek, V., Webster, J.A., Bes, M., Brun, Y., and Vernozy-Rozand, C. (1999). Recommended minimal standards for description of new staphylococcal species. Subcommittee on the taxonomy of staphylococci and streptococci of the International Committee on Systematic Bacteriology. Int. J. Syst. Bacteriol. *49*(Suppl. 2), 489–502.

Gaillot, O., Wetsch, M., Fortineau, N., and Berche, P. (2000). Evaluation of CHROMagar *Staph. aureus*, a new chromogenic medium, for isolation and presumptive identification of *Staphylococcus aureus from* human clinical specimens. J. Clin. Microbiol. *38*, 1587–1591.

Gao, J., and Stewart, G.C. (2004). Regulatory elements of the *Staphylococcus aureus* protein A (Spa) promoter. J. Bacteriol. *186*, 3738–3748.

Garcia, L.G., Lemaire, S., Kahl, B.C., Becker, K., Proctor, R.A., Denis, O., Tulkens, P.M., and Van Bambeke, F. (2013). Antibiotic activity against small-colony variants of *Staphylococcus aureus*: review of *in vitro*, animal and clinical data. J. Antimicrob. Chemother. *68*, 1455–1464.

Gardete, S., andTomasz, A. (2014). Mechanisms of vancomycin resistance in *Staphylococcus aureus*. J. Clin. Invest. *124*, 2836–2840.

Giese, B., Glowinski, F., Paprotka, K., Dittmann, S., Steiner, T., Sinha, B., and Fraunholz, M.J. (2011). Expression ofδ-toxin by *Staphylococcus aureus* mediates escape from phago-endosomes of human epithelial and endothelial cells in the presence ofβ-toxin. Cell. Microbiol. *13*, 316–329.

Geisinger, E., Adhikari, R.P., Jin, R., Ross, H.F., and Novick, R.P. (2006). Inhibition of *rot* translation by RNAIII, a key feature of *agr* function. Mol. Microbiol. *61*, 1038–1048.

Giai, C., Gonzalez, C., Ledo, C., Garofalo, A., Di Genaro, M.S., Sordelli, D.O., and Gomez, M.I. (2013). Shedding of tumor necrosis factor receptor 1 induced by protein A decrease tumor necrosis factor alpha availability and inflammation during systemic *Staphylococcus aureus* infection. Infect. Immun. *81*, 4200–4207.

Gill, S.R., Fouts, D.E., Archer, G.L., Mongodin, E.F., Deboy, R.T., Ravel, J., Paulsen, I.T., Kolonay, J.F., Brinkac, L., Beanan, M., Dodson, R.J., Daugherty, S.C., Madupu, R., Angiuoli, S.V., Durkin, A.S., Haft, D.H., et al. (2005). Insights on evolution of virulence and resistance from the complete genome analysis of an early methicillin-resistant *Staphylococcus aureus* strain and a biofilm- producing methicillin-resistant *Staphylococcus epidermidis* strain. J. Bacteriol. *187*, 2426–2438.

Gillet, Y., Issartel, B., Vanhems, P., Fournet, J.C., Lina, G., Bes, M., Vandenesch, F., Piémont, Y., Brousse, N., Floret, D., and Etienne, J. (2002). Association between *Staphylococcus aureus* strains carrying gene for Panton-Valentine leucocidin and highly lethal necrotising pneumonia in young immunocompetent patients. Lancet *359(9308)*, 53–59.

Gladstone, G.P., and van Heyningen, W.E. (1957). Staphylococcal leucocidins. Br. J. Exp. Pathol. *38*, 123–137.

Gläser, R., Becker, K., von Eiff, C., Meyer-Hoffert, U., and Harder, J. (2014). Decreased susceptibility of *Staphylococcus aureus* small-colony variants toward human antimicrobial peptides. J. Invest.Dermatol. **1**34. 2347–2350.

Glenny, A.T., and Stevens, M.F. (1935). *Staphylococcus toxins* and antitoxins. J. Pathol. Bacteriol. *40*, 201–210.

Goering, R.V., McDougal, L.K., Fosheim, G.E., Bonnstetter, K.K., Wolter, D.J., and Tenover, F.C. (2007). Epidemiologic distribution of the arginine catabolic mobile element among selected methicillin-resistant and methicillin-susceptible *Staphylococcus aureus* isolates. J. Clin. Microbiol. *45*, 1981–1984.

Gomes, F., Teixeira, P., and Oliveira, R. (2014). Mini-review: *Staphylococcus epidermidis* as the most frequent cause of nosocomial infections: old and new fighting strategies. Biofouling *30*, 131–141.

Gómez, M.I., Lee, A., Reddy, B., Muir, A., Soong, G., Pitt, A., Cheung, A., and Prince, A. (2004). *Staphylococcus aureus* protein A induces airway epithelial inflammatory responses by activating TNFR1. Nat. Med. *10*, 842–848.

Gonzalez, A., Bischoff, T., Tallent, S., Sheke, G., Ostrowsky, B., Edmond, M.B., and Wenzel, R.P. (2003). Antibiotic resistance in the community. J. Hosp. Infect. *55*, 156–157.

Gresham, H.D., Lowrance, J.H., Caver, T.E., Wilson, B.S., Cheung, A.L., and Lindberg, F.P. (2000). Survival of *Staphylococcus aureus* inside neutrophils contributes to infection. J. Immunol. *164*, 3713–3722.

Grosz, M., Kolter, J., Paprotka, K., Winkler, A.C., Schäfer, D., Chatterjee, S.S., Geiger, T., Wolz, C., Ohlsen, K., Otto, M., *et al.* (2014). Cytoplasmic replication of *Staphylococcus aureus* upon phagosomal escape triggered by phenol-soluble modulin*α*. Cell. Microbiol. *16*, 451–465.

Grov, A., Myklestad, B., and Oeding, P. (1964). Immunochemical studies on antigen preparations from *Staphylococcus aureus*. 1. Isolation and chemical characterization of antigen A. Acta Pathol. Microbiol. Scand. *61*, 588–596.

Gupta, K., Hooton, T.M., Wobbe, C.L., and Stamm, W.E. (1999). The prevalence of antimicrobial resistance among uropathogens causing acute uncomplicated cystitis in young women. Int. J. Antimicrob. Agents *11*, 305–308.

Hamill, R.J., Vann, J.M., and Proctor, R.A. (1986). Phagocytosis of *Staphylococcus aureus* by cultured bovine aortic endothelial cells: model for postadherence events in endovascular infections. Infect. Immun. *54*, 833–836.

Hanke, M.L., and Kielian, T. (2012). Deciphering mechanisms of staphylococcal biofilm evasion of host immunity. Front. Cell. Infect. Microbiol. *2*, 62–80.

Hartleib, J., Köhler, N., Dickinson, R.B., Chhatwal, G.S., Sixma, J.J., Hartford, O.M., Foster, T.J., Peters, G., Kehrel, B.E., and Herrmann, M. (2000). Protein A is the von Willebrand factor binding protein on *Staphylococcus aureus*. Blood *96*, 2149–2156.

Hawiger, J., Steckley, S., Hammond, D., Cheng, C., Timmons, S., Glick, A.D., and Des Prez, R.M. (1979). Staphylococci-induced human platelet injury mediated by protein A and immunoglobulin G Fc fragment receptor. J. Clin. Invest. *64*, 931–937.

Heilmann, C., and Peters, G. (2006). Biology and pathogenicity of *Staphylococcus epidermidis*. In Gram Positive Pathogens, 2nd edn. V.A. Fischetti, et al., Editors. 2006, ASM Press: Washington, DC, pp. 560–571.

Hermann, M., Jaconi, M.E., Dahlgren, C., Waldvogel, F.A., Stendahl, O., and Lew, D.P. (1990). Neutrophil bactericidal activity against *Staphylococcus aureus* adherent on biological surfaces. Surface-bound extracellular matrix proteins activate intracellular killing by oxygen-dependent and -independent mechanisms. J. Clin. Invest. *86*, 942–951.

Hermos, C.R., Yoong, P., and Pier, G.B. (2010). High levels of antibody to Panton–Valentine leucocidin are not associated with resistance to *Staphylococcus aureus*-associated skin and soft-tissue infection. Clin. Infect. Dis. *51*, 1138–1146.

Herold, B.C., Immergluck, L.C., Maranan, M.C., Lauderdale, D.S., Gaskin, R.E., Boyle-Vavra, S., Leitch, C.D., and Daum, R.S. (1998). Community-acquired methicillin-resistant *Staphylococcus aureus* in children with no identified predisposing risk. JAMA *279*, 593–598.

Hilmi, D., Parcina, M., Stollewerk, D., Ostrop, J., Josten, M., Meilaender, A., Zaehringer, U., Wichelhaus, T.A., Bierbaum, G., Heeg, K., *et al.* (2014). Heterogeneity of host TLR2 stimulation by *Staphylocoocus aureus* isolates. PLoS One *9*, e96416.

Hinshaw, L.B., Emerson, T.E., Taylor, F.B., Chang, A.C., Duerr, M., Peer, G.T., Flournoy, D.J., White, G.L., Kosanke, S.D., and Murray, C.K. (1992). Lethal *Staphylococcus aureus*-induced shock in primates: prevention of death with anti-TNF antibody. J. Trauma *33*, 568–573.

Hiramatsu, K., Hanaki, H., Ino, T., Yabuta, K., Oguri, T., and Tenover, F.C. (1997). Methicillin-resistant *Staphylococcus aureus* clinical strain with reduced vancomycin susceptibility. J. Antimicrob. Chemother. *40*, 135–136.

Holden, M.T., Feil, E.J., Lindsay, J.A., Peacock, S.J., Day, N.P., Enright, M.C., Foster, T.J., Moore, C.E., Hurst, L., Atkin, R., *et al.* (2004). Complete genomes of two clinical *Staphylococcus aureus* strains: evidence for the rapid evolution of virulence and drug resistance. Proc. Natl. Acad. Sci. U.S.A. *101*, 9786–9791.

House-Pompeo, K., Xu, Y., Joh, D., Speziale, P., and Höök, M. (1996). Conformational changes in the fibronectin binding MSCRAMMs are induced by ligand binding. J. Biol. Chem. *271*, 1379–1384.

Hovelius, B., and Mårdh, P.A. (1984). *Staphylococcus saprophyticus* as a common cause of urinary tract infections. Rev. Infect. Dis. *6*, 328–337.

Howden, B.P., Davies, J.K., Johnson, P.D., Stinear, T.P., and Grayson, M.L. (2010). Reduced vancomycin susceptibility in *Staphylococcus aureus*, including vancomycin-intermediate and heterogeneous vancomycin- intermediate strains: resistance mechanisms, laboratory detection, and clinical implications. Clin. Microbiol. Rev. *23*, 99–139.

Hua, L., Hilliard, J.J., Shi, Y., Tkaczyk, C., Cheng, L.I., Yu, X., Datta, V., Ren, S., Feng, H., Zinsou, R., Keller, A., O'Day, T., Du, Q., Cheng, L., Damschroder, M., Robbie, G., Suzich, J., Stover, C.K., and Sellman, B.R. (2014). Assessment of an anti-alpha-toxin monoclonal antibody for prevention and treatment of *Staphylococcus aureus*-induced pneumonia. Antimicrob Agents Chemother. *58*, 1108–1117.

Hunt, G.A., and Moses, A.J. (1958). Acute infection of mice with Smith strain of *Staphylococcus aureus*. Science *128*, 1574–1575.

Huntzinger, E., Boisset, S., Saveanu, C., Benito, Y., Geissmann, T., Namane, A., Lina, G., Etienne,

J., Ehresmann, B., Ehresmann, C., *et al.* (2005). *Staphylococcus aureus* RNAIII and the endoribonuclease III coordinately regulate *spa* gene expression. EMBO J. *24*, 824–835.

Hruz, P., Zinkernagel, A.S., Jenikova, G., Botwin, G.J., Hugot, J.P., Karin, M., Nizet, V., and Eckmann, L. (2009). NOD2 contributes to cutaneous defense against *Staphylococcus aureus* through alpha-toxin- dependent innate immune activation. Proc. Natl. Acad. Sci. U.S.A. *106*, 12873–12878.

Iandolo, J.J., Worrell, V., Groicher, K.H., Qian, Y., Tian, R., Kenton, S., Dorman, A., Ji, H., Lin, S., Loh, P., *et al.* (2002). Comparative analysis of the genomes of the temperate bacteriophages phi 11, phi 12 and phi 13 of *Staphylococcus aureus* 8325. Gene*289*, 109–118.

Inoshima, I., Inoshima, N., Wilke, G.A., Powers, M.E., Frank, K.M., Wang, Y., and Bubeck Wardenburg, J. (2011). A *Staphylococcus aureus* pore-forming toxin subverts the activity of ADAM10 to cause lethal infection in mice. Nat. Med. *17*, 1310–1314.

Jansen, K.U., Girgenti, D.Q., Scully, I.L., and Anderson, A.S. (2013). Vaccine review: "Staphyloccocus aureus vaccines: problems and prospects". Vaccine*31*, 2723–2730.

Janzon, L., Löfdahl, S., and Arvidson, S. (1989). Identification and nucleotide sequence of the delta-lysin gene, *hld*, adjacent to the accessory gene regulator (*agr*) of *Staphylococcus aureus*. Mol. Gen. Genet. *219*, 480–485.

Jayasinghe, L., Miles, G., and Bayley, H. (2006). Role of the amino latch of staphylococcal alpha-hemolysin in pore formation: a co-operative interaction between the N terminus and position 217. J. Biol. Chem. *281*, 2195–2204.

Jeffery, C.C., Sevitt, S., and Topley, E. (1949). Acute osteomyelitis due to a penicillinase-producing *Staphylococcus aureus*. Lancet *1*, 259–261.

Jensen, K. (1958). A normally occurring staphylococcal antibody in human serum. Acta Pathol. Microbiol. Scand. *44*, 421–428.

Jevons, M.P. (1961). "Celbenin" - resistant Staphylococci. Br Med J. *1*, 124–125.

Jones, J.M., Yost, S.C., and Pattee, P.A. (1987). Transfer of the conjugal tetracycline resistance transposon *Tn916* from *Streptococcus faecalis* to *Staphylococcus aureus* and identification of some insertion sites in the staphylococcal chromosome. J. Bacteriol. *169*, 2121–2131.

Julanielle, L.A. (1922). Studies of hemolytic staphylococci. Hemolytic activities, biochemical reactions, serological reactions. J. Infect. Dis. *31*, 256–284.

Kaatz, G.W., Seo, S.M., Dorman, N.J., and Lerner, S.A. (1990). Emergence of teicoplanin resistance during therapy of *Staphylococcus aureus* endocarditis. J. Infect. Dis. *162*, 103–108.

Kahl, B.C. (2014). Small colony variants (SCVs) of *Staphylococcus aureus*–a bacterial survival strategy. Infect. Genet. Evol. *21*, 515–522.

Kehoe, M., Duncan, J., Foster, T., Fairweather, N., and Dougan, G. (1983). Cloning, expression, and mapping of the *Staphylococcus aureus* alpha-hemolysin determinant in *Escherichia coli* K-12. Infect. Immun. *41*, 1105–1111.

Kiedrowski, M.R., and Horswill, A.R. (2011). New approaches for treating staphylococcal biofilm infections. Ann. N.Y. Acad. Sci. *1241*, 104–121.

Kim, H.K., Missiakas, D., and Schneewind, O. (2014). Mouse models for infectious diseases caused by *Staphylococcus aureus*. J. Immunol. Methods*410*, 88–99.

Kirby, W.M. (1944). Extraction of a highly potent penicillin inactivator from penicillin resistant Staphylococci. Science*99*, 452–453.

Klevens, R.M., Morrison, M.A., Nadle, J., Petit, S., Gershman, K., Ray, S., Harrison, L.H., Lynfield, R., Dumyati, G., Townes, J.M., *et al.* (2007). Invasive methicillin-resistant *Staphylococcus aureus* infections in the United States. JAMA *298*, 1763–1771.

Kloos, W.E., and Schleifer, K.H. (1975). Simplified scheme for routine identification of human *Staphylococcus* species. J. Clin. Microbiol. *1*, 82–88.

Knight, R.G., and Shlaes, D.M. (1983). Rapid identification of *Staphylococcus aureus* and *Streptococcus pneumoniae* from blood cultures. J. Clin. Microbiol. *17*, 97–99.

Koenig, M.G., Melly, M.A., and Rogers, D.E. (1962). Factors relating to the virulence of Staphylococci. III. Antibacterial versus antioxic immunity. J. Exp. Med. *116*, 601–610.

Koo, S.P., Bayer, A.S., Sahl, H.G., Proctor, R.A., and Yeaman, M.R. (1996). Staphylocidal action of thrombin-induced platelet microbicidal protein is not solely dependent on transmembrane potential. Infect. Immun. *64*, 1070–1074.

Kravitz, G.R., Dries, D.J., Peterson, M.L., and Schlievert, P.M. (2005). Purpura fulminans due to *Staphylococcus aureus*. Clin. Infect. Dis. *40*, 941–947.

Kuroda, M., Ohta, T., Uchiyama, I., Baba, T., Yuzawa, H., Kobayashi, I., Cui, L., Oguchi, A., Aoki, K., Nagai, Y., *et al.* (2001). Whole genome sequencing of meticillin-resistant *Staphylococcus aureus*. Lancet*357*, 1225–1240.

Kuusela, P. (1978). Fibronectin binds to *Staphylococcus aureus*. Nature. *276(5689)*, 718–720.

Kuypers, J.M., and Proctor, R.A. (1989). Reduced adherence to traumatized rat heart valves by a low-fibronectin-binding mutant of *Staphylococcus aureus*. Infect. Immun. *57*, 2306–2312.

Lachica, R.V. (1976). Simplified thermonuclease test for rapid identification of *Staphylococcus aureus* recovered on agar media. Appl. Environ. Microbiol. *32*, 633–634.

Latham, R.H., Running, K., and Stamm, W.E. (1983). Urinary tract infections in young adult women caused by *Staphylococcus saprophyticus*. JAMA *250*, 3063–3066.

Lauderdale, K.J., Boles, B.R., Cheung, A.L., and Horswill, A.R. (2009). Interconnections between Sigma B, *agr*, and proteolytic activity in *Staphylococcus aureus* biofilm maturation. Infect. Immun. *77*, 1623–1635.

Laverty, G., Gorman, S.P., and Gilmore, B.F. (2013). Biomolecular mechanisms of staphylococcal biofilm formation. Future Microbiol. *8*, 509–524.

Lechner, S., Lewis, K., and Bertram, R. (2012). *Staphylococcus aureus* persisters tolerant to bactericidal antibiotics. J. Mol. Microbiol. Biotechnol. *22*, 235–244.

Lee, C.Y., and Iandolo, J.J. (1985). Mechanism of bacteriophage conversion of lipase activity in *Staphylococcus aureus*. J. Bacteriol. *164*, 288–293.

Lina, G., Vandenesch, F., and Etienne, J. (2006). A brief history of *Staphylococcus aureus* Panton Valentine leucocidin. Antimicrobail Therapy and Vaacines: Mirobes, 1. www.antimicrobe.org/history/PVL-long.asp.

Löfdahl, S., Guss, B., Uhlén, M., Philipson, L., and Lindberg, M. (1983). Gene for staphylococcal protein A. Proc. Natl. Acad. Sci. U.S.A. *80*, 697–701.

Löffler, B., Hussain, M., Grundmeier, M., Brück, M., Holzinger, D., Varga, G., Roth, J., Kahl, B.C., Proctor, R.A., and Peters, G. (2010). *Staphylococcus aureus* Panton–Valentine leucocidin is a very potent cytotoxic factor for human neutrophils. PLOS Pathog. *6*, e1000715.

Löffler, B., Niemann, S., Ehrhardt, C., Horn, D., Lanckohr, C., Lina, G., Ludwig, S., and Peters, G. (2013). Pathogenesis of *Staphylococcus aureus* necrotizing pneumonia: the role of PVL and an influenza coinfection. Expert. Rev. Anti. Infect. Ther. *11*, 1041–1051.

Löffler, B., Niemann, S., and Peters, G. (2014). *Staphylococcus aureus* persistence in non-professional phagocytes. Int. J. Med. Microbiol. *304*, 170–176.

Mack, D., Fischer, W., Krokotsch, A., Leopold, K., Hartmann, R., Egge, H., and Laufs, R. (1996). The intercellular adhesin involved in biofilm accumulation of *Staphylococcus epidermidis* is a linear beta-1, 6-linked glucosaminoglycan: purification and structural analysis. J. Bacteriol. *178*, 175–183.

Macneal, W.J., Spence, M.J., andSlavkin, A.E. (1943). Early lesions of experimental endocarditis lenta. Am. J. Pathol. *19*, 735–749.

Madison, B.M., andBaselski, V.S. (1983). Rapid identification of *Staphylococcus aureus* in blood cultures by thermonuclease testing. J. Clin. Microbiol. *18*, 722–724.

Majerczyk, C.D., Sadykov, M.R., Luong, T.T., Lee, C., Somerville, G.A., and Sonenshein, A.L. (2008). *Staphylococcus aureus* CodY negatively regulates virulence gene expression. J. Bacteriol. *190*, 2257–2265.

Mann, P.A., Müller, A., Xiao, L., Pereira, P.M., Yang, C., Ho, Lee, S., Wang, H., Trzeciak, J., Schneeweis, J., Dos Santos, M.M., Murgolo, N., She, X., Gill, C., Balibar, C.J., Labroli, M., Su, J., Flattery. A., Sherborne. B., et al., (2013). Murgocil is a highly bioactive staphylococcal-specific inhibitor of the peptidoglycan glycosyltransferase enzyme MurG. ACS Chem Biol. *8*, 2442–2451.

Maskell, R. (1974). Importance of coagulase-negative staphylococci as pathogens in the urinary tract. Lancet *1*, 1155–1158.

McCallum, N., Berger-Bächi, B., and Senn, M.M. (2010). Regulation of antibiotic resistance in *Staphylococcus aureus*. Int. J. Med. Microbiol. *300*, 118–129.

McDougal, L.K., Steward, C.D., Killgore, G.E., Chaitram, J.M., McAllister, S.K., and Tenover, F.C. (2003). Pulsed-field gel electrophoresis typing of oxacillin-resistant *Staphylococcus aureus* isolates from the United States: establishing a national database. J. Clin. Microbiol. *41*, 5113–5120.

McNamara, P.J., Milligan-Monroe K.C., Khalili, S., and Proctor, R.A. (2000). Identification, cloning, and initial characterization of *rot*, a locus encoding a regulator of virulence factor expression in *Staphylococcus aureus*. J. Bacteriol. *182*, 3197–3203.

McTaggart, L.A., and Elliott, T.S. (1989). Is resistance to novobiocin a reliable test for confirmation of the identification of *Staphylococcus saprophyticus*? J. Med. Microbiol. *30*, 253–266.

Mediavilla, J.R., Chen, L., Mathema, B., and Kreiswirth, B.N. (2012). Global epidemiology of community- associated methicillin resistant *Staphylococcus aureus* (CA-MRSA). Curr. Opin. Microbiol. *15*, 588–595.

Miles, A.A., Williams, R.E.O., and Clayton-Cooper, B. (1944). The carriage of *Staphylococcus (pyogenes) aureus* in man and its relationship to wound infection. J. Pathol. Bacteriol. *56*, 513–524.

Miller, L.G., Perdreau-Remington, F., Rieg, G., Mehdi, S., Perlroth, J., Bayer, A.S., Tang, A.W., Phung, T.O., and Spellberg, B. (2005). Necrotizing fasciitis caused by community-associated methicillin-resistant *Staphylococcus aureus* in Los Angeles. N. Engl. J. Med. *352*, 1445–1453.

Miller, L.G., and Diep, B.A. (2008). Clinical practice: colonization, fomites, and virulence: rethinking the pathogenesis of community-associated methicillin-resistant *Staphylococcus aureus* infection. Clin. Infect. Dis. *46*, 752–760.

Mitchell, B.G., Collignon, P.J., McCann, R., Wilkinson, I.J., and Wells, A. (2014). A major reduction in hospital-onset *Staphylococcus aureus* bacteremia in Australia-12 years of progress: an observational study. Clin. Infect. Dis. *59*, 969–975.

Mitchell, R.G. (1968). Classification of *Staphylococcus albus* strains isolated from the urinary tract. J. Clin. Pathol. *21*, 93–96.

Montgomery, C.P., Boyle-Vavra, S., and Daum, R.S. (2009). The arginine catabolic mobile element is not associated with enhanced virulence in experimental invasive disease caused by the community-associated methicillin-resistant *Staphylococcus aureus* USA300 genetic background. Infect. Immun. *77*, 2650–2656.

Morfeldt, E., Janzon, L., Arvidson, S., and Löfdahl, S. (1988). Cloning of a chromosomal locus (*exp*) which regulates the expression of several exoprotein genes in *Staphylococcus aureus*. Mol. Gen. Genet. *211*, 435–440.

Morfeldt, E., Taylor, D., von Gabain, A., and Arvidson, S. (1995). Activation of alpha-toxin translation in *Staphylococcus aureus* by the trans-encoded antisense RNA, RNAIII. EMBO J. *14*, 4569–4577.

Mosher, D.F., andGrossman, J.E. (1983). Clinical use of fibronectin. (Potential applications). Ric. Clin. Lab. *13*, 43–54.

Mosher, D.F., and Proctor, R.A. (1980). Binding and factor XIIIa-mediated cross-linking of a 27-kilodalton fragment of fibronectin to *Staphylococcus aureus*. Science *209*, 927–929.

Nedzel, A.J. (1937). Experimental endocarditis. Arch. Path. *24*, 143–200.

Nicolsen, N.C., LeCroy N., Alby, K., Martin, K.E., Laux, J., Lin, F.C., Daniels, L., Weber, D.J., and Miller, M.B. (2013). Clinical outcomes with rapid detection of methicillin-resistant and methicillin-susceptible *Staphylococcus aureus* isolates from routine blood cultures. J. Clin. Microbiol. *51*, 4126–4129.

Novick, R.P., Ross, H.F., Projan, S.J., Kornblum, J., Kreiswirth, B., and Moghazeh, S. (1993). Synthesis of staphylococcal virulence factors is controlled by a regulatory RNA molecule. EMBO J. *12*, 3967–3975.

Novick, R.P. (2003). Autoinduction and signal transduction in the regulation of staphylococcal virulence. Mol. Microbiol. *48*, 1429–1449.

Ogawa, S.K., Yurberg, E.R., Hatcher, V.B., Levitt, M.A., and Lowy, F.D. (1985). Bacterial adherence to human endothelial cells *in vitro*. Infect. Immun. *50*, 218–224.

Ogston, A. (1880). Über Abscesse. Archiv fur Klinische Chirurgie 25, 588–600.

Ogston, A. (1881). Report upon micro-organisms in surgical diseases. Br. Med. J. *1*, 3 69.b2–375.

Ogston, A. (1882). Micrococcus poisoning. J. Anat. Physiol. *171054*, 24–58.

O'Kane, G.M., Gottlieb, T., and Bradbury, R. (1998). Staphylococcal bacteraemia: the hospital or the home? A review of *Staphylococcus aureus* bacteraemia at Concord Hospital in 1993. Aust. N. Z. J. Med. *28*, 23–27.

Otto, M. (2013). Staphylococcal infections: mechanisms of biofilm maturation and detachment as critical determinants of pathogenicity. Annu. Rev. Med. *64*, 175–188.

Pang, Y.Y., Schwartz, J., Thoendel, M., Ackermann, L.W., Horswill, A.R., and Nauseef, W.M. (2010). *agr*-Pasteur, L. (1880). *De l'extension de la théorie des germes à l'étiologie de queques maladies communes.* C.R. Séances Soc. Biol. Paris. *90*, 1035–1044.

Parker, M.T. (1962). Phage typing and the epidemiology of *Staphylococcus aureus* infection. Journal of Applied Bacteriology. *25*, 389–402.

Parnowska, W. (2006). In memoriam Prof. dr. Janusz Jeljaszewicz. Polish J. of Microbiol. *55*, 167–168.

Pattee, P.A. (1981). Distribution of *Tn551* insertion sites responsible for auxotrophy on the *Staphylococcus aureus* chromosome. J. Bacteriol. *145*, 479–488.

Pattee, P.A. (1986). Chromosomal map location of the alpha-hemolysin structural gene in *Staphylococcus aureus* NCTC 8325. Infect. Immun. *54*, 593–596.

Peschel, A., and Otto, M. (2013). Phenol-soluble modulins and staphylococcal infection. Nat. Rev. Microbiol. *11*, 667–673.

Peters, G., Locci, E., and Pulverer, G. (1981). Microbial colonization of prosthetic devices. II. Scanning electron microscopy of naturally infected intravenous catheters. Zentralbl Bakteriol. Mikrobiol. Hyg. B. *173*, 293–299.

Pfeltz, R.F., Singh, V.K., Schmidt, J.L., Batten. M.A., Baranyk, C.S., Nadakavukaren, M.J., Jayaswal, R.K., and Wilkinson, B.J. (2000). Characterization of passage-selected vancomycin-resistant *Staphylococcus aureus* strains of diverse parental backgrounds. Antimicrob. Agents Chemother. *44*, 294–303.

Pichereau, S., Moran, J.J., Hayney, M.S., Shukla, S.K., Sakoulas, G., and Rose, W.E. (2012). Concentration-dependent effects of antimicrobials on *Staphylococcus aureus* toxin-mediated cytokine production from peripheral blood mononuclear cells. J. Antimicrob. Chemother. *67*, 123–129.

Pickenhahn, P., Hahn, I.F., Lenz, W., and Brandis, H. (1987). Correlation between enterotoxigenicity, tested by different ELISA- techniques, antibiotic resistance patterns and phage groups of *Staphylococcus aureus* strains. Zentralbl Bakteriol. Mikrobiol. Hyg A. *266*, 127–136.

Planet, P.J., LaRussa, S.J., Dana, A., Smith, H., Xu, A., Ryan, C., Uhlemann, A.C., Boundy, S., Goldberg, J., Narechania, A., Kulkarni, R., Ratner, A.J., Geoghegan, J.A., Kolokotronis, S.O., and Prince, A. (2013). Emergence of the epidemic methicillin-resistant *Staphylococcus aureus* strain USA300 coincides with horizontal transfer of the arginine catabolic mobile element and *speG*-mediated adaptations for survival on skin. MBio. *4*, e00889–13.

Poole-Warren, L.A., Hallett, M.D., Hone, P.W., Burden, S.H., and Farrell, P.C. (1991). Vaccination for prevention of CAPD associated staphylococcal infection: results of a prospective multicentre clinical trial. Clin. Nephrol. *35*, 198–206.

Powers, M.E., Kim, H.K., Wang, Y., and Bubeck Wardenburg, J. (2012). ADAM10 mediates vascular injury induced by *Staphylococcus aureusa*-hemolysin. J. Infect. Dis. *206*, 352–356.

Proctor, R.A. (1987). The staphylococcal fibronectin receptor: evidence for its importance in invasive infections. Rev. Infect. Dis. *9* (Suppl. 4), S335–S340.

Proctor, R.A. (2012). Is there a future for a *Staphylococcus aureus* vaccine? Vaccine *30*, 2921–2927.

Proctor, R.A., Christman, G., and Mosher, D.F. (1984). Fibronectin-induced agglutination of *Staphylococcus aureus* correlates with invasiveness. J. Lab. Clin. Med. *104*, 455–469.

Proctor, R.A., Mosher, D.F., andOlbrantz, P.J. (1982). Fibronectin binding to *Staphylococcus aureus*. J. Biol. Chem. *257*, 14788–14794.

Proctor, R.A., Prendergast, E., and Mosher, D.F. (1982). Fibronectin mediates attachment of *Staphylococcus aureus* to human neutrophils. Blood*59*, 681–687.

Proctor, R.A., and Smith, D. (2014). Role ofrapid diagnostics in selecting optimal therapy for *Staphylococcus aureus* infections. In Future Medicine, Chpt 6. pp. 95–112.

Proctor, R.A., Textor, J.A., Vann, J.M., and Mosher, D.F. (1985). Role of fibronectin in human monocyte and macrophage bactericidal activity. Infect. Immun. *47*, 629–637.

Proctor, R.A., van Langevelde, P., Kristjansson, M., Maslow, J.N., and Arbeit, R.D. (1995). Persistent and relapsing infections associated with small-colony variants of *Staphylococcus aureus*. Clin. Infect. Dis. *20*, 95– 102.

Proctor, R.A., von Eiff, C., Kahl, B.C., Becker, K., McNamara, P., Herrmann, M., and Peters, G. (2006). Small colony variants: a pathogenic form of bacteria that facilitates persistent and recurrent infections. Nat. Rev. Microbiol. *4*, 295–305.

Projan, S.J., Kornblum, J., Moghazeh, S.L., Edelman, I., Gennaro, M.L., and Novick, R.P. (1985). Comparative sequence and functional analysis of pT181 and pC221, cognate plasmid replicons from *Staphylococcus aureus*. Mol. Gen. Genet. *199*, 452–464.

Projan, S.J., Nesin, M., andDunman, P.M. (2006). Staphylococcal vaccines and immunotherapy: to dream the impossible dream? Curr. Opin. Pharmacol. *6*, 473–479.

Que, Y.A., François, P., Haefliger, J.A., Entenza, J.M., Vaudaux, P., and Moreillon, P. (2001). Reassessing the role of *Staphylococcus aureus* clumping factor and fibronectin-binding protein by expression in *Lactococcus lactis*. Infect. Immun. *69*, 6296–6302.

Ragle, B.E., and Bubeck Wardenburg, J. (2009). Anti-alpha-hemolysin monoclonal antibodies mediate protection against *Staphylococcus aureus* pneumonia. Infect. Immun. *77*, 2712–2718.

Rahman, A., Izaki, K., Kato, I., and Kamio, Y. (1991). Nucleotide sequence of leucocidin S-component gene (*lukS*) from methicillin resistant *Staphylococcus aureus*. Biochem. Biophys. Res. Commun. *181*, 138–144.

Rahman, A., Nariya, H., Izaki, K., Kato, I., and Kamio, Y. (1992). Molecular cloning and nucleotide sequence of leucocidin F-component gene (*lukF*) from methicillin resistant *Staphylococcus aureus*. Biochem. Biophys. Res. Commun. *184*, 640–646.

Rasheed, J.K., Arko, R.J., Feeley, J.C., Chandler, F.W., Thornsberry, C., Gibson, R.J., Cohen, M.L., Jeffries, C.D., and Broome, C.V. (1985). Acquired ability of *Staphylococcus aureus* to produce toxic shock-associated protein and resulting illness in a rabbit model. Infect. Immun. *47*, 598–604.

Rasigade, J.P., Sicot, N., Laurent, F., Lina, G., Vandenesch, F., and Etienne, J. (2011). A history of Panton- Valentine leucocidin (PVL)-associated infection protects against death in PVL-associated pneumonia. Vaccine. *29*, 4185–4186.

Ratner, H.B., and Stratton, C.W. (1985). Thermonuclease test for same-day identification of *Staphylococcus aureus* in blood cultures. J. Clin. Microbiol. *21*, 995–996.

Raychaudhuri, P., Li, Q., Mason, A., Mikhailova, E., Heron, A.J., and Bayley, H. (2011). Fluorinated amphiphiles control the insertion of α-hemolysin pores into lipid bilayers. Biochemistry *50*, 1599–1606.

Raz, R., Colodner, R., and Kunin, C.M. (2005). Who are you–*Staphylococcus saprophyticus*? Clin. Infect. Dis. *40*, 896–898.

Recsei, P., Kreiswirth, B., O'Reilly, M., Schlievert, P., Gruss, A., and Novick, R.P. (1986). Regulation of exoprotein gene expression in *Staphylococcus aureus* by *agr*. Mol. Gen. Genet. *202*, 58–61.

Regassa, L.B., and Betley, M.J. (1992). Alkaline pH decreases expression of the accessory gene regulator (*agr*) in *Staphylococcus aureus*. J. Bacteriol. *174*, 5095–5100.

Regassa, L.B., Novick, R.P., andBetley, M.J. (1992). Glucose and nonmaintained pH decrease expression of the accessory gene regulator (*agr*) in *Staphylococcus aureus*. Infect. Immun. *60*, 3381–3388.

Roemer, T., andBoone, C. (2013). Systems-level antimicrobial drug and drug synergy discovery. Nat. Chem. Biol. *9*, 222–231.

Roemer, T., Schneider, T., and Pinho, M.G. (2013). Auxiliary factors: a chink in the armor of MRSA resistance to β-lactam antibiotics. Curr. Opin. Microbiol. *16*, 538–548.

Rogers, D.E., andTompsett, R. (1952). The survival of staphylococci within human leucocytes. J. Exp. Med. *95*, 209–230.

Romilly, C., Chevalier, C., Marzi, S., Masquida, B., Geissmann, T., Vandenesch, F., Westhof, E., and Romby, P. (2012). Loop-loop interactions involved in antisense regulation are processed by the endoribonuclease III in *Staphylococcus aureus*. RNA Biol. *9*, 1461–1472.

Rose, W.E., Eickhoff, J.C., Shukla, S.K., Pantrangi, M., Rooijakkers, S., Cosgrove, S.E., Nizet, V., and Sakoulas, G. (2012). Elevated serum interleukin-10 at time of hospital admission is predictive of mortality in patients with *Staphylococcus aureus* bacteremia. J. Infect. Dis. *206*, 1604–1611.

Rose, W.E., Berti, A.D., Hatch, J.B., and Maki, D.G. (2013). Relationship of *in vitro* synergy and treatment outcome with daptomycin plus rifampin in patients with invasive methicillin-resistant *Staphylococcus aureus* infections. Antimicrob Agents Chemother. *57*, 3450–3452.

Rothfork, J.M., Timmins, G.S., Harris, M.N., Chen, X., Lusis, A.J., Otto, M., Cheung, A.L., and Gresham, H.D. (2004). Inactivation of a bacterial virulence pheromone by phagocyte-derived oxidants: new role for the NADPH oxidase in host defense. Proc. Natl. Acad. Sci. U.S.A. *101*, 13867–13872.

Roux, A., Todd, D.A., Velázquez, J.V., Cech, N.B., and Sonenshein, A.L. (2014). CodY-mediated regulation of the *Staphylococcus aureus* Agr system integrates nutritional and population density signals. J. Bacteriol. *196*, 1184–1196.

Rubin, R.J., Harrington, C.A., Poon, A., Dietrich, K., Greene, J.A., and Moiduddin, A. (1999). The economic impact of *Staphylococcus aureus* infection in New York City hospitals. Emerg. Infect. Dis. *5*, 9–17.

Rybak, J.M., Barber, K.E., and Rybak, M.J. (2013). Current and prospective treatments for multidrug- resistant gram-positive infections. Expert Opin. Pharmacother. *14*, 1919–1932.

Salgado-Pabón, W., and Schlievert, P.M. (2014). Models matter: the search for an effective *Staphylococcus aureus* vaccine. Nat. Rev. Microbiol. *12*, 585–591.

Samuelsen, O., Haukland, H.H., Kahl, B.C., von Eiff, C., Proctor, R.A., Ulvatne, H., Sandvik, K., and Vorland, L.H. (2005). *Staphylococcus aureus* small colony variants are resistant to the antimicrobial peptide lactoferricin B. J. Antimicrob. Chemother. *56*, 1126–1129.

Schaffer, A.C., and Lee, J.C.(2009). Staphylococcal vaccines and immunotherapies. Infect. Dis. Clin. North. Am. *23*, 153–171.

Scherr, T.D., Heim, C.E., Morrison, J.M., and Kielian, T. (2014). Hiding in plain sight: interplay between staphylococcal biofilms and host immunity. Front. Immunol. *5*, 37.

Sendi, P., and Proctor, R.A. (2009). *Staphylococcus aureus* as an intracellular pathogen: the role of small colony variants. Trends Microbiol. *17*, 54–58.

Sjödahl, J. (1977). Repetitive sequences in protein A from *Staphylococcus aureus*. Arrangement of five regions within the protein, four being highly homologous and Fc-binding. Eur. J. Biochem. *73*, 343–351.

Sjöquist, J., Forsgren, A., Gustafson, G.T., and Stålenheim, G. (1967). Biological Importance of the Fc-Region of Gamma Globulins Cold Spring Harb. Symp. Quant. Biol. *32*, 577–581.

Sjöquist, J., and Stålenheim, G. (1969). Protein A from *Staphylococcus aureus*. IX. Complement-fixing activity of protein A-IgG complexes. J. Immunol. *103*, 467–473.

Smith, C.D., and Pattee, P.A. (1967). Biochemical and genetic analysis of isoleucine and valine biosynthesis in *Staphylococcus aureus*. J. Bacteriol. *93*, 1832–1838.

Smith, I.M., and Vickers, A.B. (1960). Natural history of 338 treated and untreated patients with staphylococcal septicaemia (1936–1955). Lancet *1*, 1318–1122.

Smith, T.L., Pearson, M.L., Wilcox, K.R., Cruz, C., Lancaster, M.V., Robinson-Dunn, B., Tenover, F.C., Zervos,

M.J., Band, J.D., White, E., and Jarvis, W.R. (1999). Emergence of vancomycin resistance in *Staphylococcus aureus*. Glycopeptide-Intermediate *Staphylococcus aureus* Working Group. N. Engl. J. Med. *340*, 493–450.

Somerville, G.A., and Proctor, R.A. (2009). At the crossroads of bacterial metabolism and virulence factor synthesis in staphylococci. Microbiol. Mol. Biol. Rev. *73*, 233–248.

Spaan, A.N., Surewaard, B.G., Nijland, R., and van Strijp, J.A. (2013). Neutrophils versus *Staphylococcus aureus*: a biological tug of war. Annu Rev Microbiol. *67*, 629–650.

Speziale, P., Höök, M., Wadström, T., and Timpl, R. (1982). Binding of the basement membrane protein laminin to *Escherichia coli*. FEBS Lett. *146*, 55–58.

Spika, J.S., Verbrugh, H.A., and Verhoef, J. (1981). Protein A effect on alternative pathway complement activation and opsonization of *Staphylococcus aureus*. Infect. Immun. *34*, 455–460.

Stevens, R.A. (1927). The occurrence of *Staphylococcus aureus* infections with a scarlatiniform rash. JAMA *88*, 1957–1958.

Stewart, G.T., and Holt, R.J. (1963). Evolution of natural resistance to the newer penicillins. Br. Med. J. *1*, 308–311.

Tan, C.M., Therien, A.G., Lu, J., Lee, S.H., Caron, A., Gill, C.J., Lebeau-Jacob, C., Benton-Perdomo, L., Monteiro, J.M., et al. (2012). Restoring methicillin-resistant *Staphylococcus aureus* susceptibility to β- lactam antibiotics. Sci. Transl. Med. 4: 126ra35.

Taubler, J.H., Kapral, F.A., and Mudd, S. (1963). Role of alpha-toxin in lesion formation by *Staphylococcus aureus* on sutures subcutaneously implanted in mice. J. Bacteriol. *86*, 51–57.

Tenover, F.C., Weigel, L.M., Appelbaum, P.C., McDougal, L.K., Chaitram, J., McAllister, S., Clark, N., Killgore, G., O'Hara, C.M., Jevitt, L., Patel, J.B., and Bozdogan, B. (2004). Vancomycin-resistant *Staphylococcus aureus* isolate from a patient in Pennsylvania. Antimicrob Agents Chemother. *48*, 275–280.

Thurlow, L.R., Joshi, G.S., Clark, J.R., Spontak, J.S., Neely, C.J., Maile, R., and Richardson, A.R. (2013). Functional modularity of the arginine catabolic mobile element contributes to the success of USA300 methicillin-esistant *Staphylococcus aureus*. Cell Host Microbe. *13*, 100–107.

Todd, J., Fishaut, M., Kapral, F., and Welch, T. (1978). Toxic-shock syndrome associated with phage-group-I staphylococci. Lancet. 2, 1116–1118.

Tom, S., Galbraith, J.C., Valiquette, L., Jacobsso, n G., Collignon, P., Schønheyder, H.C., Søgaard, M., Kennedy, K.J., Knudsen, J.D., Ostergaard, C., Lyytikäinen, O., and Laupland, K.B.; The International Bacteraemia Surveillance Collaborative. (2014). Case fatality ratio and mortality rate trends of community-onset *Staphylococcus aureus* bacteraemia. Clin. Microbiol. Infect. *20*, 0630–0632.

Torres Pereira, A. (1962). Coagulase-negative strains of staphylococcus possessing antigen 51 as agents of urinary infection. J. Clin. Pathol. *15*, 252–253.

Tuchscherr, L., Medina, E., Hussain, M., Völker, W., Heitmann, V., Niemann, S., Holzinger, D., Roth, J., Proctor, R.A., Becker, K., Peters, G., and Löffler, B. (2011). *Staphylococcus aureus* phenotype switching: an effective bacterial strategy to escape host immune response and establish a chronic infection. EMBO Mol. Med. *3*, 129–141.

Uhlén, M., Guss, B., Nilsson, B., Gatenbeck, S., Philipson, L., and Lindberg, M. (1984). Complete sequence of the staphylococcal gene encoding protein A. A gene evolved through multiple duplications. J. Biol. Chem. *259*, 1695–1702.

Van de Velde, H. (1894). Etude sur le mécanisme de la virulence du staphylocoque pyogène. La Cellule *10*, 403–460.

Vandenesch, F., Kornblum, J., and Novick, R.P. (1991). A temporal signal, independent of *agr*, is required for *hla* but not *spa* transcription in *Staphylococcus aureus*. J. Bacteriol. *173*, 6313–6320.

van Dissel, J.T., van Langevelde, P., Westendorp, R.G., Kwappenberg, K., and Frölich, M. (1998). Anti-inflammatory cytokine profile and mortality in febrile patients. Lancet *351*, 950–953.

Vann, J.M., and Proctor, R.A. (1988). Cytotoxic effects of ingested *Staphylococcus aureus* on bovine endothelial cells: Role of *S. aureus* alpha-hemolysin. Microb. Pathog. *4*, 443–453.

Verhoef, J., Peterson, P., Kim, Y., Sabath, L.D., and Quie, P.G. (1977). Opsonic requirements for staphylococcal phagocytosis. Heterogeneity among strains. Immunology *33*, 191–197.

Vesga, O., Groesche, l M.C., Otten, M.F., Brar, D.W., Vann, J.M., and Proctor, R.A. (1996). *Staphylococcus aureus* small colony variants are induced by the endothelial cell intracellular milieu. J. Infect. Dis. *173*, 739–742.

von Eiff, C., Becker, K., Machka, K., Stammer, H., and Peters, G. (2001). Nasal carriage as a source of *Staphylococcus aureus* bacteremia. Study Group. N. Engl. J. Med. *344*, 11–16.

Werth, N., Beerlage, C., Rosenberger, C., Yazdi, A.S., Edelmann, M., Amr, A., Bernhardt, W., von Eiff, C., Becker, K., Schäfer, A., Pesche, A., and Kempf, V.A. (2010). Activation of hypoxia inducible factor 1 is a general phenomenon in infections with human pathogens. PLoS One *5*, e11576.

Whitener, C.J., Park, S.Y., Browne, F.A., Parent, L.J., Julian, K., Bozdogan, B., Appelbaum, P.C., Chaitram, J., Weigel, L.M., Jernigan, J., McDougal, L.K., Tenover, F.C., and Fridkin, S.K. (2004). Vancomycin-resistant *Staphylococcus aureus* in the absence of vancomycin exposure. Clin. Infect. Dis. *38*, 1049–1055.

Wilmes, M., and Sahl, H.G. (2014). Defensin-based anti-infective strategies. Int. J. Med. Microbiol. *304*, 93–99.

Wood, T.K., Knabel, S.J., and Kwan, B.W. (2013). Bacterial persister cell formation and dormancy. Appl. Environ. Microbiol. *79*, 7116–7121.

Woodin, A.M. (1959). Fractionation of a leucocidin from *Staphylococcus aureus*. Biochem J. *73*, 225–237.

Wright, J. (1936). Staphylococcal leucocidin (Neisser-Wechsberg type) and antileucociddin. Lancet *230*, 1002–1004.

Wyssokowitsch, W. (1886). Beiträge zur Lehre von der Endocarditis. Virchows Arch. Pathol. Anat. Physiol. *103*, 301–332.

Yoshida, A., Mudd, S., and Lenhart, N.A. (1963). The common protein agglutinogen of *Staphylococcus aureus*. 2. Purification, chemical characterization, and serologic comparison with Jensen's antigen. J. Immunol. *91*, 777–782.

Yoshikawa, M., Matsuda, F., Naka, M., Murofushi, E., and Tsunematsu, Y. (1974). Pleiotropic alteration of activities of several toxins and enzymes in mutants of *Staphylococcus aureus*. J. Bacteriol. *119*, 117–122.

Zhu, Y., Nandakumar, R., Sadykov, M.R., Madayiputhiya, N., Luong, T.T., Gaupp, R., Lee, C.Y., and Somerville, G.A. (2011). RpiR homologues may link *Staphylococcus aureus* RNAIII synthesis and pentose phosphate pathway regulation. J. Bacteriol. *193*, 6187–6196.

Zuo, Q.F., Cai, C.Z., Ding, H.L., Wu, Y., Yang, L.Y., Feng, Q., Yang, H.J., Wei, Z.B., Zeng, H., and Zou, Q.M. (2014). Identification of the immunodominant regions of *Staphylococcus aureus* fibronectin-binding protein A. PLoS One 9, e95338.

Clinical Significance in Humans

2

Mathias Herrmann and Mark S. Smeltzer

Abstract

This chapter is intended to provide a broad overview of the staphylococci as a cause of human infections. Several key points must be made in this regard beginning with the observation that all staphylococcal species are generally adapted to live on humans and/or other animal species as commensals. However, all also exhibit some capacity to cause infection, and therefore can all be considered opportunistic pathogens. Nevertheless, the single most prominent pathogen in the context of human infections is *Staphylococcus aureus*, which has the capacity to cause a remarkably diverse array of infections ranging from superficial skin and soft tissue abscesses to life-threatening infections including bacteraemia, endocarditis, and osteomyelitis. The prominence of *S. aureus* as a human pathogen is clearly linked to two factors, the first being that it is a commensal inhabitant of 30–50% of the population and is therefore 'available' to cause disease when given the 'opportunity', and the second being its ability to produce a diverse array of virulence factors when called upon under selective pressure from the host to do so. The staphylococci are also highly adaptable at a genetic and phenotypic level, a primary manifestation of the first being the recent emergence of methicillin-resistant strains even among isolates associated with community-acquired infections, and a primary manifestation of the second being that these isolates exhibit phenotype characteristics that threaten the definition of *S. aureus* as an opportunistic pathogen. This brings up the additional key point that the epidemiology and pathogenesis of *S. aureus* infections has changed significantly in recent years. This is reflected in reports that now distinguish between 'classic' healthcare-associated strains and those associated with community-acquired infections, with resistance to methicillin and its derivatives (e.g. oxacillin) being a common and troubling characteristic of both groups. However, the goal here is not to detail this diversity but rather to provide a broad overview of the staphylococci, with a specific emphasis on *S. aureus*, as both a commensal inhabitant of humans and as a human pathogen.

Introduction

Staphylococci are natural inhabitants of many species of animals, yet it appears they are best adapted to survive and proliferate on the skin, skin glands, and mucosae of mammals and birds (see also Chapter 3). The medical focus on their potential for pathogenicity may suggest an evolutionary adaptation towards disease development in humans, but it is clear that members of the genus *Staphylococcus* evolved from free-living organisms long before the arrival of the human species and then adapted to be commensals of their mammalian hosts including humans.

It is this commensal relationship with their hosts that characterizes the overall benign lifestyle of the staphylococci, and this is evidenced by their frequent isolation from healthy human hosts on all environmentally exposed, non-sterile organs and mucosal surfaces, most notably the skin and anterior nares. Insight into the cohabitation of colonizing staphylococci and their mammalian hosts has for a long time been restricted to the identification of the various species on such surfaces, but in the era of studies focusing on the human microbiome we have begun to understand the relative abundance of the staphylococci by comparison to other microorganisms (Sanford and Gallo, 2013). It has also become clear that colonizing staphylococci play a major role in modulation of the immune response, thus further emphasizing their importance even beyond their capacity to cause infection (Brown *et al.*, 2014).

Nevertheless, the overwhelming clinical significance of the staphylococci remains their ability to cause disease. Localized (e.g. folliculitis) and spreading pyodermas (e.g. cellulitis), generalized bacteraemia, and various forms of metastatic infection (e.g. endocarditis, septic arthritis, osteomyelitis) continue to be the most prevalent forms of disease. *S. aureus* is also a common cause of well-described toxin-mediated forms of disease, some of which (e.g. food poisoning) are uncomfortable but benign while others are mediated by the superantigen properties of specific toxins (e.g. toxin shock syndrome toxin), and can be lethal. It also appears that the staphylococci may cause some forms of immune-mediated sequelae in clinical entities such as atopic dermatitis (Leung, 2013), Wegener's granulomatosis (Tadema *et al.*, 2011), and potentially Kawasaki disease (Matsubara and Fukaya, 2007). The course and consequences of all these disease mechanisms varies widely, but in absolute numbers the burden for morbidity and mortality is attributable primarily to these classic forms of infection.

During the 1980s and 1990s, the number of community and health care associated staphylococcal infections caused by *S. aureus* and coagulase-negative species (e.g. *S. epidermidis*) increased. One contributing factor to this, particularly in the context of *S. epidermidis*, is the increased use of indwelling medical devices including intravascular catheters (Banerjee *et al.*, 1991; Steinberg *et al.*, 1996); indeed, infection due to coagulase-negative staphylococci was largely unknown prior to the widespread use of such devices (Maki, 1989). The proportion of methicillin-resistant *S. aureus* strains as causative agents of all forms of infection has also increased (Panlilio *et al.*, 1992; Speller *et al.*, 1997). This is important in that, by comparison to infections caused by methicillin-sensitive strains of *S. aureus* (MSSA), infections caused by methicillin-resistant strains (MRSA) are associated with significantly higher morbidity and mortality in essentially every clinical context (Cosgrove *et al.*, 2003, 2005; Engemann *et al.*, 2003). Nevertheless, population-based data on the epidemiology of invasive *S. aureus* infection is limited. Studies from the early 2000s describe annual incidences of approximately 30/100,000 (Jacobsson *et al.*, 2007; Laupland *et al.*, 2003), with mortality rates for invasive infections ranging from 20% (Laupland *et al.*, 2003) up to 40% in severe sepsis cases (Mylotte *et al.*, 1987). With the emergence of community associated methicillin-resistant *S. aureus* (MRSA) epidemics (Chapman *et al.*, 2005; Miller *et al.*, 2005), epidemiologic studies and large meta analyses point towards higher death rates associated with invasive *S. aureus* disease (Hanberger *et al.*, 2011). In 2007, it was found that in certain US regions death rates due to MRSA were as high or surpassed death rates due to AIDS (Klevens *et al.*, 2007), a fact recognized by the public (NBC news, 16 October 2007) but largely ignored by federal funding agencies. Indeed, while recent population-based *S. aureus* bloodstream infection studies indicate that the overall incidence remains relatively stable (26/100,000) despite major secular changes and significant regional differences between MSSA and MRSA incidence (Laupland *et al.*, 2013), the associated mortality of 3.4/100,000 (Tom *et al.*, 2014) causes a death toll comparable with malignant diseases such as cancers of liver, skin, ovaries, brain or prostate.

The good news is that a recently published study found that the overall incidence of hospital-onset invasive MRSA infection declined ~31% in the USA between 2005 and 2011, the bad news being that this decrease was not reflected in the incidence of community-onset infections, particularly those classified as hospital-associated, community-onset (HACO) infections (Dantes *et al.*, 2013). This almost certainly reflects increased vigilance in the hospital setting, but nevertheless emphasizes the need to better understand how to extend this vigilance beyond the hospital. Interestingly, a recent report also demonstrated an increased incidence of invasive MRSA infections in humans associated with their proximity to high-density livestock operations (Casey *et al.*, 2013), thus further emphasizing the complex interactions that ultimately define *S. aureus* as a human pathogen.

In contrast to developed countries, almost no epidemiologic data on population-based incidence or mortality is available from low-resource countries. In fact, in many tropical countries such as sub-Saharan Africa with a high prevalence of febrile disease due to malaria and other infectious causes, staphylococcal disease is neither diagnosed nor appropriately treated, yet the anecdotal and single-centre studies available suggest that in these countries the burden of staphylococcal disease is equal if not greater than in developed countries (Breurec *et al.*, 2011; Mulholland and Adegbola, 2005; Schaumburg *et al.*, 2011; Chen and Huang, 2014; Schaumburg *et al.*, 2014). Doubtlessly, more insight into the true incidence and disease presentation in such areas is urgently needed (Herrmann *et al.*, 2013).

Thus, the staphylococci have a major impact on the daily lives of humans globally. In this chapter, we will briefly review insight into the commensal interplay between the staphylococci and their human hosts and describe major disease syndromes caused by bacteria of this genus. The chapter is written to provide physicians and scientists active in the field of staphylococcal research with information regarding hallmarks of diagnosis, clinical presentation, and principles of therapy for major staphylococcal syndromes. For issues on antimicrobial resistance, the reader is referred to Chapter 11.

Staphylococci: a short description of the species

In order to understand the clinical significance of staphylococci in humans, it is necessary to introduce the various species from a microbiological standpoint. According to modern phylogenetic classifications, the genus *Staphylococcus* is member of the family of *Staphylococcaceae*, order *Bacillales*, class *Bacilli*. They are non-motile, round bacterial cells, approximately 1.0 μm diameter, presenting in typical micrographs as grape-like clusters owing to their multiple dividing planes. They are Gram-positive, and unlike other Gram-positive cocci of medical significance (e.g. *Streptococcus* and *Enterococcus* spp.), most staphylococcal species exhibit a strong catalase reaction. They are facultatively anaerobic, and a hallmark is their ability to grow at temperatures between 18°C and 40°C and in the presence of elevated salt concentrations (10–18%), both of which almost certainly contribute to their prevalence on the skin and exposed mucosal surfaces.

Altogether, the genus consists of approximately 40 species and subspecies, with the important distinction of 'coagulase-positive' versus 'coagulase-negative' being widely used for diagnostic workup and as a popular term for clinical distinction. This is largely due to the fact that most *S. aureus* strains are coagulase positive, with most other species being coagulase negative. However, a positive coagulase test is often observed with other staphylococcal species including those of the *S. intermedius* group (SIG), although species within this group are most commonly isolated from species other than humans. Moreover, some of the coagulase-negative species (i.e. *S. lugdunensis*, *S. schleiferi* and others) contain cell wall adhesins recognizing fibrinogen and therefore can yield a positive reaction in a rapid slide test that has generally been interpreted as 'coagulase positive' but is now considered an obsolete designation for the results of this slide agglutination test.

Thus, for practical reasons in human medicine, coagulase-positive staphylococci are synonymous with *S. aureus subsp. aureus* (in the following referred to as *S. aureus*; another subspecies, *S. aureus* subsp. *anaerobius*, is clinically irrelevant in humans), while the group of coagulase-negative staphylococci (CoNS) encompasses all remaining staphylococcal species and subspecies. Within the group of coagulase-negative staphylococci, novobiocin has been employed to distinguish members of the '*S. epidermidis* group' (novobiocin sensitive) and the '*S. saprophyticus* group' (novobiocin resistant) but in times of modern species identification methods (see below), this distinction is nowadays of little clinical use. Among the *S. epidermidis* group, *S. epidermidis* and *S. haemolyticus* are most frequently recovered from human infections. *S. lugdunensis* is of specific importance due to its ability to cause systemic infections comparable to *S. aureus* (e.g. infective endocarditis), with most of the other coagulase-negative species exhibiting less pathogenic potential in humans. *S. saprophyticus* is an exception in this regard, although its disease spectrum is largely limited to urinary tract infections, particularly in young, sexually active women.

For species identification, the clinical microbiologist employs a battery of phenotypic tests including tests for fibrinogen binding proteins (clumping factor), protein A, capsular polysaccharides, deoxyribonuclease or the heat stable thermonuclease. Additional tests employing enzyme activities, substrate metabolism, or growth conditions have been the basis for the 'classical' differentiation scheme developed by Kloos and Schleifer (1975), now integrated in commercial panels available for species identification. More recently, matrix-assisted laser desorption/ionization-time of flight (MALDI-TOF) has been validated for species identification of staphylococci. Nucleic acid-based techniques are also available; typically, genes restricted to one species (e.g. *nuc*, *coa*, *eap*, *clfA*, or *femA/B* for *S. aureus*) are employed because the sequencing of ribosomal gene clusters may not provide sufficient discrimination between staphylococcal species. Overall, preliminary staphylococcal species identification can be readily obtained with rapid tests, but definitive species identification of isolates from invasive, complex staphylococcal infections requires an experienced microbiology laboratory able to identify atypical variants or rarely encountered species. For a more detailed reference concerning distinctions between staphylococcal species, in addition to the standard textbooks on systematic bacteriology, the reader is referred to the valuable and updated web resource 'List of Prokaryotic Names with Standing in Nomenclature' (http://www.bacterio.net/).

Staphylococci: natural commensals of humans

Traditional niches of colonization and new insights from the microbiome era

S. aureus colonizes the skin and mucosae of humans, but the pattern of colonization of these various body surfaces shows distinctive areas of frequent colonization and rare encounter. In humans, the anterior nares (Kluytmans *et al.*, 1997) is the niche of preference, especially in adults, while other areas such as the axilla, the perineum, or the transition zone of the non-hairy and hairy zones may be colonized by shedding from the nares [however, in certain patient populations, these areas may also be the exclusive sites of colonization, requiring additional extra-nasal screening for reliable determination of a carrier status (McKinnell *et al.*, 2013; Schechter-Perkins *et al.*, 2011)]. In females, colonization of the vagina is also a frequent and clinically important niche. Coagulase-negative staphylococci are normal commensal skin and mucous membranes colonizers with *S. epidermidis* by far being quantitatively predominant, particularly on the skin. Other coagulase-negative species tend to have a more specific reservoir, in some instances reflected by their nomenclature, e.g. *S. capitis* for the human scalp, *S. auricularis* for the external ear and *S. saprophyticus* for the genitourinary tract.

Approximately 10–30% of humans are persistently colonized by *S. aureus*, and all humans carry coagulase-negative species (typically 10–20 different strains; Feng *et al.*, 2008). It is interesting that the status of *S. aureus* colonization varies between humans: While 20% of healthy people are persistently colonized by this microorganism, approximately 30% (with a wide range) of individuals are intermittently colonized, and approximately 20–50% of humans either are not or are only rarely colonized (Wertheim *et al.*, 2004; Williams, 1963). Intermittent carriers frequently exchange strains, while constant carriers carry the same strain over prolonged times (Wertheim *et al.*, 2005). Analyses of *S. aureus* genome abundance within individuals suggest that nasal colonization results from one 'founder' organism which subsequently multiplies in the nare(s) displaying important fluctuations in the size of the within-host population (Golubchik *et al.*, 2013), yet, the individual-specific patterns of non-carriership versus persistent carriership appear to be maintained even upon experimental inoculation of other *S. aureus* strains (van Belkum *et al.*, 2009).

In the landmark study by von Eiff and colleagues it was demonstrated that in 82% of cases of *S. aureus* bacteraemia, the pulsed-field genotype of the bacteraemic isolates was identical to the genotype of an isolate concurrently recovered from the nares of patients. Even more importantly, in the parallel section of this study, in approximately 1300 patients prospectively examined by nasal swabs, 14 patients subsequently developed *S. aureus* bacteraemia with 12/14 pulsotypes again being identical between nasal and bloodstream isolates (von Eiff *et al.*, 2001). This finding was later confirmed for other types of infections caused specifically by MRSA strains (Huang and Platt, 2003). MRSA and MSSA may differ with respect to the status of nasal colonization and the subsequent risk of infection: in a meta-analysis it was found that MRSA colonization conferred a 4-fold increased risk of infection when compared to the risk of infection associated with MSSA colonization (Safdar and Bradley, 2008).

Thus, it is clear that *S. aureus* nasal and/or extra-nasal colonization is an important risk factor for ensuing invasive infection. At the same time, such nasal colonization is not a prerequisite, as a recent study found that 50% of bacteraemia cases were not associated with a positive carrier status (Marshall and McBryde, 2014). Interestingly, while carriers acquired *S. aureus* bacteraemia more frequently than non-carriers, they had a better survival rate (Wertheim *et al.*, 2004), an observation that further emphasizes the potential importance of *S. aureus* as a commensal and its impact on the innate immune status of the carrier host.

While for *S. aureus* the association between colonization and disease has been confirmed as a true causative relationship, the ubiquitous CoNS colonization of the human skin has hitherto precluded a demonstration of such dependency for CoNS. However, mucosae and skin colonization patterns in the neonate may allow for such analyses, and some studies, albeit with somewhat equivocal results, have now been published (Bjorkqvist *et al.*, 2010; Ternes *et al.*, 2013).

With the advent of next generation sequencing techniques and the coverage of entire microbiomes, a new chapter of our understanding on colonization patterns has been opened. In 2012, the Human Microbiome Project Consortium published its comprehensive analyses on the diversity and abundance of the 'normal' microbiome in a large cohort of healthy volunteers (The Human Microbiome Project Consortium, 2012). With respect to staphylococcal colonization, these data confirmed a striking abundance of different *Staphylococcus* spp. signatures from cutaneous epithelial sites when compared with mucosal epithelia (and in contrast to *Streptococcus* spp.), and that, for *S. aureus*, the anterior nares is far more abundantly colonized by comparison to retroauricular skin. In contrast, *S. epidermidis* colonization was more equally distributed between retroauricular skin and nares (but also less abundant on mucosal surfaces). Interestingly, a recent report suggested that the presence of *S. epidermidis* on a mucosal surface limits colonization with *S. aureus* owing to the production of a specific protease (Esp) that degrades *S. aureus* surface proteins required for colonization and biofilm formation (Sugimoto *et al.*, 2013), although a subsequent report has called this competitive association into question (Fredheim *et al.*, 2014). Microbiome studies have also confirmed a strong age-dependent shift of *Staphylococcus* spp. colonization (Oh *et al.*, 2012), and modified colonization patterns in patients with primary immunodeficiencies (Oh *et al.*, 2013), thus further illustrating the complex interplay between the staphylococci and their human hosts.

Staphylococcal commensals: active players modifying the immune response

The reasons for important inter-individual differences with respect to staphylococcal colonization, and in particular the factors which govern the *S. aureus* colonization pattern, are not fully elucidated. Studies describe significant contributions of host polymorphisms such as human α-defensin and β-defensin 2 levels in nasal secretions but also mutations or polymorphisms of interleukin 4, mannose-binding lectin, toll-like receptor 2, glucocorticoid receptor gene, C-reactive protein, and HLA-DR3 [summarized in a particularly valuable review by Brown *et al.* (2014), which also covers additional aspects on the effects of 'cohabitation' between *S. aureus* and other commensals in niches such as the nares].

The still unresolved question of determinants of the immune system contributing to staphylococcal colonization, however, brings into focus the reciprocal aspect, i.e. the effect of staphylococcal colonization on the human immune system, an evolving aspect in current research and of potentially substantial clinical significance in humans. The aforementioned apparent protection of *S. aureus* carriers towards a fatal course of bloodstream infection (Wertheim *et al.*, 2004) may be a strong indicator of such protective effect of *S. aureus* colonization, yet this study does not unveil the cause–consequence relationship between *S. aureus* colonization and protective status of the host.

It is clear that a complex immune response is elicited upon staphylococcal infection. It is also clear that the elicited humoral immune response is hampered by staphylococcal protein A and other Fcγ chain-binding staphylococcal proteins, and that colonization alone has only a limited effect on the human antibody response (Burian *et al.*, 2012; Joost *et al.*, 2011). The presence of an array of staphylococcal superantigens directly activates large numbers of polyclonal T-cells, however, such contact results in T-cell anergy in response to specific antigens such as effector and memory cell development (Lussow and MacDonald, 1994). Another important immune-modulating factor of *S. aureus* is the extracellular adhesive protein (Eap) which interferes with the immune response by inhibiting neutrophil recruitment due to blockage of neutrophil integrin recognition of endothelial ICAM-1 (Chavakis *et al.*, 2002), inhibiting the psoriasiform inflammatory skin reaction (Wang *et al.*, 2010), lymphocyte proliferation and the elicited Th1 response (Lee *et al.*, 2002), the surfactant-mediated *S. aureus* lung clearing (Sever-Chroneos *et al.*, 2011), and, as recently demonstrated, the activity of neutrophil serine proteases (Stapels *et al.*, 2014) (for review of the multitude of Eap interference with the immune system please also refer to Chavakis *et al.*, 2007).

S. aureus skin colonization may result in substantially reduced IFN-γ production from $CD4^+$ T cells. In addition, a chronic repeated *S. aureus* antigen-mediated stimulus of *S. aureus*-specific proinflammatory Th1/Th17 cells may result in the abrogation of this response and an overall anti-inflammatory status, in addition to an IL-10-mediated immune suppression. Taken together, it appears that in persistent carriers the constant interaction and processing of *S. aureus* antigens by local antigen-presenting cells may result in an immune tolerance and persistent colonization while in non-carriers the promotion of Th17 expansion results in phagocyte attraction, inflammation, and bacterial elimination (Brown *et al.*, 2014).

Finally, the specific modulation of the immune system by colonizing staphylococci may be better understood when comparing commensal and 'pathogenic' staphylococcal species, although once again it should be emphasized that this distinction is not absolute. Nevertheless, in their important contribution Wanke *et al.* (2011) show that these different species differ in their ability to induce expression of antimicrobial peptides/proteins and to activate different signalling pathways in primary keratinocytes: While commensals elicit expression of various antimicrobial peptides (AMP's) including HBD-3 and RNase-7, the pathogenic *S. aureus* strains activate mitogen-activated protein kinase (MAP-K) and phosphatidylinositol 3-kinase/AKT signalling pathways and suppress NF-κB expression (an effect again abrogated by *S. epidermidis*). These data demonstrate the species-specific effect of *S. aureus* and commensal staphylococcal species on the innate immune system of human keratinocytes.

Staphylococcal disease: clinically important manifestations

Staphylococci are capable of causing a wide spectrum of suppurative (pus-forming, pyogenic) infections. Generally speaking, *S. aureus* is clinically more important than the coagulase-negative species not only because of the severity of the infections it causes but also because it is increasingly capable of causing these infections even in healthy hosts. In contrast, the coagulase-negative species typically require predisposing factors such as an immature or compromised immune system, or the implantation of a foreign body. It is important to recognize that staphylococcal infections always bear the potential for a generalized or metastatic course, with any localized infection having the potential to become invasive and lead to (occult) bacteraemia and infectious organ complications. While this is definitely the case for *S. aureus*, the analysis from the International Collaboration on Endocarditis Study group suggest that transient bacteraemia due to coagulase-negative staphylococci may also have the potential to lead to invasive infection even in the absence of foreign material (Chu *et al.*, 2004).

The course of suppurative staphylococcal disease varies largely, ranging from fulminant overwhelming disease to chronic persistent forms. In his early review, Proctor *et al.* (1995) described five cases of staphylococcal organ disease, all of them bone or joint associated, with intervals between episodes of disease of up to 54 years. All of these cases had in common the recovery of *S. aureus* isolates with a phenotypic variation termed 'small colony variants' (SCV). SCVs have been recovered from patients suffering from a large variety of staphylococcal diseases as well as from carriers, and the clinical significance of staphylococci in humans is manifold as SCV variant formation is important to all clinical aspects from diagnosis, clinical presentation, antimicrobial susceptibility, treatment and prognosis. SCVs have been identified from *S. aureus* as well as from coagulase-negative species (Bogut *et al.*, 2014; von Eiff *et al.*, 1999). Overall, they appear to represent a heterogeneous group of isolates with different underlying metabolic and genetic factors leading to the SCV phenotype. The metabolic background of the SCV phenotype is addressed in other chapters of this book; for further clinical SCV-related aspects we refer to the excellent reviews available on this topic (Garcia *et al.*, 2013; Proctor *et al.*, 2006; von Eiff, 2008).

The bone, its adjacent tissues and the associated anatomic-functional structures such as joints or intervertebral spaces can be viewed as one model organ for staphylococcal infection in the human host: Almost all principles of this disease form – with the causative organisms *S. aureus* or coagulase-negative staphylococci, the various modes of pathogen inoculation, the possibility of a nosocomial or strictly spontaneous disease onset, or many diagnostic and therapeutic principles – can be particularly well illustrated with this entity. This allows to extend many of the disease principle-derived considerations also to other forms of staphylococcal clinical disease discussed subsequently. As space for this chapter is constrained, we have therefore chosen to particularly elaborate on the principles associated with this type of staphylococcal infection; while for the following subchapters, the reader is also referred to the literature cited.

Osteomyelitis and other orthopaedic infections

Osteomyelitis can be strictly defined as any inflammatory process involving bone, and it does not necessarily have an infectious aetiology. For instance, chronic recurrent multifocal osteomyelitis (CRMO) is considered an autoinflammatory disease and occurs primarily in children and adolescents (Roderick and Ramanan, 2013; Costa-Reis and Sullivan, 2013). Having said this, the overwhelming cause of osteomyelitis is infection, and even in CRMO an infectious aetiology cannot be entirely excluded as either a direct cause of the disease or as a stimulus for the resulting inflammatory process (Cimolai, 2011).

With the exception of infections associated with indwelling orthopaedic hardware, *S. aureus* is the leading cause of all forms of orthopaedic infection including osteomyelitis, and as noted above it is a prominent cause of infections involving indwelling devices (Cierny, 2011; Rao *et al.*, 2011; Zimmerli *et al.*, 2004). The prominence of *S. aureus* is such that any empiric antibiotic regimen must be focused on agents active against this pathogen. The relationship between achievable serum levels and bone penetration is also an issue, although most antibiotics have been found to penetrate bone to at least what should be therapeutically relevant levels (Rao *et al.*, 2011). Given the predominance of *S. aureus* as a causative agent, and the continued emergence of methicillin resistance even among strains causing community-associated infections (Diekema *et al.*, 2014; Otter and French, 2010), vancomycin remains the standard for empiric therapy, but there are reports suggesting that other antibiotics that are active against MRSA, including daptomycin, may offer equal if not greater efficacy (Gould *et al.*, 2013; Moenster *et al.*, 2013). If infection with a methicillin-resistant isolate appears improbable (as a result of considerations based on regional resistance profiles and/or patient history), the penicillinase-resistant penicillins (i.e. nafcillin or oxacillin) remain the therapeutic mainstay. Rifampin also warrants specific mention as an agent that is often used in combination with other antibiotics, particularly in patients with infected orthopaedic hardware (Rao *et al.*, 2011).

In the specific context of osteomyelitis, the local delivery of antibiotics also becomes an important clinical option. This is accomplished by incorporating antibiotics into any one of several delivery matrices, the two most common being polymethylmethacrylate (PMMA) bone cement and calcium sulfate (Cierny, 2011). The goal in this case is two-fold, the first being to achieve a high concentration of antibiotic directly at the site of infection while avoiding systemic toxicity, and the second being to assist in dead-space management following surgical debridement (see below). Indeed, a deciding factor in the choice between these two delivery matrices is whether the surgeon feels that the defect created by debridement does not compromise structural stability, in which case calcium sulfate is the preferred option because it is biodegradable and can be left in place; in contrast, if reconstruction is required, PMMA is the preferred choice because it is easily removed intact and leaves a clean wound environment prior to reconstruction. However, both of these have significant limitations, and this accounts for ongoing efforts to both optimize the use of these matrices and to develop viable alternatives (Beenken *et al.*, 2012, 2014b; Parker *et al.*, 2014).

Antibiotic selection is a critical issue in the treatment of all forms of *S. aureus* infection, but in osteomyelitis it is rarely the only or even the most critical issue. Indeed, with the exception of acute haematogenous osteomyelitis, which is primarily a disease of children and adolescents, antibiotic therapy alone is rarely sufficient to resolve the infection (Peltola and Paakkonen, 2014). Estimates of the overall incidence of haematogenous osteomyelitis in children range from 1 in 5000 to 1 in 10,000, with staphylococcal osteomyelitis accounting for approximately 2.5 of every 1000 hospital admissions in the USA (Gerber *et al.*, 2009; Weichert *et al.*, 2008). The incidence is roughly twice as high in boys as in girls, presumably owing to greater physical activity and its associated microtrauma (Peltola and Paakkonen, 2014), but this is likely to change as girls become more active in this regard. It should also be noted that these incidence estimates are based on studies done in developed countries and are likely an underestimate with respect to other parts of the world (Riise *et al.*, 2008). The most common site of infection in haematogenous osteomyelitis is the growth plate of long bones, particularly those of the leg (Lew and Waldvogel, 2004; Peltola and Paakkonen, 2014).

Haematogenous osteomyelitis also occurs in adults, but it this case it usually involves the spine and is more likely to present as a chronic rather than acute infection. In fact, all forms of osteomyelitis in adults are much more likely to be diagnosed only after the transition to a chronic infection, and this greatly complicates the clinical approach to treatment. The biggest issue in this regard is that systemic antibiotic therapy, irrespective of its type or duration, is essentially guaranteed to fail in the absence of adequate surgical debridement. For this reason, a number of classification schemes have been proposed as a guide for the need of surgical intervention. One of these distinguishes between haematogenous osteomyelitis, osteomyelitis secondary to a contiguous focus of infection, which is often a dreaded consequence of bone surgery or the repair of traumatic injury, osteomyelitis secondary to vascular insufficiency (e.g. the diabetic foot), and osteomyelitis associated with an infected prosthesis (Lew and Waldvogel, 2004). With the exception of haematogenous osteomyelitis, all of these forms are much more common in adults, and more importantly they almost invariably require surgical intervention in addition to long-term systemic antibiotic therapy.

However, these age-related correlations are not absolute, and for this reason an alternative classification scheme focuses simply on the distinction between acute and chronic osteomyelitis based solely on the length of time since the infection was established. There is no definitive time frame for this transition, but a diagnosis of acute infection is generally based on diagnosis within a few weeks, while a diagnosis of chronic infection is based on diagnosis only after several months (Peltola and Paakkonen, 2014; Lew and Waldvogel, 2004). Some authors also include a subacute category which is basically defined by diagnosis at a time point between these two extremes. Despite its somewhat vague nature, the distinction between acute, subacute, and chronic osteomyelitis does recognize that any form of osteomyelitis, irrespective of its source or whether it occurs in children or adults, is severely complicated by a failure to diagnose and treat in a timely manner owing to the resulting need for surgical intervention.

Several factors contribute to the necessity for surgical debridement. Included among these is a compromised local vasculature and formation of a sequestrum of dead bone, which compromises both host defences and the ability to deliver effective concentrations of antibiotic using any form of systemic antibiotic therapy. Other factors that further compromise antibiotic therapy include formation of a biofilm, which is widely considered a defining characteristic of chronic infections (Archer *et al.*, 2011; Brady *et al.*, 2008), and the emergence of persister cells and/or small colony variants (SCVs), often in association with biofilms, which exhibit reduced antibiotic susceptibility at a cellular level owing to anaerobic respiration and an overall reduction in metabolic activity (Lechner *et al.*, 2012; Lewis, 2008; Neut *et al.*, 2007). The limitation imposed by these factors is further complicated by the fact that, even in a best case scenario, antibiotic penetration into bone, which on the surface should be sufficient, is nevertheless limited by comparison to the levels observed in serum (Rao *et al.*, 2011).

The most descriptive classification system that serves as a guide in the overall treatment approach to osteomyelitis is based on the extent of anatomic damage rather than the source of the infection or the length of time since its onset (Cierny, 2011). Specifically, the Cierny–Mader system defines four stages of disease. The first is medullary or stage I, which is an 'inside out' infection that begins in the medullary canal and does not extend into cortical bone. In this respect, medullary osteomyelitis is roughly analogous to acute, haematogenous osteomyelitis. The second is superficial or stage II osteomyelitis, which can be thought of as an 'outside in' infection that involves cortical bone but not to an extent that threatens structural stability. This form most often presents as an acute infection and is analogous to early stage secondary osteomyelitis. The third and fourth stages are localized (stage III) and diffuse (stage IV), which are invariably chronic and are distinguished from each other only by the degree of damage and its impact on structural stability. Specifically, stage III disease requires limited debridement that does not compromise structural stability, while diffuse or stage IV disease requires extensive debridement that ultimately creates a structurally unstable defect.

As a further indication of the complexity of these infections, the Cierny–Mader system also includes specific consideration of the host, with the 'A' host being defined as a patient with normal physiological and immunological function. In contrast, the 'B' host suffers from some degree of localized or systemic compromise. The final category is the 'C' host, which is a patient so severely compromised that the treatment is deemed worse than the disease. In this case, the only therapeutic option is life-long suppressive antibiotic therapy or, failing that, amputation. The overriding principle of this system is that steps should always be taken to convert a 'B' host into an 'A' host before beginning therapy. As would be expected, the likelihood that the infection can be resolved by antibiotic therapy alone is highest in a IA patient, with the likelihood of such an uncomplicated outcome falling dramatically both as the stage of the infection progresses and as the status of the host declines. Unfortunately, in a summary of 2207 osteomyelitis patients classified and treated in the context of this system, 6%, 13%, 16% and 66% of patients were diagnosed in stage I, II, III and IV respectively, with 106 of these patients being deemed C hosts. The single most common presentation was a IVB patient, with 38.9% of patients falling into this category (Cierny, 2011). The success rate as judged after 2 years was 100% in a IA patient, fell to 94% in a IB patient, and fell further to 90% in a IVB patient, with this 10% failure rate being despite the intensive medical and surgical intervention employed in the treatment of these patients.

Chronic cases of bone infection are characterized in part by devascularized, necrotic sequestra, and in this respect they share many attributes with infections associated with orthopaedic implants. Indeed, a critical issue in both cases is adequate debridement that includes removal of all dead bone and/or the offending device. For this reason, a staging scheme adapted from the Cierny–Mader system was also developed for patients suffering from implant-associated infection (McPherson *et al.*, 2002). This system defines three categories of infection (early postoperative, haematogenous and chronic) and also emphasizes the importance of both local and systemic compromise in the host. In the study of 50 patients with infected total hips that led to development of this staging system and a description of its relationship to therapeutic outcome, the most common presentation was a chronic infection in a patient with both systemic and local compromise; thus, the clinical presentation in device-related infections parallels that observed in osteomyelitis itself.

The important point made by both the Cierny–Mader and McPherson classification systems is that the need to categorize patients based on multiple parameters, and the need to evaluate these parameters in as precise a manner as possible, provides an appropriate indication of the difficulties associated with the resolution of musculoskeletal infections whether or not they involve an indwelling medical device. However, in the specific context of implant-associated infections, it is important to make two additional points. First, owing to the need to maintain structural mobility in an increasingly ageing population, it has been estimated that the need for total hip and knee arthroplasties will increase by 174% and 673% respectively by 2030, while the need for revision surgery, the most common cause of which is infection, will increase by 137% and 601% respectively (Kurtz *et al.*, 2007). It has also been estimated that the current cost for such revision surgeries in US hospitals is $566 million, with this number also projected to increase to $1.62 billion by 2020 (Kurtz *et al.*, 2012). The second important point is that, while *S. aureus* remains a primary cause of such infections, it is in the context of implant-associated orthopaedic pathogens that *S. epidermidis* also emerges as an important player. For example, Steckelberg and Osmon (1994) examined 1, 033 infections of total hip and knee replacements and found that *S. aureus* and *S. epidermidis* were roughly equivalent pathogens that were collectively responsible for over half of all infections. In fact, a relatively recent review found that *S. epidermidis* was an even more common cause of prosthetic joint infections than *S. aureus* by a factor of approximately 2 (Zimmerli *et al.*, 2004).

The fact that these different staphylococcal species are essentially equivalent from a frequency point of view does not change the fact that infections caused by *S. aureus* are particularly problematic owing to their severity. This implies that *S. aureus* possesses specific attributes that define it as a musculoskeletal pathogen. In our view, and in the context of the preceding discussion, these attributes contribute to three fundamentally important processes. The first is the ability to avoid innate host defences and survive in the bloodstream in sufficient numbers to colonize bone, cartilage and/or indwelling medical devices. The second is the interaction between *S. aureus* virulence factors and specific components of bone, including osteoblasts and osteoclasts, during the acute stage of infection, the end result of which, at least in the absence of timely intervention, is extensive osteonecrosis and reactive bone formation that compromises the local vasculature (Cassat *et al.*, 2013; Smeltzer *et al.*, 1997). The third is the ability to persist in the bone for long periods, sometimes decades. Possible reasons for this, as well as the difficulties sometimes encountered in obtaining a definitive microbiological diagnosis, include internalization by osteoblasts, sequestration within a biofilm, and the aforementioned emergence of persister cells and/or SCVs (Hamza and Li, 2014; Lewis, 2010; Nelson *et al.*, 2005; Proctor *et al.*, 2006).

The virulence factors that contribute to these various stages of infection are likely to vary, but we believe there is evidence to suggest the existence of considerable overlap. For example, phenol-soluble modulins (PSMs) have been implicated in biofilm formation, which is considered a defining characteristic of the chronic stage of infection (Archer *et al.*, 2011; Brady *et al.*, 2008). The reports describing a role for PSMs in biofilm formation are somewhat contradictory and remain to be fully defined, with some reports indicating that they inhibit biofilm formation owing to their surfactant properties (Vuong *et al.*, 2000), and others indicating that they have the capacity to form amyloid-like fibril structures that stabilize biofilms (Schwartz *et al.*, 2012). However, we do not believe these are mutually exclusive possibilities in that the contribution of PSMs to biofilm formation versus biofilm dispersal may well vary depending on circumstances and the impact of these circumstances on the structural form of PSMs. Most importantly in the context of this discussion, PSMs also provide a critical defence mechanism based on their ability to lyse neutrophils via an intracellular mechanism (Surewaard *et al.*, 2013), and it was recently shown they also lyse osteoblasts in this same fashion (Rasigade *et al.*, 2013). This osteoblast cytotoxicity was also recently shown to play a primary role in the osteonecrosis and reactive bone formation during the acute stage of infection (Cassat *et al.*, 2013). Thus, PSMs have the capacity to promote the development of osteomyelitis by limiting neutrophil phagocytosis, the bone remodelling that characterizes acute infection, and formation of a biofilm during the chronic stage of infection. Similarly, Protein A was previously mentioned as a primary defence against antibody-mediated clearance, but it has also been shown to promote biofilm formation (Merino *et al.*, 2009), and was recently shown to impact osteoblast differentiation (Claro *et al.*, 2011). Protein A was also recently shown to bind tumour necrosis factor 1 on pre-osteoblastic cells and induce the secretion of soluble RANKL from osteoblasts, thereby resulting in increased recruitment of osteoclasts and an overall increase in bone loss (Widaa *et al.*, 2012). Such results strongly suggest that protein A is likely to also play an important role in multiple stages of the osteomyelitis disease process.

The list of additional virulence factors implicated in various aspects of osteomyelitis and other forms of musculoskeletal infection is far too long for a comprehensive review in this chapter, but we would note that it includes both surface associated and extracellular virulence factors. However, one particularly notable example in our view is the fibronectin-binding proteins FnbA and FnbB, which have been shown to promote both biofilm formation and internalization by host cells including osteoblasts (Ahmed *et al.*, 2001; Hamza and

Li, 2014; O'Neill *et al.*, 2008; Testoni *et al.*, 2011). The surface proteins responsible for binding collagen and bone sialoprotein (BSP) have also been implicated in bone infections (Elasri *et al.*, 2002; Ryden *et al.*, 1987, 1997). It has been suggested that there may be a specific link between acute haematogenous osteomyelitis in children and the fact that newly forming bone is particularly rich in BSP. The adhesins responsible for binding these host proteins (Cna and Bbp respectively) have also been implicated as factors that, along with FnbA/FnbB, promote osteoblast internalization (Testoni *et al.*, 2011; Meghji *et al.*, 1998) also reported the existence of a 'surface-associated material' (SAM) that directly mediates osteolytic activity by stimulating osteoclastogenesis. At present, the identity of this SAM is not known although its trypsin sensitivity suggests that it consists of one or more proteins. The same group has also demonstrated that some humans have antibodies capable of neutralizing this activity; however, the occurrence of these antibodies does not appear to be correlated with the presence or absence of disease (Nair *et al.*, 2000).

A recent report also demonstrated that as-yet-unidentified components present in conditioned medium from *S. aureus* biofilms decrease osteoblast viability, inhibit osteogenic differentiation, and increase bone resorption (Sanchez *et al.*, 2013). Although these factors could be surface proteins released from intact cells, it is also likely that they include actively secreted extracellular proteins. For instance, alpha toxin has been implicated in biofilm formation (Anderson *et al.*, 2012; Caiazza and O'Toole, 2003), and inactivation of the genes encoding both alpha and gamma toxins resulted in reduced virulence in a murine model of septic arthritis (Nilsson *et al.*, 1999). However, we believe it unlikely that these are cause–and–effect relationships related to biofilm formation in that both of these are pore-forming toxins that can directly cause cell death and elicit the production of proinflammatory cytokines (Tomita and Kamio, 1997), and this is probably a primary factor in the progression and severity of disease. Interestingly, a recent report also demonstrated that *S. aureus* forms 'biofilm-like aggregates' in human synovial fluid via a protein-dependent mechanism involving both the fibronectin-binding proteins FnbA and FnbB and the fibrinogen-binding proteins ClfA and ClfB (Dastgheyb *et al.*, 2014).

If both surface-associated and extracellular virulence factors contribute to the pathogenesis of musculoskeletal infection, then it stands to reason that the regulatory elements that modulate the production of these virulence factors would also be important. Targeting these regulatory loci has long been a therapeutic goal owing to the fact that the functional redundancy between individual virulence factors makes these less attractive therapeutic targets. The number of *S. aureus* regulatory loci continues to grow (Priest *et al.*, 2012), but to date we are aware of only three of these that have been directly examined in animal models of osteomyelitis. These are the accessory gene regulator (*agr*), the staphylococcal accessory regulator (*sarA*), and the *saePQRS* regulatory system, and in all cases mutation of these loci was shown to attenuate virulence (Beenken *et al.*, 2014a; Blevins *et al.*, 2003; Cassat *et al.*, 2013; Gillaspy *et al.*, 1995). All of these loci play global regulatory roles, thus making it difficult to clearly define why these mutants are attenuated, but one common theme that seems to be emerging is that they all impact the production of extracellular proteases. Boles and Horswill (2008) demonstrated that induction of *agr* within a biofilm mediated dispersal at least in part via a protease-dependent mechanism, but the more likely explanation for the attenuation of *agr* mutants in osteomyelitis is their reduced capacity to produce extracellular toxins. In the case of *sarA* and *saeRS*, mutation of these loci has also been shown to limit the accumulation of these toxins, as well as a number of the surface proteins discussed above, owing to either reduced production or increased protease-mediated degradation (Beenken *et al.*, 2014a; Cassat *et al.*, 2013; Mrak *et al.*, 2012; Zielinska *et al.*, 2012). It should be noted in this regard that progress is being made toward the development of small molecule inhibitors of the regulatory functions of these loci, most notably *agr* and *sarA* (Gordon *et al.*, 2013; Sully *et al.*, 2014).

Skin and soft tissue infection

The most frequent clinical presentation of *S. aureus* disease is infection of the skin or its associated soft tissues. In contrast, coagulase-negative staphylococci, while clearly predominant as skin commensals, rarely contribute to clinical disease of the intact skin. *S. aureus* infections may affect different skin layers: Superficial infections of the epidermis (impetigo), involvement of the dermis (folliculitis), or the deeper dermic structures (furuncles or carbuncles). Once the borders of the skin are transgressed, deep infections of the subepithelial tissue may arise and lead to erysipelas, cellulitis, fasciitis and pyomyositis.

With the spread of community acquired MRSA (CA-MRSA) clones such as the USA300 isolate throughout the North American continent and now globally (Daum, 2007), and with the strong association of these endemic isolates with the production of specific toxins (e.g. Panton–Valentine leucocidin or PVL) and/or the production of conserved toxins including alpha toxin and phenol-soluble modulins (PSMs) at high levels (Kobayashi *et al.*, 2011; Wang *et al.*, 2007), the picture of *S. aureus* skin and soft tissue infection has drastically changed (Moran *et al.*, 2006; Qualls *et al.*, 2012; Taira *et al.*, 2009). The overall increase in incidence of skin and soft tissue infections (SSTI), the typically recurrent form of disease (Demos *et al.*, 2012), and the frequent occurrence of deep, tissue-destroying local infections (Diep *et al.*, 2008) are all indicative of the changing patterns of staphylococcal skin disease associated with the community associated methicillin-resistant clones.

Impetigo (Darmstadt and Lane, 1994) can be caused by *S. aureus* as well as by beta-haemolytic streptococci. It is a highly contagious disease, frequently observed in children and young adults. Impetigo presents as an erythematous skin with blisters containing cloudy fluids, and, after the blister ruptures, the formation of characteristic honey-coloured crusts of various sizes. Folliculitis (Feingold, 1993) describes a follicle-penetrating entity predominantly affecting the hairy parts of the skin. Superficial bumps but also deep, follicle-destroying forms can be observed. Furuncles affect the entire hair follicle, and in case of carbuncle, form a confluent suppurative infection. Facial furuncles comprise a special entity as they carry the risk of septic sinus vein thrombosis. Another entity frequently caused by staphylococci is the localized, non-abscessing phlegmon which has to be separated from classical streptococcal erysipelas, and which is frequently associated with the presence of a skin wound, e.g. ulcer. Overall, cutaneous abscesses have become particularly frequent with the spread of CA-MRSA. Therapeutic principles (Pulia *et al.*, 2014) of these skin infections include either no treatment (e.g. in mild cases of furuncles), incision/drainage (the typical and frequently employed curative measure for outpatients with localized abscess), topical antiseptics (furuncles and limited impetigo) or topical antimicrobials (not generally recommended for resistance development considerations), and – in the case of severe, rapidly progressing disease or with accompanying systemic illness, comorbidities or immunosuppression – systemic antimicrobials. Notable is the requirement for systemic antibiotics in the case of facial furuncles for the reason cited above. In cases of recurrent skin disease, chronic *S. aureus* carriership and other comorbidities should be excluded or treated.

A deep phlegmon (involving fascia/muscle) always requires surgical intervention in conjunction with systemic antimicrobials. Necrotizing fasciitis is a rapidly progressing disease (within hours) and a medical emergency; it requires immediate surgical intervention in combination with empirical broad spectrum antimicrobial therapy to reduce mortality. It is increasingly associated with CA-MRSA (Lee *et al.*, 2007), and in areas with endemic elevated CA-MRSA, the early administration of MRSA-active bactericidal antimicrobials is crucial. The disease paradoxically presenting the least superficial signs requires a high index of clinical suspicion and early imaging, e.g. by emergency ultrasound (Iverson *et al.*, 2012).

Pulmonary infection

Staphylococci are also important causes of severe pulmonary infection (Kaye *et al.*, 1990). Both community-acquired as well as hospital-associated infections are frequent. Pneumonia due to staphylococci is almost always caused by *S. aureus*; coagulase-negative staphylococci albeit regularly recovered from respiratory specimens do not play a causative role, with potential exceptions of species such as *S. intermedius* or *S. lugdunensis* which have anecdotally been associated with cases of pneumonia. As with skin and soft tissue infections, the unique properties of CA-MRSA clones have also resulted in a changing clinical presentation of staphylococcal pneumonia. Indeed, in the past, pulmonary infection has been relatively rare in healthy children or in adults, and *S. aureus* played a relatively minor role in community-acquired and – compared to other nosocomial pathogens – a comparably lower role in the nosocomial setting (Kaye *et al.*, 1990; Musher *et al.*, 1994). A frequent complication of *S. aureus* pneumonia is empyema (Bryant and Salmon, 1996). With the endemic CA-MRSA isolates, however, a new albeit still rarely observed picture of fulminant necrotizing pneumonia incurring a high fatality rate has been described even among previously healthy individuals both in the US and in Europe (Dufour *et al.*, 2002; Mongkolrattanothai *et al.*, 2003). The course of this disease is reminiscent of the courses historically described in the late nineteenth and early twentieth century, particularly during the Spanish influenza epidemics (cited in (Bradley, 2005)). The virulence factors encoded by the CA-MRSA strains and contributing to this disease course have implicated PVL (Brown *et al.*, 2009; Lina *et al.*, 1999) as well as the high level production of other toxins including alpha toxin, PSMs, and potentially other staphylococcal toxins (DeLeo *et al.*, 2010). Early diagnosis, in particular in otherwise healthy persons with severe pneumonia and sepsis syndrome, and early implementation of CA-MRSA-covering antimicrobial therapy is essential for this disease.

Bloodstream infection and infective endocarditis

Overall, and as already pointed out in the introductory remarks of this chapter, since the early 1980s *Staphylococcus* spp. are now the leading pathogen in bloodstream infections worldwide (Banerjee *et al.*, 1991; Pittet *et al.*, 1994). In the nosocomial setting, *S. aureus* is the second most common cause of bacteraemia (Kern, 2010) with an increasing incidence of antibiotic-resistant bacteria including antibiotic-resistant *S. aureus* (Ammerlaan *et al.*, 2013). It is important to distinguish community-associated bacteraemia and nosocomial/health-care-associated bacteraemia as the etiologic principles, the types of staphylococcal organisms involved, treatment and prognosis fundamentally differ. Community-associated bloodstream infections occur in patients typically without underlying conditions, and is always the result of a deep infectious focus, e.g. an abscess, endovascular infection, osteoarticular infection, or pneumonia; focus identification, however, may be a considerable challenge (Hill *et al.*, 2001). In contrast, nosocomial and healthcare-associated infections (the latter comprising bloodstream infection which have their onset in patients with antecedent medical interventions or treatments) typically have underlying conditions such as vascular catheter access, implanted foreign material, or surgical procedures. Approximately 40–60% of *S. aureus* bacteraemias are hospital-associated, and in this nosocomial setting infected central venous catheters are by far the most frequent cause for staphylococcal bacteraemia, comprising >90% of bacteraemias associated with vascular access. With respect to the microorganisms involved, the coagulase-negative staphylococci play practically no role in community-associated bloodstream infections, yet, they account for approximately 30% of nosocomial/healthcare-associated organisms recovered from blood cultures. However, only 10% – 25% of these cases represent 'true' bacteraemia, with the remainder being due to frequent blood culture contamination from the skin when the sample was obtained (Souvenir *et al.*, 1998).

In the setting of bacteraemia, *S. aureus* differs from other blood pathogens due to its propensity to cause deep-seated metastatic infections; for instance, the risk for patients with catheter-induced *S. aureus* bacteraemia to develop infective endocarditis is approximately 10% (Cabell *et al.*, 2002; Fowler *et al.*, 2003). These complications typically occur weeks to months after the initial bacteraemic episode. It is now clear that patients with a persistent focus of infection, prolonged bacteraemia, insufficient antimicrobial treatment, or an indwelling prosthetic device are at much higher risk for infective endocarditis, osteomyelitis, or other manifestations of metastatic infection. An infectious disease-guided bundle approach on diagnostics and therapy – even in cases of *S. aureus* bacteraemia with a clinically benign course – are therefore imperative for prognosis and achieving an optimal therapeutic outcome (Forsblom *et al.*, 2013; Kern, 2010; Lopez-Cortes *et al.*, 2013). For coagulase-negative staphylococci, the potential sequelae of prolonged catheter-associated bacteraemia are less manifest, but in the case of implanted foreign material, in particular endovascular prosthesis or indwelling orthopaedic hardware, there is always a risk of foreign body infection, and bacteraemia should be treated with bactericidal antimicrobials.

Infective endocarditis due to staphylococci is probably the most serious complication following overt or occult bacteraemia. Numerous bacterial and host pathomechanistic principles and factors contribute to the infectious process with the initial attachment of the microorganisms to the endothelium or the subendothelial tissue, invasion of endothelial cells, tissue factor and interleukin release by monocytes and endothelial cells, platelet activation and deposition all being instrumental in this complex disease (Bayer *et al.*, 1998; Moreillon *et al.*, 2002; Que and Moreillon, 2011). Diagnosis is straightforward upon presentation of classic signs and symptoms, but in many 'real-life' cases proves difficult, and a high level of suspicion is required to avoid any diagnostic delay. Clinically, the approach to the patient with suspected infective endocarditis has been structured around the Duke criteria (Dodds *et al.*, 1996; Durack *et al.*, 1994), the application of which is now the basis for the diagnostic workup (Baddour *et al.*, 2015). The therapeutic approach to staphylococcal endocarditis is now also greatly standardized (Boucher *et al.*, 2010; Kern, 2010) and requires an interdisciplinary medical-surgical approach (Sohail *et al.*, 2006). Overall, staphylococcal endocarditis – prior to the antimicrobial era, was an unequivocally fatal disease – still carries significant mortality, yet with timely clinical and microbiologic diagnosis and a structured therapeutic approach is associated with an overall relatively good prognosis.

Toxin-mediated disease

The complex pathogenesis of *S. aureus* diseases includes many toxins, specific examples being PVL (Loffler *et al.*, 2010), α toxin (Bubeck Wardenburg *et al.*, 2007), phenol-soluble modulins (Peschel and Otto, 2013), toxic shock syndrome toxin (see below), and a plethora of enterotoxins. However, classically 'toxin-mediated disease' is separated from invasive or suppurative infections due to its pathogenesis being exclusively caused by the secreted toxin(s). Overall, two forms of toxin-mediated disease can be distinguished, with one form caused by *in situ* toxin production from staphylococci colonizing or invading the human body, and the other caused by ingestion of a preformed toxin, the primary example being classic staphylococcal food poisoning.

Toxic shock syndrome toxin 1 (TSST-1) is the most important and best-studied disease caused by staphylococcal toxins with superantigen activity (Schlievert, 1993). TSS first came to prominence in association with menstruation and use of hyper-absorbable tampons. The disease is characterized by an acute onset with high fever, an erythematous rash, hyperaemic mucosae, hypotension to shock, and organ manifestation (e.g. hepatic and renal failure, myalgias, disorientation). Desquamation of the palms of hand and feet is also a clinical sign that can be diagnostic. During menstruation, the vaginal milieu appears to provide optimal ecological conditions to drive colonizing staphylococci into secretion of the TSST-1 (Yarwood *et al.*, 2001) whose primary effect is on the polyclonal T-cell stimulation promoting a tight contact between the Vβ chain of the T-cell receptor and the major histocompatibility class II (MHC class II) molecules on antigen-presenting cells. While menstrual TSS has been the paradigmatic disease, non-menstrual forms of TSS caused by colonizing or invading toxin-producing staphylococcal organisms associated with minor trauma, surgery, or local infections are now of at least equivalent importance. The majority of healthy adults have anti-TSST-1 antibody titres thought to be protective against the disease. Apart from the special requirements needed for optimal toxin production, requirements that were unfortunately met by the use of hyper-absorbable tampons, it is these protective antibodies which may be the reason for the overall rather rare occurrence of clinical disease even when up to 20% of staphylococci contain the gene for TSST-1 (Becker *et al.*, 2003).

In addition to the TSST-1, *S. aureus* harbours a vast number of exotoxins with superantigen activity (Salgado-Pabon *et al.*, 2014) whereas non-*S. aureus* species lack the genes for such toxins (Becker *et al.*, 2001). The non-TSST toxins are associated with atopic dermatitis, Kawasaki syndrome, nasal polyposis, and certain autoimmune disorders [cited in: (Salgado-Pabon *et al.*, 2014). Included among this group are the enterotoxins associated with staphylococcal food poisoning, although it should be noted that the vomiting and diarrhoea characteristic of food poisoning is not associated with the superantigen activity of these toxins. Indeed, the enterotoxins are heat-stable and primarily cause emesis in primates due to complex and hitherto not fully understand mechanisms (recently reviewed in Hu and Nakane, 2014). Nausea, vomiting, abdominal cramps, and diarrhoea occur 2 to 6 h after intake of such contaminated food, and may last up to 8 to 12 h. The disease is self-limiting, has an excellent prognosis and is of concern only for elderly or debilitated persons.

Conclusions

We have attempted to summarize the role of staphylococcal species in human disease. The specific emphasis was purposefully placed on *S. aureus*, but we urge the reader not to ignore the clinical significance of other staphylococcal species; indeed, while *S. epidermidis* has become widely recognized as a cause of implant-associated infection, even a cursory search of the literature will reveal that essentially every staphylococcal species has some capacity to cause most if not all of the disease syndromes discussed above. It is also important to recognize that the interaction between staphylococcal species and their human hosts does not occur in a vacuum, with many reports confirming the importance of *S. aureus*, and likely many other staphylococcal species, as zoonotic pathogens. The extreme diversity of *S. aureus* infections, the virulence factors that contribute to these infections, and the regulatory elements that control the production of these virulence factors, makes it literally impossible to be comprehensive, at least without more narrowly focusing the discussion on specific types of infection, but we hope the preceding discussion provides a starting point for those interested in more detail. We also hope to have emphasized the important role of staphylococcal species as human commensals and the fact that the epidemiology and pathogenesis of staphylococcal infections is in a constant state of flux that occasionally reveals itself in a very dramatic manner, the most recent example being the emergence of highly invasive, methicillin-resistant strains of *S. aureus* even among isolates causing community-associated infection. All things considered, it should be clear to all concerned that the staphylococci are an integral and interactive part of human biology, and it is imperative that we understand their role in this context if we are to avoid the adverse consequences of staphylococcal infection, particularly in an increasingly ageing population and, paradoxically, the medical advances that provide these important pathogens with opportunities for infection.

References

Ahmed, S., Meghji, S., Williams, R.J., Henderson, B., Brock, J.H., and Nair, S.P. (2001). *Staphylococcus aureus* fibronectin binding proteins are essential for internalization by osteoblasts but do not account for differences in intracellular levels of bacteria. Infect. Immun. *69*, 2872–2877.

Ammerlaan, H.S., Harbarth, S., Buiting, A.G., Crook, D.W., Fitzpatrick, F., Hanberger, H., Herwaldt, L.A., Van Keulen, P.H., Kluytmans, J.A., Kola, A., *et al.* (2013). Secular trends in nosocomial bloodstream infections: antibiotic-resistant bacteria increase the total burden of infection. Clin. Infect. Dis. *56*, 798–805.

Anderson, M.J., Lin, Y.C., Gillman, A.N., Parks, P.J., Schlievert, P.M., and Peterson, M.L. (2012). Alpha-toxin promotes *Staphylococcus aureus* mucosal biofilm formation. Front. Cell. Infect. Microbiol. *2*, 64.

Archer, N.K., Mazaitis, M.J., Costerton, J.W., Leid, J.G., Powers, M.E., and Shirtliff, M.E. (2011). *Staphylococcus aureus* biofilms: properties, regulation, and roles in human disease. Virulence 2, 445–459.

Baddour, L.M., Wilson, W.R., Bayer, A.S., Fowler, V.G. Jr, Tleyjeh, I.M., Rybak, M.J., Barsic, B., Lockhart, P.B., Gewitz, M.H., Levison, M.E., *et al.*; American Heart Association Committee on Rheumatic Fever, Endocarditis, and Kawasaki Disease of the Council on Cardiovascular Disease in the Young, Council on Clinical Cardiology, Council on Cardiovascular Surgery and Anesthesia, and Stroke Counciland. (2015). Infective Endocarditis in Adults: Diagnosis, Antimicrobial Therapy, and Management of Complications: A Scientific Statement for Healthcare Professionals From the American Heart Association. Circulation. *132*, 1435–1486.

Banerjee, S.N., Emori, T.G., Culver, D.H., Gaynes, R.P., Jarvis, W.R., Horan, T., Edwards, J.R., Tolson, J., Henderson, T., and Martone, W.J. (1991). Secular trends in nosocomial primary bloodstream infections in the United States, 1980–1989. National Nosocomial Infections Surveillance System. Am. J. Med. *91*, 86S-89S.

Bayer, A.S., Bolger, A.F., Taubert, K.A., Wilson, W., Steckelberg, J., Karchmer, A.W., Levison, M., Chambers, H.F., Dajani, A.S., Gewitz, M.H., and*et al.* (1998). Diagnosis and management of infective endocarditis and its complications. Circulation. *98*, 2936–2948.

Becker, K., Friedrich, A.W., Lubritz, G., Weilert, M., Peters, G., and Von Eiff, C. (2003). Prevalence of genes encoding pyrogenic toxin superantigens and exfoliative toxins among strains of *Staphylococcus aureus* isolated from blood and nasal specimens. J. Clin. Microbiol. *41*, 1434–1439.

Becker, K., Haverkamper, G., Von Eiff, C., Roth, R., and Peters, G. (2001). Survey of staphylococcal enterotoxin genes, exfoliative toxin genes, and toxic shock syndrome toxin 1 gene in non-*Staphylococcus aureus* species. Eur. J. Clin. Microbiol. Infect. Dis. *20*, 407–409.

Beenken, K.E., Bradney, L., Bellamy, W., Skinner, R.A., Mclaren, S.G., Gruenwald, M.J., Spencer, H.J., Smith, J.K., Haggard, W.O., and Smeltzer, M.S. (2012). Use of xylitol to enhance the therapeutic efficacy of polymethylmethacrylate-based antibiotic therapy in treatment of chronic osteomyelitis. *Antimicrob. Agents Chemother.* *56*, 5839–5844.

Beenken, K.E., Mrak, L.N., Zielinska, A.K., Atwood, D.N., Loughran, A.J., Griffin, L.M., Matthews, K.A., Anthony, A.M., Spencer, H.J., Skinner, R.A., and*et al.* (2014).

Impact of the functional status of *saeRS* on *in vivo* phenotypes of *Staphylococcus aureus sarA* mutants. Mol. Microbiol. *92*, 1299–1312.

Beenken, K.E., Smith, J.K., Skinner, R.A., Mclaren, S.G., Bellamy, W., Gruenwald, M.J., Spencer, H.J., Jennings, J.A., Haggard, W.O., and Smeltzer, M.S. (2014). Chitosan coating to enhance the therapeutic efficacy of calcium sulfate-based antibiotic therapy in the treatment of chronic osteomyelitis. J. Biomater. Appl. *29*, 514–523.

Björkqvist, M., Liljedahl, M., Zimmermann, J., Schollin, J., and Söderquist, B. (2010). Colonization pattern of coagulase-negative staphylococci in preterm neonates and the relation to bacteremia. Eur. J. Clin. Microbiol. Infect. Dis. *29*, 1085–1093.

Blevins, J.S., Elasri, M.O., Allmendinger, S.D., Beenken, K.E., Skinner, R.A., Thomas, J.R., and Smeltzer, M.S. (2003). Role of *sarA* in the pathogenesis of *Staphylococcus aureus* musculoskeletal infection. Infect. Immun. *71*, 516–523.

Bogut, A., Niedzwiadek, J., Koziol-Montewka, M., Strzelec-Nowak, D., Blacha, J., Mazurkiewicz, T., Marczynski, W., and Plewik, D. (2014). Characterization of *Staphylococcus epidermidis* and *Staphyloccocus warneri* small-colony variants associated with prosthetic-joint infections. J. Med. Microbiol. *63*, 176–185.

Boles, B.R., and Horswill, A.R. (2008). Agr-mediated dispersal of *Staphylococcus aureus* biofilms. PLOS Pathog. *4*, e1000052.

Boucher, H., Miller, L.G., and Razonable, R.R. (2010). Serious infections caused by methicillin-resistant *Staphylococcus aureus*. Clin. Infect. Dis. *51* (Suppl. 2), S183–S197.

Bradley, S.F. (2005). *Staphylococcus aureus* pneumonia: emergence of MRSA in the community. Semin. Respir. Crit. Care Med. *26*, 643–649.

Brady, R.A., Leid, J.G., Calhoun, J.H., Costerton, J.W., and Shirtliff, M.E. (2008). Osteomyelitis and the role of biofilms in chronic infection. FEMS Immunol. Med. Microbiol. *52*, 13–22.

Breurec, S., Zriouil, S.B., Fall, C., Boisier, P., Brisse, S., Djibo, S., Etienne, J., Fonkoua, M.C., Perrier-Gros-Claude, J.D., Pouillot, R., Ramarokoto, C.E., Randrianirina, F., Tall, A., Thiberge, J.M., Working Group On *Staphylococcus aureus*, I., Laurent, F., and Garin, B. (2011). Epidemiology of methicillin-resistant *Staphylococcus aureus* lineages in five major African towns: emergence and spread of atypical clones. Clin. Microbiol. Infect. *17*, 160–165.

Brown, A.F., Leech, J.M., Rogers, T.R., and McLoughlin, R.M. (2014). *Staphylococcus aureus* Colonization: Modulation of host immune response and impact on human vaccine design. Front. Immunol. *4*, 507.

Brown, E.L., Dumitrescu, O., Thomas, D., Badiou, C., Koers, E.M., Choudhury, P., Vazquez, V., Etienne, J., Lina, G., Vandenesch, Fand., *et al.* (2009). The Panton–Valentine leucocidin vaccine protects mice against lung and skin infections caused by *Staphylococcus aureus* USA300. Clin. Microbiol. Infect. *15*, 156–164.

Bryant, R.E., and Salmon, C.J. (1996). Pleural empyema. Clin. Infect. Dis. *22*, 747–762.

Bubeck Wardenburg, J., Bae, T., Otto, M., Deleo, F.R., and Schneewind, O. (2007). Poring over pores: alpha-hemolysin and Panton–Valentine leucocidin in *Staphylococcus aureus* pneumonia. Nat. Med. *13*, 1405–1406.

Burian, M., Grumann, D., Holtfreter, S., Wolz, C., Goerke, C., and Broker, B.M. (2012). Expression of staphylococcal superantigens during nasal colonization is not sufficient to induce a systemic neutralizing antibody response in humans. Eur. J. Clin. Microbiol. Infect. Dis. *31*, 251–256.

Cabell, C.H., Jollis, J.G., Peterson, G.E., Corey, G.R., Anderson, D.J., Sexton, D.J., Woods, C.W., Reller, L.B., Ryan, T., and Fowler, V.G. (2002). Changing patient characteristics and the effect on mortality in endocarditis. Arch. Intern. Med. *162*, 90–94.

Caiazza, N.C., and O'Toole, G.A. (2003). Alpha-toxin is required for biofilm formation by *Staphylococcus aureus*. J. Bacteriol. *185*, 3214–3217.

Casey, J.A., Curriero, F.C., Cosgrove, S.E., Nachman, K.E., and Schwartz, B.S. (2013). High-density livestock operations, crop field application of manure, and risk of community-associated methicillin-resistant *Staphylococcus aureus* infection in Pennsylvania. JAMA Intern. Med. *173*, 1980–1990.

Cassat, J.E., Hammer, N.D., Campbell, J.P., Benson, M.A., Perrien, D.S., Mrak, L.N., Smeltzer, M.S., Torres, V.J., and Skaar, E.P. (2013). A secreted bacterial protease tailors the *Staphylococcus aureus* virulence repertoire to modulate bone remodeling during osteomyelitis. Cell Host Microbe. *13*, 759–772.

Chapman, A.L., Greig, J.M., and Innes, J.A. (2005). MRSA in the community. N. Engl. J. Med. *353*, 530–532.

Chavakis, T., Hussain, M., Kanse, S.M., Peters, G., Bretzel, R.G., Flock, J.I., Herrmann, M., and Preissner, K.T. (2002). *Staphylococcus aureus* extracellular adherence protein serves as anti-inflammatory factor by inhibiting the recruitment of host leucocytes. Nat. Med. *8*, 687–693.

Chavakis, T., Preissner, K.T., and Herrmann, M. (2007). The anti-inflammatory activities of *Staphylococcus aureus*. Trends Immunol. *28*, 408–418.

Chen, C.J., and Huang, Y.C. (2014). New epidemiology of *Staphylococcus aureus* infection in Asia. Clin. Microbiol. Infect. *20*, 605–623.

Chu, V.H., Cabell, C.H., Abrutyn, E., Corey, G.R., Hoen, B., Miro, J.M., Olaison, L., Stryjewski, M.E., Pappas, P., Anstrom, K.J., Eykyn, S., Habib, G., Benito, N., Fowler, V.G., Jr. and International Collaboration On Endocarditis Merged Database Study Group. (2004). Native valve endocarditis due to coagulase-negative staphylococci: report of 99 episodes from the International Collaboration on Endocarditis Merged Database. Clin. Infect. Dis. *39*, 1527–1530.

Cierny, G. (2011). Surgical treatment of osteomyelitis. Plast. Reconstr. Surg. *127* (Suppl. 1), 190S–204S.

Cimolai, N. (2011). Chronic multifocal osteomyelitis: is infectious causation a moot point? J. Infect. Public Health *4*, 157–168.

Claro, T., Widaa, A., O'Seaghdha, M., Miajlovic, H., Foster, T.J., O'Brien, F.J.and Kerrigan, S.W. (2011). *Staphylococcus aureus* protein A binds to osteoblasts and triggers signals that weaken bone in osteomyelitis. PLoS One *6*, e18748.

Cosgrove, S.E., Qi, Y., Kaye, K.S., Harbarth, S., Karchmer, A.W., and Carmeli, Y. (2005). The impact of methicillin resistance in *Staphylococcus aureus* bacteremia on patient outcomes: mortality, length of stay, and hospital charges. Infect. Control Hosp. Epidemiol. *26*, 166–174.

Cosgrove, S.E., Sakoulas, G., Perencevich, E.N., Schwaber, M.J., Karchmer, A.W., and Carmeli, Y. (2003). Comparison of mortality associated with methicillin-resistant and methicillin-susceptible *Staphylococcus aureus* bacteremia: a meta-analysis. Clin. Infect. Dis. *36*, 53–59.

Costa-Reis, P., and Sullivan, K.E. (2013). Chronic recurrent multifocal osteomyelitis. J. Clin. Immunol. *33*, 1043–1056.

Dantes, R., Mu, Y., Belflower, R., Aragon, D., Dumyati, G., Harrison, L.H., Lessa, F.C., Lynfield, R., Nadle, J., Petit, S., Ray, S.M., Schaffner, W., Townes, J., Fridkin, S., and Emerging Infections Program–Active Bacterial Core Surveillance MRSA Surveillance Investigators. 2013. National burden of invasive methicillin-resistant *Staphylococcus aureus* infections, United States, (2011). JAMA Intern. Med. *173*, 1970–1978.

Darmstadt, G.L., and Lane, A.T. (1994). Impetigo: an overview. Pediatr. Dermatol. *11*, 293–303.

Dastgheyb, S., Parvizi, J., Shapiro, I.M., Hickok, N.J., and Otto, M. (2015). Effect of biofilms on recalcitrance of staphylococcal joint infection to antibiotic treatment. J. Infect. Dis. *211*, 641–650.

Daum, R.S. (2007). Clinical practice. Skin and soft-tissue infections caused by methicillin-resistant *Staphylococcus aureus*. N. Engl. J. Med. *357*, 380–390.

DeLeo, F.R., Otto, M., Kreiswirth, B.N., and Chambers, H.F. (2010). Community-associated meticillin-resistant *Staphylococcus aureus*. Lancet *375*, 1557–1568.

Demos, M., McLeod, M.P., and Nouri, K. (2012). Recurrent furunculosis: a review of the literature. Br. J. Dermatol. *167*, 725–732.

Diekema, D.J., Richter, S.S., Heilmann, K.P., Dohrn, C.L., Riahi, F., Tendolkar, S., Mcdanel, J.S., and Doern, G.V. (2014). Continued emergence of USA300 methicillin-resistant *Staphylococcus aureus* in the United States: results from a nationwide surveillance study. Infec.t Control Hosp. Epidemiol. *35*, 285–292.

Diep, B.A., Chambers, H.F., Graber, C.J., Szumowski, J.D., Miller, L.G., Han, L.L., Chen, J.H., Lin, F., Lin, J., Phan, T.H., Carleton, H.A., Mcdougal, L.K., Tenover, F.C., Cohen, D.E., Mayer, K.H., Sensabaugh, G.F., and Perdreau-Remington, F. (2008). Emergence of multidrug-resistant, community-associated, methicillin-resistant *Staphylococcus aureus* clone USA300 in men who have sex with men. *Ann. Intern. Med.*, 148, 249–257.

Dodds, G.A., Sexton, D.J., Durack, D.T., Bashore, T.M., Corey, G.R., and Kisslo, J. (1996). Negative predictive value of the Duke criteria for infective endocarditis. Am. J. Cardiol. *77*, 403–407.

Dufour, P., Gillet, Y., Bes, M., Lina, G., Vandenesch, F., Floret, D., Etienne, J., and Richet, H. (2002). Community-acquired methicillin-resistant *Staphylococcus aureus* infections in France: emergence of a single clone that produces Panton–Valentine leucocidin. Clin. Infect. Dis. *35*, 819–824.

Durack, D.T., Lukes, A.S., and Bright, D.K. (1994). New criteria for diagnosis of infective endocarditis: utilization of specific echocardiographic findings. Duke Endocarditis Service. Am. J. Med. *96*, 200–209.

Elasri, M.O., Thomas, J.R., Skinner, R.A., Blevins, J.S., Beenken, K.E., Nelson, C.L., and Smeltzer, M.S. (2002). *Staphylococcus aureus* collagen adhesin contributes to the pathogenesis of osteomyelitis. Bone *30*, 275–280.

Engemann, J.J., Carmeli, Y., Cosgrove, S.E., Fowler, V.G., Bronstein, M.Z., Trivette, S.L., Briggs, J.P., Sexton, D.J., and Kaye, K.S. (2003). Adverse clinical and economic outcomes attributable to methicillin resistance among patients with *Staphylococcus aureus* surgical site infection. Clin. Infect. Dis. *36*, 592–598.

Feingold, D.S. (1993). Staphylococcal and streptococcal pyodermas. Semin. Dermatol. *12*, 331–335.

Feng, Y., Chen, C.J., Su, L.H., Hu, S., Yu, J., and Chiu, C.H. (2008). Evolution and pathogenesis of *Staphylococcus aureus*: lessons learned from genotyping and comparative genomics. FEMS Microbiol. Rev. *32*, 23–37.

Forsblom, E., Ruotsalainen, E., Ollgren, Jand., and Järvinen, A. (2013). Telephone consultation cannot replace bedside infectious disease consultation in the management of *Staphylococcus aureus* bacteremia. Clin. Infect. Dis. *56*, 527–535.

Fowler, V.G., Olsen, M.K., Corey, G.R., Woods, C.W., Cabell, C.H., Reller, L.B., Cheng, A.C., Dudley, T., and Oddone, E.Z. (2003). Clinical identifiers of complicated *Staphylococcus aureus* bacteremia. Arch. Intern. Med. *163*, 2066–2072.

Fredheim, E.G., Flægstad, T., Askarian, F., and Klingenberg, C. (2015). Colonisation and interaction between *S. epidermidis* and *S. aureus* in the nose and throat of healthy adolescents. Eur. J. Clin. Microbiol. Infect. Dis. *34*, 123–129.

Garcia, L.G., Lemaire, S., Kahl, B.C., Becker, K., Proctor, R.A., Denis, O., Tulkens, P.M., and Van Bambeke, F. (2013). Antibiotic activity against small-colony variants of *Staphylococcus aureus*: review of *in vitro*, animal and clinical data. J. Antimicrob. Chemother. *68*, 1455–1464.

Gerber, J.S., Coffin, S.E., Smathers, S.A., and Zaoutis, T.E. (2009). Trends in the incidence of methicillin-resistant *Staphylococcus aureus* infection in children's hospitals in the United States. Clin. Infect. Dis. *49*, 65–71.

Gillaspy, A.F., Hickmon, S.G., Skinner, R.A., Thomas, J.R., Nelson, C.L., and Smeltzer, M.S. (1995). Role of the accessory gene regulator (*agr*) in pathogenesis of staphylococcal osteomyelitis. Infect. Immun. *63*, 3373–3380.

Golubchik, T., Batty, E.M., Miller, R.R., Farr, H., Young, B.C., Larner-Svensson, H., Fung, R., Godwin, H., Knox, K., Votintseva, A., and*et al.* (2013). Within-host evolution of *Staphylococcus aureus* during asymptomatic carriage. PLoS One *8*, e61319.

Gordon, C.P., Williams, P., and Chan, W.C. (2013). Attenuating *Staphylococcus aureus* virulence gene regulation: a medicinal chemistry perspective. J. Med. Chem. *56*, 1389–1404.

Gould, I.M., Miró, J.M., and Rybak, M.J., and(2013). Daptomycin: the role of high-dose and combination therapy for Gram-positive infections. Int. J. Antimicrob. Agents *42*, 202–210.

Hamza, T., and Li, B. (2014). Differential responses of osteoblasts and macrophages upon *Staphylococcus aureus* infection. BMC Microbiol. *14*, 207.

Hanberger, H., Walther, S., Leone, M., Barie, P.S., Rello, J., Lipman, J., Marshall, J.C., Anzueto, A., Sakr, Y., Pickkers,

P., Felleiter, P., Engoren, M., Vincent, J.L., and EPIC II Group of Investigators (2011). Increased mortality associated with methicillin-resistant *Staphylococcus aureus* (MRSA) infection in the intensive care unit: results from the EPIC II study. *Int. J. Antimicrob. Agents.38*, 331–335.

Herrmann, M., Abdullah, S., Alabi, A., Alonso, P., Friedrich, A.W., Fuhr, G., Germann, A., Kern, W.V., Kremsner, P.G., Mandomando, I., and*et al.* (2013). Staphylococcal disease in Africa: another neglected 'tropical' disease. Future Microbiol. *8*, 17–26.

Hill, P.C., Birch, M., Chambers, S., Drinkovic, D., Ellis-Pegler, R.B., Everts, R., Murdoch, D., Pottumarthy, S., Roberts, S.A., Swager, C., Taylor, S.L., Thomas, M.G., Wong, C.G., and Morris, A.J. (2001). Prospective study of 424 cases of *Staphylococcus aureus* bacteraemia: determination of factors affecting incidence and mortality. Intern. Med. J. *31*, 97–103.

Hu, D.L., and Nakane, A. (2014). Mechanisms of staphylococcal enterotoxin-induced emesis. Eur. J. Pharmacol. *722*, 95–107.

Huang, S.S., and Platt, R. (2003). Risk of methicillin-resistant *Staphylococcus aureus* infection after previous infection or colonization. Clin. Infect. Dis. *36*, 281–285.

Iverson, K., Haritos, D., Thomas, R., and Kannikeswaran, N. (2012). The effect of bedside ultrasound on diagnosis and management of soft tissue infections in a pediatric ED. Am. J. Emerg. Med. *30*, 1347–1351.

Jacobsson, G., Dashti, S., Wahlberg, Tand Andersson, R. (2007). The epidemiology of and risk factors for invasive *Staphylococcus aureus* infections in western Sweden. Scand. J. Infect. Dis. *39*, 6–13.

Joost, I., Jacob, S., Utermöhlen, O., Schubert, U., Patti, J.M., Ong, M.F., Gross, J., Justinger, C., Renno, J.H., Preissner, K.T., and*et al.* (2011). Antibody response to the extracellular adherence protein (Eap) of *Staphylococcus aureus* in healthy and infected individuals. FEMS Immunol. Med. Microbiol. *62*, 23–31.

Kaye, M.G., Fox, M.J., Bartlett, J.G., Braman, S.S., and Glassroth, J. (1990). The clinical spectrum of *Staphylococcus aureus* pulmonary infection. Chest *97*, 788–792.

Kern, W.V. (2010). Management of *Staphylococcus aureus* bacteremia and endocarditis: progresses and challenges. Curr. Opin. Infect. Dis. *23*, 346–358.

Klevens, R.M., Morrison, M.A., Nadle, J., Petit, S., Gershman, K., Ray, S., Harrison, L.H., Lynfield, R., Dumyati, G., Townes, J.M., Craig, A.S., Zell, E.R., Fosheim, G.E., Mcdougal, L.K., Carey, R.B., Fridkin, S.K., and Active Bacterial Core surveillance (ABCs) MRSA Investigators (2007). Invasive methicillin-resistant *Staphylococcus aureus* infections in the United States. JAMA. *298*, 1763–1771.

Kloos, W.E., and Schleifer, K.H. (1975). Simplified scheme for routine identification of human *Staphylococcus* species. J. Clin. Microbiol. *1*, 82–88.

Kluytmans, J., van Belkum, A., and Verbrugh, H. (1997). Nasal carriage of *Staphylococcus aureus*: epidemiology, underlying mechanisms, and associated risks. Clin. Microbiol. Rev. *10*, 505–520.

Kobayashi, S.D., Malachowa, N., Whitney, A.R., Braughton, K.R., Gardner, D.J., Long, D., Bubeck Wardenburg, J., Schneewind, O., Otto, M., and Deleo, F.R. (2011). Comparative analysis of USA300 virulence determinants in a rabbit model of skin and soft tissue infection. J. Infect. Dis. *204*, 937–941.

Kurtz, S., Ong, K., Lau, E., Mowat, F., and Halpern, M. (2007). Projections of primary and revision hip and knee arthroplasty in the United States from 2005 to 2030. J. Bone Joint Surg. Am. *89*, 780–785.

Kurtz, S.M., Lau, E., Watson, H., Schmier, J.K and Parvizi, J. (2012). Economic burden of periprosthetic joint infection in the United States. J. Arthroplasty *27* (Suppl. 8), 61–65.e1.

Laupland, K.B., Church, D.L., Mucenski, M., Sutherland, L.R., and Davies, H.D. (2003). Population-based study of the epidemiology of and the risk factors for invasive *Staphylococcus aureus* infections. J. Infect. Dis. *187*, 1452–1459.

Laupland, K.B., Lyytikäinen, O., Søgaard, M., Kennedy, K.J., Knudsen, J.D., Ostergaard, C., Galbraith, J.C., Valiquette, L., Jacobsson, G., Collignon, P., and Schønheyder, H.C. (2013). The changing epidemiology of *Staphylococcus aureus* bloodstream infection: a multinational population-based surveillance study. Clin. Microbiol. Infect. *19*, 465–471.

Lechner, S., Lewis, K., and Bertram, R. (2012). *Staphylococcus aureus* persisters tolerant to bactericidal antibiotics. J. Mol. Microbiol. Biotechnol. *22*, 235–244.

Lee, L.Y., Miyamoto, Y.J., McIntyre, B.W., Höök, M., McCrea, K.W., McDevitt, D., and Brown, E.L. (2002). The *Staphylococcus aureus* Map protein is an immunomodulator that interferes with T cell-mediated responses. J. Clin. Invest. *110*, 1461–1471.

Lee, T.C., Carrick, M.M., Scott, B.G., Hodges, J.C., and Pham, H.Q. (2007). Incidence and clinical characteristics of methicillin-resistant *Staphylococcus aureus* necrotizing fasciitis in a large urban hospital. Am. J. Surg. *194*, 809–812; discussion 812–813.

Leung, D.Y. (2013). New insights into atopic dermatitis: role of skin barrier and immune dysregulation. Allergol. Int. *62*, 151–161.

Lew, D.P., and Waldvogel, F.A. (2004). Osteomyelitis. Lancet *364*, 369–379.

Lewis, K. (2008). Multidrug tolerance of biofilms and persister cells. Curr. Top. Microbiol. Immunol. *322*, 107–131.

Lewis, K. (2010). Persister cells. Annu. Rev. Microbiol. *64*, 357–372.

Lina, G., Piémont, Y., Godail-Gamot, F., Bes, M., Peter, M.O., Gauduchon, V., Vandenesch, F., and Etienne, J. (1999). Involvement of Panton–Valentine leucocidin-producing *Staphylococcus aureus* in primary skin infections and pneumonia. Clin. Infect. Dis. *29*, 1128–1132.

Löffler, B., Hussain, M., Grundmeier, M., Brück, M., Holzinger, D., Varga, G., Roth, J., Kahl, B.C., Proctor, R.A., and Peters, G. (2010). *Staphylococcus aureus* Panton–Valentine leucocidin is a very potent cytotoxic factor for human neutrophils. PLOS Pathog. *6*, e1000715.

López-Cortés, L.E., Del Toro, M.D., Gálvez-Acebal, J., Bereciartua-Bastarrica, E., Farinas, M.C., Sanz-Franco, M., Natera, C., Corzo, J.E., Lomas, J.M., Pasquau, J., Del Arco, A., Martinez, M.P., Romero, A., Muniain, M.A., De Cueto, M., Pascual, A., Rodriguez-Bano, J., and REIPI/SAB group (2013). Impact of an evidence-based bundle intervention in the quality-of-care management and outcome of *Staphylococcus aureus* bacteremia. Clin. Infect. Dis. 57, 1225–1233.

Lussow, A.R., and MacDonald, H.R. (1994). Differential effects of superantigen-induced "anergy" on priming and effector stages of a T cell-dependent antibody response. Eur. J. Immunol. *24*, 445–449.

Maki, D.G. (1989). Risk factors for nosocomial infection in intensive care. 'Devices vs nature' and goals for the next decade. Arch. Intern. Med. *149*, 30–35.

Marshall, C., and McBryde, E. (2014). The role of *Staphylococcus aureus* carriage in the pathogenesis of bloodstream infection. BMC Res. Notes 7, 428.

Matsubara, K., and Fukaya, T. (2007). The role of superantigens of group A *Streptococcus* and *Staphylococcus aureus* in Kawasaki disease. Curr. Opin. Infect. Dis. *20*, 298–303.

McKinnell, J.A., Huang, S.S., Eells, S.J., Cui, E., and Miller, L.G. (2013). Quantifying the impact of extranasal testing of body sites for methicillin-resistant *Staphylococcus aureus* colonization at the time of hospital or intensive care unit admission. Infect. Control. Hosp. Epidemiol. *34*, 161–170.

McPherson, E.J., Woodson, C., Holtom, P., Roidis, N., Shufelt, C., and Patzakis, M. (2002). Periprosthetic total hip infection: outcomes using a staging system. Clin. Orthop. Relat. Res. *403*, 8–15.

Meghji, S., Crean, S.J., Hill, P.A., Sheikh, M., Nair, S.P., Heron, K., Henderson, B., Mawer, E.B., and Harris, M. (1998). Surface-associated protein from *Staphylococcus aureus* stimulates osteoclastogenesis: possible role in *S. aureus*-induced bone pathology. Br. J. Rheumatol., *37*, 1095–1101.

Merino, N., Toledo-Arana, A., Vergara-Irigaray, M., Valle, J., Solano, C., Calvo, E., Lopez, J.A., Foster, T.J., Penadés, J.R., and Lasa, I. (2009). Protein A-mediated multicellular behavior in *Staphylococcus aureus*. J. Bacteriol. *191*, 832–843.

Miller, L.G., Perdreau-Remington, F., Rieg, G., Mehdi, S., Perlroth, J., Bayer, A.S., Tang, A.W., Phung, T.O., and Spellberg, B. (2005). Necrotizing fasciitis caused by community-associated methicillin-resistant *Staphylococcus aureus* in Los Angeles. N. Engl. J. Med. *352*, 1445–1453.

Moenster, R.P., Linneman, T.W., Call, W.B., Kay, C.L., McEvoy, T.A., and Sanders, J.L. (2013). The potential role of newer gram-positive antibiotics in the setting of osteomyelitis of adults. J. Clin. Pharm. Ther. *38*, 89–96.

Mongkolrattanothai, K., Boyle, S., Kahana, M.D., and Daum, R.S. (2003). Severe *Staphylococcus aureus* infections caused by clonally related community-acquired methicillin-susceptible and methicillin-resistant isolates. Clin. Infect. Dis., *37*, 1050–1058.

Moran, G.J., Krishnadasan, A., Gorwitz, R.J., Fosheim, G.E., McDougal, L.K., Carey, R.B., Talan, D.A., and Talan, D.A. (2006). Methicillin-resistant *S. aureus* infections among patients in the emergency department. N. Engl. J. Med. 355, 666–674.

Moreillon, P., Que, Y.andA., and Bayer, A.S. (2002). Pathogenesis of streptococcal and staphylococcal endocarditis. Infect. Dis. Clin. North. Am. *16*, 297–318.

Mrak, L.N., Zielinska, A.K., Beenken, K.E., Mrak, I.N., Atwood, D.N., Griffin, L.M., Lee, C.Y., and Smeltzer, M.S. (2012). *saeRS.*, and *sarA* act synergistically to repress protease production and promote biofilm formation in *Staphylococcus aureus*. PLoS One 7, e38453.

Mulholland, E.K., and Adegbola, R.A. (2005). Bacterial infections–a major cause of death among children in Africa. N. Engl. J. Med. 352, 75–77.

Musher, D.M., Lamm, N., Darouiche, R.O., Young, E.J., Hamill, R.J., and Landon, G.C. (1994). The current spectrum of *Staphylococcus aureus* infection in a tertiary care hospital. Medicine 73, 186–208.

Mylotte, J.M., McDermott, C., and Spooner, J.A. (1987). Prospective study of 114 consecutive episodes of *Staphylococcus aureus* bacteremia. Rev. Infect. Dis. *9*, 891–907.

Nair, S.P., Williams, R.J., and Henderson, B. (2000). Advances in our understanding of the bone and joint pathology caused by *Staphylococcus aureus* infection. Rheumatology 39, 821–834.

Nelson, C.L., McLaren, A.C., McLaren, S.G., Johnson, J.W., and Smeltzer, M.S. (2005). Is aseptic loosening truly aseptic? Clin. Orthop. Relat. Res. *437*, 25–30.

Neut, D., van der Mei, H.C., Bulstra, S.K., and Busscher, H.J. (2007). The role of small-colony variants in failure to diagnose and treat biofilm infections in orthopedics. Acta Orthop. 78, 299–308.

Nilsson, I.M., Hartford, O., Foster, Tand., and Tarkowski, A. (1999). Alpha-toxin and gamma-toxin jointly promote *Staphylococcus aureus* virulence in murine septic arthritis. Infect. Immun. *67*, 1045–1049.

O'neill, E., Pozzi, C., Houston, P., Humphreys, H., Robinson, D.A., Loughman, A., Foster, T.J., and O'gara, J.P. (2008). A novel *Staphylococcus aureus* biofilm phenotype mediated by the fibronectin-binding proteins, FnBPA., and FnBPB.J. Bacteriol., *190*, 3835–3850.

Oh, J., Conlan, S., Polley, E.C., Segre, J.A., and Kong, H.H. (2012). Shifts in human skin and nares microbiota of healthy children and adults. Genome Med. *4*, 77.

Oh, J., Freeman, A.F., Park, M., Sokolic, R., Candotti, F., Holland, S.M., Segre, J.A., Kong, H.H., NISC Comparative Sequencing Program and Kong, H.H. (2013). The altered landscape of the human skin microbiome in patients with primary immunodeficiencies. Genome Res. 23, 2103–2114.

Otter, J.A., and French, G.L. (2010). Molecular epidemiology of community-associated meticillin-resistant *Staphylococcus aureus* in Europe. Lancet Infect. Dis. *10*, 227–239.

Panlilio, A.L., Culver, D.H., Gaynes, R.P., Banerjee, S., Henderson, T.S., Tolson, J.S., and Martone, W.J. (1992). Methicillin-resistant *Staphylococcus aureus* in U.S. hospitals, 1975–1991. Infect. Control Hosp. Epidemiol. *13*, 582–586.

Parker, A.C., Beenken, K.E., Jennings, J.A., Hittle, L., Shirtliff, M.E., Bumgardner, J.D., Smeltzer, M.S., and Haggard, W.O. (2015). Characterization of local delivery with amphotericin B., and vancomycin from modified chitosan sponges and functional biofilm prevention evaluation. J. Orthop. Res. 33, 439–447.

Peltola, H., and Pääkkönen, M. (2014). Acute osteomyelitis in children. N. Engl. J. Med. *370*, 352–360.

Peschel, A., and Otto, M. (2013). Phenol-soluble modulins and staphylococcal infection. Nat. Rev. Microbiol. *11*, 667–673.

Pittet, D., Tarara, D., and Wenzel, R.P. (1994). Nosocomial bloodstream infection in critically ill patients. Excess length of stay, extra costs, and attributable mortality. JAMA *271*, 1598–1601.

Priest, N.K., Rudkin, J.K., Feil, E.J., van den Elsen, J.M., Cheung, A., Peacock, S.J., Laabei, M., Lucks, D.A., Recker, M., and Massey, R.C. (2012). From genotype to phenotype: can systems biology be used to predict *Staphylococcus aureus* virulence? Nat. Rev. Microbiol. *10*, 791–797.

Proctor, R.A., van Langevelde, P., Kristjansson, M., Maslow, J.N., and Arbeit, R.D. (1995). Persistent and relapsing infections associated with small-colony variants of *Staphylococcus aureus*. Clin. Infect. Dis. *20*, 95–102.

Proctor, R.A., von Eiff, C., Kahl, B.C., Becker, K., McNamara, P., Herrmann, M., and Peters, G. (2006). Small colony variants: a pathogenic form of bacteria that facilitates persistent and recurrent infections. Nat. Rev. Microbiol. *4*, 295–305.

Pulia, M.S., Calderone, M.R., Meister, J.R., Santistevan, J., and May, L. (2014). Update on management of skin and soft tissue infections in the emergency department. Curr. Infect. Dis. Rep. *16*, 418.

Qualls, M.L., Mooney, M.M., Camargo, C.A., Jr., Zucconi, T., Hooper, D.C., and Pallin, D.J. (2012). Emergency department visit rates for abscess versus other skin infections during the emergence of community-associated methicillin-resistant *Staphylococcus aureus*, 1997–2007. Clin. Infect. Dis. *55*, 103–105.

Que, Y.A., and Moreillon, P. (2011). Infective endocarditis. Nat. Rev. Cardiol. *8*, 322–336.

Rao, N., Ziran, B.H., and Lipsky, B.A. (2011). Treating osteomyelitis: antibiotics and surgery. Plast. Reconstr. Surg. *127* (Suppl. 1), 177S–187S.

Rasigade, J.P., Trouillet-Assant, S., Ferry, T., Diep, B.A., Sapin, A., Lhoste, Y., Ranfaing, J., Badiou, C., Benito, Y., Bes, M., and*et al.* (2013). PSMs of hypervirulent *Staphylococcus aureus* act as intracellular toxins that kill infected osteoblasts. PLoS One *8*, e63176.

Riise, O.R., Kirkhus, E., Handeland, K.S., Flato, B., Reiseter, T., Cvancarova, M., Nakstad, B., and Wathne, K.O. (2008). Childhood osteomyelitis-incidence and differentiation from other acute onset musculoskeletal features in a population-based study. BMC Pediatr., *8*, 45.

Roderick, M.R., and Ramanan, A.V. (2013). Chronic recurrent multifocal osteomyelitis. Adv. Exp. Med. Biol. *764*, 99–107.

Ryden, C., Maxe, I., Franzen, A., Ljungh, A., Heinegard, D., and Rubin, K. (1987). Selective binding of bone matrix sialoprotein to *Staphylococcus aureus* in osteomyelitis. Lancet. 2, 515.

Rydén, C., Tung, H.S., Nikolaev, V., Engström, Aand., and Oldberg, A. (1997). *Staphylococcus aureus* causing osteomyelitis binds to a nonapeptide sequence in bone sialoprotein. Biochem. J. *327*, 825–829.

Safdar, Nand., and Bradley, E.A. (2008). The risk of infection after nasal colonization with *Staphylococcus aureus*. Am. J. Med. *121*, 310–315.

Salgado-Pabón, W., Case-Cook, L.C., and Schlievert, P.M. (2014). Molecular analysis of staphylococcal superantigens. Methods Mol. Biol. *1085*, 169–185.

Sanchez, C.J., Jr., Ward, C.L., Romano, D.R., Hurtgen, B.J., Hardy, S.K., Woodbury, R.L., Trevino, A.V., Rathbone, C.R., and Wenke, J.C. (2013). *Staphylococcus aureus* biofilms decrease osteoblast viability, inhibits osteogenic differentiation, and increases bone resorption *in vitro*. BMC Musculoskelet. Disord. *14*, 187.

Sanford, J.A., and Gallo, R.L. (2013). Functions of the skin microbiota in health and disease. Semin. Immunol. *25*, 370–377.

Schaumburg, F., Alabi, A.S., Peters, Gand., and Becker, K. (2014). New epidemiology of *Staphylococcus aureus* infection in Africa. Clin. Microbiol. Infect. *20*, 589–596.

Schaumburg, F., Köck, R., Friedrich, A.W., Soulanoudjingar, S., Ngoa, U.A., von Eiff, C., Issifou, S., Kremsner, P.G., Herrmann, M., Peters, Gand., *et al.* (2011). Population structure of *Staphylococcus aureus* from remote African Babongo Pygmies. PLOS Negl. Trop. Dis. *5*, e1150.

Schechter-Perkins, E.M., Mitchell, P.M., Murray, K.A., Rubin-Smith, J.E., Weir, S., and Gupta, K. (2011). Prevalence and predictors of nasal and extranasal staphylococcal colonization in patients presenting to the emergency department. Ann. Emerg. Med. *57*, 492–499.

Schlievert, P.M. (1993). Role of superantigens in human disease. J. Infect. Dis. *167*, 997–1002.

Schwartz, K., Syed, A.K., Stephenson, R.E., Rickard, A.H., and Boles, B.R. (2012). Functional amyloids composed of phenol soluble modulins stabilize *Staphylococcus aureus* biofilms. PLOS Pathog. *8*, e1002744.

Sever-Chroneos, Z., Krupa, A., Davis, J., Hasan, M., Yang, C.H., Szeliga, J., Herrmann, M., Hussain, M., Geisbrecht, B.V., Kobzik, L., and Chroneos, Z.C. (2011). Surfactant protein A (SP-A)-mediated clearance of *Staphylococcus aureus* involves binding of SP-A to the staphylococcal adhesin eap and the macrophage receptors SP-A receptor 210 and scavenger receptor class A. J. Biol. Chem., *286*, 4854–4870.

Smeltzer, M.S., Thomas, J.R., Hickmon, S.G., Skinner, R.A., Nelson, C.L., Griffith, D., Parr, T.R., andand Evans, R.P. (1997). Characterization of a rabbit model of staphylococcal osteomyelitis. J. Orthop. Res. *15*, 414–421.

Sohail, M.R., Martin, K.R., Wilson, W.R., Baddour, L.M., Harmsen, W.S., and Steckelberg, J.M. (2006). Medical versus surgical management of *Staphylococcus aureus* prosthetic valve endocarditis. Am. J. Med. *119*, 147–154.

Souvenir, D., Anderson, D.E., Jr., Palpant, S., Mroch, H., Askin, S., Anderson, J., Claridge, J., Eiland, J., Malone, C., Garrison, M.W., Watson, P., and Campbell, D.M. (1998). Blood cultures positive for coagulase-negative staphylococci: antisepsis, pseudobacteremia, and therapy of patients. J. Clin. Microbiol. *36*, 1923–1926.

Speller, D.C., Johnson, A.P., James, D., Marples, R.R., Charlett, A., and George, R.C. (1997). Resistance to methicillin and other antibiotics in isolates of *Staphylococcus aureus* from blood and cerebrospinal fluid, England and Wales, 1989–95. Lancet *350*, 323–325.

Stapels, D.A., Ramyar, K.X., Bischoff, M., von Köckritz-Blickwede, M., Milder, F.J., Ruyken, M., Eisenbeis, J., McWhorter, W.J., Herrmann, M., van Kessel, K.P., and*et al.* (2014). *Staphylococcus aureus* secretes a unique class of neutrophil serine protease inhibitors. Proc. Natl. Acad. Sci. U.S.A. *111*, 13187–13192.

Steckelberg, J.M., and Osmon, D.R. (1994). Prosthetic joint infections. *In:* Infections Associated with Indwelling Medical Devices. A.L., Bisno and F.A., Waldvogel (eds.) Washington D.C.: ASM Press.

Steinberg, J.P., Clark, C.C., and Hackman, B.O. (1996). Nosocomial and community-acquired *Staphylococcus aureus* bacteraemias from 1980 to 1993: impact of intravascular devices and methicillin resistance. Clin. Infect. Dis. *23*, 255–259.

Sugimoto, S., Iwamoto, T., Takada, K., Okuda, K., Tajima, A., Iwase, T., and Mizunoe, Y. (2013). *Staphylococcus epidermidis* Esp degrades specific proteins associated with *Staphylococcus aureus* biofilm formation and host-pathogen interaction. J. Bacteriol. *195*, 1645–1655.

Sully, E.K., Malachowa, N., Elmore, B.O., Alexander, S.M., Femling, J.K., Gray, B.M., DeLeo, F.R., Otto, M., Cheung, A.L., Edwards, B.S., and*et al.* (2014). Selective chemical inhibition of *agr* quorum sensing in *Staphylococcus aureus* promotes host defense with minimal impact on resistance. PLOS Pathog. *10*, e1004174.

Surewaard, B.G., de Haas, C.J., Vervoort, F., Rigby, K.M., DeLeo, F.R., Otto, M., van Strijp, J.A., and Nijland, R. (2013). Staphylococcal alpha-phenol soluble modulins contribute to neutrophil lysis after phagocytosis. Cell. Microbiol. *15*, 1427–1437.

Tadema, H., Heeringa, P., and Kallenberg, C.G. (2011). Bacterial infections in Wegener's granulomatosis: mechanisms potentially involved in autoimmune pathogenesis. Curr. Opin. Rheumatol. *23*, 366–371.

Taira, B.R., Singer, A.J., Thode, H.C., and Lee, C.C. (2009). National epidemiology of cutaneous abscesses: 1996 to 2005. Am. J. Emerg. Med. *27*, 289–292.

Ternes, Y.M., Lamaro-Cardoso, J., André, M.C., Pessoa, V.P., Vieira, M.A., Minamisava, R., Andrade, A.L., and Kipnis, A. (2013). Molecular epidemiology of coagulase-negative *Staphylococcus* carriage in neonates admitted to an intensive care unit in Brazil. BMC Infect. Dis. *13*, 572.

Testoni, F., Montanaro, L., Poggi, A., Visai, L., Campoccia, Dand., and Arciola, C.R. (2011). Internalization by osteoblasts of two *Staphylococcus aureus* clinical isolates differing in their adhesin gene pattern. Int. J. Artif. Organs *34*, 789–798.

Tom, S., Galbraith, J.C., Valiquette, L., Jacobsson, G., Collignon, P., Schonheyder, H.C., Sogaard, M., Kennedy, K.J., Knudsen, J.D., Ostergaard, C., Lyytikainen, O., and Laupland, K.B. (2014). Case fatality ratio and mortality rate trends of community-onset *Staphylococcus aureus* bacteraemia. Clin. Microbiol. Infect. *20*, O630–O632.

Tomita, T., and Kamio, Y. (1997). Molecular biology of the pore-forming cytolysins from *Staphylococcus aureus*, alpha- and gamma-hemolysins and leucocidin. Biosci. Biotechnol. Biochem. *61*, 565–572.

van Belkum, A., Verkaik, N.J., de Vogel, C.P., Boelens, H.A., Verveer, J., Nouwen, J.L., Verbrugh, H.A., and Wertheim, H.F. (2009). Reclassification of *Staphylococcus aureus* nasal carriage types. J. Infect. Dis. *199*, 1820–1826.

von Eiff, C. (2008). *Staphylococcus aureus* small colony variants: a challenge to microbiologists and clinicians. Int. J. Antimicrob. Agents *31*, 507–510.

von Eiff, C., Becker, K., Machka, K., Stammer, H., and Peters, G. (2001). Nasal carriage as a source of *Staphylococcus aureus* bacteremia. Study Group. N. Engl. J. Med. *344*, 11–16.

Von Eiff, C., Vaudaux, P., Kahl, B.C., Lew, D., Emler, S., Schmidt, A., Peters, G., and Proctor, R.A. (1999). Bloodstream infections caused by small-colony variants of coagulase-negative staphylococci following pacemaker implantation. Clin. Infect. Dis. *29*, 932–934.

Vuong, C., Saenz, H.L., Götz, F., and Otto, M. (2000). Impact of the *agr* quorum-sensing system on adherence to polystyrene in *Staphylococcus aureus*. J. Infect. Dis. *182*, 1688–1693.

Wang, H., Von Rohrscheidt, J., Roehrbein, J., Peters, T., Sindrilaru, A., Kess, D., Preissner, K.T., and Scharffetter-Kochanek, K. (2010). Extracellular adherence protein of *Staphylococcus aureus* suppresses disease by inhibiting T-cell recruitment in a mouse model of psoriasis. J. Invest. Dermatol. *130*, 743–754.

Wang, R., Braughton, K.R., Kretschmer, D., Bach, T.H., Queck, S.Y., Li, M., Kennedy, A.D., Dorward, D.W., Klebanoff, S.J., Peschel, A., *et al.* (2007). Identification of novel cytolytic peptides as key virulence determinants for community-associated MRSA. Nat. Med. *13*, 1510–1514.

Wanke, I., Steffen, H., Christ, C., Krismer, B., Götz, F., Peschel, A., Schaller, M., and Schittek, B. (2011). Skin commensals amplify the innate immune response to pathogens by activation of distinct signaling pathways. J. Invest. Dermatol. *131*, 382–390.

Weichert, S., Sharland, M., Clarke, N.M., and Faust, S.N. (2008). Acute haematogenous osteomyelitis in children: is there any evidence for how long we should treat? Curr. Opin. Infect. Dis. *21*, 258–262.

Wertheim, H.F., Melles, D.C., Vos, M.C., van Leeuwen, W., van Belkum, A., Verbrugh, H.A., and Nouwen, J.L. (2005). The role of nasal carriage in *Staphylococcus aureus* infections. Lancet Infect. Dis. *5*, 751–762.

Wertheim, H.F., Vos, M.C., Ott, A., van Belkum, A., Voss, A., Kluytmans, J.A., van Keulen, P.H., Vandenbroucke-Grauls, C.M., Meester, M.H., and Verbrugh, H.A. (2004). Risk and outcome of nosocomial *Staphylococcus aureus* bacteraemia in nasal carriers versus non-carriers. Lancet *364*, 703–705.

Widaa, A., Claro, T., Foster, T.J., O'Brien, F.J., and Kerrigan, S.W. (2012). *Staphylococcus aureus* protein A plays a critical role in mediating bone destruction and bone loss in osteomyelitis. PLoS One 7, e40586.

Williams, R.E. (1963). Healthy carriage of *Staphylococcus aureus*: its prevalence and importance. Bacteriol. Rev. *27*, 56–71.

Yarwood, J.M., McCormick, J.K., and Schlievert, P.M. (2001). Identification of a novel two-component regulatory system that acts in global regulation of virulence factors of *Staphylococcus aureus*. J. Bacteriol. *183*, 1113–1123.

Zielinska, A.K., Beenken, K.E., Mrak, L.N., Spencer, H.J., Post, G.R., Skinner, R.A., Tackett, A.J., Horswill, A.R and Smeltzer, M.S. (2012). *sarA*-mediated repression of protease production plays a key role in the pathogenesis of *Staphylococcus aureus* USA300 isolates. Mol. Microbiol. *86*, 1183–1196.

Zimmerli, W., Trampuz, A., and Ochsner, P.E. (2004). Prosthetic-joint infections. N. Engl. J. Med. *351*, 1645–1654.

Staphylococcus: Clinical Significance in Animals

3

John Dustin Loy

Abstract

The staphylococci are important pathogens of many animal species including domestic and wild mammals and birds. Many of the animal associated staphylococci have undergone extensive host adaptation to specific body sites. Like human infections, infections in animals with staphylococci frequent the skin and mucous membranes and given the opportunity can cause severe invasive disease and septicaemia. They also possess numerous virulence factors and mechanisms that enable a diverse number of clinical manifestations and disease syndromes in host species. The introduction of molecular phylogenetics into veterinary clinical microbiology has found diversity among what were thought to only be a handful of clinically important species, and significant taxonomic changes have occurred. Like human medicine, challenges are also being faced by veterinary clinicians as methicillin resistance is emerging in animal infections. This chapter describes the role of staphylococci in animal disease, important virulence factors associated with animal disease, treatment of significant veterinary infections, as well as antimicrobial resistance in staphylococci associated with animals.

Introduction

The genus *Staphylococcus* includes the catalase producing members of the Gram-positive aerobic cocci. This genus includes more than 50 species and subspecies that have been characterized, and many of these have been isolated from animals (Parte, 2014) (Table 3.1). Several of these species are also significant pathogens in domestic, wild and zoo mammals, and birds; however, the majority of staphylococci are commensals and have evolved to colonize epithelial surfaces including skin and mucous membranes (Kloos *et al.*, 1976). Each animal host species often carries distinct populations of staphylococci that are intimately associated with that species, and for some of these associations, extensive co-evolution of microbe and host has been documented (Aarestrup, 2001). Even though many staphylococci are commensal in nature, they are flexible opportunistic pathogens; hence, animals can develop infections that are not necessarily restricted to a single host or body site. Like staphylococcal opportunistic infections in humans, infections in animals typically involve disruption of the epithelium or mucous membranes where the bacteria are introduced into deeper tissues by trauma, foreign bodies, abrasions, thermal injury or via hair follicles. Hypersensitivity reactions, which alter epithelial surfaces and reduce epithelial integrity, immune disorders, or disruption of localized defences, can potentiate disease by promoting an environment more hospitable to the establishment of infection. Many of the pathogenic *Staphylococcus* species have evolved a diverse repertoire of virulence factors to facilitate persistence in tissues through adherence, immune evasion, and secreted factors that promote tissue necrosis. Subsequent to colonization of skin and mucous membranes, many staphylococci also produce virulence factors that promote severe invasive disease including sepsis and internal abscess formation that cause diseases ranging from chronic to life-threatening.

The most important and significant *Staphylococcus* species in veterinary medicine is *Staphylococcus aureus*, a highly adaptive pathogen that causes diseases in both domestic animals and wildlife. Infections in animals with *S. aureus* can be caused by host species specific 'ecovars' that have

Table 3.1 Select *Staphylococcus* spp. associated with diseases in animals or isolated primarily from animals

Staphylococcus species	Location found or disease	Host or affected species
Staphylococcus agnetis	Subclinical or mild mastitis	Cattle
Staphylococcus arlettae	Skin flora	Poultry and horses
Staphylococcus aureus	Wound, mastitis, skin infections, septicaemia, joint, diarrhoea, vaginal infections	All animals
Staphylococcus aureus subsp. *anaerobius*	Mastitis, abscesses, opportunistic	Ruminants
Staphylococcus caprae	Skin flora	Goats
Staphylococcus chromogenes	Exudative dermatitis	Swine
	Mastitis	Cattle
	Dermatitis	Others
Staphylococcus delphini	Opportunistic and skin enteritis	Horses and dolphins, mink
Staphylococcus devriesei	Skin flora, mastitis	Bovine
Staphylococcus epidermidis	Mastitis	Cattle
Staphylococcus equorom	Skin flora	Equine
Staphylococcus felis	Soft tissue infections, otitis, cystitis	Cats
Staphylococcus gallinarum	Skin flora	Poultry
Staphylococcus kloosii	Skin flora	Wild rodents
Staphylococcus lentus	Skin flora	Sheep and goats
Staphylococcus lutrae	Skin flora	Otters
Staphylococcus microti	Skin flora	Voles
Staphylococcus nepalensis	Respiratory tract	Goats
Staphylococcus hyicus	Exudative dermatitis	Pigs
	Dermatitis	Pigs, cattle, horses
	Mastitis	Cattle
Staphylococcus pseudintermedius	Pyoderma, otitis, dermatitis, opportunistic infections	Dogs and cats, other species
Staphylococcus intermedius	Skin flora/opportunist	Pigeons
Staphylococcus rostri	Nasal flora	Swine
Staphylococcus schleiferi	Pyoderma, otitis	Dogs
Staphylococcus scuri	Skin flora	Squirrels
Staphylococcus simulans	Skin	Cats
	Mastitis	Cattle
Staphylococcus stepanovicii	Flora	Wild rodents
Staphylococcus warneri	Septicaemia, meningoencephalitis	Birds, dogs

undergone extensive genome diversification and adaptation after introduction to a new host species or environmental niche (Guinane *et al.*, 2010). This is an important concept, as the species *S. aureus* in the context of veterinary medicine is a collection of diverse ecovars, some of which have undergone extensive host-specific adaptation. Molecular characterization of *S. aureus* has found many of these associations and ecovars are associated with specific clonal complexes (CC) or sequence types (ST). Other newly recognized species of staphylococci are also increasingly being recognized in veterinary diseases. Diagnostic advances have enabled techniques such as polymerase chain reaction (PCR) and DNA sequencing to become routine in veterinary clinical microbiology laboratories, and subsequently an increasing number of species that had once been lumped together by classical biochemical testing are now being identified as pathogens. Other paradigms are also shifting, where dogma once held that coagulase-negative staphylococci (CoNS) lacked pathogenicity and were merely commensals, recent research has shown that many species of CoNS can cause opportunistic and subclinical infections in animals. Similar to challenges faced in human medicine, treatment of staphylococcal infections in animals has become increasingly difficult due to the emergence and spread of antimicrobial resistance. Methicillin-resistant *S. aureus* (MRSA) in animals was first described in 1972 in cows with mastitis (Devriese *et al.*, 1972). MRSA has since emerged as a significant problem in some animals, especially in hospital acquired infections. Food safety has also become a topic of interest as livestock-associated MRSA (LA-MRSA) is increasingly recognized in swine and other food animals and the potential impacts to the food supply are an active area of inquiry. Methicillin resistance is also being observed in other clinically important *Staphylococcus* species, such as methicillin-resistant *S. pseudintermedius* (MRSP) in dogs and cats (Weese and van Duijkeren, 2010).

Staphylococci as veterinary pathogens

Most staphylococci are readily cultured on blood agar from veterinary clinical specimens and form white to orange colonies after 18 h at 37°C that are 1–3 mm in diameter (Markey *et al.*, 2013). The use of blood agar as a primary isolation medium serves a dual purpose; providing nutrients for growth and enabling the rapid evaluation of haemolysis. However, the staphylococci present a diagnostic challenge in the clinical microbiology laboratory, as they may represent contaminants, normal flora, or possible pathogens in specimens. This challenge is compounded in veterinary microbiology where each host species can carry diverse populations of many species of *Staphylococcus*, which must be assessed for possible pathogenicity. To aide in this assessment, initially staphylococci were categorized into pathogens and non-pathogens (commensals) by the production of coagulase, which causes coagulation of blood plasma *in vitro*, a property first observed in 1903 (Loeb, 1903). The coagulase test allowed for clinical laboratories to rapidly group isolates into the coagulase positive and negative staphylococci (Kloos and Schleifer, 1975). The coagulase-positive staphylococci (CoPS) represent probable pathogens and are thus the species most likely associated with a disease process in animals. If isolates are found to be coagulase positive, biochemical tests are routinely used to differentiate among the CoPS and include: Voges–Proskauer (acetoin and 2,3-butanediol production), *ortho*-nitrophenyl-β-galactoside (beta-galactosidase production), sugar fermentation (mannitol and mannose) and polymyxin B susceptibility (Markey *et al.*, 2013). However, for many members of the CoPS routine biochemical testing can be challenging to definitively interpret, as there is significant phenotypic and reservoir host overlap amongst the species. For this reason, polymerase chain reaction (PCR) assays that target the thermonuclease gene have become a routine diagnostic method for the reliable differentiation amongst several species of CoPS (Sasaki *et al.*, 2010). Coagulase-negative staphylococci (CoNS) are increasingly being recognized as possible pathogens, having been found to be associated with diseases in animals. Traditionally they were thought to be limited to commensal members of the skin flora and non-pathogens. However, the CoNS are frequently isolated from wound infections and from tissues associated with foreign bodies such as indwelling catheters in both humans and animals (Huebner and Goldmann, 1999).

Virulence factors

Although staphylococcal virulence factors are discussed in detail in other chapters of this book, a brief overview of some of the important virulence factors relevant to veterinary medicine are described.

Coagulase

Staphylococci are able to induce blood plasma clot formation, which is thought to aid pathogenesis by sequestering infected sites from leucocyte infiltration and preventing immune recognition by the host. Clot formation is incited by the secretion of free coagulase, an extracellular protein which interacts with prothrombin in the host to promote clot formation. The formation of this clot is the functional basis for the tube coagulase test (DUTHIE and LORENZ, 1952). Clot formation occurs by the conversion of fibrinogen to fibrin by the staphylothrombin complex. Staphylococci also produce a cell associated coagulase, called bound coagulase, which causes clumping that can be assayed using the slide coagulase test (Duthie, 1954; Much, 1908). Coagulase is an important virulence factor as classically, the coagulation of plasma is seen as one of the key virulence factors to assist in determining the medical importance of an isolate in the veterinary clinical laboratory. The rapid testing and association with virulence makes it a useful tool to assist in evaluating the pathogenicity *in vitro*. CoPS of veterinary importance include *S. aureus*, *S. hyicus*, *S. schleiferi* subsp. *coagulans* and the *S. intermedius* group (SIG) consisting of *S. intermedius*, *S. delphini* and *S. pseudintermedius*, all of which are significant pathogens to a variety of animal species (Fig. 3.1). A newly described member of the genus *Staphylococcus*, *Staphylococcus agnetis*, is coagulase variable (Taponen *et al.*, 2012).

Adhesins

One requirement of pathogens is that they must be able to adhere to host tissues. To facilitate adhesion, staphylococci produce adhesins on the cell surface that potentiate adherence to host surfaces, including collagens, fibronectins, and lamins that form much of the extracellular matrix in host tissues. These have been dubbed microbial surface components recognizing adhesive matrix molecules (MSCRAMMS) which promote the formation of biofilms. These adhesins and biofilms are incredibly important in the pathogenesis of many staphylococci infections in animals. MSCRAMMs are critical to maintaining infections in mammary tissue that occurs during bovine mastitis, where a constant efflux of secretions would normally disallow bacterial colonization (Foster and Höök, 1998).

Capsular polysaccharide

The presence of capsular polysaccharide in *S. aureus* was first described in 1931 (Gilbert, 1931). Capsular polysaccharides have been typed based on recognition by specific antisera and eight types were initially described, with additions to make 11 total. (Karakawa *et al.*, 1985; Sompolinsky *et al.*, 1985) *S. aureus* serotypes 1 and 2 comprise the heavily encapsulated and mucoid strains, which are less frequently encountered clinically. The majority of both human and animal isolates of *S. aureus* are non-mucoid, encapsulated and of serotype 5 or 8 (O'Riordan and Lee, 2004). These include *S. aureus* isolates from cows, rabbits, poultry, pigs, and horses (O'Riordan and Lee, 2004). Polysaccharide capsules are important because they prevent immunoglobulin binding and interfere with immune recognition and subsequent

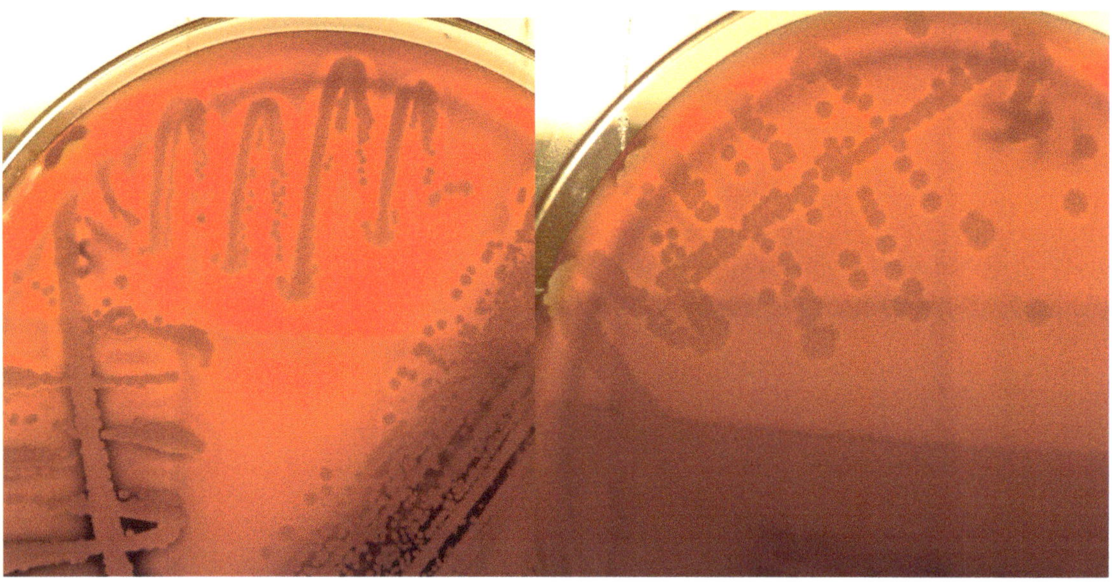

Figure 3.1 *Staphylococcus intermedius* (left) and *Staphylococcus pseudintermedius* (right). Two haemolytic and coagulase-positive members of the veterinary staphylococci.

opsonization. This enables persistent colonization in the bloodstream of the host (O'Riordan and Lee, 2004). Some capsule types have also been shown to impede phagocytic killing by neutrophils (Xu *et al.*, 1992).

Peptidoglycan and teichoic acid

Staphylococcal peptidoglycan and other cell wall components are thought to stimulate non-specific immune responses via Toll like receptors (TLR) (Schwandner *et al.*, 1999). TLR activation results in secretion of pro-inflammatory cytokines and subsequent inflammatory states in the host (Fournier and Philpott, 2005). Teichoic acids are also present in the cell walls of staphylococci. Most species of *Staphylococcus* contain a polyolphosphate teichoic acid with glycerol and/or ribitol as polyol component (Endl *et al.*, 1983). Wall teichoic acids of *S. aureus* have been shown to be critical in enabling long-term colonization of nasal passages also been shown to reduce immune recognition of peptidoglycan and assist in host immune evasion (Atilano *et al.*, 2011; Weidenmaier *et al.*, 2004). Staphylococcal lipoteichoic acid has been shown to be a critical pro-inflammatory mediator in the bovine immune response to mastitis, triggering the infiltration of neutrophils from the blood to the mammary tissue inducing somatic cell secretion into the milk (Wellnitz *et al.*, 2011). This inflammatory response is often moderated and delayed when compared with other mastitis pathogens, such as *Escherichia coli*, and is thought to enable *S. aureus* to cause chronic subclinical mastitis in infected cows (Wellnitz and Bruckmaier, 2012).

Lipase

Lipids and fatty acids are commonly present in sebaceous skin secretions and accumulate in abscesses. Additionally, mammalian milk contains significant percentages of fatty acids, with bovine milk containing 4–5% (Månsson, 2008). These compounds can disrupt bacterial membranes and cause lysis and can be toxic to many pathogenic bacteria. However, *S. aureus* produces fatty acid metabolizing enzyme (FAME) or, glycerol ester hydrolase, a lipase that can release nutrients and inhibit immune responses (MORTENSEN *et al.*, 1992). Such an enzyme is likely involved in the potentiation of abscess formation as lipase can inactivate bactericidal lipid (Shryock *et al.*, 1992). Lipase has been shown to be involved in biofilm formation and peritoneal abscess formation. Additionally, anti-lipase immune responses have been shown to protect mice from lethal challenge (Hu *et al.*, 2012). Abscess formation is critical to many veterinary infections establishing chronic infections, especially in poultry and rabbits. Lipase is also thought to be involved in the pathogenesis of bovine mastitis, enabling the bacteria to breakdown lipids in milk, allowing rapid growth in milk substrates and subsequent colonization of the mammary gland (Sutra and Poutrel, 1994).

Hyaluronidase

Hyaluronidase or hyaluronate lyase (HL), sometimes referred to as 'spreading factor', is an enzyme produced by a variety of Gram-positive bacteria associated with the skin (Linker *et al.*, 1956). The majority of *S. aureus* strains produce HL (Choudhuri and Chakrabarty, 1969) as well as *Staphylococcus hyicus* (Devriese, 1990). HL mainly degrades hyaluronic acid but some can degrade chondroitin and chondroitin sulfate to a lesser extent and at a lower degradation rate. (Hynes and Walton, 2000) These enzymes are thought to promote virulence by enabling breakdown and allowing further invasion and spread through host soft tissues, as hyaluronate is a major component of connective tissues (Hynes and Walton, 2000). Studies in mice have shown that HL appears to be important in the early stages of subcutaneous infections, where HL mutants have attenuated pathology (Makris *et al.*, 2004). *S. hyicus* is the causative agent for porcine exudative epidermitis or 'greasy pig disease' and HL may be involved the pathogenesis of this disease as HL can potentiate soft tissue infections in other species. The presence of HL is also useful clinically to differentiate *S. hyicus* from closely related species such as *S. chromogenes* (Markey *et al.*, 2013; Skalka, 1985).

Haemolysins

Alpha toxin or alpha-haemolysin (*hla*) of *S. aureus* is the most well-studied of the staphylococcal exotoxins. Alpha-haemolysin causes complete, or beta, haemolysis of mammalian red blood cells. Alpha-haemolysin requires specific receptors on susceptible cells to enable toxin binding and subsequent pore formation and cell lysis via osmotic swelling (Harshman *et al.*, 1989). Toxin binding to a large number of host cells can promote cytokine release and can promote septic shock like conditions following high levels of exposure. Alpha-haemolysin is dermonecrotic and neurotoxic and is likely involved in a variety of disease states in animals. Much of the pathophysiology of alpha-haemolysin is thought to be mediated by influx of Ca^{2+} ions through pores and can be attributed to the subsequent secondary effects of Ca^{2+} activation (Bhakdi and Tranum-Jensen, 1991). Ca^{2+} influx triggers a variety of downstream effects depending on the effected cell type and includes blood coagulation, vasoconstriction, pulmonary oedema and possibly vascular collapse (Bhakdi and Tranum-Jensen, 1991; Dinges *et al.*, 2000). Alpha-haemolysin is an important virulence factor and useful for in the clinical microbiology laboratory for evaluating growth characteristics on blood agar. Haemolysis induced by staphylococcal haemolysin is species specific, with α-haemolysin causing lysis of rabbit, sheep, and cattle red blood cells, but not horse, human or chicken (Bhakdi and Tranum-Jensen, 1991; Kenny *et al.*, 1992). Rabbit erythrocytes in particular are extremely sensitive to α-haemolysin, with sensitivity being 100-fold higher than those of other animals and more than 1000 fold more than human erythrocytes (Bhakdi *et al.*, 1984).

Beta-haemolysin (*hlb*) is a sphingomyelinase C or a type C phospholipase (Doery *et al.*, 1963). *Hlb* causes incomplete, or alpha, haemolysis of blood cells. This is also called hot-cold lysis as incubation at lower temperatures can promote further haemolysis to beta, or complete haemolysis (Glenny and Stevens, 1935). *Hlb*, like *hla*, is species specific, with sheep and cattle blood cells being affected, and horse, rabbit and chicken being unaffected (Dinges *et al.*, 2000). Differing susceptibilities of erythrocytes of different species to *hlb* are thought to be related to the percentage of sphingomyelin in the cell membranes, where sheep erythrocytes have a significantly higher concentration of sphingomyelin than rabbit erythrocytes (De Gier and Van Deenen, 1961; Doery *et al.*, 1965). The production of *hlb* is exploited for the Christie, Atkins, Munch-Petersen (CAMP) testing in the clinical veterinary microbiology laboratory, where a perpendicular streak of a β-haemolysin producing *Staphylococcus* is placed next to presumptive *Streptococcus agalactiae.*(Munch-Petersen and Christie, 1947) After an incubation period the enhancement of haemolysis where the two streaks overlap is readily apparent. *Hlb* is produced in large quantities by animal isolates of *S. aureus* (Dinges *et al.*, 2000). Although little is known about the role of *hlb* and disease, recent work suggests that β-haemolysin enhances skin colonization, perhaps by damaging keratinocytes and facilitating adherence (Katayama *et al.*, 2013).

Delta-haemolysin is a 26 amino acid polypeptide that causes complete haemolysis of a wide variety of species of erythrocytes and other cells (Freer and Birkbeck, 1982). Delta-haemolysin is slower to migrate into the blood agar than α-haemolysin forming a double zone of haemolysis (Markey *et al.*, 2013). Delta-haemolysin is under the control of the accessory gene regulatory locus (Janzon and Arvidson, 1990). Delta-haemolysin has also been associated with 'colony spreading' of *S. aureus* on agar plates (Omae *et al.*, 2012).

Leucotoxins and gamma-haemolysin

Staphylococcus aureus secretes a number of toxins that specifically target leucocytes, called leucocidal toxins or leucotoxins. One major leucotoxin is Panton–Valentine leucocidin (Luk-PV, PVL) of *S. aureus* (Panton and Valentine, 1932; Woodin, 1960). Other important members of the *Staphylococcus* leucotoxin family include γ-haemolysin, Leucocidin M, and Leucocidin E/D (Choorit *et al.*, 1995; Gravet *et al.*, 1998). Leucotoxins are bi-component toxins, called slow and fast (S and F), that induce pore formation in leucocyte cell membranes via oligomerization into a beta-barrel pore. A single *Staphylococcus* isolate can possess multiple bicomponent toxins which can form multiple combinations to increase the diversity of leucotoxins (Dinges *et al.*, 2000; Henri *et al.*, 2003). The majority of *Staphylococcus aureus* isolates from bovine mastitis have been shown to possess leucocidins genes and these have shown potent effects on bovine neutrophils *in vitro*, especially LukM (Barrio *et al.*, 2006; Fueyo *et al.*, 2005; Rainard *et al.*, 2003).

Gamma-haemolysin consists of three genes: *hlg*A, *hlg*B and *hlg*C (Cooney *et al.*, 1993). Gamma-haemolysin is inhibited by a variety of substances including agar and lipid, and thus does not cause haemolysis on blood agar (Taylor and Bernheimer, 1974; Wadström and Möllby, 1972). Almost all *S. aureus* strains express γ-haemolysin, with 99%

of human clinical isolates having *hlg* (Prevost *et al.*, 1995). Gamma-haemolysin is cytotoxic to a variety of cell types and can be lethal when injected in rabbits (Wadström and Möllby, 1972). Gamma-haemolysin has shown to be potently pro-inflammatory along with the other bi-component leucotoxins in a rabbit eye model (Siqueira *et al.*, 1997).

Enterotoxins

Staphylococcal enterotoxins (SE) are a large family of nine serotypes of heat stabile toxins that cause gastroenteritis and are potent superantigens that cause T-cell stimulation (Balaban and Rasooly, 2000). The most important of these is enterotoxin A (SEA) which is the most common SE associated with food-borne illness in humans. The mode of action is not clearly understood but it is thought to be mediated via vagal stimulation on the emesis centre of the brain and activation of local and systemic immune responses. (Argudin *et al.*, 2010) Staphylococcal enterotoxins are also commonly found in mastitis isolates from cows, although there is geographic variation in isolates with regards to prevalence and types and their role in pathogenesis is poorly understood (Larsen *et al.*, 2002). *Staphylococcus* enterotoxin C (SEC) has been shown to activate bovine neutrophils *in vitro* (Ferens *et al.*, 1998) and cows infected with *Staphylococcus* enterotoxin D (SED)-producing strains secrete SED into milk and induce SED-specific humoral and cell-mediated immune responses (Tollersrud *et al.*, 2006).

Exfolative toxins

Exfolative toxins are secreted by several species of staphylococci. Exfolative toxins are metalloproteases that target the granular and spinosus layer of the epidermis, where they prevent attachment to the spinous granulosus by specifically degrading mammalian desmoglien-1(Amagai *et al.*, 2000). Exfolative toxins are a major virulence factor involved in staphylococcal scalded skin syndrome and in a similar condition in swine called exudative epidermitis or 'greasy pig disease.' Intradermal inoculation of purified toxin induces a cleavage plane was induced in between the stratum corneum and stratum granulosum layers in swine epidermis (Sato *et al.*, 1991). In addition to exudative dermatitis in swine, exfolative toxins are also important virulence factors in canine pyoderma and soft tissue infections in other species such as horses. *S. hyicus*, *S. aureus*, *S. pseudintermedius* and some *S. chromogenes* strains have all been found to produce these toxins to varying degrees (Iyori *et al.*, 2010). Exfolative toxins appear to be highly homologous between staphylococci, suggesting some degree of horizontal gene transfer between species (Ahrens and Andresen, 2004; Andresen *et al.*, 2005; Sato *et al.*, 1994). Several distinct serotypes of exfolative toxin have been described including: SHETA, SHETB, ExhA, ExhB, ExhC, and ExhD. They all appear to be functionally related and antigenically unrelated (Andresen, 1998). All of the exfolative toxins are heat labile and some are expressed chromosomally and some extrachromosomally on plasmids (Ahrens and Andresen, 2004; Sato *et al.*, 2000).

Staphylococcus aureus infections in animals

In veterinary medicine, *S. aureus* is the predominant pathogenic species. Infections with *S. aureus* typically cause soft tissue infections that can progress to septicaemia or internal abscess formation after colonization and subsequent invasion into deeper tissues. *S. aureus* is diverse, and host specificity and adaptation to ecological niches of certain strains in specific hosts has been documented. Veterinary isolates of *S. aureus* were initially categorized as species specific ecovars and non-host specific 'biotypes' (Devriese, 1984). One of the first phenotypic observations of this distinction was the ability of ecovars isolated from ruminants to coagulate ruminant plasma where others strains isolated from other animals lacked this ability (Hajek and Marsalek, 1971). Following the advent of molecular testing these phenotypic properties were found to be associated with unique genotypes. Further evolutionary analyses of these strains revealed, for example, that strains from bovine mastitis and other livestock-associated strains are distinct from human isolates (Kapur *et al.*, 1995; Sakwinska *et al.*, 2011). However, unlike the unique and divergent lineages of many bovine isolates, canine, poultry, and rabbit isolates have been found to have much more recent ancestors in human-associated strains indicating more recent transmission events (Sasaki *et al.*, 2012; Smyth *et al.*, 2009).

Staphylococcus aureus in ruminants

S. aureus is one of the most important causes of mastitis in dairy cattle (Fig. 3.2) and small ruminants (sheep and goats). *S. aureus* isolates causing mastitis have undergone extensive adaptation to the bovine mammary gland. The majority of mastitis associated *S. aureus* strains belong to only a handful of clonal complexes or sequence types including CC97, ST151, CC130, and CC126 when tested by MLST methodologies (Peton and Le Loir, 2014). Small ruminant mastitis isolates appear to be associated with two completely different clonal complexes, ST522 and CC133 (Concepción Porrero *et al.*, 2012). The CC133 small ruminant associated clonal complex is thought to have resulted from a jump to small ruminants from humans followed by significant gene diversification, decay, and horizontal acquisition of mobile genetic elements to facilitate adaptation to the host (Guinane *et al.*, 2010). The economic impact of *S. aureus* associated mastitis is high with significant impacts on milk production and quality. In dairy cattle, it is one of the most frequent causes of mastitis (up to 44% of cases in young animals) where it typically manifests as a chronic subclinical condition that can be challenging to detect, diagnose and treat (Trinidad *et al.*, 1990; Waage *et al.*, 1999). Mastitis caused by *S. aureus* is considered a contagious mastitis, where horizontal transmission from cow to cow is critical for disease maintenance and circulation on farms (Sears and McCarthy, 2003). In modern dairies, transmission from infected to naive cows can occur through inadequately sanitized or malfunctioning milking equipment that can facilitate introduction of bacteria by retropulsion into the teat cistern (Bramley and Dodd, 1984; Sutra and Poutrel, 1994). The initial colonization of *S. aureus* in both cattle and small ruminants occurs via the mammary gland teat epithelium where adhesins and MSCRAMMS facilitate adherence to damaged epithelium (Lammers *et al.*, 1999). Phagocytosis of foreign bacteria by leucocytes is one of the most important defence mechanisms in the mammary gland; thus, leucotoxins are likely involved in mastitis infections (Craven and Williams, 1985). One barrier to clearing an *S. aureus* infection is that phagocytosis is impeded by fat globules in the milk as well as casein, making elimination of bacteria difficult (Paape and Guidry, 1977). Biofilm formation may be important in establishing or maintaining an infection because the intercellular adhesion (*ica*) locus is commonly found in isolates from cases of mastitis (Vasudevan *et al.*, 2003).

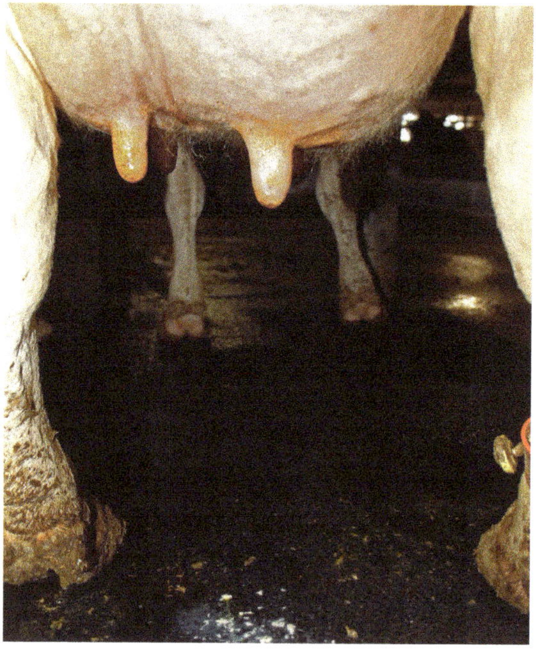

Figure 3.2 Holstein dairy cow with clinical mastitis. Note the swelling in the right rear quarter and secretion of clotted milk. Photo coutesy of Dr Conrad and Erin Spangler.

Following introduction into the teat, bacteria ascend through the mammary gland and induce chronic inflammation, stimulate somatic cell secretion into milk, and lower production by compromising secretory mammary tissue (Gudding *et al.*, 1984). Treatment regimens for mastitis focus on early identification and extended therapy if needed, with subsequent culling of unresponsive animals to reduce the risk of vertical transmission (Barkema *et al.*, 2006). Acute or peracute gangrenous mastitis, a lethal and rapidly life-threatening form of mastitis, can also be caused by infections with some strains of *S. aureus*. This syndrome is typically observed in young cows and is thought to be mediated through secretion of toxins and has been reproduced by removing circulating neutrophils by injection of anti-neutrophil antibodies (Schalm *et al.*, 1976). Alpha-haemolysin is also thought to be important in mediation of gangrenous mastitis, where damage to smooth muscle restricts blood supply to mammary tissue causing necrosis and gangrene (Bramley *et al.*, 1989; Sutra and Poutrel, 1994).

Other diseases associated with *S. aureus* in ruminants include tick pyaemia, a disease characterized by lameness and paralysis in 2- to 12-week-old lambs. This syndrome is typically associated with infestation of the tick species *Ixodes ricinus*, a species endemic to Europe. *S. aureus* has been isolated in association with this disease and the disease syndrome has been experimentally reproduced with ticks infected with *S. aureus* (Webster and Mitchell, 1986).

Staphylococcus aureus in cats and dogs

In cats and dogs, *S. aureus* is an opportunistic pathogen. The predominant *S. aureus* clonal lineage is type ST133, which is a commensal and opportunistic pathogen in cats and is only occasionally isolated from dogs (Sasaki *et al.*, 2012). It should be noted that *S. aureus* carriage or soft tissue infections in dogs are infrequent (~1% of all *Staphylococcus* infections in dogs) and when found, they are usually transient and hospital acquired or associated with extended antimicrobial therapy (Sasaki *et al.*, 2012). Many isolates from dogs are frequently misidentified as *S. aureus* based on biochemical testing, when they are much more likely to be *S. pseudintermedius* or *S. schliferi*, strains that are host adapted to canids (Sasaki *et al.*, 2012). This is important as healthcare professionals often consider pets, especially dogs, as likely reservoirs for *S. aureus*, and specifically MRSA, when it is more likely that they are acquiring the organism from humans. Thus decolonization of carrier dogs may be unwarranted as long-term colonization of dogs is rare (Weese and van Duijkeren, 2010).

Staphylococcus aureus in poultry

In birds and poultry infections with *S. aureus* cause dermatitis and subsequent bacteraemia, tenosynovitis, and osteomyelitis (Zhu and Hester, 2000). *S. aureus* is also a significant cause of 'bumble-foot' or avian pododermatitis, an infection of the subcutaneous tissue in the feet of poultry. Pododermatitis that is initiated from secondary infections of abrasions in foot pads and can progress to arthritis, internal abscesses, and septicaemia (Fig. 3.3) (Butterworth, 1999). Lameness caused by pododermatitis is a source of economic losses in the poultry industry. Pododermatitis and the arthritis–synovitis–osteomyelitis complex account for most of the outbreaks

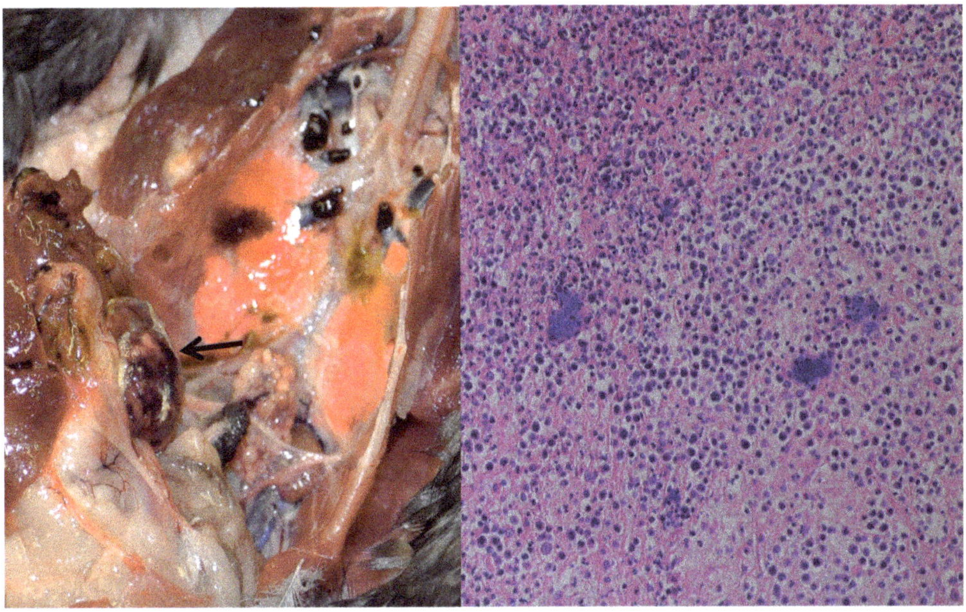

Figure 3.3 Splenitis in a duck, arrow indicates spleen (left panel). Large colonies of *Staphylococcus* are visible in splenic parenchyma (Right Panel, HE stain, 400× magnification). Courtesy of Dr Seth Harris, University of Nebraska-Lincoln, Veterinary Diagnostic Centre.

of staphylococcal related diseases in poultry (Zhu and Hester, 2000). Poor husbandry practices such as temperature fluctuation, out of feed events, transport stress, crowding, litter cleanliness and excessive humidity all contribute to this disease (Zhu and Hester, 2000). Other specific diseases associated with *S. aureus* include bacterial chondrosis with osteomyelitis or 'femoral head necrosis' that prevents growth of the femur bones in broiler chickens. This is a significant cause of lameness and mortality in chickens 4–70 days of age, with most cases occurring around 35 days (McNamee *et al.*, 2000). Staphylococcal disease mitigation strategies in poultry focus on husbandry practices including: reducing environmental stress and minimizing foot and leg trauma in facilities and are not frequently treated with antimicrobials or other drugs.

Staphylococcus aureus in pigs

In swine, *S. aureus* is associated with mostly reproductive diseases, such as sporadic mastitis, endometritis, and udder impetigo. (Everitt *et al.*, 1981; Markey *et al.*, 2013) Swine *S. aureus* isolates are primarily associated with sequence types ST9, ST398, and ST43 (Armand-Lefevre *et al.*, 2005). *S. aureus* is also associated with opportunistic infections in young pigs, such as septicaemia and pneumonia, often following infections with other agents such as viruses (Fig. 3.4). *S. aureus* in swine has recently been studied in the context as a potential reservoir for LA-MRSA which is discussed in more detail later in the chapter.

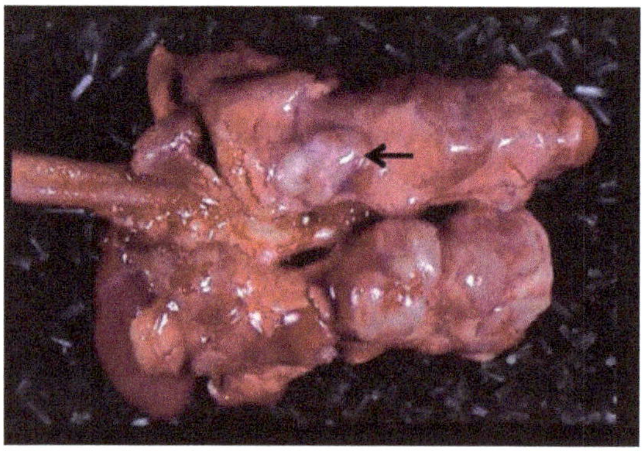

Figure 3.4 Pnemonia in a young pig caused by *Staphylococcus aureus*. Abcesses can be seen on surface of the lung (arrow). Courtesy of Dr Alan Doster, University of Nebraska-Lincoln, Veterinary Diagnostic Centre.

Staphylococcus aureus in rabbits

Rabbits are highly susceptible to infections with *S. aureus*. These infections typically manifest as pododermatitis, subcutaneous abscesses and mastitis. *S. aureus* is the most significant disease problem in some rabbit farms, and *S. aureus*-associated mastitis was found to be the most important cause of culling breeding does in one study (Segura *et al.*, 2007). Mastitis in rabbits can occur in two forms, the acute or gangrenous form, also called 'blue breast,' where tissues become cyanotic, oedematous and haemorrhagic and the chronic form causing a purulent mastitis (Corpa *et al.*, 2010). Acute mastitis spreads rapidly through rabbit farms and often causes litters of infected does to die of starvation due to agalactia. Chronic mastitis causes purulent abscess formation in mammary tissue that result in reduction in milk production and poor performance of offspring. In young kits *S. aureus* infections present clinically as an exudative epidermatitis and subsequent abscess formation in market animals can lead to condemnation or rejection at slaughter (Corpa *et al.*, 2010). Outbreaks of *S. aureus* related disease in rabbit flocks have distinct epidemiological patterns, with outbreaks categorized into low-virulence and high-virulence types. Strains characterized as low virulence occur as sporadic opportunistic infections in individual rabbits, where high-virulence strains can cause epidemics with high morbidity of chronic disease in rabbit flocks. Outbreaks of infections with high-virulence strains result in significant economic losses caused by chronically lowered production, infertility and death. High-virulence strains appear to be a clonal population that is internationally distributed and comprised of ST121 and PFGE type N2 (Vancraeynest *et al.*, 2006). ST121 strains isolated in outbreaks of high-virulence disease appear to be closely related to human skin infection associated strains, having identical MLST and *spa*-profiles indicating a recent human ancestor (Lowder *et al.*, 2009; Smyth *et al.*, 2009). Prevention of disease introduction through strict biosecurity and stock control in naive flocks is key, as treatment of infected rabbit flocks is challenging and often unrewarding (Hermans *et al.*, 2003). Treatment with tetracyclines has shown to reduce disease and mortality, but do not eliminate the disease from the flock allowing the organism to return upon cessation of therapy (Corpa *et al.*, 2010).

Staphylococcus aureus in horses

In horses, infections with *S. aureus* incite a condition called botryomycosis. Botryomycosis is a chronic granulomatous reaction that occurs in soft tissues following infections with staphylococci. Botryomycosis occurs in horses and pigs, in the head and neck muscles and the spermatic cords following castration (Markey *et al.*, 2013) and these chronic wounds can extend to adjacent muscles or underlying tissues (Mcgavin and Valentine, 2001). Pulmonary botryomycosis has been described in horses, rodents, cattle and humans (Bersoff-Matcha *et al.*, 1998; Miller *et al.*, 2001). There is typically underlying factors that predispose animals to botryomycosis, such as immune suppression or presence of a foreign nidus.

Staphylococcus intermedius group (SIG) and associated diseases

Staphylococcus intermedius was originally described as a unique species in 1976 based on phenotypic tests and G+C DNA content (Hajek, 1976). Many members of the genus *Staphylococcus* important to veterinary microbiology have been reclassified into several species groups based on 16S rRNA gene sequence analysis (Takahashi *et al.*, 1999). Some of these studies using molecular phylogeny found that *S. intermedius* is actually a collection of several species, commonly referred to as the *S. intermedius* group (SIG). The SIG contains several coagulase producing species of veterinary importance including *S. pseudintermedius*, *S. delphini* and *S. intermedius* which are involved in a variety of diseases processes such as canine and equine dermatitis. Differentiation amongst the SIG can be done using molecular methods, such PCR amplification of the phosphoacetyltransferase (*pta*) or thermonuclease (*nuc*) genes; however, many clinical microbiology laboratories do not routinely perform these tests (Bannoehr *et al.*, 2009; Sasaki *et al.*, 2010). All or nearly all canine isolates of SIG have been determined to be of a single species, *S. pseudintermedius* (Bannoehr *et al.*, 2007; Sasaki *et al.*, 2007). Many studies using phenotypic testing prior to the implementation of molecular methods refer only to *S. intermedius*, a species where the type strain was isolated from pigeons and not dogs (Devriese *et al.*, 2009). With the advent of molecular testing, all *S. intermedius* strains, including type strains, have originated from feral pigeons. Therefore, some have proposed referring to all isolates of the SIG group from canine sources as *S. pseudintermedius*, unless genomic testing is conducted and demonstrates otherwise (Devriese *et al.*, 2009). Others have proposed utilizing *Staphylococcus (pseud)intermedius* to refer to organisms in the literature that are likely to be *S. pseudintermedius*, but were phenotypically characterized as *S. intermedius* (Weese and van Duijkeren, 2010). For the purpose of this text, members of the *S. intermedius* group that are associated or isolated from dogs and other canids will be referred to as *S. pseudintermedius* (Devriese *et al.*, 2005).

Staphylococcus pseudintermedius in dogs

S. pseudintermedius is an opportunistic pathogen that causes skin infections in animals, mainly dogs but also cats (Morris *et al.*, 2006b). *S. pseudintermedius* can be isolated from many different locations in the skin of healthy dogs and cats in addition to those with inflammatory skin disorders (Abraham *et al.*, 2007). Puppies are colonized within hours of birth with *S. pseudintermedius* and colonization persists for life in some animals. Carriage of *S. pseudintermedius* in the genital mucosa in bitches prior to and immediately following parturition is high and is significantly reduced one week after whelping (Saijonmaa-Koulumies and Lloyd, 2002). *S. pseudintermedius* is the major etiologic agent involved in canine dermatitis and pyoderma (Fig. 3.5) (Bannoehr and Guardabassi, 2012). Canine pyoderma is one of the most common diseases in dogs, and is the most frequent reason for utilization of antimicrobial therapy in small animal medical practice (Baker *et al.*, 2012; Escher *et al.*, 2011; Rantala *et al.*, 2004). Pyoderma is a generalized skin infection characterized by vesicles, pustules, exudates and crust formation. The most common form of pyoderma found in dogs is a superficial pyoderma or folliculitis. Clinical features of canine pyoderma resemble those of porcine exudative epidermitis and staphylococcal scalded skin syndrome in humans. *S. pseudintermedius* is also a significant cause of canine external otitis and is the most frequent *Staphylococcus* isolated from canine urinary tract infections (Ball *et al.*, 2008; Penna *et al.*, 2010). Studies conducted on over 8,000 dogs

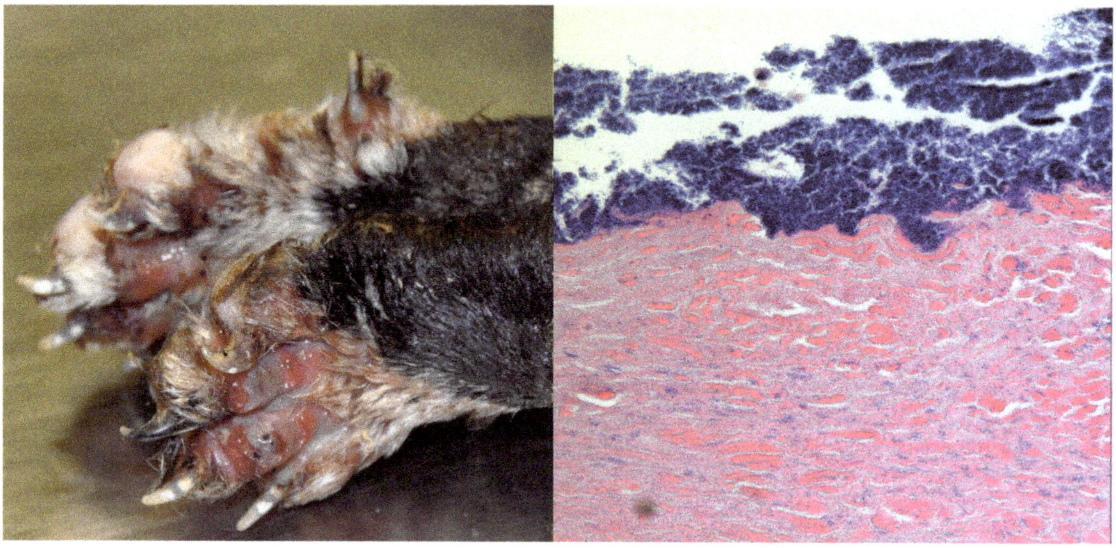

Figure 3.5 Severe pododermatitis following thermal injury (burn) in a dog (left panel). Crusts containing abundant *Staphylococcus* organisms (right panel, HE stain, 400× magnification). Courtesy of Dr Seth Harris, University of Nebraska-Lincoln, Veterinary Diagnostic Centre.

have shown that over 10% of canine urinary tract infections are caused by *Staphylococcus* spp. (Ling *et al.*, 2001) and most of these are likely *S. pseudintermedius* (Ball *et al.*, 2008). Genotyping data suggest that *S. pseudintermedius* isolates from different host species among the superfamily *Canoidea* cluster phylogenetically along with the host species (Aarestrup, 2001). Like many other *Staphylococcus* species, this indicates that *S. pseudintermedius* demonstrates a high degree of host specificity and likely co-evolution between canid host and microbe.

Other SIG staphylococci

Other SIG include *Staphylococcus delphini*, which was originally isolated from dolphins, but is also found in pigeons, horses and camels and mustelids (Guardabassi *et al.*, 2012; Sasaki *et al.*, 2007). *S. delphini* falls into two distinct groups, A and B, with the former thought to represent a divergent group that has evolved to colonize members of the *Mustalidae* (including ferrets, badgers, and minks) (Guardabassi *et al.*, 2012). Enterotoxigenic *S. delphini* has caused outbreaks of diarrhoea in mink farms (Sledge *et al.*, 2010).

Staphylococcus schleiferi

Staphylococcus schleiferi is a more recently recognized pathogen in veterinary medicine. Cases of recurrent canine pyoderma and otitis have been associated with *S. schleiferi*, both *S. schleiferi* subsp. *schleiferi* and *S. schleiferi* subsp. *coagulans* (Frank *et al.*, 2003; May *et al.*, 2005). Canine infections typically have a history of beta-lactam antimicrobial treatment, underlying allergic dermatitis and isolates are found to be resistant to oxacillin. (Cain *et al.*, 2011). Some concern over *S. schleiferi* subsp. *schleiferi* has also arisen as the potential to pose as an opportunistic zoonotic infection in susceptible humans (Davis *et al.*, 2013; Tzamalis *et al.*, 2013).

Staphylococcus hyicus

Staphylococcus hyicus can cause a significant disease in young swine called porcine exudative epidermitis (PEE), or 'greasy pig disease'. *S. hyicus* isolates associated with PEE are typically coagulase positive; however, some are coagulase negative (Markey *et al.*, 2013). Other species such as toxigenic *Staphylococcus chromogenes* are also isolated in cases of PEE (Andresen *et al.*, 2005). Porcine exudative epidermitis is characterized by a generalized exudative epidermitis in pigs of varying ages and is characterized by high morbidity and rapid onset (Wegener and Skov-Jensen, 2006) (Fig. 3.6). Exudative epidermitis can be associated with damage to the epithelium caused by fighting, parasites (e.g. mange), or other environmental factors that

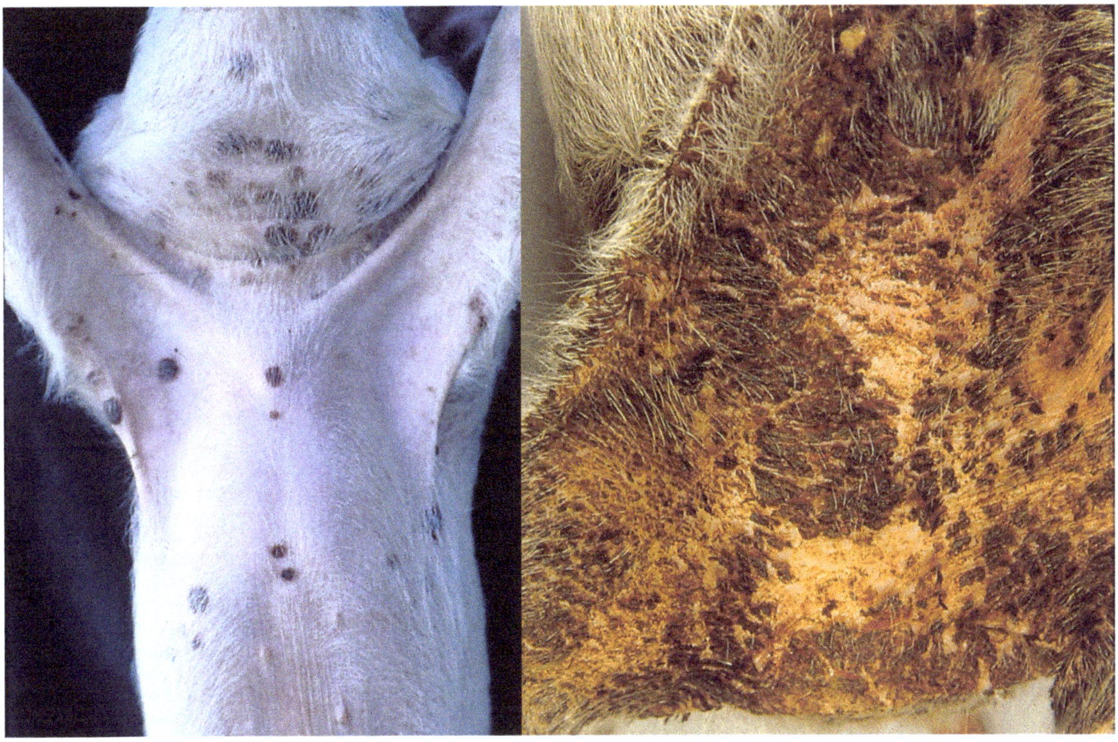

Figure 3.6 Porcine Exudative epidermitis caused by *Staphylococcus hyicus*, with multiple epidermitis plaques (left) and generalized epidermitis (right). Courtesy of Dr Bruce Brodersen and Dr Alan Doster, University of Nebraska-Lincoln, Veterinary Diagnostic Centre.

disrupt skin defences. Secretion of exfolative toxins by either species is thought to be necessary for the development of disease, and *S. hyicus* exofolative toxin has been shown to specifically cleave porcine desmonglein-1(Nishifuji *et al.*, 2005). Exudative epidermitis can have significant impact on swine production, as loss of significant amounts of epithelium integrity can predispose animals to secondary infections with other pathogens and cause severe dehydration from fluid loss. Infections with these bacteria also cause sporadic joint and urinary tract infections in young pigs and have been associated with ear tip necrosis syndrome (Park *et al.*, 2013a). *S. hyicus* has been associated with infections in cattle secondary to mange infestation (Hazarika *et al.*, 1991) and in cases of conjunctivitis in poultry (Cheville *et al.*, 1988). Autogenous vaccines generated from toxigenic strains of *S. hyicus* have shown some efficacy against exudative epidermitis when given to sows prior to farrowing (Sieverding, 1993).

Coagulase-negative staphylococci

Several members of the CoNS are associated with pathogenic processes such as canine skin infections, and for this reason, there is a great deal of ongoing research into the virulence and pathogenicity of these bacteria (Davis *et al.*, 2013; Rainard *et al.*, 1990). For example, CoNS are an emerging cause of subclinical mastitis in dairy cattle and have been isolated in over 25% of mammary gland infections in cows (Chaffer *et al.*, 1999). The number of CoNS associated mastitis is likely to continue to increase as many farms have eliminated *S. aureus* from their herds. The majority of these CoNS infections are caused by *S. chromogenes*, *S. simulans* or *S. epidermidis* (Pyörälä and Taponen, 2009; Thorberg *et al.*, 2009). There are also a large number of CoNS that are isolated with some frequency from a variety of animals that have uncertain significance and may represent normal skin flora in some species (Table 3.1). Therefore great care must be taken when evaluating cultures of *Staphylococcus*, especially the CoNS, in the context of species of origin, clinical history, quantity and site of isolation.

Treatment of veterinary staphylococci infections

Treatment of veterinary infections caused by staphylococci has some challenges that are unique when compared with treating human infections. For example, pharmacokinetic data are not available for most host–pathogen species combinations and calculation of antimicrobial susceptibility breakpoints instead relies on human-derived data. Many of these drugs that are most frequently used are also not explicitly approved and labelled for animal infections, so their usage is extra-label. It should also be noted that antimicrobial therapy in food- and milk-producing animals in the USA is heavily regulated by the Food and Drug Administration and similar organizations in other countries, due to concerns over antimicrobial residues in food. Therefore, many choices are limited to approved therapies and extra-label usage is explicitly restricted for some drugs due to concerns over the emergence of antimicrobial resistance. Many of these antimicrobials, especially the systemically administered agents, have established withdrawal periods to prevent antibiotic residues from being present in meat and milk, which must be strictly adhered to. The regulations of antimicrobial usage in food and milk producing animals are extremely complex and subject to continual changes by governing bodies, which extend beyond the scope of this chapter. However, research conducted into efficacy and common therapy protocols for common veterinary infections are described below.

Bovine mastitis caused by *S. aureus* is one of the more challenging veterinary infections to treat clinically. There are several antibiotics that have been utilized for treatment, and many *S. aureus* isolates appear to be susceptible *in vitro*. However, *in vivo* cure rates are low, and frequently many of these drugs, which include sulfonamides, penicillins, aminoglycosides, and cephalosporins do not directly penetrate mammary tissue (Barkema *et al.*, 2006). Another challenge is to achieve therapeutic antimicrobial concentrations in chronic infections, where microabcesses and extensive fibrosis have developed, and preventing efficient drug distribution. Therefore, most mastitis treatments involve intramammary infusion of antimicrobials directly through the teat canal. Frequently these are administered after the lactation period has ceased when the cows are 'dry', or immediately prior to calving where they can be most effective and not be flushed out by milking. For treatment of mastitis there are only two approved drug classes for intramammary use in the United States, the beta-lactams (amoxicillin, ceftiofur, cephapirin, cloxicillin, hetacillin, and penicillin) and lincosamides (pirilimycin) (Oliveira and Ruegg, 2014). There are no systemically administered drugs approved, but some are used extralabel in the USA (Oliveira and Ruegg, 2014). Efficacy of these treatments varies, especially in lactating animals and several studies have directly compared treatments. One study did not demonstrate any differences between penicillin-novobiocin, penicillin-streptomycin, cephapirin, tilmicosin, and cephalonium antimicrobials, all of which were significantly more effective than the spontaneous cure rate and approached 100% cure in heifers when administered prior to calving (Owens *et al.*, 2001). The susceptibility patterns of the infecting organism can make a significant difference in cure rates. For example, when clinical mastitis cases were found to be caused by penicillin resistant *S. aureus*, less than a third of cases were cured with systemic and intramammary amoxicillin–clavulanic acid or systemic spiramycin (Taponen *et al.*, 2003). Other active compounds such as cefuroxime has been shown to be more effective than cloxacillin (52.4% vs. 12.5% cured) (Wraight, 2003) and lincomycin/neomycin combination have been shown to be more effective than an ampilcillin/cloxacillin combination (41.2% vs. 15.4% cured) (Deluyker *et al.*, 1999). Treatment of dairy animals should follow treatment protocols provided by their veterinary professional to ensure prudent use of antimicrobials, adherence to current regulations, and minimize the risk of residues in milk (AABP 2015)

Therapy for pigs infected with *S. hyicus* or other exfoliative toxin producing species can also be challenging. Many of these isolates have been found to be resistant to penicillins, which were previously considered the first therapy of choice (Park *et al.*, 2013b). A Canadian study examining treatment of *S. hyicus* found the majority of veterinarians preferred to treat these infectious with topical therapy of novobiocin in conjunction with a systemic antimicrobial treatment (penicillin G, ceftiofur, and trimethoprim/sulfadoxine) (Park *et al.*, 2013b). However, many veterinarians consider 'greasy pig disease' a husbandry issue, and offer management recommendations such as clipping of needle teeth to reduce skin abrasions, reducing humidity in barns and improving overall hygiene instead of antimicrobial therapy (Park *et al.*, 2013b).

In companion animals and pets there is a larger number of treatment options for infections with staphylococci, of which the majority are skin and integumentary infections. However, this can often lead to confusion as to which therapy is optimum. To address this The Antimicrobial Guidelines Working Group of the International Society for Companion Animal Infectious Diseases has published guidelines for therapy of superficial bacterial folliculitis in dogs. Superficial bacterial folliculitis is the most commonly treated disorder in small animal clinics, often with antimicrobials that are also used in human medicine (Hillier *et al.*, 2014). The working group strongly recommends the use of culture and susceptibility testing to guide informed therapy. In addition, topical antimicrobials such as chlorhexidine based scrubs and benzoyl peroxide are often helpful as high concentrations in topical products are less likely to lead to resistance. For systemic antimicrobial treatments, prudent drug selection is categorized into tiers. First tier antimicrobials for *S. psudeintermedius* pyoderma include clindamycin or lincomycin, first generation cephalosporins, amoxicillin–clavulanate, and potentiated sulfonamides if regional susceptibility is known (Hillier *et al.*, 2014). Second tier drugs are to be used when first tier drugs are not appropriate, such as if the isolate is resistant and if susceptibility of the isolate has been determined. These include the tetracyclines, fluoroquinolones, and aminoglycosides. A consensus could not be reached by the committee on the use of third generation cephalosporins, so these are considered either first or second tier. The use of linezolid, teicoplanin, and vancomycin, are considered third tier and their usage is strongly discouraged as they are to be reserved for the treatment of serious MRSA infections in humans.

Antimicrobial resistance in veterinary staphylococci

In veterinary medicine, a significant hurdle to treating *Staphylococcus* infections is the emergence of antimicrobial resistance. Animals having methicillin-resistant *S. aureus* (MRSA) infections were first described in cows with mastitis (Devriese *et al.*, 1972). Treatment of mastitis accounts for the majority of antibiotic drugs that are used to treat animals on dairy farms (Oliveira and Ruegg, 2014). MRSA has since emerged as a significant problem in humans and companion animals, especially in hospital acquired infections. Antibiotic resistance in animal MRSA strains, like those in humans, are mediated by the presence of the *mecA* gene, which alters the penicillin binding protein to have a lower affinity for beta-lactam antimicrobials (Beck *et al.*, 1986; Kwon *et al.*, 2006; Wielders *et al.*, 2002). The *mecA* gene is carried on a large mobile genetic element called staphylococcal cassette chromosome *mec* (SCC*mec*), which also carries site-specific recombinases responsible for the mobility of SCC*mec* (Kwon *et al.*, 2005).

Livestock-associated MRSA

Livestock-associated MRSA (LA-MRSA) strains have increasingly been isolated from food-producing animals, even in the absence of treatment with beta-lactam antimicrobials. The majority of isolates are associated with the ST398 clonal complex. ST398 strains were first isolated from animals and workers in pig farms, but have subsequently been isolated from poultry and horses (Nemati *et al.*, 2008; Van den Eede *et al.*, 2009). LA-MRSA appears to be a potential zoonotic agent, as strains seem to have high rates of colonization in at-risk populations, such as pig farmers, where the prevalence was found to be over 760 times the rate in the normal European population (Voss *et al.*, 2005). However, follow up studies in other countries have shown lower prevalence among pigs and at-risk populations (Smith *et al.*, 2013). LA-MRSA has been found in retail meat samples (e.g. pork, beef and poultry), although the origin of these organisms is unclear (Weese *et al.*, 2010). Importantly, some studies examining MRSA prevalence in beef cattle at slaughter were unable to isolate any MRSA

(Weese *et al.*, 2012), suggesting that contamination likely occurs post-harvest. In contrast, high rates of carriage have been found in veal calves at slaughter (Tenhagen *et al.*, 2014), which may correspond to a higher rate of LA-MRSA colonization in veal farm workers and veal calves observed in the Netherlands (Graveland *et al.*, 2010). LA-MRSA have also been found in bulk tank milk samples from dairies, and rarely isolates of human community acquired-MRSA (CA-MRSA) (USA100) are isolated from bulk tank milk (Haran *et al.*, 2012; Kreausukon *et al.*, 2012). Although emergence of MRSA in livestock is often attributed to widespread microbial usage, epidemiological studies have been unable to establish this association. For example, no difference in MRSA prevalence has been found in pork reared conventionally or alternatively (without antibiotic usage) in US pork products (O'Brien *et al.*, 2012). Other studies examining the prevalence of antimicrobial resistant strains of *S. hyicus* and *S. aureus* found no significant difference between farms that were antibiotic-free and conventional commercial farms that used antibiotics (Park *et al.*, 2013b). Research examining the sources and risks of foodborne LA-MRSA is an active area, but currently causative factors and mitigation methods remain a challenge to identify.

MRSA in horses

Equine MRSA cases have been documented and are usually associated with veterinary hospitals and surgery site infections, and outbreaks among patients have followed exposure to human carriers (reverse zoonosis) (Seguin *et al.*, 1999). Veterinary hospital acquired MRSA isolates are characterized as related to human origin clonal complexes ST8 and ST254 in addition to LA-MRSA ST398. Importantly, ST398 strains thought to have come from a foal in a veterinary hospital have caused human infections (van Duijkeren *et al.*, 2011). In cats and dogs, MRSA can be found in healthy animals at low levels (Abraham *et al.*, 2007; Weese and van Duijkeren, 2010). All MRSA isolates in cats appear to be highly related (Morris *et al.*, 2006a). No differences in signalment, clinical signs, or outcomes were observed in cats infected with MRSA versus MSSA for a variety of types of infections (Morris *et al.*, 2006a). MRSA colonization in dogs has been associated with hospital visitation and contact with children; however, its presence is thought to be transient as *S. aureus* is not a commensal in dogs and colonization is transient (Lefebvre *et al.*, 2009).

Methicillin-resistant *Staphylococcus pseudintermedius*

Other antibiotic resistant members of the genus *Staphylococcus* include methicillin-resistant *S. pseudintermedius* (MRSP). MRSP carriage rates in dogs appear to vary from 1% to 30%, depending on the population and study (Weese and van Duijkeren, 2010). Colonization rates with *S. pseudintermedius* in humans are low; however, they are commonly isolated from dog bite wounds and present a risk for opportunistic infections, especially in veterinary hospitals and clinics (Weese and van Duijkeren, 2010). However, the incidence of MRSP in the general human population is very low and the risk is small with proper hygiene and care should be taken to explain the differences in risk to owners with pets diagnosed with MRSP infections. Methicillin-resistant *Staphylococcus pseudintermedius* are also emerging causes of post-surgical infections in dogs making treatment challenging (van Duijkeren *et al.*, 2008).

Conclusions and future trends

The staphylococci remain important pathogens in nearly all animal species, including companion and production animals and humans. Since the first staphylococci were described in the late nineteenth century, research has made tremendous strides in increasing understanding of these remarkably flexible pathogens and the role they play in disease. As new technologies such as next generation sequencing and new epidemiological tools enable rapid data acquisition, new findings will only come more rapidly, as we are only beginning to understand disease dynamics and virulence factors for a variety of staphylococcal infections. Research into new interventions, including novel vaccines and non-traditional antimicrobials will be extremely important as the prevalence of resistance to traditional therapies increases. New therapies and disease mitigation strategies are critical to maintaining a healthy livestock population and a safe food supply, especially as worldwide demand for protein sources is greatly increasing to feed a growing human population. The human–animal bond with companion animals is also increasingly being recognized as beneficial for both humans and animals across the globe. Subsequently, the nexus of humans and animals will also be the forefront of zoonotic disease. Thus, new strategies will need to be developed to mitigate staphylococcal diseases that impact both production and companion animals and the humans they contact.

Web resources

List of current prokaryotic names with standing in the nomenclature (LPSN) for the genus *Staphylococcus:* http://www.bacterio.net/staphylococcus.html

References

AABP, American Association of Bovine Practitioners (2015). Guidelines for Therapy of Clinical Mastitis in Lactating Dairy Cows.

Aarestrup, F.M. (2001). Comparative ribotyping of *Staphylococcus intermedius* isolated from members of the *Canoidea* gives possible evidence for host-specificity and co-evolution of bacteria and hosts. Int. J. Syst. Evol. Microbiol. *51*, 1343–1347.

Abraham, J.L., Morris, D.O., Griffeth, G.C., Shofer, F.S., and Rankin, S.C. (2007). Surveillance of healthy cats and cats with inflammatory skin disease for colonization of the skin by methicillin-resistant coagulase-positive staphylococci and *Staphylococcus schleiferi* subsp. *schleiferi*. Vet. Dermatol. *18*, 252–259.

Ahrens, P., and Andresen, L.O. (2004). Cloning and sequence analysis of genes encoding *Staphylococcus hyicus* exfoliative toxin types A, B, C, and D. J. Bacteriol. *186*, 1833–1837.

Amagai, M., Matsuyoshi, N., Wang, Z.H., Andl, C., and Stanley, J.R. (2000). Toxin in bullous impetigo and staphylococcal scalded-skin syndrome targets desmoglein 1. Nat. Med. *6*, 1275–1277.

Andresen, L.O. (1998). Differentiation and distribution of three types of exfoliative toxin produced by *Staphylococcus hyicus* from pigs with exudative epidermitis. FEMS Immunol. Med. Microbiol. *20*, 301–310.

Andresen, L.O., Ahrens, P., Daugaard, L., and Bille-Hansen, V. (2005). Exudative epidermitis in pigs caused by toxigenic *Staphylococcus chromogenes*. Vet. Microbiol. *105*, 291–300.

Argudín, M.Á., Mendoza, M.C., and Rodicio, M.R. (2010). Food poisoning and *Staphylococcus aureus* enterotoxins. Toxins 2, 1751–1773.

Armand-Lefevre, L., Ruimy, R., and Andremont, A. (2005). Clonal comparison of *Staphylococcus aureus* isolates from healthy pig farmers, human controls, and pigs. Emerg. Infect. Dis. *11*, 711–714.

Atilano, M.L., Yates, J., Glittenberg, M., Filipe, S.R., and Ligoxygakis, P. (2011). Wall teichoic acids of *Staphylococcus aureus* limit recognition by the *Drosophila* peptidoglycan recognition protein-SA to promote pathogenicity. PLOS Pathog *7*, e1002421.

Baker, S.A., Van-Balen, J., Lu, B., Hillier, A., and Hoet, A.E. (2012). Antimicrobial drug use in dogs prior to admission to a veterinary teaching hospital. J. Am. Vet. Med. Assoc. *241*, 210–217.

Balaban, N., and Rasooly, A. (2000). Staphylococcal enterotoxins. Int. J. Food Microbiol. *61*, 1–10.

Ball, K.R., Rubin, J.E., Chirino-Trejo, M., and Dowling, P.M. (2008). Antimicrobial resistance and prevalence of canine uropathogens at the Western College of Veterinary Medicine Veterinary Teaching Hospital, 2002–2007. Can Vet. J. *49*, 985–990.

Bannoehr, J., Ben Zakour, N.L., Waller, A.S., Guardabassi, L., Thoday, K.L., van den Broek, A.H.M., and Fitzgerald, J.R. (2007). Population Genetic Structure of the *Staphylococcus intermedius* Group: Insights into agr Diversification and the Emergence of Methicillin-Resistant Strains. J. Bacteriol. *189*, 8685–8692.

Bannoehr, J., Franco, A., Iurescia, M., Battisti, A., and Fitzgerald, J.R. (2009). Molecular diagnostic identification of Staphylococcus pseudintermedius. J. Clin. Microbiol. *47*, 469–471.

Bannoehr, J., and Guardabassi, L. (2012). *Staphylococcus pseudintermedius* in the dog: taxonomy, diagnostics, ecology, epidemiology and pathogenicity. Veterinary Dermatology 23, 253-e252.

Barkema, H.W., Schukken, Y.H., and Zadoks, R.N. (2006). Invited Review: The role of cow, pathogen, and treatment regimen in the therapeutic success of bovine *Staphylococcus aureus* mastitis. J. Dairy Sci. *89*, 1877–1895.

Barrio, M.B., Rainard, P., and Prévost, G. (2006). LukM/LukF'-PV is the most active *Staphylococcus aureus* leucotoxin on bovine neutrophils. Microbes Infect. *8*, 2068–2074.

Beck, W.D., Berger-Bächi, B., and Kayser, F.H. (1986). Additional DNA in methicillin-resistant *Staphylococcus aureus* and molecular cloning of mec-specific DNA. J. Bacteriol. *165*, 373–378.

Bersoff-Matcha, S.J., Roper, C.C., Liapis, H., and Little, J.R. (1998). Primary pulmonary botryomycosis: case report and review. Clin. Infect. Dis. *26*, 620–624.

Bhakdi, S., Muhly, M., and Füssle, R. (1984). Correlation between toxin binding and hemolytic activity in membrane damage by staphylococcal alpha-toxin. Infect. Immun. *46*, 318–323.

Bhakdi, S., and Tranum-Jensen, J. (1991). Alpha-toxin of *Staphylococcus aureus*. Microbiol. Rev. *55*, 733–751.

Bramley, A.J., and Dodd, F.H. (1984). Reviews of the progress of dairy science: mastitis control–progress and prospects. J. Dairy Res. *51*, 481–512.

Bramley, A.J., Patel, A.H., O'Reilly, M., Foster, R., and Foster, T.J. (1989). Roles of alpha-toxin and beta-toxin in virulence of *Staphylococcus aureus* for the mouse mammary gland. Infect. Immun. *57*, 2489–2494.

Butterworth, A. (1999). Infectious components of broiler lameness: a review. World Poultry Sci. J. *55*, 327–352.

Cain, C.L., Morris, D.O., and Rankin, S.C. (2011). Clinical characterization of *Staphylococcus schleiferi* infections and identification of risk factors for acquisition of oxacillin-resistant strains in dogs: 225 cases (2003–2009). J. Am. Vet. Med. Assoc. *239*, 1566–1573.

Chaffer, M., Leitner, G., Winkler, M., Glickman, A., Krifucks, O., Ezra, E., and Saran, A. (1999). Coagulase-negative staphylococci and mammary gland infections in cows. Zentralblatt Veterinarmedizin Reihe B *46*, 707–712.

Cheville, N.F., Tappe, J., Ackermann, M., and Jensen, A. (1988). Acute Fibrinopurulent blepharitis and conjunctivitis associated with *Staphylococcus hyicus*, *Escherichia coli*, and *Streptococcus* sp. in chickens and turkeys. Vet. Path. Online *25*, 369–375.

Choorit, W., Kaneko, J., Muramoto, K., and Kamio, Y. (1995). Existence of a new protein component with the same function as the LukF component of leucocidin or gamma-hemolysin and its gene in *Staphylococcus aureus* P83. FEBS Lett. *357*, 260–264.

Choudhuri, K.K., and Chakrabarty, A.N. (1969). Hyaluronate lyase activity of staphylococci. Ind. J. Exp. Biol. *7*, 183–185.

Concepción Porrero, M., Hasman, H., Vela, A.I., Fernández-Garayzábal, J.F., Domínguez, L., and Aarestrup, F.M. (2012). Clonal diversity of *Staphylococcus aureus* originating from the small ruminants goats and sheep. Vet. Microbiol. *156*, 157–161.

Cooney, J., Kienle, Z., Foster, T.J., and O'Toole, P.W. (1993). The gamma-hemolysin locus of *Staphylococcus aureus* comprises three linked genes, two of which are identical to the genes for the F and S components of leucocidin. Infect. Immun. *61*, 768–771.

Corpa, J., Hermans, K., and Haesebrouck, F. (2010). Main pathologies associated with *Staphylococcus aureus* infections in rabbits: a review. World Rabbit Science *17*, 115–125.

Craven, N., and Williams, M.R. (1985). Defences of the bovine mammary gland against infection and prospects for their enhancement. Vet. Immunol. Immunopathol. *10*, 71–127.

Davis, M.F., Cain, C.L., Brazil, A.M., and Rankin, S.C. (2013). Two coagulase-negative staphylococci emerging as potential zoonotic pathogens: wolves in sheep's clothing? Front. Microbiol. *4*, 123.

De Gier, J., and Van Deenen, L.L.M. (1961). Some lipid characteristics of red cell membranes of various animal species. Biochim. Biophys. Acta *49*, 286–296.

Deluyker, H.A., Chester, S.T., and Van Oye, S.N. (1999). A multilocation clinical trial in lactating dairy cows affected with clinical mastitis to compare the efficacy of treatment with intramammary infusions of a lincomycin/neomycin combination with an ampicillin/cloxacillin combination. J. Vet. Pharmacol. Ther. *22*, 274–282.

Devriese, L.A. (1984). A simplified system for biotyping *Staphylococcus aureus* strains isolated from animal species. J. Appl. Bacteriol. *56*, 215–220.

Devriese, L.A. (1990). Staphylococci in healthy and diseased animals. Soc. Appl. Bacteriol. Symp. Ser. *19*, 71S–80S.

Devriese, L.A., Hermans, K., Baele, M., and Haesebrouck, F. (2009). *Staphylococcus pseudintermedius* versus *Staphylococcus intermedius*. Vet. Microbiol. *133*, 206–207.

Devriese, L.A., Van Damme, L.R., and Fameree, L. (1972). Methicillin (Cloxacillin)-Resistant *Staphylococcus aureus* strains Isolated from Bovine Mastitis Cases. Zentralblatt für Veterinärmedizin Reihe B *19*, 598–605.

Devriese, L.A., Vancanneyt, M., Baele, M., Vaneechoutte, M., De Graef, E., Snauwaert, C., Cleenwerck, I., Dawyndt, P., Swings, J., Decostere, A., *et al.* (2005). *Staphylococcus pseudintermedius* sp. nov., a coagulase-positive species from animals. Int. J. Syst. Evol. Microbiol. *55*, 1569–1573.

Dinges, M.M., Orwin, P.M., and Schlievert, P.M. (2000). Exotoxins of *Staphylococcus aureus*. Clin. Microbiol. Rev. *13*, 16–34.

Doery, H.M., Magnusson, B.J., Cheyne, I.M., and Sulasekharam, J. (1963). A phospholipase in staphylococcal toxin which hydrolyses sphingomyelin. Nature *198*, 1091–1092.

Doery, H.M., Magnusson, B.J., Gulasekharam, J., and Pearson, J.E. (1965). The properties of phospholipase enzymes in staphylococcal toxins. J. Gen. Microbiol. *40*, 283–296.

Duthie, E.S. (1954). Evidence for two forms of staphylococcal coagulase. J. Gen. Microbiol. *10*, 427–436.

Duthie, E.S., and Lorenz, L.L. (1952). Staphylococcal coagulase; mode of action and antigenicity. J. Gen. Microbiol. *6*, 95–107.

Endl, J., Seidl, H.P., Fiedler, F., and Schleifer, K.H. (1983). Chemical composition and structure of cell wall teichoic acids of staphylococci. Arch. Microbiol. *135*, 215–223.

Escher, M., Vanni, M., Intorre, L., Caprioli, A., Tognetti, R., and Scavia, G. (2011). Use of antimicrobials in companion animal practice: a retrospective study in a veterinary teaching hospital in Italy. J. Antimicrob. Chemother. *66*, 920–927.

Everitt, J.I., Fetter, A.W., Kenney, R.M., and Della-Ferra, M.A. (1981). Porcine Necrotizing Staphylococcal Endometritis. Vet. Pathol. Online *18*, 125–127.

Ferens, W.A., Davis, W.C., Hamilton, M.J., Park, Y.H., Deobald, C.F., Fox, L., and Bohach, G. (1998). Activation of bovine lymphocyte subpopulations by staphylococcal enterotoxin C. Infect. Immun. *66*, 573–580.

Foster, T.J., and Höök, M. (1998). Surface protein adhesins of *Staphylococcus aureus*. Trends Microbiol. *6*, 484–488.

Fournier, B., and Philpott, D.J. (2005). Recognition of *Staphylococcus aureus* by the innate immune system. Clin. Microbiol. Rev. *18*, 521–540.

Frank, L.A., Kania, S.A., Hnilica, K.A., Wilkes, R.P., and Bemis, D.A. (2003). Isolation of *Staphylococcus schleiferi* from dogs with pyoderma. J. Am. Vet. Med. Assoc. *222*, 451–454.

Freer, J.H., and Birkbeck, T.H. (1982). Possible conformation of delta-lysin, a membrane-damaging peptide of *Staphylococcus aureus*. J. Theor. Biol. *94*, 535–540.

Fueyo, J.M., Mendoza, M.C., Rodicio, M.R., Muñiz, J., Alvarez, M.A., and Martín, M.C. (2005). Cytotoxin and Pyrogenic Toxin Superantigen Gene Profiles of *Staphylococcus aureus* Associated with Subclinical Mastitis in Dairy Cows and Relationships with Macrorestriction Genomic Profiles. Clin. Micro. *43*, 1278–1284.

Gilbert, I. (1931). Dissociation in an Encapsulated Staphylococcus. J. Bacteriol. *21*, 157–160.

Glenny, A.T., and Stevens, M.F. (1935). Staphylococcus toxins and antitoxins. J. Pathol. Bacteriol. *40*, 201–210.

Graveland, H., Wagenaar, J.A., Heesterbeek, H., Mevius, D., van Duijkeren, E., and Heederik, D. (2010). Methicillin resistant *Staphylococcus aureus* ST398 in veal calf farming: human MRSA carriage related with animal antimicrobial usage and farm hygiene. PLoS One *5*, e10990.

Gravet, A., Colin, D.A., Keller, D., Girardot, R., Monteil, H., Prévost, G., and Giradot, R. (1998). Characterization of a novel structural member, LukE-LukD, of the bi-component staphylococcal leucotoxins family. FEBS Lett. *436*, 202–208.

Guardabassi, L., Schmidt, K.R., Petersen, T.S., Espinosa-Gongora, C., Moodley, A., Agersø, Y., and Olsen, J.E. (2012). Mustelidae are natural hosts of Staphylococcus delphini group A. Vet. Microbiol. *159*, 351–353.

Gudding, R., McDonald, J.S., and Cheville, N.F. (1984). Pathogenesis of *Staphylococcus aureus* mastitis: bacteriologic, histologic, and ultrastructural pathologic findings. Am. J. Vet. Res *45*, 2525–2531.

Guinane, C.M., Ben Zakour, N.L., Tormo-Mas, M.A., Weinert, L.A., Lowder, B.V., Cartwright, R.A., Smyth, D.S., Smyth, C.J., Lindsay, J.A., Gould, K.A., *et al.* (2010). Evolutionary genomics of *Staphylococcus aureus* reveals insights into the origin and molecular basis of ruminant host adaptation. Genome Biol. Evol. *2*, 454–466.

Hajek, V. (1976). *Staphylococcus intermedius*, a New Species Isolated from Animals. Int. J. Syst Bacteriol. *26*, 401–408.

Hajek, V., and Marsalek, E. (1971). [The differentiation of pathogenic staphylococci and a suggestion for their taxonomic classification]. Zentralblatt fur Bakteriologie, Parasitenkunde, Infektionskrankheiten und Hygiene Erste Abteilung Originale Reihe A: Medizinische Mikrobiologie und Parasitologie *217*, 176–182.

Haran, K.P., Godden, S.M., Boxrud, D., Jawahir, S., Bender, J.B., and Sreevatsan, S. (2012). Prevalence and characterization of *Staphylococcus aureus*, including methicillin-resistant *Staphylococcus aureus*, isolated from bulk tank milk from Minnesota dairy farms. J. Clin. Microbiol. *50*, 688–695.

Harshman, S., Boquet, P., Duflot, E., Alouf, J.E., Montecucco, C., and Papini, E. (1989). Staphylococcal alpha-toxin: a study of membrane penetration and pore formation. J. Biol. Chem. *264*, 14978–14984.

Hazarika, R.A., Mahanta, P.N., Dutta, G.N., and Devriese, L.A. (1991). Cutaneous infection associated with *Staphylococcus hyicus* in cattle. Res. Vet. Sci. *50*, 374–375.

Henri, M., Alain, G., Mauro Dalla, S., Gianfranco, M., Manuela, C., Didier, A.C., Stephen, B., Gilles, P.v., Lamin Baba, M., and Sandra, W. (2003). Staphylococcal bicomponent leucotoxins, mechanism of action, impact on cells and contribution to virulence. In Pore-forming Peptides and Protein Toxins (CRC Press), pp. 3–26.

Hermans, K., Devriese, L.A., and Haesebrouck, F. (2003). Rabbit staphylococcosis: difficult solutions for serious problems. Vet. Microbiol. *91*, 57–64.

Hillier, A., Lloyd, D.H., Weese, J.S., Blondeau, J.M., Boothe, D., Breitschwerdt, E., Guardabassi, L., Papich, M.G., Rankin, S., Turnidge, J.D., *et al.* (2014). Guidelines for the diagnosis and antimicrobial therapy of canine superficial bacterial folliculitis (Antimicrobial Guidelines Working Group of the International Society for Companion Animal Infectious Diseases). Vet. Dermatol. *25*, 163–175, e142–163.

Hu, C., Xiong, N., Zhang, Y., Rayner, S., and Chen, S. (2012). Functional characterization of lipase in the pathogenesis of *Staphylococcus aureus*. Biochem. Biophys. Res. Commun. *419*, 617–620.

Huebner, J., and Goldmann, D.A. (1999). Coagulase-negative staphylococci: role as pathogens. Annu. Rev. Med. *50*, 223–236.

Hynes, W.L., and Walton, S.L. (2000). Hyaluronidases of Gram-positive bacteria. FEMS Microbiol. Lett. *183*, 201–207.

Iyori, K., Hisatsune, J., Kawakami, T., Shibata, S., Murayama, N., Ide, K., Nagata, M., Fukata, T., Iwasaki, T., Oshima, K., *et al.* (2010). Identification of a novel *Staphylococcus pseudintermedius* exfoliative toxin gene and its prevalence in isolates from canines with pyoderma and healthy dogs. FEMS Microbiol. Lett. *312*, 169–175.

Janzon, L., and Arvidson, S. (1990). The role of the delta-lysin gene (hld) in the regulation of virulence genes by the accessory gene regulator (agr) in *Staphylococcus aureus*. EMBO J. *9*, 1391–1399.

Kapur, V., Sischo, W.M., Greer, R.S., Whittam, T.S., and Musser, J.M. (1995). Molecular population genetic analysis of *Staphylococcus aureus* recovered from cows. J. Clin. Microbiol. *33*, 376–380.

Karakawa, W.W., Fournier, J.M., Vann, W.F., Arbeit, R., Schneerson, R.S., and Robbins, J.B. (1985). Method for the serological typing of the capsular polysaccharides of *Staphylococcus aureus*. J. Clin. Microbiol. *22*, 445–447.

Katayama, Y., Baba, T., Sekine, M., Fukuda, M., and Hiramatsu, K. (2013). Beta-hemolysin promotes skin colonization by *Staphylococcus aureus*. J. Bacteriol. *195*, 1194–1203.

Kenny, K., Bastida, F.D., and Norcross, N.L. (1992). Secretion of alpha-hemolysin by bovine mammary isolates of *Staphylococcus aureus*. Can. J. Vet. Res. *56*, 265–268.

Kloos, W.E., and Schleifer, K.H. (1975). Simplified scheme for routine identification of human *Staphylococcus* species. J. Clin. Microbiol. *1*, 82–88.

Kloos, W.E., Zimmerman, R.J., and Smith, R.F. (1976). Preliminary studies on the characterization and distribution of *Staphylococcus* and *Micrococcus* species on animal skin. Appl. Environ. Microbiol. *31*, 53–59.

Kreausukon, K., Fetsch, A., Kraushaar, B., Alt, K., Müller, K., Krömker, V., Zessin, K.H., Käsbohrer, A., and Tenhagen, B.A. (2012). Prevalence, antimicrobial resistance, and molecular characterization of methicillin-resistant *Staphylococcus aureus* from bulk tank milk of dairy herds. J. Dairy Sci. *95*, 4382–4388.

Kwon, N.H., Park, K.T., Jung, W.K., Youn, H.Y., Lee, Y., Kim, S.H., Bae, W., Lim, J.Y., Kim, J.Y., Kim, J.M., *et al.* (2006). Characteristics of methicillin resistant *Staphylococcus aureus* isolated from chicken meat and hospitalized dogs in Korea and their epidemiological relatedness. Vet. Microbiol. *117*, 304–312.

Kwon, N.H., Park, K.T., Moon, J.S., Jung, W.K., Kim, S.H., Kim, J.M., Hong, S.K., Koo, H.C., Joo, Y.S., and Park, Y.H. (2005). Staphylococcal cassette chromosome mec (SCCmec) characterization and molecular analysis for methicillin-resistant *Staphylococcus aureus* and novel SCCmec subtype IVg isolated from bovine milk in Korea. J. Antimicrob. Chemother. *56*, 624–632.

Lammers, A., Nuijten, P.J., and Smith, H.E. (1999). The fibronectin binding proteins of *Staphylococcus aureus* are required for adhesion to and invasion of bovine mammary gland cells. FEMS Microbiol. Lett. *180*, 103–109.

Larsen, H.D., Aarestrup, F.M., and Jensen, N.E. (2002). Geographical variation in the presence of genes encoding superantigenic exotoxins and β-hemolysin among *Staphylococcus aureus* isolated from bovine mastitis in Europe and USA. Vet. Microbiol. *85*, 61–67.

Lefebvre, S.L., Reid-Smith, R.J., Waltner-Toews, D., and Weese, J.S. (2009). Incidence of acquisition of methicillin-resistant *Staphylococcus aureus, Clostridium difficile*, and other health-care-associated pathogens by dogs that participate in animal-assisted interventions. J. Am. Vet. Med. Assoc. *234*, 1404–1417.

Ling, G.V., Norris, C.R., Franti, C.E., Eisele, P.H., Johnson, D.L., Ruby, A.L., and Jang, S.S. (2001). Interrelations of organism prevalence, specimen collection method, and host age, sex, and breed among 8, 354 canine urinary tract infections (1969–1995). J. Vet. Intern. Med.*15*, 341–347.

Linker, A., Meyer, K., and Hoffman, P. (1956). The production of unsaturated uronides by bacterial hyaluronidases. J. Biol. Chem. *219*, 13–25.

Loeb, L. (1903). The influence of certain bacteria on the coagulation of the blood. J. Med. Res. *10*, 407–419.

Lowder, B.V., Guinane, C.M., Ben Zakour, N.L., Weinert, L.A., Conway-Morris, A., Cartwright, R.A., Simpson, A.J., Rambaut, A., Nübel, U., and Fitzgerald, J.R. (2009). Recent human-to-poultry host jump, adaptation, and pandemic spread of *Staphylococcus aureus*. Proc. Natl. Acad. Sci. U.S.A. *106*, 19545–19550.

Makris, G., Wright, J.D., Ingham, E., and Holland, K.T. (2004). The hyaluronate lyase of *Staphylococcus aureus* – a virulence factor? Microbiology *150* 46, 2005–2013.

Månsson, H.L. (2008). Fatty acids in bovine milk fat. Food Nutr. Res. 52.

Markey, B., Leonard, F., Archambault, M., Cullinan, A., and Maguire, D. (2013). *Staphylococcus* species. In Veterinary Clinical Microbiology (Edinburgh: Mosby Elsevier).

May, E.R., Hnilica, K.A., Frank, L.A., Jones, R.D., and Bemis, D.A. (2005). Isolation of *Staphylococcus schleiferi* from healthy dogs and dogs with otitis, pyoderma, or both. J. Am. Vet. Med. Assoc. 227, 928–931.

Mcgavin, D., and Valentine, B. (2001). Muscle. In Thomson's Special Veterinary Pathology (St. Louis, MO: Mosby), p. 475.

McNamee, P.T., Smyth, J.A., and Smyth, J.A. (2000). Bacterial chondronecrosis with osteomyelitis ('femoral head necrosis') of broiler chickens: A review. Avian Pathol. *29*, 253–270.

Miller, M.A., Fales, W.H., Tyler, J.W., and Suedmeyer, W.K. (2001). Pulmonary botryomycosis in a Scottish highland steer. J. Vet. Diagn. Invest. *13*, 74–76.

Morris, D.O., Mauldin, E.A., O'Shea, K., Shofer, F.S., and Rankin, S.C. (2006a). Clinical, microbiological, and molecular characterization of methicillin-resistant *Staphylococcus aureus* infections of cats. Am. J. Vet. Res. *67*, 1421–1425.

Morris, D.O., Rook, K.A., Shofer, F.S., and Rankin, S.C. (2006b). Screening of *Staphylococcus aureus, Staphylococcus intermedius*, and *Staphylococcus schleiferi* isolates obtained from small companion animals for antimicrobial resistance: a retrospective review of 749 isolates (2003–04). Vet. Dermatol. *17*, 332–337.

Mortensen, J.E., Shryock, T.R., and Kapral, F.A. (1992). Modification of bactericidal fatty acids by an enzyme of *Staphylococcus aureus*. J. Med. Microbiol. *36*, 293–298.

Much, H. (1908). Uber eine Vorstufe des Fibrinfermentes in Kulturen von *Staphylococcus aureus*. Biochem. Z. *14*, 143–155.

Munch-Petersen, E., and Christie, R. (1947). On the effect of the interaction of staphylococcal β toxin and group-B streptococcal substance on red blood corpuscles and its use as a test for the identification of *Streptococcus agalactiae*. J. Pathol. Bacteriol. *59*, 367–371.

Nemati, M., Hermans, K., Lipinska, U., Denis, O., Deplano, A., Struelens, M., Devriese, L.A., Pasmans, F., and Haesebrouck, F. (2008). Antimicrobial Resistance of Old and Recent *Staphylococcus aureus* isolates from poultry: first detection of livestock-associated methicillin-resistant strain ST398. Antimicrob. Agents Chemother. *52*, 3817–3819.

Nishifuji, K., Fudaba, Y., Yamaguchi, T., Iwasaki, T., Sugai, M., and Amagai, M. (2005). Cloning of swine desmoglein 1 and its direct proteolysis by *Staphylococcus hyicus* exfoliative toxins isolated from pigs with exudative epidermitis. Vet. Dermatol. *16*, 315–323.

O'Brien, A.M., Hanson, B.M., Farina, S.A., Wu, J.Y., Simmering, J.E., Wardyn, S.E., Forshey, B.M., Kulick, M.E., Wallinga, D.B., and Smith, T.C. (2012). MRSA in conventional and alternative retail pork products. PLoS One *7*, e30092.

O'Riordan, K., and Lee, J.C. (2004). *Staphylococcus aureus* capsular polysaccharides. Clin. Microbiol. Rev. *17*, 218–234.

Oliveira, L., and Ruegg, P.L. (2014). Treatments of clinical mastitis occurring in cows on 51 large dairy herds in Wisconsin. J. Dairy Sci. *97*, 5426–5436.

Omae, Y., Sekimizu, K., and Kaito, C. (2012). Inhibition of colony-spreading activity of *Staphylococcus aureus* by secretion of δ-hemolysin. J. Biol. Chem. *287*, 15570–15579.

Owens, W.E., Nickerson, S.C., Boddie, R.L., Tomita, G.M., and Ray, C.H. (2001). Prevalence of mastitis in dairy heifers and effectiveness of antibiotic therapy. J. Dairy Sci. *84*, 814–817.

Paape, M.J., and Guidry, A.J. (1977). Effect of fat and casein on intracellular killing of *Staphylococcus aureus* by milk leucocytes. Proc. Soc. Exp. Biol. Med. *155*, 588–593.

Panton, P.N., and Valentine, F.C.O. (1932). STAPHYLOCOCCAL TOXIN. The Lancet *219*, 506–508.

Park, J., Friendship, R.M., Poljak, Z., DeLay, J., Slavic, D., and Dewey, C.E. (2013a). An investigation of ear necrosis in pigs. Can. Vet. J. *54*, 491–495.

Park, J., Friendship, R.M., Poljak, Z., Weese, J.S., and Dewey, C.E. (2013b). An investigation of exudative epidermitis (greasy pig disease) and antimicrobial resistance patterns of *Staphylococcus hyicus* and *Staphylococcus aureus* isolated from clinical cases. Can Vet. J. *54*, 139–144.

Parte, A.C. (2014). Genus *Staphylococcus*. In bacterionet, A.C. Parte, ed. (LPSN), pp. List of prokaryotic names with standing in nomenclature.

Penna, B., Varges, R., Medeiros, L., Martins, G.M., Martins, R.R., and Lilenbaum, W. (2010). Species distribution and antimicrobial susceptibility of staphylococci isolated from canine otitis externa. Vet. Dermatol. *21*, 292–296.

Peton, V., and Le Loir, Y. (2014). *Staphylococcus aureus* in veterinary medicine. Infection, genetics and evolution: journal of molecular epidemiology and evolutionary genetics in infectious diseases *21*, 602–615.

Prevost, G., Couppie, P., Prevost, P., Gayet, S., Petiau, P., Cribier, B., Monteil, H., and Piemont, Y. (1995). Epidemiological data on *Staphylococcus aureus* strains producing synergohymenotropic toxins. J. Med. Microbiol. *42*, 237–245.

Pyörälä, S., and Taponen, S. (2009). Coagulase-negative staphylococci-emerging mastitis pathogens. Vet. Microbiol. *134*, 3–8.

Rainard, P., Corrales, J.-C., Barrio, M.B., Cochard, T., and Poutrel, B. (2003). Leucotoxic activities of *Staphylococcus aureus* strains Isolated from cows, ewes, and goats with mastitis: importance of LukM/LukF′-PV leucotoxin. Clin. Diagn. Lab. Immunol. *10*, 272–277.

Rainard, P., Ducelliez, M., and Poutrel, B. (1990). The contribution of mammary infections by coagulase-negative staphylococci to the herd bulk milk somatic cell count. Vet. Res. Commun. *14*, 193–198.

Rantala, M., Hölsö, K., Lillas, A., Huovinen, P., and Kaartinen, L. (2004). Survey of condition-based prescribing of antimicrobial drugs for dogs at a veterinary teaching hospital. Vet. Rec. *155*, 259–262.

Saijonmaa-Koulumies, L.E., and Lloyd, D.H. (2002). Colonization of neonatal puppies by *Staphylococcus intermedius*. Vet. Dermatol. *13*, 123–130.

Sakwinska, O., Giddey, M., Moreillon, M., Morisset, D., Waldvogel, A., and Moreillon, P. (2011). *Staphylococcus aureus* host range and human-bovine host shift. Appl. Environ. Microbiol. *77*, 5908–5915.

Sasaki, T., Kikuchi, K., Tanaka, Y., Takahashi, N., Kamata, S., and Hiramatsu, K. (2007). Reclassification of phenotypically identified *staphylococcus intermedius* strains. J. Clin. Microbiol. *45*, 2770–2778.

Sasaki, T., Tsubakishita, S., Tanaka, Y., Ohtsuka, M., Hongo, I., Fukata, T., Kabeya, H., Maruyama, S., and Hiramatsu, K. (2012). Population genetic structures of *Staphylococcus aureus* isolates from cats and dogs in Japan. J. Clin. Microbiol. *50*, 2152–2155.

Sasaki, T., Tsubakishita, S., Tanaka, Y., Sakusabe, A., Ohtsuka, M., Hirotaki, S., Kawakami, T., Fukata, T., and Hiramatsu, K. (2010). Multiplex-PCR method for species identification of coagulase-positive staphylococci. J. Clin. Microbiol. *48*, 765–769.

Sato, H., Matsumori, Y., Tanabe, T., Saito, H., Shimizu, A., and Kawano, J. (1994). A new type of staphylococcal exfoliative toxin from a *Staphylococcus aureus* strain isolated from a horse with phlegmon. Infect. Immun. *62*, 3780–3785.

Sato, H., Tanabe, T., Kuramoto, M., Tanaka, K., Hashimoto, T., and Saito, H. (1991). Isolation of exfoliative toxin from *Staphylococcus hyicus* subsp. *hyicus* and its exfoliative activity in the piglet. Vet. Microbiol. *27*, 263–275.

Sato, H., Watanabe, T., Higuchi, K., Teruya, K., Ohtake, A., Murata, Y., Saito, H., Aizawa, C., Danbara, H., and Maehara, N. (2000). Chromosomal and extrachromosomal synthesis of exfoliative toxin from *Staphylococcus hyicus*. J. Bacteriol. *182*, 4096–4100.

Schalm, O.W., Lasmanis, J., and Jain, N.C. (1976). Conversion of chronic staphylococcal mastitis to acute gangrenous mastitis after neutropenia in blood and bone marrow produced by an equine anti-bovine leucocyte serum. Am. J. Vet. Res. *37*, 885–890.

Schwandner, R., Dziarski, R., Wesche, H., Rothe, M., and Kirschning, C.J. (1999). Peptidoglycan- and lipoteichoic acid-induced cell activation is mediated by toll-like receptor 2. J. Biol. Chem. *274*, 17406–17409.

Sears, P.M., and McCarthy, K.K. (2003). Management and treatment of staphylococcal mastitis. Vet. Clin. North. Am. Food. Anim. Pract. *19*, 171–185, vii.

Seguin, J.C., Walker, R.D., Caron, J.P., Kloos, W.E., George, C.G., Hollis, R.J., Jones, R.N., and Pfaller, M.A. (1999). Methicillin-Resistant *Staphylococcus aureus* Outbreak in a Veterinary Teaching Hospital: Potential Human-to-Animal Transmission. J. Clin. Microbiol. *37*, 1459–1463.

Segura, P., Martinez, J., Peris, B., Selva, L., Viana, D., Penades, J.R., and Corpa, J.M. (2007). Staphylococcal infections in rabbit does on two industrial farms. Vet. Rec. *160*, 869–872.

Shryock, T.R., Dye, E.S., and Kapral, F.A. (1992). The accumulation of bactericidal lipids in staphylococcal abscesses. J. Med. Microbiol. *36*, 332–336.

Sieverding, E. (1993). Use of an autogenous vaccine against exudative epidermitis in suckled piglets. Tierarztliche-Umschau *48*, 2.

Siqueira, J.A., Speeg-Schatz, C., Freitas, F.I.S., Sahel, J., Monteil, H., and Prévost, G. (1997). Channel-Forming Leucotoxins from *Staphylococcus aureus* cause Severe Inflammatory Reactions in a Rabbit Eye Model. J. Med. Micrbiol. *46*, 486–494.

Skalka, B. (1985). [Hyaluronidase test in the diagnosis of staphylococci.] Vet. Med. *30*, 373–378.

Sledge, D.G., Danieu, P.K., Bolin, C.A., Bolin, S.R., Lim, A., Anderson, B.C., and Kiupel, M. (2010). Outbreak of Neonatal Diarrhea in Farmed Mink Kits (Mustella vison) Associated With Enterotoxigenic *Staphylococcus delphini*. Vet. Pathol. *47*, 751–757.

Smith, T.C., Gebreyes, W.A., Abley, M.J., Harper, A.L., Forshey, B.M., Male, M.J., Martin, H.W., Molla, B.Z., Sreevatsan, S., Thakur, S., *et al.* (2013). Methicillin-resistant *Staphylococcus aureus* in pigs and farm workers on conventional and antibiotic-free swine farms in the USA. PLoS One *8*, e63704.

Smyth, D.S., Feil, E.J., Meaney, W.J., Hartigan, P.J., Tollersrud, T., Fitzgerald, J.R., Enright, M.C., and Smyth, C.J. (2009). Molecular genetic typing reveals further insights into the diversity of animal-associated *Staphylococcus aureus*. J. Med. Microbiol. *58*, 1343–1353.

Sompolinsky, D., Samra, Z., Karakawa, W.W., Vann, W.F., Schneerson, R., and Malik, Z. (1985). Encapsulation and capsular types in isolates of *Staphylococcus aureus* from different sources and relationship to phage types. J. Clin. Microbiol. *22*, 828–834.

Sutra, L., and Poutrel, B. (1994). Virulence factors involved in the pathogenesis of bovine intramammary infections due to *Staphylococcus aureus*. J. Med. Microbiol. *40*, 79–89.

Takahashi, T., Satoh, I., and Kikuchi, N. (1999). NOTE: Phylogenetic relationships of 38 taxa of the genus

Staphylococcus based on 16S rRNA gene sequence analysis. In.t J. Syst. Bacteriol. *49*, 725–728.
Taponen, S., Jantunen, A., Pyörälä, E., and Pyörälä, S. (2003). Efficacy of targeted 5-day combined parenteral and intramammary treatment of clinical mastitis caused by penicillin-susceptible or penicillin-resistant *Staphylococcus aureus*. Acta Vet. Scand. *44*, 53–62.
Taponen, S., Supré, K., Piessens, V., Van Coillie, E., De Vliegher, S., and Koort, J.M.K. (2012). *Staphylococcus agnetis* sp. nov., a coagulase-variable species from bovine subclinical and mild clinical mastitis. Int. J. Syst. Evol. Microbiol. *62*, 61–65.
Taylor, A.G., and Bernheimer, A.W. (1974). Further characterization of staphylococcal gamma-hemolysin. Infect. Immun. *10*, 54–59.
Tenhagen, B.-A., Vossenkuhl, B., Käsbohrer, A., Alt, K., Kraushaar, B., Guerra, B., Schroeter, A., and Fetsch, A. (2014). Methicillin-resistant *Staphylococcus aureus* in cattle food chains – Prevalence, diversity, and antimicrobial resistance in Germany. J. Anim. Sci. *92*, 2741–2751.
Thorberg, B.M., Danielsson-Tham, M.L., Emanuelson, U., and Persson Waller, K. (2009). Bovine subclinical mastitis caused by different types of coagulase-negative staphylococci. J. Dairy Sci. *92*, 4962–4970.
Tollersrud, T., Kampen, A.H., and Kenny, K. (2006). *Staphylococcus aureus* enterotoxin D is secreted in milk and stimulates specific antibody responses in cows in the course of experimental intramammary infection. Infect. Immun. *74*, 3507–3512.
Trinidad, P., Nickerson, S.C., and Alley, T.K. (1990). Prevalence of intramammary infection and teat canal colonization in unbred and primigravid dairy heifers. J. Dairy Sci. 73, 107–114.
Tzamalis, A., Chalvatzis, N., Anastasopoulos, E., Tzetzi, D., and Dimitrakos, S. (2013). Acute postoperative *Staphylococcus schleiferi* endophthalmitis following uncomplicated cataract surgery: first report in the literature. Eur. J. Ophthalmol. *23*, 427–430.
Van den Eede, A., Martens, A., Lipinska, U., Struelens, M., Deplano, A., Denis, O., Haesebrouck, F., Gasthuys, F., and Hermans, K. (2009). High occurrence of methicillin-resistant *Staphylococcus aureus* ST398 in equine nasal samples. Vet. Microbiol. *133*, 138–144.
van Duijkeren, E., Houwers, D.J., Schoormans, A., Broekhuizen-Stins, M.J., Ikawaty, R., Fluit, A.C., and Wagenaar, J.A. (2008). Transmission of methicillin-resistant *Staphylococcus intermedius* between humans and animals. Vet. Microbiol. *128*, 213–215.
van Duijkeren, E., Ten Horn, L., Wagenaar, J.A., de Bruijn, M., Laarhoven, L., Verstappen, K., de Weerd, W., Meessen, N., and Duim, B. (2011). Suspected horse-to-human transmission of MRSA ST398. Emerg. Infect. Dis. *17*, 1137–1139.
Vancraeynest, D., Haesebrouck, F., Deplano, A., Denis, O., Godard, C., Wildemauwe, C., and Hermans, K. (2006). International Dissemination of a High Virulence Rabbit *Staphylococcus aureus* Clone. J. Vet. Med., Series B 53, 418–422.
Vasudevan, P., Nair, M.K.M., Annamalai, T., and Venkitanarayanan, K.S. (2003). Phenotypic and genotypic characterization of bovine mastitis isolates of *Staphylococcus aureus* for biofilm formation. Vet. Microbiol. *92*, 179–185.
Voss, A., Loeffen, F., Bakker, J., Klaassen, C., and Wulf, M. (2005). Methicillin-resistant *Staphylococcus aureus* in pig farming. Emerg. Infect. Dis. *11*, 1965–1966.
Waage, S., Mørk, T., Røros, A., Aasland, D., Hunshamar, A., and Odegaard, S.A. (1999). Bacteria associated with clinical mastitis in dairy heifers. J. Dairy Sci. *82*, 712–719.
Wadström, T., and Möllby, R. (1972). Some biological properties of purified staphylococcal haemolysins. Toxicon *10*, 511–519.
Webster, K.A., and Mitchell, G.B. (1986). Experimental production of tick pyaemia. Vet. Rec. *119*, 186–187.
Weese, J.S., Avery, B.P., and Reid-Smith, R.J. (2010). Detection and quantification of methicillin-resistant *Staphylococcus aureus* (MRSA) clones in retail meat products. Lett. Appl. Microbiol. *51*, 338–342.
Weese, J.S., Hannon, S.J., Booker, C.W., Gow, S., Avery, B.P., and Reid-Smith, R.J. (2012). The prevalence of methicillin-resistant *Staphylococcus aureus* colonization in feedlot cattle. Zoonoses Public Health 59, 144–147.
Weese, J.S., and van Duijkeren, E. (2010). Methicillin-resistant *Staphylococcus aureus* and *Staphylococcus pseudintermedius* in veterinary medicine. Vet. Microbiol. *140*, 418–429.
Wegener, H., and Skov-Jensen, E. (2006). Exudative epidermitis. In Diseases of Swine, B. Straw, J. Zimmerman, S. D'Allaire, and D. Taylor, eds. (Ames, IA: Blackwell Publishing), pp. 675–679.
Weidenmaier, C., Kokai-Kun, J.F., Kristian, S.A., Chanturiya, T., Kalbacher, H., Gross, M., Nicholson, G., Neumeister, B., Mond, J.J., and Peschel, A. (2004). Role of teichoic acids in *Staphylococcus aureus* nasal colonization, a major risk factor in nosocomial infections. Nat. Med. *10*, 243–245.
Wellnitz, O., Arnold, E.T., and Bruckmaier, R.M. (2011). Lipopolysaccharide and lipoteichoic acid induce different immune responses in the bovine mammary gland. J. Dairy Sci. *94*, 5405–5412.
Wellnitz, O., and Bruckmaier, R.M. (2012). The innate immune response of the bovine mammary gland to bacterial infection. Vet. J. *192*, 148–152.
Wielders, C.L., Fluit, A.C., Brisse, S., Verhoef, J., and Schmitz, F.J. (2002). mecA gene is widely disseminated in *Staphylococcus aureus* population. J. Clin. Microbiol. *40*, 3970–3975.
Woodin, A.M. (1960). Purification of the two components of leucocidin from *Staphylococcus aureus*. Biochem. J. *75*, 158–165.
Wraight, M.D. (2003). A comparative efficacy trial between cefuroxime and cloxacillin as intramammary treatments for clinical mastitis in lactating cows on commercial dairy farms. N. Z. Vet. J. *51*, 26–32.
Xu, S., Arbeit, R.D., and Lee, J.C. (1992). Phagocytic killing of encapsulated and microencapsulated *Staphylococcus aureus* by human polymorphonuclear leucocytes. Infect. Immun. *60*, 1358–1362.
Zhu, X.Y., and Hester, P.Y. (2000). The etiology of staphylococcosis. Avian Poultry Biol. Rev. *11*, 97–112.

Staphylococcal Variation and Evolution

Jodi A. Lindsay

4

Abstract

A staphylococcal genome carries all of the information responsible for behaviour of that cell. The genome sequence of each staphylococcal isolate can vary substantially from other isolates, even those that cause similar disease. And *S. aureus* populations as a whole are not stagnant, but constantly evolving. The population structure of *S. aureus* groups isolates into independent lineages, each with unique combinations of surface, regulatory and immune evasion genes. Individual isolates also acquire and lose mobile genetic elements (MGEs) encoding virulence, resistance and host-adaptation factors. Transfer of DNA within populations is controlled by molecular machinery that partially dictates how the species evolves. Selection, such as by antimicrobials or host specificity, leads to the emergence and spread of successful clones in particular environments. Other staphylococcal species also show evidence of lineages and MGEs that shape populations. Improving technology for sequencing and comparing genomes is allowing analysis of thousands of isolates and will ensure this is a rapidly growing area of research. Investigation of the diversity and selection of staphylococcal genomic variation leads to greater understanding of physiology, host–pathogen interactions, epidemiology and the development of better therapeutic and public health strategies.

Introduction

New technologies such as whole genome sequencing (WGS) have revealed the genomes of thousands of staphylococci and revolutionized our understanding of how genomes and populations vary and evolve (Lindsay, 2010). The first *S. aureus* genome sequences were published in 2001 (Kuroda *et al.*, 2001) and the progress made since is breathtaking. More is known about *S. aureus* than the other species, and therefore this chapter will focus on *S. aureus* but also includes a section on progress with other species. The overall aim of the chapter is to describe the genetic variation seen in large population studies, describe at the molecular level how genetic variation arises, how variation might be selected for in successful populations leading to the evolution of the populations we see and the expansion of successful clones. A section on technology is included to explain briefly how it works, how to interpret data critically, how to investigate your own genes of interest, and how the field may change in the future.

The *S. aureus* genome

The first *S. aureus* whole genome sequences revealed a relatively small chromosomal genome, less than 3000 kb and fewer than 3000 genes (Kuroda *et al.*, 2001). The G+C content is low, around 32%, and this does not fluctuate very much, suggesting little transfer of DNA from species with differing content. Many isolates also carry plasmids (Holden and Lindsay, 2008). The *S. aureus* genome defines the metabolic and physiological pathways used by *S. aureus*, including those for nutrient uptake, utilization, synthesis, protein production, regulation, cell replication, synthesis of structural components and more, as well as revealing an extensive array of genes predicted to play a role in virulence or interact with the host (Holden and Lindsay, 2008). Despite the enormous amount of information we have gleaned from reading the genome, the function of around 30% of predicted coding regions is still unknown.

Variation between isolates

A remarkable feature of *S. aureus* populations is the extensive amount of diversity between isolates. Isolates differ in three major ways: Firstly, they can have single nucleotide polymorphisms (SNPs) potentially changing a single base pair, or other minor variations such as small replications, insertions or deletions. Secondly, whole genes or clusters of genes may vary substantially or be missing between isolates. Finally, mobile genetic elements (MGEs), discrete pieces of DNA that encode their own integration or replication mechanisms, move into and out of genomes as well as between isolates.

Lineages

Overall, the population structure of *S. aureus* is composed of lineages of closely related isolates, with about a dozen lineages prevalent in humans, while animal isolates belong to their own distinct lineages (Lindsay *et al.*, 2006; Sung *et al.*, 2008). Two isolates belonging to the same lineage have highly conserved genomes, except for the MGEs (Holden and Lindsay, 2008; Lindsay, 2010). Their genomes will have a conserved gene order, and they will share many SNPs, gene variants and deletions. An early example is the genome sequences of two isolates from the same lineage, clonal complex (CC)1, MSSA476, a methicillin-sensitive isolate from an osteomyelitis patient in the UK, and MW2, a representative of the community-associated MRSA clone from the USA causing severe skin and soft tissue infection. The two genomes only differed in 285 SNPs within functional coding regions, excluding those found on MGEs (Holden *et al.*, 2004). This is remarkably stable considering the geographical differences and the different types of disease caused by these bacteria, suggesting that *S. aureus* core genomes within lineages are inherently stable. Subsequent whole genome sequencing projects have confirmed the relatively conserved genomes of staphylococci belonging to the same lineage (Harris *et al.*, 2010; Holden *et al.*, 2013).

Isolates from different lineages are much more distinct. They differ substantially in variation and deletions of gene fragments, whole genes or operons in the non-MGE genome. These core-variable or lineage-associated genes are often in coding regions associated with surface proteins, immune interaction and gene regulation (Lindsay *et al.*, 2006; McCarthy and Lindsay, 2010; McCarthy and Lindsay, 2013). Hundreds of these 'core variable' genes are present and include those genes encoding protein A, coagulase, capsule type, fibrinogen binding proteins, collagen binding protein, immunodominant antigen IsaB, haemagglutinin-like protein, mammalian protein binding MapW, large surface protein Ebh, serine-aspartate repeat surface proteins, accumulation associated protein, proteases, superantigens, lipoproteins, leucocidin LukDE, hyaluronate lyase, lantibiotics, accessory gene regulators and staphylococcal accessory regulators SarH2 and SarH3 (Holden and Lindsay, 2008; McCarthy and Lindsay, 2010). This is interesting, because these lineage-associated genes are exactly those thought to be important in host–pathogen interaction, yet most human *S. aureus* lineages seem equally capable of colonization or causing similar types of disease (Lindsay *et al.*, 2006). If a gene is missing or substantially variant in *S. aureus* populations, is it essential for disease? Importantly, such an individual protein may be discarded as a good vaccine or drug target. However, it may not be that simple. Different *S. aureus* lineages may have related genes that perform the same or similar functions (gene redundancy), and as a group can still be useful targets (McCarthy and Lindsay, 2010, 2013).

The dominant lineages of *S. aureus* in humans are labelled clonal complex (CC)1, CC5, CC8, CC12, CC15, CC22, CC25, CC30, CC45, CC51 (Day *et al.*, 2001; Lindsay *et al.*, 2006). They are called clonal complex based on the multilocus sequence typing (MLST) system that groups isolates together if they share identical sequences in five of seven housekeeping genes (Enright *et al.*, 2000). Minor lineages are also common in population studies and the reference database of MLST types list thousands of such lineages or individual sequence types (ST) that share seven of seven genes (www.mlst.net). Isolates can be assigned to a lineage using a variety of methods (Stefani *et al.*, 2012). Briefly, the whole genome sequence of an isolate will easily reveal an isolate's lineage, as will the pattern of lineage-associated genes detected by multi-strain microarray (Lindsay *et al.*, 2006). MLST is a relatively simple method, and the spa-typing method, where the genome sequence of only one gene, the protein A (*spa*) gene, is compared to a reference library of sequences and assigned to a lineage is even easier (Harmsen *et al.*, 2003) (www.

spaserver.ridom.de). A simple method to identify the major lineages is by a PCR test for the lineage-specific restriction-enzyme loci *hsdS* (Cockfield *et al.*, 2007).

Although the different lineages are very unique from each other, many of the gene variants are shared between lineages. This suggests recombination between ancestral lineages followed by selection of particular combinations. For example, there are two dominant capsule types, type 5 found in CC5, CC8, CC22 and CC25, and type 8 found in CC1, CC12, CC15, CC30, CC45 and CC51. There are three major types of accessory gene regulator, with type I found in CC8, CC22, CC25 and CC45, type II in CC5, CC12 and CC15, and type III in CC1 and CC30. Collagen binding protein (cna) is found only in CC1, CC12, CC22, CC45, CC51 (Lindsay *et al.*, 2006) Therefore, it seems there has been recombination in the evolutionary history of *S. aureus*, where these lineage-associated genes exchanged, and certain combinations were selected for and these particular populations expanded.

Surprisingly, non-human lineages often share lineage-associated genes with human lineages (Sung *et al.*, 2008). For example, the dominant lineages in cows include CC151, CC133 and CC97, while the major lineage associated with pigs is CC398. While we might have expected animal isolates to carry their own surface proteins binding to distinct animal host-proteins this does not seem to be the case. For example, CC151 is capsule type 8 and agr II, CC133 is capsule type 8 and agr I, and CC97 and CC398 are capsule type 5 and agr I (Sung *et al.*, 2008). Again, it suggests common ancestors of both human and animal lineages, recombination, and selection of successful combinations in the different hosts.

Mobile genetic elements (MGEs)

Individual *S. aureus* isolates can carry a wide variety of different MGEs, and these vary substantially even between isolates belonging to the same lineage (Lindsay *et al.*, 2006). The major types of MGE are bacteriophage, *S. aureus* pathogenicity islands (SaPIs), plasmids, staphylococcal cassette chromosomes (SCC) and transposons (Lindsay and Holden, 2006; Lindsay, 2008). MGEs encode a large proportion of resistance genes and many important toxins, virulence and host-adaptation factors. MGEs account for about 10–20% of the genetic content of each *S. aureus* isolate. In this section, the MGEs are described, and in the next, the movement of MGEs is described.

Bacteriophage in *S. aureus* genomes are temperate phage that have inserted into specific sites in the chromosome. Eight φ families have been described, based on the sequence of the integrase gene that determines the integration site (McCarthy *et al.*, 2012c). Each prophage is about 45 kb. Mosaicism is common in *S. aureus* prophage, and the sequence of each is homologous to a series of fragments derived from a variety of other phage. Many prophage encode genes that may benefit the host, such as immune-evasion pathways or toxins such as enterotoxin A or Panton–Valentine leucocidin. All clinical isolates carry at least one prophage, and some can carry up to four.

SaPIs are about 15 kb long and also encode site specific integrases (Lindsay *et al.*, 1998; Lindsay, 2008). Five families of SaPI have been described based on the sequences of the integrases, and they also have mosaic genomes. SaPIs encode toxins such as toxic shock syndrome toxins, enterotoxins B and C, host immune evasion factors and occasionally resistance genes. Many *S. aureus* isolates carry one SaPI.

Plasmids are circular pieces of DNA that replicate autonomously. They are found in many *S. aureus* and there are three major types (McCarthy and Lindsay, 2012). Small plasmids that replicate by the rolling circle method may have only a few genes, and contain only a few thousand base pairs, but frequently carry antibiotic resistance genes. Type II plasmids are much bigger, often 20–50 kb, and also frequently carry resistance genes as well as heavy metal resistance genes and some toxin genes. Smaller plasmids and transposons can insert, and mosaicism is common. Type III plasmids are similar to Type II but additionally encode the *tra* genes necessary for conjugative transfer.

SCC elements integrate into a single *orfX* gene site in the *S. aureus* chromosome and frequently carry the *mecA* gene, and in this case are called SCC*mec*. Other resistances, transposons and plasmids can also insert into SCC*mec* elements. At least 11 major types have been described, based on the *ccr* genes that catalyse excision and integration of the element, as well as other markers (Ito *et al.*, 2004; Shore and Coleman, 2013; www.sccmec.org).

Mosaicism is common, and size can vary from a few kilobases up to 70 kb.

Transposons encode a transposase gene, and in *S. aureus*, often encode resistance genes. Some isolates can carry more than one copy of the same transposon (Lindsay, 2008). Insertion sequences (IS), literally just encoding a single transposase gene are also found and these can flank regions such as resistance genes, potentially catalysing their movement. However, many IS seem stable in *S. aureus*, although in *S. epidermidis* they can move and insert into genes altering gene function or expression (Ziebuhr *et al.*, 1999). The genomic islands alpha and beta described in *S. aureus* encoded Type I restriction modification systems and multiple proteases and exotoxins, however, they appear to be very lineage specific and very stable in the genome, and therefore fall outside the definition of an MGE (Lindsay, 2008).

Variation: mechanisms

The major mechanisms controlling genomic variation are horizontal gene transfer (HGT) of DNA from another organism, deletion, transposition movement, duplication, rearrangement, inversion, and point mutation. HGT of DNA from one bacterium to another is a rapid strategy for bacteria to adapt to new environments (Lindsay, 2014). The major mechanisms of HGT in bacteria are transformation, conjugation and transduction, but in *S. aureus*, transduction is the major method.

Transduction requires bacteriophage, which are viruses that infect microbes. Prophage, bacteriophage genomes that insert into the bacterial chromosome and lie dormant, are extremely common in *S. aureus*. Transduction involves excision of the prophage, replication of the phage genome, production of phage particles, packaging of DNA into the phage heads of the particles, lysis of the host cell and release of virulent phage particles. The phage particles bind to receptors on other *S. aureus*, probably glycosylated *S. aureus* wall teichoic acid, which is species specific (Xia *et al.*, 2011), and inject the phage genome into the new recipient cell, where most of the phage undergo a lytic cycle leading to production of more virulent phage particles. However, a small proportion of phage genomes will integrate site-specifically into the chromosome and resume the dormant or lysogenic phase. SaPIs are MGEs that can take over the packaging step of particular phage to ensure that their own genomes are preferentially packaged into phage heads and transferred horizontally (Tormo-Mas *et al.*, 2010).

Generalized transduction occurs when the phage particles package bacterial DNA instead of phage DNA. The phage particle produced is not virulent, but it can still deliver foreign DNA to a new host (Quiles-Puchalt *et al.*, 2014). Once inside the new host, the DNA needs to integrate into the chromosome or replicate on its own, or it will be lost once the bacterium goes through a cycle of replication. Plasmids replicate themselves. Phage, SaPI and transposons encode integrases or transposases which catalyse the integration step into the chromosome. Other pieces of DNA must find a way to integrate, usually by homologous recombination. This recombination is catalysed by *recA*, the gene involved in DNA mismatch repair (Bayles *et al.*, 1994). In *S. aureus*, it appears to be very inefficient, certainly compared to other laboratory bacteria such as *E. coli*, making *S. aureus* particularly difficult to genetically manipulate.

Transduction is widely used in the laboratory to move DNA fragments of 45 kb or less (the capacity that can be carried in a phage head) between *S. aureus* isolates and the process is efficient, especially when selecting for transfer with a selectable resistance marker. The triggers for induction of bacteriophage are known to include antibiotics at sub-MIC concentrations and stresses such as UV light and mitomycin C (Goerke *et al.*, 2006; Mesak *et al.*, 2008). Recent studies have also shown that during colonization *in vivo* in a piglet animal model, HGT due to transduction occurs at extremely high frequencies, resulting in detectable transfer within hours, even without resistance marker selection (McCarthy *et al.*, 2014). Generation of strains with a wide variety of combinations of transferred MGEs was detected, showing that HGT occurs at much higher frequency than SNP generation and at higher frequency *in vivo* than *in vitro*.

Conjugation is very efficient in transferring DNA between *S. aureus*, and about 5% of isolates carry MGEs with the necessary *tra* genes for transfer (McCarthy and Lindsay, 2012). The *tra* genes encode proteins for a pore or pilus that forms between the donor and recipient bacteria, allowing the DNA to transfer. DNA can transfer between species and genera, such as the transfer of vancomycin resistance from enterococci into *S. aureus*. Transformation, in contrast, is inefficient in *S. aureus*. Transformation involves competence proteins that form a pore in the cell wall allowing free DNA to be taken into the cell. This occurs at high frequency when the system is switched on in *E. coli*, streptococci and many other bacteria. However, through a combination of environmental, mutational and RNA inhibition pathways competence genes are transcribed at extremely low frequency in *S. aureus* (Morikawa *et al*., 2012), suggesting it is not a common HGT mechanism.

There are two major types of barriers that prevent bacteria from taking up foreign DNA, which may be a strategy to protect the cell from lytic phage (Lindsay, 2014). The CRISPR system is found in many bacteria, including some *S. epidermidis* (Marraffini and Sontheimer, 2008; Conlan *et al*., 2012), and a few unusual *S. aureus* isolates (Holt *et al*., 2011; Golding *et al*., 2012; Kinnevey *et al*., 2013). The system recognizes small regions of DNA in specific MGEs and targets them by binding with complementary RNA molecules, triggering a specific cutting of the DNA at that site (Makarova *et al*., 2011). Much more common in *S. aureus* are the restriction modification (RM) systems that recognize and digest foreign DNA based on whether it has been modified at specific sequences (Lindsay, 2014). All isolates carry a type I system and most carry a type IV system, with some isolates carrying additional type I and II systems on MGEs.

The type I RM system is a restriction enzyme complex that recognizes a specific DNA sequence occurring hundreds of times in the chromosome, binds to the DNA sequence, slides along the DNA until it meets another enzyme and digests the DNA (Murray, 2000; Roberts *et al*., 2013). In order to not digest its own DNA, the cell also produces a simpler modification enzyme complex that targets the exact same sequence and methylates the DNA, protecting it from recognition and digestion by the restriction enzyme (Waldron and Lindsay, 2006; Roberts *et al*., 2013). Foreign DNA that is not modified is therefore recognized and digested precisely because it is not modified. The type I RM found in all *S. aureus* is lineage specific, and the two different copies of the *hsdS* gene that encodes the enzyme complex subunit that determines the sequence specificity of the system are highly conserved between isolates of the same lineage (Waldron and Lindsay, 2006). This means that DNA is much more readily transferred between isolates of the same lineage than between isolates of different lineages, leading to relatively independent evolution of the lineages and suggesting it is responsible for the *S. aureus* population structure (Lindsay, 2008). The target sites of many common type I RMs have now been identified. Interestingly, some large conjugative plasmids have adapted to evade the type I RM system by reducing the number of target sites, allowing movement of resistance plasmids between clinically important lineages (Roberts *et al*., 2013).

The type IV RM system found in *S. aureus* consists of a single enzyme, a restriction enzyme that recognizes and digests only modified DNA (Corvaglia *et al*., 2010; Xu *et al*., 2011). *S. aureus* does not produce the corresponding modification enzyme and cannot modify its own DNA. The specifically modified DNA that is targeted by the type IV RM comes from *E. coli*. This explains why plasmids from *E. coli* are difficult to transfer into *S. aureus*, and indeed the most useful laboratory strain for moving DNA into *S. aureus* is RN4220 that has mutations in both type I and IV RM restriction enzymes.

Type II RM systems are the ones exploited by molecular biology labs that recognize unmodified DNA at a specific genome sequence and digest it exactly at this site, often generating sticky ends. Type I and II systems are found on some MGEs such as SCC*mec*, phage and plasmids (Holden *et al*., 2004; Dempsey *et al*., 2005). Presumably they protect the bacterial host from phage lysis in a synergistic relationship that also protects the MGE's own genome.

Transposons are known to move within genomes, but whole genome projects suggest that most are relatively stable. Gene duplication occurs in specific regions of the genome, particularly in the genomic islands alpha and beta that encode multiple variants of superantigen-like genes or proteases and enterotoxin genes, respectively (Lindsay, 2008). In both cases, the close similarity of the genes suggests repeated duplication followed by variation or adaptation and rearrangements. Interestingly, the sequences are highly conserved within lineages, indicating that rearrangement predominantly occurred around the time of lineage emergence and/or expansion, but is currently stable. Another example of duplication is in the sigma S gene (*sigS*) that controls transcription of the competence pathway (Morikawa *et al.*, 2013). Inversion is relatively uncommon in *S. aureus*, as is rearrangement of the core genome. However, the sequences of phage, SaPIs and plasmids show high levels of mosaicism, which is evidence of very frequent recombination, at least within related types of element. Recently we have observed recombination of phage genomes during experimental colonization of piglets, indicating this could occur at extremely high frequency (McCarthy *et al.*, 2014).

Single nucleotide polymorphisms, or other minor variations in the genome, are thought to occur at random during the chromosomal replication process and then are repaired using the intrinsic DNA repair mechanisms (Bayles *et al.*, 1994; Mesak *et al.*, 2008). SNPs can be synonymous or non-synonymous, depending on whether they are predicted to cause an amino acid change and potentially alter the function of a protein. Occasionally, a SNP will have a beneficial effect on a cell in a particular environment and be selected for. It is estimated from the large epidemiology projects that in the core genome, one SNP per genome occurs and is selected for every 8 weeks (Harris *et al.*, 2010; Uhlemann *et al.*, 2014). This means that a patient that is colonized may have a few SNP variants in the isolates, which becomes important when conducting epidemiological studies based on WGS of the core genome and tracing the movement of isolates (Harris *et al.*, 2013). Within lineages over time, a vast variety of SNPs have accumulated as the lineage expands. For example, in a recent study of the lineage CC22, 192 isolates were sequenced and over 8000 SNPs identified, allowing a comprehensive tree of the evolutionary history of this lineage to be constructed (Holden *et al.*, 2013). Similar trees have also been constructed for successful MRSA clones including ST239, CC30, USA300 CC8, CC398, ST225 (Harris *et al.*, 2010; McAdam *et al.*, 2012; Uhlemann *et al.*, 2012, 2014; Price *et al.*, 2012; Nubel *et al.*, 2010). These trees are extraordinarily powerful for understanding how populations have arisen, spread over time and geographically, and how they might have adapted to new environments.

Selection

Evolution occurs when DNA varies and the variant isolate is subsequently selected because it has an advantage. *S. aureus* are able to grow in a wide variety of environments, such as in the human nose, throat, groin, armpit and gut, as well as evading the human immune response. They can also colonize a range of mammalian and bird species, and survive in high salt, anaerobic conditions and chemically defined media. Perhaps our most powerful tools for understanding selection of *S. aureus* populations so far have been the large scale epidemiological studies of successful MRSA clones (Harris *et al.*, 2010; McAdam *et al.*, 2012; Uhlemann *et al.*, 2012; Price *et al.*, 2012; Nubel *et al.*, 2010). In this way, we can compare genomes of isolates selected in different environments and identify changes consistent with that environment compared to control isolates that are not selected. These studies often show that lineages are stable, MGEs vary substantially, and that fixation of an MGE in a particular environment suggests it plays a role in adaptation (Lindsay, 2010).

Methicillin-resistant *S. aureus* are *S. aureus* that have acquired a SCC*mec* element carrying the *mecA* gene encoding resistance to all β-lactam antibiotics (Ito *et al.*, 2004). SCC*mec* have been found in most lineages, but only a handful of specific clones have become successful and spread. Some rare *S. aureus* lineages have become much more prevalent once SCC*mec* was acquired (CC80, CC398). SCC*mec* is a relatively stable MGE and moves at lower frequency than other MGEs, so successful MRSA clones are often described by their lineage and SCC*mec* type (Shore and Coleman, 2013). There are three main environments and epidemiological behaviours of MRSA, those that have adapted to healthcare, community and livestock (Lindsay, 2010). Interestingly, some of the original large SCC*mec* types have a fitness cost on the cell, but must survive because the ability to resist β-lactam antibiotics overrides this cost (Ender *et al.*, 2004; Knight *et al.*, 2013)

Healthcare-associated (HA)-MRSA is associated with health care exposure, and these isolates can spread rapidly in this environment (Lindsay, 2013). The reservoir of infecting MRSA is the patient's own colonizing flora (McCarthy *et al.*, 2012a). The major lineages involved are CC22, CC30, CC5, CC8, CC45 and ST239 (which was formed by a recombination event between CC30 and CC8). Each country or geographical region is usually dominated by only a few clones, but some of these clones are extremely successful in hospitals throughout the world, especially CC22 and ST239 (Harris *et al.*, 2010; Holden *et al.*, 2013). HA-MRSA cause a variety of infections, predominantly in immunocompromised patients or those that have had their skin barrier breached by wounds, surgery or catheters. Isolates are often multi-drug resistant, which is likely to be beneficial in the healthcare setting as a high proportion of colonized patients receive antibiotics. Interestingly, most HA-MRSA have multiple resistances, but fully drug resistant isolates that have accumulated all resistances are not seen (Knight *et al.*, 2012). Recently it has been suggested that HA-MRSA are successful in healthcare settings because of two key steps: acquisition of *mecA* as well as SNP mutations conferring fluoroquinolone resistance (Knight *et al.*, 2012; Holden *et al.*, 2013). Other types of MRSA and methicillin susceptible isolates tend to be sensitive to fluoroquinolones (Knight *et al.*, 2012). Importantly, there is evidence that a decrease in fluoroquinolone prescribing leads to a decrease in incidence of HA-MRSA, suggesting resistance to this class of drugs is particularly important in selection of HA-MRSA (Knight *et al.*, 2012).

Community-associated (CA-) MRSA did not arise from HA-MRSA but acquired SCC*mec* independently. These isolates typically belong to lineages CC8 (including USA300), CC1, CC59, CC30, CC80 and CC93, and again only one or two types tend to dominate in each geographical area (Lindsay, 2010; Chua *et al.*, 2011; Stefani *et al.*, 2012). CA-MRSA infections are particularly prevalent in the USA and Asia and often present as severe skin and soft tissue infection in healthy patients, so the spectrum of disease caused by these bacteria is different to HA-MRSA. A common feature of all CA-MRSA is carriage of a prophage of the φ2 family that has a copy of the Panton–Valentine leucocidin genes. Although the role of these genes is somewhat controversial (Otto, 2013), especially in animal models, their ability to kill human neutrophils potentially describes the unique presentation of disease (Loffler *et al.*, 2010).

Livestock-associated (LA-) MRSA was first described in 2005, predominantly belong to lineages CC398 and CC9, and also evolved independently (van Cleef *et al.*, 2011; McCarthy *et al.*, 2012b). LA-MRSA is prevalent in mainland Europe, Asia, and increasing in North America. LA-MRSA are currently associated with intensive pig and veal-calf farming, where they colonize the animals at high bacterial load but cause little infection. Humans that come into contact with the animals, such as farm workers and vets, become colonized and can subsequently become infected. LA-MRSA do not appear to colonize humans as efficiently, and farmers that go on holiday normally clear their carriage, and isolates do not spread well in human hospitals (van Cleef *et al.*, 2011; Wassenberg *et al.*, 2011). LA-MRSA are resistant to β-lactams via SCC*mec* and also to tetracycline due to carriage of resistance plasmids and/or transposons (McCarthy *et al.*, 2012b). In contrast, tetracycline resistance in human isolates is relatively uncommon. Tetracyclines and penicillins are widely used in agriculture, and presumably have selected for these isolates, as *S. aureus* is otherwise an uncommon colonizer of pigs.

Nearly all isolates of *S. aureus* and MRSA from humans carry prophage from the φ3 family integrated into the β-haemolysin toxin gene. These phage encode at least two human specific immune escape proteins, chemotaxis inhibitory protein (*chips*), staphylococcal complement inhibitor (*scin*) and staphylokinase (*sak*) (van Wamel *et al.*, 2006). In contrast, less than 20% of *S. aureus* from animals carry this phage (Sung and Lindsay, 2008). In the major LA-MRSA CC398 clone, φ3 is rare, but uncommon isolates of CC398 lineage that can spread from human to human are positive for φ3 bacteriophage. Several other phage and plasmids also show some association with epidemiological spread, and this holds true for isolates from different continents, although the specific genes responsible are currently unknown (McCarthy *et al.*, 2012b).

S. aureus isolates from chickens and birds mostly belong to the CC5 lineage and appear to have jumped from humans to poultry in recent years, which is associated with several MGEs that may play a role in host adaptation (Lowder *et al.*, 2009). The bovine and ovine lineage CC133 is associated with a SaPI that has a unique von-Willebrand-binding protein that coagulates ruminent plasma (Guinane *et al.*, 2010).

An important question is whether colonizing and infecting isolates of *S. aureus* might have different genomes, indicating selection during the course of infection. The data suggests that there is little evidence that colonizing and infecting isolates are different, that colonizing isolates of humans are the reservoir of infecting isolates, and any changes arising during infection have been difficult attribute to selection (Lindsay *et al.*, 2006). However, if the patient was treated with antibiotics, especially for chronic infection, selection is well documented due to SNPs that increase resistance to vancomycin, daptomycin, rifampicin and β-lactams (Mwangi *et al.*, 2007; Howden *et al.*, 2008). During chronic infection, such as in cystic fibrosis patients, movement of MGEs is also documented (Goerke *et al.*, 2004), and recent studies suggest *in vivo* colonization models also detect high levels of MGE transfer (McCarthy *et al.*, 2014). While the movement and accumulation of MGEs may contribute to selection in the host, this is yet to be proven experimentally. Studies of human-adapted isolates are particularly difficult due to the ethics of using resistant and toxigenic isolates in human volunteers.

Other staphylococcal species

S. epidermidis is the most common species of staphylococci colonizing humans, and a common cause of infections in immunocompromised patients, particularly those with prosthetic devices. *S. epidermidis* isolates cluster into two major groups, and the virulent isolates all belong to one group. Considering these groups, there is still a lot of variation between isolates, and *S. epidermidis* also carry a high proportion of mobile genetic elements including resistance genes, and some isolates carry CRISPR genes (Conlan *et al.*, 2012). There is some suggestion that certain genes are associated with disease, although there is clearly further explanations for the difference between commensals and pathogens (Conlan *et al.*, 2012; Thomas *et al.*, 2014).

S. haemolyticus is a human pathogen that is often associated with resistance. WGS has clustered isolates into one major group and the remaining minority of isolates having very diverse genomes. The major group has been divided into clades, and variation in many core genes between clades is detected. MGEs are common, as are RM systems, and expansion of the major group is associated with resistance (Cavanagh *et al.*, 2014).

S. pseudintermedius is a pathogen of dogs, and of increasing concern because methicillin resistance has emerged quickly as well as resistant to all classes of antibiotics used in veterinary medicine. This species has a lineage structure and there is an established MLST typing system. Fifteen isolates have now been sequenced, and multidrug resistance is due to a family of related transposons. A variety of other MGEs are also detected, including phage, SaPIs and SCCmec. RM systems are found in all isolates, and CRISPR in a small proportion (McCarthy *et al.*, 2015).

Other species of staphylococci have been sequenced including *S. lugdunensis* (Heilbronner *et al.*, 2011), *S. pseudintermedius* (Ben Zakour *et al.*, 2011) and *S. saprophyticus* (Kuroda *et al.*, 2005). For these species only a small number of WGS are currently available, so the population structure within each species is less clear. Overall, MGEs are common features of all the genomes, suggesting variation and evolution is highly likely. One significant exception is *S. carnosus*, a meat starter bacterium that is not pathogenic, which is notably free of mobile MGEs (Rosenstein and Gotz, 2010).

Intriguingly, sequence of multiple staphylococcal species reveals evidence of large scale inversion and replacements in the genomes, particularly near the origin of replication, which may have contributed to the emergence of each of the species (Holden and Lindsay, 2008). Each of the species has presumably adapted to their own niches, and a variety of factors may have been selected for such as specific attachment mechanisms (Kuroda *et al.*, 2005), or bacterial sensing and adapting systems (Shah *et al.*, 2013).

Technology

Sequencing methods

When the first whole genome sequences of *S. aureus* was published, high-quality DNA from a laboratory culture was fractionated, cloned into a library of *E. coli* vectors, sequenced by Sanger sequencing, and the resulting fragments aligned based on where sequence overlapped. Typically a project would generate hundreds of large and small contigs of aligned sequence, and the next step would involve painstakingly slow step-by-step sequencing across the 'gaps' between the contigs to 'close' the sequence into a single circle. The genome was annotated, and a manuscript for each of these reference genomes was published and the sequence and strains deposited in a public databases and repositories (Holden and Lindsay, 2008).

The arrival of next generation sequencing (NGS) made it possible to sequence short fragments at very high frequency, without the cloning step and at much lower cost. Rather than sequencing a whole genome to completion and closing the gaps, the fragments of sequence are aligned to one of the reference genomes, preferably from the same lineage. The gaps, which are often repetitive regions, such as ribosomes or transposons or repeat regions within genes, are not closed. Unfortunately, this has a downside for *S. aureus* projects where MGEs can vary so much, as the reference genome is unlikely to have the same MGE complement as the isolate being sequenced, and novel MGE data often remains unanalysed (Lindsay, 2014). Nevertheless, NGS has allowed the large scale sequencing projects that have revolutionized our understanding of core genome variation, and particularly epidemiology and how successful clones of bacteria have spread (Harris *et al.*, 2010).

New sequencing technologies allow new approaches. For example, the Pacific Biosciences system allows large fragments of genome to be sequenced at once and has great potential for studying MGEs. It also allows modification of DNA to be identified, although the method and hardware are expensive (Clark *et al.*, 2012). Nanopore sequencing technologies allow long reads of sequence to be performed with minimal hardware (Laszlo *et al.*, 2014), and the sequencing of genomes from a single cell is now feasible (Blainey, 2013). Increasingly rapid sequencing is showing potential as a diagnostic tool in the clinical microbiology setting (Torok and Peacock, 2012).

Internet resources for the non-bioinformatician

With the increasing accessibility of staphylococcal genomes comes increased demand for tools to analyse them. Staphylococcal genomes vary widely and a book chapter can only summarize the major features. Yet searching genomes for yourself can be very rewarding, especially if you plan to focus on your own genes of interest. Bioinformatics, including the study of genetic variation and sequences is a specialty area of research in its own right, but many of its resources are available online for all to use.

Thousands of staphylococci have now been sequenced, and many more will be in the next few years, using a variety of technologies. The most useful genomes are those that were sequenced to completion and represent the major lineages. Useful examples (with their lineage, GenBank accession number and reference) are MRSA N315, CC5, BA000018 (Kuroda *et al.*, 2001); MRSA252, CC30, BX571856 (Holden *et al.*, 2004); 8325 (the ancestor of 8325-4 and RN4220), CC8, CP000253; MW2, CC1, BA000033 (Baba *et al.*, 2002); USA300 FPR3757, CC8, CP000255 (Diep *et al.*, 2006); HO 5096 0412, CC22, HE681097 (Koser, 2012); TW20, ST239, FN433596 (Holden *et al.*, 2010); and S0385, CC398, AM9900992 (Schijffelen *et al.*, 2010). A more extensive list can be found in McCarthy and Lindsay (2010). These individual genomes can be accessed from GenBank using the nucleotide search option at http://www.ncbi.nlm.nih.gov/nuccore or from the *S. aureus* page (http://www.ncbi.nlm.nih.gov/genome/154). Caution should be used when viewing GenBank's automated phylogenetic trees however, as these are skewed by including MGE data (Lindsay *et al.*, 2006). The individual genome sequences can be viewed in a variety of customizable ways, and by choosing basic features such as gene, RNA and CDC features, the fully annotated completed genome is shown, listing the predicted open reading frames, predicted or known functions, and other regions of interest. A quirk of the sequence deposition process is that only the depositor can modify the annotation, so by definition, a whole genome annotation should never be considered up-to-the-minute. The actual nucleotide sequence itself can also be viewed, or specific regions of the sequence.

To compare genomes from different bacteria, a useful tool is the Artemis Comparison Tool (ACT) developed by the Wellcome Trust Sanger Centre (Carver *et al.*, 2005). The software can be downloaded from their website, or a user friendly web-based option is available at www.webact.org. ACT aligns the sequences of large genomes, allowing conserved and variant regions between isolates to be quickly and easily visualized.

To investigate a specific gene, it is beneficial to start with one of the whole genomes listed above, or to search with the gene option http://www.ncbi.nlm.nih.gov/gene along with the organism's name. For example, the coagulase gene in isolate MRSA252 is annotated as gene number SAR0222 (from the whole genome sequence at http://www.ncbi.nlm.nih.gov/nuccore/NC_002952.2), and links to its predicted protein at http://www.ncbi.nlm.nih.gov/protein/49240592. From this page, further links are available to features about the protein or its gene. It is also possible to search for similar genes in other bacteria using the BLAST search option, either by nucleotide or predicted amino acid sequence, with the possibility of narrowing down the databases that are searched to species level or to individual isolates. The results of the BLAST search will identify which other genomes carry the gene, and how variable the sequence is.

There is increased interest in developing rapid whole genome sequencing for clinical diagnostic application, and for the key features of the genome to be summarized for the non-expert automatically (Torok and Peacock, 2012; www.staphopia.com). Aspects such as resistance, toxin, lineage profile and strain relatedness could be useful for guiding patient management and infection control strategy (Harris *et al.*, 2013).

Conclusions and future

Understanding how staphylococcal genomes vary has focussed attention on the adaptability of staphylococci and on MGEs as drivers of adaptation and evolution. In particular, antimicrobial resistance and host-associated genes and mutations appear to play a key role. These are potential targets for therapeutic interventions, although staphylococci often exhibit genome redundancy and carry multiple genes with similar function. The ability of staphylococci to rapidly lose and acquire key genes by carrying them on MGEs also makes these particularly challenging targets for vaccines and drugs. Nevertheless, these challenges can be understood and overcome to design effective therapeutics.

Future studies will establish the function of many genes on MGEs associated with selection, as well as proving their role in adaptation using experimental approaches. Greater understanding of how variation arises and is selected for will also identify novel approaches to combating infection. Rapid diagnostics using WGS that are inexpensive and reliable will allow tailored management of antibiotics and other therapeutics to patients, optimizing treatment and reducing the risk of resistance selection.

We also expect staphylococcal populations to continue to evolve and evade any new challenges we throw at them. Resistance to new antibiotics emerges almost immediately they are introduced, and adaptation to new host groups or environments has to be expected. Emerging clones must be fit enough to survive in competitive environments with other commensals and pathogens as well as evading host immunity. Accumulation of different MGEs could lead to *S. aureus* that can colonize animals and humans, infect healthy and immunocompromised hosts, cause food poisoning, toxic shock syndrome, skin and soft tissue infections and can resist the full spectrum of antimicrobials. Each new successful clone teaches us more about the remarkable ability of staphylococci to affect our health, agriculture, pets, food, and environment, and prediction, recognition and containment of emerging clones with new epidemic features are important goals.

References

Baba, T., Takeuchi, F., Kuroda, M., Yuzawa, H., Aoki, K., Oguchi, A., Nagai, Y., Iwama, N., Asano, K., Naimi, T., *et al.* (2002). Genome and virulence determinants of high virulence community-acquired MRSA. Lancet *359*, 1819–1827.

Bayles, K.W., Brunskill, E.W., Iandolo, J.J., Hruska, L.L., Huang, S., Pattee, P.A., Smiley, B.K., and Yasbin, R.E. (1994). A genetic and molecular characterization of the recA gene from *Staphylococcus aureus*. Gene *147*, 13–20.

Ben Zakour, N.L., Bannoehr, J., van den Broek, A.H., Thoday, K.L., and Fitzgerald, J.R. (2011). Complete genome sequence of the canine pathogen Staphylococcus pseudintermedius. J. Bacteriol. *193*, 2363–2364.

Blainey, P.C. (2013). The future is now: single-cell genomics of bacteria and archaea. FEMS Microbiol. Rev. *37*, 407–427.

Cavanagh, J.P., Hjerde, E., Holden, M.T., Kahlke, T., Klingenberg, C., Flaegstad, T., Parkhill, J., Bentley, S.D., and Sollid, J.U. (2014). Whole genome sequencing reveals clonal expansion of multiresistant *Staphylococcus haemolyticus* in European hospitals. J. Antimicrob. Chemother. *69*, 2920–2927.

Chua, K.Y., Seemann, T., Harrison, P.F., Monagle, S., Korman, T.M., Johnson, P.D., Coombs, G.W., Howden, B.O., Davies, J.K., Howden, B.P., *et al.* (2011). The dominant Australian community-acquired methicillin-resistant *Staphylococcus aureus* clone ST93-IV [2B] is highly virulent and genetically distinct. PLoS One *6*, e25887.

Clark, T.A., Murray, I.A., Morgan, R.D., Kislyuk, A.O., Spittle, K.E., Boitano, M., Fomenkov, A., Roberts, R.J., and Korlach, J. (2012). Characterization of DNA methyltransferase specificities using single-molecule, real-time DNA sequencing. Nucleic Acids Res. *40*, e29.

Cockfield, J.D., Pathak, S., Edgeworth, J.D., and Lindsay, J.A. (2007). Rapid determination of hospital-acquired meticillin-resistant *Staphylococcus aureus* lineages. J. Med. Microbiol. *56*, 614–619.

Conlan, S., Mijares, L.A., NISC Comparative Sequencing Program, Becker, J., Blakesley, R.W., Bouffard, G.G., Brooks, S., Coleman, H., Gupta, J., Gurson, N., *et al.* (2012). *Staphylococcus epidermidis* pan-genome sequence analysis reveals diversity of skin commensal and hospital infection-associated isolates. Genome Biol. *13*, R64.

Corvaglia, A.R., François, P., Hernandez, D., Perron, K., Linder, P., and Schrenzel, J. (2010). A type III-like restriction endonuclease functions as a major barrier to horizontal gene transfer in clinical *Staphylococcus aureus* strains. Proc. Natl. Acad. Sci. U.S.A. *107*, 11954–11958.

Day, N.P., Moore, C.E., Enright, M.C., Berendt, A.R., Smith, J.M., Murphy, M.F., Peacock, S.J., Spratt, B.G., and Feil, E.J. (2001). A link between virulence and ecological abundance in natural populations of *Staphylococcus aureus*. Science *292*, 114–116.

Dempsey, R.M., Carroll, D., Kong, H., Higgins, L., Keane, C.T., and Coleman, D.C. (2005). Sau42I, a BcgI-like restriction-modification system encoded by the *Staphylococcus aureus* quadruple-converting phage Phi42. Microbiology *151*, 1301–1311.

Diep, B.A., Gill, S.R., Chang, R.F., Phan, T.H., Chen, J.H., Davidson, M.G., Lin, F., Lin, J., Carleton, H.A., Mongodin, E.F., *et al.* (2006). Complete genome sequence of USA300, an epidemic clone of community-acquired meticillin-resistant *Staphylococcus aureus*. Lancet. *367*, 731–739.

Ender, M., McCallum, N., Adhikari, R., and Berger-Bächi, B. (2004). Fitness cost of SCCmec and methicillin resistance levels in *Staphylococcus aureus*. Antimicrob. Agents Chemother. *48*, 2295–2297.

Enright, M.C., Day, N.P., Davies, C.E., Peacock, S.J., and Spratt, B.G. (2000). Multilocus sequence typing for characterization of methicillin-resistant and methicillin-susceptible clones of *Staphylococcus aureus*. J. Clin. Microbiol. *38*, 1008–1015.

Goerke, C., Köller, J., and Wolz, C. (2006). Ciprofloxacin and trimethoprim cause phage induction and virulence modulation in *Staphylococcus aureus*. Antimicrob. Agents Chemother. *50*, 171–177.

Goerke, C., Matias y Papenberg, S., Dasbach, S., Dietz, K., Ziebach, R., Kahl, B.C., and Wolz, C. (2004). Increased frequency of genomic alterations in *Staphylococcus aureus* during chronic infection is in part due to phage mobilization. J. Infect. Dis. *189*, 724–734.

Guinane, C.M., Ben Zakour, N.L., Tormo-Mas, M.A., Weinert, L.A., Lowder, B.V., Cartwright, R.A., Smyth, D.S., Smyth, C.J., Lindsay, J.A., Gould, K.A., *et al.* (2010). Evolutionary genomics of *Staphylococcus aureus* reveals insights into the origin and molecular basis of ruminant host adaptation. Genome Biol. Evol. *2*, 454–466.

Harmsen, D., Claus, H., Witte, W., Rothgänger, J., Claus, H., Turnwald, D., and Vogel, U. (2003). Typing of methicillin-resistant *Staphylococcus aureus* in a university hospital setting by using novel software for *spa* repeat determination and database management. J. Clin. Microbiol. *41*, 5442–5448.

Heilbronner, S., Holden, M.T., van Tonder, A., Geoghegan, J.A., Foster, T.J., Parkhill, J., and Bentley, S.D. (2011). Genome sequence of Staphylococcus lugdunensis N920143 allows identification of putative colonization and virulence factors. FEMS Microbiol. Lett. *322*, 60–67.

Holden, M.T., Feil, E.J., Lindsay, J.A., Peacock, S.J., Day, N.P., Enright, M.C., Foster, T.J., Moore, C.E., Hurst, L., Atkin, R., *et al.* (2004). Complete genomes of two clinical *Staphylococcus aureus* strains: evidence for the rapid evolution of virulence and drug resistance. Proc. Natl. Acad. Sci. U.S.A. *101*, 9786–9791.

Holden, M.T., Hsu, L.Y., Kurt, K., Weinert, L.A., Mather, A.E., Harris, S.R., Strommenger, B., Layer, F., Witte, W., de Lencastre, H., *et al.* (2013). A genomic portrait of the emergence, evolution and global spread of a methicillin resistant *Staphylococcus aureus* pandemic. Genome Res. *23*, 653–664.

Holden, M.T., Lindsay, J.A., Corton, C., Quail, M.A., Cockfield, J.D., Pathak, S., Batra, R., Parkhill, J., Bentley, S.D., and Edgeworth, J.D. (2010). Genome sequence of a recently emerged, highly transmissible, multi-antibiotic- and antiseptic-resistant variant of methicillin-resistant *Staphylococcus aureus*, sequence type 239 (TW). J. Bacteriol. *192*, 888–892.

Holden, M.T.G., and Lindsay, J.A. (2008). Whole genomes: sequence, microarray and systems biology. In Staphylococcus: Molecular Genetics, J.A. Lindsay, ed. (Caister Academic Press, Norfolk, UK) pp. 1–28.

Howden, B.P., Stinear, T.P., Allen, D.L., Johnson, P.D., Ward, P.B., and Davies, J.K. (2008). Genomic analysis reveals a point mutation in the two-component sensor gene *graS* that leads to intermediate vancomycin resistance in clinical *Staphylococcus aureus*. Antimicrob. Agents. Chemother. *52*, 3755–3762.

Ito, T., Ma, X.X., Takeuchi, F., Okuma, K., Yuzawa, H., and Hiramatsu, K. (2004). Novel type V staphylococcal cassette chromosome mec driven by a novel cassette chromosome recombinase, ccrC. Antimicrob. Agents Chemother. *48*, 2637–2651.

Knight, G.M., Budd, E.L., and Lindsay, J.A. (2013). Large mobile genetic elements carrying resistance genes that do not confer a fitness burden in healthcare-associated methicillin-resistant *Staphylococcus aureus* (HA-MRSA). Microbiology *159*, 1661–1672.

Knight, G.M., Budd, E.L., Whitney, L., Thornley, A., Al-Ghusein, H., Planche, T., and Lindsay, J.A. (2012). Shift in dominant hospital-associated methicillin-resistant *Staphylococcus aureus* (HA-MRSA) clones over time. J. Antimicrob. Chemother. *67*, 2514–2522.

Kuroda, M., Ohta, T., Uchiyama, I., Baba, T., Yuzawa, H., Kobayashi, I., Cui, L., Oguchi, A., Aoki, K., Nagai, Y., *et al.* (2001). Whole genome sequencing of meticillin-resistant *Staphylococcus aureus*. Lancet *357*, 1225–1240.

Kuroda, M., Yamashita, A., Hirakawa, H., Kumano, M., Morikawa, K., Higashide, M., Maruyama, A., Inose, Y., Matoba, K., Toh, H., *et al.* (2005). Whole genome sequence of Staphylococcus saprophyticus reveals the pathogenesis of uncomplicated urinary tract infection. Proc. Natl. Acad. Sci. U.S.A. *102*, 13272–13277.

Laszlo, A.H., Derrington, I.M., Ross, B.C., Brinkerhoff, H., Adey, A., Nova, I.C., Craig, J.M., Langford, K.W., Samson, J.M., Daza, R., *et al.* (2014). Decoding long nanopore sequencing reads of natural DNA. Nat. Biotechnol. *32*, 829–833.

Lindsay, J.A. (2008). *S. aureus* evolution: lineages and mobile genetic elements. In Staphylococcus: Molecular

Genetics, J.A. Lindsay, ed. (Caister Academic Press, Norfolk, UK), pp. 45–69.

Lindsay, J.A. (2010). Genomic variation and evolution of *Staphylococcus aureus*. Int. J. Med. Microbiol. *300*, 98–103.

Lindsay, J.A. (2013). Hospital-associated MRSA and antibiotic resistance-what have we learned from genomics? Int. J. Med. Microbiol. *303*, 318–323.

Lindsay, J.A. (2014). *Staphylococcus aureus* genomics and the impact of horizontal gene transfer. Int. J. Med. Microbiol. *304*, 103–109.

Lindsay, J.A., Moore, C.E., Day, N.P., Peacock, S.J., Witney, A.A., Stabler, R.A., Husain, S.E., Butcher, P.D., and Hinds, J. (2006). Microarrays reveal that each of the ten dominant lineages of *Staphylococcus aureus* has a unique combination of surface-associated and regulatory genes. J. Bacteriol. *188*, 669–676.

Lindsay, J.A., Ruzin, A., Ross, H.F., Kurepina, N., and Novick, R.P. (1998). The gene for toxic shock toxin is carried by a family of mobile pathogenicity islands in *Staphylococcus aureus*. Mol. Microbiol. *29*, 527–543.

Löffler, B., Hussain, M., Grundmeier, M., Brück, M., Holzinger, D., Varga, G., Roth, J., Kahl, B.C., Proctor, R.A., and Peters, G. (2010). *Staphylococcus aureus* Panton–Valentine leucocidin is a very potent cytotoxic factor for human neutrophils. PLOS Pathog. *6*, e1000715.

Lowder, B.V., Guinane, C.M., Ben Zakour, N.L., Weinert, L.A., Conway-Morris, A., Cartwright, R.A., Simpson, A.J., Rambaut, A., Nübel, U., and Fitzgerald, J.R. (2009). Recent human-to-poultry host jump, adaptation, and pandemic spread of *Staphylococcus aureus*. Proc. Natl. Acad. Sci. U.S.A. *106*, 19545–19550.

Marraffini, L.A., and Sontheimer, E.J. (2008). CRISPR interference limits horizontal gene transfer in staphylococci by targeting DNA. Science *322*, 1843–1845.

McAdam, P.R., Templeton, K.E., Edwards, G.F., Holden, M.T., Feil, E.J., Aanensen, D.M., Bargawi, H.J., Spratt, B.G., Bentley, S.D., Parkhill, J., *et al.* (2012). Molecular tracing of the emergence, adaptation, and transmission of hospital-associated methicillin-resistant *Staphylococcus aureus*. Proc. Natl. Acad. Sci. U.S.A. *109*, 9107–9112.

McCarthy A.J., Breathnach, A.S., and Lindsay, J.A. (2012a). Detection of mobile genetic element variation between colonizing and infecting hospital-associated methicillin-resistant *Staphylococcus aureus* isolates. J. Clin. Microbiol. *50*, 1073–1075.

McCarthy A.J., and Lindsay, J.A. (2010). Genetic variation in *Staphylococcus aureus* surface and immune evasion genes is lineage associated: implications for vaccine design and host-pathogen interactions. BMC Microbiol. *10*, 173.

McCarthy, A.J., and Lindsay, J.A. (2012). The distribution of plasmids that carry virulence and resistance genes in *Staphylococcus aureus* is lineage associated. BMC Microbiol. *12*, 104.

McCarthy, A.J., and Lindsay, J.A. (2013). *Staphylococcus aureus* innate immune evasion is lineage-specific: a bioinfomatics study. Infect. Genet. Evol. *19*, 7–14.

McCarthy, A.J., Loeffler, A., Witney, A.A., Gould, K.A., Lloyd, D.H., and Lindsay, J.A. (2014). Extensive horizontal gene transfer during *Staphylococcus aureus* co-colonization *in vivo*. Genome Biol. Evol. *6*, 2697–2708.

McCarthy, A.J., Harrison, E.M., Stanczak-Mrozek, K., Leggett, B., Waller, A., Holmes, M.A., Lloyd, D.H., Lindsay, J.A., and Loeffler, A. (2015). Genomic insights into the rapid emergence and evolution of MDR in Staphylococcus pseudintermedius. J. Antimicrob. Chemother. *70*, 997–1007.

McCarthy, A.J., van Wamel, W., Vandendriessche, S., Larsen, J., Denis, O., Garcia-Graells, C., Uhlemann, A.C., Lowy, F.D., Skov, R., and Lindsay, J.A. (2012b). *Staphylococcus aureus* CC398 clade associated with human-to-human transmission. Appl. Environ. Microbiol. *78*, 8845–8848.

McCarthy, A.J., Witney, A.A., and Lindsay, J.A. (2012c). *Staphylococcus aureus* temperate bacteriophage: carriage and horizontal gene transfer is lineage associated. Front. Cell. Infect. Microbiol. *2*, 6.

Mesak, L.R., Miao, V., and Davies, J. (2008). Effects of subinhibitory concentrations of antibiotics on SOS and DNA repair gene expression in *Staphylococcus aureus*. Antimicrob. Agents Chemother. *52*, 3394–3397.

Morikawa, K., Takemura, A.J., Inose, Y., Tsai, M., Nguyen Thi le, T., Ohta, T., and Msadek, T. (2012). Expression of a cryptic secondary sigma factor gene unveils natural competence for DNA transformation in *Staphylococcus aureus*. PLOS Pathog. *8*, e1003003.

Murray, N.E. (2000). Type I restriction systems: sophisticated molecular machines (a legacy of Bertani and Weigle). Microbiol. Mol. Biol. Rev. *64*, 412–434.

Mwangi, M.M., Wu, S.W., Zhou, Y., Sieradzki, K., de Lencastre, H., Richardson, P., Bruce, D., Rubin, E., Myers, E., Siggia, E.D., *et al.* (2007). Tracking the *in vivo* evolution of multidrug resistance in *Staphylococcus aureus* by whole-genome sequencing. Proc. Natl. Acad. Sci. U.S.A. *104*, 9451–9456.

Nübel, U., Dordel, J., Kurt, K., Strommenger, B., Westh, H., Shukla, S.K., Zemlicková, H., Leblois, R., Wirth, T., Jombart, T., *et al.* (2010). A timescale for evolution, population expansion, and spatial spread of an emerging clone of methicillin-resistant *Staphylococcus aureus*. PLOS Pathog. *6*, e1000855.

Otto, M. (2013). Community-associated MRSA: what makes them special? Int. J. Med. Microbiol. *303*, 324–330.

Price, L.B., Stegger, M., Hasman, H., Aziz, M., Larsen, J., Andersen, P.S., Pearson, T., Waters, A.E., Foster, J.T., Schupp, J., *et al.* (2012). *Staphylococcus aureus* CC398: host adaptation and emergence of methicillin resistance in livestock. MBio 3, e00305–11.

Quiles-Puchalt, N., Martínez-Rubio, R., Ram, G., Lasa, I., and Penadés, J.R. (2014). Unravelling bacteriophage ϕ11 requirements for packaging and transfer of mobile genetic elements in *Staphylococcus aureus*. Mol. Microbiol. *91*, 423–437.

Roberts, G.A., Houston, P.J., White, J.H., Chen, K., Stephanou, A.S., Cooper, L.P., Dryden, D.T., and Lindsay, J.A. (2013). Impact of target site distribution for Type I restriction enzymes on the evolution of methicillin-resistant *Staphylococcus aureus* (MRSA) populations. Nucleic Acids Res. *41*, 7472–7484.

Rosenstein, R., and Götz, F. (2010). Genomic differences between the food-grade Staphylococcus carnosus and pathogenic staphylococcal species. Int. J. Med. Microbiol. *300*, 104–108.

Schijffelen, M.J., Boel, C.H., van Strijp, J.A., and Fluit, A.C. (2010). Whole genome analysis of a livestock-associated methicillin-resistant *Staphylococcus aureus* ST398 isolate from a case of human endocarditis. BMC Genomics *11*, 376.

Shah, N., Gaupp, R., Moriyama, H., Eskridge, K.M., Moriyama, E.N., and Somerville, G.A. (2013). Reductive evolution and the loss of PDC/PAS domains from the genus Staphylococcus. BMC Genomics *14*, 524.

Stefani, S., Chung, D.R., Lindsay, J.A., Friedrich, A.W., Kearns, A.M., Westh, H., and Mackenzie, F.M. (2012). Meticillin-resistant *Staphylococcus aureus* (MRSA): global epidemiology and harmonisation of typing methods. Int. J. Antimicrob. Agents *39*, 273–282.

Shore, A.C., and Coleman, D.C. (2013). Staphylococcal cassette chromosome mec: recent advances and new insights. Int. J. Med. Microbiol. *303*, 350–359.

Sung, J.M., Lloyd, D.H., and Lindsay, J.A. (2008). *Staphylococcus aureus* host specificity: comparative genomics of human versus animal isolates by multi-strain microarray. Microbiology *154*, 1949–1959.

Thomas, J.C., Zhang, L., and Robinson, D.A. (2014). Differing lifestyles of *Staphylococus epidermidis* as revealed through Bayesian clustering of multilocus sequence types. Infect. Genet. Evol. *22*, 257–264.

Tormo-Más, M.A., Mir, I., Shrestha, A., Tallent, S.M., Campoy, S., Lasa, I., Barbé, J., Novick, R.P., Christie, G.E., and Penadés, J.R. (2010). Moonlighting bacteriophage proteins derepress staphylococcal pathogenicity islands. Nature. *465*, 779–782.

Török, M.E., and Peacock, S.J. (2012). Rapid whole-genome sequencing of bacterial pathogens in the clinical microbiology laboratory–pipe dream or reality? J. Antimicrob. Chemother. *67*, 2307–2308.

Uhlemann, A.C., Kennedy, A.D., Martens, C., Porcella, S.F., Deleo, F.R., and Lowy, F.D. (2012). Toward an understanding of the evolution of *Staphylococcus aureus* strain USA300 during colonization in community households. Genome Biol. Evol. *4*, 1275–1285.

Uhlemann, A.C., Dordel, J., Knox, J.R., Raven, K.E., Parkhill, J., Holden, M.T., Peacock, S.J., and Lowy, F.D. (2014). Molecular tracing of the emergence, diversification, and transmission of S. aureus sequence type 8 in a New York community. Proc. Natl. Acad. Sci. U.S.A. *111*, 6738–6743.

van Cleef, B.A., Graveland, H., Haenen, A.P., van de Giessen, A.W., Heederik, D., Wagenaar, J.A., and Kluytmans, J.A. (2011). Persistence of livestock-associated methicillin-resistant *Staphylococcus aureus* in field workers after short-term occupational exposure to pigs and veal calves. J. Clin. Microbiol. *49*, 1030–1033.

van Cleef, B.A., Monnet, D.L., Voss, A., Krziwanek, K., Allerberger, F., Struelens, M., Zemlickova, H., Skov, R.L., Vuopio-Varkila, J., Cuny, C., *et al.* (2011). Livestock-associated methicillin-resistant *Staphylococcus aureus* in humans, Europe. Emerg. Infect. Dis. *17*, 502–505.

van Wamel, W.J., Rooijakkers, S.H., Ruyken, M., van Kessel, K.P., and van Strijp, J.A. (2006). The innate immune modulators staphylococcal complement inhibitor and chemotaxis inhibitory protein of *Staphylococcus aureus* are located on beta-hemolysin-converting bacteriophages. J. Bacteriol. *188*, 1310–1315.

Waldron, D.E., and Lindsay, J.A. (2006). Sau1: a novel lineage-specific type I restriction-modification system that blocks horizontal gene transfer into *Staphylococcus aureus* and between *S. aureus* isolates of different lineages. J. Bacteriol. *188*, 5578–5585.

Wassenberg, M.W., Bootsma, M.C., Troelstra, A., Kluytmans, J.A., and Bonten, M.J. (2011). Transmissibility of livestock-associated methicillin-resistant *Staphylococcus aureus* (ST398) in Dutch hospitals. Clin. Microbiol. Infect. *17*, 316–319.

Xia, G., Corrigan, R.M., Winstel, V., Goerke, C., Gründling, A., and Peschel, A. (2011). Wall teichoic Acid-dependent adsorption of staphylococcal siphovirus and myovirus. J. Bacteriol. *193*, 4006–4009.

Xu, S.Y., Corvaglia, A.R., Chan, S.H., Zheng, Y., and Linder, P. (2011). A type IV modification-dependent restriction enzyme SauUSI from *Staphylococcus aureus* subsp. aureus USA300. Nucleic Acids Res. *39*, 5597–5610.

Ziebuhr, W., Krimmer, V., Rachid, S., Lössner, I., Götz, F., and Hacker, J. (1999). A novel mechanism of phase variation of virulence in *Staphylococcus epidermidis*: evidence for control of the polysaccharide intercellular adhesin synthesis by alternating insertion and excision of the insertion sequence element IS256. Mol. Microbiol. *32*, 345–356.

Staphylococcal Virulence Factors

5

Patrick M. Schlievert

Abstract

Staphylococcus aureus is a multi-dimensional pathogen that colonizes and infects only animals, including humans, with immune systems. The organism causes large numbers of infections worldwide, both relatively benign and highly serious, through its production of colonization factors and three tiers of virulence factors which interfere with immune function. Secreted superantigens and exfoliative toxins systemically interfere with immune function through dysregulation of cells and opening immune barriers, respectively. Secreted cytotoxins kill immune cells influxing into local areas of infection. At the same time, some microbial colonization factors sequester the organism in walled off sites, the hallmark of *S. aureus* infections, and in biofilms. Finally, if the immune system somehow bypasses these first two tiers of microbial resistance, *S. aureus* has capsules, other polysaccharides, and pigment as last lines of defence. This review discusses these virulence factors from the presumed staphylococcal point of view.

Introduction

Staphylococcus aureus is a Gram-positive bacterium that is a major cause of benign and serious infections worldwide, through its production of a myriad of cell-surface and secreted virulence factors. The cell-surface virulence factors include capsules, polysaccharide intercellular adhesion (PIA), pigments, microbial surface components recognizing adhesive matrix molecules (MSCRAAMs), and surface factor promoting resistance to oxidative killing (SOK). The secreted virulence factors include exotoxin superantigens (Sags), exfoliative toxins and other serine proteases, cytotoxins, glycerol ester hydrolase (lipase), nucleases, and hyaluronidase. *S. aureus* has such a large array of virulence factors that the organism can cause nearly any type of infection and at almost any body site. The hallmark infection of *S. aureus* is a walled-off furuncle or soft tissue abscess that depends first on host colonization by the organism and then sequestration from the host immune system. *S. aureus* only colonizes and infects animals that have immune systems and thus critically depends on subversion of immune cells and molecules. In walled-off infections, upon nutrient depletion, the organism also depends on spread to adjacent tissues through production of exoenzymes, where the pathogenic process can begin anew.

S. aureus colonization depends on both cell wall-associated teichoic acids and MSCRAAMs. Subsequently, as I envisage *S. aureus* pathogenesis, there are three tiers of microbial defence against the host immune system. These will be discussed in this order in this review: The first line of microbial defence is production of families of exotoxins, called superantigens (SAgs) and exfoliative toxins (ETs), which systemically interfere with immune system function (SAgs) or open host barriers to infection (ETs) (Fig. 5.1). The second line of microbial defence against the host immune system is local production of cytotoxins that kill immune cells influxing into infection areas. These toxins are primarily produced as a function of nutrient limitation within furuncles and abscesses, as the microbe must spread to persist in disease causation. Exoenzymes are also produced at this time to degrade host tissue to provide nutrients sources for new growth. Most closely associated with the microbe, and comprising the third and last line of defence against the host immune system, are capsules, PIA,

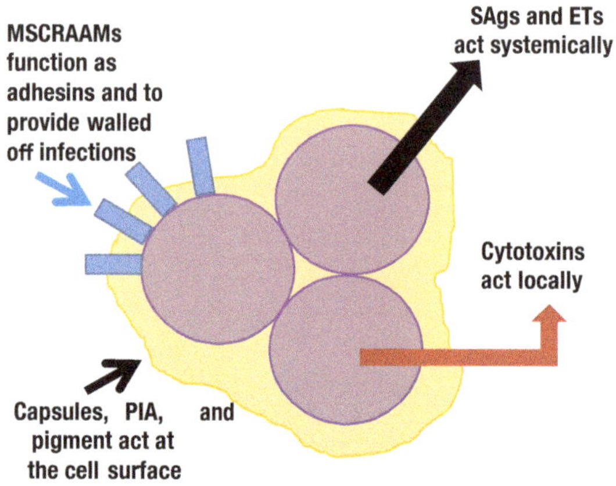

Figure 5.1 Three tiers of virulence factors facilitate disease causation and resistance in *S. aureus* to the host immune system. Systemic secreted toxins, such as SAgs and ETs dysregulate the immune system and open protective barriers at a distance. Cytotoxins kill immune cells influxing locally. Capsules, PIA, and pigment form the last line of staphylococcal defence against the host immune system.

pigment, and cell surface factors that resist phagocytosis and sequester the microbe.

In 2007, the Centers for Disease Control and Prevention (CDC) and their collaborators published that *S. aureus* is the most significant cause of serious infectious diseases and infectious disease deaths in the USA (Klevens *et al.*, 2007). *S. aureus* causes relatively manageable furuncles and soft tissue abscesses but commonly also causes life-threatening infections such as infective endocarditis (as many as 40,000 cases/year), pneumonia (as many as 70,000 cases/year), sepsis (number two cause behind coagulase-negative staphylococci), osteomyelitis, and toxic shock syndrome (TSS; 5000–35,000 cases/year) (Bergdoll *et al.*, 1981; Bohach *et al.*, 1990; CDC, 1999; Herold *et al.*, 1998; Lowy, 1998; McCormick *et al.*, 2001; Navarre and Schneewind, 1999; Que and Moreillon, 2011; Schlievert *et al.*, 1981; Schlievert *et al.*, 2010; Xiong *et al.*, 2009). Estimates suggest that 30–40% of the human population are asymptomatically colonized at any given time on one or more of their mucosal or skin surfaces; up to 70% of people may be transiently colonized (Lowy, 1998; Schlievert *et al.*, 2010). These colonization sites allow the organism to persist in the human host and transition into disease production when conditions become favourable. Many conditions that facilitate transition are known, for example host immunosuppression, damaged skin and mucosal surfaces, and pre-existent respiratory viral infections (Lowy, 1998; MacDonald *et al.*, 1987). The major goal of this review is to present new information on cell-surface virulence factors and exotoxins, and their roles in human infections.

Colonization: teichoic acids and microbial surface components recognizing adhesive matrix molecules (MSCRAAMs)

The first requirement for *S. aureus* to cause disease is its ability to colonize the host. The organism does so through its production of teichoic acids and MSCRAAMs, many of which are adhesins that allow binding to host tissue, but many also participate in sequestering the organism from the human immune system (Foster and Hook, 1998). Protein A is unique in its anti-phagocytic ability by binding to the Fc of IgG and thereby turning potentially protective antibodies to the microbes' resistance to clearance (Lofdahl *et al.*, 1983; Pauli *et al.*, 2014).

S. aureus is multi-dimensional in its production of colonization virulence factors. Initial attachment of the organism to host tissues likely includes cell wall teichoic acids, with stronger attachment dependent on MSCRAAMs. Teichoic acids are anionic glycopolymers of about 40 polyribitol phosphates with 1,5-phosphodiester bonds. Some of the C4 hydroxyls are substituted with α-O or β-O N-acetyl glucosamine and C2 hydroxyls substituted with D-alanines (Koc *et al.*, 2015). Teichoic acids are present in the cell wall throughout cell growth. These molecules also facilitate cell division and may participate in antimicrobial resistance.

MSCRAAMs are typically produced during the exponential phase of staphylococcal growth and are primarily attached to the bacterial surface by an enzymatic sortase-anchoring mechanism (Foster and Hook, 1998). Production of many *S. aureus* virulence factors is influenced strongly or weakly by multiple global regulatory systems. In general, these regulatory systems assure the microbes' production of MSCRAAMs during active growth, wherein the microbe has likely just spread to or landed on a new body site (Novick, 2003). In contrast, many other virulence factors are either produced in the post-exponential growth phase when MSCRAAM genes are no longer being transcribed, as in production of toxic shock syndrome toxin-1 (TSST-1) and many cytotoxins and exoenzymes, or are produced independent of global regulatory systems, as in the majority of SAg exotoxins (Novick, 2003).

The surface MSCRAAMs allow the microbe to adhere to the extracellular matrix. These virulence factors include protein A, clumping factors A and B (ClfA and ClfB), iron surface determinants (Isds), and fibronectin (FnBPA and B), fibrinogen, collagen (Cna), elastin, and von Willebrand factor binding proteins. Because of redundancy, it is difficult to determine which of these factors are the most critical in host binding. In addition to host binding, these factors also lead to sequestration of *S. aureus* in furuncles and abscesses as part of the third line of defence against the host. MSCRAAMs also vary among *S. aureus* isolates whether or not they are encoded, and additionally, multiple allelic forms may be produced that presumably increase or decrease the ability to attach to host tissues.

Cell wall anchored MSCRAAMs have an N-terminal, cleavable signal peptide and a C-terminal region that contains an LPXTG motif (with cleavage between the T and G by sortase and attachment of the T to peptidoglycan) and hydrophobic membrane spanning domain, followed by hydrophilic tail; elastin-binding protein (EbpS) is not LPXTG anchored (Foster and Hook, 1998; Schneewind *et al.*, 1995; Schneewind *et al.*, 1992). The ability of MSCRAAMs to bind to host tissues is usually found in the amino terminal region after signal peptide cleavage.

Staphylococcal protein A (SpA) is the prototypical LPXTG-anchored protein. This molecule functions as a B cell SAg and binds to the Fc of IgG molecules to inhibit opsonphagocytosis (Pauli *et al.*, 2014). SpA has a molecular weight of approximately 45,000 Da, with ability to bind to 4–5 IgG Fcs (Lofdahl *et al.*, 1983). The protein may be attached to the bacterial surface as a 45,000 molecular weight molecule, or may be secreted as proteins of 15,000, 30,000 and 45,000 molecular weights, all with ability to bind IgG Fcs. Recent studies, suggest that the presence of protein A on the surface of *S. aureus* skews the B cell response of humans to itself such that there is minimal protective response to other virulence factors, resulting in recurrent infections (Pauli *et al.*, 2014). It was also suggested that inactivation of the B cell superantigenicity function of protein A could lead to greater vaccine efficacy against *S. aureus*, but this was unfortunately not observed (Rauch *et al.*, 2014).

Fibronectin binding through FnBPA and B is a common trait of *S. aureus*; these proteins appear to be bi-functional with the ability also to bind fibrinogen (Foster and Hook, 1998; Greene *et al.*, 1995; Jonsson *et al.*, 1991). These proteins contribute to adherence of *S. aureus* in plasma clots and coated catheters, thus contributing both to attachment and avoidance of the host immune system. In addition, they contribute to the well-known foreign body response, wherein foreign bodies in humans predispose significantly to *S. aureus* infections. FnBPA is made up of an A domain of amino acids, connected to two B domains, followed by a C domain, four D domains, a wall-spanning region and a membrane-spanning region. The nearly identical ligand binding domain of both FnBPA and B resides in the D region; these molecules bind to the amino terminus of fibronectin.

The collagen-binding protein (Cna) mediates *S. aureus* binding to collagen substrates (Foster and Hook, 1998; Switalski *et al.*, 1989). Cna appears to be produced by only 30–50% of *S. aureus* strains. Cna has a large A domain (55,000 molecular weight) followed by 1–4 repeats of a 20,000 molecular weight B domain and then wall and membrane spanning domains. The A domain contains the collagen-binding region.

There are two fibrinogen binding proteins (ClfA and B) that may be produced by *S. aureus* (Entenza *et al.*, 2000; Foster and Hook, 1998; Moreillon *et al.*, 1995; Vaudaux *et al.*, 1995; Walsh *et al.*, 2008). These molecules mediate binding to plasma clots and biomaterials coated with plasma, and are responsible for the clumping phenotype of the organism. Thus, like FnBPs, Clf proteins are both adhesions and molecules that sequester the microbe from the host immune system. Clf proteins are organized similar to other MSCRAAMs except for the presence of an R region; this domain may function as a stalk to ensure presentation of the A domain to ligand occurs. The A domain of ClfA binds to fibrinogen and is the primary binding site to platelets; the carboxyl terminus of the γ-chain of fibrinogen binds to ClfA. There are other *S. aureus* surface proteins related to ClfA and B; these are referred to as Sdr proteins, based on SD repeats (SdrC, SdrD, and SdrE).

Recent studies have identified other cell-surface factors that may be important in virulence. Two of these merit special mention: the Giant Staphylococcal Surface Protein (GSSP) and SOK. The hallmark of *S. aureus* infections is their ability to become walled off in clots as a function of free coagulase and clumping factors. Indeed, we have recently shown in rabbits that further clumping of *S. aureus*, as mediated by IgG antibodies to surface factors of the organism, enhance virulence. Presumably, once nutrients are used up within a clot region, the organism will need to spread to adjacent tissue. The GSSP is a 1.1 million molecular weight protein, appearing possibly as a 'porcupine quill' that may facilitate spread by preventing clots from forming (Cheng *et al.*, 2014; Walker *et al.*, 2013).

SOK has recently been partially characterized as a surface protein that enhances resistance to oxidative killing; SOK does not appear anchored to the staphylococcal surface by the LPXTG motif (Malachowa *et al.*, 2011). Deletion mutagenesis of SOK increased sensitivity of *S. aureus* to killing by polymorphonuclear leucocytes, affected levels of detectable reactive oxygen species following phagocytosis, and increased bacterial susceptibility to oxidative killing. SOK expression also dramatically reduced staphylococcal virulence in an infective endocarditis model.

Superantigens (SAgs) and exfoliative toxins (ETs): systemic defence against the human immune system

SAgs

Two important families of exotoxins have been conclusively shown to cause human diseases, including SAgs, also known as toxic shock syndrome toxin-1 (TSST-1), staphylococcal enterotoxins (SEs) and related toxins (staphylococcal enterotoxin-like; SE*l*SAgs, and exfoliative toxins (ETs). As noted above, the major effect of SAgs in the host is systemic inactivation of the immune system.

SAgs are a large family of low molecular weight exotoxins, formerly referred to as pyrogenic toxins based on their being the most potent pyrogens known. They are secreted by all human pathogenic *S. aureus* and β-haemolytic streptococci, with secretion dependent on a cleavable signal peptide. Some SAgs are produced during exponential phase growth, specifically, those that are produced in high amounts (mg/ml amounts in biofilms), such as TSST-1 and SEs B and C. Importantly, these SAgs cause serious human TSS illnesses. In contrast to these SAgs, others are expressed in ng/ml or pg/ml amounts. Typically, these are produced during the exponential growth phase and may be important in facilitating immune system inactivation to permit colonization. Occasionally, these low-abundance SAgs may be produced in high amounts, such as SEA, wherein the SAg causes TSS. The mechanism controlling the switch to high-level synthesis remains unknown. SAgs have molecular weights ranging from 19,000 to 30,000 Da (McCormick *et al.*, 2001). The proteins are heat stable, resistant to proteolysis and acids, and highly resistant to desiccation (Dinges *et al.*, 2000; McCormick, 2006; McCormick *et al.*, 2001). Their toxicity and unusual stability have resulted in staphylococcal enterotoxin SAgs (serotypes A-E) being categorized as select agents of bioterrorism.

S. aureus strains secrete from 1 to 23 of at least 24 serologically distinct SAgs (Schlievert *et al.*, 2008). The only *S. aureus* strain we are aware of that does not secrete SAgs is NCTC 8325–4 and its variant strains (i.e. RN4220, RN 6390, and RN450). *S. aureus* SAgs include TSST-1, SEs (serotypes A, B_n, C_n, D, E and G); and the *SEl* SAgs, serotypes H-X (Bohach *et al.*, 1990; Dinges *et al.*, 2000; Lina *et al.*, 2004; McCormick, 2006; McCormick *et al.*, 2001). There is no SEF, as the name was retired during the renaming of staphylococcal pyrogenic exotoxin C (PE C) and SEF as TSST-1 in 1984 (Bergdoll and Schlievert, 1984).

The SEs are defined by emetic activity when ingested or when given orally to monkeys or selected other species (Lina *et al.*, 2004). TSST-1 was originally thought to have emetic activity when purified by Bergdoll *et al.* (1981), but that activity was later shown to result from SE contamination. TSST-1 lacks emetic activity and lacks the cystine loop structure thought to be important for emetic activity of SEs (Hovde *et al.*, 1994). TSST-ovine is biologically inactive when tested against human lymphocytes, but is active against lymphocytes from sheep (Lee *et al.*, 1992). The SE*l* proteins either lack emetic activity or have not been tested (Harris *et al.*, 1993; Lina *et al.*, 2004; Orwin *et al.*, 2001, 2002, 2003). With the exception of *SEl*-X, which is encoded on the core chromosome, other SAgs are encoded on variable DNA elements (McCormick *et al.*, 2001; Wilson *et al.*, 2011). SEA is encoded by the *sea* gene located on a bacteriophage (Betley *et al.*, 1984), SED is encoded by the *sed* gene found on a plasmid (Bayles and Iandolo, 1989), but most staphylococcal SAgs are encoded by genes on *S. aureus* pathogenicity islands (SaPIs) (McCormick *et al.*, 2001).

Like other exotoxins and cell-surface virulence factors, SAgs are under complex regulatory control, including global regulators such as *agr*, *sae*, *rot*, and *srr*. It is not our intent to discuss these DNA regulatory elements. Readers are encouraged to read Chapter 12 and important reviews that discuss global regulatory pathways in detail (Novick, 2003; Pragman and Schlievert, 2004; Yarwood and Schlievert, 2003).

S. aureus are unusually successful pathogens. It is likely that SAgs arose and were selected for because they interfere with normal immune function, increasing the survival and transmission chances for *S. aureus*. Approximately 20% of humans cannot make neutralizing antibodies to any given SAg (Parsonnet *et al.*, 2005), and because of this nearly all humans cannot produce antibodies to one or more SAgs. It appears that *S. aureus* strains may be selected for in humans based on SAgs, wherein lack of ability to produce antibodies may determine strain selection.

SAgs were initially known by their described biological activities (Bohach *et al.*, 1990; Watson, 1960) and SEs were known as enterotoxins due to their causation of staphylococcal food poisoning (Bergdoll, 1988; McCormick *et al.*, 2003a). For many years, the Sags, TSST-1, SEs, and SE*l* proteins were known as pyrogenic toxins, since potent pyrogenicity was a shared activity of all of the proteins (Bohach *et al.*, 1990). Additionally, all of the proteins shared the ability to enhance susceptibility to lethal shock caused by lipopolysaccharide (LPS) by up to 10^6-fold (Kim and Watson, 1970; Schlievert and Watson, 1978). These became the defining properties of the family. It had also been known since the 1970s that pyrogenic toxins were potent non-specific stimulators of T lymphocyte proliferation, dependent on antigen-presenting cells, and that this property was important in human illness (Schlievert, 1983; Schlievert *et al.*, 1977, 1979, 1980b, 1981).

The SAg term was coined by Marrack and Kappler (1990a) to emphasize the novel way pyrogenic toxins activate T cells. SAgs cross-bridge the variable regions of the β-chain of T cell receptors (Vβ-TCR) with major histocompatibility complex (MHC) II molecules on antigen-presenting cells in a relatively non-specific manner, causing high level proliferation of T cells and activation of antigen-presenting cells, such as macrophages (Kotzin *et al.*, 1993; Li *et al.*, 1999; McCormick *et al.*, 2001). SAgs stimulate up to 50% of T cells, inducing what is referred to as a 'cytokine storm' that defines many symptoms of TSS. It is also possible for strong conventional antigens to appear as weak SAgs, so it is necessary to demonstrate that new putative SAgs have defining activities.

Numerous structural and mutational analyses have provided key information regarding the three-dimensional structure and host cell interactions of SAgs. Studies that have examined the crystal structure of SAgs determined that these simple proteins contain a conserved overall structure made of two major protein domains: an amino-terminal oligosaccharide/oligonucleotide binding (O/B) fold, comprising a β-barrel, and a carboxy-terminal β-grasp domain made of anti-parallel β-strands, with domains connected by a central, diagonal α-helix (Brosnahan and Schlievert, 2011; Mitchell *et al.*, 2000). Based on small variations in this common core structure, SAgs can be categorized into five major groups, with staphylococcal SAgs falling into four of the five groups.

Group I SAgs are characterized by TSST-1, but this group also includes TSST-ovine and the recently described SE*l*-X (Acharya *et al.*, 1994; McCormick *et al.*, 2001; Prasad *et al.*, 1993; Wilson *et al.*, 2011). Group I SAgs have unique primary amino acid sequences. They have only the basic core structure, lacking the emetic cystine loop of SEs and the extra loop of 15 amino acids in group V SAgs (McCormick *et al.*, 2001). Group I SAgs contain only one MHC II binding site, the low-affinity MHC II binding site in their O/B folds that interacts with the α-chain of MHC II molecules (Kim *et al.*, 1994). The Vβ-TCR binding site of TSST-1 is known; this binding site is located on the top back of the molecule in the standard view, in a groove formed between the O/B fold and β-grasp domains (McCormick *et al.*, 2003b; Moza *et al.*, 2007).

The group II SAgs are characterized primarily by SEBn and SECn (Table 5.1) (Hovde *et al.*, 1994; McCormick *et al.*, 2001). These SAgs contain the basic core structure plus a cystine loop with between 10 to 19 amino acids separating the cysteine residues (Hovde *et al.*, 1994; Schlievert *et al.*, 2000). The cystine loop appears required for emetic activity, but its presence does not guarantee emesis. Similar to group I SAgs, group II SAgs contain only one MHC II site, the low-affinity, α-chain MHC II binding site in the O/B fold, and this interaction does not depend on interaction with the antigenic peptide within the MHC II peptide-binding groove (Brosnahan and Schlievert, 2011; Jardetzky *et al.*, 1994; Papageorgiou *et al.*, 1999). Interestingly and opposite of TSST-1, the Vβ-TCR binding site of the group II SAgs is located on the top front (Brosnahan and Schlievert, 2011; Fields *et al.*, 1996; Leder *et al.*, 1998; Li *et al.*, 1998).

Group III SAgs include SEA, SED, and SEE, and others (Table 5.1) (McCormick *et al.*, 2001). The group III SAgs contain cystine loops and have emetic activity. The loop of the group III SAgs is always nine amino acids long (McCormick *et al.*, 2001). Importantly, the group III SAgs contain both the low-affinity α-chain MHC II binding site on their O/B folds and also a second, high-affinity site, referred to as a Zn^{2+} dependent MHC II binding site in their β-grasp domains (McCormick *et al.*, 2001; Petersson *et al.*, 2002). The high affinity Zn^{2+} site interacts with the β-chain of MHC II molecules (McCormick *et al.*, 2001; Petersson *et al.*, 2002). The presence of two MHC II sites allows these SAgs to crossbridge MHC II molecules on adjacent antigen-presenting cells and increase activity by 10- to 100-fold (Kozono *et al.*, 1995). Group III SAgs bind Vβ-TCRs in the comparable site as group II SAgs (i.e. top front in the groove between O/B fold and β-grasp domains) (Antonsson *et al.*, 1997; Kozono *et al.*, 1995).

The Group IV SAgs are produced by group A streptococci but not *S. aureus* (McCormick *et al.*, 2001). Group V SAgs are the last described group. These include SE*l*-K to M, SE*l*-Q, and SE*l*-V (Orwin *et al.*, 2003; Orwin *et al.*, 2001; Orwin *et al.*, 2002). Group V SAgs contain both the low and high affinity MHC II binding sites but lack the emetic cystine loop. The Group V SAgs have an additional 15 amino acid insert that is unique; the insert exists between the third α-helix and β-strand 8, hence being given the name the α3-β8 loop. This loop appears to be critical for the specificity of the interaction of the SAgs with their corresponding Vβ-TCRs (Gunther *et al.*, 2007). Group V SAgs have no known association with specific human diseases.

In addition to the presence of up to two MHC II binding sites, a site for Vβ-TCR binding, and a cystine loop required for emesis, SAgs also contain a host cell binding site, which is referred to as the dodecapeptide binding region (Arad *et al.*, 2000; Brosnahan *et al.*, 2008; Brosnahan and Schlievert, 2011). This site is important for interaction with epithelial cells (Brosnahan *et al.*, 2008), possibly endothelial cells and adipocytes, and the immune co-stimulatory molecules CD40 (Spaulding *et al.*, 2012a) and CD28 (Arad *et al.*, 2011). It has been known since the 1980s that TSST-1 has the ability to bind to both epithelial and endothelial cells (Kushnaryov *et al.*, 1984a,b, 1989; Lee *et al.*, 1991b); however, the importance of these interactions has only recently been appreciated. Studies indicate the interaction of SAgs with epithelial, endothelial and adipose cells most likely initiates many disease processes (Brosnahan and Schlievert, 2011). The dodecapeptide binding region is fairly

Table 5.1 Major groupings of *S. aureus* superantigens

Group	Representative superantigens	Characteristic features
I	TSST-1, SE-like X	Distinct sequence; low affinity MHCII site
II	SEB, SEC, SEG	Presence of cystine loop of 10–19 amino acids; low affinity MHC II site
III	SEA, SED, SEE, SE/H, SE/J	Presence of cystine loop of exactly 9 amino acids; low and high affinity MHC II sites
IV	None	Present in Group A streptococci
V	SE/I, SE/K, SE/L, SE/P, SE/Q	No cystine loop; low- and high -ffinity MHC II sites; extra 15 amino acid insertion

conserved among SAgs, with greatest difference being between TSST-1 and other SAgs (Arad *et al.*, 2000; Brosnahan *et al.*, 2008; Brosnahan and Schlievert, 2011). This difference in the dodecapeptide binding region has been suggested to explain the greater mucosal epithelium penetration of TSST-1 relative to other SAgs (Brosnahan *et al.*, 2008; Brosnahan and Schlievert, 2011; Hamad *et al.*, 1997), and thus TSST-1's unique association with menstrual, vaginal TSS (Schlievert *et al.*, 2000). The dodecapeptide region is located at the base of the central, diagonal α-helix (Brosnahan and Schlievert, 2011).

Clinically, SAgs are classically defined by their abilities to cause TSS, they have been associated with many other illnesses (Table 5.2). All serious illnesses caused by *S. aureus* have significant vascular-cardiovascular effects on the host, including capillary leak syndromes, in the form of TSS, sepsis, infective endocarditis, pneumonia, and osteomyelitis. Since SAgs are produced by all pathogenic *S. aureus*, these toxins should be considered as contributory to all illnesses. It is noteworthy that we have hypothesized that *S. aureus* appears to need at least one major SAg and one major cytotoxin (discussed later) to cause disease (Schlievert *et al.*, 2010). The following is a brief discussion of SAg association with human diseases.

TSS including menstrual TSS (mTSS) and non-menstrual TSS

mTSS, which is a potentially life-threatening illness, is the most recognized of the SAg diseases. As originally described, TSS was most often seen in women during their menstrual periods, hence, is referred to mTSS (Davis *et al.*, 1980; Shands *et al.*, 1980). Importantly, mTSS refers only to the association with menstruation and does not refer to the site of *S. aureus* isolation; approximately 95% of mTSS cases are menstrual, with vaginal colonization, whereas the remaining 5% are menstrual, non-vaginal colonization, occurring with almost any other type of infection. Most vaginal staphylococcal mTSS cases are associated with tampon use (Davis *et al.*, 1980; Shands *et al.*, 1980), with higher incidence associated with higher tampon absorbency (Davis *et al.*, 1980; Osterholm *et al.*, 1982; Shands *et al.*, 1980). The incidence of mTSS peaked in the early 1980s when the incidence was as high as 10/100,000 women (Osterholm *et al.*, 1982). The incidence of this illness is lower today, at approximately 1–3/100,000 women (DeVries *et al.*, 2011; Gaventa *et al.*, 1989; Schlievert *et al.*, 2007b; Schlievert *et al.*, 2004); it is alarming that the illness is considerably less-recognized today due to lack of media attention, and the disease is occurring in younger individuals (typically 12–15 years

Table 5.2 Disease association with *S. aureus* SAgs

Human disease	SAg association
Menstrual TSS	TSST-1
Non-menstrual TSS	TSST-1, SEB, SEC
Purpura fulminans	TSST-1, SEB, SEC
Extreme pyrexia	Any
RED syndrome	Any
Anaphylactic TSS-like	TSST-1
Kawasaki-like syndrome	TSST-1, SEB, SEC
Neonatal exanthematous disease	TSST-1
Food poisoning	Any SE
Nasal polyposis	Any
Perineal erythema	Any
Desquamative inflammatory vaginitis	Any
Sudden infant death syndrome	Any
Type II diabetes	Any

of age). mTSS is defined as occurring within 2–3 days of onset of menstruation, during menstruation, and within 2–3 days after menstruation. *S. aureus* reaches its highest numbers on days 2–3 of menstruation which precedes by one day the peak onset of TSS (days 3–4) (Schlievert *et al.*, 2007b). *S. aureus* counts vaginally as measured by numbers on tampons may exceed 10^{11} per tampon (Schlievert *et al.*, 2007b). TSST-1 amounts as high as 100 μg may be present on soiled tampons (Schlievert *et al.*).

TSS is defined by the following criteria: high fever, hypotension, erythematous (scarlet fever-like) rash, peeling of the skin often during recovery, and any three multi-organ components. Unfortunately, TSS symptoms are often mistaken as flu-like symptoms, including vomiting and diarrhoea (Davis *et al.*, 1980; Schlievert, 2007; Shands *et al.*, 1980; Todd *et al.*, 1978). It is now recognized that often one of the defining criteria is not present and this illness is considered probable TSS (Reingold *et al.*, 1982). The defining criterion most often missing is the scarlet fever-like rash that occurs in association with pre-existent delayed hypersensitivity to SAgs (John *et al.*, 2009; Schlievert *et al.*, 1979). However, cases of probable TSS have been described in which each defining criterion has been absent. If more than one criterion is absent and other causes ruled out, the illness can be considered toxin-mediated disease (Parsonnet, 1998).

The majority of TSS patients do not meet the full criteria for TSS that were established to perform epidemiologic studies by the CDC (Shands *et al.*, 1980). Additionally, it is now recognized that the majority of TSS cases are not menstrual and vaginal-associated. Multiple additional non-menstrual categories of staphylococcal TSS are recognized, including the more commonly seen post-respiratory viral pneumonia-associated TSS (MacDonald *et al.*, 1987). This illness may progress exceptionally rapidly, being lethal in a matter of hours, and in children this illness is associated with an extremely high case-fatality rate (up to 90%) (MacDonald *et al.*, 1987). Other categories of staphylococcal TSS include: (1) illness associated with any type of soft tissue infection; (2) recalcitrant erythematous desquamating syndrome, an unrelenting TSS in AIDS patients resulting in death after as many as 70 days of illness (Cone *et al.*, 1992); (3) purpura fulminans, a rapidly progressing TSS illness associated with clotting abnormalities and purpuric rash (Adem *et al.*, 2005; Kravitz, 2005); (4) extreme pyrexia syndrome, an illness with a 100% case-fatality rate and fevers in excess of 108°F (Assimacopoulos *et al.*, 2009); and (5) anaphylactic TSS, an acute illness that has a 100% case-fatality rate and high cardiac eosinophilia. This latter illness may be a severe form of chronic TSS in patients who develop atopic dermatitis rashes instead of the characteristic scarlet fever-like rash. Importantly, each year new categories of staphylococcal TSS are identified, including most recently a re-discovered, severe enterocolitis TSS illness (Kotler *et al.*, 2007; Lin *et al.*, 2010).

Nearly all mTSS is caused by TSST-1 producing *S. aureus* strains (Bergdoll *et al.*, 1981; Schlievert *et al.*, 1981; Schlievert *et al.*, 2004). Rare mTSS cases appear to be associated with production of SEG and SE*I*-I (Jarraud *et al.*, 1999). The reason for the high association of TSST-1 with mTSS is not completely clear but likely depends on at least three factors: (1) TSST-1 is produced in high concentrations relative to the majority of other Sags; (2) TSST-1 has greater mucosal surface penetrating ability than other SAgs (Hamad *et al.*, 1997; Schlievert *et al.*, 2000); and (3) large numbers of mucosal *S. aureus* strains produce TSST-1 (Kuehnert *et al.*, 2006). It is important to recognize, that when first identified as causing mTSS, *S. aureus* strains highly associated with illness belonged to bacteriophage types 29/52 complex (Altemeier *et al.*, 1981, 1982). These strains with the ability to produce TSST-1 emerged as major clones in 1972 (Altemeier *et al.*, 1981, 1982), such that today essentially 100% of these organisms produce TSST-1. As many as 25% of people colonized with *S. aureus* on mucosal surfaces, may be colonized with TSST-1 producing strains (Kuehnert *et al.*, 2006). We recognize these strains today by other, newer typing mechanisms, including pulsed-field gel electrophoresis as primarily type USA200 and staphylococcal protein A type and/or multilocus sequence type as clonal complex 30. USA200 strains appear highly adapted to mucosal surface colonization, with infections resulting from TSST-1 production and/or microbial spread from those surfaces. For example, >95% of USA200 strains have a mutation in the α-toxin gene that greatly reduces production of the cytotoxin (Lin *et al.*, 2011; O'Reilly *et al.*, 1990; Schlievert *et al.*, 1982). α-toxin is highly toxic to humans, and is the most inflammatory protein produced by *S. aureus* strains (Lin *et al.*, 2011). While α-toxin is critical for production of furuncles and soft tissue abscesses originating from skin infections, production of high levels of this cytotoxin would be expected to result in highly lethal infections if produced on mucosal surfaces; thus, it is likely that the reduction in α-toxin production by USA200 strains has allowed them to colonize mucosal surfaces effectively. This mucosal niche location of USA200 strains, with consequent diseases originating from those mucosal surfaces, has been underappreciated, as evidenced by a recent publication suggesting that USA200 strains are less virulent than historic skin-adapted strains referred to as bacteriophage type 80/81 (Deleo *et al.*, 2011).

The environmental factors that allow USA200 strains to produce TSST-1 have been established; these are similar for most other SAgs. These conditions include growth in complex media containing animal protein with low glucose (glucose functions as a catabolite repressor of exotoxin production), neutral pH (as expected vaginally during menstruation), temperature of 37–40°C, and oxygen balanced with CO_2 (Schlievert and Blomster, 1983; Yarwood and Schlievert, 2000). As noted previously, blood components, and more specifically haemoglobin peptides, negatively affect production of TSST-1 (Schlievert *et al.*, 2007a). The introduction of oxygen into the human vagina, a typically anaerobic environment, by tampons is now considered the major reason for the tampon association with mTSS (Hill *et al.*, 2005; Schlievert and Blomster, 1983; Yarwood and Schlievert, 2000). The introduction of oxygen would also explain the major association of mTSS with higher absorbency tampons in that those tampons introduce more oxygen. Studies of regulation of TSST-1 production by oxygen led to the identification of a global regulator of TSST-1 production called staphylococcal respiratory response (Srr) A/B (Yarwood *et al.*, 2001). This important two-component system functions as a repressor of TSST-1 (and other exotoxins) production when oxygen levels are low. The repressor function in anaerobic conditions appears to be dominant over all other global regulators of exotoxin production (Yarwood *et al.*, 2001). One other factor has been identified that increases TSST-1 production, the surfactant pluronic L-92 which was present in one tampon in the early 1980s (Melish *et al.*, 1989; Schlievert, 1996a). It is hypothesized that pluronic L-92 alters staphylococcal two-component system signalling that up-regulates TSST-1 production by as much as 8-fold. It has also been suggested that tampons composed of all cotton reduce TSST-1 production, compared to tampons composed of cotton-rayon blends or all rayon (Tierno and Hanna, 1989). These studies are refuted by multiple, carefully performed studies that fail to find a reduction in TSST-1 production by all cotton tampons (Fischetti *et al.*, 1989; Parsonnet *et al.*, 1996; Schlievert, 1995, 1996b).

S. aureus typically colonizes the vaginal mucosal surface, resulting in TSST-1 production and penetration through the mucosa. TSST-1 producing *S. aureus* strains accomplish this feat by a mechanism called 'outside-in signalling, ' where initial interactions with epithelial cells promote TSST-1 penetration and recruitment and stimulation of immune cells (Brosnahan and Schlievert, 2011). Porcine vaginal *ex vivo* models, which nearly completely mimic human vaginal mucosa, have shown that TSST-1 alone is able to penetrate the vaginal mucosa (Davis *et al.*, 2003; Squier *et al.*, 2008), but low levels of the cytotoxin α-toxin greatly enhance penetration.

It appears that TSST-1, alone and as enhanced by cytotoxin production, penetrates the vaginal mucosa through stimulation of chemokine production by epithelial cells (Brosnahan *et al.*, 2009; Brosnahan and Schlievert, 2011; Peterson, 2005). The combined effects of TSST-1 and cytotoxin-induced inflammation and cytolysin toxicity likely contribute to destabilization of the stratified squamous epithelial barrier, allowing TSST-1 access to the deeper layers of the mucosal epithelium where the SAg can directly interact with epithelial cells close to the basement membrane (Brosnahan and Schlievert, 2011). Approximately 10^4 TSST-1 receptor sites per cell have been demonstrated on primary human epithelial cells and immortalized human vaginal epithelial cells (Kushnaryov *et al.*, 1984b; Peterson, 2005). TSST-1 induces the production of pro-inflammatory chemokines IL-8 (CCL8) and MIP-3α (CCL20) in human vaginal epithelial cells *in vitro* by a mechanism dependent on signalling via ADAM17 and epithelial growth factor receptor (Breshears *et al.*, 2012; Peterson, 2005). These chemokines attract neutrophils and other immune cells, including T cells and macrophages, to infection sites. In line with this, immune cell recruitment to the subepithelial mucosa has been shown to occur in the *ex vivo* porcine vaginal model (Davis *et al.*, 2003; Peterson, 2005). Hence, the concerted action of TSST-1 and low levels of cytotoxins, such as α-toxin, results in mucosal epithelium inflammation and increased permeability, followed by TSST-1 penetration and induction of chemokines by metabolically active epithelial cells, and finally recruitment of immune cells to the submucosa. This outside-in signalling mechanism provides TSST-1 accessibility to a sufficient pool of T cells and macrophages to elicit the cytokine storm characteristic of mTSS; specifically, IL-1β and TNF-α (produced by macrophages) and TNF-β, IL-2 and IFN-γ (produced by T cells).

Once TSS-inducing SAgs are produced and T cells and macrophages become activated to secrete a cytokine storm, the cascade of visible events seen in TSS begins. Fever depends on SAg and/or cytokine stimulation of the hypothalamic fever response control centre (Schlievert *et al.*, 1978; Schlievert and Watson, 1978, 1979). The most severe symptom associated with TSS is hypotension, which may progress to shock and death, resulting from capillary leakage. SAgs induce vascular injury in part by the combined effect of toxin-induced systemic release of vasoactive mediators, such as TNF-α and TNF-β (Kotzin *et al.*, 1993; McCormick *et al.*, 2001), synergy with other molecules such as LPS (Dinges and Schlievert, 2001; Schlievert, 1982; Stone and Schlievert, 1987), and direct toxic interaction with the vascular endothelium (Kushnaryov *et al.*, 1989; Lee *et al.*, 1991b). The identity of the endothelial cell receptor for TSST-1 is currently unknown and under investigation. However, it is clear that fluid replacement to offset capillary leak is required for management of TSS cases in humans (Davis *et al.*, 1980; Shands *et al.*, 1980) and rabbit models (Lee *et al.*, 1991a).

An interesting and potentially important property of SAgs is their ability to enhance the lethality of LPS by up to 10^6-fold. This mechanism depends on synergistic TNF production in the presence of both SAgs and LPS (Dinges and Schlievert, 2001), and the impaired LPS clearance function of the liver in the presence of SAgs (Schlievert *et al.*, 1980a). Although the LPS enhancement mechanism is not universally accepted as contributing to TSS, it is intriguing for many reasons: Humans and rabbits are approximately equally susceptible to TSST-1 and TSS, and both have high numbers of LPS-containing Gram-negative intestinal and vaginal flora (Schlievert, 2009). In contrast, mice are approximately 10^{11} more resistant to TSST-1 lethality on a per gram basis than rabbits and humans (Schlievert, 2009), and mice are less colonized by LPS-containing Gram-negative bacteria. Rabbits at 8 months of age are approximately 1000 times more susceptible to TSST-1 than young rabbits, and this corresponds to the same time as rabbits become 1000 times more susceptible to the lethal effects of LPS.

The ability of *S. aureus* to cause mTSS depends also on the immune status and genetics of colonized persons. It has been long appreciated that women with low levels or no antibodies to TSST-1 are sero-susceptible to mTSS, whereas those individuals with antibodies appear protected (Bergdoll *et al.*, 1981; Kansal *et al.*, 2007; Parsonnet *et al.*, 2005; Vergeront *et al.*, 1983). There is an age-dependent appearance of antibodies to TSST-1 in humans, with 80% of humans having antibodies to the SAg by 12 years of age (Parsonnet *et al.*, 2005; Vergeront *et al.*, 1983). The remaining 20% of sero-susceptible individuals are among those who develop TSS. Importantly, these 20% appear not to be able to produce protective antibodies to TSST-1, and thus remain susceptible to TSS recurrences (Osterholm *et al.*, 1982). The failure to develop antibodies in sero-susceptible women likely results from the TSS cytokine storm preventing B cell function (Kotb *et al.*, 2002; Schlievert, 1983). The same lack of antibody production has been seen in approximately 50% of rabbits tested an model of human TSS. In contrast, 100% of rabbits respond with production of neutralizing antibodies when challenged with a non-toxic mutant of TSST-1. These studies indicate that the failure to develop antibodies is not due to genetic non-responsiveness, but rather due to native TSST-1 effects on the immune system.

As noted above, non-menstrual staphylococcal TSS occurs in association with nearly any type of staphylococcal infection. As with mTSS, not all SAgs are equally associated with non-menstrual TSS. Studies indicate that 50% of non-menstrual TSS is caused by USA200 and related strains producing TSST-1 (Schlievert and Kim, 1991; Schlievert *et al.*, 2004). The remaining 50% of strains nearly always produce the SAgs SEB or SEC (Schlievert and Kim, 1991; Schlievert *et al.*, 2004). The reason for the association of these three SAgs with most TSS cases, whether mTSS or non-menstrual TSS, is their high-level of production (up to 15 mg/ml in biofilms).

While TSST-1 is restricted to USA200 and related strains of *S. aureus*, SEB and SEC may be produced by USA200 and other clonal groups. It has been shown that amounts of SAgs as low as 0.1 μg may induce TSS symptoms in humans (Giantonio *et al.*, 1997).

Nearly one-half of TSS cases are associated with TSST-1 production. In the United States, the majority, but not all, of these TSST-1 producing organisms are MSSA; however, in other countries, TSST-1 positive MRSA strains appear more common (Kim *et al.*, 2007; Nagao *et al.*, 2009; Robert *et al.*, 2011). Given their increased presence on mucosal surfaces today as opposed to 1980 (Kuehnert *et al.*, 2006), it is possible that TSST-1 positive MRSA strains will continue to increase in numbers in the United States. USA400 strains all produce SEB or SEC and were the initially identified causes of CA-MRSA necrotizing (haemorrhagic) pneumonia and sepsis. Pneumonia and sepsis do not preclude patients from also simultaneously having TSS, the symptoms of which are usually present. USA400 strains were more recently displaced in many, but not all, regions of the USA by USA300 CA-MRSA with ability to cause necrotizing pneumonia with TSS-like symptoms (2003; Francis *et al.*, 2005). These strains do not produce TSST-1, SEB, or SEC (Diep *et al.*, 2006; Francis *et al.*, 2005); however, the strains produce a newly recognized SAg, SE-like X, which has been linked to necrotizing pneumonia in rabbit model studies (Wilson *et al.*, 2011). Additionally, these strains appear to produce a deletion derivative of TSST-1, as associated with extreme pyrexia syndrome (Assimacopoulos *et al.*, 2009).

SAgs clearly cause TSS. This statement is supported by the association of SAgs and causative bacteria with human illnesses. Additionally, SAgs cause TSS symptoms in animal models. For example, studies have shown that administration of SAgs in subcutaneously implanted miniature osmotic pumps duplicate the symptoms of TSS in rabbits (Lee and Schlievert, 1989; Parsonnet *et al.*, 1987). Mice are highly resistant to SAgs unless the liver is damaged first with D-galactosamine (Silverstein, 2004), which causes liver necrosis, a feature not seen in cases of human TSS. SAgs also cause lethal TSS in rabbits as applied intra-pulmonary (Strandberg *et al.*, 2010). Studies with use of isogenic strains that differ only in production of SAgs (de Azavedo *et al.*, 1985), and use of active and passive immunization against specific SAgs to protect rabbits, conclusively establish that SAgs cause TSS (Buonpane *et al.*, 2007; Spaulding *et al.*, 2012a; Strandberg *et al.*, 2010). Finally, SAgs directly injected into humans cause TSS symptoms (Giantonio *et al.*, 1997).

Staphylococcal SAg food poisoning

The emetic cystine loop in SEs has been discussed. Not all SAgs are emetic, but SEs, including most commonly SEA, cause 24–48 h episodes of retching, vomiting, and diarrhoea every 15–30 min, without fever after human or non-human primate ingestion of nanogram quantities of SEs (Bergdoll, 1988; McCormick *et al.*, 2003a). The lack of fever with staphylococcal food poisoning likely results from combinations of non-pyrogenic quantities required to cause emesis and failure of SEs to exhibit high mucosa penetration. Emetic activity has been shown in studies to be independent of Superantigenicity (Harris *et al.*, 1993; Schlievert *et al.*, 2000). Through their emetic activity, SEs are primary causes of toxin-mediated food-borne illness, and the second leading cause of food-borne illness overall (Le Loir *et al.*, 2003). Domestic cats and non-human primates have been used extensively in studies of the SE causation of food poisoning; a house musk shrew model is a newly developed animal model used to investigate SE emetic activity (Hu *et al.*, 2003; Hu *et al.*, 1999).

While much of the mechanism of the ability of SEs to induce food poisoning remains unknown, recent studies, to examine the effects of SEA on human intestinal epithelial cells, demonstrated that SEA induces increases in the intracellular calcium concentration of these cells (Hu *et al.*, 2005). Using the house musk shrew model, Hu *et al.*, (2007) demonstrated that SEA induces the release of serotonin in the intestine to cause emesis. Other work suggests that the vagus nerve is involved, where SEs stimulate the vagus nerve, thereby activating the sympathetic nervous system (Sugiyama and Hayama, 1965). For further information on mechanisms and outbreaks of staphylococcal food poisoning, readers are referred to other reviews (Bohach *et al.*, 1990; Hennekinne *et al.*, 2012).

Staphylococcal food poisoning is a self-limiting illness that is rarely if ever fatal. However, this incapacitating activity of SEs may have been primarily responsible for SEs, such as SEB, to be listed as select agents of bioterrorism. It is noteworthy that during the 1950s and 1960s, the United States stockpiled tons of SEB yearly as its major bioweapon. Additionally, because it is difficult to denature SAgs and unnecessary to weaponize them to be taken orally or intra-pulmonary, SEs pose a significant hazard by these two routes. Likewise, there is no evidence to indicate that humans can be vaccinated against this activity.

Staphylococcal pneumonia

There are an estimated 70,000 cases of *S. aureus* pneumonia in the USA each year. Because all pathogenic *S. aureus* strains produce high levels of SAgs, these toxins contribute to severe pneumonia, as demonstrated in animal models of human disease (Buonpane *et al.*, 2007; Spaulding *et al.*, 2012a; Strandberg *et al.*, 2010). Even though pneumonia is an illness that describes the primary infection site, the illness does not exclude pneumonia-associated TSS, since TSS is defined as a collection of symptoms rather than a body site of infection.

In 1987, MacDonald *et al.* (1987) described post-influenza TSS associated with TSST-1 and SEB. The illness does not require influenza virus infection in that cases of pneumonia-associated TSS occurs in association with many other upper respiratory viral infections, and even asthma. This illness occurs each year usually during the winter months, throughout the world. In the MacDonald *et al.*, study, there was a 90% case fatality rate in children, all associated with TSST-1 producing USA200 strains. One strain from the sole surviving child was a USA400 strain producing SEB. I have tested large numbers of other strains from children with post-influenza TSS, and the majority of *S. aureus* strains belong to the USA200 clonal group and produce TSST-1 (unpublished data).

In 1999, the CDC and colleagues published on four fatal cases of *S. aureus* necrotizing (haemorrhagic) pneumonia in children associated with the recently emergent USA400 clonal group of CA-MRSA (CDC, 1999). Two of these isolates produced SEB and two produced SEC, the expected SAgs produced by USA400 strains. Subsequent studies have shown that these two SAgs are nearly always present in USA400 CA-MRSA strains (Fey *et al.*, 2003).

A rabbit model of necrotizing pneumonia conclusively showed that SAgs are critical for the development of lethal pulmonary illness associated with staphylococcal infection (Strandberg *et al.*, 2010). Purified SEB and SEC, when installed intrapulmonary, induce haemorrhagic lung tissue, respiratory distress, and lethal TSS (Strandberg *et al.*, 2010). Intrabroncheal administration of CA-MRSA strains of the USA200 (TSST-1 positive) and USA400 (SEB and SEC positive) clonal types in rabbits induces similar lung pathology and lethality as individual administration of SAgs (Strandberg *et al.*, 2010). The newly discovered SAg SE-like X (Wilson *et al.*, 2011), encoded by at least USA300, and as well USA100 and USA400 clonal types, appears critical for the development of necrotizing pneumonia and lethal TSS in rabbits. Furthermore, vaccination against TSST-1, SEB, or SEC provides complete protection against highly lethal doses of *S. aureus* strains producing the corresponding SAgs (Strandberg *et al.*, 2010).

Staphylococcal infective endocarditis

Infective endocarditis is a serious life-threatening infection of the heart endothelium characterized by the formation of 'cauliflower-like' vegetations on heart valves composed of microorganisms and host factors (Spaulding *et al.*, 2012b). *S. aureus* is a major cause of infective endocarditis worldwide (Bashore *et al.*, 2006). Infective endocarditis occurs most often at sites of pre-existing heart damage, usually involving valves. Damaged sites become coated with platelets and other host factors that in regions of turbulence, allow *S. aureus* to adhere and grow. As many as 50% of infective endocarditis patients may succumb to the infection, and up to one-half of patients may have strokes and metastatic abscesses. Of the patients who die, many are deaths are associated with valve failure, strokes, and sepsis associated with metastatic abscesses.

Recent studies have shown that as many as 90% of infective endocarditis isolates of *S. aureus* belong to the USA200 clonal group, all of which produce TSST-1, among other virulence factors (Nienaber *et al.*, 2011; Xiong *et al.*, 2009). Pragman *et al.* showed that TSST-1 is critical for development of infective endocarditis as tested in rabbits with use of isogenic strains, one strain positive for TSST-1 (TSST-1^+) and the other strain negative for TSST-1 (TSST-1^-) (Pragman *et al.*, 2004). Only one rabbit administered the TSST-1^- strains developed vegetations, containing only 100 colony-forming units. In contrast, the TSST-1^+ strain caused nearly 7 logs more colony-forming units in cardiac vegetations. In a recent study using passive neutralization of SEC in CA-MRSA USA400 strain MW2, administration of Vβ-TCRs specific for SEC prevented development of cardiac vegetations (Mattis *et al.*, 2013). Finally, active immunization studies targeting SAgs and cytotoxins effectively prevented USA200 *S. aureus* infective endocarditis (Spaulding *et al.*, 2012a).

There are at least two possible explanations for SAg roles in infective endocarditis. Staphylococci initially colonize damaged endothelium to initiate infective endocarditis. SAgs may then interact with host cells to prevent endothelium wound healing through direct effects on endothelial cells (Lee *et al.*, 1991b). Support for this hypothesis comes from a study showing that TSST-1 is toxic to porcine aortic endothelial cells (Lee *et al.*, 1991b). In addition, SAgs may also systemically dysregulate the immune system and prevent clearance of the growing *S. aureus* valve infections.

Both staphylococcal pneumonia and infective endocarditis patients often develop sepsis as a consequence of their infections, and significant numbers of patients develop sepsis following surgery. It is likely that many systemic effects in these patients result from SAg activities on the host since the infecting strains are likely to produce these toxins. Investigators have shown that certain SAgs, notably SEA, TSST-1, and SEC, are over-represented in sepsis cases compared to bacteria isolated from non-sepsis patients (Becker *et al.*, 2003; Dauwalder *et al.*, 2006; Ferry *et al.*, 2005). However, it is unlikely that SAgs are directly produced in the bloodstream of patients, since haemoglobin peptides inhibit SAg production (Schlievert *et al.*, 2007a). It is more likely that SAgs are produced in focal sites of infection protected from haemoglobin peptides and then secreted into the bloodstream.

Atopic dermatitis (AD)

Atopic dermatitis (AD), is a chronic relapsing, highly pruritic, inflammatory skin disease that is often the prelude to development of food allergy, asthma and allergic rhinitis (Boguniewicz and Leung; Guttman-Yassky *et al.*). *S. aureus* skin infections exacerbate skin disease in patients with AD and modify the host response to environmental allergens and viral pathogens (Kawakami *et al.*, 2009; Laouini *et al.*, 2003). Recent studies suggest that host–pathogen interactions stemming from the production of *S. aureus* derived virulence factors, such as SAgs and α-toxin, contribute greatly to the skin inflammation seen in AD (Brauweiler, 2013; Skov *et al.*, 2000).

SAgs and α-toxin activate cells involved in the pathogenesis of AD (Schlievert *et al.*, 2010). Due to their highly potent immunostimulatory activity, much of the focus has been on the role of SAgs (Kotzin *et al.*, 1993). Support for a role of SAgs in AD include the following: (1) the severity of AD correlates with number of SAg-secreting *S. aureus* colonizing the skin (Boguniewicz and Leung, 2010); (2) most AD patients make IgE antibodies directed against SAgs found on their skin (Leung *et al.*, 1993); (3) SEB, applied to the skin, can induce eczematoid skin changes (Skov *et al.*, 2000); (4) SAgs have been demonstrated to induce T cell expression of the skin homing receptor via stimulation of IL-12 production (Leung *et al.*, 1995); (5) after stimulation by SEB, T regulatory cells lose their immunosuppressive activity, suggesting a novel mechanism by which SAgs could augment T-cell activation in patients with AD (Ou *et al.*, 2004); (6) SAgs induce corticosteroid resistance suggesting that several mechanisms exist by which SAg increase AD severity (Leung *et al.*, 2009); and (7) I developed 3–4 months of AD after accidentally splashing highly pure SEA on my face (unpublished data).

Type II diabetes

Recently, Vu *et al.* (2013, 2015) showed that the purified SAg TSST-1has the ability to induce clinical findings of type II diabetes in rabbits, including chronic inflammation, elevated blood glucose, and insulin-resistance. Their data show that *S. aureus* and its SAgs are at least one cause of type II diabetes. Clearly, this area should be a major research direction in the future.

Exfoliative toxins (ETs) A and B

Two serologically distinct exfoliative toxins have been described associated with human disease-causing strains of *S. aureus*, ETA and ETB (Ladhani *et al.*, 1999). These serine protease exotoxins cause staphylococcal scalded skin syndrome (SSSS), usually seen in newborns and characterized by intraepidermal separation of layers of the skin at the desmosomes (Melish *et al.*, 1972). The illness begins with a generalized erythema, often near the mouth and spreading over the entire body in the course of a few days. When the skin is lightly rubbed, the epidermal layer wrinkles irreversibly, giving rise to the characteristic positive Nikolsky sign (Melish *et al.*, 1972). Later, large flaccid sterile bullae appear that lead to separation through the stratum granulosum layer. The entire sequence from disease onset to recovery takes 7–10 days. There is no permanent scarring on areas of the skin, and the causative toxins are not themselves lethal to the host; as many as 5% of affected neonates may succumb due to production of SAgs and other factors of *S. aureus* (Ladhani, 2001; Ladhani *et al.*, 1999).

In 1970, Melish and Glasgow provided an animal model for the study of SSSS, the new-born mouse model, and were thus able to fulfil Koch's postulates for the role of both *S. aureus* and toxin in the illness (Melish and Glasgow, 1970). ETA was purified by several investigators during the early 1970s, and ETB was purified in 1974 (Ladhani, 2001; Ladhani *et al.*, 1999). Both genes, designated *etA* and *etB*, have been cloned and the nucleotide sequences determined. *etA* is chromosomal and its protein is encoded by a bacteriophage gene and is the more commonly seen of the two toxins. ETB is encoded by a gene present on a large plasmid. Both genes encode ETs with a signal peptide removed upon secretion. ETA and ETB are both made as proteins of 280 amino acids, of which the first 38 constitute signal peptides. The mature proteins have molecular weights of 27,000 Da. These two toxins are serologically distinct and share only 45% sequence similarity. Antibodies against the individual toxins is neutralizing. ETA has a pI of approximately 7.0, while ETB has a pI of approximately 4.

Both ETs are produced in large amounts in standard cultivation media; we use beef heart medium without added CO_2 (Vath *et al.*, 1999; Vath *et al.*, 1997). Clones such as *S. aureus* RN4220 containing ET genes make 10 to 20 times more toxin than clinical isolates in which yields of 10–20 mg/l are easily achievable. We find that the simplest method to make large quantities of ETA and ETB are to precipitate toxins made in RN4220 with 4 volumes of ethanol and subject the resolubilized toxins to successive isoelectric focusing in pH gradients of 3–10 and then a narrower range. Toxins obtained using our approach are homogeneous in SDS-PAGE.

Although ETs have definitively been established to cause SSSS for many years, the molecular mechanism of action of the toxins has only recently been tentatively elucidated. Lillibridge *et al.* (1972) reported that injection of ET into the skin of newborn mice resulted in positive Nikolsky signs by 25 min post injection. At 20 min post injection, light-density oval bubbles in the region of the stratum granulosum widened and lost their oval appearance. By 25 min, the desmosomes, which are specialized thickenings of cell membranes that hold cells together, began to split and separate. At 150 min post-injection, all desmosomes along the region of cleavage were split; the cleavage plane was in the stratum granulosum layer. These researchers proposed that ET may activate proteolytic enzymes contained within the ovoid bubbles to cleave the desmosomes.

Lymphocyte mitogenic activity of ET was described (Monday *et al.*, 1999; Rago *et al.*, 2000a) and that the toxin primarily stimulated T-cells to proliferate. Several other laboratories confirmed the findings that ET was mitogenic, culminating in the report by Marrack and Kappler (1990b) that ETs are SAgs. We have also noted the very strong mitogenic activity associated with ETA (Monday *et al.*, 1999; Rago *et al.*, 2000a). In contrast, a recent study by Fleischer *et al.* (1995) suggests that ETA is not mitogenic; rather, its apparent superantigenic activity results from contamination with SEs. Further studies are needed to clarify this issue. If the toxin is shown to be superantigenic, it is possible that the separation of layers of the skin results in part from capillary leakage and pressure as a result of fluid accumulation.

Recently we solved the three dimensional structures of both ETA and ETB (Vath *et al.*, 1997, 1999). Both structures are consistent with chymotrypsin-like serine proteases that cleave substrates after acidic residues, often glutamic acid. In both ETs, the studies confirmed the importance of the catalytic triad of amino acids (His72, Asp120, and Ser195 in ETA, and His65, Asp114, and Ser186 in ETB) as important in the active sites. The structural studies suggested the active site of the enzymes may be very small and partially blocked by a loop, permitting only very small peptides or unique proteins to be susceptible to proteolysis. With this in mind, our group tested a variety of small peptides containing glutamic acid residues for their ability to be cleaved by ETs. Both α- and β-melanocyte-stimulating hormone peptides were cleaved by ETA, after the Glu residue; thus, it was shown for the first time that ETs are serine proteases (Rago *et al.*, 2000b). Subsequently, it was shown that desmoglein I is also cleaved by ETs, suggesting that this activity accounts for the clinical illness SSSS (Amagai *et al.*, 2000).

Cytotoxins: the second line of defence against the human immune system

S. aureus produces a myriad of cytotoxins, including alpha, beta, gamma (and related leucocidins), delta (and related phenol soluble modulins) and a new cytotoxin, referred to as epsilon. The majority of these cytotoxins are encoded in the core genome, and at least one major cytotoxin appears required for *S. aureus* strains to cause human disease, with most strains producing multiple cytotoxins.

α-Toxin

Perhaps the most-studied staphylococcal cytotoxin is α-toxin. There are many articles that summarize early data related to α-toxin that will not be discussed in this review, but are worth noting due to their importance (Bernheimer and Schwartz, 1963, 1965; Berube and Bubeck Wardenburg, 2013; Bhakdi *et al.*, 1996; Bhakdi and Tranum-Jensen, 1991). α-Toxin is encoded by the hla gene in the chromosome of at least 80% of *S. aureus* strains. Major mucosal strains of *S. aureus*, including USA200 strains, have a mutation in hla that reduces production of the toxin by 50–100 fold (Lin *et al.*, 2011; O'Reilly *et al.*, 1990). This reduction in α-toxin may afford these strains the ability to colonize mucosal surfaces without producing too much toxin that the host is killed. α-Toxin is cytotoxic and one of the most inflammatory molecules produced by *S. aureus*, and is typically required for furuncle and abscess formation. The toxin is relatively invariant and is produced as a mature protein of approximately 33,000 Da molecular weight (Tweten *et al.*, 1983). Electron microscopy and structural studies indicate α-toxin forms homoheptamer structures within host cell membranes and pores of approximately 10 nm are formed as a result of oligomerization (Gouaux *et al.*, 1994; Song *et al.*, 1996). The toxin has a cap that is exposed to the aqueous environment, a rim that is adjacent to the host cell plasma membrane, and a stem that creates the transmembrane pore. Like other secreted virulence factors, α-toxin production is under complex control by multiple global regulatory mechanisms (Novick, 2003).

The ability of α-toxin to bind to host cells has been studied since the early to mid-1900s. Originally, the toxin was referred to as a haemolysin based on its exquisite ability to lyse rabbit erythrocytes (Bernheimer and Schwartz, 1963; Bhakdi and Tranum-Jensen, 1991; Cassidy and Harshman, 1973). Interestingly, human erythrocytes are much less sensitive to the lytic activity of the toxin, but human epidermal cells and other cell types can be killed (Bhakdi and Tranum-Jensen, 1991). Thus, it has long been recognized that α-toxin causes the dermonecrosis associated with furuncles and abscesses, whether superficial or the result of systemic infection. More recently, it was shown that an important host cell receptor for α-toxin is A disintegrin and metalloprotease (ADAM) 10 (Inoshima *et al.*, 2011). This was shown by the differential susceptibility of rabbit and human erythrocytes to α-toxin, combined with immunoprecipitation of ADAM10 with α-toxin.

ADAM10 is responsible for cleaving the ectodomain of a large number of host proteins, including members of the cadherins, epithelial growth factor, syndecans, cytokines, and many others (Berube and Bubeck Wardenburg, 2013). Many of the extracellular and intracellular cleavage products are important signalling molecules, perhaps the most important of which lead to high inflammation. While low levels of inflammation may be important in early events in innate immunity, uncontrolled inflammation, as occurs in the presence of α-toxin, opens skin and mucosal barriers and high inflammation in lung tissue can lead to lethal pneumonia. α-toxin has been shown in murine models to be important in necrotizing pneumonia, and vaccination against the toxin in multiple models provides at least partial protection (Bubeck Wardenburg *et al.*, 2007; Bubeck Wardenburg and Schneewind, 2008; Kennedy *et al.*, 2010).

Relative to epithelial cells, including skin, pulmonary and vaginal, the key target of α-toxin is E-cadherin, an important substrate for ADAM10 (Anderson *et al.*, 2012; Bubeck Wardenburg *et al.*, 2007; Bubeck Wardenburg and Schneewind, 2008; Kennedy *et al.*, 2010). Cleavage of E-cadherin disrupts epithelial barrier function, resulting in inflammatory oedema and cell necrosis.

β-Toxin

β-Toxin is classically referred to as the hot–cold haemolysin, in which incubation of erythrocytes at 37°C prepares them for lysis at 4°C. This artefact, combined with the demonstration that the majority of *S. aureus* strains have bacteriophage inactivation of the β-toxin gene, led investigators to believe the toxin has little role in human disease. More recent data, however, refute those claims. For example, nearly all USA200 strains of *S. aureus* that concomitantly produce TSST-1 but minimal α-toxin, produce high levels of β-toxin. Additionally, it has been shown that all clonal groups of *S. aureus*, despite the presence of the β-toxin-inactivating bacteriophage, have colony variants that produce the toxin, particularly *in vivo* (Salgado-Pabon *et al.*, 2014). *In vivo*, as tested in a rabbit model of infective endocarditis, β-toxin increases the severity of disease (Salgado-Pabon *et al.*, 2014).

β-Toxin is a neutral sphingomyelinase as shown recently by Huseby *et al.* (2010). This sphingomyelinase activity is critical for erythrocyte lysis, but more importantly leads to killing of immune cells. It should also be considered that sphingomyelin cleavage products are also likely to have immunosuppressive effects. In their description of the structure of β-toxin, the Huseby group showed that the toxin also has biofilm ligase activity, with capacity to covalently link extracellular DNA, as is postulated to be in biofilms to proteins (Huseby *et al.*, 2010). While the catalytic site of the sphingomyelinase activity is known, the amino acid residues that contribute to biofilm ligase activity remain unknown.

γ-Toxin and leucocidins

Readers are encouraged to read the recent review by DuMont and Torres related to these cytotoxins (DuMont and Torres, 2014). Among the most recently studied cytotoxins are γ-toxin and related leucocidins. These cytotoxins are commonly referred to as bi-component cytotoxins in which two different proteins oligomerize to form pores (unlike the homoheptamer pore-forming cytotoxin α-toxin). The two different proteins are known as S-class (slow-eluting) and F-class (fast-eluting) from chromatographic columns. Individual *S. aureus* strains may produce up to five bio-component cytotoxins: (1) γ-toxin AB, (2) γ-toxin BC, (3) leucocidin ED, (4) leucocidin S-PV/F-PV [also known as Panton–Valentine (PV) leucocidin] and (5) leucocidin AB (also known as leucocidin GH). Nearly all clinical isolates of *S. aureus* have the genes for γ-toxin. Similarly, approximately 90% have leucocidin ED genes, 100% have leucocidin AB genes, but only 5% have the PV leucocidin genes. When methicillin-resistant *S. aureus* isolates of USA300 and USA400 clonal groups were examined, 80–85% have the PV leucocidin genes; this toxin has been proposed to be important but not required for production of necrotizing pneumonia (Bubeck Wardenburg *et al.*, 2007, 2008; Diep *et al.*, 2008; Labandeira-Rey *et al.*, 2007). Based on structure–function studies, the bi-component cytotoxins are considered octamer pore formers (Jayasinghe and Bayley, 2005; Yamashita *et al.*, 2011), with each subunit having a molecular weight of approximately 34,000 Da. The S-class components appear to target host cell receptors first, followed by F-class proteins.

The bi-component toxins were originally described as haemolysins. However, like α- and β-toxin effects in humans and other animals, it is unlikely that haemolysis is of importance; this investigator believes that humans who succumb to *S. aureus* do not succumb to systemic haemolysis. In contrast, the more likely important effects of these bi-component toxins are inactivation of immune cells. Even in this regard, there is great variability of immune cell killing, both across animal species and within humans, by the five members of this family of cytotoxins. Unfortunately, there has not been a thorough side-by-side examination of all of these toxins to determine which have greater activity and against which cell types are they most active.

Multiple protein receptors have been identified for these bi-component cytotoxins. Leucocidin ED targets the chemokine receptor CCR5, and this targeting leads to death of human lymphocytes, macrophages, and dendritic cells; however, the toxin also kills PMNs by a CCR5-independent mechanism (Alonzo *et al.*, 2013). PV leucocidin appears to target the chemokine receptors C5aR and C5L2 found on PMNs and macrophages (Spaan *et al.*, 2013). It is considered likely that γ-toxin AB and BC will use the same chemokine receptors as leucocidin ED and PV leucocidin, based on sequence identity.

The receptor for leucocidin AB appears to be different than the above. It has been shown that the β-integrin Mac-1 (CD11b/CD18) is the receptor for this toxin on phagocytic cells (DuMont *et al.*, 2013).

It is my experience that doses of α-toxin as low as 0.1 μg/kg intravenously in Dutch-belted rabbits is uniformly lethal in 1 h. This effect may result from blockade of nerve repolarization (Bhakdi and Tranum-Jensen, 1991). Interestingly, the same effect is not observed in New Zealand white rabbits (unpublished data). In tests of γ-toxin AB, γ-toxin BC, and PV leucocidin, these toxins, although capable of causing dermonecrosis and vascular necrosis like α-toxin, are not lethal to Dutch-belted or New Zealand white rabbits in doses as high as 100 μg/kg.

Delta toxin and phenol soluble modulins (PSMs)

δ-Toxin has long been known as a haemolysin and cytotoxin for nearly any cell type, with activity apparently dependent on pore-formation or surfactant properties (Cheung *et al.*, 2014; Dinges *et al.*, 2000). The proteins may also form fibrillar, amyloid structures that may be important in biofilm formation (Schwartz and Boles, 2013). More recently, with the discovery of the *agr* global regulator system, it has been shown that δ-toxin is encoded by the gene giving rise to the RNAIII regulatory molecule (Novick, 2003); thus δ-toxin production depends on *agr*. δ-Toxin is a low molecular weight cytotoxin of 26 amino acids.

The term PSM was coined by the Klebanoff group as a series of three peptides PSMα, PSMβ, and PSMγ (δ-toxin) (Mehlin *et al.*, 1999). PSMs can be classified according to their lengths with PSMα variants having sizes of 20–25 amino acids and PSMβs having lengths of 43–45 amino acids (Cheung *et al.*, 2014). All PSMs form an amphipathic α-helix. Their host cell receptor appears to be FPR2 (Kretschmer *et al.*, 2010). Many studies have been done recently that show that PSMs contribute importantly to *S. aureus* diseases as studies in animal models; readers are referred to the following review (Cheung *et al.*, 2014).

ε-Cytotoxin

Recently, we have described a new cytotoxin that kills human keratinocytes and is both cytotoxic and pro-inflammatory in rabbit skin; we have designated this secreted cytotoxin ε-cytotoxin (Merriman *et al.*, 2015). This toxin was discovered during a survey of secreted proteins produced by surgical site infection *S. aureus*. The protein is produced by the majority of *S. aureus* strains as a 16,000 Da molecular weight protein with no sequence similarity to other proteins. Future studies will provide further characterization of the protein, both functionally and in regard to its importance in infections.

Third line of *S. aureus* defence against the human immune system

S. aureus strains that cause human diseases have the ability to produce capsules, make pigments, and produce polysaccharide intercellular adhesion.

Capsular polysaccharides

The function of capsules is to enhance virulence by making *S. aureus* resistant to phagocytosis (reviewed in O'Riordan and Lee, 2004). Serotyping studies indicate that the majority of human isolates of *S. aureus* produce capsule types 5 (25%) and 8 (50%). Strains that do not react with antisera to capsule types 1, 2, 5 and 8 may be referred to as non-typable. The structures of the capsules of these four types are known.

Capsule types 5 and 8 are most prevalent in clinical isolates. Strains that produce these capsules do not appear mucoid and the capsules are essentially micro-capsules, unlike the larger capsules associated with types 1 and 2. Production of types 5 and 8 capsules is amplified under iron limitation and growth on solid medium, and is inhibited by anaerobiasis and alkaline conditions, and at least for type 5 capsule CO_2. Both of the capsules are produced *in vivo* in animal models of human diseases, but production is a function of body site of infection. For example, capsule can be detected in strains from infective endocarditis vegetations (Arbeit and Nelles, 1987; Lee *et al.*, 1993), but capsule is not detected in microbes from lung or nasal polyp tissue (O'Riordan and Lee, 2004).

There have been a large number of studies performed to assess the role of capsule types 5 and 8 in virulence. While there have been conflicting data, there are ample data to suggest that these two capsule types interfere with phagocytosis *in vivo* and in this way enhance colonization and increase lethality approximately 10-fold in animal models [extensively reviewed in O'Riordan and Lee (2004)]. There have been clinical studies done in humans in attempt to vaccinate patients against these capsules; unfortunately, these studies indicate conjugate capsule vaccines are not highly immunogenic, and patients do not appear protected from infection.

Polysaccharide intercellular adhesin (PIA)

The formation of biofilms by *S. aureus* is an important aspect of many infections, most notably those associated with indwelling medical devices, infective endocarditis, and osteomyelitis. Other infections, such as vaginal infections in the presence of tampons leading to TSS may be viewed as short-term biofilms since mucosal epithelial tissue sloughs in periods of 8–12 h. Additionally, tampons are changed with biofilm removal in 4- to 8-h periods. The major components of biofilms are polysaccharides, such as PIA, extracellular DNA, and various proteins, dependent on bacterial strain. PIA is a poly *N*-acetyl-glucosamine which is synthesized by 4 genes in an operon, referred to as *icaADBC* (see review by Cue *et al.*, 2012). Additionally, another gene designated *icaR* functions as a repressor. PIA production is induced by high salt, high glucose, and ethanol. Although produced by *S. aureus* strains, PIA appears to be more highly produced in *S. epidermidis* strains in association with biofilms.

Pigment

Carotinoid pigment production defines the golden colour of *S. aureus*, though *S. aureus* strains may vary from dark brown to white; for example, the majority if not all of USA200 methicillin-resistant *S. aureus* strains from human infections lack any detectable pigment. Isogenic mutants with disrupted ability to produce carotinoid pigments are more easily susceptible to oxidative killing, have impaired neutrophil survival, and are less pathogenic in a mouse subcutaneous model of infection, compared to wild-type organisms (Liu *et al.*, 2005; Liu and Nizet, 2009).

References

Acharya, K.R., Passalacqua, E.F., Jones, E.Y., Harlos, K., Stuart, D.I., Brehm, R.D., and Tranter, H.S. (1994). Structural basis of superantigen action inferred from crystal structure of toxic-shock syndrome toxin-1. Nature *367*, 94–97.

Adem, P.V., Montgomery, C.P., Husain, A.N., Koogler, T.K., Arangelovich, V., Humilier, M., Boyle-Vavra, S., and Daum, R.S. (2005). *Staphylococcus aureus* sepsis and the Waterhouse–Friderichsen syndrome in children. N. Engl. J. Med. 353, 1245–1251.

Alonzo, F., Kozhaya, L., Rawlings, S.A., Reyes-Robles, T., DuMont, A.L., Myszka, D.G., Landau, N.R., Unutmaz, D., and Torres, V.J. (2013). CCR5 is a receptor for *Staphylococcus aureus* leucotoxin ED. Nature *493*, 51–55.

Altemeier, W.A., Lewis, S., Schlievert, P.M., and Bjornson, H.S. (1981). Studies of the staphylococcal causation of toxic shock syndrome. Surg. Gynecol. Obstet. *153*, 481–485.

Altemeier, W.A., Lewis, S.A., Schlievert, P.M., Bergdoll, M.S., Bjornson, H.S., Staneck, J.L., and Crass, B.A. (1982). *Staphylococcus aureus* associated with toxic shock syndrome: phage typing and toxin capability testing. Ann. Intern. Med. *96* 4, 978–982.

Amagai, M., Matsuyoshi, N., Wang, Z.H., Andl, C., and Stanley, J.R. (2000). Toxin in bullous impetigo and staphylococcal scalded-skin syndrome targets desmoglein 1. Nat. Med. *6*, 1275–1277.

Anderson, M.J., Lin, Y.C., Gillman, A.N., Parks, P.J., Schlievert, P.M., and Peterson, M.L. (2012). Alpha-toxin promotes *Staphylococcus aureus* mucosal biofilm formation. Front. Cell. Infect. Microbiol. 2, 64.

Antonsson, P., Wingren, A.G., Hansson, J., Kalland, T., Varga, M., and Dohlsten, M. (1997). Functional characterization of the interaction between the superantigen staphylococcal enterotoxin A and the TCR. J. Immunol. *158*, 4245–4251.

Arad, G., Levy, R., Hillman, D., and Kaempfer, R. (2000). Superantigen antagonist protects against lethal shock and defines a new domain for T-cell activation. Nat. Med. *6*, 414–421.

Arad, G., Levy, R., Nasie, I., Hillman, D., Rotfogel, Z., Barash, U., Supper, E., Shpilka, T., Minis, A., and Kaempfer, R. (2011). Binding of superantigen toxins into the CD28 homodimer interface is essential for induction of cytokine genes that mediate lethal shock. PLOS Biol. *9*, e1001149.

Arbeit, R.D., and Nelles, M.J. (1987). Capsular polysaccharide antigenemia in rats with experimental endocarditis due to *Staphylococcus aureus*. J. Infect. Dis. *155*, 242–246.

Assimacopoulos, A.P., Strandberg, K.L., Rotschafer, J.H., and Schlievert, P.M. (2009). Extreme pyrexia and rapid death due to *Staphylococcus aureus* infection: analysis of 2 cases. Clin. Infect. Dis. *48*, 612–614.

Bashore, T.M., Cabell, C., and Fowler, V. (2006). Update on infective endocarditis. Curr. Probl. Cardiol. *31*, 274–352.

Bayles, K.W., and Iandolo, J.J. (1989). Genetic and molecular analyses of the gene encoding staphylococcal enterotoxin D. J. Bacteriol. *171*, 4799–4806.

Becker, K., Friedrich, A.W., Lubritz, G., Weilert, M., Peters, G., and Von Eiff, C. (2003). Prevalence of genes encoding pyrogenic toxin superantigens and exfoliative toxins among strains of *Staphylococcus aureus* isolated from blood and nasal specimens. J. Clin. Microbiol. *41*, 1434–1439.

Bergdoll, M.S. (1988). Monkey feeding test for staphylococcal enterotoxin. Methods Enzymol. *165*, 324–333.

Bergdoll, M.S., Crass, B.A., Reiser, R.F., Robbins, R.N., and Davis, J.P. (1981). A new staphylococcal enterotoxin, enterotoxin F, associated with toxic-shock-syndrome *Staphylococcus aureus* isolates. Lancet *1*, 1017–1021.

Bergdoll, M.S., and Schlievert, P.M. (1984). Toxic-shock syndrome toxin. Lancet *ii*, 691.

Bernheimer, A.W., and Schwartz, L.L. (1963). Isolation and composition of staphylococcal alpha toxin. J. Gen. Microbiol. *30*, 455–468.

Bernheimer, A.W., and Schwartz, L.L. (1965). Effect of staphylococcal and other bacterial toxins on platelets *in vitro*. J. Pathol. Bacteriol. *89*, 209–223.

Berube, B.J., and Bubeck Wardenburg, J. (2013). *Staphylococcus aureus* α-toxin: nearly a century of intrigue. Toxins *5*, 1140–1166.

Betley, M.J., Löfdahl, S., Kreiswirth, B.N., Bergdoll, M.S., and Novick, R.P. (1984). Staphylococcal enterotoxin A gene is associated with a variable genetic element. Proc. Natl. Acad. Sci. U.S.A. *81*, 5179–5183.

Bhakdi, S., Bayley, H., Valeva, A., Walev, I., Walker, B., Kehoe, M., and Palmer, M. (1996). Staphylococcal alpha-toxin, streptolysin-O, and Escherichia coli hemolysin: prototypes of pore-forming bacterial cytolysins. Arch. Microbiol. *165*, 73–79.

Bhakdi, S., and Tranum-Jensen, J. (1991). Alpha-toxin of *Staphylococcus aureus*. Microbiol. Rev. *55*, 733–751.

Boguniewicz, M., and Leung, D.Y. (2011). Atopic dermatitis: a disease of altered skin barrier and immune dysregulation. Immunol. Rev. *242*, 233–246.

Boguniewicz, M., and Leung, D.Y. (2010). Recent insights into atopic dermatitis and implications for management of infectious complications. J. Allergy. Clin. Immunol. *125*, 4–13.

Bohach, G.A., Fast, D.J., Nelson, R.D., and Schlievert, P.M. (1990). Staphylococcal and streptococcal pyrogenic toxins involved in toxic shock syndrome and related illnesses. Crit. Rev. Microbiol. *17*, 251–272.

Brauweiler, A.M., Bin, L., Kim, B.E., Oyoshi, M.K., Geha, R.S., Goleva, E., and Leung, D.Y.M. (2013). Filaggrin dependent secretion of sphingomyelinase protects against staphylococcal alpha toxin induced keratinocyte death. J. Allegy Clin. Immunol. *131*, 421–427.

Breshears, L.M., Schlievert, P.M., and Peterson, M.L. (2012). A disintegrin and metalloproteinase 17 (ADAM17) and epidermal growth factor receptor (EGFR) signaling drive the epithelial response to *Staphylococcus aureus* toxic shock syndrome toxin-1 (TSST-1). J. Biol. Chem. *287*, 32578–32587.

Brosnahan, A.J., Mantz, M.J., Squier, C.A., Peterson, M.L., and Schlievert, P.M. (2009). Cytolysins augment superantigen penetration of stratified mucosa. J. Immunol. *182*, 2364–2373.

Brosnahan, A.J., Schaefers, M.M., Amundson, W.H., Mantz, M.J., Squier, C.A., Peterson, M.L., and Schlievert, P.M. (2008). Novel toxic shock syndrome toxin-1 amino acids required for biological activity. Biochemistry *47*, 12995–13003.

Brosnahan, A.J., and Schlievert, P.M. (2011). Gram-positive bacterial superantigen outside-in signaling causes toxic shock syndrome. FEBS J. *278*, 4649–4667.

Bubeck Wardenburg, J., Bae, T., Otto, M., Deleo, F.R., and Schneewind, O. (2007). Poring over pores: alpha-hemolysin and Panton–Valentine leucocidin in *Staphylococcus aureus* pneumonia. Nat. Med. *13*, 1405–1406.

Bubeck Wardenburg, J., Palazzolo-Ballance, A.M., Otto, M., Schneewind, O., and DeLeo, F.R. (2008). Panton–Valentine leucocidin is not a virulence determinant in murine models of community-associated methicillin-resistant *Staphylococcus aureus* disease. J. Infect. Dis. *198*, 1166–1170.

Bubeck Wardenburg, J., and Schneewind, O. (2008). Vaccine protection against *Staphylococcus aureus* pneumonia. J. Exp. Med. *205*, 287–294.

Buonpane, R.A., Churchill, H.R., Moza, B., Sundberg, E.J., Peterson, M.L., Schlievert, P.M., and Kranz, D.M. (2007). Neutralization of staphylococcal enterotoxin B by soluble, high-affinity receptor antagonists. Nat. Med. *13*, 725–729.

Cassidy, P.S., and Harshman, S. (1973). The binding of staphylococcal 125I-alpha-toxin (B) to erythrocytes. J. Biol. Chem. *248*, 5545–5546.

Centers for Disease Control and Prevention (CDC). (1999). From the Centers for Disease Control and Prevention. Four pediatric deaths from community-acquired methicillin-resistant *Staphylococcus aureus*– Minnesota and North Dakota, 1997–1999. J.A.M.A. *282*, 1123–1125.

Centers for Disease Control and Prevention (CDC). (2003). Methicillin-resistant *Staphylococcus aureus* infections in correctional facilities--Georgia, California, and Texas, 2001–2003. MMWR. Morb. Mortal. Wkly Rep. *52*, 992–996.

Cheng, A.G., Missiakas, D., and Schneewind, O. (2014). The giant protein Ebh is a determinant of *Staphylococcus aureus* cell size and complement resistance. J. Bacteriol. *196*, 971–981.

Cheung, G.Y., Joo, H.S., Chatterjee, S.S., and Otto, M. (2014). Phenol-soluble modulins–critical determinants of staphylococcal virulence. FEMS Microbiol. Rev. *38*, 698–719.

Cone, L.A., Woodard, D.R., Byrd, R.G., Schulz, K., Kopp, S.M., and Schlievert, P.M. (1992). A recalcitrant, erythematous, desquamating disorder associated with toxin-producing staphylococci in patients with AIDS. J. Infect. Dis. *165*, 638–643.

Cue, D., Lei, M.G., and Lee, C.Y. (2012). Genetic regulation of the intercellular adhesion locus in staphylococci. Front. Cell. Infect. Microbiol. *2*, 38.

Dauwalder, O., Thomas, D., Ferry, T., Debard, A.L., Badiou, C., Vandenesch, F., Etienne, J., Lina, G., and Monneret, G. (2006). Comparative inflammatory properties of staphylococcal superantigenic enterotoxins SEA and SEG: implications for septic shock. J. Leucocyte Biol. *80*, 753–758.

Davis, C.C., Kremer, M.J., Schlievert, P.M., and Squier, C.A. (2003). Penetration of toxic shock syndrome toxin-1 across porcine vaginal mucosa ex vivo: permeability characteristics, toxin distribution, and tissue damage. Am. J. Obstet. Gynecol. *189*, 1785–1791.

Davis, J.P., Chesney, P.J., Wand, P.J., and LaVenture, M. (1980). Toxic-shock syndrome: epidemiologic features, recurrence, risk factors, and prevention. N. Engl. J. Med. *303*, 1429–1435.

de Azavedo, J.C., Foster, T.J., Hartigan, P.J., Arbuthnott, J.P., O'Reilly, M., Kreiswirth, B.N., and Novick, R.P. (1985). Expression of the cloned toxic shock syndrome toxin 1 gene (tst) *in vivo* with a rabbit uterine model. Infect. Immun. *50*, 304–309.

DeLeo, F.R., Kennedy, A.D., Chen, L., Bubeck Wardenburg, J., Kobayashi, S.D., Mathema, B., Braughton, K.R., Whitney, A.R., Villaruz, A.E., Martens, C.A., *et al.* (2011). Molecular differentiation of historic phage-type 80/81 and contemporary epidemic *Staphylococcus aureus*. Proc. Natl. Acad. Sci. U.S.A. *108*, 18091–18096.

DeVries, A.S., Lesher, L., Schlievert, P.M., Rogers, T., Villaume, L.G., Danila, R., and Lynfield, R. (2011). Staphylococcal toxic shock syndrome 2000–2006: epidemiology, clinical features, and molecular characteristics. PLoS One *6*, e22997.

Diep, B.A., Gill, S.R., Chang, R.F., Phan, T.H., Chen, J.H., Davidson, M.G., Lin, F., Lin, J., Carleton, H.A., Mongodin, E.F., *et al.* (2006). Complete genome sequence of USA300, an epidemic clone of community-acquired meticillin-resistant *Staphylococcus aureus*. Lancet *367*, 731–739.

Diep, B.A., Palazzolo-Ballance, A.M., Tattevin, P., Basuino, L., Braughton, K.R., Whitney, A.R., Chen, L., Kreiswirth, B.N., Otto, M., DeLeo, F.R., *et al.* (2008). Contribution of Panton–Valentine leucocidin in community-associated methicillin-resistant *Staphylococcus aureus* pathogenesis. PloS ONE 3, e3198.

Dinges, M.M., Orwin, P.M., and Schlievert, P.M. (2000). Exotoxins of *Staphylococcus aureus*. Clin. Microbiol. Rev. *13*, 16–34.

Dinges, M.M., and Schlievert, P.M. (2001). Comparative analysis of lipopolysaccharide-induced tumor necrosis factor alpha activity in serum and lethality in mice and rabbits pretreated with the staphylococcal superantigen toxic shock syndrome toxin 1. Infect. Immun. *69*, 7169–7172.

DuMont, A.L., and Torres, V.J. (2014). Cell targeting by the *Staphylococcus aureus* pore-forming toxins: it's not just about lipids. Trends. Microbiol. *22*, 21–27.

DuMont, A.L., Yoong, P., Day, C.J., Alonzo, F., McDonald, W.H., Jennings, M.P., and Torres, V.J. (2013). *Staphylococcus aureus* LukAB cytotoxin kills human neutrophils by targeting the CD11b subunit of the integrin Mac-1. Proc. Natl. Acad. Sci. U.S.A. *110*, 10794–10799.

Entenza, J.M., Foster, T.J., Ni Eidhin, D., Vaudaux, P., Francioli, P., and Moreillon, P. (2000). Contribution of clumping factor B to pathogenesis of experimental endocarditis due to *Staphylococcus aureus*. Infect. Immun. *68*, 5443–5446.

Ferry, T., Thomas, D., Genestier, A.L., Bes, M., Lina, G., Vandenesch, F., and Etienne, J. (2005). Comparative prevalence of superantigen genes in *Staphylococcus aureus* isolates causing sepsis with and without septic shock. Clin. Infect. Dis. *41*, 771–777.

Fey, P.D., Saïd-Salim, B., Rupp, M.E., Hinrichs, S.H., Boxrud, D.J., Davis, C.C., Kreiswirth, B.N., and Schlievert, P.M. (2003). Comparative molecular analysis of community- or hospital-acquired methicillin-resistant *Staphylococcus aureus*. Antimicrob. Agents Chemother. *47*, 196–203.

Fields, B.A., Malchiodi, E.L., Li, H., Ysern, X., Stauffacher, C.V., Schlievert, P.M., Karjalainen, K., and Mariuzza, R.A. (1996). Crystal structure of a T-cell receptor beta-chain complexed with a superantigen. Nature *384*, 188–192.

Fischetti, V.A., Chapman, F., Kakani, R., James, J., Grun, E., and Zabriskie, J.B. (1989). Role of air in growth and production of toxic shock syndrome toxin 1 by *Staphylococcus aureus* in experimental cotton and rayon tampons. Rev. Infect. Dis. *11* (Suppl. 1), S176–S181.

Fleischer, B., Gerlach, D., Fuhrmann, A., and Schmidt, K.H. (1995). Superantigens and pseudosuperantigens of gram-positive cocci. Med. Microbiol. Immunol. *184*, 1–8.

Foster, T.J., and Höök, M. (1998). Surface protein adhesins of *Staphylococcus aureus*. Trends. Microbiol. *6*, 484–488.

Francis, J.S., Doherty, M.C., Lopatin, U., Johnston, C.P., Sinha, G., Ross, T., Cai, M., Hansel, N.N., Perl, T., Ticehurst, J.R., *et al.* (2005). Severe community-onset pneumonia in healthy adults caused by methicillin-resistant *Staphylococcus aureus* carrying the Panton–Valentine leucocidin genes. Clin. Infect. Dis. *40*, 100–107.

Gaventa, S., Reingold, A.L., Hightower, A.W., Broome, C.V., Schwartz, B., Hoppe, C., Harwell, J., Lefkowitz, L.K., Makintubee, S., and Cundiff, D.R. (1989). Active surveillance for toxic shock syndrome in the United States, 1986. Rev. Infect. Dis. *11* (Suppl. 1), S28–S34.

Giantonio, B.J., Alpaugh, R.K., Schultz, J., McAleer, C., Newton, D.W., Shannon, B., Guedez, Y., Kotb, M., Vitek, L., Persson, R., *et al.* (1997). Superantigen-based immunotherapy: a phase I trial of PNU-214565, a monoclonal antibody-staphylococcal enterotoxin A recombinant fusion protein, in advanced pancreatic and colorectal cancer. J. Clin. Oncol. *15*, 1994–2007.

Gouaux, J.E., Braha, O., Hobaugh, M.R., Song, L., Cheley, S., Shustak, C., and Bayley, H. (1994). Subunit stoichiometry of staphylococcal alpha-hemolysin in crystals and on membranes: a heptameric transmembrane pore. Proc. Natl. Acad. Sci. U.S.A. *91*, 12828–12831.

Greene, C., McDevitt, D., Francois, P., Vaudaux, P.E., Lew, D.P., and Foster, T.J. (1995). Adhesion properties of mutants of *Staphylococcus aureus* defective in fibronectin-binding proteins and studies on the expression of fnb genes. Mol. Microbiol. *17*, 1143–1152.

Günther, S., Varma, A.K., Moza, B., Kasper, K.J., Wyatt, A.W., Zhu, P., Rahman, A.K., Li, Y., Mariuzza, R.A., McCormick, J.K., *et al.* (2007). A novel loop domain in superantigens extends their T cell receptor recognition site. J. Mol. Biol. *371*, 210–221.

Guttman-Yassky, E., Nograles, K.E., and Krueger, J.G. (2011). Contrasting pathogenesis of atopic dermatitis and psoriasis–part I: clinical and pathologic concepts. J. Allergy. Clin. Immunol. *127*, 1110–1118.

Hamad, A.R., Marrack, P., and Kappler, J.W. (1997). Transcytosis of staphylococcal superantigen toxins. J. Exp. Med. *185*, 1447–1454.

Harris, T.O., Grossman, D., Kappler, J.W., Marrack, P., Rich, R.R., and Betley, M.J. (1993). Lack of complete correlation between emetic and T-cell-stimulatory activities of staphylococcal enterotoxins. Infect. Immun. *61*, 3175–3183.

Hennekinne, J.A., De Buyser, M.L., and Dragacci, S. (2012). *Staphylococcus aureus* and its food poisoning toxins: characterization and outbreak investigation. FEMS Microbiol. Rev. *36*, 815–836.

Herold, B.C., Immergluck, L.C., Maranan, M.C., Lauderdale, D.S., Gaskin, R.E., Boyle-Vavra, S., Leitch, C.D., and Daum, R.S. (1998). Community-acquired methicillin-resistant *Staphylococcus aureus* in children with no identified predisposing risk. J.A.M.A. *279*, 593–598.

Hill, D.R., Brunner, M.E., Schmitz, D.C., Davis, C.C., Flood, J.A., Schlievert, P.M., Wang-Weigand, S.Z., and Osborn, T.W. (2005). *In vivo* assessment of human vaginal oxygen and carbon dioxide levels during and post menses. J. Appl. Physiol. *99*, 1582–1591.

Hovde, C.J., Marr, J.C., Hoffmann, M.L., Hackett, S.P., Chi, Y.I., Crum, K.K., Stevens, D.L., Stauffacher, C.V., and Bohach, G.A. (1994). Investigation of the role of the disulphide bond in the activity and structure of staphylococcal enterotoxin C1. Mol. Microbiol. *13*, 897–909.

Hu, D.L., Omoe, K., Shimoda, Y., Nakane, A., and Shinagawa, K. (2003). Induction of emetic response to staphylococcal enterotoxins in the house musk shrew (Suncus murinus). Infect. Immun. *71*, 567–570.

Hu, D.L., Omoe, K., Shimura, H., Ono, K., Sugii, S., and Shinagawa, K. (1999). Emesis in the shrew mouse (Suncus murinus) induced by peroral and intraperitoneal administration of staphylococcal enterotoxin A. J. Food Prot. *62*, 1350–1353.

Hu, D.L., Suga, S., Omoe, K., Abe, Y., Shinagawa, K., Wakui, M., and Nakane, A. (2005). Staphylococcal enterotoxin A modulates intracellular Ca2+ signal pathway in human intestinal epithelial cells. FEBS Lett. *579*, 4407–4412.

Hu, D.L., Zhu, G., Mori, F., Omoe, K., Okada, M., Wakabayashi, K., Kaneko, S., Shinagawa, K., and Nakane, A. (2007). Staphylococcal enterotoxin induces emesis through increasing serotonin release in intestine and it is downregulated by cannabinoid receptor 1. Cell. Microbiol. *9*, 2267–2277.

Huseby, M., Shi, K., Brown, C.K., Digre, J., Mengistu, F., Seo, K.S., Bohach, G.A., Schlievert, P.M., Ohlendorf, D.H., and Earhart, C.A. (2007). Structure and biological activities of beta toxin from *Staphylococcus aureus*. J. Bacteriol. *189*, 8719–8726.

Huseby, M.J., Kruse, A.C., Digre, J., Kohler, P.L., Vocke, J.A., Mann, E.E., Bayles, K.W., Bohach, G.A., Schlievert, P.M., Ohlendorf, D.H., *et al.* (2010). Beta toxin catalyzes formation of nucleoprotein matrix in staphylococcal biofilms. Proc. Natl. Acad. Sci. U.S.A. *107*, 14407–14412.

Inoshima, I., Inoshima, N., Wilke, G.A., Powers, M.E., Frank, K.M., Wang, Y., and Bubeck Wardenburg, J. (2011). A *Staphylococcus aureus* pore-forming toxin subverts the activity of ADAM10 to cause lethal infection in mice. Nat. Med. *17*, 1310–1314.

Jardetzky, T.S., Brown, J.H., Gorga, J.C., Stern, L.J., Urban, R.G., Chi, Y.I., Stauffacher, C., Strominger, J.L., and Wiley, D.C. (1994). Three-dimensional structure of a human class II histocompatibility molecule complexed with superantigen. Nature *368*, 711–718.

Jarraud, S., Cozon, G., Vandenesch, F., Bes, M., Etienne, J., and Lina, G. (1999). Involvement of enterotoxins G and I in staphylococcal toxic shock syndrome and staphylococcal scarlet fever. J. Clin. Microbiol. *37*, 2446–2449.

Jayasinghe, L., and Bayley, H. (2005). The leucocidin pore: evidence for an octamer with four LukF subunits and four LukS subunits alternating around a central axis. Protein Sci. *14*, 2550–2561.

John, C.C., Niermann, M., Sharon, B., Peterson, M.L., Kranz, D.M., and Schlievert, P.M. (2009). Staphylococcal toxic shock syndrome erythroderma is associated with superantigenicity and hypersensitivity. Clin. Infect. Dis. *49*, 1893–1896.

Jonsson, K., Signas, C., Muller, H.P., and Lindberg, M. (1991). Two different genes encode fibronectin binding proteins in *Staphylococcus aureus*. The complete nucleotide sequence and characterization of the second gene. Eur. J. Biochem. *202*, 1041–1048.

Kansal, R., Davis, C., Hansmann, M., Seymour, J., Parsonnet, J., Modern, P., Gilbert, S., and Kotb, M. (2007). Structural and functional properties of antibodies to the superantigen TSST-1 and their relationship to menstrual toxic shock syndrome. J. Clin. Immunol. *27*, 327–338.

Kawakami, Y., Tomimori, Y., Yumoto, K., Hasegawa, S., Ando, T., Tagaya, Y., Crotty, S., and Kawakami, T. (2009). Inhibition of NK cell activity by IL-17 allows vaccinia virus to induce severe skin lesions in a mouse model of eczema vaccinatum. J. Exp. Med. *206*, 1219–1225.

Kennedy, A.D., Bubeck Wardenburg, J., Gardner, D.J., Long, D., Whitney, A.R., Braughton, K.R., Schneewind, O., and DeLeo, F.R. (2010). Targeting of alpha-hemolysin by active or passive immunization decreases severity of USA300 skin infection in a mouse model. J. Infect. Dis. *202*, 1050–1058.

Kim, J., Urban, R.G., Strominger, J.L., and Wiley, D.C. (1994). Toxic shock syndrome toxin-1 complexed with a class II major histocompatibility molecule HLA-DR1. Science *266*, 1870–1874.

Kim, J.S., Kim, H.S., Song, W., Cho, H.C., Lee, K.M., and Kim, E.C. (2007). Molecular epidemiology of methicillin-resistant *Staphylococcus aureus* isolates with toxic shock syndrome toxin and staphylococcal enterotoxin C genes. Korean J. Lab. Med. *27*, 118–123.

Kim, Y.B., and Watson, D.W. (1970). A purified group A streptococcal pyrogenic exotoxin. Physiochemical and biological properties including the enhancement of susceptibility to endotoxin lethal shock. J. Exp. Med. *131*, 611–622.

Klevens, R.M., Morrison, M.A., Nadle, J., Petit, S., Gershman, K., Ray, S., Harrison, L.H., Lynfield, R., Dumyati, G., Townes, J.M., *et al.* (2007). Invasive methicillin-resistant *Staphylococcus aureus* infections in the United States. J.A.M.A. *298*, 1763–1771.

Koç, C., Gerlach, D., Beck, S., Peschel, A., Xia, G., and Stehle, T. (2015). Structural and enzymatic analysis of TarM glycosyltransferase from *Staphylococcus aureus* reveals an oligomeric protein specific for the glycosylation of wall teichoic acid. J. Biol. Chem. *290*, 9874–9885.

Kotb, M., Norrby-Teglund, A., McGeer, A., El-Sherbini, H., Dorak, M.T., Khurshid, A., Green, K., Peeples, J., Wade, J., Thomson, G., *et al.* (2002). An immunogenetic and molecular basis for differences in outcomes of invasive group A streptococcal infections. Nat. Med. *8*, 1398–1404.

Kotler, D.P., Sandkovsky, U., Schlievert, P.M., and Sordillo, E.M. (2007). Toxic shock-like syndrome associated with staphylococcal enterocolitis in an HIV-infected man. Clin. Infect. Dis. *44*, e121–3.

Kotzin, B.L., Leung, D.Y., Kappler, J., and Marrack, P. (1993). Superantigens and their potential role in human disease. Adv. Immunol. *54*, 99–166.

Kozono, H., Parker, D., White, J., Marrack, P., and Kappler, J. (1995). Multiple binding sites for bacterial superantigens on soluble class II MHC molecules. Immunity *3*, 187–196.

Kravitz, G.R., Dries, D.J., Peterson, M.L., and Schlievert, P.M. (2005). Purpura fulminans due to *Staphylococcus aureus*. Clin. Infect. Dis. *40*, 941–947.

Kretschmer, D., Gleske, A.K., Rautenberg, M., Wang, R., Köberle, M., Bohn, E., Schöneberg, T., Rabiet, M.J., Boulay, F., Klebanoff, S.J., *et al.* (2010). Human formyl peptide receptor 2 senses highly pathogenic *Staphylococcus aureus*. Cell Host Microbe *7*, 463–473.

Kuehnert, M.J., Kruszon-Moran, D., Hill, H.A., McQuillan, G., McAllister, S.K., Fosheim, G., McDougal, L.K., Chaitram, J., Jensen, B., Fridkin, S.K., *et al.* (2006). Prevalence of *Staphylococcus aureus* nasal colonization in the United States, 2001–2002. J. Infect. Dis. *193*, 172–179.

Kushnaryov, V.M., Bergdoll, M.S., MacDonald, H.S., Vellinga, J., and Reiser, R. (1984a). Study of staphylococcal toxic shock syndrome toxin in human epithelial cell culture. J. Infect. Dis. *150*, 535–545.

Kushnaryov, V.M., MacDonald, H.S., Reiser, R., and Bergdoll, M.S. (1984b). Staphylococcal toxic shock toxin specifically binds to cultured human epithelial cells and is rapidly internalized. Infect. Immun. *45*, 566–571.

Kushnaryov, V.M., MacDonald, H.S., Reiser, R.F., and Bergdoll, M.S. (1989). Reaction of toxic shock syndrome toxin 1 with endothelium of human umbilical cord vein. Rev. Infect. Dis. *11* (Suppl. 1), S282–S287.

Labandeira-Rey, M., Couzon, F., Boisset, S., Brown, E.L., Bes, M., Benito, Y., Barbu, E.M., Vazquez, V., Hook, M., Etienne, J., *et al.* (2007). *Staphylococcus aureus* Panton–Valentine leucocidin causes necrotizing pneumonia. Science *315*, 1130–1133.

Ladhani, S. (2001). Recent developments in staphylococcal scalded skin syndrome. Clin. Microbiol. Infect. *7*, 301–307.

Ladhani, S., Joannou, C.L., Lochrie, D.P., Evans, R.W., and Poston, S.M. (1999). Clinical, microbial, and biochemical aspects of the exfoliative toxins causing staphylococcal scalded-skin syndrome. Clin. Microbiol. Rev. *12*, 224–242.

Laouini, D., Kawamoto, S., Yalcindag, A., Bryce, P., Mizoguchi, E., Oettgen, H., and Geha, R.S. (2003). Epicutaneous sensitization with superantigen induces allergic skin inflammation. J. Allergy. Clin. Immunol. *112*, 981–987.

Le Loir, Y., Baron, F., and Gautier, M. (2003). *Staphylococcus aureus* and food poisoning. Genet. Mol. Res. *2*, 63–76.

Leder, L., Llera, A., Lavoie, P.M., Lebedeva, M.I., Li, H., Sékaly, R.P., Bohach, G.A., Gahr, P.J., Schlievert, P.M., Karjalainen, K., *et al.* (1998). A mutational analysis of the binding of staphylococcal enterotoxins B and C3 to the T cell receptor beta chain and major histocompatibility complex class II. J. Exp. Med. *187*, 823–833.

Lee, J.C., Takeda, S., Livolsi, P.J., and Paoletti, L.C. (1993). Effects of *in vitro* and *in vivo* growth conditions on expression of type 8 capsular polysaccharide by *Staphylococcus aureus*. Infect. Immun. *61*, 1853–1858.

Lee, P.K., Deringer, J.R., Kreiswirth, B.N., Novick, R.P., and Schlievert, P.M. (1991a). Fluid replacement protection of rabbits challenged subcutaneous with toxic shock syndrome toxins. Infect. Immun. *59*, 879–884.

Lee, P.K., Kreiswirth, B.N., Deringer, J.R., Projan, S.J., Eisner, W., Smith, B.L., Carlson, E., Novick, R.P., and Schlievert, P.M. (1992). Nucleotide sequences and biologic properties of toxic shock syndrome toxin 1 from ovine- and bovine-associated *Staphylococcus aureus*. J. Infect. Dis. *165*, 1056–1063.

Lee, P.K., and Schlievert, P.M. (1989). Quantification and toxicity of group A streptococcal pyrogenic exotoxins in an animal model of toxic shock syndrome-like illness. J. Clin. Microbiol. *27*, 1890–1892.

Lee, P.K., Vercellotti, G.M., Deringer, J.R., and Schlievert, P.M. (1991b). Effects of staphylococcal toxic shock syndrome toxin 1 on aortic endothelial cells. J. Infect. Dis. *164*, 711–719.

Leung, D.Y., Gately, M., Trumble, A., Ferguson-Darnell, B., Schlievert, P.M., and Picker, L.J. (1995). Bacterial superantigens induce T cell expression of the skin-selective homing receptor, the cutaneous lymphocyte-associated antigen, via stimulation of interleukin 12 production. J. Exp. Med. *181*, 747–753.

Leung, D.Y., Hanifin, J.M., Pariser, D.M., Barber, K.A., Langley, R.G., Schlievert, P.M., Abrams, B., and Hultsch, T. (2009). Effects of pimecrolimus cream 1% in the treatment of patients with atopic dermatitis who demonstrate a clinical insensitivity to topical corticosteroids: a randomized, multicentre vehicle-controlled trial. Br. J. Dermatol. *161*, 435–443.

Leung, D.Y., Harbeck, R., Bina, P., Reiser, R.F., Yang, E., Norris, D.A., Hanifin, J.M., and Sampson, H.A. (1993). Presence of IgE antibodies to staphylococcal exotoxins on the skin of patients with atopic dermatitis. Evidence for a new group of allergens. J. Clin. Invest. *92*, 1374–1380.

Li, H., Llera, A., Malchiodi, E.L., and Mariuzza, R.A. (1999). The structural basis of T cell activation by superantigens. Annu. Rev. Immunol. *17*, 435–466.

Li, H., Llera, A., Tsuchiya, D., Leder, L., Ysern, X., Schlievert, P.M., Karjalainen, K., and Mariuzza, R.A. (1998). Three-dimensional structure of the complex between a T cell receptor beta chain and the superantigen staphylococcal enterotoxin B. Immunity *9*, 807–816.

Lillibridge, C.B., Melish, M.E., and Glasgow, L.A. (1972). Site of action of exfoliative toxin in the staphylococcal scaled-skin syndrome. Pediatrics *50*, 728–738.

Lin, Y.C., Anderson, M.J., Kohler, P.L., Strandberg, K.L., Olson, M.E., Horswill, A.R., Schlievert, P.M., and Peterson, M.L. (2011). Proinflammatory exoprotein characterization of toxic shock syndrome *Staphylococcus aureus*. Biochemistry *50*, 7157–7167.

Lin, Z., Kotler, D.P., Schlievert, P.M., and Sordillo, E.M. (2010). Staphylococcal enterocolitis: forgotten but not gone? Dig. Dis. Sci. *55*, 1200–1207.

Lina, G., Bohach, G.A., Nair, S.P., Hiramatsu, K., Jouvin-Marche, E., Mariuzza, R., and International Nomenclature Committee for Staphylococcal Superantigens. (2004). Standard nomenclature for the superantigens expressed by Staphylococcus. J. Infect. Dis. *189*, 2334–2336.

Liu, G.Y., Essex, A., Buchanan, J.T., Datta, V., Hoffman, H.M., Bastian, J.F., Fierer, J., and Nizet, V. (2005). *Staphylococcus aureus* golden pigment impairs neutrophil killing and promotes virulence through its antioxidant activity. J. Exp. Med. *202*, 209–215.

Liu, G.Y., and Nizet, V. (2009). Color me bad: microbial pigments as virulence factors. Trends. Microbiol. *17*, 406–413.

Löfdahl, S., Guss, B., Uhlén, M., Philipson, L., and Lindberg, M. (1983). Gene for staphylococcal protein A. Proc. Natl. Acad. Sci. U.S.A. *80*, 697–701.

Lowy, F.D. (1998). *Staphylococcus aureus* infections. N. Engl. J. Med. *339*, 520–532.

MacDonald, K.L., Osterholm, M.T., Hedberg, C.W., Schrock, C.G., Peterson, G.F., Jentzen, J.M., Leonard, S.A., and Schlievert, P.M. (1987). Toxic shock syndrome. A newly recognized complication of influenza and influenzalike illness. JAMA *257*, 1053–1058.

Malachowa, N., Kohler, P.L., Schlievert, P.M., Chuang, O.N., Dunny, G.M., Kobayashi, S.D., Miedzobrodzki, J., Bohach, G.A., and Seo, K.S. (2011). Characterization of a *Staphylococcus aureus* surface virulence factor that promotes resistance to oxidative killing and infectious endocarditis. Infect. Immun. *79*, 342–352.

Marrack, P., and Kappler, J. (1990a). The staphylococcal enterotoxins and their relatives. Science *248*, 1066.

Marrack, P., and Kappler, J. (1990b). The staphylococcal enterotoxins and their relatives. Science *248*, 705–711.

Mattis, D.M., Spaulding, A.R., Chuang-Smith, O.N., Sundberg, E.J., Schlievert, P.M., and Kranz, D.M. (2013). Engineering a soluble high-affinity receptor domain that neutralizes staphylococcal enterotoxin C in rabbit models of disease. Protein Eng. Des. Sel. *26*, 133–142.

McCormick, J., M.L Peterson, and P.M Schlievert (2006). Toxins and superantigens of group A streptococci. In Gram Positive Pathogens, R.N. V.A. Fischetti, J.J. Ferretti, D.A. Portnoy, and J.I. Rood, eds. (Washington, DC: ASM Press), pp. 47–58.

McCormick, J.K., Bohach, G.A., and Schlievert, P.M. (2003a). Pyrogenic, lethal, and emetic properties of superantigens in rabbits and primates. Methods Mol. Biol. *214*, 245–253.

McCormick, J.K., Tripp, T.J., Llera, A.S., Sundberg, E.J., Dinges, M.M., Mariuzza, R.A., and Schlievert, P.M. (2003b). Functional analysis of the TCR binding domain of toxic shock syndrome toxin-1 predicts further diversity in MHC class II/superantigen/TCR ternary complexes. J. Immunol. *171*, 1385–1392.

McCormick, J.K., Yarwood, J.M., and Schlievert, P.M. (2001). Toxic shock syndrome and bacterial superantigens: an update. Annu. Rev. Microbiol. *55*, 77–104.

Mehlin, C., Headley, C.M., and Klebanoff, S.J. (1999). An inflammatory polypeptide complex from Staphylococcus epidermidis: isolation and characterization. J. Exp. Med. *189*, 907–918.

Melish, M.E., and Glasgow, L.A. (1970). The staphylococcal scalded-skin syndrome. N. Engl. J. Med. *282*, 1114–1119.

Melish, M.E., Glasgow, L.A., and Turner, M.D. (1972). The staphylococcal scalded-skin syndrome: isolation and partial characterization of the exfoliative toxin. J. Infect. Dis. *125*, 129–140.

Melish, M.E., Murata, S., Fukunaga, C., Frogner, K., and McKissick, C. (1989). Vaginal tampon model for toxic shock syndrome. Rev. Infect. Dis. *11* (Suppl. 1), S238-S246.

Merriman, J.A., Klingelhutz, A.J., Diekema, D.J., Leung, D.Y., and Schlievert, P.M. (2015). Novel *Staphylococcus aureus* secreted protein alters keratinocyte proliferation and elicits a pro-inflammatory response *in vitro* and *in vivo*. Biochemistry *54*, 4855–4862.

Mitchell, D.T., Levitt, D.G., Schlievert, P.M., and Ohlendorf, D.H. (2000). Structural evidence for the evolution of pyrogenic toxin superantigens. J. Mol. Evol. *51*, 520–531.

Monday, S.R., Vath, G.M., Ferens, W.A., Deobald, C., Rago, J.V., Gahr, P.J., Monie, D.D., Iandolo, J.J., Chapes, S.K., Davis, W.C., *et al.* (1999). Unique superantigen activity of staphylococcal exfoliative toxins. J. Immunol. *162*, 4550–4559.

Moreillon, P., Entenza, J.M., Francioli, P., McDevitt, D., Foster, T.J., François, P., and Vaudaux, P. (1995). Role of *Staphylococcus aureus* coagulase and clumping factor in pathogenesis of experimental endocarditis. Infect. Immun. *63*, 4738–4743.

Moza, B., Varma, A.K., Buonpane, R.A., Zhu, P., Herfst, C.A., Nicholson, M.J., Wilbuer, A.K., Seth, N.P., Wucherpfennig, K.W., McCormick, J.K., *et al.* (2007). Structural basis of T-cell specificity and activation by the bacterial superantigen TSST-1. EMBO J. *26*, 1187–1197.

Nagao, M., Okamoto, A., Yamada, K., Hasegawa, T., Hasegawa, Y., and Ohta, M. (2009). Variations in amount of TSST-1 produced by clinical methicillin resistant *Staphylococcus aureus* (MRSA) isolates and allelic variation in accessory gene regulator (agr) locus. BMC Microbiol. 9, 52.

Navarre, W.W., and Schneewind, O. (1999). Surface proteins of gram-positive bacteria and mechanisms of their targeting to the cell wall envelope. Microbiol. Mol. Biol. Rev. *63*, 174–229.

Nienaber, J.J., Sharma Kuinkel, B.K., Clarke-Pearson, M., Lamlertthon, S., Park, L., Rude, T.H., Barriere, S., Woods, C.W., Chu, V.H., Marin, M., *et al.* (2011). Methicillin-susceptible *Staphylococcus aureus* endocarditis isolates are associated with clonal complex 30 genotype and a distinct repertoire of enterotoxins and adhesins. J. Infect. Dis. *204*, 704–713.

Novick, R.P. (2003). Autoinduction and signal transduction in the regulation of staphylococcal virulence. Mol. Microbiol. *48*, 1429–1449.

O'Reilly, M., Kreiswirth, B., and Foster, T.J. (1990). Cryptic alpha-toxin gene in toxic shock syndrome and septicaemia strains of *Staphylococcus aureus*. Mol. Microbiol. *4*, 1947–1955.

O'Riordan, K., and Lee, J.C. (2004). *Staphylococcus aureus* capsular polysaccharides. Clin. Microbiol. Rev. *17*, 218–234.

Orwin, P.M., Fitzgerald, J.R., Leung, D.Y., Gutierrez, J.A., Bohach, G.A., and Schlievert, P.M. (2003). Characterization of *Staphylococcus aureus* enterotoxin L. Infect. Immun. *71*, 2916–2919.

Orwin, P.M., Leung, D.Y., Donahue, H.L., Novick, R.P., and Schlievert, P.M. (2001). Biochemical and biological properties of Staphylococcal enterotoxin K. Infect. Immun. *69*, 360–366.

Orwin, P.M., Leung, D.Y., Tripp, T.J., Bohach, G.A., Earhart, C.A., Ohlendorf, D.H., and Schlievert, P.M. (2002). Characterization of a novel staphylococcal enterotoxin-like superantigen, a member of the group V subfamily of pyrogenic toxins. Biochemistry *41*, 14033–14040.

Osterholm, M.T., Davis, J.P., Gibson, R.W., Mandel, J.S., Wintermeyer, L.A., Helms, C.M., Forfang, J.C., Rondeau, J., and Vergeront, J.M. (1982). Tri-state toxic-state syndrome study. I. Epidemiologic findings. J. Infect. Dis. *145*, 431–440.

Ou, L.S., Goleva, E., Hall, C., and Leung, D.Y. (2004). T regulatory cells in atopic dermatitis and subversion of their activity by superantigens. J. Allergy Clin. Immunol. *113*, 756–763.

Papageorgiou, A.C., Collins, C.M., Gutman, D.M., Kline, J.B., O'Brien, S.M., Tranter, H.S., and Acharya, K.R. (1999). Structural basis for the recognition of superantigen streptococcal pyrogenic exotoxin A (SpeA1) by MHC class II molecules and T-cell receptors. EMBO J. *18*, 9–21.

Parsonnet, J. (1998). Case definition of staphylococcal TSS: a proposed revision incorporating laboratory findings. International Congress and Symposium Series *229*, 15.

Parsonnet, J., Gillis, Z.A., Richter, A.G., and Pier, G.B. (1987). A rabbit model of toxic shock syndrome that uses a constant, subcutaneous infusion of toxic shock syndrome toxin 1. Infect. Immun. *55*, 1070–1076.

Parsonnet, J., Hansmann, M.A., Delaney, M.L., Modern, P.A., Dubois, A.M., Wieland-Alter, W., Wissemann, K.W., Wild, J.E., Jones, M.B., Seymour, J.L., *et al.* (2005). Prevalence of toxic shock syndrome toxin 1-producing *Staphylococcus aureus* and the presence of antibodies to this superantigen in menstruating women. J. Clin. Microbiol. *43*, 4628–4634.

Parsonnet, J., Modern, P.A., and Giacobbe, K.D. (1996). Effect of tampon composition on production of toxic shock syndrome toxin-1 by *Staphylococcus aureus in vitro*. J. Infect. Dis. *173*, 98–103.

Pauli, N.T., Kim, H.K., Falugi, F., Huang, M., Dulac, J., Henry Dunand, C., Zheng, N.Y., Kaur, K., Andrews, S.F., Huang, Y., *et al.* (2014). *Staphylococcus aureus* infection induces protein A-mediated immune evasion in humans. J. Exp. Med. *211*, 2331–2339.

Peterson, M.L., Ault, K., Kremer, M.J., Klingelhutz, A.J., Davis, C.C., Squier, C.A., and Schlievert, P.M. (2005). The innate immune system is activated by stimulation of vaginal epithelial cells with *Staphylococcus aureus* and toxic shock syndrome toxin 1. Infect. Immun. *73*, 2164–2174.

Petersson, K., Thunnissen, M., Forsberg, G., and Walse, B. (2002). Crystal structure of a SEA variant in complex with MHC class II reveals the ability of SEA to cross-link MHC molecules. Structure *10*, 1619–1626.

Pragman, A.A., and Schlievert, P.M. (2004). Virulence regulation in *Staphylococcus aureus*: the need for *in vivo* analysis of virulence factor regulation. FEMS Immunol. Med. Microbiol. *42*, 147–154.

Pragman, A.A., Yarwood, J.M., Tripp, T.J., and Schlievert, P.M. (2004). Characterization of virulence factor regulation by SrrAB, a two-component system in *Staphylococcus aureus*. J. Bacteriol. *186*, 2430–2438.

Prasad, G.S., Earhart, C.A., Murray, D.L., Novick, R.P., Schlievert, P.M., and Ohlendorf, D.H. (1993). Structure of toxic shock syndrome toxin 1. Biochemistry *32*, 13761–13766.

Que, Y.A., and Moreillon, P. (2011). Infective endocarditis. Nat. Rev. Cardiol. *8*, 322–336.

Rago, J.V., Vath, G.M., Bohach, G.A., Ohlendorf, D.H., and Schlievert, P.M. (2000a). Mutational analysis of the superantigen staphylococcal exfoliative toxin A (ETA). J. Immunol. *164*, 2207–2213.

Rago, J.V., Vath, G.M., Tripp, T.J., Bohach, G.A., Ohlendorf, D.H., and Schlievert, P.M. (2000b). Staphylococcal exfoliative toxins cleave alpha- and beta-melanocyte-stimulating hormones. Infect. Immun. *68*, 2366–2368.

Rauch, S., Gough, P., Kim, H.K., Schneewind, O., and Missiakas, D. (2014). Vaccine protection of leucopenic mice against *Staphylococcus aureus* bloodstream infection. Infect. Immun. *82*, 4889–4898.

Reingold, A.L., Hargrett, N.T., Dan, B.B., Shands, K.N., Strickland, B.Y., and Broome, C.V. (1982). Nonmenstrual toxic shock syndrome: a review of 130 cases. Ann. Intern. Med. *96* 11, 871–874.

Robert, J., Tristan, A., Cavalie, L., Decousser, J.W., Bes, M., Etienne, J., and Laurent, F. (2011). Panton–Valentine leucocidin-positive and toxic shock syndrome toxin 1-positive methicillin-resistant *Staphylococcus aureus*: a French multicenter prospective study in 2008. Antimicrob. Agents Chemother. *55*, 1734–1739.

Salgado-Pabón, W., Herrera, A., Vu, B.G., Stach, C.S., Merriman, J.A., Spaulding, A.R., and Schlievert, P.M. (2014). *Staphylococcus aureus* β-toxin production is common in strains with the β-toxin gene inactivated by bacteriophage. J. Infect. Dis. *210*, 784–792.

Schlievert, P.M. (1982). Enhancement of host susceptibility to lethal endotoxin shock by staphylococcal pyrogenic exotoxin type C. Infect. Immun. *36*, 123–128.

Schlievert, P.M. (1983). Alteration of immune function by staphylococcal pyrogenic exotoxin type C: possible role in toxic-shock syndrome. J. Infect. Dis. *147*, 391–398.

Schlievert, P.M. (1995). Comparison of cotton and cotton/rayon tampons for effect on production of toxic shock syndrome toxin. J. Infect. Dis. *172*, 1112–1114.

Schlievert, P.M. (1996a). Effect of Merocel vaginal sponge on growth of *Staphylococcus aureus* and production of toxic shock syndrome-associated toxins. J. Am. Coll. Surg. *183*, 19–24.

Schlievert, P.M. (1996b). Tampons and toxic shock syndrome. Med. J. Aust. *164*, 635–636.

Schlievert, P.M. (2009). Cytolysins, superantigens, and pneumonia due to community-associated methicillin-resistant *Staphylococcus aureus*. J. Infect. Dis. *200*, 676–678.

Schlievert, P.M., and Bohach, G.A. (2007). Staphylococcal and streptococcal superantigens: an update. In Superantigens Molecular Basis for Their Role in Human Diseases, M. Kotb, and Fraser, J.D., ed. (Washington, DC: ASM Press), pp. 21–36.

Schlievert, P.M., Bettin, K.M., and Watson, D.W. (1977). Purification and characterization of group A streptococcal pyrogenic exotoxin type C. Infect. Immun. *16*, 673–679.

Schlievert, P.M., Bettin, K.M., and Watson, D.W. (1978). Effect of antipyretics on group A streptococcal pyrogenic exotoxin fever production and ability to enhance lethal endotoxin shock. Proc. Soc. Exp. Biol. Med. *157*, 472–475.

Schlievert, P.M., Bettin, K.M., and Watson, D.W. (1979). Reinterpretation of the Dick test: role of group A streptococcal pyrogenic exotoxin. Infect. Immun. *26*, 467–472.

Schlievert, P.M., Bettin, K.M., and Watson, D.W. (1980a). Inhibition of ribonucleic acid synthesis by group A streptococcal pyrogenic exotoxin. Infect. Immun. *27*, 542–548.

Schlievert, P.M., and Blomster, D.A. (1983). Production of staphylococcal pyrogenic exotoxin type C: influence of physical and chemical factors. J. Infect. Dis. *147*, 236–242.

Schlievert, P.M., Case, L.C., Nemeth, K.A., Davis, C.C., Sun, Y., Qin, W., Wang, F., Brosnahan, A.J., Mleziva, J.A., Peterson, M.L., *et al.* (2007a). Alpha and beta chains of hemoglobin inhibit production of *Staphylococcus aureus* exotoxins. Biochemistry *46*, 14349–14358.

Schlievert, P.M., Case, L.C., Strandberg, K.L., Abrams, B.B., and Leung, D.Y. (2008). Superantigen profile of *Staphylococcus aureus* isolates from patients with steroid-resistant atopic dermatitis. Clin. Infect. Dis. *46*, 1562–1567.

Schlievert, P.M., Case, L.C., Strandberg, K.L., Tripp, T.J., Lin, Y.C., and Peterson, M.L. (2007b). Vaginal *Staphylococcus aureus* superantigen profile shift from 1980 and 1981 to 2003, 2004, and 2005. J. Clin. Microbiol. *45*, 2704–2707.

Schlievert, P.M., Jablonski, L.M., Roggiani, M., Sadler, I., Callantine, S., Mitchell, D.T., Ohlendorf, D.H., and Bohach, G.A. (2000). Pyrogenic toxin superantigen site specificity in toxic shock syndrome and food poisoning in animals. Infect. Immun. *68*, 3630–3634.

Schlievert, P.M., and Kim, M.H. (1991). Reporting of toxic shock syndrome *Staphylococcus aureus* in 1982 to 1990. J. Infect. Dis. *164*, 1245–1246.

Schlievert, P.M., Nemeth, K.A., Davis, C.C., Peterson, M.L., and Jones, B.E. (2010). *Staphylococcus aureus* exotoxins are present *in vivo* in tampons. Clin. Vaccine. Immunol. *17*, 722–727.

Schlievert, P.M., Osterholm, M.T., Kelly, J.A., and Nishimura, R.D. (1982). Toxin and enzyme characterization of *Staphylococcus aureus* isolates from patients with and without toxic shock syndrome. Ann. Intern. Med. *96* 5, 937–940.

Schlievert, P.M., Schoettle, D.J., and Watson, D.W. (1980b). Ganglioside and monosaccharide inhibition of nonspecific lymphocyte mitogenicity by group A streptococcal pyrogenic exotoxins. Infect. Immun. *27*, 276–279.

Schlievert, P.M., Shands, K.N., Dan, B.B., Schmid, G.P., and Nishimura, R.D. (1981). Identification and characterization of an exotoxin from *Staphylococcus aureus* associated with toxic-shock syndrome. J. Infect. Dis. *143*, 509–516.

Schlievert, P.M., Strandberg, K.L., Lin, Y.C., Peterson, M.L., and Leung, D.Y. (2010). Secreted virulence factor comparison between methicillin-resistant and methicillin-sensitive *Staphylococcus aureus*, and its relevance to atopic dermatitis. J. Clin. Allergy Immunol. *125*, 39–49.

Schlievert, P.M., Tripp, T.J., and Peterson, M.L. (2004). Reemergence of staphylococcal toxic shock syndrome in Minneapolis-St. Paul, Minnesota, during the 2000–2003 surveillance period. J. Clin. Microbiol. *42*, 2875–2876.

Schlievert, P.M., and Watson, D.W. (1978). Group A streptococcal pyrogenic exotoxin: pyrogenicity, alteration of blood-brain barrier, and separation of sites for pyrogenicity and enhancement of lethal endotoxin shock. Infect. Immun. *21*, 753–763.

Schlievert, P.M., and Watson, D.W. (1979). Biogenic amine involvement in pyrogenicity and enhancement of lethal endotoxin shock by group A streptococcal pyrogenic exotoxin. Proc. Soc. Exp. Biol. Med. *162*, 269–274.

Schneewind, O., Fowler, A., and Faull, K.F. (1995). Structure of the cell wall anchor of surface proteins in *Staphylococcus aureus*. Science *268*, 103–106.

Schneewind, O., Model, P., and Fischetti, V.A. (1992). Sorting of protein A to the staphylococcal cell wall. Cell *70*, 267–281.

Schwartz, K., and Boles, B.R. (2013). Microbial amyloids – functions and interactions within the host. Curr. Opin. Microbiol. *16*, 93–99.

Shands, K.N., Schmid, G.P., Dan, B.B., Blum, D., Guidotti, R.J., Hargrett, N.T., Anderson, R.L., Hill, D.L., Broome, C.V., Band, J.D., *et al.* (1980). Toxic-shock syndrome in menstruating women: association with tampon use and *Staphylococcus aureus* and clinical features in 52 cases. N. Engl. J. Med. *303*, 1436–1442.

Silverstein, R. (2004). D-galactosamine lethality model: scope and limitations. J. Endotoxin Res. *10*, 147–162.

Skov, L., Olsen, J.V., Giorno, R., Schlievert, P.M., Baadsgaard, O., and Leung, D.Y. (2000). Application of Staphylococcal enterotoxin B on normal and atopic skin induces up-regulation of T cells by a superantigen-mediated mechanism. J. Allergy Clin. Immunol. *105*, 820–826.

Song, L., Hobaugh, M.R., Shustak, C., Cheley, S., Bayley, H., and Gouaux, J.E. (1996). Structure of staphylococcal alpha-hemolysin, a heptameric transmembrane pore. Science *274*, 1859–1866.

Spaan, A.N., Henry, T., van Rooijen, W.J., Perret, M., Badiou, C., Aerts, P.C., Kemmink, J., de Haas, C.J., van Kessel, K.P., Vandenesch, F., *et al.* (2013). The staphylococcal toxin Panton–Valentine Leucocidin targets human C5a receptors. Cell Host Microbe *13*, 584–594.

Spaulding, A.R., Lin, Y.C., Merriman, J.A., Brosnahan, A.J., Peterson, M.L., and Schlievert, P.M. (2012a). Immunity to *Staphylococcus aureus* secreted proteins protects rabbits from serious illnesses. Vaccine *30*, 5099–5109.

Spaulding, A.R., Satterwhite, E.A., Lin, Y.-C., Chuang-Smith, O.N., Frank, K.L., Merriman, J.A., Schaefers, M.M., Yarwood, J.M., Peterson, M.L., and Schlievert, P.M. (2012b). Differential ability of *Staphylococcus aureus* to cause infective endocarditis and lethal sepsis in rabbits. Front. Cell. Infect. Microbiol. *2*, 18.

Squier, C.A., Mantz, M.J., Schlievert, P.M., and Davis, C.C. (2008). Porcine vagina ex vivo as a model for studying permeability and pathogenesis in mucosa. J. Pharm. Sci. *97*, 9–21.

Stone, R.L., and Schlievert, P.M. (1987). Evidence for the involvement of endotoxin in toxic shock syndrome. J. Infect. Dis. *155*, 682–689.

Strandberg, K.L., Rotschafer, J.H., Vetter, S.M., Buonpane, R.A., Kranz, D.M., and Schlievert, P.M. (2010). Staphylococcal superantigens cause lethal pulmonary disease in rabbits. J. Infect. Dis. *202*, 1690–1697.

Sugiyama, H., and Hayama, T. (1965). Abdominal viscera as site of emetic action for staphylococcal enterotoxin in the monkey. J. Infect. Dis. *115*, 330–336.

Switalski, L.M., Speziale, P., and Höök, M. (1989). Isolation and characterization of a putative collagen receptor from *Staphylococcus aureus* strain Cowan 1. J. Biol. Chem. *264*, 21080–21086.

Tierno, P.M., and Hanna, B.A. (1989). Ecology of toxic shock syndrome: amplification of toxic shock syndrome toxin 1 by materials of medical interest. Rev. Infect. Dis. *11* (Suppl. 1), S182–S186.

Todd, J., Fishaut, M., Kapral, F., and Welch, T. (1978). Toxic-shock syndrome associated with phage-group-I Staphylococci. Lancet *2*, 1116–1118.

Tweten, R.K., Christianson, K.K., and Iandolo, J.J. (1983). Transport and processing of staphylococcal alpha-toxin. J. Bacteriol. *156*, 524–528.

Vath, G.M., Earhart, C.A., Monie, D.D., Iandolo, J.J., Schlievert, P.M., and Ohlendorf, D.H. (1999). The crystal structure of exfoliative toxin B: a superantigen with enzymatic activity. Biochemistry *38*, 10239–10246.

Vath, G.M., Earhart, C.A., Rago, J.V., Kim, M.H., Bohach, G.A., Schlievert, P.M., and Ohlendorf, D.H. (1997). The structure of the superantigen exfoliative toxin A suggests a novel regulation as a serine protease. Biochemistry *36*, 1559–1566.

Vaudaux, P.E., Francois, P., Proctor, R.A., McDevitt, D., Foster, T.J., Albrecht, R.M., Lew, D.P., Wabers, H., and Cooper, S.L. (1995). Use of adhesion-defective mutants of *Staphylococcus aureus* to define the role of specific plasma proteins in promoting bacterial adhesion to canine arteriovenous shunts. Infect. Immun. *63*, 585–590.

Vergeront, J.M., Stolz, S.J., Crass, B.A., Nelson, D.B., Davis, J.P., and Bergdoll, M.S. (1983). Prevalence of serum antibody to staphylococcal enterotoxin F among Wisconsin residents: implications for toxic-shock syndrome. J. Infect. Dis. *148*, 692–698.

Vu, B.G., Gourronc, F.A., Bernlohr, D.A., Schlievert, P.M., and Klingelhutz, A.J. (2013). Staphylococcal superantigens stimulate immortalized human adipocytes to produce chemokines. PLoS One *8*, e77988.

Vu, B.G., Stach, C.S., Kulhankova, K., Salgado-Pabon, W., Klingelhutz, A.J., and Schlievert, P.M. (2015). Chronic superantigen exposure induces systemic inflammation, elevated bloodstream endotoxin, and abnormal glucose tolerance in rabbits: possible role in diabetes. mBio *6*, e02554.

Walker, J.N., Crosby, H.A., Spaulding, A.R., Salgado-Pabón, W., Malone, C.L., Rosenthal, C.B., Schlievert, P.M., Boyd, J.M., and Horswill, A.R. (2013). The *Staphylococcus aureus* ArlRS two-component system is a novel regulator of agglutination and pathogenesis. PLOS Pathog. *9*, e1003819.

Walsh, E.J., Miajlovic, H., Gorkun, O.V., and Foster, T.J. (2008). Identification of the *Staphylococcus aureus* MSCRAMM clumping factor B (ClfB) binding site in the alphaC-domain of human fibrinogen. Microbiology *154* 12, 550–558.

Watson, D.W. (1960). Host-parasite factors in group A streptococcal infections. Pyrogenic and other effects of immunologic distinct exotoxins related to scarlet fever toxins. J. Exp. Med. *111*, 255–284.

Wilson, G.J., Seo, K.S., Cartwright, R.A., Connelley, T., Chuang-Smith, O.N., Merriman, J.A., Guinane, C.M., Park, J.Y., Bohach, G.A., Schlievert, P.M., *et al.* (2011). A novel core genome-encoded superantigen contributes to lethality of community-associated MRSA necrotizing pneumonia. PLOS Pathog. *7*, e1002271.

Xiong, Y.Q., Fowler, V.G., Yeaman, M.R., Perdreau-Remington, F., Kreiswirth, B.N., and Bayer, A.S. (2009). Phenotypic and genotypic characteristics of persistent methicillin-resistant *Staphylococcus aureus* bacteremia *in vitro* and in an experimental endocarditis model. J. Infect. Dis. *199*, 201–208.

Yamashita, K., Kawai, Y., Tanaka, Y., Hirano, N., Kaneko, J., Tomita, N., Ohta, M., Kamio, Y., Yao, M., and Tanaka, I. (2011). Crystal structure of the octameric pore of staphylococcal γ-hemolysin reveals the β-barrel pore formation mechanism by two components. Proc. Natl. Acad. Sci. U.S.A. *108*, 17314–17319.

Yarwood, J.M., McCormick, J.K., and Schlievert, P.M. (2001). Identification of a novel two-component regulatory system that acts in global regulation of virulence factors of *Staphylococcus aureus*. J. Bacteriol. *183*, 1113–1123.

Yarwood, J.M., and Schlievert, P.M. (2000). Oxygen and carbon dioxide regulation of toxic shock syndrome toxin 1 production by *Staphylococcus aureus* MN8. J. Clin. Microbiol. *38*, 1797–1803.

Yarwood, J.M., and Schlievert, P.M. (2003). Quorum sensing in Staphylococcus infections. J. Clin. Invest. *112*, 1620–1625.

Staphylococcus aureus Metabolism and Physiology

6

Greg A. Somerville

Abstract

All living organisms require endogenous and/or exogenous sources of biosynthetic precursors (e.g. nucleic acids, amino acids, carbohydrates, etc.) for polymerizing and assembly reactions that make DNA, RNA, proteins, and everything else required for life. In *Staphylococcus aureus*, the *de novo* synthesis of biosynthetic precursors requires thirteen biosynthetic intermediates (e.g. glucose-6-phosphate and oxaloacetate) and the coordinated action of numerous enzymes. The life cycle of *S. aureus* is such that its environment is in a near constant state of flux. This flux has important implications because enzyme activity is altered in response to co-factor, substrate, and product availability, pH, temperature, post-translational modifications, and many other environmental and nutritional factors. In other words, changes in the environment precipitate changes in enzymatic activity, which alters the availability of biosynthetic precursors and the ability to synthesize cellular components. Understanding how environmental changes alter metabolism, begins with an understanding of metabolism.

Introduction

As stated by Dr Jodi Lindsay in Chapter 4, the first *Staphylococcus aureus* genomes were published in 2001 (Kuroda *et al.*, 2001) and since that time the pace of newly sequenced staphylococcal genomes has been breath-taking. Unfortunately, there are well-known limitations to whole genome sequencing projects; namely, there is little or no experimental validation of the *majority* of proteins encoded within a genome. This is further complicated by the fact that these genomes are usually not curated by experts, but are 'machine' curated; hence, functionality is determined by the transfer of knowledge from one genome to the next. That being said, these genomes have greatly assisted in our understanding of *S. aureus* metabolism and physiology (e.g. Shah *et al.*, 2013; Somerville *et al.*, 2003a). This wealth of genome data has shed light on observations such as the niacin requirement for growth (Knight, 1937). This observation is likely related to the absence of NAD biosynthetic genes encoding quinolinate phosphoribosyl transferase and quinolinate synthetase from most *S. aureus* genomes. (In genomes where these genes do appear, their homology is moderate and the enzymes have not been experimentally validated.) In contrast, the genome data have provided only minimal insight into the observations indicating *S. aureus* is auxtrophic for several amino acids (e.g. Emmett and Kloos, 1975; Gladstone, 1937). One thing is clear from all the genome sequencing data, *S. aureus* metabolic capabilities appear to be very robust (http://biocyc.org/).

Because an understanding of staphylococcal metabolism is predicated on the reproducible cultivation of the bacterium, I will begin this chapter with a brief overview of the important considerations regarding bacterial cultivation. This will be followed by a discussion of bacterial growth phases and the differences in metabolism, and finally, a more in-depth review of important *S. aureus* metabolism.

Cultivation of *Staphylococcus aureus*

Historically, bacteria have been classified by many morphological and physiological characteristics, such as cytological features, antibiotic resistance traits, nutritional requirements, metabolic properties, and their requirement/response to oxygen. On this latter characteristic, *S. aureus* is defined as a facultative anaerobe, which means that it can respire in the presence of an electron acceptor (i.e. oxygen, organics, nitrate, and nitrite) and ferment carbohydrates in the absence of one. Although these classifications do not always stand the test of time; as examples, *Pseudomonas aeruginosa* and *Bacillus subtilis* were originally classified as strict aerobes, the classification of *S. aureus* as a facultative anaerobe is accurate. Being a facultative anaerobe makes a bacterium highly adaptable and is another reason that *S. aureus* is a metabolically robust bacterium. Because *S. aureus* can grow under a wide range of environmental conditions, it is imperative to understand that minor/subtle changes in cultivation conditions can produce profound changes in its metabolism.

Cultivation of bacteria, fungi, viruses, eukaryotic cells, etc. requires consideration of three essential parameters: cell type/strain, cultivation medium, and cultivation conditions (e.g. temperature and percentage of oxygen). In the case of *S. aureus,* strain usage varies greatly. Some scientists choose to work with the ever evolving contemporary strains (e.g. USA300 lineage), while others work with historical strains such as COL and NCTC 8325 (Dyke *et al.,* 1966; Novick, 1967). This heterogeneous approach of having many investigators studying many different strains is good in the sense that it provides a population overview of *S. aureus* metabolism, physiology, and regulation, but it comes at a cost; namely, variation within a population generates 'noise' that can obscure the functions and relationships between proteins, genes, and regulators (Cassat *et al.,* 2006; Somerville *et al.,* 2003b).

The second essential parameter that must be considered during *S. aureus* cultivation is the cultivation medium. *S. aureus* will not grow in minimal medium, as stated, it needs supplements for cultivation (Gladstone, 1937). For this reason, many groups have developed chemically defined media that facilitate growth, synthesis of virulence determinants, metabolism, or whatever they have chosen to study (e.g. Miller and Fung, 1973). Because scientists study a heterogeneous collection of *S. aureus* strains with diverse nutritional needs, there are a large number of formulations of chemically defined media for *S. aureus* cultivation. Unfortunately, the inconsistent nutritional requirements of *S. aureus* make choosing the correct media formulation difficult, which makes comparison of data between labs challenging if not impossible. Because there is no 'best' chemically defined medium for cultivating *S. aureus,* many laboratories choose to grow *S. aureus* in complex media (Novick, 1991; Sadykov *et al.,* 2010). Complex media have the advantage of promoting growth of most *S. aureus* stains to a high biomass; however, they complicate interpretation of metabolic studies due to unknown genetic and metabolic (i.e. feedback and feedforward) regulation of metabolic pathways. As with strain selection, staphylococcal researchers have chosen to use diverse media for their experiments, all but assuring that data generated in different laboratories are incomparable (Neidhardt, 2006).

Cultivation conditions are the third critical parameter to consider when growing *S. aureus* (Fig. 6.1). Since *S. aureus* is a facultative anaerobe, subtle differences in cultivation conditions can dramatically alter the bacterial metabolic state (Ledala *et al.,* 2014), which will change the concentrations of biosynthetic intermediates and precursors, affecting the polymerizing and assembly reactions that generate cellular components, including virulence determinants (Richardson *et al.,* 2015). The sensitivity of *S. aureus* to changes in cultivation conditions is the reason these conditions must be explicitly stated in protocols and manuscripts. Without this information, it is impossible to accurately reproduce experiments and to integrate data into our global understanding of *S. aureus.*

Staphylococcal growth phases and changes in metabolism

Cultivation of bacteria results in distinct growth phases; specifically, the lag, exponential, stationary, and death phases (Fig. 6.1). The lag phase is the 'quiescent' period before bacteria commence rapid growth, and is characterized by an increase in ribosome synthesis, DNA replication, and cell mass. After the bacterial initiation mass has been achieved, they begin dividing by binary fission, leading to exponential growth. Once bacteria cease to grow exponentially, they are said to enter into the stationary phase of growth. Defining this as the stationary phase is somewhat misleading, as the biomass continues to increase, albeit, at a reduced

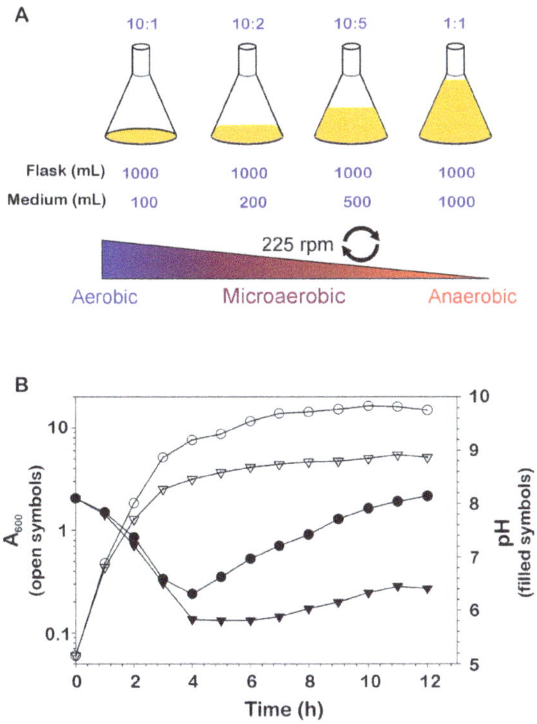

Figure 6.1 The availability of oxygen dramatically affects *S. aureus* biomass generation and pH of the cultivation medium. Panel A shows the transition from aerobic growth to anaerobic growth as the flask-to-medium ratio is altered. Panel B demonstrates effect of altering the flask-to-medium ratio on *S. aureus* cultivation. The circles represent the growth and pH profiles of *S. aureus* grown with a flask-to-medium ration of 10:1, while the inverted triangles represent *S. aureus* grown using a flask-to-medium ration of 1:1.

rate. For this reason, the period between the exponential growth phase and the cessation of growth is often referred to as the post-exponential growth phase, and the stationary phase is the period when growth and death are equal or when cell division has stopped. The *S. aureus* death phase is an active process involving the synthesis of pore forming lysins, which dictates that carbon and energy are required for this process (Somerville *et al.*, 2002). Importantly, these growth phases not only define the life cycle of *S. aureus*, but they also define different metabolic states.

Inaccessibility of metabolic precursors, or an imbalance in intracellular precursor pools, will necessitate alterations in bacterial metabolism to establish or maintain sufficient precursors to permit the rapid growth that occurs during the exponential growth phase. During *in vitro* cultivation, variation in the availability of nutrients is more often due to lot-to-lot variation in the manufacturing of growth medium, differences in autoclaving times and/or temperatures that lead to differences in pyrolysis of medium components, bacterial depletion of nutrients, and differences in medium formulation, particularly in chemically defined media. Often media variations are subtle, but because bacteria have evolved to respond to changes in their environment, these subtle differences can alter bacterial metabolism. In short, the life cycle of *S. aureus*, or any bacterium, involves a gradient of metabolic states; however, for simplicity, it is easiest to view *S. aureus* as having two distinct metabolic states.

During aerobic cultivation of *S. aureus*, the first metabolic state occurs in the exponential growth phase and involves the incomplete oxidation of rapidly catabolizable metabolites that generate pyruvate (e.g. carbohydrates, serine). Pyruvate can be used to produce acetyl-phosphate, which is used in substrate level phosphorylation to generate ATP and acetic acid (this occurs via the phosphotransacetylase/acetate kinase pathway) (Sadykov *et al.*, 2013; Somerville *et al.*, 2003b). In addition, to a rapidly catabolizable carbon source(s), the bacteria must have sufficient biosynthetic precursors (e.g. amino acids, nucleic acids) for macromolecular synthesis to occur. In a complex medium, biosynthetic precursors are usually present in abundance; hence, biosynthesis of precursors is restricted to those whose concentrations are in limiting supply. As mentioned in the preceding paragraph, differences in media formulation and/or preparation will strongly influence the availability of biosynthetic precursors; therefore, when examining *S. aureus* metabolism, it is essential to consider that some differences in metabolism may simply be due to unintended alteration of metabolite availability. This becomes important when biosynthetic intermediates of the tricarboxylic acid (TCA) cycle are in limiting supply, but glucose or other rapidly catabolizable are present in abundance. Under these conditions, TCA cycle activity will increase to produce the necessary biosynthetic intermediates; however, increased TCA cycle activity is a hallmark for the second distinct metabolic state found in *S. aureus* (Zhu *et al.*, 2009).

The second metabolic state occurs upon de-repression of the TCA cycle when readily catabolizable carbon source(s) and/or glutamate are depleted from the cultivation medium. In addition to the depletion of nutrients, induction of the TCA cycle also requires the presence of an electron acceptor (e.g. oxygen, nitrate, or fumarate) that can be used in the electron transport chain to facilitate the oxidation of reduced dinucleotides. Without an electron acceptor, reduced dinucleotides accumulate and TCA cycle activity is inhibited until the redox balance is restored. Obviously, this metabolic state marks the transition from substrate level phosphorylation to oxidative phosphorylation as the primary source of ATP. To fuel the TCA cycle, the acetate (In culture media with a pH above 5.5, acetic acid is present as its conjugate base acetate.) produced during the exponential growth phase is catabolized (Somerville *et al.*, 2003b). Because *S. aureus* lacks a glyoxylate shunt that would by-pass the decarboxylation steps of the TCA cycle, for every two carbons that enter into the TCA cycle as acetyl-CoA, two carbons are lost as CO_2. This is a problem if carbons are simultaneously withdrawn as biosynthetic intermediates (i.e. α-ketoglutarate, succinyl-CoA, and oxaloacetate) to produce biosynthetic precursors (e.g. glutamate, aspartate). The withdrawal of C_4 and C_5 intermediates will lead to a shortage of carbon to keep the TCA cycle going; however, this scenario is avoided through anaplerotic reactions. In *S. aureus*, anaplerotic carbons are usually derived from amino acids, which necessitates deamination reactions and results in the accumulation of ammonia in the cultivation medium, another hallmark of the second metabolic state. Combined, the depletion of incompletely oxidized metabolites and the accumulation of ammonia cause an alkalization of the cultivation medium. In summary, the second metabolic state in *S. aureus* occurs during the post-exponential growth phase and is dependent upon TCA cycle activity, electron transport, and catabolism of non-preferred carbon sources.

Staphylococcus aureus metabolism

Metabolism is the sum of energy-generating catabolism (breaking down) and energy-requiring anabolism (building up). Metabolism can also be divided into central, intermediary, and, in eukaryotes, peripheral metabolism. Central metabolism (i.e. the glycolytic and pentose phosphate pathways and the TCA cycle) is both catabolic and anabolic and provides *S. aureus* with the 13 canonical biosynthetic intermediates (Table 6.1) by which it makes all biosynthetic precursors (Somerville and Proctor, 2009). To produce these 13 biosynthetic intermediates, *S. aureus* will typically use carbohydrates when they are available; however, many macromolecules (e.g. amino acids, nucleic acids) can be catabolized to produce these intermediates. For the sake of brevity and because it has been the most extensively studied, this chapter will focus on carbohydrate catabolism.

In *S. aureus*, carbohydrates are frequently translocated into the cytoplasm by the phosphotransferase system (PTS) (Kundig *et al.*, 1964), a process known as group translocation. The PTS uses three enzymes to transfer the high-energy phosphate from phosphoenolpyruvate (PEP) to the sugar being translocated, enzyme I (EI), enzyme II (EII) and a histidine-containing phosphocarrier protein (HPr) (Kundig *et al.*, 1964). EI and HPr are soluble cytoplasmic proteins involved in the transport of all PTS translocated sugars, while specificity for a particular sugar is provided by the EII components (domains EIIA, EIIB, and EIIC).

Table 6.1 Central metabolic pathways and their biosynthetic intermediates

Glycolysis/gluconeogenesis	Pentose phosphate pathway	Tricarboxylic acid cycle
Glucose-6-phosphate	Ribose-5-phosphate	Oxaloacetate
Fructose-6-phosphate	Sedoheptulose-7-phophate	α-Ketoglutarate
Dihydroxyacetone phosphate	Erythrose-4-phosphate	Succinate
Glyceraldehyde-3-phosphate		
Phosphoenolpyruvate		
Pyruvate		
Acetyl-coenzyme A		

S. aureus has at least 15 EII components encoded within its genome, highlighting *S. aureus*' versatility in utilizing carbohydrates (Gill *et al.*, 2005). The process of translocating a sugar initiates with PEP transferring its phosphate to a histidine on EI, which is then transferred to another histidine on HPr. The phosphorylated HPr transfers the high-energy phosphate to EIIA, which then transfers it to the sugar. Transfer of phosphate from EIIA to the sugar has two primary consequences: first, the phosphorylated sugar is no longer recognized by the EII permease, preventing efflux of the sugar out of the cytoplasm; and, second, the phosphorylated sugar is activated for catabolism.

Glycolysis

Once translocated into the cytoplasm, most sugars and sugar alcohols will enter into glycolysis after conversion into fructose-6-phosphate. The enzyme responsible for this conversion will depend on the sugar; as examples, activated mannitol (i.e. mannitol-1-phosphate) is processed into fructose-6-phosphate by mannitol-1-phosphate 5-dehydrogenase, whereas activated glucose (i.e. glucose-6-phosphate) is processed by glucose-6-phosphate isomerase. Non-activated sugars, specifically, those not transported via the PTS, also follow this pathway, but they must be activated by a hexokinase or glucokinase in the case of glucose. Fructose-6-phosphate is converted into fructose-1, 6-bisphosphate by 6-phosphofructokinase, a step that requires ATP and represents the committed step in glycolysis. In the next step of glycolysis, fructose bisphosphate aldolase and triosephosphate isomerase convert fructose-1, 6-bisphosphate into dihydroxyacetone phosphate and glyceraldehyde-3-phosphate. The enzymatic conversion of glyceraldehyde-3-phosphate into glycerate 1, 3-bisphosphate by glyceraldehyde-3-phosphate dehydrogenase produces reduced nicotinamide adenine dinucleotide (NADH), which can be used in some biosynthetic reactions, to drive lactate dehydrogenase, used in the electron transport chain, etc. Phosphoglycerate kinase converts glycerate 1, 3-bisphosphate into 3-phosphoglycerate and in the process produces an ATP. The isomerase phosphoglycerate mutase transforms 3-phosphoglycerate into 2-phosphoglycerate, which is the substrate for enolase. Enolase produces PEP from 2-phosphoglycerate. PEP is a versatile phosphodonor that can be used to generate a second ATP and pyruvate via the pyruvate kinase; it can be used in chorismate biosynthesis; initiate group translocation through the PTS, etc.

While glycolysis is an oxygen-independent pathway, the catabolic fate of pyruvate is heavily dependent upon the availability of oxygen. Under microaerobic or anaerobic growth, pyruvate is primarily reduced to lactic acid (Kendall *et al.*, 1930; Krebs, 1937), whereas during aerobic growth pyruvate undergoes oxidative decarboxylation to produce acetyl-CoA (Gardner and Lascelles, 1962). Whether lactic acid or acetic acid is produced, both organic acids will accumulate in the cultivation medium and lead to its acidification (Fig. 6.1) (Ledala *et al.*, 2014). Generation of acetyl-CoA from pyruvate is primarily driven by the pyruvate dehydrogenase complex (Gardner and Lascelles, 1962), while lactic acid is produced by the lactate dehydrogenases (Garrard and Lascelles, 1968; Stockland and San Clemente, 1969). Regarding acetyl-CoA, it is usually converted into acetyl-phosphate by phosphotransacetylase, which can be used as a substrate for acetate kinase in substrate-level phosphorylation to generate ATP and acetate (Somerville *et al.*, 2003b). Whether lactate or acetate accumulation in the culture medium, both can be used as carbon sources if the growth conditions are favourable; namely, having an electron acceptor to facilitate oxidation of reduced dinucleotides.

Pentose phosphate pathway

In addition to glycolysis, *S. aureus* also catabolizes carbohydrates through the pentose phosphate pathway. Activated carbohydrates (e.g. glucose-6-phosphate, fructose-6-phosphate, and gluconate-6-phosphate) are shunted into the pentose phosphate pathway to produce three of the 13 biosynthetic intermediates; specifically, ribose-5-phosphate, sedoheptulose-7-phosphate, and erythrose-4-phosphate (Dickens, 1938; Warburg *et al.*, 1935). In addition to generating biosynthetic intermediates, the pentose phosphate pathway also reduces $NADP^+$ to $NADPH/H^+$ that can be used an electron donor in biosynthesis. The generation of $NADPH/H^+$ occurs in the oxidative branch of the pentose phosphate pathway, which involves the oxidation and decarboxylation of glucose-6-phosphate. The committed step in this pathway occurs during the glucose-6-phosphate dehydrogenase catalysed oxidation of glucose-6-phosphate to 6-phospho-D-glucono-1, 5-lactone. The next reaction is the 6-phosphogluconolactonase-mediated decarboxylation of 6-phospho-D-glucono-1, 5-lactone into the pentose sugar ribulose-5-phosphate. Ribulose-5-phosphate is converted to the furanose ring structure ribose-5-phosphate through the action of the ribose-5-phosphate isomerase (RpiA) (Zhu *et al.*, 2011). Ribulose-phosphate-3-epimerase produces the stereoisomer of ribulose-5-phosphate, namely, xylulose-5-phosphate. Transketolase transfers a C_2 moiety from xylulose-5-phosphate to ribose-5-phosphate or erythrose-4-phosphate to generate sedheptulose-7-phosphate or fructose-6-phosphate, respectively. Collectively, glycolysis and the pentose phosphate pathway catabolize the majority of glucose consumed by *S. aureus* (Blumenthal, 1972). This catabolism is aided by the fact that the TCA cycle is carbon catabolite repressed when glucose and a source of glutamate are present in the cultivation medium (Hartmann *et al.*, 2013, 2014) (see Chapter 12).

Tricarboxylic acid cycle

The TCA cycle is believed to have evolved from two amino acid biosynthetic pathways, one oxidative pathway and one reductive pathway. In addition, the TCA cycle was complete in ancestral Proteobacteria prior to the divergence of low G+C and high G+C Gram-positive bacteria (Melendez-Hevia *et al.*, 1996; Huynen *et al.*, 1999). The TCA cycle serves three primary functions:

1. As with other central metabolic pathways, the TCA cycle supplies biosynthetic intermediates (Table 6.1).
2. It generates reduced NADPH, NADH, $FADH_2$, and menaquinol, which serve as electron donors in the electron transport chain.
3. It produces energy in the form of ATP (much more is produced during oxidative phosphorylation from the electron donors listed previously).

Carbon enters into the TCA cycle in the form of acetyl-CoA in a condensation reaction with oxaloacetate. Because the TCA cycle is catabolite repressed during the exponential growth phase, very little carbon flows into the TCA cycle and acetyl-CoA is shunted into the phosphotransacetylase/acetate kinase (Pta/AckA) pathway, which causes acetate to accumulate in the cultivation fluid. This means that in the post-exponential growth phase, *S. aureus* has an abundant supply of acetate to feed into the TCA cycle. To get acetate into the TCA cycle, the acetate must be converted into acetyl-CoA, either by reversing the Pta/AckA pathway or through the action of acetyl-CoA synthetase.

The first enzymatic reaction of the TCA cycle is an aldol condensation by citrate synthase that transfers two carbons from acetyl-CoA to oxaloacetate and generates citric acid. *S. aureus* appears to lack a citrate lyase that could reverse the citrate synthase catalysed condensation reaction; hence, this condensation reaction is the first committed step in the TCA cycle. As mentioned in Chapter 9, citrate is not only important for TCA cycle function but also the synthesis of staphylococcal siderophores. The second enzymatic reaction in the TCA cycle is the aconitase-mediated reversible isomerization of citric acid to isocitric acid via a *cis*-aconitate intermediate. Because the activity of aconitase is influenced by the availability of iron, oxidative stress, substrate availability, and other environmental stresses it is an excellent bellwether for the overall status of the bacterium (Somerville and Proctor, 2009). The genome of *S. aureus* encodes for a single aconitase (*acnA*/*citB*); hence, any disruption of aconitase activity (e.g. iron-limitation) will cause a metabolic block in the TCA cycle that limits carbon low through the TCA cycle. When the environment is conducive to aconitase activity, it produces isocitric acid which is the substrate for isocitrate dehydrogenase. Isocitrate dehydrogenase catalyses the irreversible decarboxylation of isocitrate to α-ketoglutarate and CO_2. This reaction requires $NADP^+$ as a co-factor to accept a hydrogen anion, which facilitates the generation of the intermediate oxalosuccinate (Prasad *et al.*, 2013). Subsequent to the formation of oxalosuccinate, the carboxyl group is deprotonated, facilitating the release of CO_2 and the formation of α-ketoglutarate. The α-ketoglutarate dehydrogenase complex [i.e. oxoglutarate dehydrogenase (E1), dihydrolipoyl succinyltransferase

(E2), and dihydrolipoyl dehydrogenase (E3)] catalyses the second decarboxylation step in the TCA cycle. To facilitate formation of succinyl-CoA from α-ketoglutarate, the α-ketoglutarate dehydrogenase complex requires thiamine pyrophosphate, CoA, FAD, NAD^+ and lipoate (an organosulfur molecule containing a ditholane ring). Processing of α-ketoglutarate requires three key steps: (1) decarboxylation, (2) reduction of NAD^+ to NADH, and (3) formation of a thioester bond between succinate and CoA. Once succinyl-CoA has been produced, it can used in biosynthesis (e.g. to make porphyrins) or it can be processed to make ATP, succinate, and free CoA (i.e. CoASH). This latter process requires the enzymes of the succinyl-CoA synthetase (aka, succinyl-CoA ligase); specifically, the succinyl-CoA synthetase alpha (SucD) and beta (SucC) subunits. Succinate can be oxidized to fumarate by the succinate dehydrogenase complex, which consists of three subunits: (1) succinate dehydrogenase flavoprotein subunit (SdhA); (2) succinate dehydrogenase iron–sulfur subunit (SdhB); and (3) succinate dehydrogenase cytochrome b-558 (SdhC). This latter subunit, SdhC, is membrane bound and serves to link the TCA cycle to the respiratory chain. Concomitant with the production of fumarate is the reduction of FAD to $FADH_2$, which transfers its electrons to menaquinone to generate menaquinol and fuel electron transport. Fumarate is reversibly converted to malate by fumarate hydratase (aka, fumarase) via a hydration/dehydration mechanism. Malate can be oxidized to oxaloacetate by malate dehydrogenase (aka, malic enzyme), a process that also reduces NAD^+ to NADH. Alternatively, the malate–quinone oxidoreductase (Mqo1) can oxidize malate to oxaloacetate by coupling it with the reduction of menaquinone to menaquinol. Interestingly, many of the *S. aureus* genomes are annotated as having two malate–quinone oxidoreductases (i.e. Mqo1 and Mqo2). Instead of having two malate–quinone oxidoreductases, these annotations highlight the need for human curation and experimental validation of protein functions, because the Mqo2 is actually a lactate–quinone oxidoreductase (Fuller *et al.*, 2011). Irrespective of the enzyme, the generation of oxaloacetate is significant because it allows for a new round of acetyl-CoA incorporation into the TCA cycle and it is the precursor for gluconeogenesis.

Gluconeogenesis

Upon the depletion of glucose from the cultivation medium, gluconeogenesis is needed to generate seven of the thirteen biosynthetic precursors (Table 6.1). As stated, depletion of glucose from the cultivation medium coincides with de-repression of the TCA cycle, a necessary step for gluconeogenesis because of the need for oxaloacetate. Gluconeogenesis begins with oxaloacetate being decarboxylated into pyruvate or PEP by pyruvate carboxylase or phosphoenolpyruvate carboxykinase (PckA), respectively (Michalik *et al.*, 2012). If pyruvate is produced, then pyruvate kinase (PykA) transfers a phosphate from ATP to pyruvate to generate PEP (Zoraghi *et al.*, 2010). The remainder of gluconeogenesis is essentially the reverse of glycolysis, except that the gluconeogenic glyceraldehyde-3-phosphate dehydrogenase (GapB) is different from the glycolytic enzyme (GapA) (Purves *et al.*, 2010). In *S. aureus*, gluconeogenesis is important during chronic infections such as osteomyelitis in which nutrient availability appears to be limited (Szafranska *et al.*, 2014); hence, this pathway could be an attractive therapeutic target for antimicrobials directed at persistent infections.

Amino sugar biosynthesis

Amino sugars (e.g. *N*-acetyl-glucosamine) are used in polymerizing reactions to make exopolysaccharides [i.e. capsule and polysaccharide intracellular adhesin (PIA), discussed in Chapter 5] and cell wall peptidoglycan (discussed in Chapter 8). Biosynthesis of amino sugars begins with the transfer of an amino group from glutamine to the glycolytic intermediate fructose-6-phosphate (Table 6.1) to generate glucosamine-6-phosphate, a process catalysed by glucosamine-fructose-6-phosphate aminotransferase (GlmS) (Komatsuzawa *et al.*, 2004). The GlmS catalysed transamination of fructose-6-phosphate represents the committed step in *N*-acetyl-glucosamine biosynthesis, an important step because this directs carbon away from glycolysis. This importance is highlighted by the fact that *glmS* mRNA stability and translation initiation are regulated by a *glmS*-riboswitch that responds to changes in the intracellular concentration of glucosamine-6-phosphate (Lunse *et al.*, 2011; and see Chapter 12). Glucosamine-6-phosphate is isomerized to glucosamine-1-phosphate via phosphoglucosamine mutase (GlmM). Interestingly, GlmM (initially identified as FemD (Jolly *et al.*, 1997)) was found to be important for methicillin resistance in the historical *S. aureus* strain COL (Jolly *et al.*, 1997) and the prototypical laboratory strain NCTC 8325 (Glanzmann *et al.*, 1999). The function of GlmM in mediating methicillin resistance is likely a reflection of the fact that UDP-activated *N*-acetyl-glucosamine is a biosynthetic precursor for *N*-acetylmuramic acid and *N*-acetylmannosamine, which are major constituents of peptidoglycan and wall teichoic acid (Chapter 8). Glucosamine-1-phosphate is acetylated and activated by *N*-acetylglucosamine-1-phosphate uridyltransferase/glucosamine-1-phosphate acetyltransferase (GlmU) to produce UDP-*N*-acetyl-glucosamine. UDP-*N*-acetyl-glucosamine is the biosynthetic precursor used in the synthesis of PIA; hence, its synthesis by GlmU has been proposed as a potential therapeutic target for treating biofilm associated infections (Burton *et al.*, 2006). While formation of sugar moieties in amino sugar biosynthesis is directly linked to glycolysis/gluconeogenesis, the amino group(s) is/are supplied by amino acids.

Amino acid synthesis

As mentioned, *S. aureus* is frequently described as an amino acid auxotroph; however, auxotrophic *S. aureus* strains often revert to a prototrophic state (Miller and Fung, 1973; Emmett and Kloos, 1975; Gladstone, 1937). The ability to revert to a prototrophic state implies that all of the genetic information necessary to encode for the biosynthesis of these amino acids is present within the genome. This is important because *S. aureus* must adapt to host/environmental niches where nutrients may be limiting and the ability to synthesize amino acids can determine whether a bacterium survives or dies. Central to amino acid biosynthesis are the amino acids glutamate and glutamine because they serve as the primary amino donors in biosynthetic reactions (Reitzer, 2003). Synthesis of glutamate usually involves one of two enzymes: glutamate dehydrogenase or glutamine oxoglutarate aminotransferase (aka glutamate synthase). Glutamate dehydrogenase drives ammonia assimilation by catalysing the reductive amination of α-ketoglutarate, while glutamate synthase converts glutamine and α-ketoglutarate into two molecules of glutamate by a transamidation reaction (Tempest *et al.*, 1970). Glutamine is synthesized from glutamate, ATP, and ammonia in a process catalysed by glutamine synthetase. Alternatively, glutamine can be synthesized from glutamyl charged tRNA through the action of a tRNA-dependent amidotransferase (i.e. aspartyl/glutamyl-tRNA amidotransferase; GatCAB) to generate a glutaminyl-tRNA (Ito and Yokoyama, 2010). While glutamate and glutamine are both important, glutamate is the nitrogen donor for most amino acids, with the exceptions being asparagine and tryptophan, and histidine, which uses both glutamate and glutamine (Reitzer, 2003).

The lack of experimental validation for many amino acid biosynthetic pathways in *S. aureus* hinders an exhaustive analysis of these pathways; hence, this section will focus on examples of amino acid biosynthesis. Amino acid biosynthesis can be as simple as a transamination of one of the thirteen biosynthetic intermediates of central metabolism (Table 6.1). As an example, alanine can be synthesized from pyruvate and glutamate by the enzyme alanine aminotransferase. In addition, amino acids can be synthesized by the transamination of another amino acid, such as asparagine. In many bacteria, asparagine is synthesized by asparagine synthetase from aspartate, glutamine, and ATP; however, *S. aureus* lacks an asparagine synthetase. Despite the absence of asparagine synthetase, *S. aureus* is not an asparagine auxotroph because it uses asparaginyl-tRNA synthetase and the GatCAB amidotransferase to produce asparaginyl-tRNA (Mladenova *et al.*, 2014). For some amino acids, biosynthesis is a much more complex process involving multiple enzymes or enzyme complexes with several layers of regulation. For example, methionine biosynthesis requires the proteins encoded within the *met* operon; specifically, *metI* (cystathionine γ-synthase), *metC* (cystathionine β-lyase), *metF* (methylenetetrahydrofolate reductase), *metE* (methionine synthase), and *mdh* (metal-dependent hydrolase) (Schoenfelder *et al.*, 2013). These enzymes are part of the transsulfuration pathway of methionine biosynthesis (Rodionov *et al.*, 2004). Synthesis of methionine begins with aspartate being phosphorylated by aspartate kinase, an ATP requiring process. Aspartate semialdehyde dehydrogenase utilizes the oxidation of NADPH to drive the conversion of aspartyl-4-phosphate into aspartate semialdehyde, which is further processed into homoserine by homoserine dehydrogenase. Homoserine is acetylated by homoserine O-acetyltransferase and acetyl-CoA. Sulfate assimilation is likely accomplished by cystathionine γ-synthase and cystathionine β-lyase when cysteine is incorporated with O-acetyl-homoserine to generate cystathionine, which is cleaved to yield homocysteine. Alternatively, hydrogen sulfide can be used for sulfate assimilation; however, the gene coding for O-acetylhomoserine aminocarboxypropyltransferase (*metY*) appears to be missing in the *S. aureus* genome. Irrespective of the pathway to generate homocysteine, a methyl group is transferred from 5-methyltetrahydrofolate to homocysteine by methionine synthase to generate methionine. Interestingly, methionine biosynthesis in *S. aureus* is regulated by a T-box riboswitch mechanism as opposed to an S-adenosylmethionine-binding S-box riboswitch, which is found in the closely related *Bacillus* spp. (Schoenfelder *et al.*, 2013).

Nucleic acid synthesis

Most research involving nucleic acid 'biosynthesis' in *S. aureus* relates to the effects of antibiotics on cell physiology and metabolism (e.g. Jordan and Inniss, 1959; discussed in Chapter 11); hence, there is little *Staphylococcus*-specific literature to review. Unfortunately, this creates a gap in our knowledge of staphylococcal metabolism that can only be filled by analogy to the model organisms *Escherichia coli* and *Bacillus subtilis*. For this reason, readers are referred to reviews of *B. subtilis* nucleic acid biosynthesis (e.g. Switzer *et al.*, 2002). That being said and in keeping with the theme of this chapter, I want to point out how nucleic acid biosynthesis relies on one of the thirteen canonical biosynthetic intermediates (Table 6.1); specifically, the pentose phosphate pathway intermediate ribose-5-phophate. Both pyrimidine and purine biosynthesis rely on 5-phospho-α-D-ribose 1-diphosphate (PRPP) as the source of the ribosyl group. Synthesis of PRPP is accomplished by the enzyme ribose-phosphate pyrophosphokinase (aka, PRPP synthetase) from ribose-5-phophate and ATP. Not only is PRPP used in nucleotide biosynthesis, but it is also essential for tryptophan and histidine biosynthesis.

Conclusion

Although *S. aureus* lacks much of the metabolic overlap/redundancy found in Gram-negative bacteria, it is still a metabolically robust bacterium. Nowhere is this robustness more apparent than in the metabolic adaptations that permit *S. aureus* to survive in the presence of sublethal concentrations of antibiotics (Gaupp *et al.*, 2015; Nelson *et al.*, 2007; Zhang *et al.*, 2011). These adaptations often involve changes in central metabolism, which alters the availability of the thirteen biosynthetic intermediates used to make all macromolecules in and on the bacterium. If the available concentration of a biosynthetic intermediate drops below that necessary for the synthesis of precursors (e.g. amino acids, nucleic acids), then metabolic blocks are created within the bacterium and growth will be impaired. To circumvent these metabolic blocks and continue dividing, *S. aureus* has evolved or acquired virulence factors that liberate nutrients from host cells, transporters to take up exogenous nutrients, and salvage and alternative metabolic pathways to process these nutrients. This ability to adapt to environmental challenges also explains why regulation of virulence determinants and metabolism are intimately linked (Richardson *et al.*, 2015). Importantly, when studying *S. aureus* in the laboratory, the interconnectedness of metabolism and regulation means that fervent attention to bacterial cultivation must be practised.

Web resources

BioCyc Database Collection: http://biocyc.org/. This website is an outstanding and comprehensive resource of metabolic pathways found in genome sequenced organisms.
KEGG: Kyoto Encyclopedia of Genes and Genomes: http://www.genome.jp/kegg/. A multi-purpose resource for exploring gene functions and relationships across a wide breadth of organisms.
BRENDA The Comprehensive Enzyme Information System: http://www.brenda-enzymes.org/index.php. This is a well-organized collection of information regarding enzyme functions.
Sigma-Aldrich Company enzymatic assay library: http://www.sigmaaldrich.com/life-science/metabolomics/enzyme-explorer/learning-center/assay-library.html. This is a good starting point for those wishing to assay the enzymatic activity of common enzymes.

References

Blumenthal, H.J. (1972). Glucose catabolism in staphylococci. In: Cohen, J.O. (ed.) The Staphylococci. New York: Wiley-Interscience.

Burton, E., Gawande, P.V., Yakandawala, N., Lovetri, K., Zhanel, G.G., Romeo, T., Friesen, A.D., and Madhyastha, S. (2006). Antibiofilm activity of GlmU enzyme inhibitors against catheter-associated uropathogens. Antimicrob. Agents Chemother. *50*, 1835–1840.

Cassat, J., Dunman, P.M., Murphy, E., Projan, S.J., Beenken, K.E., Palm, K.J., Yang, S.J., Rice, K.C., Bayles, K.W., and Smeltzer, M.S. (2006). Transcriptional profiling of a *Staphylococcus aureus* clinical isolate and its isogenic *agr* and *sarA* mutants reveals global differences in comparison to the laboratory strain RN6390. Microbiology *152*, 3075–3090.

Dickens, F. (1938). Oxidation of phosphohexonate and pentose phosphoric acids by yeast enzymes: Oxidation of phosphohexonate. II. Oxidation of pentose phosphoric acids. Biochem. J. *32*, 1626–1644.

Dyke, K. G., Jevons, M. P., and Parker, M. T. 1966. Penicillinase production and intrinsic resistance to penicillins in *Staphylococcus aureus*. Lancet *1*, 835–838.

Emmett, M., and Kloos, W.E. (1975). Amino acid requirements of staphylococci isolated from human skin. Can. J. Microbiol. *21*, 729–733.

Fuller, J.R., Vitko, N.P., Perkowski, E.F., Scott, E., Khatri, D., Spontak, J.S., Thurlow, L.R., and Richardson, A.R. (2011). Identification of a lactate-quinone oxidoreductase in *Staphylococcus aureus* that is essential for virulence. Front. Cell. Infect. Microbiol. *1*, 19.

Gardner, J. F., and Lascelles, J. 1962. The requirement for acetate of a streptomycin-resistant strain of *Staphylococcus aureus*. J. Gen. Microbiol. 29, 157–164.

Garrard, W., andand Lascelles, J. (1968). Regulation of *Staphylococcus aureus* lactate dehydrogenase. J. Bacteriol. *95*, 152–156.

Gaupp, R., Lei, S., Reed, J.M., Peisker, H., Boyle-Vavra, S., Bayer, A.S., Bischoff, M., Herrmann, M., Daum, R.S., Powers, R., and Somerville, G.A. (2015). *Staphylococcus aureus* metabolic adaptations during the transition from a daptomycin susceptibility phenotype to a daptomycin nonsusceptibility phenotype. Antimicrob. Agents Chemother. *59*, 4226–4238.

Gill, S.R., Fouts, D.E., Archer, G.L., Mongodin, E.F., Deboy, R.T., Ravel, J., Paulsen, I.T., Kolonay, J.F., Brinkac, L., Beanan, M., Dodson, R.J., Daugherty, S.C., Madupu, R., Angiuoli, S.V., Durkin, A.S., Haft, D.H., Vamathevan, J., Khouri, H., Utterback, T., Lee, C., Dimitrov, G., Jiang, L., Qin, H., Weidman, J., Tran, K., Kang, K., Hance, I.R., Nelson, K.E., and Fraser, C.M. (2005). Insights on evolution of virulence and resistance from the complete genome analysis of an early methicillin-resistant *Staphylococcus aureus* strain and a biofilm-producing methicillin-resistant *Staphylococcus epidermidis* strain. J. Bacteriol. *187*, 2426–2438.

Gladstone, G.P. (1937). The nutrition of *Staphylococcus aureus*; nitrogen requirements. Br. J. Exp. Pathol., *18*, 322–333.

Glanzmann, P., Gustafson, J., Komatsuzawa, H., Ohta, K., and Berger-Bächi, B. (1999). *glmM* operon and methicillin-resistant *glmM* suppressor mutants in *Staphylococcus aureus*. Antimicrob. Agents Chemother., *43*, 240–245.

Hartmann, T., Baronian, G., Nippe, N., Voss, M., Schulthess, B., Wolz, C., Eisenbeis, J., Schmidt-Hohagen, K., Gaupp, R., Sunderkötter, C., Beisswenger, C., Bals, R., Somerville, G.A., Herrmann, M., Molle, V., and Bischoff, M. (2014). The catabolite control protein E (CcpE) affects virulence determinant production and pathogenesis of *Staphylococcus aureus*. J. Biol. Chem. *289*, 29701–29711.

Hartmann, T., Zhang, B., Baronian, G., Schulthess, B., Homerova, D., Grubmüller, S., Kutzner, E., Gaupp, R., Bertram, R., Powers, R., Eisenreich, W., Kormanec, J., Herrmann, M., Molle, V., Somerville, G.A., and Bischoff, M. (2013). Catabolite control protein E (CcpE) is a LysR-type transcriptional regulator of tricarboxylic acid cycle activity in *Staphylococcus aureus*. J. Biol. Chem. *288*, 36116–36128.

Huynen, M.A., Dandekar, T., and Bork, P. (1999). Variation and evolution of the citric-acid cycle: a genomic perspective. Trends Microbiol. *7*, 281–291.

Ito, T., and Yokoyama, S. (2010). Two enzymes bound to one transfer RNA assume alternative conformations for consecutive reactions. Nature, *467*, 612–616.

Jolly, L., Wu, S., Van Heijenoort, J., De Lencastre, H., Mengin-Lecreulx, D., and Tomasz, A. (1997). The *femR315* gene from *Staphylococcus aureus*, the interruption of which results in reduced methicillin resistance, encodes a phosphoglucosamine mutase. J. Bacteriol. *179*, 5321–5325.

Jordan, D.C., and Inniss, W.E. (1959). Selective inhibition of ribonucleic acid synthesis in *Staphylococcus aureus* by vancomycin. Nature *184*(Suppl. 24), 1894–1895.

Kendall, A.I., Friedemann, T.E., and Ishikawa, M. (1930). Quantitative observations on the chemical activity of "resting" *Staphylococcus aureus*. J. Infect. Dis., 47, 223–228.

Knight, B.C. (1937). The nutrition of *Staphylococcus aureus*; nicotinic acid and vitamin B(1). Biochem. J. *31*, 731–737.

Komatsuzawa, H., Fujiwara, T., Nishi, H., Yamada, S., Ohara, M., Mccallum, N., Berger-Bächi, B., and Sugai, M. (2004). The gate controlling cell wall synthesis in *Staphylococcus aureus*. Mol. Microbiol. *53*, 1221–1231.

Krebs, H.A. (1937). Dismutation of pyruvic acid in *Gonococcus* and *Staphylococcus*. Biochem. J. *31*, 661–671.

Kundig, W., Ghosh, S., and Roseman, S. (1964). Phosphate bound to histidine in a protein as an intermediate in a novel phospho-transferase system. Proc. Natl. Acad. Sci. U.S.A, *52*, 1067–1074.

Kuroda, M., Ohta, T., Uchiyama, I., Baba, T., Yuzawa, H., Kobayashi, I., Cui, L., Oguchi, A., Aoki, K., Nagai, Y. *et al.* (2001). Whole genome sequencing of meticillin-resistant *Staphylococcus aureus*. Lancet *357*, 1225–1240.

Ledala, N., Zhang, B., Seravalli, J., Powers, R., and Somerville, G.A. (2014). Influence of iron and aeration on *Staphylococcus aureus* growth, metabolism, and transcription. J. Bacteriol. *196*, 2178–2189.

Lünse, C.E., Schmidt, M.S., Wittmann, V., and Mayer, G. (2011). Carba-sugars activate the *glmS*-riboswitch of *Staphylococcus aureus*. ACS Chem. Biol. *6*, 675–678.

Meléndez-Hevia, E., Waddell, T.G., and Cascante, M. (1996). The puzzle of the Krebs citric acid cycle: assembling the pieces of chemically feasible reactions, and opportunism in the design of metabolic pathways during evolution. J. Mol. Evol. *43*, 293–303.

Michalik, S., Bernhardt, J., Otto, A., Moche, M., Becher, D., Meyer, H., Lalk, M., Schurmann, C., Schlüter, R., Kock, H., *et al.* (2012). Life and death of proteins: a case study of glucose-starved *Staphylococcus aureus*. Mol. Cell. Proteomics *11*, 558–570.

Miller, R.D., and Fung, D.Y. (1973). Amino acid requirements for the production of enterotoxin B by *Staphylococcus aureus* S-6 in a chemically defined medium. Appl. Microbiol. 25, 800–806.

Mladenova, S.R., Stein, K.R., Bartlett, L., and Sheppard, K. (2014). Relaxed tRNA specificity of the *Staphylococcus aureus* aspartyl-tRNA synthetase enables RNA-dependent asparagine biosynthesis. FEBS Lett. *588*, 1808–1812.

Neidhardt, F.C. (2006). Apples, oranges and unknown fruit. Nat. Rev. Microbiol. *4*, 876.

Nelson, J. L., Rice, K. C., Slater, S. R., Fox, P. M., Archer, G. L., Bayles, K. W., Fey, P. D., Kreiswirth, B. N., and Somerville, G. A. 2007. Vancomycin-intermediate *Staphylococcus aureus* strains have impaired acetate catabolism: implications for polysaccharide intercellular adhesin synthesis and autolysis. Antimicrob. Agents Chemother. *51*, 616–622.

Novick, R. (1967). Properties of a cryptic high-frequency transducing phage in *Staphylococcus aureus*. Virology *33*, 155–166.

Novick, R.P. (1991). Genetic systems in staphylococci. *Methods in Enzymology.* San Diego: Academic Press, Inc.

Prasad, U.V., Vasu, D., Kumar, Y.N., Kumar, P.S., Yeswanth, S., Swarupa, V., Phaneendra, B.V., Chaudhary, A., and Sarma, P.V. (2013). Cloning, expression and characterization of NADP-dependent isocitrate dehydrogenase from *Staphylococcus aureus*. Appl. Biochem. Biotechnol. *169*, 862–869.

Purves, J., Cockayne, A., Moody, P.C., and Morrissey, J.A. (2010). Comparison of the regulation, metabolic functions, and roles in virulence of the glyceraldehyde-3-phosphate dehydrogenase homologues *gapA* and *gapB* in *Staphylococcus aureus*. Infect. Immun. *78*, 5223–5232.

Reitzer, L. (2003). Nitrogen assimilation and global regulation in *Escherichia coli*. Annu. Rev. Microbiol. *57*, 155–176.

Richardson, A.R., Somerville, G.A., and Sonenshein, A.L. (2015). Regulating the intersection of metabolism and pathogenesis in Gram-positive bacteria. Microbiol. Spectr. 3.

Rodionov, D.A., Vitreschak, A.G., Mironov, A.A., and Gelfand, M.S. (2004). Comparative genomics of the methionine metabolism in Gram-positive bacteria: a variety of regulatory systems. Nucleic Acids Res. *32*, 3340–3353.

Sadykov, M.R., Thomas, V.C., Marshall, D.D., Wenstrom, C.J., Moormeier, D.E., Widhelm, T.J., Nuxoll, A.S., Powers, R., and Bayles, K.W. (2013). Inactivation of the Pta-AckA pathway causes cell death in *Staphylococcus aureus*. J. Bacteriol. *195*, 3035–3044.

Sadykov, M.R., Zhang, B., Halouska, S., Nelson, J.L., Kreimer, L.W., Zhu, Y., Powers, R., and Somerville, G.A. (2010). Using NMR metabolomics to investigate tricarboxylic acid cycle dependent signal transduction in *Staphylococcus epidermidis*. J. Biol. Chem. *285*, 36616–36624.

Schoenfelder, S.M., Marincola, G., Geiger, T., Goerke, C., Wolz, C., and Ziebuhr, W. (2013). Methionine biosynthesis in *Staphylococcus aureus* is tightly controlled by a hierarchical network involving an initiator tRNA-specific T-box riboswitch. PLoS Pathog., *9*, e1003606.

Shah, N., Gaupp, R., Moriyama, H., Eskridge, K. M., Moriyama, E. N., and Somerville, G. A. 2013. Reductive evolution and the loss of PDC/PAS domains from the genus *Staphylococcus*. BMC Genomics *14*, 524.

Somerville, G. A., Chaussee, M. S., Morgan, C. I., Fitzgerald, J. R., Dorward, D. W., Reitzer, L. J., and Musser, J. M. 2002. *Staphylococcus aureus* aconitase inactivation unexpectedly inhibits post-exponential-phase growth and enhances stationary-phase survival. Infect. Immun. *70*, 6373–6382.

Somerville, G.A., Cockayne, A., Dürr, M., Peschel, A., Otto, M., and Musser, J.M. (2003). Synthesis and deformylation of *Staphylococcus aureus* delta-toxin are linked to tricarboxylic acid cycle activity. J. Bacteriol. *185*, 6686–6694.

Somerville, G.A., and Proctor, R.A. (2009). At the crossroads of bacterial metabolism and virulence factor synthesis in Staphylococci. Microbiol. Mol. Biol. Rev. *73*, 233–248.

Somerville, G.A., Saïd-Salim, B., Wickman, J.M., Raffel, S.J., Kreiswirth, B.N., and Musser, J.M. (2003). Correlation of acetate catabolism and growth yield in *Staphylococcus aureus*: Implications for host-pathogen interactions. Infect. Immun. *71*, 4724–4732.

Stockland, A.E., and San Clemente, C.L. (1969). Multiple forms of lactate dehydrogenase in *Staphylococcus aureus*. J. Bacteriol. *100*, 347–353.

Switzer, R.L., Zalkin, H., and Saxild, H.H. (2002). Purine, pyrimidine, and pyridine nucleotide metabolism. In: Sonenshein, A.L., Hoch, J.A., and Losick, R. (eds.) *Bacillus subtilis* and Its Closest Relatives. Washington, D.C.: ASM Press.

Szafranska, A.K., Oxley, A.P., Chaves-Moreno, D., Horst, S.A., Rosslenbroich, S., Peters, G., Goldmann, O., Rohde, M., Sinha, B., Pieper, D.H., *et al.* (2014). High-resolution transcriptomic analysis of the adaptive response of *Staphylococcus aureus* during acute and chronic phases of osteomyelitis. MBio, *5*.

Tempest, D.W., Meers, J.L., and Brown, C.M. (1970). Synthesis of glutamate in *Aerobacter aerogenes* by a hitherto unknown route. Biochem. J. *117*, 405–407.

Warburg, O., Christian, W., and Griese, A. (1935). Hydrogen-transferring coenzyme, its composition and mode of action. Biochem. Z. *282*, 157–205.

Zhang, B., Halouska, S., Schiaffo, C.E., Sadykov, M.R., Somerville, G.A., and Powers, R. (2011). NMR analysis of a stress response metabolic signaling network. J. Proteome Res. *10*, 3743–3754.

Zhu, Y., Nandakumar, R., Sadykov, M.R., Madayiputhiya, N., Luong, T.T., Gaupp, R., Lee, C.Y., and Somerville, G.A. (2011). RpiR homologues may link *Staphylococcus aureus* RNAIII synthesis and pentose phosphate pathway regulation. J. Bacteriol. *193*, 6187–6196.

Zhu, Y., Xiong, Y.Q., Sadykov, M.R., Fey, P.D., Lei, M.G., Lee, C.Y., Bayer, A.S., and Somerville, G.A. (2009). Tricarboxylic acid cycle-dependent attenuation of *Staphylococcus aureus in vivo* virulence by selective inhibition of amino acid transport. Infect. Immun. *77*, 4256–4264.

Zoraghi, R., See, R.H., Gong, H., Lian, T., Swayze, R., Finlay, B.B., Brunham, R.C., Mcmaster, W.R., and Reiner, N.E. (2010). Functional analysis, overexpression, and kinetic characterization of pyruvate kinase from methicillin-resistant *Staphylococcus aureus*. Biochemistry *49*, 7733–7747.

Physiological Proteomics of *Staphylococcus aureus*: from the Protein Inventory to Stress Physiology and *in vivo* Adaptation

7

Susanne Engelmann, Stephan Fuchs and Michael Hecker

Abstract

Here we present the different facets of proteomics turning it into a valuable tool for a more comprehensive understanding of metabolism and stress and starvation responses of *Staphylococcus aureus*. Relying on a combination of gel-based and gel-free proteomics procedures, new fields in staphylococci physiology such as phosphoproteomics, protein damage or proteolysis on a proteome-wide scale are addressed. Finally, we show that stress/starvation proteomics signature libraries are essential diagnostic tools for the analysis of the physiological state of *S. aureus* cells under both *in vitro* and *in vivo* conditions.

Introduction

In 2001, a new era of *Staphylococcus aureus* research was opened by the publication of the first two genome sequences of this pathogen (Kuroda *et al.*, 2001). With the completion of 54 genome sequences and more than 4000 sequencing projects currently under way, *S. aureus* belongs to the top ten of sequenced bacteria (BioProject; http://www.ncbi.nlm.nih.gov/bioproject). Despite this impressive number of complete genome sequences, the pan-genome of *S. aureus* has not yet been closed indicating that the genomic diversity of *S. aureus* is very high (Tettelin *et al.*, 2008) (unpublished data). Genome sequences alone do not explain life but they do provide the essential basis for all the other OMICs technologies in general and for proteomics in particular. These in turn are what are required to understand how the 'blueprint of life' controls and directs all physiological processes of a living cell in its interaction with its environment. With these techniques at hand, a completely new view of virulence and physiology of *S. aureus* has become possible.

Because proteins are the 'workhorses' of cellular life, it is the proteome that translates the genome sequence into cellular function. In contrast to the genome, the proteome is significantly more dynamic and allows the rapid adaptation of a bacterium to changing conditions. To understand bacterial physiology in molecular detail and in its whole complexity, the proteome in all its variants has to be analysed. Ultimately, all living processes depend on the number and nature of the active protein molecules available. While the quantity of a protein depends on its synthesis rate and stability, the activity of a protein can be additionally influenced by posttranslational modifications, protein–protein interactions and interactions with metabolites. Shortly after or even during synthesis, specific amino acid residues of a protein become chemically modified, which may alter physical and chemical properties such as folding and stability. Proteins can also work together to achieve a particular function, and they often associate to form stable protein complexes.

The accumulated knowledge on the state of all the proteins of *S. aureus* under both defined laboratory conditions and during infection, in combination with its genome information and metabolites, have brought us a quantum leap forward in understanding the physiology of *S. aureus*, and this has revolutionized *S. aureus* research since 2001.

The protein inventory of *S. aureus*

Proteomics is defined as the analysis of the whole set of proteins (= proteome) of a given cell or organism. Because of their low complexity, bacteria are an attractive model in which to carry out these analyses. From the genome sequences of *S. aureus*, it was predicted that *S. aureus* codes for at least 2400 proteins, with 75% of these proteins being conserved amongst *S. aureus*. Using bioinformatic tools, the subcellular localization of each protein was predicted (Zhou *et al.*, 2008), and it was found that about 30% of the *S. aureus* proteome is localized outside the cytoplasm (Kusch and Engelmann, 2014; Sibbald *et al.*, 2006). However, there are still questions that need be clarified: (i) Under which conditions and in what quantities are these proteins produced? (ii) Which proteins interact with each other? (iii) How stable are the individual proteins; and (iv) Are they modified or (v) immunogenic?

A very expensive but simple method to separate and identify proteins of complex crude extracts is the two-dimensional (2-D) polyacrylamide gel electrophoresis (PAGE) technique combined with mass spectrometry (MS) (Fig. 7.1). This technique allows the separation of more than one thousand proteins in a 20 × 20 cm area and was originally invented by O'Farrel (1975) and Klose (1975). With this technique, 521 cytosolic *S. aureus* proteins have been identified that cover about 30% of the entire cytoplasmic proteome predicted for *S. aureus* strain COL and 41% of the cytoplasmic proteome predicted for the analytical proteomic window of pI 4 to 7 (Fuchs *et al.*, 2013; Becher *et al.*, 2009). Among these proteins, most belong to metabolism: 42% of the identified proteins are involved in biosynthetic and metabolic pathways, 13% in gene expression and its regulation, 6% in protein fate, and 2% in transport processes (Fuchs *et al.*, 2013). By expanding the pI range up to 11 the total number of identified cytosolic proteins was increased to almost 700 (Becher *et al.*, 2009). Moreover, 66 extracellular proteins were identified by analysing secreted proteins in different *S. aureus* strains and these represent about 50% of the extracellular proteins predicted to be encoded within *S. aureus* genomes (Ziebandt *et al.*, 2001, 2004, 2010; Kusch and Engelmann, 2014; Rogasch *et al.*, 2006; Becher *et al.*, 2009). Unfortunately, essential limitations of this technique cannot be ignored: many proteins such as hydrophobic proteins (i.e. most membrane proteins), proteins with extreme pI values and molecular weights, and very low abundancy proteins escape detection by 2-D PAGE based proteomics. Hence, during the last decade, proteomics based on 2-D gels has been more and more replaced by mass spectrometry (MS) of peptide mixtures generated using mild ionization methods and online liquid chromatography (LC-MS/MS). This approach provides a completely new quality in proteomic research. Protein or peptide mixtures may be prefractionated, either by one-dimensional (1-D) PAGE (GeLC-MS/MS) or liquid chromatography (2-D LC-MS/MS), before the peptides are identified by ultra-performance liquid chromatography (UPLC) coupled MS/MS analysis (Fig. 7.1). When applied to *S. aureus*, the GeLC-MS/MS approach led to the identification of more than 1426 cytosolic proteins, covering 79% of the cytosolic proteins encoded by the genome of *S. aureus* strain COL. Among them, are more than 500 new proteins most of which are of low abundance, very acidic, very alkaline, very large, or very small. Moreover, 64% of the membrane proteins, 95% of the lipoproteins, 95% of the sortase anchored proteins, 100% of other cell wall associated proteins, and 79% of the extracellular proteins encoded by the genome of *S. aureus* COL have been identified in this way. This corresponds

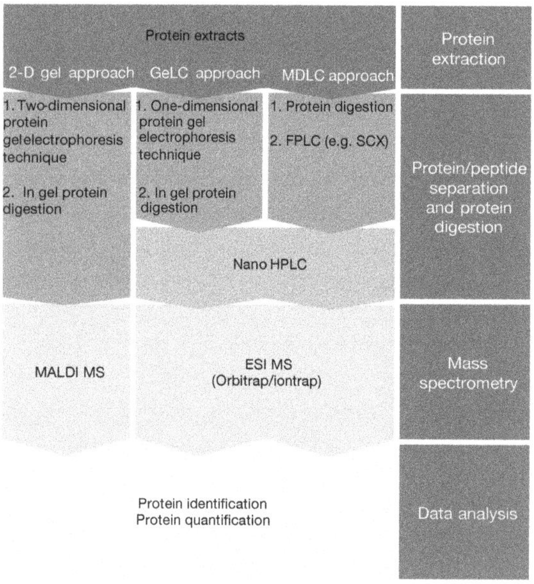

Figure 7.1 Typical workflow of proteome analyses.

to 76% coverage of the predicted proteome of *S. aureus* strain COL (Otto *et al.*, 2014). In total, 1450 proteins were quantified in this study, resulting in a coverage of approximately three-quarters of the expressed proteins (Becher *et al.*, 2009).

The identification of the protein inventory of *S. aureus* is certainly a substantial achievement; however, it is just the beginning of a long road towards the elucidation of the molecular mechanisms governing the life of this pathogen. The next step is to unravel how these different proteins once released from ribosomes organize cellular life.

Protein fate from birth to death on a proteome-wide scale: protein phosphorylation, protein protection, protein damage and proteolysis

Once formed, proteins exist for only a certain period of time and are then degraded and recycled by the cell's machinery through the processes of protein turnover. Abnormal and misfolded proteins are degraded more rapidly than fully active proteins. Additionally, proteins can be modified so as to regulate their activity, for example, by phosphorylation or to protect them from damage, for example by thiolation. All these processes are highly dynamic within a bacterial cell, highlighting the need to not only detect all these proteins, but also to address all the different events that might befall proteins during their life as part of the native proteome.

Protein phosphorylation at a proteome-wide scale

Phosphorylation events on proteins during growth and stress/starvation can represent crucial regulatory processes inside the bacterial cell. In particular, the activity of regulatory proteins is frequently modified by phosphorylation and dephosphorylation events. Serine, threonine and tyrosine phosphorylated proteins in *S. aureus* have been analysed using different proteomic approaches. By separating *in vivo* ^{32}P-labelled protein extracts on 2-D gels followed by MS, 11 phosphoproteins were identified. Phosphorylation of these proteins was primarily catalysed by serine/threonine kinase activity and to a lesser extent by tyrosine kinase activity (Lomas-Lopez *et al.*, 2007). Very recently, the number of identified serine, threonine, and tyrosine phosphorylated proteins has been significantly increased using 2-D gel analysis with a phosphosensitive stain (Pro-Q Diamond) in combination with the gel-free titanium dioxide based enrichment of phosphopeptides. As a result, 103 putative phosphorylated proteins with 68 different phosphorylation sites were found in the soluble protein fraction; among them, 38 serine, 19 threonine and 11 tyrosine target residues were identified. Additionally, eight proteins phosphorylated on arginine residues have been identified (Bäsell *et al.*, 2014).

In both studies, several proteins related to central metabolic pathways and to the virulence of *S. aureus* were phosphorylated. Among them are the virulence regulator SarA, various glycolytic enzymes such as fructose-bisphosphate aldolase (FbaA) and triose phosphate isomerase (TpiA), and the elastin-binding protein EbpS (Lomas-Lopez *et al.*, 2007; Bäsell *et al.*, 2014). Of note, more than one phosphorylation site has been identified for some of these proteins. Three phosphorylation events on highly conserved positions have been detected in FbaA. Two distinct phosphorylation sites have been mapped for SarA, which is a target of the kinases PknB and SA0077 (Didier *et al.*, 2010). Both phosphorylation sites correlate to those of the virulence regulator MrgA, which is also a member of the SarA family of regulatory proteins in *S. aureus*. These phosphorylation events on MrgA have been shown to be important for its regulatory activity; hence, these results imply that the same may be true for SarA phosphorylation (Truong-Bolduc and Hooper, 2010; Bäsell *et al.*, 2014).

Since phosphorylation events might modify the activity of proteins, and thus fine tune metabolic and virulence activity directly by the activation and inactivation of key enzymes or indirectly by affecting global gene expression, changes in the protein phosphorylation patterns in response to growth conditions may reflect physiological changes taking place inside the bacterial cell. Differential analysis of phosphorylation patterns in response to changing growth conditions (e.g. glucose starvation and osmotic, oxidative, and nitrosative stress) using 2-D gels showed significant changes in phosphorylation for ten proteins in *S. aureus*. Most changes were found in response to glucose starvation and nitrosative stress. Among them are changes in the phosphorylation level of phosphorylated isoforms of DeoB, Pgm, SACOL1588, Pyk, and SACOL1483 under glucose starvation. Moreover, newly appearing phosphorylated isoforms of FdaB and HchA were detected under nitrosative stress conditions (Bäsell *et al.*, 2014). It seems very likely that phosphorylation of these proteins might be a rapid and reversible way for modulating metabolic activity – in particular glycolytic activity – under these conditions. However, to understand the physiological consequences, the role of these phosphorylation events in modulating the respective protein's activity needs be studied in greater detail.

Protein protection and damage by oxidative stress

Protein damage caused by oxidative stress includes carbonylation, oxidation of methionine to methionine sulfoxide, and the formation of non-native disulfide bridges or over oxidation of cysteine residues. Protein thiol groups are very sensitive to oxidation, particularly under oxidative stress conditions, and this modulates the activity of a growing number of proteins (Hondorp and Matthews, 2004; Jakob *et al.*, 1999; Weber *et al.*, 2004; Zheng *et al.*, 1998). Thiol groups of cysteinyl residues in proteins fulfil important and versatile cellular functions: They are involved in enzymatic catalysis, in metal coordination, and in determining the structure of proteins by formation of disulfide bounds. Because of their strong nucleophilic character they are easily oxidizable by reactive oxygen species (ROS) and reactive nitrogen species (RNS), leading to disulfides, to sulfenic, sulfinic, or sulfonic acids, mixed disulfides with low molecular weight (LMW) thiols (S-thiolations), and S-nitrosylations (Barford, 2004; Demple, 2002). Reversible S-thiolations, such as disulfides, are thought to protect protein thiol residues from irreversible oxidation to sulfinic and sulfonic acid, thereby preserving the structure and function of the respective proteins. They can be detected by using the fluorescence thiol-modification assay. Briefly, reversibly oxidized thiols are selectively labelled with a fluorescent dye. The protein mixture can then be separated by 2-D PAGE and the fluorescence/protein ratio can be calculated for each individual protein (Hochgräfe *et al.*, 2005). In growing *S. aureus* cells, most of protein thiols were found in the reduced state. In the presence of diamide, a strong increase of reversibly oxidized thiols has been observed in a variety of metabolic enzymes. By contrast, hydrogen peroxide mainly induced reversible oxidation of enzymes containing cysteine residues within their active site. Interestingly, paraquat did not significantly increase the thiol-oxidation of proteins in *S. aureus* (Wolf *et al.*, 2008).

S-thiolation is frequently achieved by the formation of mixed disulfides with LMW-thiols, which are small molecules with diverse functions inside the cell. They serve numerous roles including as detoxifiers of ROS and RNS or xenobiotics, or as buffer reagents to maintain the cellular redox status within the cell. The most common LMW is gluthathione; however, many Gram-positive bacteria including *S. aureus* lack glutathione (Newton *et al.*, 1996). It is noticeable that both *Bacillus subtilis* and *S. aureus* respond to disulfide stress with an induced expression of cysteine biosynthesis (Leichert *et al.*, 2003; Wolf *et al.*, 2008). *In vivo* labelling of proteins with [^{35}S]-cysteine and non-reducing 2-D PAGE analyses revealed that a large subset of proteins previously identified as having redox-sensitive thiols are modified by cysteine in response to treatment with the thiol-specific oxidant diamide, which is possibly accompanied by a higher consumption of cysteine in these bacteria (Hochgräfe *et al.*, 2007; Pöther *et al.*, 2009). By means of multidimensional shotgun proteomics, the sites of S-cysteinylation for at least seven proteins were identified in *S. aureus* (Pöther *et al.*, 2009). This cysteinylation can be regarded as protection against an irreversible oxidation to sulfonic acid, which may lead to a complete inactivation of the protein, as described for glyceraldehyde dehydrogenase in *S. aureus* (Weber *et al.*, 2004). In addition to cysteine, bacillithiol (Cys-GlcN-malate, BSH) the major LMW-thiol in low GC Gram-positive bacteria, has a significant role in protein protection against oxidation in *S. aureus*. Using HPLC, varying levels of bacillithiol have been identified in different *S. aureus* strains. Furthermore, it has been shown that BSH-producing *S. aureus* strains had higher resistance against the antibiotic fosfomycin and the oxidant hypochlorite but not against hydrogen peroxide or diamide. In addition, a higher bacterial load of BSH-producing strains was detected in human blood as well as in human upper-airway epithelial cells and murine macrophages revealing a potential role of BSH in protection of *S. aureus* during infection (Pöther *et al.*, 2013; Posada *et al.*, 2014).

Proteolysis at a proteome-wide scale and the role of Clp proteins

The lifetime of proteins is limited and is terminated by their proteolytic degradation. The individual stability of bacterial proteins can vary considerably. The major system involved in protein degradation in bacteria is the Clp machinery. The ATP dependent Clp proteases are highly conserved and widely distributed among eubacteria. The main substrates for the Clp protease are damaged and misfolded proteins. In non-growing cells, there is an enhanced degradation of vegetative proteins required for growth and reproduction. Using an *in vivo* pulse-chase labelling approach with amino acids containing stable isotopes combined with a GeLC-MS/MS approach, protein degradation in the human pathogen *S. aureus* during glucose starvation was analysed. Altogether, about 1400 proteins were identified and the protein stability profiles of approximately 900 proteins were monitored in a wild-type and an isogenic *clpP* mutant. The results showed that (i) proteolysis mainly affected proteins with vegetative functions as well as anabolic and selected catabolic enzymes, whereas the level of tricarboxylic acid (TCA) cycle and gluconeogenesis enzymes increased; (ii) most proteins were prone to aggregation in the *clpP* mutant; (iii) the absence of ClpP correlated with protein denaturation and oxidative stress responses, deregulation of virulence factors, and repression of CodY activity. These results provide clear evidence that mainly inactive proteins, probably excluded from functional complexes, become amenable to proteolytic attack (Michalik *et al.*, 2012; Michalik *et al.*, 2009). This process is not restricted to *S. aureus*, but is a fundamental cellular process found in all living cells to regain nutrients from unemployed proteins under conditions of nutrient starvation and to guarantee protein homeostasis.

It became obvious that proteolysis in *S. aureus* was very selective and depends on active ClpP for most of proteins. By *in vivo* expression of a proteolytically inactive form of ClpP that retained but did not degrade substrates translocated into its proteolytic chamber, about 70 proteins were identified as potential targets of the protease in two different *S. aureus* strains. Most of the ClpP substrates belong to the functional categories: energy metabolism, protein fate and synthesis, regulatory function, biosynthesis of co-factors, prosthetic groups and carriers, and DNA metabolism and repair. Among them are regulatory proteins such as CtsR, Spx, McsB, MecA, PerR and HrcA as well as proteins involved in DNA damage repair (RecA, UvrA, and UvrB), protein synthesis and protein fate (RpoB, Tuf, DnaJ, DnaK, GroEL, Tig, HslU, SecA, ClpX) and in major metabolic pathways such as GuaB, GuaA, MetK, NrdE, NdrF, Pnp (Feng *et al.*, 2013). In order to be degraded by ClpP, substrates first interact with a Clp ATPase component which leads to their being unfolded and subsequently translocated into the ClpP proteolytic chamber (Martin *et al.*, 2005). ClpP dependent degradation of proteins is a costly process and the energy required is provided from ATP hydrolysis by the Clp ATPases. ClpP associates with different Clp ATPases which confer distinct substrate specificities to the Clp protease complexes (for review see Kirstein *et al.*, 2009). In *S. aureus*, ClpP forms active Clp protease complexes either with ClpX or ClpC (Frees *et al.*, 2004). By using a ClpC variant that formed stable complexes with its substrates but was no longer able to bind and hydrolyse ATP, ClpC substrates have been trapped and co-purified and were subsequently analysed by GeLC-MS. A total of 103 *S. aureus* proteins were identified as potential substrates of ClpC and these belong to several functional categories, such as amino acid metabolism, cell envelope turnover, energy metabolism, protein synthesis, and protein fate, indicating substantial effects of ClpC on cell physiology and survival (Graham *et al.*, 2013). Moreover, regulatory proteins such as AgrA, CcpA, RsbW, CodY, HprK, RecA, and Rex were found to be associated with ClpC (Graham *et al.*, 2013), which might explain the observed global effects of *clpC* deletion on gene transcription in *S. aureus* (Chatterjee *et al.*, 2009; Luong *et al.*, 2011). Feng and co-workers (Feng *et al.*, 2013) further examined whether the two Clp protease complexes ClpCP and ClpXP have distinct substrate specificities. They showed that ClpC itself is the most abundant substrate that is exclusively degraded by the ClpCP complex. In addition, they identified CtsR, UvrA, RecA, CodY FtsZ, HrcA, and GlmS as ClpCP specific substrates. Among the most abundant substrates captured only by the ClpXP protease complex, were Sle1 and Spx. Less abundant substrates included Tig and FtsA (Feng *et al.*, 2013).

Interestingly, the stability of some *S. aureus* proteins remained unaffected by the activity of ClpP indicating that proteolytic systems apart from the ClpP complexes might be involved in their degradation (Michalik *et al.*, 2012). The proteases FtsH, SACOL0396 or ClpYQ are considered to be probable candidates; however, detailed studies on their proteolytic activities and possible targets in *S. aureus* have not yet been provided.

Global-scale quantification of absolute protein amounts

To modulate biological processes in systems biology, it is essential to know not only the relative amounts of proteins for a single cell but also their absolute concentrations. To determine absolute protein concentrations, there are different approaches available. In *S. aureus*, absolute concentrations for cytoplasmic proteins were initially determined for a selected set of proteins by applying a targeted mass spectrometry approach with isotopically labelled internal standard peptides. In a second step, known concentrations for these proteins were used to calibrate 2-D protein gels, which subsequently allowed the calculation of absolute amounts for all detected protein spots in the gel. In this way, the concentrations of the majority of metabolic enzymes were determined. Altogether, 503 proteins have been quantified and these range from 1 to 25,000 molecule numbers per cell. In addition, stoichiometric information for selected proteins known to form protein complexes such as OdhA/OdhB, SucA/SucB and PdhA/PdhB/PdhC was determined.

Ribosomal proteins were detected at 70 to 10,000 molecules per cell. Most proteins present in high concentrations are involved in energy metabolism, while low-abundance proteins are mainly involved in DNA replication and repair. Among the glycolytic enzymes glyceraldehyde dehydrogenase (GapA1) and enolase (Eno) are the enzymes with the highest copy numbers in exponentially growing cells. With the exception of glucokinase, phosphofructokinase, and pyruvate kinase, the concentration of glycolytic enzymes was diminished under glucose-limiting conditions. At the same time, the copy number of enzymes belonging to the TCA cycle increased significantly (Maass *et al.*, 2011). Very similar observations were made using relative protein quantification methods based on 2-D gels and metabolite analyses; specifically, under glucose excess, glycolytic activity is very high and glucose is mainly catabolized to acetate (Kohler *et al.*, 2003, 2005; Liebeke *et al.*, 2011).

As expected, transcriptional regulatory proteins are present in rather low copy numbers per cell. Among the identified regulatory proteins, CcpA and CodY are the most abundant, while GapR, Rex, and the response regulators SaeR and SrrA are present in 6- to 10-fold lower concentrations in growing *S. aureus* cultures. In glucose starved cells, a significant increase in SrrA (13-fold), SaeR (2.7-fold), and CodY (2-fold) molecules was observed (Maass *et al.*, 2011). Importantly, combining DNA binding affinity data, post-translational modification events, and concentrations of potential interaction partners (i.e. proteins or metabolites) that might influence the activity of these regulatory proteins, transcriptional activity of genes belonging to the regulons can be modelled. This is mandatory for an advanced understanding of the function of global regulatory networks and metabolism and provides the essential basis for integrated systems biology models for the different environments that *S. aureus* might encounter during colonization and infection.

S. aureus proteomic signatures: towards an understanding of *S. aureus* adaptation to the human host during colonization and infection

Proteomic signatures of *S. aureus* in response to infection related stimuli

During colonization and infection of host tissues, *S. aureus* encounters a number of growth-limiting factors and life-threatening host defence mechanisms. Among these are low oxygen levels, oxidative stress, and nutrient limitations; hence, adaptation of *S. aureus* to these hostile habitats is characterized by a multi-signal response that is crucial for survival of the bacterium within the host. Understanding this very complex process is a considerable challenge and an essential prerequisite for the development of new strategies to combat infections with antibiotic resistant strains, which are already a serious problem in hospitals and nursing homes.

Several proteomic studies have provided valuable insight into the diverse signal transduction network responsible for *S. aureus* adaptations in gene expression under *in vivo* conditions that are relevant to infections (Wolf *et al.*, 2008; Reiss *et al.*, 2012; Fuchs *et al.*, 2007, 2010, 2013; Hochgräfe *et al.*, 2008; Hessling *et al.*, 2013; Kohler *et al.*, 2003, 2005, 2008; Islam *et al.*, 2014; Liebeke *et al.*, 2011; Resch *et al.*, 2006). Each condition induced and repressed the synthesis/amount of a definite set of proteins, which is also defined as proteomic signature and expresses the cell's state under the specific conditions (VanBogelen *et al.*, 1999) and can be used as biomarkers to interpret the proteome pattern obtained under more complex *in vivo* conditions. The highly sensitive 2-D gel electrophoresis technique, using radioactively pulse-labelled protein extracts in combination with mass spectrometry to identify proteins with an altered synthesis pattern in response to changing environmental conditions, has been widely used to visualize rapid stress and starvation responses (Engelmann and Hecker, 2008; Hecker *et al.*, 2009, 2010; Fuchs *et al.*, 2013).

To distinguish between stress/starvation-specific and more general (e.g. growth rate dependent) effects, a global approach for inter-experimental data analyses was initiated for *S. aureus* by using data from different adaptation experiments including oxidative stress (i.e. H_2O_2, diamide, and paraquat), nitrosative stress (NO), oxygen limitation (i.e. fermentation and nitrate respiration), elevated temperatures, and subinhibitory concentrations of antibiotics such as mupirocin and puromycin. To facilitate differentiating between these stress/stimuli responses, an online resource called *Aureolib* (http://www.aureolib.de) was established containing 4692 time-dependent protein synthesis profiles for 521 cytosolic proteins of *S. aureus* strain COL exposed to the nine infection-related stress and starvation stimuli (Fuchs *et al.*, 2013). When comparing all these protein synthesis profiles, it became clear that the synthesis of every protein was affected by at least one of the stimuli and 226 protein synthesis profiles showed induction ratios of 2.5-fold or higher under at least one of the tested conditions. For instance, 73 protein synthesis profiles showed induction during diamide stress, 48 during hydrogen peroxide stress, 47 during nitrogen oxide stress, and 44 during fermentation (Fuchs *et al.*, 2013). Moreover, stress and starvation specific marker proteins were identified (Fig. 7.2). As an example, specific marker proteins were identified during oxidative stress, which represents one of the challenges *S. aureus* has to cope with under *in vivo* conditions. For this analysis, cells were exposed to different oxidative stresses, such as superoxide (paraquate), peroxide (H_2O_2) and nitric oxide stress and, in addition, the thiol stress response induced by diamide was also analysed. Exposure to H_2O_2, diamide, and nitric oxide caused transient changes of the global protein synthesis pattern. Detailed studies of the expression kinetics of H_2O_2-induced proteins, revealed that adaptation of *S. aureus* to this stimulus proceeds in three different phases: (i) acute phase, (ii) recovery phase, and (iii) tolerance phase (Fuchs *et al.*, 2013; Wolf *et al.*, 2008). Remarkably, only two proteins (SACOL0815 and SACOL2722) were differentially synthesized in response to hydrogen peroxide, paraquate, nitric oxide and diamide indicating significantly different physiological consequences for the cells exposed to these various oxidants (Fuchs *et al.*, 2013; Hochgräfe *et al.*, 2008; Wolf *et al.*, 2008). Most of the induced proteins are shared by diamide and nitric oxide stress (17 proteins), followed by diamide and paraquate (14 proteins), and paraquat and nitric oxide (13 proteins). Eight marker proteins are shared by these three oxidative stress stimuli. Remarkably, the response to hydrogen peroxide has the least similarities to the other oxidative stress stimuli (http://www.aureolib.de).

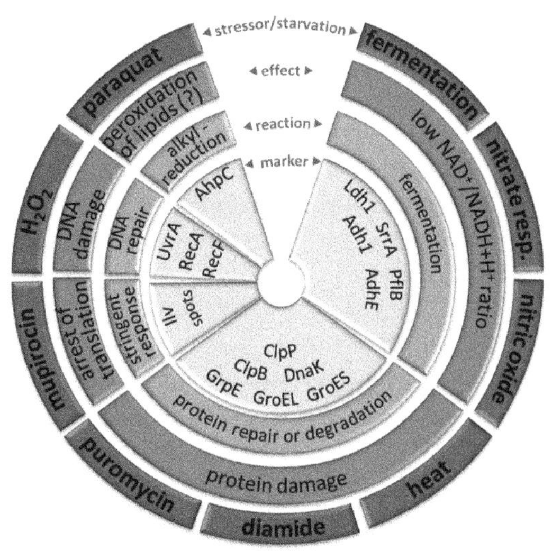

Figure 7.2 *S. aureus* marker proteins for different growth stimuli.

One crucial adaptive strategy for survival of bacteria under growth-restricting conditions is the stringent response, which has been shown to be inducible in *S. aureus* in response to mupirocin (Reiß *et al.*, 2011). Being a global response to nutritional deprivation, growth arrest-induced repression of stringent controlled proteins should also be observed under other growth restricting conditions. In fact, 53 of the proteins with reduced signal intensities after mupirocin treatment were also found to be repressed under at least one additional stress condition. The most prominent functional categories were nucleotide biosynthesis, energy metabolism, and translation. Remarkably, signal intensities of seven protein spots were reduced following treatment with five additional stress stimuli. These are AcnA, CarB, PurH, PurQ, PyrF and SACOL0430 which are mainly involved in nucleotide biosynthesis. About 50% of the protein spots repressed by mupirocin were found altered during heat stress (28 of 212 protein spots repressed by heat) and diamide stress (24 of 158 protein spots repressed by diamide). Although to a lesser extent, NO stress (19 of 99 protein spots repressed by NO) and paraquat stress (14 of 37 protein spots repressed by paraquat) proteomic profiles also resembled that of mupirocin treated bacteria (Fuchs *et al.*, 2013; Wolf *et al.*, 2008; Reiss *et al.*, 2012).

Clustering of proteins according to their synthesis profiles revealed only in rare cases a grouping of proteins controlled by the same regulator, suggesting that synthesis of most proteins is affected by multiple factors acting on both transcriptional and post transcriptional processes. Consequently, for a comprehensive understanding of cell physiology, proteomics under certain conditions represents the method of choice. However, for deciphering regulatory networks active under given environmental conditions, data on transcriptional and translational activities as well as information on the stability of both transcripts and proteins are required. Hence, only a combination of proteomics and transcriptomics will provide the entire information to successfully unravel signal transduction and regulatory mechanisms involved in gene expression in *S. aureus* under *in vitro* and *in vivo* conditions (Fuchs *et al.*, 2013).

From stimulons to regulons to regulatory networks in *S. aureus*

Comparison of protein synthesis patterns before and after exposure to stimuli allows a rapid assignment of proteins to stimulons, which suggests these proteins might be involved in adaptation to that stimulus. The next step is to dissect the stimulons into their individual regulons. The regulon structure can be defined by comparing protein synthesis profiles of wild type and mutant strains lacking regulatory proteins of interest.

In *S. aureus*, proteomics was used to unravel the regulatory network involved in adaptation to anaerobic conditions (Fuchs *et al.*, 2007, 2010, 2013; Kohler *et al.*, 2003, 2008; Makhlin *et al.*, 2007; Pagels *et al.*, 2010; Schlag *et al.*, 2008) (Fig. 7.3). The redox-sensitive protein Rex, which affects the transcription of more than 30 genes, was identified as a master regulator for anaerobic gene transcription in *S. aureus* (Pagels *et al.*, 2010). Being a transcriptional repressor, Rex blocks transcription of many genes active under anaerobic conditions, e.g. genes encoding enzymes for fermentation reactions, the arginine deiminase pathway, and nitrate/nitrite respiration. At high NADH levels, Rex is inactivated and no longer able to interact with the operator region of the Rex-regulated genes. Using a bio-informatics approach and DNA binding assays, Rex was shown to be directly involved in transcription of at least 19 genes. As shown by a proteomics comparison between a wild-type and a *rex* mutant strain, derepression of these genes is in most cases necessary but not sufficient for gene activation. The usually weak promoters of genes controlled by oxygen limitation require, in addition to Rex

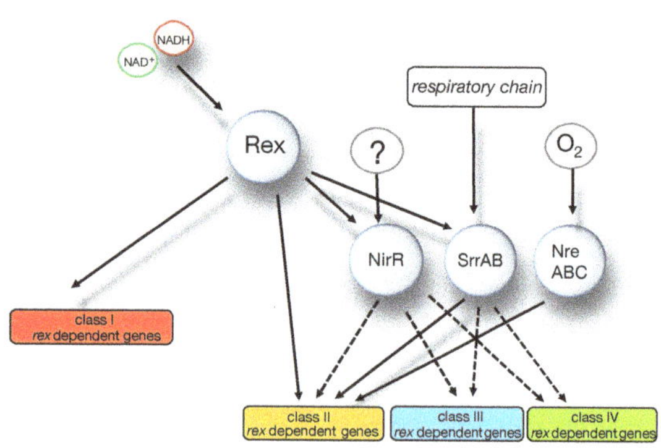

Figure 7.3 The Rex-modulon in *S. aureus*.

derepression, the binding of an activator, very likely a phosphorylated response regulator. Interesting candidates are the response regulator SrrB but also NreC or NirR. Genes encoding the two component system SrrAB and the transcriptional regulator NirR are among the Rex dependently regulated genes in *S. aureus*. This might also explain the observed indirect effects on gene expression of Rex in *S. aureus* under anaerobic conditions (Pagels *et al.*, 2010). The activity of the regulatory system NreABC has been shown to be directly influenced by oxygen (Kamps *et al.*, 2004). For SrrAB activation, a menadione dependent regulatory mechanism as described for ArcAB in *Escherichia coli* has been discussed (Kohler *et al.*, 2008) (Fig. 7.3).

Response of *S. aureus* to antibiotics

Treating *S. aureus* with antibiotics provides valuable insights into the physiological consequences of specifically impairing metabolic processes and functions, such as protein- and cell wall biosynthesis, membrane functions, replication and transcription. As an example, addition of mupirocin blocks charging of isoleucyl-tRNA with isoleucine, thus, inhibits protein biosynthesis. *S. aureus* specifically adapts to mupirocin with the increased expression of genes involved in isoleucine biosynthesis (*ilvD, ilvB, ilvC, ilvA2*) and isoleucyl-tRNA charging (*ileS*). The induction of these genes, suggests that branched chain amino acid starvation is potentially affected by the binding of uncharged Ile-tRNA to a cis-regulatory consensus motif, the T-Box that has been identified in front of these genes. In addition, it has been shown that the GTP level decreases in the presence of mupirocin and this may have considerable effects on the global gene expression; specifically, it has been associated with the stringent response in *S. aureus* which is triggered by the accumulation of uncharged tRNAs (Reiss *et al.*, 2012). The protein pattern induced by puromycin is very distinct from that of mupirocin, although both antibiotics terminate translation. This clearly shows that the physiological consequences for a cell in the presence of puromycin are different from those in the presence of mupirocin (Fig. 7.4). Puromycin mimics the 3′ end of aminoacyl-tRNAs and is incorporated into the nascent peptide chain. As consequence, ongoing protein biosynthesis is interrupted and dysfunctional peptides accumulate in the cell. *S. aureus* responds to puromycin treatment with the induction of the protein stress stimulon including Clp proteins and chaperons (Fuchs *et al.*, 2010). As expected, the cell wall active antibiotic vancomycin specifically affected the expression of genes belonging to the VraSR regulon in *S. aureus* (Hessling *et al.*, 2013). VraSR constitutes a positive regulator of cell wall peptidoglycan synthesis and mediates beta-lactam and glycopeptide resistance in *S. aureus* (Kuroda *et al.*, 2003).

These examples highlight the ways in which proteomic signatures may be used to discriminate the mode of action of antibiotics, even if they act on the same cellular process but at different steps, as has already been successfully done in *B. subtilis* (Bandow *et al.*, 2003; Brötz-Oesterhelt *et al.*, 2005) (Fig. 7.4). Proteome analyses of stationary cells of a methicillin-resistant *S. aureus* isolate exposed to an acyldepsipeptide (ADEP) derivative, ADEP4, revealed that hyperactive ClpP acts as a non-specific protease. ADPE4 activated ClpP facilitates degrading more than 400 proteins, induces self-digestion of persister cells, and eradicates *S. aureus* biofilm in an infection animal model (Conlon *et al.*, 2013).

Proteomics has also been used to unravel molecular mechanisms involved in antibiotic resistance in *S. aureus*. This has been done for daptomycin and glycopeptide resistance. Resistant and susceptible *S. aureus* strains were subjected to both transcriptome and proteome analyses and the transcript and protein patterns obtained were compared to identify molecular targets involved in resistance to these antibiotics (Fischer *et al.*, 2011; Scherl *et al.*, 2006). Daptomycin resistant strains are characterized by changes in metabolic functions, in the activity of global regulators, and in the biogenesis of cell envelope components. The most interesting observation was the increased expression of the essential two component system, YycFGH, which has been shown to be a key regulator for genes involved in cell membrane lipid homeostasis, cell wall metabolism and biofilm formation. It was suggested that differential expression of this regulator might contribute to modification of the net surface charge via both *mprF* and *dlt* pathways; hence, daptomycin binding may affect the initial attachment during biofilm formation (Fischer *et al.*, 2011). For glycopeptide resistant isolates, proteins involved in cell wall biosynthesis and metabolism were found at elevated levels. The VraSR regulatory system obviously contributes to the upregulation of these

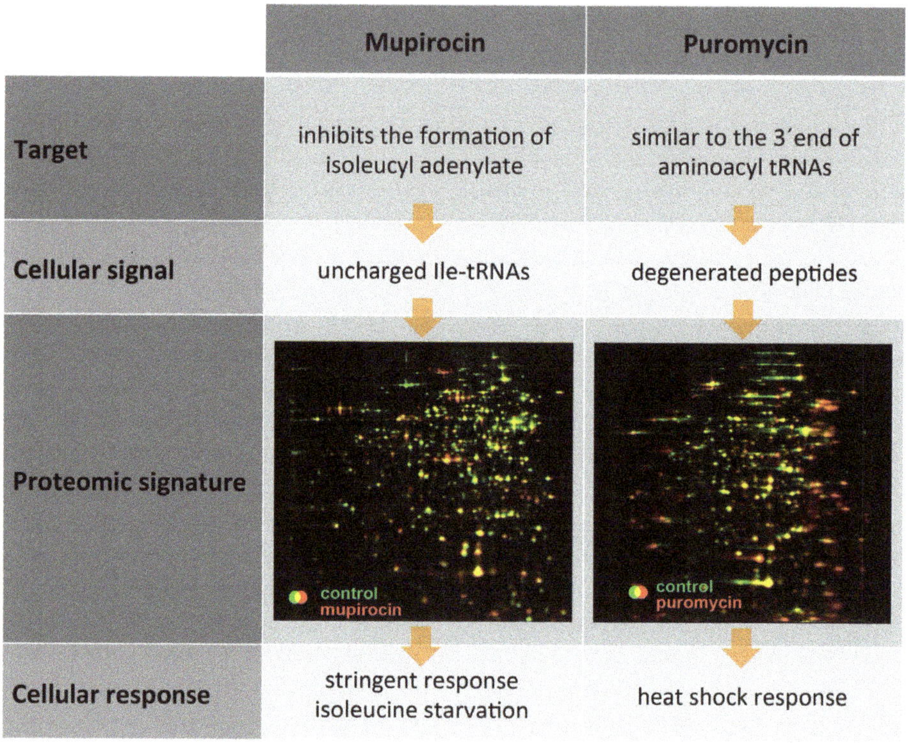

Figure 7.4 Proteomic signatures in response to the translational inhibitors mupirocin and puromycin in *S. aureus*.

proteins. Finally, proteins involved in stress protection such as the proteinases CtpA, MsrA and MsrB as well as the regulator MsrR were overproduced in the resistant isolates (Scherl *et al.*, 2006).

Adaptation of *S. aureus* to *in vivo* conditions

Coming from defined *in vitro* conditions, proteomics is more and more used to probe the adaptation of this pathogen to conditions that will normally confront it during host colonization and infection. This is even more challenging since these conditions are highly complex, the material is frequently limited, and a significant number of non-staphylococcal proteins that interfere with protein identification are present. In an initial approach, adaptation of *S. aureus* after internalization by two human lung epithelial cells lines (i.e. S9 and A549), and human embryonic kidney cells (HEK 293) was analysed (Pförtner *et al.*, 2014; Schmidt *et al.*, 2010; Surmann *et al.*, 2014). For these experiments, *gfp* tagged *S. aureus* strains were used for fluorescence activated cell sorting (Pförtner *et al.*, 2013); however, *S. aureus* cells coated with gold or ferric oxide-core nanoparticles combined with magnetic separation have also been successfully applied (Depke *et al.*, 2014). In each case, two million sorted bacterial cells were used for highly sensitivity MS analyses. Altogether, 1443 *S. aureus* proteins were identified and quantitatively monitored during the time course of internalization in the three different cell lines. After internalization, the growth rate of *S. aureus* was significantly reduced. This is accompanied by reduced levels of proteins involved in protein and purine biosynthesis, ribonucleotide reduction, and cold shock response. At the same, time proteins involved in arginine and lysine biosynthesis, terminal oxidases, and some stress proteins became more prominent. In general, the protein patterns of *S. aureus* internalized by epithelial cell lines S9 and A549 were more similar to each other than to those of HEK293-internalized *S. aureus* (Surmann *et al.*, 2014). The alternative sigma factor, SigB, might have a decisive role in *S. aureus* adaptation to the intracellular milieu, as it has been shown to be activated under these conditions

(Surmann *et al.*, 2014; Pförtner *et al.*, 2014). In S9 cells, loss of SigB retarded *S. aureus* replication and resulted in profound changes in the *S. aureus* proteome. Specifically, the amount of proteins involved in resistance to antibiotics and oxidative stress and in protein quality control was affected (Pförtner *et al.*, 2014).

In an ambitious work presented by Attia and co-workers, secreted *S. aureus* proteins were identified in host cell tissues from *S. aureus* infected mice (Attia *et al.*, 2013). The *S. aureus* abscess proteome of wild type and neutropenic mice was analysed to highlight both the *S. aureus*-specific host response and bacterial virulence factors expressed under these conditions. The identification of the very low abundant *S. aureus* proteins within an overwhelming amount of mouse proteins is like searching for a needle in a haystack. Using abscessed and lesioned tissues, 315 mouse proteins and 22 staphylococcal proteins were detectable. The amount of 40 mouse proteins was decreased in lesions from neutropenic mice, compared to neutrophil-rich abscesses demonstrating the impact of neutrophils on the staphylococcal abscess proteome. Among the staphylococcal proteins, thirteen were more abundant in neutropenic mice, and included known virulence factors such as LukD, LukE, LukS, HlgB, HlgC, MapN/Eap, and Ssp/Emp, but also proteins involved in iron acquisition (IsdA) and glycolysis (Adh1, Ldh, Eno, GapA). Moreover, Usp2 was found at elevated levels upon neutrophil depletion; however, it is not essential for bacterial survival in a systemic model of infection (Attia *et al.*, 2013). Under *in vitro* conditions, Usp2 belongs to the group of proteins that are more generally induced in response to different growth stimuli such as oxidative stress, fermentation, heat stress and mupirocin; however, its precise function is unknown (Fuchs *et al.*, 2013). In the study presented by Attia and co-workers, higher abundance of bacterial proteins can be caused by both higher expression rates and/or by the multiplying effects of higher bacterial loads. These two factors could not be distinguished in this study (Attia *et al.*, 2013).

Concluding remarks

With the information presented in this review, it should be clear that an intelligent use of proteomics can provide tremendous insight into the life of a large proportion of the proteins in *S. aureus*, including those involved in synthesis, localization, concentration, modification, damage, repair, and degradation. Since proteins are the main players of life, fundamental cellular processes under differing growth conditions can be described on the basis of these data. This is not only possible for *S. aureus* grown under laboratory conditions, but also for *S. aureus* isolated under more relevant *in vivo* conditions. By integrating *in vitro* and *in vivo* data, a new window of staphylococcal virulence and physiology has been opened, significantly enhancing our understanding of how *S. aureus* infects its host. This is of special importance, as there is currently a serious problem in hospital settings associated with an increasing number of nosocomial infections with multiresistant *S. aureus* isolates that result in alarmingly high mortality rates of infected patients. The increased antibiotic resistance and omnipresent nosocomial infections, highlights the urgent demand for new treatment strategies. The comprehensive proteomic view of *S. aureus* physiology and virulence may help to solve this problem by identifying completely new therapeutic targets therapeutic.

Acknowledgements

We thank all co-workers and students for their excellent work on proteome analyses. We are grateful to Robert S. Jack for critical reading the manuscript. This work was supported by grants from the Bildungsministerium für Bildung und Forschung (GenomikPlus, InfectionGenomik, ERA-Net) and the Deutsche Forschungsgeminschaft (GK212, TRR34, FOR585).

References

Attia, A.S., Cassat, J.E., Aranmolate, S.O., Zimmerman, L.J., Boyd, K.L., and Skaar, E.P. (2013). Analysis of the *Staphylococcus aureus* abscess proteome identifies antimicrobial host proteins and bacterial stress responses at the host-pathogen interface. Pathog. Dis. *69*, 36–48.

Bandow, J.E., Brötz, H., Leichert, L.I., Labischinski, H., and Hecker, M. (2003). Proteomic approach to understanding antibiotic action. Antimicrob. Agents Chemother. *47*, 948–955.

Barford, D. (2004). The role of cysteine residues as redox-sensitive regulatory switches. Curr. Opin. Struct. Biol. *14*, 679–686.

Bäsell, K., Otto, A., Junker, S., Zühlke, D., Rappen, G.M., Schmidt, S., Hentschker, C., Macek, B., Ohlsen, K., Hecker, M., *et al.* (2014). The phosphoproteome and its physiological dynamics in *Staphylococcus aureus*. Int. J. Med. Microbiol. *304*, 121–132.

Becher, D., Hempel, K., Sievers, S., Zühlke, D., Pané-Farré, J., Otto, A., Fuchs, S., Albrecht, D., Bernhardt, J., Engelmann, S., *et al.* (2009). A proteomic view of an important human pathogen – towards the quantification of the entire *Staphylococcus aureus* proteome. PLoS One *4*, e8176.

Brötz-Oesterhelt, H., Bandow, J.E., and Labischinski, H. (2005). Bacterial proteomics and its role in antibacterial drug discovery. Mass Spectrom. Rev. *24*, 549–565.

Chatterjee, I., Schmitt, S., Batzilla, C.F., Engelmann, S., Keller, A., Ring, M.W., Kautenburger, R., Ziebuhr, W., Hecker, M., Preissner, K.T., *et al.* (2009). *Staphylococcus aureus* ClpC ATPase is a late growth phase effector of metabolism and persistence. Proteomics *9*, 1152–1176.

Conlon, B.P., Nakayasu, E.S., Fleck, L.E., LaFleur, M.D., Isabella, V.M., Coleman, K., Leonard, S.N., Smith, R.D., Adkins, J.N., and Lewis, K. (2013). Activated ClpP kills persisters and eradicates a chronic biofilm infection. Nature *503*, 365–370.

Demple, B. (2002). Signal transduction by nitric oxide in cellular stress responses. Mol. Cell. Biochem. *234–235*, 11–18.

Depke, M., Surmann, K., Hildebrandt, P., Jehmlich, N., Michalik, S., Stanca, S.E., Fritzsche, W., Völker, U., and Schmidt, F. (2014). Labeling of the pathogenic bacterium *Staphylococcus aureus* with gold or ferric oxide-core nanoparticles highlights new capabilities for investigation of host-pathogen interactions. Cytomet. A: J. Internat. Soc. Anal. Cytol. *85*, 140–150.

Didier, J.P., Cozzone, A.J., and Duclos, B. (2010). Phosphorylation of the virulence regulator SarA modulates its ability to bind DNA in *Staphylococcus aureus*. FEMS Microbiol. Lett. *306*, 30–36.

Engelmann, S., and Hecker, M. (2008). Proteomic analysis to investigate regulatory networks in *Staphylococcus aureus*. Methods Mol. Biol. *431*, 25–45.

Feng, J., Michalik, S., Varming, A.N., Andersen, J.H., Albrecht, D., Jelsbak, L., Krieger, S., Ohlsen, K., Hecker, M., Gerth, U., *et al.* (2013). Trapping and proteomic identification of cellular substrates of the ClpP protease in *Staphylococcus aureus*. J. Proteome Res. *12*, 547–558.

Fischer, A., Yang, S.J., Bayer, A.S., Vaezzadeh, A.R., Herzig, S., Stenz, L., Girard, M., Sakoulas, G., Scherl, A., Yeaman, M.R., *et al.* (2011). Daptomycin resistance mechanisms in clinically derived *Staphylococcus aureus* strains assessed by a combined transcriptomics and proteomics approach. J. Antimicrob. Chemother. *66*, 1696–1711.

Frees, D., Chastanet, A., Qazi, S., Sørensen, K., Hill, P., Msadek, T., and Ingmer, H. (2004). Clp ATPases are required for stress tolerance, intracellular replication and biofilm formation in *Staphylococcus aureus*. Mol. Microbiol. *54*, 1445–1462.

Fuchs, S., Mehlan, H., Kusch, H., Teumer, A., Zühlke, D., Berth, M., Wolf, C., Dandekar, T., Hecker, M., Engelmann, S., *et al.* (2010). Protecs, a comprehensive and powerful storage and analysis system for OMICS data, applied for profiling the anaerobiosis response of *Staphylococcus aureus* COL. Proteomics *10*, 2982–3000.

Fuchs, S., Pané-Farré, J., Kohler, C., Hecker, M., and Engelmann, S. (2007). Anaerobic gene expression in *Staphylococcus aureus*. J. Bacteriol. *189*, 4275–4289.

Fuchs, S., Zühlke, D., Pané-Farré, J., Kusch, H., Wolf, C., Reiß, S., Binh, l.e.T.N., Albrecht, D., Riedel, K., Hecker, M., *et al.* (2013). Aureolib - a proteome signature library: towards an understanding of *Staphylococcus aureus* pathophysiology. PLoS One *8*, e70669.

Graham, J.W., Lei, M.G., and Lee, C.Y. (2013). Trapping and identification of cellular substrates of the *Staphylococcus aureus* ClpC chaperone. J. Bacteriol. *195*, 4506–4516.

Hecker, M., Becher, D., Fuchs, S., and Engelmann, S. (2010). A proteomic view of cell physiology and virulence of *Staphylococcus aureus*. Int. J. Med. Microbiol. *300*, 76–87.

Hecker, M., Reder, A., Fuchs, S., Pagels, M., and Engelmann, S. (2009). Physiological proteomics and stress/starvation responses in *Bacillus subtilis* and *Staphylococcus aureus*. Res. Microbiol. *160*, 245–258.

Hessling, B., Bonn, F., Otto, A., Herbst, F.A., Rappen, G.M., Bernhardt, J., Hecker, M., and Becher, D. (2013). Global proteome analysis of vancomycin stress in *Staphylococcus aureus*. Int. J. Med. Microbiol. *303*, 624–634.

Hochgräfe, F., Mostertz, J., Albrecht, D., and Hecker, M. (2005). Fluorescence thiol modification assay: oxidatively modified proteins in *Bacillus subtilis*. Mol. Microbiol. *58*, 409–425.

Hochgräfe, F., Mostertz, J., Pöther, D.C., Becher, D., Helmann, J.D., and Hecker, M. (2007). S-cysteinylation is a general mechanism for thiol protection of *Bacillus subtilis* proteins after oxidative stress. J. Biol. Chem. *282*, 25981–25985.

Hochgräfe, F., Wolf, C., Fuchs, S., Liebeke, M., Lalk, M., Engelmann, S., and Hecker, M. (2008). Nitric oxide stress induces different responses but mediates comparable protein thiol protection in *Bacillus subtilis* and *Staphylococcus aureus*. J. Bacteriol. *190*, 4997–5008.

Hondorp, E.R., and Matthews, R.G. (2004). Oxidative stress inactivates cobalamin-independent methionine synthase (MetE) in *Escherichia coli*. PLOS Biol. *2*, e336.

Islam, N., Kim, Y., Ross, J.M., and Marten, M.R. (2014). Proteomic analysis of *Staphylococcus aureus* biofilm cells grown under physiologically relevant fluid shear stress conditions. Proteome Sci. *12*, 21.

Jakob, U., Muse, W., Eser, M., and Bardwell, J.C. (1999). Chaperone activity with a redox switch. Cell *96*, 341–352.

Kamps, A., Achebach, S., Fedtke, I., Unden, G., and Götz, F. (2004). Staphylococcal NreB: an O(2)-sensing histidine protein kinase with an O(2)-labile iron-sulphur cluster of the FNR type. Mol. Microbiol. *52*, 713–723.

Kirstein, J., Molière, N., Dougan, D.A., and Turgay, K. (2009). Adapting the machine: adaptor proteins for Hsp100/Clp and AAA+ proteases. Nat. Rev. Microbiol. *7*, 589–599.

Klose, J. (1975). Protein mapping by combined isoelectric focusing and electrophoresis of mouse tissues. A novel approach to testing for induced point mutations in mammals. Humangenetik *26*, 231–243.

Kohler, C., von Eiff, C., Liebeke, M., McNamara, P.J., Lalk, M., Proctor, R.A., Hecker, M., and Engelmann, S. (2008). A defect in menadione biosynthesis induces global changes in gene expression in *Staphylococcus aureus*. J. Bacteriol. *190*, 6351–6364.

Kohler, C., von Eiff, C., Peters, G., Proctor, R.A., Hecker, M., and Engelmann, S. (2003). Physiological characterization of a heme-deficient mutant of *Staphylococcus aureus* by a proteomic approach. J. Bacteriol. *185*, 6928–6937.

Kohler, C., Wolff, S., Albrecht, D., Fuchs, S., Becher, D., Büttner, K., Engelmann, S., and Hecker, M. (2005). Proteome analyses of *Staphylococcus aureus* in growing and non-growing cells: a physiological approach. Int. J. Med. Microbiol. *295*, 547–565.

Kuroda, M., Kuroda, H., Oshima, T., Takeuchi, F., Mori, H., and Hiramatsu, K. (2003). Two-component system VraSR positively modulates the regulation of cell wall biosynthesis pathway in *Staphylococcus aureus*. Mol. Microbiol. *49*, 807–821.

Kuroda, M., Ohta, T., Uchiyama, I., Baba, T., Yuzawa, H., Kobayashi, I., Cui, L., Oguchi, A., Aoki, K., Nagai, Y., *et al.* (2001). Whole genome sequencing of meticillin-resistant *Staphylococcus aureus*. Lancet 357, 1225–1240.

Kusch, H., and Engelmann, S. (2014). Secrets of the secretome in *Staphylococcus aureus*. Int. J. Med. Microbiol. *304*, 133–141.

Leichert, L.I., Scharf, C., and Hecker, M. (2003). Global characterization of disulfide stress in *Bacillus subtilis*. J. Bacteriol. *185*, 1967–1975.

Liebeke, M., Dörries, K., Zühlke, D., Bernhardt, J., Fuchs, S., Pané-Farré, J., Engelmann, S., Völker, U., Bode, R., Dandekar, T., *et al.* (2011). A metabolomics and proteomics study of the adaptation of *Staphylococcus aureus* to glucose starvation. Mol. Biosyst. *7*, 1241–1253.

Lomas-Lopez, R., Paracuellos, P., Riberty, M., Cozzone, A.J., and Duclos, B. (2007). Several enzymes of the central metabolism are phosphorylated in *Staphylococcus aureus*. FEMS Microbiol. Lett. *272*, 35–42.

Luong, T.T., Sau, K., Roux, C., Sau, S., Dunman, P.M., and Lee, C.Y. (2011). *Staphylococcus aureus* ClpC divergently regulates capsule via *sae* and *codY* in strain Newman but activates capsule via *codY* in strain UAMS-1 and in strain Newman with repaired *saeS*. J. Bacteriol. *193*, 686–694.

Maass, S., Sievers, S., Zühlke, D., Kuzinski, J., Sappa, P.K., Muntel, J., Hessling, B., Bernhardt, J., Sietmann, R., Völker, U., *et al.* (2011). Efficient, global-scale quantification of absolute protein amounts by integration of targeted mass spectrometry and two-dimensional gel-based proteomics. Anal. Chem. *83*, 2677–2684.

Makhlin, J., Kofman, T., Borovok, I., Kohler, C., Engelmann, S., Cohen, G., and Aharonowitz, Y. (2007). *Staphylococcus aureus* ArcR controls expression of the arginine deiminase operon. J. Bacteriol. *189*, 5976–5986.

Martin, A., Baker, T.A., and Sauer, R.T. (2005). Rebuilt AAA + motors reveal operating principles for ATP-fuelled machines. Nature *437*, 1115–1120.

Michalik, S., Bernhardt, J., Otto, A., Moche, M., Becher, D., Meyer, H., Lalk, M., Schurmann, C., Schlüter, R., Kock, H., *et al.* (2012). Life and death of proteins: a case study of glucose-starved *Staphylococcus aureus*. Mol. Cell. Proteomics *11*, 558–570.

Michalik, S., Liebeke, M., Zühlke, D., Lalk, M., Bernhardt, J., Gerth, U., and Hecker, M. (2009). Proteolysis during long-term glucose starvation in *Staphylococcus aureus* COL. Proteomics *9*, 4468–4477.

Newton, G.L., Arnold, K., Price, M.S., Sherrill, C., Delcardayre, S.B., Aharonowitz, Y., Cohen, G., Davies, J., Fahey, R.C., and Davis, C. (1996). Distribution of thiols in microorganisms: mycothiol is a major thiol in most actinomycetes. J. Bacteriol. *178*, 1990–1995.

O'Farrell, P.H. (1975). High resolution two-dimensional electrophoresis of proteins. J. Biol. Chem. *250*, 4007–4021.

Otto, A., van Dijl, J.M., Hecker, M., and Becher, D. (2014). The *Staphylococcus aureus* proteome. Int. J. Med. Microbiol. *304*, 110–120.

Pagels, M., Fuchs, S., Pané-Farré, J., Kohler, C., Menschner, L., Hecker, M., McNamarra, P.J., Bauer, M.C., von Wachenfeldt, C., Liebeke, M., *et al.* (2010). Redox sensing by a Rex-family repressor is involved in the regulation of anaerobic gene expression in *Staphylococcus aureus*. Mol. Microbiol. *76*, 1142–1161.

Pförtner, H., Burian, M.S., Michalik, S., Depke, M., Hildebrandt, P., Dhople, V.M., Pané-Farré, J., Hecker, M., Schmidt, F., and Völker, U. (2014). Activation of the alternative sigma factor SigB of *Staphylococcus aureus* following internalization by epithelial cells - an *in vivo* proteomics perspective. Int. J. Med. Microbiol. *304*, 177–187.

Pförtner, H., Wagner, J., Surmann, K., Hildebrandt, P., Ernst, S., Bernhardt, J., Schurmann, C., Gutjahr, M., Depke, M., Jehmlich, U., *et al.* (2013). A proteomics workflow for quantitative and time-resolved analysis of adaptation reactions of internalized bacteria. Methods *61*, 244–250.

Posada, A.C., Kolar, S.L., Dusi, R.G., Francois, P., Roberts, A.A., Hamilton, C.J., Liu, G.Y., and Cheung, A. (2014). Importance of bacillithiol in the oxidative stress response of *Staphylococcus aureus*. Infect. Immun. *82*, 316–332.

Pöther, D.C., Gierok, P., Harms, M., Mostertz, J., Hochgräfe, F., Antelmann, H., Hamilton, C.J., Borovok, I., Lalk, M., Aharonowitz, Y., *et al.* (2013). Distribution and infection-related functions of bacillithiol in *Staphylococcus aureus*. Int. J. Med. Microbiol. *303*, 114–123.

Pöther, D.C., Liebeke, M., Hochgräfe, F., Antelmann, H., Becher, D., Lalk, M., Lindequist, U., Borovok, I., Cohen, G., Aharonowitz, Y., *et al.* (2009). Diamide triggers mainly S Thiolations in the cytoplasmic proteomes of *Bacillus subtilis* and *Staphylococcus aureus*. J. Bacteriol. *191*, 7520–7530.

Reiss, S., Pané-Farré, J., Fuchs, S., François, P., Liebeke, M., Schrenzel, J., Lindequist, U., Lalk, M., Wolz, C., Hecker, M., *et al.* (2012). Global analysis of the *Staphylococcus aureus* response to mupirocin. Antimicrob. Agents Chemother. *56*, 787–804.

Resch, A., Leicht, S., Saric, M., Pasztor, L., Jakob, A., Götz, F., and Nordheim, A. (2006). Comparative proteome analysis of *Staphylococcus aureus* biofilm and planktonic cells and correlation with transcriptome profiling. Proteomics *6*, 1867–1877.

Rogasch, K., Rühmling, V., Pané-Farré, J., Höper, D., Weinberg, C., Fuchs, S., Schmudde, M., Bröker, B.M., Wolz, C., Hecker, M., *et al.* (2006). Influence of the two-component system SaeRS on global gene expression in two different *Staphylococcus aureus* strains. J. Bacteriol. *188*, 7742–7758.

Scherl, A., Francois, P., Charbonnier, Y., Deshusses, J.M., Koessler, T., Huyghe, A., Bento, M., Stahl-Zeng, J., Fischer, A., Masselot, A., *et al.* (2006). Exploring glycopeptide-resistance in *Staphylococcus aureus*: a combined proteomics and transcriptomics approach for the identification of resistance-related markers. BMC Genomics 7, 296.

Schlag, S., Fuchs, S., Nerz, C., Gaupp, R., Engelmann, S., Liebeke, M., Lalk, M., Hecker, M., and Götz, F. (2008). Characterization of the oxygen-responsive NreABC regulon of *Staphylococcus aureus*. J. Bacteriol. *190*, 7847–7858.

Schmidt, F., Scharf, S.S., Hildebrandt, P., Burian, M., Bernhardt, J., Dhople, V., Kalinka, J., Gutjahr, M., Hammer, E., and Völker, U. (2010). Time-resolved quantitative proteome profiling of host-pathogen interactions: the response of *Staphylococcus aureus* RN1HG to internalisation by human airway epithelial cells. Proteomics *10*, 2801–2811.

Sibbald, M.J., Ziebandt, A.K., Engelmann, S., Hecker, M., de Jong, A., Harmsen, H.J., Raangs, G.C., Stokroos, I., Arends, J.P., Dubois, J.Y., *et al.* (2006). Mapping the pathways to staphylococcal pathogenesis by comparative secretomics. Microbiol. Mol. Biol. Rev. *70*, 755–788.

Surmann, K., Michalik, S., Hildebrandt, P., Gierok, P., Depke, M., Brinkmann, L., Bernhardt, J., Salazar, M.G., Sun, Z., Shteynberg, D., *et al.* (2014). Comparative proteome analysis reveals conserved and specific adaptation patterns of *Staphylococcus aureus* after internalization by different types of human non-professional phagocytic host cells. Front. Microbiol. *5*, 392.

Tettelin, H., Riley, D., Cattuto, C., and Medini, D. (2008). Comparative genomics: the bacterial pan-genome. Curr. Opin. Microbiol. *11*, 472–477.

Truong-Bolduc, Q.C., and Hooper, D.C. (2010). Phosphorylation of MgrA and its effect on expression of the NorA and NorB efflux pumps of *Staphylococcus aureus*. J. Bacteriol. *192*, 2525–2534.

VanBogelen, R.A., Schiller, E.E., Thomas, J.D., and Neidhardt, F.C. (1999). Diagnosis of cellular states of microbial organisms using proteomics. Electrophoresis *20*, 2149–2159.

Weber, H., Engelmann, S., Becher, D., and Hecker, M. (2004). Oxidative stress triggers thiol oxidation in the glyceraldehyde-3-phosphate dehydrogenase of *Staphylococcus aureus*. Mol. Microbiol. *52*, 133–140.

Wolf, C., Hochgräfe, F., Kusch, H., Albrecht, D., Hecker, M., and Engelmann, S. (2008). Proteomic analysis of antioxidant strategies of *Staphylococcus aureus*: diverse responses to different oxidants. Proteomics *8*, 3139–3153.

Zheng, M., Aslund, F., and Storz, G. (1998). Activation of the OxyR transcription factor by reversible disulfide bond formation. Science *279*, 1718–1721.

Zhou, M., Boekhorst, J., Francke, C., and Siezen, R.J. (2008). LocateP: genome-scale subcellular-location predictor for bacterial proteins. BMC Bioinf. *9*, 173.

Ziebandt, A.K., Becher, D., Ohlsen, K., Hacker, J., Hecker, M., and Engelmann, S. (2004). The influence of *agr* and sigmaB in growth phase dependent regulation of virulence factors in *Staphylococcus aureus*. Proteomics *4*, 3034–3047.

Ziebandt, A.K., Kusch, H., Degner, M., Jaglitz, S., Sibbald, M.J., Arends, J.P., Chlebowicz, M.A., Albrecht, D., Pantucek, R., Doskar, J., et al. (2010). Proteomics uncovers extreme heterogeneity in the *Staphylococcus aureus* exoproteome due to genomic plasticity and variant gene regulation. Proteomics *10*, 1634–1644.

Ziebandt, A.K., Weber, H., Rudolph, J., Schmid, R., Höper, D., Engelmann, S., and Hecker, M. (2001). Extracellular proteins of *Staphylococcus aureus* and the role of SarA and sigma B. Proteomics *1*, 480–493.

Cell Wall Assembly and Physiology

8

Angelika Gründling

Abstract

Staphylococcus aureus has a typical Gram-positive cell wall, which is composed of a thick peptidoglycan layer, cell surface proteins, teichoic acids and carbohydrate polymers. The functions of these different cell wall components are diverse: the peptidoglycan layer protects bacteria from osmotic lysis, cell surface proteins are required for adhesion, biofilm formation and immune evasion, and teichoic acids help to protect bacteria from the action of cationic antimicrobial peptides. Proper assembly of the different cell wall structures is not only essential for the pathogenesis of *S. aureus* but also for bacterial growth, and several antimicrobial agents target the synthesis of key cell wall components. In this chapter, the synthesis pathways and functions of the diverse cell wall components of *S. aureus* will be summarized, along with the action of important cell wall targeting antibiotics and the corresponding bacterial resistance mechanisms.

Introduction

Staphylococcus aureus is a Gram-positive bacterium that belongs to the phylum Firmicutes. A detailed picture of the overall cell wall architecture and its composition in *S. aureus* has emerged from a large number of biochemical, genetic and electron microscopy studies. Recent work using modern cryo-electron microscopy techniques in particular has provided new insight into the overall cell wall architecture (Matias and Beveridge, 2006, 2007) (Fig. 8.1). The cytoplasmic content is contained with a lipid bilayer, which in turn is surrounded

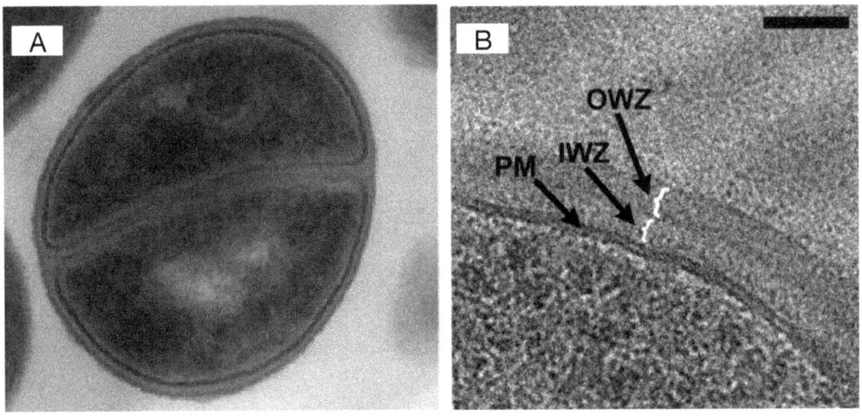

Figure 8.1 Electron microscopy images of *S. aureus.* (A) Thin section EM images of a dividing *S. aureus* cell using conventional embedding methods. (B) EM image showing the *S. aureus* cell envelope using a frozen-hydrated section method, where the plasma membrane (PM) is seen surrounded by a bipartite wall with a low-density inner wall zone (IWZ) and a higher density outer wall zone (OWZ). The bar shown represents 50 nm. The image shown in panel B was adapted with permission from Matias and Beveridge (2006). The publisher for this copyrighted material is the American Society for Microbiology.

by the cell wall. The *S. aureus* cell wall, as is typical for low G+C Gram-positive bacteria, is composed of a thick peptidoglycan layer, proteins, teichoic acids and carbohydrate polymers (Fig. 8.2). There are two types of teichoic acid (TA): wall teichoic acid (WTA) and lipoteichoic acid (LTA). Both are zwitterionic cell wall polymers, but they are synthesized through separate pathways. In the majority of *S. aureus* strains, WTA is a poly-ribitolphosphate polymer that is polymerized on a linker unit on undecaprenyl-phosphate (C_{55}-P) within the cytoplasm if the cell. WTA is transported across the membrane and subsequently covalently linked to the peptidoglycan layer (Brown *et al.*, 2013; Weidenmaier and Peschel, 2008; Xia and Peschel, 2008). LTA is a poly-glycerolphosphate chain that is linked via a glycolipid anchor to the bacterial membrane (Fischer, 1990, 1994; Percy and Gründling, 2014; Reichmann and Gründling, 2011) (Fig. 8.2). *S. aureus* synthesizes two extracellular carbohydrate polymers in addition to teichoic acids. One of these is the capsular polysaccharide, which is covalently linked to the peptidoglycan layer (Chan *et al.*, 2014; O'Riordan and Lee, 2004). The exact composition of capsular polysaccharide varies between different strains, allowing their classification into different serotypes. The second extracellular carbohydrate polymer is a poly-*N*-acetylglucosamine (PNAG) polymer (Cue *et al.*, 2012). How this polymer is retained within the cell wall is currently not known. Lastly, different proteins are embedded within the membrane or the cell wall and fulfil important functions ranging from nutrient uptake to adhesion. The three classes of proteins found in the cell wall are lipoproteins and covalently and non-covalently attached proteins (Fig. 8.2). The polymers and

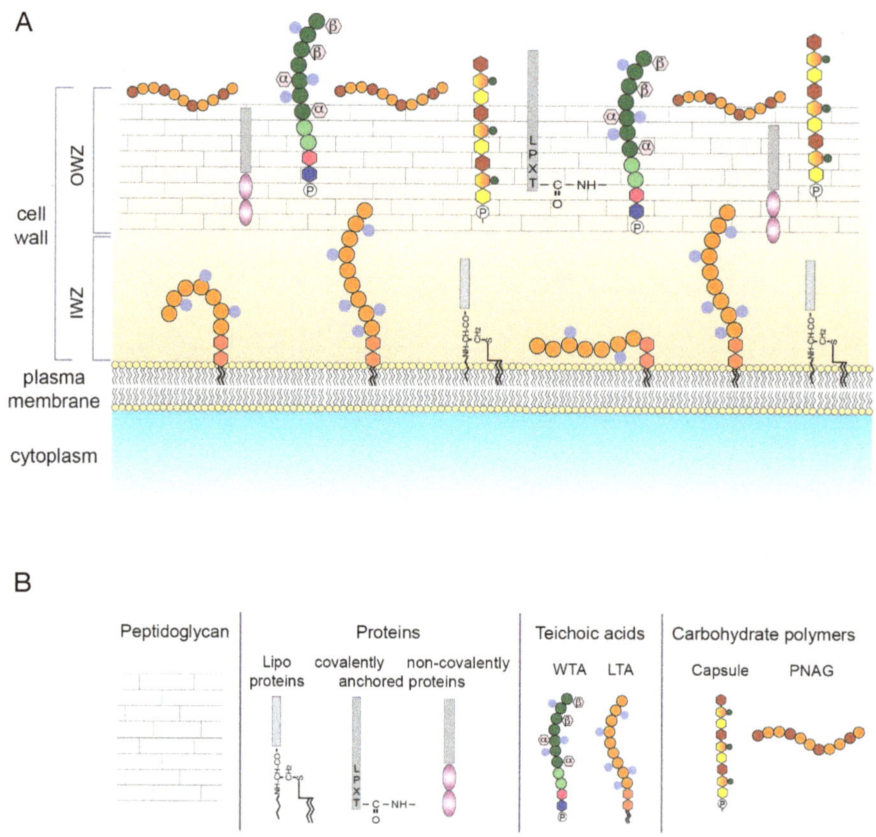

Figure 8.2 *Staphylococcus aureus cell wall.* (A) Schematic representation of the *S. aureus* cell wall. The cytoplasm of the cell is surrounded by a lipid bilayer or plasma membrane followed by the cell wall. The cell wall has a bipartite structure with an inner wall zone (IWZ) and outer wall zone (OWZ). (B) Depiction of major cell wall components in *S. aureus.* These are peptidoglycan, membrane anchored-, covalently anchored- and non-covalently anchored cell wall proteins, the teichoic acids WTA and LTA as well as the capsule and PNAG exopolysaccharides.

proteins found within the cell wall are not uniformly distributed within the cell envelope. This has been highlighted in recent cryo-electron microscopy images, which revealed that the cell wall of *S. aureus* has two distinct zones, an inner wall zone (IWZ) with LTA as one of the major constituents and an outer wall zone (OWZ), which has been suggested to be composed of the peptidoglycan cell wall polymer as well as peptidoglycan associated proteins and polymers such WTA (Matias and Beveridge, 2006, 2007) (Figs. 8.1 and 8.2).

In this chapter, the chemical structures, the synthesis pathways and main enzymes involved in the production of the diverse cell wall structures will be summarized along with the key physiological functions of these components. A brief description of the composition and synthesis of the plasma membrane will also be provided, as this lipid barrier, which separates the cell cytoplasm and cell wall, has an important function during the production of precursors of the diverse cell wall polymers. In the final section, the action of the main classes of cell wall targeting antibiotics will be discussed, along with common resistance mechanisms observed in *S. aureus*.

The majority of the proteins and enzymes required for the synthesis of the different cell wall components are conserved among the different *S. aureus* strains. However, some strain-specific differences are observed and these will be highlighted throughout this chapter. For consistency and ease of reference, the gene locus tag numbers for the methicillin-resistant *S. aureus* (MRSA) strain USA300_FR3757 will be used throughout (Diep *et al.*, 2006). This nomenclature was chosen as this strain is of clinical relevance and closely related to the USA300 strain LAC, which is used in many laboratories for genetic studies as a large sequence-defined transposon mutant library has been assembled (Fey *et al.*, 2013).

S. aureus membrane lipids and their synthesis

The *S. aureus* plasma membrane is composed of glycerolipids, which are typical of bacterial membranes. Membrane lipid homeostasis and the ability to adjust the lipid composition in response to changing environmental conditions are essential for bacterial survival. In *S. aureus*, specific membrane lipids together with other cell wall components contribute to the ability of this organism to resist killing by cationic antimicrobial peptides (CAMPs) and the recently introduced antimicrobial daptomycin.

S. aureus contains four major membrane lipids during growth: phosphatidylglycerol (PG; around 50% of the total membrane), diacylglycerol (DAG; 24%), lysyl-PG (10%) and the glycolipid diglucosyldiacylglycerol (Glc_2-DAG; 8%) (Fischer, 1994; Koch *et al.*, 1984) (Fig. 8.3A). In addition, approximately 6% of the membrane consists of LTA, which is anchored to the membrane by the glycolipid Glc_2-DAG (Fischer, 1994; Koch *et al.*, 1984). In contrast to the other membrane lipids, the LTA polymer is present only within the outer layer of the membrane bilayer. Details of its synthesis and function will be summarized in a later section. In addition to these four major lipids, a small amount of cardiolipin (also referred to as bisphosphatidylglycerol; ~1% of the total membrane lipid composition) is present in exponentially growing *S. aureus* cells. However, its concentration increases drastically when cells enter the stationary phase and under other stress conditions (Femling *et al.*, 2005; Koprivnjak *et al.*, 2011; Short and White, 1971).

The membrane lipids in *S. aureus* are produced by the acylation of the hydroxyl groups at the C1 and C2 positions of sn-glycerol-3 phosphate, yielding phosphatidic acid (PA) (Zhang and Rock, 2008). The majority of fatty acids or acyl chains required for the production of lipids are synthesized through the type II fatty acid synthesis pathway (FASII), although fatty acids used for lipid biosynthesis can also be taken up by *S. aureus* from the environment (Parsons *et al.*, 2014). The end products of the FASII pathway are fully formed acyl chains linked to the acyl carrier protein (ACP) to form acyl-ACP. Acyl-ACP is then used for the synthesis of the phosphatidic acid (PA) (Fig. 8.3B). The presumed *S. aureus* membrane lipid biosynthesis enzymes along with their corresponding activity and USA300 FPR3757 locus tag numbers are listed in Table 8.1. In Gram-positive bacteria, including in *S. aureus*, PA is synthesized by the enzymes PlsX, PlsY and PlsC (Fig. 8.3B and Table 8.1). PlsX is an acyl-ACP phosphotransacylase that produces acylphosphate (acyl-P) from acyl-ACP (Fig. 8.3B). Using acyl-P, the glycerolphosphate acyltransferase PlsY acylates the hydroxyl group at the C1 position

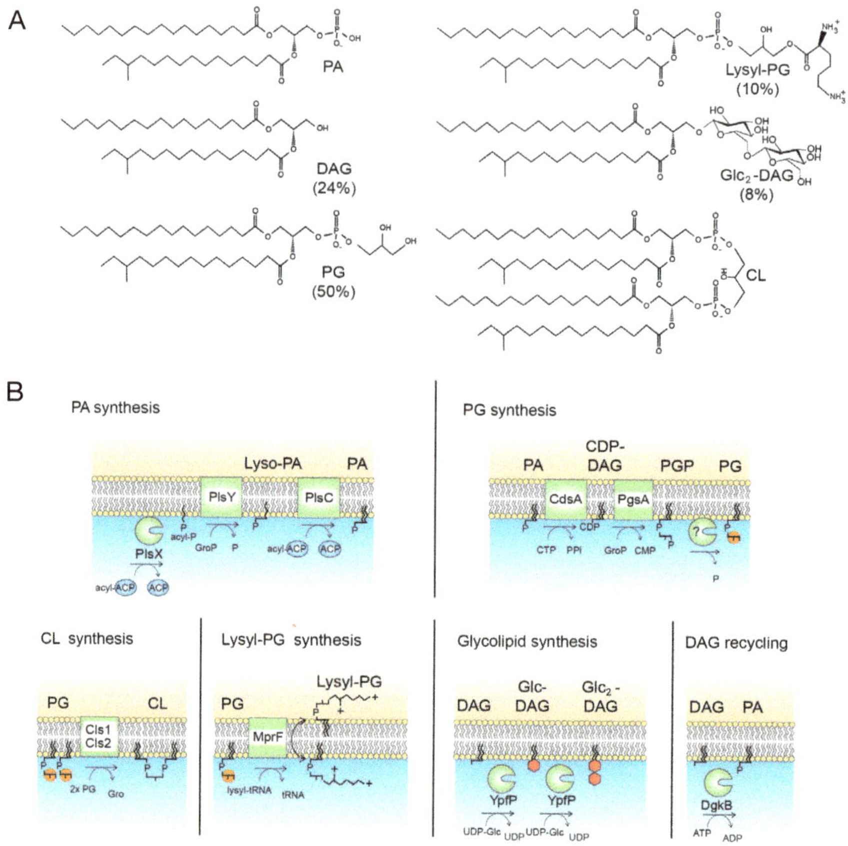

Figure 8.3 The *S. aureus* membrane. (A) Chemical structures of the *S. aureus* membrane lipids PA, DAG, PG, Lysyl-PG, Glc_2-DAG, CL with% of membrane lipids in growing cells given in parenthesis. (B) Schematic representation of the *S. aureus* enzymes involved in PA, PG, CL, Lysyl-PG, Glc_2-DAG synthesis and DAG recycling pathway.

Table 8.1 *S. aureus* lipid synthesis proteins

Name	USA300_FPR3757 locus tag	Function
PlsX	SAUSA300_1122	Acyl-ACP phosphotransacylase
PlsY	SAUSA300_1249	Glycerol-3-phosphate acyltransferase
PlsC	SAUSA300_1673	1-Acyl-sn-glycerol-3-phosphate acyltransferase
CdsA	SAUSA300_1154	Phosphatidate cytidylyltransferase
PgsA	SAUSA300_1176	CDP-diacylglycerol-glycerol-3-phosphate 3-phosphatidyltransferase
Cls1	SAUSA300_1216	Cardiolipin synthase
Cls2	SAUSA300_2044	Cardiolipin synthase
MprF (FmtC)	SAUSA300_1255	Phosphatidylglycerol lysyltransferase
YpfP	SAUSA300_0918	Processive diacylglycerol β-glucosyltransferase
DgkB	SAUSA300_1879	Diacylglycerol kinase

of sn-glycerolphosphate to produce lysophosphatidic acid (LPA) (Lu *et al.*, 2006). In a final step, the acyltransferase PlsC acylates LPA on the hydroxyl group at the C2 position to yield PA (Lu *et al.*, 2006) (Fig. 8.3B). The acyl chains at the C1 and C2 positions differ in their structure. Straight and longer-chain C16, C18 or C20 fatty acids are preferentially linked at position 1, whereas branched iso- or anteiso-C15 fatty acids or shorter chain C14 fatty acids are attached at the C2 position (Fischer, 1994).

PA is present only in very small amounts in the membrane of *S. aureus,* but it is the key intermediate and required for the synthesis of other membrane lipids. PG is produced from PA by the sequential action of the phosphatidate cytidyltransferase CdsA, the phosphatidylglycerol phosphate synthase PgsA, and one or more unidentified enzymes with phosphatidylglycerol phosphate phosphatase activity (Fig. 8.3B and Table 8.1). PG can then be used as the substrate to form other cell wall components. The glycerolphosphate head group of PG is used for the production of LTA, converting PG to DAG, which is subsequently converted to PA by the essential DAG kinase DgkB (Fig. 8.3B and Table 8.1) (Jerga *et al.,* 2007; Miller *et al.,* 2008). DAG can also be used as a substrate for the synthesis of glycolipids. The processive glycosyltransferase YpfP uses the nucleotide-activated sugar UDP-glucose (UDP-Glc) initially for the synthesis of the monosaccharide-containing glycolipid Glc-DAG. However, only small amounts of Glc-DAG are present within the membrane, as this lipid is subsequently converted by YpfP to the disaccharide-containing glycolipid Glc_2-DAG by the addition of a second saccharide (Gründling and Schneewind, 2007a; Jorasch *et al.,* 2000; Kiriukhin *et al.,* 2001) (Fig. 8.3B and Table 8.1).

As observed in many other bacteria, entry of *S. aureus* into stationary phase is accompanied by the conversion of most of the PG to cardiolipin (CL) (Short and White, 1971). The CL concentration increases to around 30% in stationary phase (Short and White, 1971). Similarly, phagocytosis of *S. aureus* by human neutrophils induces the production of CL from PG (Femling *et al.,* 2005). CL is produced by the condensation of two molecules of PG to yield one molecule of CL and glycerol (Short and White, 1972) (Fig. 8.3B). This reaction is catalysed by CL synthases, of which *S. aureus* produces two, Cls1 and Cls2 (Koprivnjak *et al.,* 2011) (Fig. 8.3B and Table 8.1). While both enzymes are required for the accumulation of CL in stationary phase, disruption of *cls2* has a more pronounced effect on CL production than disruption of *cls1* (Koprivnjak *et al.,* 2011).

In *S. aureus,* PG can also be further converted to lysyl-PG by the membrane protein MprF (Peschel *et al.,* 2001). MprF has two functional domains, a synthase and a lipid flippase domain (Ernst *et al.,* 2009). The C-terminal domain contains the enzymatic activity that transfers L-lysine derived from lysyl-tRNA onto the head group of PG on the cytoplasmic side of the membrane (Fig. 8.3B and Table 8.1). The resulting lysyl-PG is then transported to the outer leaflet of the membrane by the flippase domain present in the N-terminal part of MprF (Ernst *et al.,* 2009). This enzyme was named MprF, for multiple peptide resistance factor, as *S. aureus mprF* mutants are very sensitive to the action of cationic antimicrobial peptides (CAMPs) (Peschel *et al.,* 2001). The lysyl-PG synthesis domain and flippase domain are essential for CAMP resistance and also resistance to the antibiotic daptomycin, which will be discussed in more detail in the final section of this chapter (Ernst *et al.,* 2009) (see also review by Bayer *et al.,* 2013).

S. aureus peptidoglycan structure, synthesis and function

The bacterial peptidoglycan layer determines the shape of the bacterium and protects bacteria from lysis as a result of the high internal osmotic pressure. This macromolecule is made up of glycan strands to which short peptides, so called stem peptides, are attached. Cross-linking of the stem peptides of neighbouring glycan strands by another peptide, a pentaglycine cross-bridge in the case of *S. aureus,* connects the strands to form the large three-dimensional structure of peptidoglycan (Fig. 8.4A). The glycan strands are composed of disaccharide units consisting of *N*-acetylmuramic acid (MurNAc) and *N*-acetylglucosamine (GlcNAc) joined by a β1–4 linkage. The stem peptides are attached to the lactyl moiety of the MurNAc units (for review see Navarre and Schneewind, 1999) (Fig. 8.4A). In *S. aureus,* the stem peptide is a pentapeptide with the sequence L-Ala-D-iGln-L-Lys-D-Ala-D-Ala (for review see Navarre and Schneewind, 1999). The stem peptides of neighbouring glycan strands are cross-linked via a pentaglycine cross-bridge that is produced on the L-Lys in position 3 of one strand and then connected to the D-Ala in position 4 of the neighbouring stem peptide (Fig. 8.4A).

The peptidoglycan synthesis pathway is summarized in Fig. 8.4B and key enzymes involved in

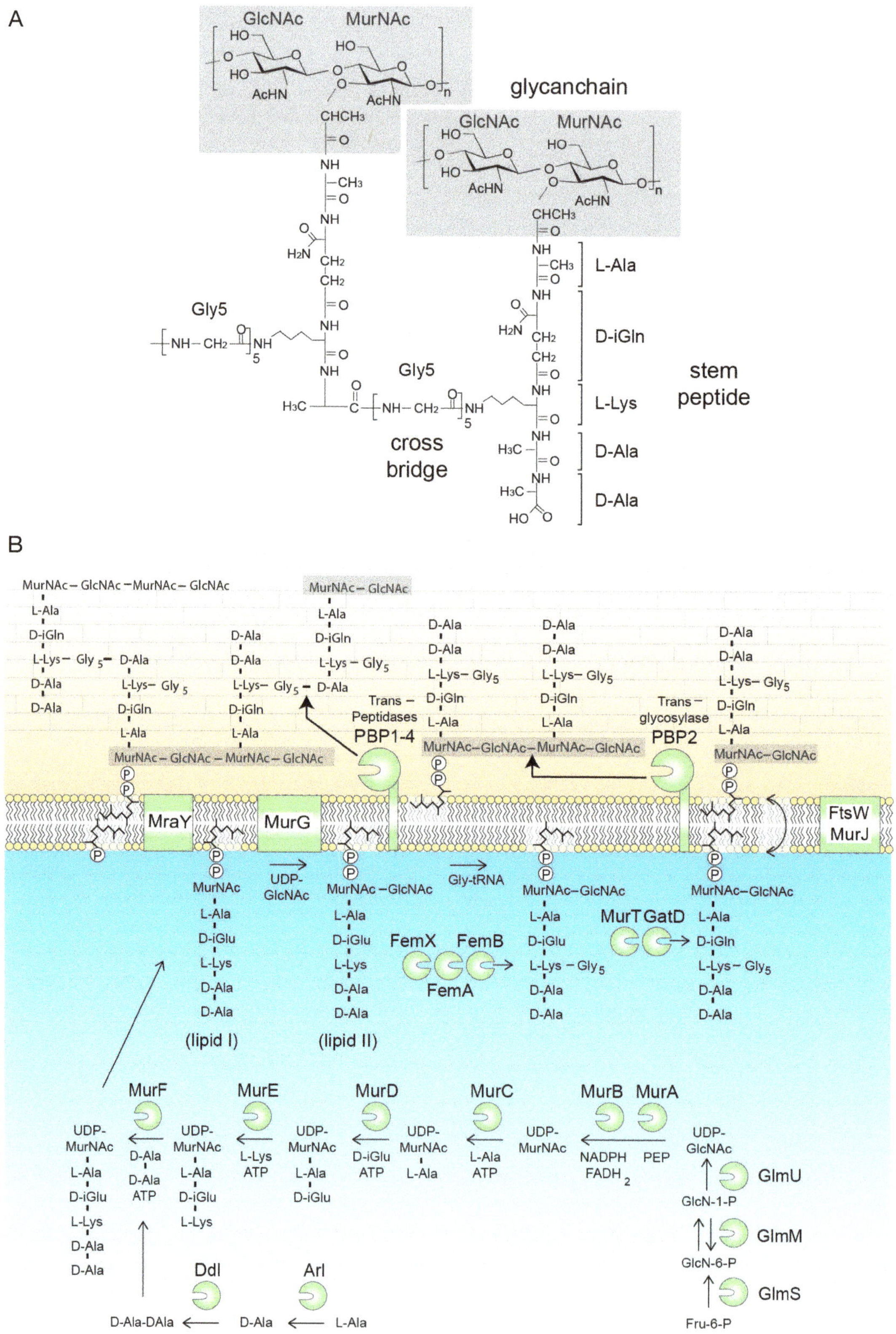

Figure 8.4 *S. aureus* peptidoglycan structure and synthesis pathway. (A) Chemical structure of the *S. aureus* peptidoglycan. (B) *S. aureus* enzymes involved in the peptidoglycan synthesis and schematic representation of its synthesis pathway.

the synthesis of this polymer are listed in Table 8.2. Peptidoglycan synthesis begins in the cytoplasm of the cell with the production of UDP-GlcNAc from fructose-6-phosphate by the sequential activity of GlmS, GlmM and GlmU (Fig. 8.4B and Table 8.2). Next, UDP-GlcNAc is converted to UDP-MurNAc by the enzymes MurA and MurB. Subsequently the amino acids L-Ala, D-iGlu, L-Lys and the dipeptide D-Ala-D-Ala are added on the lactyl moiety of the MurNAc residue by the sequential actions of the enzymes MurC, MurD, MurE, and MurF, respectively, forming the so-called Park's nucleotide [(Park, 1952a,b,c), reviewed in (Lovering *et al.*, 2012; Navarre and Schneewind, 1999)]. The D-Ala-D-Ala dipeptide is synthesized by the combined action of the D-Ala racemase Arl, which converts L-Ala to D-Ala, and the D-Ala ligase Ddl, which produces the dipeptide D-Ala-D-Ala (see review (Navarre and Schneewind, 1999) and references therein). The UDP-MurNAc pentapeptide is then transferred by MraY to undecaprenyl-phosphate (C_{55}-P) to give the membrane-linked intermediate lipid I. Next, MurG uses UDP-GlcNAc as substrate and links the GlcNAc sugar to MurNAc, forming the membrane intermediate lipid II (see reviews (Lovering *et al.*, 2012; Navarre and Schneewind, 1999) and references therein). In *S. aureus*, the L-iGlu residue within this lipid II precursor is further amidated

Table 8.2 *S. aureus* peptidoglycan synthesis proteins

Name	USA300_FPR3757 locus tag	Function
GlmS	SAUSA300_2104	Glucosamine-6-phosphate synthase
GlmM (FemD)	SAUSA300_2111	phosphoglucosamine mutase
GlmU	SAUSA300_0477	Bifunctional: UDP-N-acetylglucosamine pyrophosphorylase and glucosamine-1-phosphate N-acetyltransferase
MurA	SAUSA300_2055	UDP-N-acetylglucosamine 1 carboxyvinyltransferase 1
MurB	SAUSA300_0722	UDP-N-acetylenolpyruvoylglucosamine reductase
MurC	SAUSA300_1686	UDP-N-acetylmuramate alanine ligase
MurD	SAUSA300_1077	UDP-N-acetylmuramoylalanine D-glutamate ligase
MurE	SAUSA300_0919	UDP-N-acetylmuramoylalanyl-D-glutamate L-lysine ligase
MurF	SAUSA300_2038	UDP-N-acetylmuramoyl-tripeptide-D-alanyl-D-alanine ligase
Arl	SAUSA300_2027	Alanine racemase
Ddl	SAUSA300_2039	D-Alanyl-alanine synthetase
MraY	SAUSA300_1076	Phospho-N-acetylmuramoyl-pentapeptide-transferase
MurG	SAUSA300_1311	UDP-N-acetylglucosamine-N-acetylmuramyl-(pentapeptide) pyrophosphoryl-undecaprenol N-acetylglucosamine transferase
FemX	SAUSA300_2214	Aminoacyltransferase (Gly1)
FemA	SAUSA300_1269	Aminoacyltransferase (Gly2 and Gly3)
FemB	SAUSA300_1270	Aminoacyltransferase (Gly4 and Gly5)
MurT	SAUSA300_1873	Murine ligase family protein
GatD	SAUSA300_1872	Glutamine amidotransferase-like protein
FtsW/RodA	SAUSA300_2040/SAUSA300_1013	Putative lipid II flippase
MurJ	SAUSA300_1700	Putative lipid II flippase
PBP1	SAUSA300_1075	Penicillin-binding protein 1 (transpeptidase)
PBP2	SAUSA300_1341	Penicillin-binding protein 2 (transpeptidase and transglycosylase)
PBP3	SAUSA300_1512	Penicillin-binding protein 3 (transpeptidase)
PBP4	SAUSA300_0629	Penicillin-binding protein 4 (transpeptidase)
PBP2A (PBP2′)	SAUSA300_0032	Penicillin-binding protein 2A (transpeptidase)
OatA	SAUSA300_2504	O-acetyltransferase

to L-iGln by MurT, a member of the murine ligase family, and the glutamine amidotransferase-like protein GatD (Figueiredo *et al.*, 2012; Munch *et al.*, 2012). Without this amidation, *S. aureus* has a severe growth defect and decreased resistance to β-lactam antibiotics and lysozyme (Figueiredo *et al.*, 2012). At this stage, the pentaglycines are added to the ε-amino group of the lysine residue of the stem peptide by the peptidyltransferases FemX, FemA, and FemB, all of which use glycyl-tRNAs as the glycine donor (Maidhof *et al.*, 1991; Rohrer *et al.*, 1999; Schneider *et al.*, 2004; Stranden *et al.*, 1997). Addition of the first glycine by FemX is an essential process in *S. aureus.* FemA adds glycine residues 2 and 3 and FemB adds glycine residues 4 and 5 (Rohrer *et al.*, 1999; Schneider *et al.*, 2004). A *femAB* double mutant, which contains only a single glycine in the cross-bridge, is viable, although its growth is impaired. Bacteria become more sensitive to methicillin and other cell wall active antibiotics as they display reduced peptidoglycan cross-linking (Stranden *et al.*, 1997). Next, the lipid-linked peptidoglycan intermediate is transported to the outer leaflet of the membrane by a lipid II flippase. Two different types of membrane proteins, FtsW/RodA and MurJ, have been suggested to fulfil this function in *Escherichia coli* (Mohammadi *et al.*, 2011; Ruiz, 2008; Sham *et al.*, 2014). *S. aureus* encodes two proteins with homology to FtsW/RodA, (SAUSA300_2040 and SAUSA300_1013) and one protein with homology to MurJ (SAUSA300_1700) (Table 8.2). After the lipid II peptidoglycan precursors have been transported to the outside of the membrane, they are polymerized by the penicillin binding proteins. *S. aureus* has four penicillin binding proteins, PBP1 through PBP4, which have distinct functions. The high molecular weight PBP2 contains a transglycosylase domain and catalyses the glycan chain formation. This enzyme cleaves the disaccharide stem peptide unit from lipid II and attaches the MurNAc residue via a β1–4 linkage to a GlcNAc residue in the growing peptidoglycan layer, thereby expending the glycan strand by a disaccharide repeating unit. All four PBPs have a transpeptidase domain and perform the peptide cross-linking reaction, in which the terminal Gly residue of the pentaglycine cross-bridge is linked to the D-Ala in position 4 of a neighbouring stem peptide. As part of this reaction, the PBP enzymes cleave the terminal D-Ala and form a covalent intermediate with the carbonyl group of the D-Ala at position four. This intermediate is resolved through a nucleophilic attack by the terminal amino group of the pentaglycine residues of a neighbouring peptide chain, resulting in the formation of a new peptide bond. The β-lactam ring in β-lactam antibiotics mimics the D-alanyl-D-alanine bond within the stem peptide (Tipper and Strominger, 1965). PBPs bind and hydrolyse the β-lactam ring to form a covalent intermediate, which blocks their transpeptidase activity (Kozarich and Strominger, 1978). Although the peptidoglycan in *S. aureus* is highly cross-linked, not all peptides are cross-linked with other peptidoglycan strands. It is thought that the ratio of cross-linked versus free pentapeptide chains is important for the proper functioning of the peptidoglycan layer (Labischinski and Maidhof, 1994). Furthermore, during the peptidoglycan synthesis process, proteins are covalently linked to the pentaglycine cross-bridge by the sortase enzymes (described later in this chapter), which also prevent further cross-linking of these peptides. The peptidoglycan polymer undergoes additional modifications after its synthesis. The hydroxyl group at the C6 position of MurNAc is partially O-acetylated by the O-acetyl transferase enzyme OatA (Bera *et al.*, 2005) or used as attachment site for the cell wall polymer wall teichoic acid (Araki and Ito, 1989). Both of these modifications have important functional consequences, as the O-acetylation is important for resistance to the peptidoglycan-hydrolysing host enzyme lysozyme and staphylococci become even more sensitive to lysozyme in the absence of both WTA and the O-acetylation (Bera *et al.*, 2005, 2007). Furthermore, WTA contributes to the resistance to methicillin in *S. aureus* (Campbell *et al.*, 2011; Maki *et al.*, 1994).

Cell wall proteins in *S. aureus*

There are many proteins that function in the staphylococcal cell wall. The three main mechanisms by with protein are retained within the cell wall are: attachment of an N-terminal lipid anchor, thereby retaining the protein in the membrane (as in the case of lipoproteins), sortase-catalysed covalent attachment to peptidoglycan and non-covalent attachment to a cell wall component through a dedicated cell wall binding domain.

Lipoprotein synthesis and function

Lipoproteins are synthesized with a characteristic N-terminal type II signal peptide. The signal peptide has a typical tripartite structure, with a short basic n-region, followed by a hydrophobic h-region and a c-region containing the recognition site for the signal peptidase. In case of lipoproteins, the c-region is recognized by the type II signal peptidase and contains specific amino acids also called a lipo-box motif at amino acid position −3 to +1 with respect to the cleavage site. Specifically, this lipo-box motif is L_{-3} $[A/S/T]_{-2}$ $[G/A]_{-1}$ followed by an invariant cysteine (C_{+1}) residue immediately after the signal peptidase cleavage site at position +1 (Fig. 8.5A) (Hutchings *et al.*, 2009). Most lipoproteins are transported across the membrane by the general secretory (Sec) machinery. Following transport across the membrane, the cysteine residue is modified with a lipid moiety that allows the protein to be retained within the membrane. Once the protein is lipidated at the invariant cysteine residue at position +1, the type II signal peptidase cleaves the N-terminal signal peptide; in the absence of lipidation the signal sequence is not cleaved. The lipoprotein synthesis pathway has been best characterized in *E. coli* (for reviews see Hayashi and Wu, 1990; Nakayama *et al.*, 2012). In this organism, three enzymes that act sequentially are required for the lipidation and production of the mature triacylated lipoprotein (Hayashi and Wu, 1990; Nakayama *et al.*, 2012). Lgt, the preprolipoprotein diacylglycerol transferase, adds diacylglycerol (DAG), derived from an anionic membrane lipid such as phosphatidylglycerol (PG) to the cysteine residue at position +1 (Hayashi and Wu, 1990; Nakayama *et al.*, 2012). Next, the type II signal peptidase Lsp cleaves the signal peptide, after which the third enzyme, Lnt, transfers an acyl group, also derived from a membrane lipid, to the newly generated α-amino group of the N-terminal *S*-diacylglycerol cysteine residue; thereby, generating the final triacylated lipoprotein (reviewed in Nakayama *et al.*, 2012).

In contrast to Gram-negative bacteria, in which lipoproteins are always triacylated, lipoproteins in Gram-positive bacteria can be either diacylated or triacylated (Nakayama *et al.*, 2012). In *S. aureus* Lgt and Lsp fulfil the same function as their *E. coli*

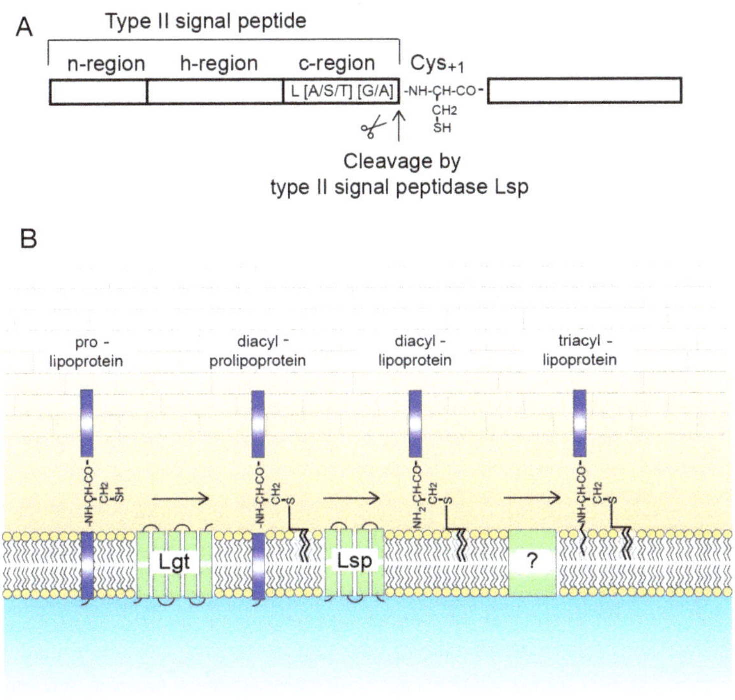

Figure 8.5 *S. aureus* lipoproteins. (A) Schematic representation of pre-lipoprotein with characteristic features of the N-terminal type II signal peptide indicated (B) *S. aureus* enzymes involved in lipoprotein synthesis and schematic representation of lipoprotein synthesis pathways.

orthologues: Lgt transfers the DAG lipid onto the conserved cysteine residue of the lipo-box and Lsp subsequently cleaves the N-terminal signal peptide (Bubeck Wardenburg *et al.*, 2006; Hutchings *et al.*, 2009; Stoll *et al.*, 2005) (Fig. 8.5B and Table 8.3). In contrast to *E. coli*, no orthologue of Lnt is present in *S. aureus* (Hutchings *et al.*, 2009; Stoll *et al.*, 2005). This finding, in conjunction with initial experimental evidence in *S. aureus* and other Gram-positive bacteria, lead to the assumption that all lipoproteins in *S. aureus* are diacylated (Tawaratsumida *et al.*, 2009). However, it has now been demonstrated that lipoproteins in *S. aureus* can be triacylated, although the enzyme responsible for the addition of the third acyl chain is unknown (Asanuma *et al.*, 2011; Kurokawa *et al.*, 2009, 2012). Furthermore, lipoproteins were thought to be either diacylated or triacylated in a strain specific fashion. Interestingly, this is not the case in *S. aureus*, where a lipoprotein can be either diacylated or triacylated, depending on the growth conditions. Lipoproteins in *S. aureus* are usually triacylated, but are diacylated under acidic growth conditions in the post-exponential growth phase. In addition, other stress conditions, such as a high temperatures and high salt concentrations, accelerate the accumulation of diacylated lipoproteins (Kurokawa *et al.*, 2012).

Lipoproteins perform their function at the interface between the membrane and the cell wall. Based on genome analysis, it has been suggested that around 2% of all *S. aureus* proteins are lipoproteins, the majority of which have not been functionally characterized (Babu *et al.*, 2006). These proteins probably play a critical role in a wide range of physiological processes and may be of particular importance in nutrient uptake, including iron acquisition (Maresso and Schneewind, 2006). However, the *lgt* and *lsp* genes are not essential for the growth of *S. aureus* (Bae *et al.*, 2004; Bubeck Wardenburg *et al.*, 2006; Stoll *et al.*, 2005). Growth of an *S. aureus lgt* mutant is only minimally affected in rich medium, however a growth defect is observed under nutrient or iron limiting conditions and within macrophages (Schmaler *et al.*, 2009; Stoll *et al.*, 2005). Bacterial lipoproteins are now well-established pathogen-associated molecular patterns (PAMPs) and known to activate an immune response through the mammalian toll-like receptor 2 (TLR2) (Aliprantis *et al.*, 1999). Of all staphylococcal cell wall components, lipoproteins appear to be the most potent activators of TLR2 (Hashimoto *et al.*, 2006a,b). Several other cell wall components, including LTA and peptidoglycan fragments, have been implicated as PAMPs for TLR2, but as the concentration of these components necessary to trigger activation are rather high, the physiological relevance of TLR2 activation has been questioned (Zähringer *et al.*, 2008). TLR2-deficient mice show a more severe disease outcome, a higher bacterial load and decreased inflammation upon infection with *S. aureus*, highlighting that TLR2 is an important receptor for the detection of *S. aureus* during infections (Kielian *et al.*, 2005; Miller *et al.*, 2006; Schmaler *et al.*, 2009; Sun *et al.*, 2006; Takeuchi *et al.*, 2000). Furthermore, infection of mice with an *lgt* mutant strain, which is unable to produce lipidated proteins, resulted in a decrease in TLR2-dependent production of cytokines compared to an infection with a wild-type strain (Schmaler *et al.*, 2009; Stoll *et al.*, 2005); highlighting, that lipoproteins are important for generating an innate immune response through TLR2 during infection. As lipoproteins in Gram-positive bacteria are embedded within the cell wall, they are not easily accessible for recognition by TLRs. Recent studies on *S. aureus* have highlighted that cell wall degradation in the acidic phagosome environment is important for stimulation of an immune response through the TLR2 receptor (Ip *et al.*, 2010; Müller *et al.*, 2010).

Table 8.3 *S. aureus* enzymes involved in lipoprotein production and covalent anchoring of cell wall proteins

Name	USA300_FPR3757 locus tag	Function
Lgt	SAUSA300_0744	Prolipoprotein diacylglycerol transferase
LspA	SAUSA300_1089	Type II signal peptidase
SpsB	SAUSA300_0868	Type I signal peptidase
SrtA	SAUSA300_2467	Sortase A
SrtB	SAUSA300_1034	Sortase B

Covalently and non-covalently anchored cell wall proteins and their function

In *S. aureus* most secreted proteins as well as covalently and non-covalently attached cell wall proteins are transported across the membrane through the Sec pathway. In contrast to lipoproteins, which contain a type II signal peptide at their N-terminus, these proteins contain an N-terminal type I signal peptide. Type I signal peptides also have a tripartite structure, with an N-terminal basic n-region, a hydrophobic h-region and a c-region, which contains the signal peptidase recognition site for the type I signal peptidase. Commonly found amino acids at positions −3 to −1, with respect to the signal peptidase cleavage site, are AXA, where X can be any amino acid (Fig. 8.6A). Two signal peptidases, SpsA and SpsB, are encoded in the genome of *S. aureus*; however, only one, SpsB, is catalytically active and is essential for growth (Cregg *et al.*, 1996). Proteins that contain only an N-terminal type I signal peptide are secreted into the supernatant; covalent or non-covalent attachment to the cell wall requires additional sequence motifs. Distinct binding domains mediate non-covalent attachment of proteins to the cell wall. In the case of the major *S. aureus* autolysin Atl, binding is mediated by repeat (R) domains, which belong to the family of bacterial SH3 (SH3b) domains, while other proteins contain one or multiple LysM domains for cell wall binding (Fig. 8.6A). SH3b domains are thought to bind to LTA, while LysM domains bind to the glycan strands within the peptidoglycan polymer (Götz *et al.*, 2014; Mesnage *et al.*, 2014; Steen *et al.*, 2003; Zoll *et al.*, 2012). LysM domains are also often found in proteins containing an additional cell wall hydrolase domain. Therefore, many of these non-covalently attached cell wall proteins are required for the remodelling of the peptidoglycan layer, cell division and septation (Vollmer *et al.*, 2008).

The major autolysin Atl is one of the best-characterized *S. aureus* enzymes belonging to this family of non-covalently anchored cell wall proteins (see review by Götz *et al.*, 2014). Atl localizes specifically to the cell division site and is required for cell septation and an *S. aureus atl* mutant is viable but forms large cell clusters (Baba and Schneewind, 1998; Biswas *et al.*, 2006; Schlag *et al.*, 2010; Yamada *et al.*, 1996). Atl is also an important adhesion molecule that is required for attachment to surfaces and biofilm formation (Biswas *et al.*, 2006). In addition, Atl can also interact with eukaryotic extracellular matrix glyoproteins through its cell wall anchoring domains and therefore is also an important adhesion and virulence factor (Kohler *et al.*, 2014; see also review by Götz *et al.*, 2014).

Additional important adhesion molecules in *S. aureus* are surface proteins that are covalently linked to the peptidoglycan. Covalently anchored cell wall proteins contain a specific C-terminal sorting signal, which is composed of the amino acid sequence LPXTG (where X is any amino acid), followed by a hydrophobic domain and a string of positively charged amino acids (Fischetti *et al.*, 1990; Schneewind *et al.*, 1993) (Fig. 8.6A). After secretion, these proteins are initially retained in the membrane through their hydrophobic domain and positively charged C-terminal tail (Fig. 8.6B). The LPXTG motif, which is now exposed on the outside of the membrane, is recognized by the sortase enzyme SrtA (Mazmanian *et al.*, 1999). SrtA is a membrane-anchored transpeptidase, which cleaves the peptide bond between the threonine (T) and glycine (G) residue in the LPXTG motif, generating an enzyme-surface protein intermediate (Fig. 8.6B). Next, the protein is transferred onto the pentaglycine cross-bridge of lipid II, the peptidoglycan synthesis intermediate described above (Fig. 8.6B). Finally, the surface proteins-lipid II peptidoglycan intermediate is incorporated into the peptidoglycan layer through the activity of the penicillin binding proteins (PBPs) [see reviews (Navarre and Schneewind, 1999; Schneewind and Missiakas, 2014) (Fig. 8.6B). *S. aureus* encodes a second sortase enzyme, SrtB, although only one substrate for this sortase is known, a protein named IsdC (Mazmanian *et al.*, 2003) (Table 8.3). SrtB acts similar to SrtA, but recognizes an NPQTN motif instead of the LPXTG motif recognized by SrtA (Table 8.3). The SrtB enzyme cleave between the T and N within the motif but rather than transferring the protein onto lipid II the enzyme transfers it onto assembled peptidoglycan. This results in a more buried location of the IsdC proteins in the cell wall, as compared to SrtA-anchored cell surface proteins (Marraffini and Schneewind, 2005).

Sortase-anchored surface proteins can be readily identified based on the presence of the C-terminal sorting motif. They have been grouped

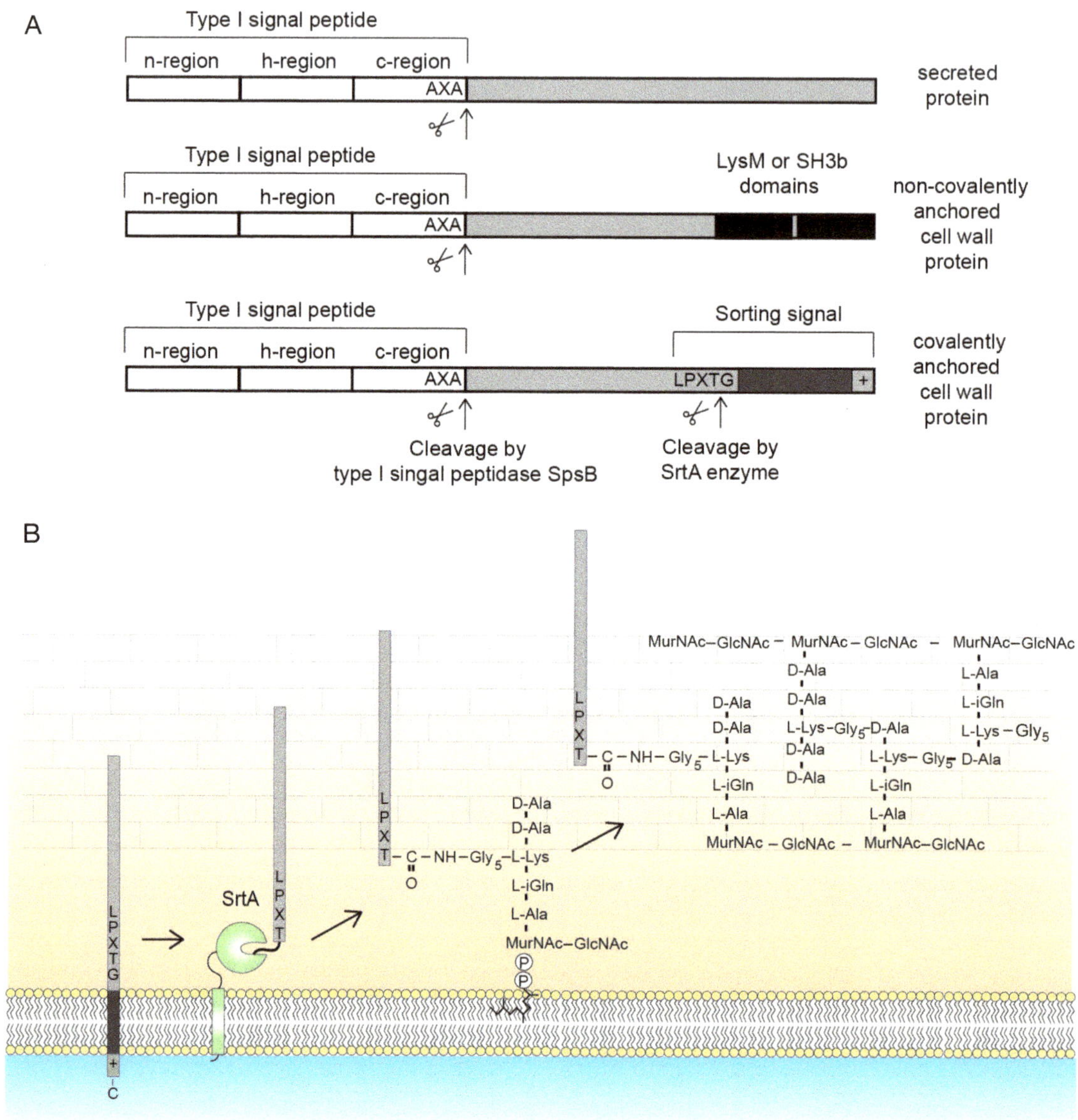

Figure 8.6 Cell wall proteins. (A) Schematic representation of secreted, non-covalently and covalently attached cell wall proteins. These proteins are produced with a characteristic N-terminal type I signal peptide. Non-covalently attached proteins contain extra SH3b or LysM cell wall binding domains, while covalently attached cell wall proteins contain a C-terminal cell wall sorting signal. (B) Schematic representation of the sortase enzyme mediated process of anchoring cell wall proteins to the peptidoglycan layer.

based on their function, and more recently, based on specific protein motifs that have been defined by a structure–function analysis (Foster *et al.*, 2014). While the precise repertoire of sortase-anchored proteins varies between different *S. aureus* strains, more than 20 sortase substrates can be expressed in an individual strain (McCarthy and Lindsay, 2010) (Table 8.4). Several of these sortase-anchored proteins, including FnBPA, SasF, AadA and protein A, are present in all *S. aureus* strains, while others such as CnaA, SasK, SdrE, are present in only a few strains (McCarthy and Lindsay, 2010). Thus, not all surface proteins are required for the growth and pathogenesis of *S. aureus*, likely owing to the fact that several of the proteins have redundant functions. The expression

Table 8.4 *S. aureus* sortase anchored cell surface proteins

Name	USA300_FPR3757 locus tag	Function
AdsA (SasH)	SAUSA300_0025	Adenosine synthase A
ClfA[1]	SAUSA300_0772	Clumping factor A
ClfB[1]	SAUSA300_2565	Clumping factor B
CnaA[1]	Not present	Collagen adhesin
FnBPA[1]	SAUSA300_2441	Fibronectin binding protein A
FnBPB[1]	SAUSA300_2440	Fibronectin binding protein B
IsdA[2]	SAUSA300_1029	Iron-regulated surface determinants A
IsdB[2] (SasJ)	SAUSA300_1028	Iron-regulated surface determinants B
IsdH[2]	SAUSA300_1677	Iron-regulated surface determinants H
SasB	SAUSA300_2109	*Staphylococcus aureus* protein B
SasC	SAUSA300_1702	*Staphylococcus aureus* protein C
SasD	SAUSA300_0136	*Staphylococcus aureus* protein D
SasF	SAUSA300_2581	*Staphylococcus aureus* protein F
SasG[3]	SAUSA300_2436	*Staphylococcus aureus* protein G
SasK	Not present	*Staphylococcus aureus* protein K
SasL	Not present	*Staphylococcus aureus* protein L
SasX	Not present	*Staphylococcus aureus* protein X
SdrC[1]	SAUSA300_0546	Serine aspartate repeat protein C
SdrD[1]	SAUSA300_0547	Serine aspartate repeat protein D
SdrE[1]	SAUSA300_0548	Serine aspartate repeat protein E
Spa[4]	SAUSA300_0113	*Staphylococcus aureus* protein A
SraP	SAUSA300_2589	Serine-rich adhesin for platelets

[1]Microbial surface component recognizing adhesive matrix molecule (MSCAMM) family protein.
[2]Near transporter (NEAT) motif family protein.
[3]G5-E repeat family protein (annotated as described in Foster *et al.*, 2014).
[4]Three-helix bundle family.

of the sortase-anchored proteins also varies depending on the growth conditions. Some proteins, such as the iron-regulated surface proteins IsdA, IsdB and IsdH, which are involved in haem uptake, are expressed only under iron-limiting conditions (Mazmanian *et al.*, 2003). Several other sortase-anchored proteins are present primarily during exponential phase growth, whereas others are present only in the stationary phase (Bischoff *et al.*, 2004; McAleese *et al.*, 2001).

Surface proteins can be located at specific sites within the cell wall. As discussed above, the SrtB substrate IsdC is buried within the cell wall, whereas SrtA substrates appear to be deposited either at the division site or the cell poles, depending on the nature of their N-terminal signal peptide (DeDent *et al.*, 2008). Several of the sortase-anchored proteins are produced with a longer signal peptide containing a YSIRK-G/S motif within the N-terminal n-region. Proteins with such a signal peptide are primarily deposited at the cell division site, while proteins without this motif are deposited at the poles (DeDent *et al.*, 2008).

Many sortase-anchored proteins are involved in the adhesion of *S. aureus* to host components, including cell matrix proteins such as fibrinogen, fibronectin or collagen and are therefore, essential for the colonization process. For example, the clumping factor proteins ClfA and ClfB bind to immobilized fibrinogen and allow the bacterium to adhere to desquamated epithelial cells through their fibrinogen binding ability. Similarly, the fibronectin-binding proteins FnBPA and FnBPB bind to fibrinogen and fibronectin and are involved in the adhesion to the extracellular matrix. The ligands, binding mechanism and function of sortase-anchored surface proteins have been summarized in a recent review (Foster *et al.*, 2014).

Several sortase-anchored proteins also have important immune-evasion functions. FnBPs promote the internalization of *S. aureus* into host cells, in addition to their function in adherence to the extracellular matrix. As a result, the internalized bacteria are protected from recognition by extracellular host defence molecules and this facilitates the formation of slow growing small colony variants (SCVs) that can resist antibiotic therapy (Peacock *et al.*, 1999; Sendi and Proctor, 2009; Thwaites and Gant, 2011). Other sortase-anchored proteins, including SdrE and ClfA, can interact with components of the host's complement system. They contribute to immune evasion by promoting either the destruction of a complement component or by preventing interactions between complement components (Hair *et al.*, 2008; Sharp *et al.*, 2012). Perhaps one of the best-known immune evasion proteins is protein A, encoded by the *spa* gene [see review (Foster *et al.*, 2014). Protein A interacts with the Fc region of IgG antibodies, orienting antibodies in such a way that they cannot interact with other components of the immune system, and preventing the activation of the classical complement pathway. Taken together, many sortase-anchored cell wall proteins are important virulence factors, as they are required for the adhesion to, colonization and immune evasion of the host. The role of many of these proteins in virulence has been confirmed experimentally in different animal models (reviewed in Foster *et al.*, 2014). As they are exposed to the immune system, recombinant sortase-anchored cell wall proteins are currently also extensively pursued as vaccine candidates to combat infections with *S. aureus*. However, vaccination with the IsdB protein, which was protective in an animal model of infection, did not protect patients from serious infection (Bagnoli *et al.*, 2012; Fowler *et al.*, 2013). The reasons for this failure are currently under investigation. Possibly superior protection can be achieved using a vaccine composed of multiple components (Anderson *et al.*, 2012; Bagnoli *et al.*, 2012; Jansen *et al.*, 2013; Stranger-Jones *et al.*, 2006).

Teichoic acids in *S. aureus*

Teichoic acids, named after the Greek word *teichos*, meaning 'wall', were first detected by Baddiley and co-workers in 1958 (Armstrong *et al.*, 1958, 1959). A large structural diversity amongst teichoic acids was recognized soon after their discovery and the term teichoic acids was subsequently used to describe all polymers in the bacterial cell wall, membrane and capsule that contain glycerolphosphate (GroP) or ribitolphosphate (RboP) residues (Baddiley, 1972). The term teichoic acid now collectively refers to the two Gram-positive cell wall polymers, wall teichoic acid (WTA) and lipoteichoic acid (LTA), and is no longer associated with both Gram-negative and Gram-positive bacterial polymers. *S. aureus* produces two types of teichoic acids, the lipid-linked glycerolphosphate LTA polymer and the peptidoglycan-linked ribitolphosphate WTA polymer. While both of these polymers are collectively referred to as teichoic acids, they have distinct synthesis pathways and functions (Brown *et al.*, 2013; Percy and Gründling, 2014; Santa Maria *et al.*, 2014).

Lipoteichoic acid synthesis and function

LTA is a zwitterionic polymer found in the cell wall of many Gram-positive bacteria. One of the best-studied and most widespread forms of LTA is a 1–3-linked glycerolphosphate polymer, modified with D-Ala residues and tethered to the membrane via a glycolipid anchor (Fischer, 1990, 1994; Percy and Gründling, 2014) (Fig. 8.7A). This type of LTA, referred to as type I LTA, is also produced by *S. aureus* (Fischer, 1994). In *S. aureus*, the LTA polymer has an average chain length of 25 GroP units, although the individual chains vary in length between 4 and 30 repeating units (Fischer, 1994). Under normal growth conditions, 70–80% of the GroP units are modified at the hydroxyl group at the C2 position with D-Ala residues, but the amount of modification varies depending on the growth conditions (Fischer, 1994).

LTA synthesis is intimately linked with lipid biosynthesis and turnover (Koch *et al.*, 1984) and all currently known enzymes involved in its synthesis are listed in Table 8.5. The LTA lipid anchor in *S. aureus* is glycosyl (β1–6) glycosyl (β1–3) DAG (Glc_2-DAG), which is generated in the cytoplasm by the transfer of two glucose moieties from the nucleotide-activated sugar UDP-glucose (UDP-Glc) to DAG (Fig. 8.7B). UDP-Glc is produced by the α-phosphoglucomutase PgcA, which converts

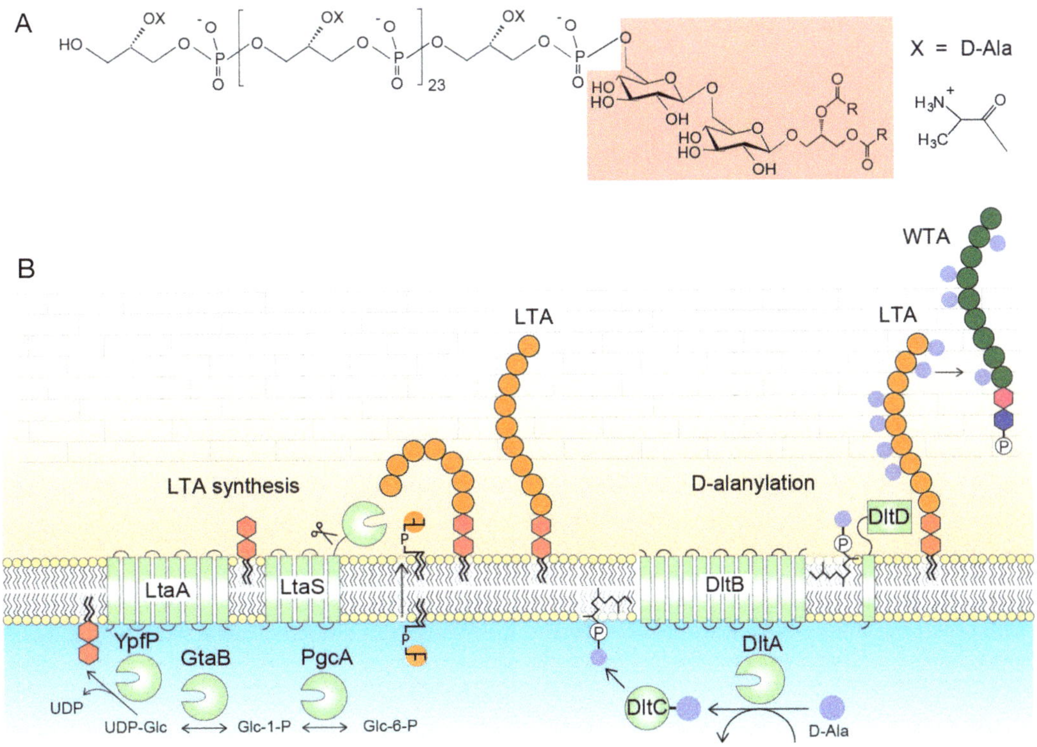

Figure 8.7 *S. aureus* lipoteichoic acid (LTA) structure and synthesis pathway. (A) Chemical structure of the *S. aureus* LTA polymer. R indicates the fatty acids in the glycolipid anchor and X indicates the position of D-Ala modifications on the GroP repeating units. (B) *S. aureus* enzymes involved in LTA synthesis and its modification with D-Ala esters and schematic representation of the different steps required for its synthesis.

Table 8.5 *S. aureus* lipoteichoic acid (LTA) synthesis proteins

Name	USA300_FPR3757 locus tag	Function
PgcA	SAUSA300_2433	α-Phosphoglucomutase
GtaB	SAUSA300_2439	UTP-glucose-1-phosphate uridylyltransferase
YpfP	SAUSA300_0918	Processive diacylglycerol β-glucosyltransferase
LtaA	SAUSA300_0917	Putative glycolipid permease
LtaS	SAUSA300_0703	Lipoteichoic acid synthase
DltA	SAUSA300_0835	D-Alanine-D-alanyl carrier protein ligase
DltB	SAUSA300_0836	D-Alanine transfer protein DltB
DltC	SAUSA300_0837	D-Alanyl carrier protein
DltD	SAUSA300_0838	D-Alanine transfer protein DltD

glucose-6-phosphate (Glc-6-P) to glucose-1-phosphate (Glc-1-P), which is subsequently converted to UDP-Glc by the UTP:α-glucose-1-phosphate uridyltransferase GtaB [(Gründling and Schneewind, 2007a) and references therein] (Fig. 8.7B and Table 8.5). In the last step, the processive glycosyltransferase YpfP produces the glycolipids Glc-DAG and Glc_2-DAG by the sequential transfers of two glucose moieties from UDP-Glc to DAG (Jorasch *et al.*, 2000; Jorasch *et al.*, 1998; Kiriukhin *et al.*, 2001). In *S. aureus* mutants lacking *pgcA*, *gtaB* or *ypfP*, no glycolipids are produced and the LTA backbone, which is still synthesized, is anchored to the membrane lipid DAG (Gründling and

Schneewind, 2007a; Kiriukhin *et al.*, 2001). After synthesis in the cytoplasm, the Glc_2-DAG glycolipid needs to be transported to the outer leaflet of the membrane to serve as LTA lipid anchor. The *S. aureus* protein LtaA, a member of a major facilitator superfamily clan, the gene of which is encoded in an operon with *ypfP*, has been implicated as a glycolipid flippase (Gründling and Schneewind, 2007a).

In contrast to the Glc_2-DAG, the poly-glycerolphosphate chain is produced on the outside of the cell, where the GroP subunits are derived from the head group of the membrane lipid phosphatidylglycerol (PG) and added to the tip of the growing chain (Karatsa-Dodgson *et al.*, 2010; Koch *et al.*, 1984; Taron *et al.*, 1983). The enzyme responsible for the synthesis of the polyglycerolphosphate backbone is the LTA synthase enzyme LtaS, a membrane protein with a large extracellular enzymatic domain (Gründling and Schneewind, 2007b; Lu *et al.*, 2009) (Fig. 8.7B and Table 8.5). In the absence of this enzyme *S. aureus* is unable to produce the LTA backbone and bacteria show severe growth and morphological defects, indicating the importance of LTA in the physiology of the bacterium.

The GroP residues of the LTA backbone, as well as the RboP residues of the WTA backbone, are modified with D-alanine (D-Ala) residues. The D-Ala residues are initially introduced into LTA and from there can be transferred onto WTA. The proteins required for this D-alanylation process are encoded in the *dltABCD* operon (Heaton and Neuhaus, 1992; Neuhaus and Baddiley, 2003). The D-alanylation process starts in the cytoplasm of the cell, where the D-alanine-D-alanyl carrier protein ligase DltA activates D-alanine using ATP to form a high-energy D-alanyl-AMP intermediate (Neuhaus and Baddiley, 2003) and then transfers this onto the D-alanine carrier protein DltC, a small protein with sequence and structural homology to acyl carrier proteins (ACPs) (Volkman *et al.*, 2001) (Fig. 8.7B and Table 8.5). Different models have been proposed for the function of DltB and DltD and how the D-alanines are transferred from DltC across the membrane and onto LTA (Debabov *et al.*, 2000; Neuhaus and Baddiley, 2003; Perego *et al.*, 1995). Recent work has provided experimental support for a model originally proposed by Fischer and colleagues (Perego *et al.*, 1995; Reichmann *et al.*, 2013) in which the multi-membrane spanning protein DltB transfers the D-Ala from DltC onto undecaprenyl-phosphate (C_{55}-P) to form the membrane intermediate D-Ala-P-C_{55} and subsequently transfers this intermediate across the membrane (Fig. 8.7B). In this model, DltD functions on the outside of the cell and aids the transfer of the D-Ala from the D-Ala-P-C_{55} intermediate onto the LTA backbone (Fig. 8.7B). However, it should be stressed that the proposed activities for DltB and DltD have not yet been experimentally verified.

The level of D-Ala modifications on LTA is influenced by a number of environmental factors. A decrease in D-Ala modifications is seen with an increase in growth temperature, NaCl concentration or pH (for review see Fischer, 1994; Neuhaus and Baddiley, 2003). The modification of TAs with D-alanines, which decreases the negative charge of the polymers, is a critical factor contribution to the resistance of *S. aureus* towards cationic antimicrobial peptides. *S. aureus dlt* mutants are more susceptible to bacteriocins, cationic host defence molecules as well as other antimicrobial host enzymes such as the human group II phospholipase A2 and lysozyme (Collins *et al.*, 2002; Herbert *et al.*, 2007; Koprivnjak *et al.*, 2002; Peschel *et al.*, 1999; Peschel and Sahl, 2006). Therefore, D-Ala modifications on TAs are also important for the full virulence of *S. aureus* (Collins *et al.*, 2002). Expression of the *dlt* operon is controlled by the three-component system ApsXRS (also named GraXRS) in conjunction with the ABC transporter VraFG (Falord *et al.*, 2012; Herbert *et al.*, 2007; Li *et al.*, 2007a). It is thought that binding of antimicrobial peptides to the sensor histidine kinase ApsS (or GraS) and the ABC transporter VraFG activates the three-component system, resulting in increased transcription of the *dlt* operon and D-alanine modifications on TAs (Falord *et al.*, 2012; Li *et al.*, 2007a,b). It is also of note, that transcription of *mprF*, which encodes for the enzyme responsible for the lysinylation of PG, is also controlled by the AspXRS three-component system. This further increases the net positive charge within the cell envelope and contributes to antimicrobial peptide resistance (Li *et al.*, 2007a). Transcription of the *dlt* operon is additionally controlled by regulatory systems that respond to other cell-envelope stresses. Transcription of the *dlt* operon is repressed in *S. aureus* by ArlSR in response to an increase in the

Mg^{2+} concentration (Jordan *et al.*, 2008; Koprivnjak *et al.*, 2006). The complex regulatory control of the *dlt* operon highlights that a fine-tuning of the amount of D-Ala modifications on TAs is critical for the maintenance of the bacterial physiology.

Together with WTA, LTA is required for maintaining an adequate homeostasis of cations within the cell wall and helps to regulate the activity of cell wall hydrolases (Fischer, 1994; Neuhaus and Baddiley, 2003). LTA synthesis is essential for the growth of *S. aureus* under standard laboratory growth conditions. *S. aureus ltaS* mutant strains, which completely lack the ability to produce the LTA polymer, can only grow under osmotically stabilizing conditions achieved by the addition of NaCl or sucrose to the medium, after acquisition of compensatory mutations or, for some strains, by lowering the growth temperature (Corrigan *et al.*, 2011; Gründling and Schneewind, 2007b; Oku *et al.*, 2009). In *S. aureus ypfP* mutants, which are unable to produce glycolipids, the LTA polymer is linked to the membrane lipid DAG (Gründling and Schneewind, 2007a; Kiriukhin *et al.*, 2001). These mutant bacteria display growth defects and the bacteria are enlarged, forming misshapen cocci (Kiriukhin *et al.*, 2001). Furthermore, bacteria with a reduced amount of LTA do not form biofilms and have a reduced ability to penetrate the blood–brain barrier (Fedtke *et al.*, 2007; Sheen *et al.*, 2010). LTA has also been identified as a ligand for scavenger receptors on macrophages (Greenberg *et al.*, 1996) and this interaction is thought to aid the uptake and clearance of *S. aureus* cells; consistent with this, scavenger receptor-deficient mice are more susceptible to infection than wild-type mice (Thomas *et al.*, 2000) (see also review by Weidenmaier and Peschel, 2008). Nonetheless, because of the multiple important functions of LTA for bacterial growth and for pathogenesis, the enzymes involved in its synthesis are pursued as potential targets for the development of new antimicrobials (Pasquina *et al.*, 2013; Richter *et al.*, 2013).

Wall teichoic acid synthesis and function

WTA in *S. aureus* is a 1–5 linked RboP polymer with approximately 40 repeating units. An exception to this are strains of the phylogenetically isolated *S. aureus* ST395 linage that produce a 1–3-linked glycerolphosphate WTA polymer, which is decorated with *N*-acetyl-D-galactosamine residues (Winstel *et al.*, 2013, 2014a). However, as most *S. aureus* strains produce an RboP WTA polymer, its synthesis and function will be summarized in this section. WTA is covalently attached to the C6 position of the *N*-acetylmuramic acid residue of peptidoglycan via the complex – $(glycerolphosphate)_2$ *N*-acetylmannosamine-β(1–4)-*N*-acetylglucosamine linker unit [or short $(GroP)_2$-ManNAc-β(1–4)-GlcNAc)] (Fig. 8.8A). The enzymes responsible for WTA synthesis are often referred to as *tar* (teichoic acid ribitol) genes and are listed in Table 8.6. The enzymes catalysing the first steps required for the formation of the linker unit (TarO, TarA and TarB) have functions identical to the corresponding enzymes in the synthesis pathway of a glycerolphosphate WTA polymer and are therefore sometimes also referred to as Tag (teichoic acid glycerol) proteins. The RboP polymer is synthesized in the cytoplasm of the cell on an undecaprenyl-phosphate (C_{55}-P) lipid carrier molecule (Fig. 8.8B), the same lipid carrier used in peptidoglycan synthesis (see also reviews by Brown *et al.*, 2013; Xia and Peschel, 2008). TarO catalyses the first step in the WTA synthesis process, the transfer of GlcNAc-1-P from UDP-GlcNAc to C_{55}-P (Fig. 8.8B). Next, TarA transfers ManNAc, using UDP-ManNAc as substrate, to the C4 hydroxyl group of GlcNAc, forming the ManNAc β(1–4) GlcNAc disaccharide lipid-linked intermediate. The UDP-ManNAc is produced from UDP-GlcNAc by the mannose-6-phosphate isomerase ManA. Two *manA* genes with USA300_FPR3757 locus tags SAUSA300_2096 and SAUSA300_2577 are present in the *S. aureus* genome (Table 8.6). In addition, the CapP enzyme encoded in the capsule operon, and described in a later section, has the same enzymatic activity. As the next step in the WTA synthesis pathways, TarB transfers a GroP unit to the C4 position of ManNAc and TarF subsequently transfers a second GroP unit to form the $(GroP)_2$-ManNAc-β(1–4)-GlcNAc–lipid linked membrane intermediate (Fig. 8.8B). The GroP units are derived from CDP-GroP, which are produced by TarD from CTP and GroP. The WTA backbone chain is produced by an RboP polymerase and two such polymerases, TarL and TarK, are present in *S. aureus*, where TarL appears to be the main polymerase (Meredith *et al.*, 2008). These polymerases add approximately 40 RboP

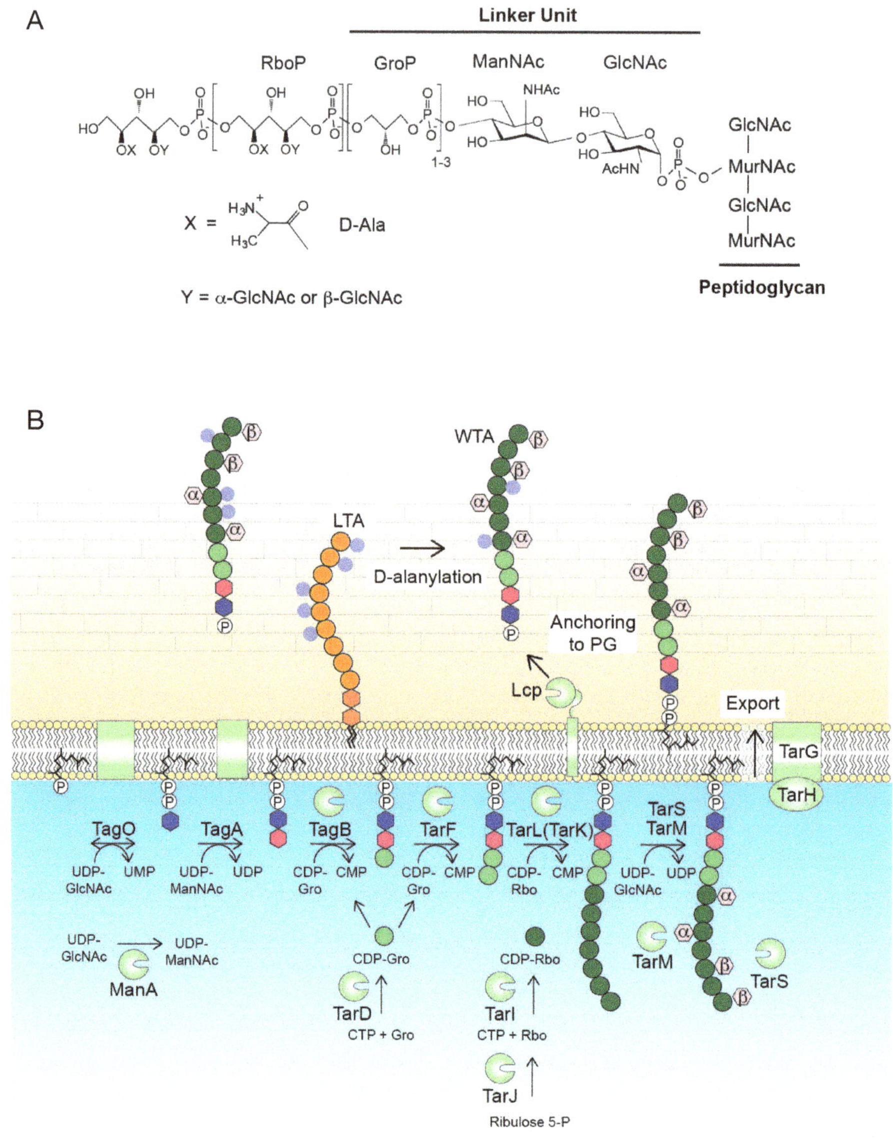

Figure 8.8 *S. aureus* wall teichoic acid (WTA) structure and synthesis pathway. (A) Chemical structure of the *S. aureus* WTA polymer. The ribitolphosphate repeating units can me modified with D-Ala or sugar residues, indicated by X and Y in the structure. (B) *S. aureus* enzymes involved in WTA synthesis and schematic representation of the different steps in the synthesis pathway.

units onto the membrane-anchored linker unit to complete the production of the full-length polymer. TarL and TarK use CDP-RboP as substrate, which is produced from ribulose-5-phosphate in a two-step process by the sequential activity of TarJ and TarI. While several of the genes required for WTA synthesis are scattered around the genome, the majority of the *tar* genes are encoded in two genomic regions. TarI, TarJ and TarL are encoded in the *tarIJL* operon that is duplicated in *S. aureus;* these duplicated genes are found immediately upstream of the *traIJL* operon and are denoted *tarI'J'K*. TarI'/TarJ' do not appear to be functional enzymes in *S. aureus*, but TarK is a functional RboP polymerase (Meredith *et al.*, 2008; Pereira *et al.*, 2008). The significance of the duplication of this operon remains unclear.

The WTA polymer is modified with α- and β-linked GlcNAc residues prior to translocation across the membrane. α-linked GlcNAc residues

Table 8.6 *S. aureus* wall teichoic acid (WTA) synthesis proteins

Name	USA300_FPR3757 locus tag	Function
TarO (TagO)	SAUSA300_0731	Undecaprenyl-phosphate N-acetylglucosaminyl-1-phosphate transferase
TarA (TagA)	SAUSA300_0623	N-Acetylmannosaminyltransferase
TarH	SAUSA300_0624	Teichoic acid export: ATP-binding protein
TarG	SAUSA300_0625	Teichoic acid export: ABC transporter protein
TarB (TagB)	SAUSA300_0626	Glycosyl/glycerophosphate transferase
TarX	SAUSA300_0627	Putative glycosyltransferase
TarD	SAUSA300_0628	Glycerol-3-phosphate cytidylyltransferase
TarI'	SAUSA300_0245	D-Ribitol-5-phosphate cytidylyltransferases (non functional)
TarJ'	SAUSA300_0246	Ribitol-5-phosphate dehydrogenase (non functional)
TarK	SAUSA300_0247	Ribitol-phosphate polymerase
TarF	SAUSA300_0248	CDP-glycerol glycerophosphotransferase
TarI	SAUSA300_0249	D-Ribitol-5-phosphate cytidylyltransferases
TarJ	SAUSA300_0250	Ribitol-5-phosphate dehydrogenase
TarL	SAUSA300_0251	Ribitol-phosphate polymerase
TarS	SAUSA300_0252	Glycosyltransferase (β-GlcNAc residues)
TarM	SAUSA300_0939	Glycosyltransferase (α-GlcNAc residues)
ManA	SAUSA300_2096 SAUSA300_2577	Mannose-6-phosphate isomerase
LcpB	SAUSA300_0958	LytR-CpsA-Psr family protein (transfer of undecaprenyl-pyrophosphate linked precursor onto peptidoglycan)
LcpA/MsrR	SAUSA300_1257	LytR-CpsA-Psr family protein
LcpC	SAUSA300_2259	LytR-CpsA-Psr family protein

are introduced by the glycosyltransferase TarM, whereas β-linked GlcNAc residues are introduced by the glycosyltransferase TarS (Brown *et al.*, 2013; Xia *et al.*, 2010); in both cases, the nucleotide-activated sugar UDP-GlcNAc is used as a substrate (Fig. 8.8B). Interestingly, individual WTA chains are either not modified or heavily modified with sugars that share the same stereochemistry, suggesting that TarM and TarS are highly processive enzymes [see review (Brown *et al.*, 2013). The discovery of the enzymes involved in the glycosylation process and creation of *S. aureus* mutants lacking the specific glycosyl-groups provided fundamental insight into the role of these modifications (reviewed inWinstel *et al.*, 2014b). *S. aureus* strains lacking this modifications on their WTA become resistant to phage infection, highlighting that WTA has a function as a phage receptor (Xia *et al.*, 2011; Xia *et al.*, 2010). Furthermore, TarS-mediated β-glycosylation is essential for methicillin resistance of *S. aureus* (Brown et al., 2012); however, the exact mechanism responsible for this phenotype is currently not known.

Following RboP polymerization and modification of the polymer with GlcNAc residues, the C_{55}-P-linked WTA polymer is transported from the cytosolic phase of the membrane to the outer leaflet of the membrane by the ABC (ATP-binding cassette) transporter TarGH. Finally, the WTA polymer is transferred from the C_{55}-P-linked membrane intermediate onto the C6 hydroxyl group of the MurNAc unit of peptidoglycan. The *S. aureus* enzymes responsible for this attachment step are LcpA (also referred to as MsrR), LcpB, and LcpC, (Table 8.6) and they all belong to the family of LytR-CpsA-Psr (LCP) proteins (Chan *et al.*, 2013; Dengler *et al.*, 2012; Kawai *et al.*, 2011; Over *et al.*, 2011). Deletion of *lcpA* or *lcpB* reduces the WTA content in *S. aureus* cells to around 60%, whereas deletion of *lcpC* has only a minor effect on WTA levels (Chan *et al.*, 2013; Dengler *et al.*, 2012). A complete absence of WTA in the cell wall is only observed in the absence of all three Lcp enzymes, revealing a functional redundancy of these enzymes. As discussed later, these enzymes, and in particular LcpC, are also required for the attachment of the capsular polysaccharide to the peptidoglycan polymer. Lastly, WTA is modified on the outside of the cell with D-Ala residues. The Dlt proteins, discussed earlier in the section describing LTA, are essential for this process and in their absence neither LTA nor WTA are modified with D-Ala residues. However, it is thought that WTA is not directly or only very inefficiently modified by the action of the Dlt proteins. Instead, D-alanylated LTA is likely the donor for the D-Ala residues on WTA. This is based on pulse-chase experiments using radiolabelled alanine, which revealed that the radioactivity was first recovered in the LTA fraction. During the chase phase, the radioactivity decreased in the LTA fraction and a concomitant increase in radioactivity was seen in the WTA fraction (Haas *et al.*, 1984). Furthermore, in the absence of LTA (but in the presence of an intact *dlt* operon) the D-Ala content on WTA is drastically reduced (Reichmann *et al.*, 2013).

WTA has a wide range of functions in the cell. Together with LTA, it is thought to bind cations and contribute to ion retention within the envelope and to the regulation and cellular localization of cell wall hydrolytic enzymes. WTA has also been implicated in the coordination of peptidoglycan synthesis (Atilano *et al.*, 2010). In *S. aureus* the PBP peptidoglycan synthesis enzymes are localized at the cell division site. In the absence of WTA, PBP4, which is responsible for the high degree of cross-linking of the staphylococcal peptidoglycan, is miss-localized and becomes dispersed throughout the membrane (Atilano *et al.*, 2010). In addition, WTA is required for the resistance of *S. aureus* to lysozyme (Bera *et al.*, 2007) and β-lactam antibiotics (Brown *et al.*, 2012; Campbell *et al.*, 2011; Maki *et al.*, 1994); specifically, methicillin-resistant *S. aureus* (MRSA) strains that lack WTA or β-linked GlcNAc (but not α-linked GlcNAc) are very sensitive to β-lactam antibiotics, despite the expression of the methicillin insensitive penicillin binding protein PBP2A (Brown *et al.*, 2012; Campbell *et al.*, 2011; Maki *et al.*, 1994).

Several studies have also indicated an important role for WTA in the adherence of *S. aureus* to host cells (Weidenmaier and Peschel, 2008). A *S. aureus* mutant lacking WTA shows a defect in a cotton rat nasal colonization model owing to an inability to bind to nasal epithelial cells (Weidenmaier *et al.*, 2004). Furthermore, WTA also contributes to the adhesion to endothelial cells (Weidenmaier *et al.*, 2005b). *In vitro* and *in vivo* data suggest that WTA contributes to a similar extent to the binding to epithelial and endothelial cells as cell surface proteins (Weidenmaier *et al.*, 2008). D-Ala modifications on the WTA are important components in this interaction, as *dlt* mutants display a significant reduction in adhesion to host cells and reduced virulence in animal models of nasal colonization and endocarditis (Weidenmaier *et al.*, 2004, 2005a).

Although it was initially thought that WTA is essential for the growth of *S. aureus*, *S. aureus* strains completely lacking WTA can be obtained either by inactivating *tarO* or *tarA*, which encode for the first two enzymes in the biosynthetic pathway of WTA [see review (Sewell and Brown, 2014)]. A strain without WTA shows morphological and cell division defects, an increased sensitivity to cell wall acting antimicrobials and decreased virulence. Somewhat paradoxically, genes that encode enzymes acting at later stages in the biosynthetic process, including *tarB*, *tarF*, *tarG* and *tarH*, cannot be deleted. It is thought that deletion of these genes leads to the accumulation of C_{55}-P-linked membrane intermediates and a concomitant depletion of the cellular C_{55}-P pool; deletion of *tarO* and *tarA* does not lead to the formation of these intermediates. The C_{55}-P membrane carrier lipid is also required for peptidoglycan synthesis and its depletion in *tarB, tarF, tarG and tarH* mutants will block, or severely decrease, peptidoglycan synthesis and lead to a growth arrest. Consistent with this idea, late acting WTA synthesis enzymes that cannot be deleted in wild type strains can be readily inactivated in *tarO* or *tarA* mutant strains as these mutant strains will not produce any C_{55}-P linked WTA intermediates. Taken together, the WTA biosynthesis genes are now grouped into dispensable early acting genes *tarO* and *tarA* and conditional essential late acting genes. These unique synthetic viable genetic interactions among WTA synthesis genes are currently exploited in pathway-specific cell-based inhibitor screens (Swoboda *et al.*, 2009). These screens have already led to the identification of several promising small molecule inhibitors, many of which target the ABC transporter TarG (reviewed in Sewell and Brown, 2014). WTA synthesis enzymes are currently amongst the most promising targets for the development of new antimicrobials working against *S. aureus*, based on their essential function for growth, virulence and resistance to cell β-lactam antibiotics, as well as the ability to set up pathway-specific inhibitor screens.

Exopolysaccharides in *S. aureus*

Capsular polysaccharide synthesis and function

S. aureus produces two major exopolysaccharides: the capsule and a poly-*N*-acetylglucosamine (PNAG) polymer. The production of a capsular polysaccharide in *S. aureus* was first described by Gilbert (1931). The initial methods for the detection of capsular polysaccharides were insensitive and could only detect highly encapsulated strains, which appeared mucoid on plates. Therefore it was initially assumed that only a few *S. aureus* strains produce capsular material. Only when more sensitive detection methods were developed, did it become apparent that most *S. aureus* strains produce a capsular polysaccharide (Karakawa and Vann, 1982). Such encapsulated but non-mucoid *S. aureus* strains are sometimes referred to as micro-capsulated. Strains have been classified into different serotypes based on their capsular polysaccharide, using a scheme originally developed by Karakawa and Vann (Karakawa and Vann, 1982). In a subsequent large typing study, 11 different serotypes were identified (Sompolinsky *et al.*, 1985) and now up to 13 serotypes have been described (von Eiff *et al.*, 2007). The originally identified highly capsulated strain M and Smith diffuse were assigned serotype 1 and 2, respectively, whereas serotype 5 (CP5) and serotype 8 (CP8) strain are the most abundant isolates from humans; these latter two serotypes are also isolated from a range of different animal species (see review (O'Riordan and Lee, 2004) and references therein). The chemical structures of the capsular polysaccharides of the prototype CP5 strain Reynold and the prototype CP8 strain Becker were first investigated over 20 years ago and shown to be composed of 2-acetamido-2-deoxy-D-mannuronic acid (D-ManNAcA) and two 2-acetamido-2, 6-dideoxy galactose residues with D and L configurations (D-FucNAc and L-FucNAc), with further serotype-specific O-acetylation of some of these sugars (Fournier *et al.*, 1984; Moreau *et al.*, 1990). The originally proposed chemical structures have recently been revised and it is now thought that the CP5 capsular polysaccharide is a polymer consisting of →4)-β-D-ManNAcA-(1→4)-α-L-FucNAc(3OAc)-(1→3)-β-D-FucNAc-(1→, whereas CP8 capsular polysaccharide is a →3)-β-D-ManNAcA(4OAc)-(1→3)-α-L-FucNAc-(1→3)-α-D-FucNAc-(1→ polymer (Jones, 2005) (Fig. 8.9A).

Sixteen genes, *cap5A* through *cap5P* and *cap8A* through *cap8P*, are required for CP5 and CP8 synthesis, respectively (Fig. 8.9B and Table 8.7). All genes are closely linked in an approximately 17.5 kb region and transcribed in the same orientation. The main promoter used for expression of these genes is located upstream of the first gene, but additional weaker internal promoters are present in front of several other genes (see review by O'Riordan and Lee, 2004). As CP5 and CP8 are composed of the same repeating trisaccharide units (D-ManNAcA, L-FucNAc and D-FucNAc), 12 of the 16 capsular genes encode proteins with more than 97% identity and these likely possess identical enzymatic activities. The remaining four genes, coding for CapH, CapI, CapJ and CapK, are serotype-specific (O'Riordan and Lee, 2004; Sau *et al.*, 1997). With the exception of CapC, a function for all other proteins encoded in the capsular biosynthesis gene operon has been proposed and is summarized in Fig. 8.9B (Chan *et al.*, 2014; O'Riordan and Lee, 2004); however, many of these functions still await experimental verification.

UDP-D-GlcNAc is used as precursor for the production of all three sugars in the capsular polysaccharide, D-ManNAcA, L-FucNAc and D-FucNAc (Fig. 8.9B). It is thought that UDP-D-GlcNAc is converted to UDP-D-FucNAc by the sequential activity of the dehydrogenase CapD and the reductase CapN. UDP-L-FucNAc is produced from UDP-D-GlcNAc by the bifunctional dehydrogenase epimerase CapE, the reductase CapF and epimerase CapG (Kneidinger *et al.*, 2003). Based on additional experimental evidence it is thought that UDP-D-ManNAcA is synthesized from UDP-GlcNAc by the epimerase CapP and the dehydrogenase CapO (Kiser *et al.*, 1999; O'Riordan and Lee, 2004). CapP, which converts UDP-GlcNAc to UDP-ManNAc; however, this enzyme is not

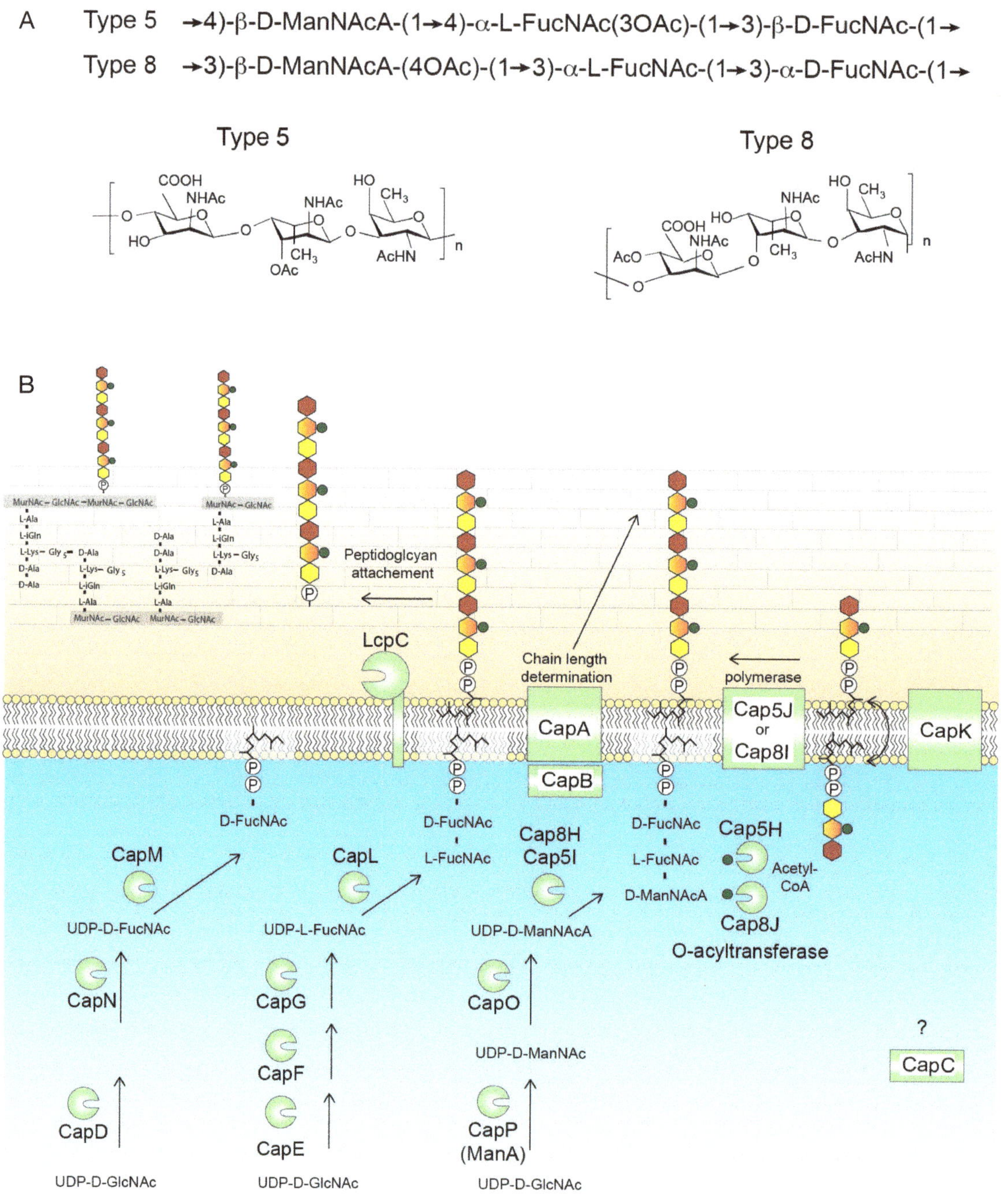

Figure 8.9 *S. aureus* capsular polysaccharide structure and synthesis pathway (A) Chemical structure of the *S. aureus* type 5 (CP5) and type 8 (CP8) capsular polysaccharides. (B) *S. aureus* enzymes involved in CP5 and CP8 capsular polysaccharides synthesis and schematic representation of the different steps involved in the synthesis process.

essential for capsular biosynthesis, as additional genes at different chromosomal locations, code for proteins with the same enzymatic activity (Kiser *et al.*, 1999). In the type 5 capsule, L-FucNAc is further acetylated, likely by the O-acetyltransferase Cap5H, whereas in the type 8 capsule D-ManNAcA

Table 8.7 *S. aureus* type 5 capsular polysaccharide (CP5) synthesis proteins

Name	USA300_FPR3757 locus tag	Function
Cap5A	SAUSA300_0152	Chain length determinant protein
Cap5B	SAUSA300_0153	Chain length determinant protein
Cap5C	SAUSA300_0154	Unknown function: hydrolase
Cap5D	SAUSA300_0155	4,6-Dehydratase for synthesis of UDP-D-FrucNAc
Cap5E	SAUSA300_0156	4,6-Dehydratase and 3,5-epimerase for synthesis of UDP-L-FrucNAc
Cap5F	SAUSA300_0157	Reductase for synthesis of UDP-L-FrucNAc
Cap5G	SAUSA300_0158	2-Epimerase for synthesis of UDP-L-FrucNAc
Cap5H	SAUSA300_0159	O-Acetyltransferase
Cap5I	SAUSA300_0160	Glycosyltransferase (transfer of D-ManNAcA onto lipid intermediate)
Cap5J	SAUSA300_0161	Capsule polymerization
Cap5K	SAUSA300_0162	Transporter
Cap5L	SAUSA300_0163	Glycosyltransfer (transfer of L-FrucNAc onto lipid intermediate)
Cap5M	SAUSA300_0164	Glycosyltransfer (transfer of D-FrucNAc onto lipid intermediate)
Cap5N	SAUSA300_0165	4-Reductase (for synthesis of UDP-D-FrucNAc)
Cap5O	SAUSA300_0166	Dehydrogenase (for synthesis of UDP-ManNAcA)
Cap5P	SAUSA300_0167	2-Epimerase (for synthesis of UDP-ManNAcA)
LcpC	SAUSA300_2259	LytR-CpsA-Psr family protein (transfer of capsular polysaccharide from undecaprenol intermediate onto peptidoglycan)

is acetylated by the O-acetyltransferase Cap8J. Next, it is predicted that the glycosyltransferases CapM, CapL and Cap5I or Cap8H transfer the nucleotide-activated sugars D-FucNAc, L-FucNAc and D-ManNAcA, respectively, onto undecaprenyl-phosphate (C_{55}-P) to form a membrane-linked intermediate. CapK is thought to aid in the transfer of this intermediate from the inner to the outer leaflet of the membrane. Next, Cap5J and Cap8I are thought to aid in the polymerization of the capsular material, with CapA and CapB serving as chain length determinants (O'Riordan and Lee, 2004). The polymerized capsular material is covalently attached to the peptidoglycan layer by enzymes of the LytR-CpsA-Psr (LCP) family, members of which can catalyse the transfer of different undecaprenol-linked saccharides onto the C6 hydroxyl group of MurNAc in the peptidoglycan polymer (Chan *et al.*, 2014). As already discussed in the WTA section, three LCP enzymes, LcpA (also named MsrR), LcpB and LcpC, are encoded in the genome of *S. aureus* outside of the capsular biosynthesis operon. LcpC is the key enzyme required for the attachment of the capsular polysaccharide to peptidoglycan (Chan *et al.*, 2014).

The expression of CP5 and CP8 can vary depending on the growth conditions, and increases when cultures are well-aerated and in the post-exponential growth phase; the latter of which is consistent with the observation that its production is positively controlled by the *agr* quorum sensing system (see review by O'Riordan and Lee, 2004) and references therein). Also, significant amounts of capsular polysaccharide are produced when bacteria are grown on solid medium under iron-limiting conditions or in the presence of a high NaCl concentration (see review by O'Riordan and Lee, 2004, and references therein). However, even within a population of cells, there can be large variations between individual cells in the amount of capsular polysaccharide that they produce (Poutrel *et al.*, 1997).

The requirement of the capsular polysaccharide during infection depends on the infection model used. Highly encapsulated strains are more resistant to opsonophagocytic killing *in vitro* (Lee *et al.*, 1987; Peterson *et al.*, 1978; Verbrugh *et al.*, 1982; Xu *et al.*, 1992) and more virulent in a mouse model of infection using an intraperitoneal rout of inoculation compared to isogenic non-capsulated or micro-capsulated strains (Albus *et al.*, 1991). In contrast, capsulated *S. aureus* strains have been reported to have a reduced virulence in a catheter-induced endocarditis model, where bacterial adhesion to the damaged heart valve is important (Baddour *et al.*, 1992; Nemeth and Lee, 1995), suggesting that the presence of large amounts of capsular polysaccharide may mask adhesion proteins (Kuypers and Proctor, 1989; Moreillon *et al.*, 1995). Capsular polysaccharides are also pursued as vaccine candidates, as immunization with capsular polysaccharide provides protection against subsequent challenge with a highly or micro-capsulated *S. aureus* strains producing the same type of capsule (but not against a mutant lacking a capsule) and the protection correlated with the level of capsular antibodies (Lee *et al.*, 1988).

Poly-*N*-acetylglucosamine (PNAG) polymer synthesis and function

Formation of biofilms requires that the bacteria aggregate. In many cases, exopolysaccharides are the intercellular adhesion components that allow the cells to adhere to each other. The major exopolysaccharide in *S. aureus* that acts as an intercellular adhesion molecule is the poly-*N*-acetylglucosamine (PNAG) polymer. This polymer was first characterized in *Staphylococcus epidermidis*, where it was named polysaccharide intercellular adhesion (PIA) polymer and shown to be an important constituent of the biofilm matrix (Mack *et al.*, 1992, 1994, 1996). As a result, much of the information about PIA/PNAG polymer production and function stems from research on *S. epidermidis*. However, much of this information will likely also apply to *S. aureus*.

The chemical structure of the PIA/PNAG was first described for the exopolysaccharides isolated from *S. epidermidis*, where it was shown to be a β(1–6)-linked *N*-acetylglucosamine (GlcNAc) polymer (Mack *et al.*, 1996). In *S. epidermidis*, about 15–20% of the sugar residues are de-*N*-acetylated and therefore positively charged (Mack *et al.*, 1996). The distribution of these de-*N*-acetylated sugars is roughly even along the polymer (Mack *et al.*, 1996). The sugar residues can also be modified with ester-linked succinate residues, introducing negative charges on the polymer (Mack *et al.*, 1996). These positive and negative charges provide the polymer with a zwitterionic character, which is of functional importance for the adhesive properties of the polymer (Rohde *et al.*, 2010).

The chemical structure of the PNAG polymer produced by the *S. aureus* strain MN8m, a mucoid derivative of strain MN8, was characterized in detail by NMR spectroscopy and chemical analysis (Joyce *et al.*, 2003) (Fig. 8.10A). This strain overproduces the PNAG polymer due to a mutation in the promoter region of the *ica* operon, resulting in increased transcription of the operon and increased biofilm formation (Jefferson *et al.*, 2003). As described in more detail below, the *ica* operon codes for the enzymes required for PNAG production. It was shown that the PNAG polymer in *S. aureus* has the same β(1–6) linked *N*-acetylglucosamine structure as observed in the *S. epidermidis* polymer (Joyce *et al.*, 2003). A portion of the hydroxyl groups at the C3 and C4 positions in GlcNAc are O-succinylated and the *S. aureus* and *S. epidermidis* PNAG polymers contain comparable amounts of this modification

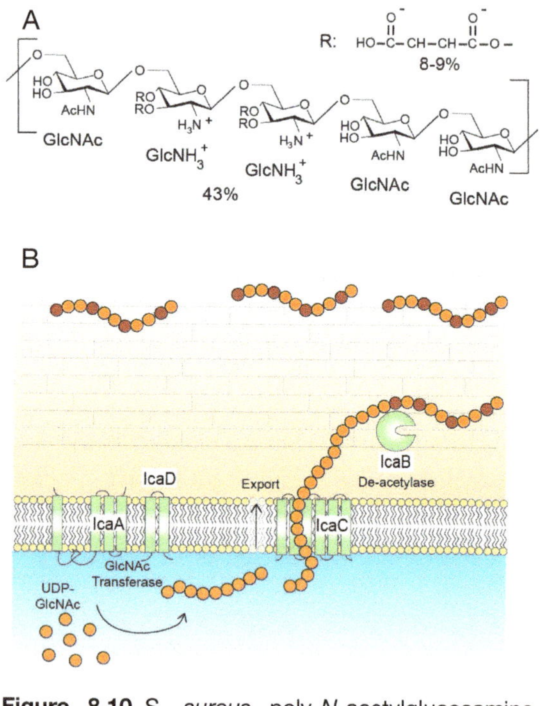

Figure 8.10 *S. aureus* poly-*N*-acetylglucosamine (PNAG) structure and synthesis pathway. (A) Chemical structure of the *S. aureus* PNAG polymer. Approximately 43% of the repeating units are de-*N*-acetylated and 8–9% of the sugar residues are modified with succinate moieties at positions indicated with R. (B) *S. aureus* enzymes involved in PNAG production and schematic representation of the different steps required for the synthesis of this exopolysaccharide polymer.

with 8–9% and 6%, respectively (Sadovskaya *et al.*, 2005) (Fig. 8.10A). However, the proportion of de-*N*-acetylated glucosamine residues is higher in *S. aureus* and found to be approximately 43% as compared with the 15–20% observed in *S. epidermidis* (Joyce *et al.*, 2003; Sadovskaya *et al.*, 2005). Size-exclusion chromatography indicated that the *S. aureus* PNAG polymer has an apparent molecular weight of 346,000 Da (Joyce *et al.*, 2003), corresponding to a polymer with around 1500 GlcNAc residues. However, the determination of the polymer length based on its migration during size exclusion chromatography likely results in a drastic overestimation of the polymer length, as the purified polymer can form aggregates. For *S. epidermidis,* the polymer length has been estimated to be approximately 130 GlcNAc residues, based on a quantification of the ratio of the amount of reducing (terminal) sugars and the total amount of sugars (Mack *et al.*, 1996). This would correspond to a polymer with an approximate molecular weight of 30,000 Da, which is several-fold shorter than estimated for the *S. aureus* polymer.

The enzymes required for PNAG synthesis are encoded in the *icaADBC* operon, which was identified in *S. epidermidis* in a transposon mutagenesis screen for isolates that are unable to form biofilms (Gerke *et al.*, 1998; Heilmann *et al.*, 1996). Close orthologues of the *ica* genes are found in other staphylococcal species, including *S. aureus* (Table 8.8) and even several Gram-negative bacteria, such as pathogenic *E. coli* and *Yersinia pestis* strains (see review (Rohde *et al.*, 2010) and references there in). Hence, the functions of PNAG extend beyond biofilm formation in staphylococcal species.

All four Ica proteins, IcaA, IcaD, IcaB and IcaC, are required for the production of active PNAG (Cramton *et al.*, 1999; Gerke *et al.*, 1998; Heilmann *et al.*, 1996). IcaA encodes an *N*-acetylglucosamine transferase, which uses UDP-GlcNAc as substrate and the current view is that it functions together with IcaD to form GlcNAc polymers of approximately 20 residues (Cue *et al.*, 2012; Gerke *et al.*, 1998) (Fig. 8.10B). The transmembrane protein IcaC has been suggested to be involved in the transport of the polymer across the membrane and the elongation of the polymer to its final length (Cue *et al.*, 2012; Gerke *et al.*, 1998). IcaB is a secreted protein with similarity to chitin deacetylases and is responsible for the de-acetylation of PNAG (Heilmann *et al.*, 1996; Vuong *et al.*, 2004). The genetic determinants responsible for the O-succinylation are currently not known. Although IcaB is not essential for the PNAG polymer production, the de-*N*-acetylation step and introduction of positively charged residues is essential for the production of a functionally active PNAG polymer. An *S. epidermidis icaB* mutant strain is defective in biofilm formation and less virulent in an animal model of foreign material associated infection (Vuong *et al.*, 2004). The de-*N*-acetylation step, which provides a net positive charge to the polymer, is also important for the intercellular adhesive properties and attachment of the polymer to the bacterial cell. How the PNAG polymer is attached to the bacterial cell wall is still unclear. It was suggested that the PNAG polymer interacts with negatively charged teichoic acids (TAs), but a recent study showed that WTA is dispensable for the binding of the PNAG polymer to the cell as the PNAG levels on the cell and the interaction strength between PNAG and the bacterial cell surface were not affected in an *S. aureus* mutant lacking WTA (Vergara-Irigaray *et al.*, 2008).

Expression of the *ica* genes is regulated by the transcriptional regulator IcaR that belongs to the TetR family of transcription regulators. The gene encoding IcaR is located upstream of *icaA* but is transcribed divergently (Cramton *et al.*, 1999). It binds to the promoter region of the *ica* operon and acts as a transcriptional repressor (Conlon *et al.*,

Table 8.8 *S. aureus* poly-N-acetyl-glucosamine (PNAG) synthesis proteins

Name	USA300_FPR3757 locus tag	Function
IcaA	SAUSA300_2600	*N*-Acetylglycosyltransferase (PNAG polymerase works together with IcaD)
IcaB	SAUSA300_2601	PNAG deacetylase
IcaC	SAUSA300_2602	PNAG inner membrane transporter
IcaD	None	*N*-Acetylglycosyltransferase (PNAG polymerase works with IcaA)
IcaR	SAUSA300_2599	Transcription repressor

2002; Jefferson *et al.*, 2003; Jefferson *et al.*, 2004). Furthermore, expression of *icaR* itself is regulated by an intricate post-transcriptional regulatory mechanism (Ruiz de los Mozos *et al.*, 2013). In addition to IcaR, several other regulatory factors can influence the transcription of the *ica* operon (reviewed in Cue *et al.*, 2012). These factors included Sar (staphylococcal accessory regulator) family proteins, the tecoplanin-associated locus transcription factor TcaR, Rbf, which also regulates biofilm formation, the global regulatory alternative sigma factor σ^B and the SrrAB two component system, which is responsive to anaerobiosis (see review by Cue *et al.*, 2012, and references therein). The signals controlling PNAG production within the host have not been clearly defined, but several environmental factors have been shown to induce PNAG production, such as an increase in temperature, anaerobiosis, high osmolarity, ethanol and glucose. However, significant strain-to-strain variations are observed.

The PNAG polymer was initially shown to be an important biofilm matrix component and essential for biofilm formation in *S. epidermidis* (Mack *et al.*, 1992, 1994, 1996) and it has since become clear that the PNAG polymer is also important for biofilm formation in *S. aureus* under conditions in which the organism forms a protein-independent biofilm (Cramton *et al.*, 1999). Furthermore, PNAG polymer production has also been implicated in protecting *S. aureus* from killing by neutrophils under anaerobic growth conditions (Ulrich *et al.*, 2007). Lastly, the PNAG polymer is also currently pursued as a potential vaccine candidate.

Cell wall synthesis inhibitors and resistance mechanisms

Action and resistance to methicillin and other β-lactam antibiotics in *S. aureus*

Staphylococcus aureus has four PBP transpeptidases (PBP1–4), all of which can be inhibited by β-lactam antibiotics through an irreversible acylation of the active site serine residue. By inhibiting the peptide cross-linking step of peptidoglycan synthesis, lactam antibiotics induce cell lysis and cell death. Methicillin-resistant *S. aureus* (MRSA) strains express PBP2A (also called PBP2′, encoded by the *mecA* gene), an additional PBP with transpeptidase activity (Hartman and Tomasz, 1984). The first MRSA strain was isolated in 1960 in the UK (Jevons, 1961) and MRSA strains have been a worldwide epidemic since the 1970s. The antibiotic methicillin itself is no longer produced or in use, but the name MRSA has persisted and MRSA strains are now regarded as resistant to virtually all β-lactam antibiotics including the β-lactam antibiotics that are currently used for susceptibility testing, oxacillin or cefoxitin. The *mecA* gene is located on the staphylococcal cassette chromosome (SCC) element, a 32 to 60 kb chromosomal locus that is acquired through horizontal gene transfer (Hiramatsu *et al.*, 2013; Ito *et al.*, 1999; Katayama *et al.*, 2000). The SCC elements are flanked by characteristic conserved direct repeats and are inserted in proximity of the origin of replication at the 3′ end of *orfX*. The insertion site within the direct repeats is recognized by the cassette chromosome recombinase (CCR), which mediates precise insertion and excision of the SCC element (Katayama *et al.*, 2000). The type of *mec* and *ccr* genes define the SSC*mec* type, of which there are currently eleven (International Working Group on the Classification of Staphylococcal Cassette Chromosome Elements (IWG-SCC), 2009). Besides the *mec* and *ccr* genes, SSC*mec* elements contain additional genes that code for proteins with diverse functions, including resistance to fusaric acid (Holden *et al.*, 2004) or resistance to mercury (Chongtrakool *et al.*, 2006), enterotoxins (Baba *et al.*, 2002), the type-1 capsule synthesis machinery (Luong *et al.*, 2002), an additional PBP4 protein, or an arginine catabolic mobile genetic element (ACME) [(Diep *et al.*, 2006; see also review (Hiramatsu *et al.*, 2013). Type IV and V SSC*mec* elements are often found in the more recently emerged community-associated MRSA (CA-MRSA) strains and are also frequently found in other methicillin-resistant *Staphylococcus* species such as methicillin-resistant *Staphylococcus epidermidis* (MRSE) strains (see review (Hiramatsu *et al.*, 2013) and references therein). Ten *SSCmec* types contain the *mecA* gene, whereas the type XI *SSCmec* element contains a distinct *mec* gene now referred to as *mecC*. This gene was first described in 2007 in an epidemiological study on bovine mastitis caused by the *S. aureus* strain LGA251 and was initially referred to as $mecA_{LGA251}$ (Graveland *et al.*, 2011). It was renamed *mecC* when it was discovered that there was only 69% similarity between this gene and *mecA* (Ito *et al.*, 2012; Paterson *et al.*, 2014). A retrospective analysis of *S. aureus* collections in the UK and Denmark indicated that *mecC* is present in isolates from cattle and humans, with some isolates dating back to 1975 (Garcia-Alvarez *et al.*, 2011). Due to the low similarity to *mecA*, common PCR and protein-based detection methods failed to detect the presence of the *mecC* gene, but new modified PCR tests that can detect *mecC* have now been developed (see review by Paterson *et al.*, 2014, and references therein). Furthermore, the penicillin binding proteins encoded by *mecA* and *mecC* have notably different properties (Kim *et al.*, 2012). Both proteins have a low affinity for cefoxitin, but MecC has relatively high affinity for oxacillin, whereas PBP2A has a low affinity for this antibiotic (Kim *et al.*, 2012). Hence, strains that encode *mecA* strains are resistant to both oxacillin and cefoxitin, whereas *mecC* strains are resistant only to cefoxitin but not to oxacillin, as determined by commercially available resistance profiling systems (Cartwright *et al.*, 2013). MRSA strains that contain *mecC* have so far been reported only in Western Europe, and their prevalence is quite low, with the highest prevalence reported in Demark, where 2.8% of the MRSA strains were positive for *mecC* in 2011 (Paterson *et al.*, 2014; Petersen *et al.*, 2013). Two prevalent *S. aureus* lineages, CC130 and ST425, contain the *mecC* gene. These lineages were originally regarded as animal-associated lineages, suggesting a spread from animals to humans (Garcia-Alvarez *et al.*, 2011; Paterson *et al.*, 2014). Furthermore, the *mecC* gene has now also been detected in several other *Staphylococcus* spp., including *Staphylococcus stepanovic*, *Staphylococcus xylosus* and *Staphylococcus scirui* (see review by Paterson *et al.*, 2014, and references therein).

Vancomycin action and resistance mechanisms in *S. aureus*

Vancomycin is a glycopeptide antibiotic that is used for the treatment of infections with Gram-positive bacteria. It binds with high affinity to the terminal D-alanyl-D-alanine residues of the peptidoglycan precursor molecules lipid II, thereby preventing the synthesis of peptidoglycan. With the increase in infections by MRSA strains, vancomycin has been used increasingly in several countries to treat MRSA infections (Levine, 2006). The first MRSA strain with decreased susceptibility to vancomycin was reported in 1997 from Japan (Hiramatsu *et al.*, 1997). This strain showed a moderate resistance to vancomycin with a minimum inhibitory concentration (MIC) in the range of 3–8 μg/ml. Strains with such low level of vancomycin resistance are now commonly referred to as vancomycin intermediate-resistant *S. aureus* (VISA) strains. Within a population of VISA cells, the bacteria also display a heterogeneous resistance profile as often only a small fraction of cells show increased resistance (Hiramatsu, 2001). Nonetheless, despite their low level of resistance, treatment failure is often seen in patients infected with VISA strains (Fridkin *et al.*, 2003; Linares, 2001). The genetic basis for the increased vancomycin resistance of VISA strains are varied and the mechanism of their resistance is not completely understood (see review by Gardete and Tomasz, 2014). Frequently observed mutations in clinical VISA isolates are changes in either *walKR* (also referred to as *yycGF*) or *vraSR*, which encode two-component systems, *yvqF*, a regulator of the *vraSR* operon, or *rpoB*, which encodes the β subunit of RNA polymerase (see review (Gardete and Tomasz, 2014) and references therein). The two-component systems control the transcription of genes involved in the biosynthesis or metabolism of the staphylococcal cell wall; VISA strains frequently have a thicker cell wall than wild type cells, show aberrant septation and often have a decreased rate of autolysis (see review by Gardete and Tomasz, 2014). For VISA strains to continue peptidoglycan synthesis in the presence of vancomycin, the lipid II precursor must somehow be protected. It has been speculated that in VISA strains the amount of pentapeptides in the mature cell wall that still contain the terminal D-Ala-D-Ala residues in the stem peptide is increased, possibly owing to a decrease in peptidoglycan cross-linking. This increase in D-alanyl-D-alanine targets in the mature cell wall may serve as binding sites for vancomycin and thereby protecting the lipid II precursors. Indeed, an increase in vancomycin binding sites was observed in the cell walls of several VISA strains; however, the amount of vancomycin that bound was not proportional to the MIC (Pereira *et al.*, 2007). A more careful binding analysis suggested that the rate with which the antibiotic reaches the cell division site, where the lipid II precursor is located, is slowed in VISA strains and correlates with their resistance (Pereira *et al.*, 2007).

The first *S. aureus* strains with high-level resistance to vancomycin (VRSA strains), with MIC values for vancomycin >100 μg/ml, were isolated in 2002 (Sievert *et al.*, 2008). These stains acquired a plasmid from *Enterococcus faecalis* containing the transposon Tn1546 through horizontal gene transfer. The *van* genes present on this transposon code for enzymes that catalyse the formation of lipid II molecules in which the terminal D-Ala-D-Ala residues are replaced by D-Ala-D-Lac residues. Vancomycin has greatly decreased affinity for this precursor and peptidoglycan synthesis can continue even in the presence of high concentrations of vancomycin. Interestingly, as this modified peptidoglycan precursor cannot be used as a substrate by the *mecA*-encoded alternative penicillin-binding proteins PBP2, VRSA strains will be sensitive to β-lactam antibiotics in the presence of vancomycin (Severin *et al.*, 2004).

Daptomycin action and resistance mechanisms in *S. aureus*

Daptomycin (DAP) is a recently introduced antibiotic that is used in the treatment of Gram-positive infections. Its use for the treatment of skin and soft tissue infections with Gram-positive bacteria was approved in the USA in 2003 and in 2006, its use was approved for the treatment of *S. aureus* bacteraemia and right–sided endocarditis. DAP is a cyclic lipopeptide that is naturally produced by *Streptomyces roseosporus* using non-ribosomal peptide synthetases (Robbel and Marahiel, 2010). Although DAP is an anionic molecule, it is only active in the presence of Ca^{2+}, which makes the Ca^{2+}–DAP complex comparable to a cationic peptide in charge and mechanism of action (for review see Bayer *et al.*, 2013; Humphries *et al.*, 2013). Ca^{2+} induces a conformational change in DAP that leads to oligomerization and subsequent insertion of the antibiotic into the bacterial membrane through the interaction with the negatively charged membrane lipids phosphatidylglycerol (PG) and cardiolipin (CL) (Jung *et al.*, 2004; Muraih *et al.*, 2011; Straus and Hancock, 2006), which ultimately leads to membrane depolarization, ion leakage and cell death.

Since the introduction of DAP, *S. aureus* strains with an increased MIC of > 1 μg/ml, referred to as daptomycin non-susceptible (DAP-NS) strains, have been isolated from patients treated with this antimicrobial [see reviews (Humphries *et al.*, 2013; Moise *et al.*, 2009)]. However, surveillance studies indicate that the incidence of *S. aureus* strains that are no longer susceptible to daptomycin is in general very low [see review (Moise *et al.*, 2009)]. These DAP-NS strains display alterations in the cytoplasmic membrane and the cell wall. Most frequently DAP-NS strains contain mutations in *mprF*, which encodes the multiple peptide resistance factor MprF (see reviews by Bayer *et al.*, 2013; Humphries *et al.*, 2013). MprF was discovered as a resistance factor for cationic antimicrobial peptides, and is responsible for converting the negatively charged membrane lipid PG into the positively charged lysyl-PG and its subsequent transport to the outer leaflet of the membrane (Peschel *et al.*, 2001; Ernst *et al.*, 2009). The mutations observed in the DAP-NS strains are gain-of-function mutations that result in either increased levels of lysyl-PG synthesis or increased transport of lysyl-PG to the outer leaflet of the membrane, thereby increasing the positive surface charge of the membrane (Jones *et al.*, 2008; see also reviews by Bayer *et al.*, 2013; Humphries *et al.*, 2013). DAP-NS strains with mutations in *pgsA* and *cls2*, genes involved in PG and CL lipid synthesis, respectively, have also been discovered (Peleg *et al.*, 2012). These mutations provide resistance to DAP through a combination of factors including a charge-repulsion of the Ca^{2+}–DAP complex from the more positively charged membrane and cell wall. Consistent with this hypothesis, is the recent observation that an increase in transcription of the *dlt* operon can also improve DAP resistance. An increase in the transcription of the *dlt* operon, will likely lead to an increase in the positive surface charge of the cell wall through the alanylation of LTA and WTA. However, a strict correlation between cell wall charge and DAP resistance is not observed and therefore, additional factors must play a role. These have been suggested to include a decrease in the PG or CL content in the membrane [see review (Bayer *et al.*, 2013)]. In addition, changes in the membrane fluidity have been associated with an increase in DAP resistance and as well as a thickening of the cell wall (see reviews by Bayer *et al.*, 2013; Humphries *et al.*, 2013).

As the Ca^{2+}–DAP complex functions similarly to cationic antimicrobial peptides, a cross-resistance of DAP-NS strains to positively charged host defence peptides has been observed (Mishra *et al.*, 2011). Furthermore, a correlation between a pre-exposure of strains to vancomycin and occurrence of vancomycin intermediate resistance strains and daptomycin resistance has been reported [(Mwangi *et al.*, 2007) and see also reviews (Bayer *et al.*, 2013; Humphries *et al.*, 2013; Moise *et al.*, 2009)]. Over 50% of the isolated VISA strains show an MIC >1 μg/ml for DAP, which is regarded as non-susceptible (see reviews by Humphries *et al.*, 2013; Moise *et al.*, 2009). The mechanism for the decreased susceptibility of these VISA strains to DAP is not understood; however, the resistance appears to involve mutations in genes other than *mprF* (Pillai *et al.*, 2007). Upon serial passage of *S. aureus* in the presence of a sublethal concentration of DAP, *mprF* mutations are usually obtained first. However, these strains subsequently also acquire mutations in *walKR* (also named *yycGF*) and *rpoB* or *rpoC* (Friedman *et al.*, 2006; Patel *et al.*, 2011); mutations in several of these genes have also been detected in VISA strains. While the modes of action of DAP and vancomycin are very different, cell wall changes observed in VISA strains can potentially also reduce the efficacy of DAP, which will need to be considered when choosing antibiotic treatment options. While antibiotics such as DAP that target the membrane of the bacterium can be a potent antibiotic, it is clear that *S. aureus* can also adapt by making far-reaching changes in its membrane and cell wall leading to a decreased susceptibility to DAP.

Conclusions

As presented in this chapter, the *S. aureus* cell wall is composed of a variety of different structures. While the synthesis pathways of the different components have been discussed in isolation, their production proceeds in a highly coordinated manner. Many of the proteins required for the synthesis of the different cell wall components have now been identified, but further characterization or experimental verification of their function is required for a large number of them to fully understand their function. As highlighted throughout this chapter, the essential function for growth, virulence and ease of accessibility on the outside of the cell makes cell wall proteins, cell wall polymers and the enzymes required for their synthesis interesting targets for the development of novel anti-infective strategies and as vaccine targets.

Acknowledgements

I would like to thank Christiaan van Ooij for his time and help with the English language editing of this chapter and Marko Susnik for his help in the preparation of the image with the chemical structures of the capsular polysaccharides. The research in the laboratory of A.G. is currently supported by the European Research Council grant 260371 and the Wellcome Trust grant 100289.

References

Albus, A., Arbeit, R.D., and Lee, J.C. (1991). Virulence of *Staphylococcus aureus* mutants altered in type 5 capsule production. Infect. Immun. *59*, 1008–1014.

Aliprantis, A.O., Yang, R.B., Mark, M.R., Suggett, S., Devaux, B., Radolf, J.D., Klimpel, G.R., Godowski, P., and Zychlinsky, A. (1999). Cell activation and apoptosis by bacterial lipoproteins through toll-like receptor-2. Science *285*, 736–739.

Anderson, A.S., Miller, A.A., Donald, R.G., Scully, I.L., Nanra, J.S., Cooper, D., and Jansen, K.U. (2012). Development of a multicomponent *Staphylococcus aureus* vaccine designed to counter multiple bacterial virulence factors. Hum. Vaccin. Immunother. *8*, 1585–1594.

Araki, Y., and Ito, E. (1989). Linkage units in cell walls of gram-positive bacteria. Crit. Rev. Microbiol. *17*, 121–135.

Armstrong, J.J., Baddiley, J., Buchanan, J.G., Davision, A.L., Kelemen, M.V., and Neuhaus, F.C. (1958). Isolation and structure of ribitol phosphate derivatives (teichoic acids) from bacterial cell walls. J. Chem. Soc. *1958*, 4344–4354.

Armstrong, J.J., Baddiley, J., Buchanan, J.G., Davision, A.L., Kelemen, M.V., and Neuhaus, F.C. (1959). Composition of teichoic acids from a number of bacterial walls. Nature *184*, 247–248.

Asanuma, M., Kurokawa, K., Ichikawa, R., Ryu, K.H., Chae, J.H., Dohmae, N., Lee, B.L., and Nakayama, H. (2011). Structural evidence of α-aminoacylated lipoproteins of *Staphylococcus aureus*. FEBS J. *278*, 716–728.

Atilano, M.L., Pereira, P.M., Yates, J., Reed, P., Veiga, H., Pinho, M.G., and Filipe, S.R. (2010). Teichoic acids are temporal and spatial regulators of peptidoglycan cross-linking in *Staphylococcus aureus*. Proc. Natl. Acad. Sci. U.S.A. *107*, 18991–18996.

Baba, T., and Schneewind, O. (1998). Targeting of muralytic enzymes to the cell division site of Gram-positive bacteria: repeat domains direct autolysin to the equatorial surface ring of *Staphylococcus aureus*. EMBO J. *17*, 4639–4646.

Baba, T., Takeuchi, F., Kuroda, M., Yuzawa, H., Aoki, K., Oguchi, A., Nagai, Y., Iwama, N., Asano, K., Naimi, T., *et al.* (2002). Genome and virulence determinants of high virulence community-acquired MRSA. Lancet *359*, 1819–1827.

Babu, M.M., Priya, M.L., Selvan, A.T., Madera, M., Gough, J., Aravind, L., and Sankaran, K. (2006). A database of bacterial lipoproteins (DOLOP) with functional assignments to predicted lipoproteins. J. Bacteriol. *188*, 2761–2773.

Baddiley, J. (1972). Teichoic acids in cell walls and membranes of bacteria. Essays Biochem. *8*, 35–77.

Baddour, L.M., Lowrance, C., Albus, A., Lowrance, J.H., Anderson, S.K., and Lee, J.C. (1992). *Staphylococcus aureus* microcapsule expression attenuates bacterial virulence in a rat model of experimental endocarditis. The J. Inect. Dis. *165*, 749–753.

Bae, T., Banger, A.K., Wallace, A., Glass, E.M., Aslund, F., Schneewind, O., and Missiakas, D.M. (2004). *Staphylococcus aureus* virulence genes identified by bursa aurealis mutagenesis and nematode killing. Proc. Natl. Acad. Sci. U.S.A. *101*, 12312–12317.

Bagnoli, F., Bertholet, S., and Grandi, G. (2012). Inferring reasons for the failure of *Staphylococcus aureus* vaccines in clinical trials. Front. Cell. Infect. Microbiol. *2*, 16.

Bayer, A.S., Schneider, T., and Sahl, H.G. (2013). Mechanisms of daptomycin resistance in *Staphylococcus aureus*: role of the cell membrane and cell wall. Ann. N.Y. Acad. Sci. *1277*, 139–158.

Bera, A., Biswas, R., Herbert, S., Kulauzovic, E., Weidenmaier, C., Peschel, A., and Götz, F. (2007). Influence of wall teichoic acid on lysozyme resistance in *Staphylococcus aureus*. J. Bacteriol. *189*, 280–283.

Bera, A., Herbert, S., Jakob, A., Vollmer, W., and Götz, F. (2005). Why are pathogenic staphylococci so lysozyme resistant? The peptidoglycan O-acetyltransferase OatA is the major determinant for lysozyme resistance of *Staphylococcus aureus*. Mol. Microbiol. *55*, 778–787.

Bischoff, M., Dunman, P., Kormanec, J., Macapagal, D., Murphy, E., Mounts, W., Berger-Bächi, B., and Projan, S. (2004). Microarray-based analysis of the *Staphylococcus aureus* sigmaB regulon. J. Bacteriol. *186*, 4085–4099.

Biswas, R., Voggu, L., Simon, U.K., Hentschel, P., Thumm, G., and Götz, F. (2006). Activity of the major staphylococcal autolysin Atl. FEMS Microbiol. Lett. *259*, 260–268.

Brown, S., Santa Maria, J.P., and Walker, S. (2013). Wall teichoic acids of gram-positive bacteria. Annu. Rev. Microbiol. *67*, 313–336.

Brown, S., Xia, G., Luhachack, L.G., Campbell, J., Meredith, T.C., Chen, C., Winstel, V., Gekeler, C., Irazoqui, J.E., Peschel, A., *et al.* (2012). Methicillin resistance in *Staphylococcus aureus* requires glycosylated wall teichoic acids. Proc. Natl. Acad. Sci. U.S.A. *109*, 18909–18914.

Bubeck Wardenburg, J., Williams, W.A., and Missiakas, D. (2006). Host defenses against *Staphylococcus aureus* infection require recognition of bacterial lipoproteins. Proc. Natl. Acad. Sci. U.S.A. *103*, 13831–13836.

Campbell, J., Singh, A.K., Santa Maria, J.P., Jr., Kim, Y., Brown, S., Swoboda, J.G., Mylonakis, E., Wilkinson, B.J., and Walker, S. (2011). Synthetic lethal compound combinations reveal a fundamental connection between wall teichoic acid and peptidoglycan biosyntheses in *Staphylococcus aureus*. ACS Chem. Biol. *6*, 106–116.

Cartwright, E.J., Paterson, G.K., Raven, K.E., Harrison, E.M., Gouliouris, T., Kearns, A., Pichon, B., Edwards, G., Skov, R.L., Larsen, A.R., *et al.* (2013). Use of Vitek 2 antimicrobial susceptibility profile to identify *mecC* in methicillin-resistant *Staphylococcus aureus*. J. Clin. Microbiol. *51*, 2732–2734.

Chan, Y.G., Frankel, M.B., Dengler, V., Schneewind, O., and Missiakas, D. (2013). *Staphylococcus aureus* mutants lacking the LytR-CpsA-Psr family of enzymes release cell wall teichoic acids into the extracellular medium. J. Bacteriol. *195*, 4650–4659.

Chan, Y.G., Kim, H.K., Schneewind, O., and Missiakas, D. (2014). The capsular polysaccharide of *Staphylococcus aureus* is attached to peptidoglycan by the LytR-CpsA-Psr (LCP) family of enzymes. J. Biol. Chem. *289*, 15680–15690.

Chongtrakool, P., Ito, T., Ma, X.X., Kondo, Y., Trakulsomboon, S., Tiensasitorn, C., Jamklang, M., Chavalit, T., Song, J.H., and Hiramatsu, K. (2006). Staphylococcal cassette chromosome mec (SCCmec) typing of methicillin-resistant *Staphylococcus aureus* strains isolated in 11 Asian countries: a proposal for a new nomenclature for SCCmec elements. Antimicrob. Agents Chemother. *50*, 1001–1012.

Collins, L.V., Kristian, S.A., Weidenmaier, C., Faigle, M., Van Kessel, K.P., Van Strijp, J.A., Götz, F., Neumeister, B., and Peschel, A. (2002). *Staphylococcus aureus* strains lacking D-alanine modifications of teichoic acids are highly susceptible to human neutrophil killing and are virulence attenuated in mice. J. Infect. Dis. *186*, 214–219.

Conlon, K.M., Humphreys, H., and O'Gara, J.P. (2002). *icaR* encodes a transcriptional repressor involved in environmental regulation of ica operon expression and biofilm formation in *Staphylococcus epidermidis*. J. Bacteriol. *184*, 4400–4408.

Corrigan, R.M., Abbott, J.C., Burhenne, H., Kaever, V., and Gründling, A. (2011). c-di-AMP is a new second messenger in *Staphylococcus aureus* with a role in controlling cell size and envelope stress. PLOS Pathog. *7*, e1002217.

Cramton, S.E., Gerke, C., Schnell, N.F., Nichols, W.W., and Götz, F. (1999). The intercellular adhesion (*ica*) locus is present in *Staphylococcus aureus* and is required for biofilm formation. Infect. Immun. *67*, 5427–5433.

Cregg, K.M., Wilding, I., and Black, M.T. (1996). Molecular cloning and expression of the *spsB* gene encoding an essential type I signal peptidase from *Staphylococcus aureus*. J. Bacteriol. *178*, 5712–5718.

Cue, D., Lei, M.G., and Lee, C.Y. (2012). Genetic regulation of the intercellular adhesion locus in staphylococci. Front. Cell. Infect. Microbiol. *2*, 38.

Debabov, D.V., Kiriukhin, M.Y., and Neuhaus, F.C. (2000). Biosynthesis of lipoteichoic acid in *Lactobacillus rhamnosus*: role of DltD in D-alanylation. J. Bacteriol. *182*, 2855–2864.

DeDent, A., Bae, T., Missiakas, D.M., and Schneewind, O. (2008). Signal peptides direct surface proteins to two distinct envelope locations of *Staphylococcus aureus*. EMBO J. *27*, 2656–2668.

Dengler, V., Meier, P.S., Heusser, R., Kupferschmied, P., Fazekas, J., Friebe, S., Staufer, S.B., Majcherczyk, P.A., Moreillon, P., Berger-Bächi, B., *et al.* (2012). Deletion of hypothetical wall teichoic acid ligases in *Staphylococcus aureus* activates the cell wall stress response. FEMS Microbiol. Lett. *333*, 109–120.

Diep, B.A., Gill, S.R., Chang, R.F., Phan, T.H., Chen, J.H., Davidson, M.G., Lin, F., Lin, J., Carleton, H.A., Mongodin, E.F., *et al.* (2006). Complete genome sequence of USA300, an epidemic clone of community-acquired meticillin-resistant *Staphylococcus aureus*. Lancet *367*, 731–739.

Ernst, C.M., Staubitz, P., Mishra, N.N., Yang, S.J., Hornig, G., Kalbacher, H., Bayer, A.S., Kraus, D., and Peschel, A. (2009). The bacterial defensin resistance protein MprF consists of separable domains for lipid lysinylation and antimicrobial peptide repulsion. PLOS Pathog. *5*, e1000660.

Falord, M., Karimova, G., Hiron, A., and Msadek, T. (2012). GraXSR proteins interact with the VraFG ABC transporter to form a five-component system required for cationic antimicrobial peptide sensing and resistance in *Staphylococcus aureus*. Antimicrob. Agents Chemother. *56*, 1047–1058.

Fedtke, I., Mader, D., Kohler, T., Moll, H., Nicholson, G., Biswas, R., Henseler, K., Götz, F., Zähringer, U., and Peschel, A. (2007). A *Staphylococcus aureus ypfP* mutant with strongly reduced lipoteichoic acid (LTA) content: LTA governs bacterial surface properties and autolysin activity. Mol. Microbiol. 65, 1078–1091.

Femling, J.K., Nauseef, W.M., and Weiss, J.P. (2005). Synergy between extracellular group IIA phospholipase A2 and phagocyte NADPH oxidase in digestion of phospholipids of *Staphylococcus aureus* ingested by human neutrophils. J. Immunol. *175*, 4653–4661.

Fey, P.D., Endres, J.L., Yajjala, V.K., Widhelm, T.J., Boissy, R.J., Bose, J.L., and Bayles, K.W. (2013). A genetic resource for rapid and comprehensive phenotype screening of nonessential *Staphylococcus aureus* genes. MBio *4*, e00537–00512.

Figueiredo, T.A., Sobral, R.G., Ludovice, A.M., Almeida, J.M., Bui, N.K., Vollmer, W., de Lencastre, H., and Tomasz, A. (2012). Identification of genetic determinants and enzymes involved with the amidation of glutamic acid residues in the peptidoglycan of *Staphylococcus aureus*. PLOS Pathog. *8*, e1002508.

Fischer, W. (1990). Bacterial phosphoglycolipids and lipoteichoic acids. In Handbook of Lipid Research. D. Hanahan, ed. (New York, Plenum Press), 123–234.

Fischer, W. (1994). Lipoteichoic acid and lipids in the membrane of *Staphylococcus aureus*. Med. Microbiol. Immunol. *183*, 61–76.

Fischetti, V.A., Pancholi, V., and Schneewind, O. (1990). Conservation of a hexapeptide sequence in the anchor region of surface proteins from gram-positive cocci. Mol. Microbiol. *4*, 1603–1605.

Foster, T.J., Geoghegan, J.A., Ganesh, V.K., and Höök, M. (2014). Adhesion, invasion and evasion: the many functions of the surface proteins of *Staphylococcus aureus*. Nat. Rev. Microbiol. *12*, 49–62.

Fournier, J.M., Vann, W.F., and Karakawa, W.W. (1984). Purification and characterization of *Staphylococcus aureus* type 8 capsular polysaccharide. Infect. Immun. *45*, 87–93.

Fowler, V.G., Allen, K.B., Moreira, E.D., Moustafa, M., Isgro, F., Boucher, H.W., Corey, G.R., Carmeli, Y., Betts, R., Hartzel, J.S., *et al.* (2013). Effect of an investigational vaccine for preventing *Staphylococcus aureus* infections after cardiothoracic surgery: a randomized trial. J. Am. Med. Assoc. *309*, 1368–1378.

Fridkin, S.K., Hageman, J., McDougal, L.K., Mohammed, J., Jarvis, W.R., Perl, T.M., Tenover, F.C., and Vancomycin-Intermediate *Staphylococcus aureus* Epidemiology Study Group. (2003). Epidemiological and microbiological characterization of infections caused by *Staphylococcus aureus* with reduced susceptibility to vancomycin, United States, 1997–2001. Clin. Infect. Dis. *36*, 429–439.

Friedman, L., Alder, J.D., and Silverman, J.A. (2006). Genetic changes that correlate with reduced susceptibility to daptomycin in *Staphylococcus aureus*. Antimicrob. Agents Chemother. *50*, 2137–2145.

Garcia-Alvarez, L., Holden, M.T., Lindsay, H., Webb, C.R., Brown, D.F., Curran, M.D., Walpole, E., Brooks, K., Pickard, D.J., Teale, C., *et al.* (2011). Meticillin-resistant *Staphylococcus aureus* with a novel mecA homologue in human and bovine populations in the UK and Denmark: a descriptive study. Lancet *11*, 595–603.

Gardete, S., and Tomasz, A. (2014). Mechanisms of vancomycin resistance in *Staphylococcus aureus*. J. Clin. Invest. *124*, 2836–2840.

Gerke, C., Kraft, A., Sussmuth, R., Schweitzer, O., and Götz, F. (1998). Characterization of the N-acetylglucosaminyltransferase activity involved in the biosynthesis of the *Staphylococcus epidermidis* polysaccharide intercellular adhesin. J. Biol. Chem. *273*, 18586–18593.

Gilbert, I. (1931). Dissociation in an Encapsulated Staphylococcus. J. Bacteriol. *21*, 157–160.

Götz, F., Heilmann, C., and Stehle, T. (2014). Functional and structural analysis of the major amidase (Atl) in Staphylococcus. Int. J. Med. Microbiol. *304*, 156–163.

Graveland, H., Duim, B., van Duijkeren, E., Heederik, D., and Wagenaar, J.A. (2011). Livestock-associated methicillin-resistant *Staphylococcus aureus* in animals and humans. Int. J. Med. Microbiol. *301*, 630–634.

Greenberg, J.W., Fischer, W., and Joiner, K.A. (1996). Influence of lipoteichoic acid structure on recognition by the macrophage scavenger receptor. Infect. Immun. *64*, 3318–3325.

Gründling, A., and Schneewind, O. (2007a). Genes required for glycolipid synthesis and lipoteichoic acid anchoring in *Staphylococcus aureus*. J. Bacteriol. *189*, 2521–2530.

Gründling, A., and Schneewind, O. (2007b). Synthesis of glycerol phosphate lipoteichoic acid in *Staphylococcus aureus*. Proc. Natl. Acad. Sci. U.S.A. *104*, 8478–8483.

Haas, H., Koch, U., and Fischer, W. (1984). Alanyl turnover from lipoteichoic acid to teichoic acid in *Staphylococcus aureus*. FEMS Microbiol. Lett. *21*, 27–31.

Hair, P.S., Ward, M.D., Semmes, O.J., Foster, T.J., and Cunnion, K.M. (2008). *Staphylococcus aureus* clumping factor A binds to complement regulator factor I and increases factor I cleavage of C3b. J. Infect. Dis. *198*, 125–133.

Hartman, B.J., and Tomasz, A. (1984). Low-affinity penicillin-binding protein associated with beta-lactam resistance in *Staphylococcus aureus*. J. Bacteriol. *158*, 513–516.

Hashimoto, M., Tawaratsumida, K., Kariya, H., Aoyama, K., Tamura, T., and Suda, Y. (2006a). Lipoprotein is a predominant Toll-like receptor 2 ligand in *Staphylococcus aureus* cell wall components. Int. Immunol. *18*, 355–362.

Hashimoto, M., Tawaratsumida, K., Kariya, H., Kiyohara, A., Suda, Y., Krikae, F., Kirikae, T., and Götz, F. (2006b). Not lipoteichoic acid but lipoproteins appear to be the dominant immunobiologically active compounds in *Staphylococcus aureus*. J. Immunol. *177*, 3162–3169.

Hayashi, S., and Wu, H.C. (1990). Lipoproteins in bacteria. J. Bioenerg. Biomembr. *22*, 451–471.

Heaton, M.P., and Neuhaus, F.C. (1992). Biosynthesis of D-alanyl-lipoteichoic acid: cloning, nucleotide sequence, and expression of the *Lactobacillus casei* gene for the D-alanine-activating enzyme. J. Bacteriol. *174*, 4707–4717.

Heilmann, C., Schweitzer, O., Gerke, C., Vanittanakom, N., Mack, D., and Götz, F. (1996). Molecular basis of intercellular adhesion in the biofilm-forming *Staphylococcus epidermidis*. Mol. Microbiol. *20*, 1083–1091.

Herbert, S., Bera, A., Nerz, C., Kraus, D., Peschel, A., Goerke, C., Meehl, M., Cheung, A., and Götz, F. (2007). Molecular basis of resistance to muramidase and cationic antimicrobial peptide activity of lysozyme in staphylococci. PLOS Pathog. *3*, e102.

Hiramatsu, K. (2001). Vancomycin-resistant *Staphylococcus aureus*: a new model of antibiotic resistance. Lancet Infect. Dis. *1*, 147–155.

Hiramatsu, K., Hanaki, H., Ino, T., Yabuta, K., Oguri, T., and Tenover, F.C. (1997). Methicillin-resistant *Staphylococcus aureus* clinical strain with reduced vancomycin susceptibility. J. Antimicrob. Chemother. *40*, 135–136.

Hiramatsu, K., Ito, T., Tsubakishita, S., Sasaki, T., Takeuchi, F., Morimoto, Y., Katayama, Y., Matsuo, M., Kuwahara-Arai, K., Hishinuma, T., *et al.* (2013). Genomic Basis for Methicillin Resistance in *Staphylococcus aureus*. Infect. Chemother. *45*, 117–136.

Holden, M.T., Feil, E.J., Lindsay, J.A., Peacock, S.J., Day, N.P., Enright, M.C., Foster, T.J., Moore, C.E., Hurst, L., Atkin, R., *et al.* (2004). Complete genomes of two clinical *Staphylococcus aureus* strains: evidence for the rapid evolution of virulence and drug resistance. Proc. Natl. Acad. Sci. U.S.A. *101*, 9786–9791.

Humphries, R.M., Pollett, S., and Sakoulas, G. (2013). A current perspective on daptomycin for the clinical microbiologist. Clin. Microbiol. Rev. *26*, 759–780.

Hutchings, M.I., Palmer, T., Harrington, D.J., and Sutcliffe, I.C. (2009). Lipoprotein biogenesis in Gram-positive bacteria: knowing when to hold 'em, knowing when to fold 'em. Trends Microbiol. *17*, 13–21.

International Working Group on the Classification of Staphylococcal Cassette Chromosome Elements (IWG-SCC). (2009). Classification of staphylococcal cassette chromosome mec (SCCmec): guidelines for reporting novel SCCmec elements. Antimicrob. Agents Chemother. *53*, 4961–4967.

Ip, W.K., Sokolovska, A., Charriere, G.M., Boyer, L., Dejardin, S., Cappillino, M.P., Yantosca, L.M., Takahashi, K., Moore, K.J., Lacy-Hulbert, A., *et al.* (2010). Phagocytosis and phagosome acidification are required for pathogen processing and MyD88-dependent responses to *Staphylococcus aureus*. J. Immunol. *184*, 7071–7081.

Ito, T., Hiramatsu, K., Tomasz, A., de Lencastre, H., Perreten, V., Holden, M.T., Coleman, D.C., Goering, R., Giffard, P.M., Skov, R.L., *et al.* (2012). Guidelines for reporting novel mecA gene homologues. Antimicrob. Agents Chemother. *56*, 4997–4999.

Ito, T., Katayama, Y., and Hiramatsu, K. (1999). Cloning and nucleotide sequence determination of the entire mec DNA of pre-methicillin-resistant *Staphylococcus aureus* N315. Antimicrob. Agents Chemother. *43*, 1449–1458.

Jansen, K.U., Girgenti, D.Q., Scully, I.L., and Anderson, A.S. (2013). Vaccine review: "*Staphylococcus aureus* vaccines: problems and prospects". Vaccine *31*, 2723–2730.

Jefferson, K.K., Cramton, S.E., Götz, F., and Pier, G.B. (2003). Identification of a 5-nucleotide sequence that controls expression of the ica locus in *Staphylococcus aureus* and characterization of the DNA-binding properties of IcaR. Mol. Microbiol. *48*, 889–899.

Jefferson, K.K., Pier, D.B., Goldmann, D.A., and Pier, G.B. (2004). The teicoplanin-associated locus regulator (TcaR) and the intercellular adhesin locus regulator (IcaR) are transcriptional inhibitors of the ica locus in *Staphylococcus aureus*. J. Bacteriol. *186*, 2449–2456.

Jerga, A., Lu, Y.J., Schujman, G.E., de Mendoza, D., and Rock, C.O. (2007). Identification of a soluble diacylglycerol kinase required for lipoteichoic acid production in Bacillus subtilis. J. Biol. Chem. *282*, 21738–21745.

Jevons, M.P. (1961). "Celbenin"-resistant staphylococci. Br. Med. J. *1*, 124–125.

Jones, C., (2005). Revised structures for the capsular polysaccharides from *Staphylococcus aureus* Types 5 and 8, components of novel glycoconjugate vaccines. Carbohydr. Res. *340*, 1097–1106.

Jones, T., Yeaman, M.R., Sakoulas, G., Yang, S.J., Proctor, R.A., Sahl, H.G., Schrenzel, J., Xiong, Y.Q., and Bayer, A.S. (2008). Failures in clinical treatment of *Staphylococcus aureus* Infection with daptomycin are associated with alterations in surface charge, membrane phospholipid asymmetry, and drug binding. Antimicrob. Agents Chemother. *52*, 269–278.

Jorasch, P., Warnecke, D.C., Lindner, B., Zahringer, U., and Heinz, E. (2000). Novel processive and nonprocessive glycosyltransferases from *Staphylococcus aureus* and *Arabidopsis thaliana* synthesize glycoglycerolipids, glycophospholipids, glycosphingolipids and glycosylsterols. Eur. J. Biochem. *267*, 3770–3783.

Jorasch, P., Wolter, F.P., Zahringer, U., and Heinz, E. (1998). A UDP glucosyltransferase from *Bacillus subtilis* successively transfers up to four glucose residues to 1,2-diacylglycerol: expression of *ypfP* in *Escherichia coli* and structural analysis of its reaction products. Mol. Microbiol. 29, 419–430.

Jordan, S., Hutchings, M.I., and Mascher, T. (2008). Cell envelope stress response in Gram-positive bacteria. FEMS Microbiol. Rev. 32, 107–146.

Joyce, J.G., Abeygunawardana, C., Xu, Q., Cook, J.C., Hepler, R., Przysiecki, C.T., Grimm, K.M., Roper, K., Ip, C.C., Cope, L., *et al.* (2003). Isolation, structural characterization, and immunological evaluation of a high-molecular-weight exopolysaccharide from *Staphylococcus aureus*. Carbohydr. Res. *338*, 903–922.

Jung, D., Rozek, A., Okon, M., and Hancock, R.E. (2004). Structural transitions as determinants of the action of the calcium-dependent antibiotic daptomycin. Chem. Biol. *11*, 949–957.

Karakawa, W.W., and Vann, W.F. (1982). Capsular polysaccharides of *Staphylococcus aureus*. Semin. Infect. Dis. *4*, 285–293.

Karatsa-Dodgson, M., Wörmann, M.E., and Gründling, A. (2010). *In vitro* analysis of the *Staphylococcus aureus* lipoteichoic acid synthase enzyme using fluorescently labeled lipids. J. Bacteriol. *192*, 5341–5349.

Katayama, Y., Ito, T., and Hiramatsu, K. (2000). A new class of genetic element, staphylococcus cassette chromosome mec, encodes methicillin resistance in *Staphylococcus aureus*. Antimicrob. Agents Chemother. *44*, 1549–1555.

Kawai, Y., Marles-Wright, J., Cleverley, R.M., Emmins, R., Ishikawa, S., Kuwano, M., Heinz, N., Bui, N.K., Hoyland, C.N., Ogasawara, N., *et al.* (2011). A widespread family of bacterial cell wall assembly proteins. EMBO. J. *30*, 4931–4941.

Kielian, T., Haney, A., Mayes, P.M., Garg, S., and Esen, N. (2005). Toll-like receptor 2 modulates the proinflammatory milieu in *Staphylococcus aureus*-induced brain abscess. Infect. Immun. *73*, 7428–7435.

Kim, C., Milheiriço, C., Gardete, S., Holmes, M.A., Holden, M.T., de Lencastre, H., and Tomasz, A. (2012). Properties of a novel PBP2A protein homolog from *Staphylococcus aureus* strain LGA251 and its contribution to the β-lactam-resistant phenotype. J. Biol. Chem. *287*, 36854–36863.

Kiriukhin, M.Y., Debabov, D.V., Shinabarger, D.L., and Neuhaus, F.C. (2001). Biosynthesis of the glycolipid anchor in lipoteichoic acid of *Staphylococcus aureus* RN4220: role of YpfP, the diglucosyldiacylglycerol synthase. J. Bacteriol. *183*, 3506–3514.

Kiser, K.B., Bhasin, N., Deng, L., and Lee, J.C. (1999). *Staphylococcus aureus* cap5P encodes a UDP-N-acetylglucosamine 2-epimerase with functional redundancy. J. Bacteriol. *181*, 4818–4824.

Kneidinger, B., O'Riordan, K., Li, J., Brisson, J.R., Lee, J.C., and Lam, J.S. (2003). Three highly conserved proteins catalyze the conversion of UDP-N-acetyl-D-glucosamine to precursors for the biosynthesis of O antigen in *Pseudomonas aeruginosa* O11 and capsule in *Staphylococcus aureus* type 5. Implications for the UDP-N-acetyl-L-fucosamine biosynthetic pathway. J. Biol. Chem. *278*, 3615–3627.

Koch, H.U., Haas, R., and Fischer, W. (1984). The role of lipoteichoic acid biosynthesis in membrane lipid metabolism of growing *Staphylococcus aureus*. Eur. J. Biochem. *138*, 357–363.

Kohler, T.P., Gisch, N., Binsker, U., Schlag, M., Darm, K., Völker, U., Zähringer, U., and Hammerschmidt, S. (2014). Repeating structures of the major staphylococcal autolysin are essential for the interaction with human thrombospondin 1 and vitronectin. J. Biol. Chem. *289*, 4070–4082.

Koprivnjak, T., Mlakar, V., Swanson, L., Fournier, B., Peschel, A., and Weiss, J.P. (2006). Cation-induced transcriptional regulation of the dlt operon of *Staphylococcus aureus*. J. Bacteriol. *188*, 3622–3630.

Koprivnjak, T., Peschel, A., Gelb, M.H., Liang, N.S., and Weiss, J.P. (2002). Role of charge properties of bacterial envelope in bactericidal action of human group IIA phospholipase A2 against *Staphylococcus aureus*. J. Biol. Chem. *277*, 47636–47644.

Koprivnjak, T., Zhang, D., Ernst, C.M., Peschel, A., Nauseef, W.M., and Weiss, J.P. (2011). Characterization of *Staphylococcus aureus* cardiolipin synthases 1 and 2 and their contribution to accumulation of cardiolipin in stationary phase and within phagocytes. J. Bacteriol. *193*, 4134–4142.

Kozarich, J.W., and Strominger, J.L. (1978). A membrane enzyme from *Staphylococcus aureus* which catalyzes transpeptidase, carboxypeptidase, and penicillinase activities. J. Biol. Chem. *253*, 1272–1278.

Kurokawa, K., Kim, M.S., Ichikawa, R., Ryu, K.H., Dohmae, N., Nakayama, H., and Lee, B.L. (2012). Environment-mediated accumulation of diacyl lipoproteins over their triacyl counterparts in *Staphylococcus aureus*. J. Bacteriol. *194*, 3299–3306.

Kurokawa, K., Lee, H., Roh, K.B., Asanuma, M., Kim, Y.S., Nakayama, H., Shiratsuchi, A., Choi, Y., Takeuchi, O., Kang, H.J., *et al.* (2009). The Triacylated ATP Binding Cluster Transporter Substrate-binding Lipoprotein of *Staphylococcus aureus* Functions as a Native Ligand for Toll-like Receptor 2. J. Biol. Chem. *284*, 8406–8411.

Kuypers, J.M., and Proctor, R.A. (1989). Reduced adherence to traumatized rat heart valves by a low-fibronectin-binding mutant of *Staphylococcus aureus*. Infect. Immun. *57*, 2306–2312.

Labischinski, H., and Maidhof, H. (1994). Bacterial peptidoglycan: overview and evolving concepts. In Bacterial cell wall, J.-M. Ghuysen, and R. Hakenbeck, eds. (Amsterdam, Elsevier Science BV), 23–38.

Lee, J.C., Betley, M.J., Hopkins, C.A., Perez, N.E., and Pier, G.B. (1987). Virulence studies, in mice, of transposon-induced mutants of *Staphylococcus aureus* differing in capsule size. J. Infect. Dis. *156*, 741–750.

Lee, J.C., Perez, N.E., Hopkins, C.A., and Pier, G.B. (1988). Purified capsular polysaccharide-induced immunity to *Staphylococcus aureus* infection. J. Infect. Dis. *157*, 723–730.

Levine, D.P. (2006). Vancomycin: a history. Clin. Infect. Dis. *42* (Suppl. 1), S5–S12.

Li, M., Cha, D.J., Lai, Y., Villaruz, A.E., Sturdevant, D.E., and Otto, M. (2007a). The antimicrobial peptide-sensing system aps of *Staphylococcus aureus*. Mol. Microbiol. *66*, 1136–1147.

Li, M., Lai, Y., Villaruz, A.E., Cha, D.J., Sturdevant, D.E., and Otto, M. (2007b). Gram-positive three-component antimicrobial peptide-sensing system. Proc. Natl. Acad. Sci. U.S.A. *104*, 9469–9474.

Linares, J. (2001). The VISA/GISA problem: therapeutic implications. Clin. Microbiol. Infect. 7 Suppl. *4*, 8–15.

Lovering, A.L., Safadi, S.S., and Strynadka, N.C. (2012). Structural perspective of peptidoglycan biosynthesis and assembly. Annu. Rev. Biochem. *81*, 451–478.

Lu, D., Wörmann, M.E., Zhang, X., Schneewind, O., Gründling, A., and Freemont, P.S. (2009). Structure-based mechanism of lipoteichoic acid synthesis by *Staphylococcus aureus* LtaS. Proc. Natl. Acad. Sci. U.S.A. *106*, 1584–1589.

Lu, Y.J., Zhang, Y.M., Grimes, K.D., Qi, J., Lee, R.E., and Rock, C.O. (2006). Acyl-phosphates initiate membrane phospholipid synthesis in Gram-positive pathogens. Mol. Cell. *23*, 765–772.

Luong, T.T., Ouyang, S., Bush, K., and Lee, C.Y. (2002). Type 1 capsule genes of *Staphylococcus aureus* are carried in a staphylococcal cassette chromosome genetic element. J. Bacteriol. *184*, 3623–3629.

Mack, D., Fischer, W., Krokotsch, A., Leopold, K., Hartmann, R., Egge, H., and Laufs, R. (1996). The intercellular adhesin involved in biofilm accumulation of *Staphylococcus epidermidis* is a linear beta-1, 6-linked glucosaminoglycan: purification and structural analysis. J. Bacteriol. *178*, 175–183.

Mack, D., Nedelmann, M., Krokotsch, A., Schwarzkopf, A., Heesemann, J., and Laufs, R. (1994). Characterization of transposon mutants of biofilm-producing *Staphylococcus epidermidis* impaired in the accumulative phase of biofilm production: genetic identification of a hexosamine-containing polysaccharide intercellular adhesin. Infect. Immun. *62*, 3244–3253.

Mack, D., Siemssen, N., and Laufs, R. (1992). Parallel induction by glucose of adherence and a polysaccharide antigen specific for plastic-adherent *Staphylococcus epidermidis*: evidence for functional relation to intercellular adhesion. Infect. Immun. *60*, 2048–2057.

Maidhof, H., Reinicke, B., Blumel, P., Berger-Bächi, B., and Labischinski, H. (1991). *femA*, which encodes a factor essential for expression of methicillin resistance, affects glycine content of peptidoglycan in methicillin-resistant and methicillin-susceptible *Staphylococcus aureus* strains. J. Bacteriol. *173*, 3507–3513.

Maki, H., Yamaguchi, T., and Murakami, K. (1994). Cloning and characterization of a gene affecting the methicillin resistance level and the autolysis rate in *Staphylococcus aureus*. J. Bacteriol. *176*, 4993–5000.

Maresso, A.W., and Schneewind, O. (2006). Iron acquisition and transport in *Staphylococcus aureus*. Biometals *19*, 193–203.

Marraffini, L.A., and Schneewind, O. (2005). Anchor structure of staphylococcal surface proteins. V. Anchor structure of the sortase B substrate IsdC. J. Biol. Chem. *280*, 16263–16271.

Matias, V.R., and Beveridge, T.J. (2006). Native cell wall organization shown by cryo-electron microscopy confirms the existence of a periplasmic space in *Staphylococcus aureus*. J. Bacteriol. *188*, 1011–1021.

Matias, V.R., and Beveridge, T.J. (2007). Cryo-electron microscopy of cell division in *Staphylococcus aureus* reveals a mid-zone between nascent cross walls. Mol. Microbiol. *64*, 195–206.

Mazmanian, S.K., Liu, G., Ton-That, H., and Schneewind, O. (1999). *Staphylococcus aureus* sortase, an enzyme that anchors surface proteins to the cell wall. Science *285*, 760–763.

Mazmanian, S.K., Skaar, E.P., Gaspar, A.H., Humayun, M., Gornicki, P., Jelenska, J., Joachmiak, A., Missiakas, D.M., and Schneewind, O. (2003). Passage of heme-iron across the envelope of *Staphylococcus aureus*. Science *299*, 906–909.

McAleese, F.M., Walsh, E.J., Sieprawska, M., Potempa, J., and Foster, T.J. (2001). Loss of clumping factor B fibrinogen binding activity by *Staphylococcus aureus* involves cessation of transcription, shedding and cleavage by metalloprotease. J. Biol. Chem. *276*, 29969–29978.

McCarthy, A.J., and Lindsay, J.A. (2010). Genetic variation in *Staphylococcus aureus* surface and immune evasion genes is lineage associated: implications for vaccine design and host-pathogen interactions. BMC Microbiol. *10*, 173.

Meredith, T.C., Swoboda, J.G., and Walker, S. (2008). Late-stage polyribitol phosphate wall teichoic acid biosynthesis in *Staphylococcus aureus*. J. Bacteriol. *190*, 3046–3056.

Mesnage, S., Dellarole, M., Baxter, N.J., Rouget, J.B., Dimitrov, J.D., Wang, N., Fujimoto, Y., Hounslow, A.M., Lacroix-Desmazes, S., Fukase, K., *et al.* (2014). Molecular basis for bacterial peptidoglycan recognition by LysM domains. Nat. Commun. *5*, 4269.

Miller, D.J., Jerga, A., Rock, C.O., and White, S.W. (2008). Analysis of the *Staphylococcus aureus* DgkB structure reveals a common catalytic mechanism for the soluble diacylglycerol kinases. Structure *16*, 1036–1046.

Miller, L.S., O'Connell, R.M., Gutierrez, M.A., Pietras, E.M., Shahangian, A., Gross, C.E., Thirumala, A., Cheung, A.L., Cheng, G., and Modlin, R.L. (2006). MyD88 mediates neutrophil recruitment initiated by IL-1R but not TLR2 activation in immunity against *Staphylococcus aureus*. Immunity *24*, 79–91.

Mishra, N.N., McKinnell, J., Yeaman, M.R., Rubio, A., Nast, C.C., Chen, L., Kreiswirth, B.N., and Bayer, A.S. (2011). *In vitro* cross-resistance to daptomycin and host defense cationic antimicrobial peptides in clinical methicillin-resistant *Staphylococcus aureus* isolates. Antimicrob. Agents Chemother. *55*, 4012–4018.

Mohammadi, T., van Dam, V., Sijbrandi, R., Vernet, T., Zapun, A., Bouhss, A., Diepeveen-de Bruin, M., Nguyen-Distèche, M., de Kruijff, B., and Breukink, E. (2011). Identification of FtsW as a transporter of lipid-linked cell wall precursors across the membrane. EMBO J. *30*, 1425–1432.

Moise, P.A., North, D., Steenbergen, J.N., and Sakoulas, G. (2009). Susceptibility relationship between vancomycin and daptomycin in *Staphylococcus aureus*: facts and assumptions. Lancet Infect. Dis. *9*, 617–624.

Moreau, M., Richards, J.C., Fournier, J.M., Byrd, R.A., Karakawa, W.W., and Vann, W.F. (1990). Structure of the type 5 capsular polysaccharide of *Staphylococcus aureus*. Carbohydr. Res. *201*, 285–297.

Moreillon, P., Entenza, J.M., Francioli, P., McDevitt, D., Foster, T.J., François, P., and Vaudaux, P. (1995). Role of *Staphylococcus aureus* coagulase and clumping factor in pathogenesis of experimental endocarditis. Infect. Immun. *63*, 4738–4743.

Müller, P., Müller-Anstett, M., Wagener, J., Gao, Q., Kaesler, S., Schaller, M., Biedermann, T., and Götz, F. (2010). The *Staphylococcus aureus* lipoprotein SitC colocalizes with Toll-like receptor 2 (TLR2) in murine keratinocytes and elicits intracellular TLR2 accumulation. Infect. Immun. *78*, 4243–4250.

Münch, D., Roemer, T., Lee, S.H., Engeser, M., Sahl, H.G., and Schneider, T. (2012). Identification and *in vitro* analysis of the GatD/MurT enzyme-complex catalyzing lipid II amidation in *Staphylococcus aureus*. PLOS Pathog. *8*, e1002509.

Muraih, J.K., Pearson, A., Silverman, J., and Palmer, M. (2011). Oligomerization of daptomycin on membranes. Biochim. Biophys. Acta *1808*, 1154–1160.

Mwangi, M.M., Wu, S.W., Zhou, Y., Sieradzki, K., de Lencastre, H., Richardson, P., Bruce, D., Rubin, E., Myers, E., Siggia, E.D., *et al.* (2007). Tracking the *in vivo* evolution of multidrug resistance in *Staphylococcus aureus* by whole-genome sequencing. Proc. Natl. Acad. Sci. U.S.A. *104*, 9451–9456.

Nakayama, H., Kurokawa, K., and Lee, B.L. (2012). Lipoproteins in bacteria: structures and biosynthetic pathways. FEBS J. *279*, 4247–4268.

Navarre, W.W., and Schneewind, O. (1999). Surface proteins of gram-positive bacteria and mechanisms of their targeting to the cell wall envelope. Microbiol. Mol. Biol. Rev. *63*, 174–229.

Nemeth, J., and Lee, J.C. (1995). Antibodies to capsular polysaccharides are not protective against experimental *Staphylococcus aureus* endocarditis. Infect. Immun. *63*, 375–380.

Neuhaus, F.C., and Baddiley, J. (2003). A continuum of anionic charge: structures and functions of D-alanyl-teichoic acids in gram-positive bacteria. Microbiol. Mol. Biol. Rev. *67*, 686–723.

O'Riordan, K., and Lee, J.C. (2004). *Staphylococcus aureus* capsular polysaccharides. Clin. Microbiol. Rev. *17*, 218–234.

Oku, Y., Kurokawa, K., Matsuo, M., Yamada, S., Lee, B.L., and Sekimizu, K. (2009). Pleiotropic roles of polyglycerolphosphate synthase of lipoteichoic acid in growth of *Staphylococcus aureus* cells. J. Bacteriol. *191*, 141–151.

Over, B., Heusser, R., McCallum, N., Schulthess, B., Kupferschmied, P., Gaiani, J.M., Sifri, C.D., Berger-Bächi, B., and Stutzmann Meier, P. (2011). LytR-CpsA-Psr proteins in *Staphylococcus aureus* display partial functional redundancy and the deletion of all three severely impairs septum placement and cell separation. FEMS Microbiol. Lett. *320*, 142–151.

Park, J.T. (1952a). Uridine-5′-pyrophosphate derivatives. II. A structure common to three derivatives. J. Biol. Chem. *194*, 885–895.

Park, J.T. (1952b). Uridine-5′-pyrophosphate derivatives. II. Isolation from *Staphylococcus aureus*. J. Biol. Chem. *194*, 877–884.

Park, J.T. (1952c). Uridine-5′-pyrophosphate derivatives. III. Amino acid-containing derivatives. J. Biol. Chem. *194*, 897–904.

Parsons, J.B., Broussard, T.C., Bose, J.L., Rosch, J.W., Jackson, P., Subramanian, C., and Rock, C.O. (2014). Identification of a two-component fatty acid kinase responsible for host fatty acid incorporation by *Staphylococcus aureus*. Proc. Natl. Acad. Sci. U.S.A. *111*, 10532–10537.

Pasquina, L.W., Santa Maria, J.P., and Walker, S. (2013). Teichoic acid biosynthesis as an antibiotic target. Curr. Opin. Microbiol. *16*, 531–537.

Patel, D., Husain, M., Vidaillac, C., Steed, M.E., Rybak, M.J., Seo, S.M., and Kaatz, G.W. (2011). Mechanisms of in-vitro-selected daptomycin-non-susceptibility in *Staphylococcus aureus*. Int. J. Antimicrob. Agents *38*, 442–446.

Paterson, G.K., Harrison, E.M., and Holmes, M.A. (2014). The emergence of mecC methicillin-resistant *Staphylococcus aureus*. Trends Microbiol. *22*, 42–47.

Peacock, S.J., Foster, T.J., Cameron, B.J., and Berendt, A.R. (1999). Bacterial fibronectin-binding proteins and endothelial cell surface fibronectin mediate adherence of *Staphylococcus aureus* to resting human endothelial cells. Microbiology *145*, 3477–3486.

Peleg, A.Y., Miyakis, S., Ward, D.V., Earl, A.M., Rubio, A., Cameron, D.R., Pillai, S., Moellering, R.C., and Eliopoulos, G.M. (2012). Whole genome characterization of the mechanisms of daptomycin resistance in clinical and laboratory derived isolates of *Staphylococcus aureus*. PLoS One *7*, e28316.

Percy, M.G., and Gründling, A. (2014). Lipoteichoic acid synthesis and function in gram-positive bacteria. Annu. Rev. Microbiol. *68*, 81–100.

Perego, M., Glaser, P., Minutello, A., Strauch, M.A., Leopold, K., and Fischer, W. (1995). Incorporation of D-alanine into lipoteichoic acid and wall teichoic acid in *Bacillus subtilis*. Identification of genes and regulation. J. Biol. Chem. *270*, 15598–15606.

Pereira, M.P., D'Elia, M.A., Troczynska, J., and Brown, E.D. (2008). Duplication of teichoic acid biosynthetic genes in *Staphylococcus aureus* leads to functionally redundant poly(ribitol phosphate) polymerases. J. Bacteriol. *190*, 5642–5649.

Pereira, P.M., Filipe, S.R., Tomasz, A., and Pinho, M.G. (2007). Fluorescence ratio imaging microscopy shows decreased access of vancomycin to cell wall synthetic sites in vancomycin-resistant *Staphylococcus aureus*. Antimicrob. Agents Chemother. *51*, 3627–3633.

Peschel, A., Jack, R.W., Otto, M., Collins, L.V., Staubitz, P., Nicholson, G., Kalbacher, H., Nieuwenhuizen, W.F., Jung, G., Tarkowski, A., *et al.* (2001). *Staphylococcus aureus* resistance to human defensins and evasion of neutrophil killing via the novel virulence factor MprF is based on modification of membrane lipids with l-lysine. J. Exp. Med. *193*, 1067–1076.

Peschel, A., Otto, M., Jack, R.W., Kalbacher, H., Jung, G., and Götz, F. (1999). Inactivation of the dlt operon in *Staphylococcus aureus* confers sensitivity to defensins, protegrins, and other antimicrobial peptides. J. Biol. Chem. *274*, 8405–8410.

Peschel, A., and Sahl, H.G. (2006). The co-evolution of host cationic antimicrobial peptides and microbial resistance. Nat. Rev. Microbiol. *4*, 529–536.

Petersen, A., Stegger, M., Heltberg, O., Christensen, J., Zeuthen, A., Knudsen, L.K., Urth, T., Sorum, M., Schouls, L., Larsen, J., *et al.* (2013). Epidemiology of methicillin-resistant *Staphylococcus aureus* carrying the novel *mecC* gene in Denmark corroborates a zoonotic reservoir with transmission to humans. Clin. Microbiol. Infect. *19*, E16–E22.

Peterson, P.K., Wilkinson, B.J., Kim, Y., Schmeling, D., and Quie, P.G. (1978). Influence of encapsulation on staphylococcal opsonization and phagocytosis by human polymorphonuclear leucocytes. Infect. Immun. *19*, 943–949.

Pillai, S.K., Gold, H.S., Sakoulas, G., Wennersten, C., Moellering, R.C., and Eliopoulos, G.M. (2007). Daptomycin nonsusceptibility in *Staphylococcus aureus* with reduced vancomycin susceptibility is independent of alterations in MprF. Antimicrob. Agents Chemother. *51*, 2223–2225.

Poutrel, B., Rainard, P., and Sarradin, P. (1997). Heterogeneity of cell-associated CP5 expression on *Staphylococcus aureus* strains demonstrated by flow cytometry. Clin. Diagn. Lab. Immunol. *4*, 275–278.

Reichmann, N.T., Cassona, C.P., and Gründling, A. (2013). Revised mechanism of D-alanine incorporation into cell wall polymers in Gram-positive bacteria. Microbiology *159*, 1868–1877.

Reichmann, N.T., and Gründling, A. (2011). Location, synthesis and function of glycolipids and polyglycerolphosphate lipoteichoic acid in Gram-positive bacteria of the phylum Firmicutes. FEMS Microbiol. Lett. *319*, 97–105.

Richter, S.G., Elli, D., Kim, H.K., Hendrickx, A.P., Sorg, J.A., Schneewind, O., and Missiakas, D. (2013). Small molecule inhibitor of lipoteichoic acid synthesis is an antibiotic for Gram-positive bacteria. Proc. Natl. Acad. Sci. U.S.A. *110*, 3531–3536.

Robbel, L., and Marahiel, M.A. (2010). Daptomycin, a bacterial lipopeptide synthesized by a nonribosomal machinery. J. Biol. Chem. *285*, 27501–27508.

Rohde, H., Frankenberger, S., Zähringer, U., and Mack, D. (2010). Structure, function and contribution of polysaccharide intercellular adhesin (PIA) to *Staphylococcus epidermidis* biofilm formation and pathogenesis of biomaterial-associated infections. Eur. J. Cell Biol. *89*, 103–111.

Rohrer, S., Ehlert, K., Tschierske, M., Labischinski, H., and Berger-Bächi, B. (1999). The essential *Staphylococcus aureus* gene *fmhB* is involved in the first step of peptidoglycan pentaglycine interpeptide formation. Proc. Natl. Acad. Sci. U.S.A. *96*, 9351–9356.

Ruiz de los Mozos, I., Vergara-Irigaray, M., Segura, V., Villanueva, M., Bitarte, N., Saramago, M., Domingues, S., Arraiano, C.M., Fechter, P., Romby, P., *et al.* (2013). Base pairing interaction between 5'- and 3'-UTRs controls icaR mRNA translation in *Staphylococcus aureus*. PLOS Genet. *9*, e1004001.

Ruiz, N. (2008). Bioinformatics identification of MurJ (MviN) as the peptidoglycan lipid II flippase in *Escherichia coli*. Proc. Natl. Acad. Sci. U.S.A. *105*, 15553–15557.

Sadovskaya, I., Vinogradov, E., Flahaut, S., Kogan, G., and Jabbouri, S. (2005). Extracellular carbohydrate-containing polymers of a model biofilm-producing strain, *Staphylococcus epidermidis* RP62A. Infect. Immun. *73*, 3007–3017.

Santa Maria, J.P., Jr., Sadaka, A., Moussa, S.H., Brown, S., Zhang, Y.J., Rubin, E.J., Gilmore, M.S., and Walker, S. (2014). Compound-gene interaction mapping reveals distinct roles for *Staphylococcus aureus* teichoic acids. Proc. Natl. Acad. Sci. U.S.A. *111*, 12510–12515.

Sau, S., Bhasin, N., Wann, E.R., Lee, J.C., Foster, T.J., and Lee, C.Y. (1997). The *Staphylococcus aureus* allelic genetic loci for serotype 5 and 8 capsule expression contain the type-specific genes flanked by common genes. Microbiology *143*, 2395–2405.

Schlag, M., Biswas, R., Krismer, B., Kohler, T., Zoll, S., Yu, W., Schwarz, H., Peschel, A., and Götz, F. (2010). Role of staphylococcal wall teichoic acid in targeting the major autolysin Atl. Mol. Microbiol. *75*, 864–873.

Schmaler, M., Jann, N.J., Ferracin, F., Landolt, L.Z., Biswas, L., Götz, F., and Landmann, R. (2009). Lipoproteins in *Staphylococcus aureus* mediate inflammation by TLR2 and iron-dependent growth *in vivo*. J. Immunol. *182*, 7110–7118.

Schneewind, O., Mihaylova-Petkov, D., and Model, P. (1993). Cell wall sorting signals in surface proteins of gram-positive bacteria. EMBO J. *12*, 4803–4811.

Schneewind, O., and Missiakas, D. (2014). Sec-secretion and sortase-mediated anchoring of proteins in Gram-positive bacteria. Biochim. Biophys. Acta *1843*, 1687–1697.

Schneider, T., Senn, M.M., Berger-Bächi, B., Tossi, A., Sahl, H.G., and Wiedemann, I. (2004). *In vitro* assembly of a complete, pentaglycine interpeptide bridge containing cell wall precursor (lipid II-Gly5) of *Staphylococcus aureus*. Mol. Microbiol. *53*, 675–685.

Sendi, P., and Proctor, R.A. (2009). *Staphylococcus aureus* as an intracellular pathogen: the role of small colony variants. Trends Microbiol. *17*, 54–58.

Severin, A., Wu, S.W., Tabei, K., and Tomasz, A. (2004). Penicillin-binding protein 2 is essential for expression of high-level vancomycin resistance and cell wall synthesis in vancomycin-resistant *Staphylococcus aureus* carrying the enterococcal vanA gene complex. Antimicrob. Agents Chemother. *48*, 4566–4573.

Sewell, E.W., and Brown, E.D. (2014). Taking aim at wall teichoic acid synthesis: new biology and new leads for antibiotics. J. Antibiot. *67*, 43–51.

Sham, L.T., Butler, E.K., Lebar, M.D., Kahne, D., Bernhardt, T.G., and Ruiz, N. (2014). Bacterial cell wall. MurJ is the flippase of lipid-linked precursors for peptidoglycan biogenesis. Science *345*, 220–222.

Sharp, J.A., Echague, C.G., Hair, P.S., Ward, M.D., Nyalwidhe, J.O., Geoghegan, J.A., Foster, T.J., and Cunnion, K.M. (2012). *Staphylococcus aureus* surface protein SdrE binds complement regulator factor H as an immune evasion tactic. PLoS One *7*, e38407.

Sheen, T.R., Ebrahimi, C.M., Hiemstra, I.H., Barlow, S.B., Peschel, A., and Doran, K.S. (2010). Penetration of the blood-brain barrier by *Staphylococcus aureus*: contribution of membrane-anchored lipoteichoic acid. J. Mol. Med. *88*, 633–639.

Short, S.A., and White, D.C. (1971). Metabolism of phosphatidylglycerol, lysylphosphatidylglycerol, and cardiolipin of *Staphylococcus aureus*. J. Bacteriol. *108*, 219–226.

Short, S.A., and White, D.C. (1972). Biosynthesis of cardiolipin from phosphatidylglycerol in *Staphylococcus aureus*. J. Bacteriol. *109*, 820–826.

Sievert, D.M., Rudrik, J.T., Patel, J.B., McDonald, L.C., Wilkins, M.J., and Hageman, J.C. (2008). Vancomycin-resistant *Staphylococcus aureus* in the United States, 2002–2006. Clin. Infect. Dis. *46*, 668–674.

Sompolinsky, D., Samra, Z., Karakawa, W.W., Vann, W.F., Schneerson, R., and Malik, Z. (1985). Encapsulation and capsular types in isolates of *Staphylococcus aureus* from different sources and relationship to phage types. J. Clin. Microbiol. *22*, 828–834.

Steen, A., Buist, G., Leenhouts, K.J., El Khattabi, M., Grijpstra, F., Zomer, A.L., Venema, G., Kuipers, O.P., and Kok, J. (2003). Cell wall attachment of a widely distributed peptidoglycan binding domain is hindered by cell wall constituents. J. Biol. Chem. *278*, 23874–23881.

Stoll, H., Dengjel, J., Nerz, C., and Götz, F. (2005). *Staphylococcus aureus* deficient in lipidation of prelipoproteins is attenuated in growth and immune activation. Infect. Immun. *73*, 2411–2423.

Stranden, A.M., Ehlert, K., Labischinski, H., and Berger-Bächi, B. (1997). Cell wall monoglycine cross-bridges and methicillin hypersusceptibility in a *femAB* null mutant of methicillin-resistant *Staphylococcus aureus*. J. Bacteriol. 179, 9–16.

Stranger-Jones, Y.K., Bae, T., and Schneewind, O. (2006). Vaccine assembly from surface proteins of *Staphylococcus aureus*. Proc. Natl. Acad. Sci. U.S.A. *103*, 16942–16947.

Straus, S.K., and Hancock, R.E. (2006). Mode of action of the new antibiotic for Gram-positive pathogens daptomycin: comparison with cationic antimicrobial peptides and lipopeptides. Biochim. Biophys. Acta *1758*, 1215–1223.

Sun, Y., Hise, A.G., Kalsow, C.M., and Pearlman, E. (2006). *Staphylococcus aureus*-induced corneal inflammation is dependent on Toll-like receptor 2 and myeloid differentiation factor 88. Infect. Immun. *74*, 5325–5332.

Swoboda, J.G., Meredith, T.C., Campbell, J., Brown, S., Suzuki, T., Bollenbach, T., Malhowski, A.J., Kishony, R., Gilmore, M.S., and Walker, S. (2009). Discovery of a small molecule that blocks wall teichoic acid biosynthesis in *Staphylococcus aureus*. ACS Chem. Biol. *4*, 875–883.

Takeuchi, O., Hoshino, K., and Akira, S. (2000). Cutting edge: TLR2-deficient and MyD88-deficient mice are highly susceptible to *Staphylococcus aureus* infection. J. Immunol. *165*, 5392–5396.

Taron, D.J., Childs, W.C., and Neuhaus, F.C. (1983). Biosynthesis of D-alanyl-lipoteichoic acid: role of diglyceride kinase in the synthesis of phosphatidylglycerol for chain elongation. J. Bacteriol. *154*, 1110–1116.

Tawaratsumida, K., Furuyashiki, M., Katsumoto, M., Fujimoto, Y., Fukase, K., Suda, Y., and Hashimoto, M. (2009). Characterization of N-terminal structure of TLR2-activating lipoprotein in *Staphylococcus aureus*. J. Biol. Chem. *284*, 9147–9152.

Thomas, C.A., Li, Y., Kodama, T., Suzuki, H., Silverstein, S.C., and El Khoury, J. (2000). Protection from lethal gram-positive infection by macrophage scavenger receptor-dependent phagocytosis. J. Exp. Med. *191*, 147–156.

Thwaites, G.E., and Gant, V. (2011). Are bloodstream leucocytes Trojan Horses for the metastasis of *Staphylococcus aureus*? Nat. Rev. Microbiol. *9*, 215–222.

Tipper, D.J., and Strominger, J.L. (1965). Mechanism of action of penicillins: a proposal based on their structural similarity to acyl-D-alanyl-D-alanine. Proc. Natl. Acad. Sci. U.S.A. *54*, 1133–1141.

Ulrich, M., Bastian, M., Cramton, S.E., Ziegler, K., Pragman, A.A., Bragonzi, A., Memmi, G., Wolz, C., Schlievert, P.M., Cheung, A., *et al.* (2007). The staphylococcal respiratory response regulator SrrAB induces *ica* gene transcription and polysaccharide intercellular adhesin expression, protecting *Staphylococcus aureus* from neutrophil killing under anaerobic growth conditions. Mol. Microbiol. *65*, 1276–1287.

Verbrugh, H.A., Peterson, P.K., Nguyen, B.Y., Sisson, S.P., and Kim, Y. (1982). Opsonization of encapsulated *Staphylococcus aureus*: the role of specific antibody and complement. J. Immunol. *129*, 1681–1687.

Vergara-Irigaray, M., Maira-Litrán, T., Merino, N., Pier, G.B., Penadés, J.R., and Lasa, I. (2008). Wall teichoic acids are dispensable for anchoring the PNAG exopolysaccharide to the *Staphylococcus aureus* cell surface. Microbiology *154*, 865–877.

Volkman, B.F., Zhang, Q., Debabov, D.V., Rivera, E., Kresheck, G.C., and Neuhaus, F.C. (2001). Biosynthesis of D-alanyl-lipoteichoic acid: the tertiary structure of apo-D-alanyl carrier protein. Biochemistry *40*, 7964–7972.

Vollmer, W., Joris, B., Charlier, P., and Foster, S. (2008). Bacterial peptidoglycan (murein) hydrolases. FEMS Microbiol. Rev. *32*, 259–286.

von Eiff, C., Taylor, K.L., Mellmann, A., Fattom, A.I., Friedrich, A.W., Peters, G., and Becker, K. (2007). Distribution of capsular and surface polysaccharide serotypes of *Staphylococcus aureus*. Diagn. Microbiol. Infect. Dis. *58*, 297–302.

Vuong, C., Kocianova, S., Voyich, J.M., Yao, Y., Fischer, E.R., DeLeo, F.R., and Otto, M. (2004). A crucial role for exopolysaccharide modification in bacterial biofilm formation, immune evasion, and virulence. J. Biol. Chem. *279*, 54881–54886.

Weidenmaier, C., Kokai-Kun, J.F., Kristian, S.A., Chanturiya, T., Kalbacher, H., Gross, M., Nicholson, G., Neumeister, B., Mond, J.J., and Peschel, A. (2004). Role of teichoic acids in *Staphylococcus aureus* nasal colonization, a major risk factor in nosocomial infections. Nat. Med. *10*, 243–245.

Weidenmaier, C., Kokai-Kun, J.F., Kulauzovic, E., Kohler, T., Thumm, G., Stoll, H., Götz, F., and Peschel, A. (2008). Differential roles of sortase-anchored surface proteins and wall teichoic acid in *Staphylococcus aureus* nasal colonization. Int. J. Med. Microbiol. *298*, 505–513.

Weidenmaier, C., and Peschel, A. (2008). Teichoic acids and related cell wall glycopolymers in Gram-positive physiology and host interactions. Nat. Rev. Microbiol. *6*, 276–287.

Weidenmaier, C., Peschel, A., Kempf, V.A., Lucindo, N., Yeaman, M.R., and Bayer, A.S. (2005a). DltABCD- and MprF-mediated cell envelope modifications of *Staphylococcus aureus* confer resistance to platelet microbicidal proteins and contribute to virulence in a rabbit endocarditis model. Infect. Immun. *73*, 8033–8038.

Weidenmaier, C., Peschel, A., Xiong, Y.Q., Kristian, S.A., Dietz, K., Yeaman, M.R., and Bayer, A.S. (2005b). Lack of wall teichoic acids in *Staphylococcus aureus* leads to reduced interactions with endothelial cells and to attenuated virulence in a rabbit model of endocarditis. J. Infect. Dis. *191*, 1771–1777.

Winstel, V., Liang, C., Sanchez-Carballo, P., Steglich, M., Munar, M., Bröker, B.M., Penadés, J.R., Nübel, U., Holst, O., Dandekar, T., *et al.* (2013). Wall teichoic acid structure governs horizontal gene transfer between major bacterial pathogens. Nat. Commun. *4*, 2345.

Winstel, V., Sanchez-Carballo, P., Holst, O., Xia, G., and Peschel, A. (2014a). Biosynthesis of the unique wall teichoic acid of *Staphylococcus aureus* lineage ST395. MBio 5, e00869.

Winstel, V., Xia, G., and Peschel, A. (2014b). Pathways and roles of wall teichoic acid glycosylation in *Staphylococcus aureus*. Int. J. Med. Microbiol. *304*, 215–221.

Xia, G., Corrigan, R.M., Winstel, V., Goerke, C., Gründling, A., and Peschel, A. (2011). Wall teichoic Acid-dependent adsorption of staphylococcal siphovirus and myovirus. J. Bacteriol. *193*, 4006–4009.

Xia, G., Maier, L., Sanchez-Carballo, P., Li, M., Otto, M., Holst, O., and Peschel, A. (2010). Glycosylation of wall teichoic acid in *Staphylococcus aureus* by TarM. J. Biol. Chem. *285*, 13405–13415.

Xia, G., and Peschel, A. (2008). Toward the pathway of S. aureus WTA biosynthesis. Chem. Biol. *15*, 95–96.

Xu, S., Arbeit, R.D., and Lee, J.C. (1992). Phagocytic killing of encapsulated and microencapsulated *Staphylococcus aureus* by human polymorphonuclear leucocytes. Infect. Immun. *60*, 1358–1362.

Yamada, S., Sugai, M., Komatsuzawa, H., Nakashima, S., Oshida, T., Matsumoto, A., and Suginaka, H. (1996). An autolysin ring associated with cell separation of *Staphylococcus aureus*. J. Bacteriol. *178*, 1565–1571.

Zähringer, U., Lindner, B., Inamura, S., Heine, H., and Alexander, C. (2008). TLR2 - promiscuous or specific? A critical re-evaluation of a receptor expressing apparent broad specificity. Immunobiology *213*, 205–224.

Zhang, Y.M., and Rock, C.O. (2008). Membrane lipid homeostasis in bacteria. Nat. Rev. Microbiol. *6*, 222–233.

Zoll, S., Schlag, M., Shkumatov, A.V., Rautenberg, M., Svergun, D.I., Götz, F., and Stehle, T. (2012). Ligand-binding properties and conformational dynamics of autolysin repeat domains in staphylococcal cell wall recognition. J. Bacteriol. *194*, 3789–3802.

Transition Metal Ion Homeostasis

9

Jessica R. Sheldon, Ronald S. Flannagan, Mélissa Hannauer and David E. Heinrichs

Abstract

The six 3*d*-block transition metals – iron, manganese, zinc, nickel, cobalt and copper – are biologically relevant to both prokaryotes and eukaryotes alike, serving as essential micronutrients while at the same time being potentially toxic in excess. As such, intricate mechanisms exist to control the availability of metal ions within the human host and in bacteria, in an effort to maintain homeostatic concentrations of these elements. Perturbations to metal ion availability within the host can impact both overall health, as well as susceptibility to infectious diseases, as bacteria, such as the staphylococci, express a plethora of acquisition systems that allow for the extraction of transition metals from the environment to fulfil their nutritional requirements. The host counters this effort to acquire metals with immune mechanisms that not only actively sequester transition metals, rendering them unavailable to support microbial growth, but also that harness the destructive redox potential of copper and zinc to intoxicate invading bacteria; these processes are collectively termed nutritional immunity. Importantly, exploitation of metal acquisition pathways in the staphylococci may represent a viable option for the realization of novel antimicrobials or vaccine based therapies that are desperately needed to combat important human staphylococcal pathogens such as *S. aureus*.

Introduction

Biologically relevant transition metals include the six 3*d*-block elements manganese, iron, cobalt, nickel, copper and zinc, which serve as essential micronutrients in eukaryotes and prokaryotes alike (Bleackley and Macgillivray, 2011). Nearly half of all enzymes require metal to function, highlighting the indispensability of transition metals to cellular physiology, as a whole (Andreini *et al.*, 2008; Waldron and Robinson, 2009; Waldron *et al.*, 2009). Transition metals within metalloproteins provide structure, exploit their redox potential to facilitate electron transport, and serve as enzymatic cofactors (Gaupp *et al.*, 2012; Maret, 2010).

The very characteristics engendering transition metals with utility as biological catalysts, namely their interchangeable redox states, and affinity for the functional groups of proteins, renders them potentially toxic to the cell. Redox active metals, such as Fe and Cu, are capable of generating damaging reactive oxygen species (ROS) through Fenton chemistry (Gutteridge *et al.*, 1982; Imlay and Linn, 1988). Highly competitive metals, defined by their propensity to displace another metal from the active site of an enzyme, can result in mismetallation, and potentially inactivation, of enzymatic targets. The natural stability of metal ion complexes is given by the Irving–Williams series: $Mn^{2+} < Fe^{2+} < Co^{2+} < Ni^{2+} < Cu^{2+} > Zn^{2+}$, and is a reflection of metal competitiveness (Braymer and Giedroc, 2014; Irving and Williams, 1948). Stringent control over metal ion concentrations within the cell is required both to prevent undesired and/or uncontrolled redox reactions, as well as to prevent metal displacement from proteins by competitive metals such as Cu. Accordingly, the bioavailability of these metals, referring to the proportion of their total cellular content that is readily exchangeable with metabolites, is roughly in the inverse of the Irving–Williams series (Lisher and Giedroc, 2013).

The effort to promote correct metallation of essential proteins and avoid unfavourable redox reactions is further confounded in pathogenic bacteria, which have to contend with host-imposed restrictions on extracellular metal concentrations. Through the process of nutritional immunity, host sequestration strategies constrain the availability of essential metals to curtail bacterial proliferation (Kochan, 1973; Weinberg, 1974, 1975). This process effectively precipitates what is referred to as a 'fight over metals', whereby invading bacteria elaborate mechanisms of metal acquisition to subvert metal-withholding within the host (Hood and Skaar, 2012).

The concerted mechanisms governing intracellular metal ion concentration in the face of changing or limiting environmental conditions is referred to as transition metal homeostasis (Lisher and Giedroc, 2013). Serving to maximize bacterial cell viability and propagation, metal homeostasis is coordinated by metal sensor proteins, which respond to prevailing metal concentrations by regulating expression of genes encoding metal acquisition systems, efflux pumps, and sequestration proteins. Herein, the mechanisms employed by the staphylococci to maintain Fe, Mn, Zi, Ni, Co, and Cu homeostasis, along with the means by which the host limits or exploits the availability of these essential transition metals to promote immunity, are discussed.

Iron

Iron chemistry

Iron, arguably the most versatile of the transition metals as a catalytic cofactor, is an essential micronutrient for almost all living organisms, with the notable exceptions of *Lactobacillus* spp., and the obligate intracellular parasites *Borrelia burgdorferi*, and *Treponema pallidum* (Archibald, 1983; Posey and Gherardini, 2000; Weinberg, 1997). Due to its unfilled *d* atomic orbitals, iron possesses two readily interchangeable stable oxidation states, Fe^{2+} (ferrous) and Fe^{3+} (ferric), and can adopt both high and low spin conformations depending on the ligand bound. Iron exhibits octahedral coordination geometry, with six potential binding sites along three axes of symmetry. Flexibility exists in both the bond length and angle formed with the central Fe ion, as determined by the oxidation state of the ion and the ligand environment (Griffiths, 1999). All of the aforementioned properties influence the ease with which electrons are gained or lost and, accordingly, iron possesses an extensive range of redox potentials. These unique electronic characteristics make iron an ideal cofactor for metalloproteins involved in electron transfer and oxidation reactions, where iron is most often in complex with the sulfhydryl groups of cysteine residues in the form of iron–sulfur clusters, or in the tetrapyrrole ring of haem (Evans *et al.*, 1999). Correspondingly, iron is involved in several key metabolic pathways in bacteria, including photosynthesis, nitrogen fixation, amino acid synthesis, the tricarboxylic acid (TCA) cycle, electron transport, and cellular respiration (Table 9.1).

As with other redox-active metals, the very characteristics that make iron a favourable catalyst are also capable of promoting the production of intoxicating intracellular ROS. ROS, such as superoxide anions ($\cdot O_2^-$), hydrogen peroxide (H_2O_2), and hydroxyl radicals ($\cdot OH$), can damage nucleic acids, proteins, and lipids and, without mitigation, compromise bacterial survival (Cabiscol *et al.*, 2000; Imlay, 2002; Imlay and Linn, 1988; McManus and Josephy, 1995). ROS are produced through the iron-catalysed Haber–Weiss reaction, a two-step reaction that makes use of Fenton chemistry (Fenton, 1894; Haber and Weiss, 1932):

$$Fe^{3+} + \cdot O_2^- \rightarrow Fe^{2+} + O_2 \text{ (slow reaction)}$$

$$Fe^{2+} + H_2O_2 \rightarrow Fe^{3+} + \cdot OH + OH^- \text{ (Fenton reaction; fast)}$$

Net reaction:

$$\cdot O_2^- + H_2O_2 \rightarrow O_2 + \cdot OH + OH^-$$

Both $\cdot O_2^-$ and H_2O_2 are produced during normal aerobic metabolism, and host immune cells produce the former as a defence against invading microbes. The staphylococci, as with all aerobic organisms, require means by which ROS are detoxified, as well as mechanisms to sequester free iron within the cell, and these mechanisms are discussed in later sections of this chapter.

Bioavailability of iron

Although iron is one of the most abundant elements, constituting approximately 5% of the Earth's crust, its bioavailability is limited by its insolubility under physiological conditions. In the presence of oxygen at neutral pH, iron forms ferric oxyhydroxide precipitates, reducing the concentration of free iron to 10^{-8}–10^{-9} M (Ratledge and Dover, 2000). While this concentration is already orders of magnitude less than what is required to support microbial growth (~ 10^{-6} M) free iron within the host is further depleted (up to 10^{-18} M) through binding to host glycoproteins such as transferrin and lactoferrin, coordination with haem in haemoproteins, or sequestration within storage proteins such as ferritin and haemosiderin (Bullen, 1981; Ratledge, 2007). Sequestration effectively reduces the potential for Fenton chemistry and also severely limits the availability of this essential trace element to invading microbes. As such, the staphylococci have developed numerous strategies to access both sequestered iron and trace amounts of free iron through: (i) the elaboration and/or appropriation of low molecular weight, high-affinity iron-binding molecules, known as siderophores, that are capable of stripping iron from transferrin; (ii) the expression of haem-iron acquisition systems which facilitate the extraction of iron from haem/haemoproteins; and (iii) the uptake of free inorganic iron. In the following sections, we discuss the mechanisms of iron transport employed by the staphylococci, the means by which iron is liberated from siderophore or haem complexes, how it is safely housed in intracellular storage proteins or functional groups, and how these processes are regulated to maintain iron homeostasis.

Acquisition of iron from transferrin and lactoferrin

The transferrins are a family of high-affinity, Fe^{3+}-binding glycoproteins that serve to maintain iron homeostasis within the host. Serum transferrin controls free iron concentrations within the blood by mediating uptake from sites of absorption and storage, and shuttling it to sites of utilization (Evans *et al.*, 1999). Lactoferrin is not responsible for iron transport, but instead maintains bacteriostatic concentrations of iron in bodily secretions such as saliva, tears, mucous, and milk. During an infection, degranulation of neutrophils enhances localized release of lactoferrin, further reducing the pool of iron available to invading pathogens (Ganz, 2009). The role of these glycoproteins in host–pathogen interactions is discussed in-depth in the section entitled 'Regulation of extracellular iron availability: transferrin and lactoferrin'. While both transferrin and lactoferrin possess two high-affinity binding sites for Fe^{3+}, and negligible affinity for Fe^{2+}, lactoferrin is distinct from transferrin in that it releases iron at a much lower pH (pH 3.0 versus 5.5, respectively) (Baker and Baker, 2005). Both glycoproteins bind Fe^{3+} tightly, but reversibly, with a stability constant as high as 10^{23} M^{-1} at physiological pH (Aisen *et al.*, 1978). To exploit transferrin and lactoferrin as iron sources, bacteria require high-affinity acquisition systems.

Many pathogenic bacteria have developed sophisticated means of utilizing transferrin and lactoferrin as iron sources through the elaboration of specific cell surface receptors. Characterization of transferrin and lactoferrin-binding proteins has been performed primarily in Gram-negative bacteria that are incapable of producing their own siderophores, specifically *Neisseria* and *Pasteurella* spp. (Biswas and Sparling, 1995; Cornelissen *et al.*, 1992; Ogunnariwo and Schryvers, 1990; Ogunnariwo *et al.*, 1991; Shouldice *et al.*, 2003). While sporadic reports of transferrin-binding proteins in Gram-positive bacteria, including in the staphylococci, exist (Hartford *et al.*, 1993; Modun *et al.*, 1994, 1998, 2000), they lack mechanistic confirmation, and no further investigations into the presence of a functional transferrin receptor in the staphylococci have been reported. Rather, it has been consistently shown that siderophores are indispensable for the acquisition of transferrin-iron among the staphylococci (Beasley *et al.*, 2009, 2011; Park *et al.*, 2005).

Staphylococcal siderophores

Siderophores are low molecular weight (typically less than 1 kDa), high-affinity, secreted scavengers of ferric iron, capable of divesting host glycoproteins, such as transferrin and lactoferrin, of iron. With stability constants as high as 10^{52} M^{-1} (Carrano and Raymond, 1979), siderophores are classified based on the functional groups involved in iron coordination, broadly including catechol, hydroxamate, and α-hydroxycarboxylate types. Two α-hydroxycarboxylate-type siderophores have been characterized in the staphylococci, staphyloferrin A (SA) and staphyloferrin B (SB) (Drechsel *et al.*, 1993; Haag *et al.*, 1994; Konetschny-Rapp *et al.*, 1990; Meiwes *et al.*, 1990), synthesized by enzymes

Table 9.1 Selected examples of important metalloproteins in eukaryotes and prokaryotes

Metal	Protein	Biochemical function	Prokaryotic/eukaryotic	Reference
Co				
	Methionyl aminopeptidase	Removes N-terminal methionine from proteins or peptides greater than three residues in length	Eukaryotic/prokaryotic (staphylococci)	Oefner *et al*. (2003)
Cu				
	Ceruloplasmin	Abundant serum protein. Ferric iron formed through its copper-dependent ferroxidase activity is picked up by circulating transferrin	Eukaryotic	Gitlin and Janeway (1960)
	Cytochrome *c* oxidase	The terminal enzyme in the electron transport chain. Transfers electrons to molecular oxygen. Generates proton electrochemical gradient needed for ATP synthesis	Eukaryotic/prokaryotic	Solioz *et al*. (2010), Wende and Wainio (1960)
Fe				
	Transferrin	Serum glycoprotein mobilizes ferric iron to cells/tissues	Eukaryotic	Aisen and Listowsky (1980)
	Lactoferrin	Ferric iron binding glycoprotein in external secretions	Eukaryotic	Aisen and Listowsky (1980)
	Deoxyhypusine hydroxylase	Catalyses the formation of hypusine, the unique post-translational modification found on eukaryotic translation initiation factor 5A (eIF5A)	Eukaryotic	Kim *et al*. (2006)
	α-Ketoglutarate-dependent dioxygenase	Repair of alkylated DNA and/or RNA by catalysing oxidative demethylation	Eukaryotic/prokaryotic	Duncan *et al*. (2002), Shen *et al*. (2014)
	ScdA	Repair of Fe–S clusters damaged by NO or H_2O_2	Prokaryotic (staphylococci)	Overton *et al*. (2008)
Fe-S				
	Aconitase	Catalyses the isomerization of citrate to isocitrate in the TCA cycle	Eukaryotic/Prokaryotic (staphylococci)	Walden (2002)
	DNA polymerase (Pol α, δ, ε)	Replication of double-stranded DNA	Eukaryotic	Netz *et al*. (2012)
	Dihydroxy-acid dehydratase (IlvD)	Biosynthesis of branched-chain amino acids isoleucine and valine	Prokaryotic (staphylococci)	Duan *et al*. (2009)

Haem				
	Haemoglobin	Erythrocyte protein required for blood gas exchange	Eukaryotic	Evans *et al*. (1999)
	Lactoperoxidase	Catalyses the oxidation of halides and thiocyanate to form toxic antimicrobial compounds.	Eukaryotic	Singh *et al*. (2008)
	Cytochrome *c*	Small, water-soluble component of the electron transport chain	Eukaryotic/prokaryotic (staphylococci)	Le Brun *et al*. (2000), Fisher *et al*. (1973)
Mn				
	Superoxide dismutase	Dismutates (detoxifies) superoxide anion (O_2^-) to O_2 and H_2O_2	Eukaryotic/prokaryotic	Barra *et al*. (1987), Clements *et al*. (1999)
	Arginase	Catalyses the hydrolysis of arginine to urea and ornithine	Eukaryotic/prokaryotic (staphylococci)	Kanyo *et al*. (1996)
	Ribonucleotide reductase (NrdF)	Subunit of Class Ib ribonucleotide reductase that catalyses the formation of deoxyribonucleotides from ribonucleotides under low-iron conditions	Prokaryotic (staphylococci)	Andrews (2011), Martin and Imlay (2011)
	Conserved virulence factor A (CvfA)	Hydrolyses phosphodiester linkage of 2′,3′-cyclic AMP/GMP and 3′-terminal RNA	Prokaryotic (staphylococci)	Nagata *et al*. (2008)
Ni				
	Urease	Catalyses the hydrolysis of urea into CO_2 and ammonia	Prokaryotic (staphylococci)	Grüter *et al*. (1992)
Zn				
	Early endosome antigen 1 (EEA1)	An adaptor protein that localizes to early endosomes and is involved in endosomal trafficking	Eukaryotic	Mishra *et al*. (2010)
	Transcription factor IIIA (TFIIIA)	Facilitates the initiation and maintenance of gene transcription	Eukaryotic	Freedman *et al*. (1988)
	SufU	Catalyses the transfer of sulfur in the biogenesis of Fe–S clusters	Prokaryotic (staphylococci)	Selbach *et al*. (2014)
	Fusidic acid resistance protein (FusB)	Rescues elongation factor G function in the presence of fusidic acid	Prokaryotic (staphylococci)	Cox *et al*. (2012)

encoded in the loci *sfaABCD* (Beasley *et al.*, 2009; Cotton *et al.*, 2009) and *sbnABCDEFGHI* (Cheung *et al.*, 2009; Dale *et al.*, 2004a, b), respectively. SA is composed of two citrate molecules condensed with a central D-ornithine; synthesis of the SA molecule *in vitro* has confirmed that the SfaD and SfaB synthetases are responsible for this activity (Cotton *et al.*, 2009). More recently, the ferric-SA crystal structure was solved, demonstrating how the carboxyl groups from the citrates coordinate the iron atom (Grigg *et al.*, 2010a). SB is composed of one molecule each of citrate and α-ketoglutarate (α-KG), and two molecules of L-2,3-diaminopropionate (L-Dap), one of which is decarboxylated to form 1, 2-diaminoethane (Cheung *et al.*, 2009; Cotton *et al.*, 2009; Drechsel *et al.*, 1993). Both α-KG and L-Dap can be derived enzymatically from the activity of SbnA and SbnB (Kobylarz *et al.*, 2014), while citrate, synthesized from oxaloacetate (OAA) and acetyl-CoA, can be supplied by either SbnG or CitZ (Cheung *et al.*, 2012; Sheldon *et al.*, 2014). The SbnCEF synthetases condense the substrate molecules into the final structure (Cheung *et al.*, 2009).

The vast majority of coagulase-positive and coagulase-negative staphylococci (CoNS) encode for the production of SA, with the notable exception of *S. lugdunensis*, which bears a large deletion in the *sfa* locus (Beasley and Heinrichs, 2010; Beasley *et al.*, 2009; Brozyna *et al.*, 2014). In contrast, despite early reports describing SB production by the CoNS (Haag *et al.*, 1994; Lindsay *et al.*, 1994; Meiwes *et al.*, 1990), relatively few CoNS genomes (e.g. *S. arlettae*, *S. equorum*) possess the SB biosynthetic locus, and instead its production seems to be largely limited to *S. aureus* and other coagulase-positive staphylococci. A third, structurally uncharacterized siderophore, named aureochelin, was reported in *S. aureus*, however no further studies have been done to elucidate its structure (Courcol *et al.*, 1997). Deletion of the loci for both SA and SB production is sufficient to abolish *S. aureus* growth in iron-restricted media containing serum or transferrin (Beasley *et al.*, 2009, 2011) suggesting, at the very least, that these are the predominant siderophores produced by *S. aureus*.

Siderophore synthesis in the staphylococci is highly iron-regulated, with consensus sequences for the ferric uptake regulator, Fur, located in the intergenic region between *sfaD* and *sfaABC* (Beasley *et al.*, 2009), as well as upstream of *sbnABCDEFGHI* (Dale *et al.*, 2004a). Accordingly, and as discussed further in 'Fur – global transcriptional regulator of iron homeostasis', siderophore production is derepressed under iron-deplete growth conditions, and aids in the acquisition of iron to fulfil metabolic requirements. Production of SA is also strongly linked to central metabolism, as citrate for incorporation into this siderophore is derived from the TCA cycle (CitZ) (Sheldon *et al.*, 2014). SA production is severely hindered under iron-restriction in glucose-containing media, conditions under which TCA cycle activity are strongly repressed and metabolism favours fermentative metabolism (Collins and Lascelles, 1962; Friedman *et al.*, 2006; Strasters and Winkler, 1963). Given that concentrations of glucose in human serum (approximately 5 mM, up to 11.1 mM in diabetics) are sufficient to repress SA production *in vitro* (American Diabetes Association, 2013; Psychogios *et al.*, 2011), it is conceivable that SA is not the primary siderophore employed in instances of staphylococcal bacteraemia. Conversely, *sbnA-I* encodes all the necessary enzymes for the formation of SB, independent of TCA cycle activity, likely ensuring iron acquisition from serum transferrin transpires in the absence of SA production (Sheldon *et al.*, 2014). Given that the production of SB is limited primarily to the more pathogenic coagulase-positive staphylococci, it may provide a fitness advantage to these bacteria *in vivo*; however the specific host niches wherein SA and SB are expressed have yet to be investigated.

Once secreted, both SA and SB bind Fe^{3+}, either having removed it from transferrin or lactoferrin, or solubilized it from ferric hydroxide precipitates in the extracellular milieu. The Fe–SA and Fe–SB complexes are then transported back into the cell by way of the dedicated uptake systems, HtsABC and SirABC, respectively (Beasley *et al.*, 2009; Dale *et al.*, 2004b; Heinrichs *et al.*, 1999) (Fig. 9.1). These transporters possess the hallmark components of ATP-binding cassette (ABC) superfamily proteins, where HtsA and SirA are membrane-anchored, substrate-binding lipoproteins, and HtsBC and SirBC comprise the associated integral transmembrane proteins required for translocation across the cytoplasmic membrane (Davidson *et al.*, 2008). Energy for ferric-siderophore uptake by both transporters is provided by the genetically unlinked, and promiscuous ATPase, FhuC (Speziali *et al.*,

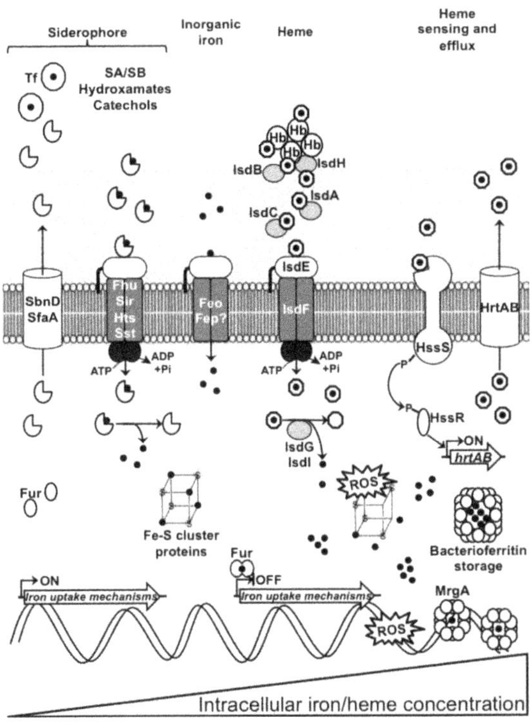

Figure 9.1 Overview of iron homeostasis in *S. aureus*. During iron starvation, high-affinity iron acquisition mechanisms are maximally expressed. As intracellular iron increases, the acquisition systems are repressed and mechanisms are evoked that protect against ROS from iron-dependent Fenton chemistry. More detail for what is depicted is described in the text. Abbreviations: SA, staphyloferrin A; SB, staphyloferrin B; Hb, haemoglobin; Hm, haem; ROS, reactive oxygen species.

2006). Both *htsABC* and *sirABC* are located adjacent to, and are divergently transcribed from, the biosynthetic loci of their cognate siderophores and, like the biosynthetic loci, are demonstrably iron-regulated (Beasley *et al.*, 2009; Dale *et al.*, 2004b).

HtsABC and SirABC bind and transport SA and SB, respectively, with exquisite specificity (Beasley *et al.*, 2009; Grigg *et al.*, 2010a,b). Both HtsA and SirA belong to the cluster A-II (class III) substrate binding proteins (SBP), where members of this group specifically coordinate siderophores in a bi-lobate receptor structure consisting of two α/β domains linked by a central α-helix (Berntsson *et al.*, 2010). The specificity of these SBPs for their siderophores has been attributed to a lack of conservation between SirA and HtsA in the key ionic contacts used in siderophore coordination (Cooper *et al.*, 2014), differential orientation of the siderophore within the binding pocket (the long axis of Fe-SB runs parallel to the protein domain interface of SirA, whereas Fe-SA is oriented perpendicularly in HtsA), and distinct conformational changes induced in the receptors upon substrate binding (Grigg *et al.*, 2010a,b). The latter factor may also contribute to the high-affinity of these receptors for their siderophores, by expanding the protein-substrate contact. Both HtsA and SirA possess affinities for their respective ferric-siderophores in the low nanomolar range, and it has been suggested that the high-affinity, high specificity nature of these SBP allows the staphylococci to outcompete other organisms attempting to usurp Fe-SA and Fe-SB to meet their own iron requirements (Grigg *et al.*, 2010a,b).

As previously discussed, all staphylococci are predicted to synthesize at least one siderophore to facilitate iron acquisition from transferrin and/or lactoferrin. The noteworthy exception to this is *S. lugdunensis*, an atypically pathogenic CoNS capable of causing afflictions ranging from skin and soft tissue infections (SSTIs) to severe endocarditis (Anguera *et al.*, 2005; Böcher *et al.*, 2009). Surprisingly, while unable to synthesize siderophores, *S. lugdunensis* expresses all of the genes necessary for the uptake of SA and SB (Brozyna *et al.*, 2014), as well as hydroxamate and catechol-type siderophores. The ability of *S. lugdunensis* to pirate siderophores from other staphylococci highlights an important mechanism that many bacteria use to compete for limited amounts of iron from within complex microbial communities.

Xenosiderophores

As mentioned above, in addition to utilizing endogenously produced siderophores, many bacteria are also capable of appropriating siderophores produced by other microorganisms, termed xenosiderophores (Fig. 9.1). Although unable to synthesize hydroxamate-type siderophores (Courcol *et al.*, 1997), *S. aureus* has long been reported to scavenge iron using desferrioxamine B, a hydroxamate siderophore produced by *Streptomyces pilosus*, and its mesylate salt, Desferal, a widely used drug for the treatment of iron overload diseases (Brock and Ng, 1983; Heinrichs *et al.*, 1999; Sebulsky *et al.*, 2000). Early investigations revealed that, similar to *Bacillus subtilis*, this attribute was not limited to the uptake of desferrioxamine B, but also included other hydroxamate siderophores including ferrichrome (from the fungus *Ustilago sphaerogena*), coprogen (from *Neurospora crassa*), and aerobactin (produced by some members of the Enterobacteriaceae), as well as catechol siderophores such as enterobactin (from Gram-negative bacteria such as *Escherichia coli*) (Schneider and Hantke, 1993; Sebulsky *et al.*, 2000). To exploit these xenosiderophores as an iron source, the staphylococci require additional uptake systems specific to hydroxamate and/or catechol-type siderophores.

The use of hydroxamate siderophores by the staphylococci has been attributed to an ABC transport system, termed Fhu (for ferric hydroxamate uptake). Conserved components of this iron-regulated transporter include two integral transmembrane proteins, FhuB and FhuG, and the FhuC ATPase (Sebulsky *et al.*, 2000; Speziali *et al.*, 2006). Ferric-hydroxamate binding in *S. aureus* is facilitated by two paralogous and partially redundant SBPs, FhuD1 and FhuD2 (Sebulsky *et al.*, 2003, 2004; Sebulsky and Heinrichs, 2001), where FhuD2 recognizes common hydroxamate features allowing for the accommodation of a wide variety of these siderophores (Podkowa *et al.*, 2014). FhuD2 has also been identified as a staphylococcal virulence factor, with a potential role in dissemination, since its disruption yielded a strain with reduced staphylococcal burden in the blood and kidneys of infected mice (Cheng *et al.*, 2009; Mishra *et al.*, 2012). Recent structural characterization of apo- and holo-FhuD2 (in complex with ferrichrome or ferrioxamine B) revealed that, similar to other cluster A-II SBP (Berntsson *et al.*, 2010), the protein undergoes minimal conformational changes upon siderophore binding, an important facet of vaccine targets where well-ordered epitopes are desired (Mariotti *et al.*, 2013; Podkowa *et al.*, 2014; Sebulsky *et al.*, 2003). Notably, FhuD2 has indeed attracted attention as a potential anti-staphylococcal vaccine target (Mariotti *et al.*, 2013; Mishra *et al.*, 2012), and immunization of mice with FhuD2 conferred resistance to lethal challenge by *S. aureus* in a sepsis model. The protective effects of immunization were attributed not only to opsonophagocytic killing mediated by functional antibodies derived against FhuD2, but also to the inhibition of bacterial proliferation through blocking FhuD2 iron acquisition (Mishra *et al.*, 2012).

The catechol family of siderophores are also recognized by receptors expressed by the staphylococci (Beasley *et al.*, 2011; Miethke and Marahiel, 2007). Moreover, in the presence of serum, catecholamine stress hormones have also been shown to stimulate the growth of bacteria, including members of the CoNS (Freestone *et al.*, 2000, 2008; Lyte *et al.*, 2003; Neal *et al.*, 2001). This phenomenon has been attributed to the ability of these hormones, including adrenaline, noradrenaline, and dopamine, to promote the reduction of transferrin-bound Fe^{3+} to Fe^{2+}, eliciting an exchange of iron from transferrin to catecholamine (Sandrini *et al.*, 2010). The ability of *S. aureus* to utilize Fe-bound catecholamines, as well as catechol siderophores, such as enterobactin, is afforded by the expression of the widely conserved staphylococcal siderphore transport system, SstABCD (Beasley *et al.*, 2011), where SstA and SstB are the transmembrane domains, SstC is an ATPase, and SstD is the catechol/catecholamine-binding lipoprotein (Beasley *et al.*, 2011; Morrissey *et al.*, 2000). SstD binds catechol siderophores and catecholamine hormones with micromolar affinity, in contrast to the much higher affinity observed for HtsA and SirA for SA and SB. The reduced affinity of SstD for its ligands may reflect a sacrifice of ligand affinity for greater ligand diversity. Indeed, the Sst transport system enhances *S. aureus* growth in the presence of serum and catechols or catecholamines, but only in the absence of SA and SB production which, when present, mask the function of SstABCD (Beasley *et al.*, 2011). However, inactivation of *sst* does reduce bacterial burden in murine organs during a systemic model of infection, indicating it does play a role in iron acquisition *in vivo* (Beasley *et al.*, 2011). This suggests that, in conjunction with siderophores, catechols/catecholamines may be at sufficient concentrations to subvert the bacteriostatic effects of transferrin and/or lactoferrin and allow for proliferation and dissemination to occur within the host; this may be particularly true in the gastrointestinal tract, where catecholamine stress hormones, lactoferrin, and bacteria all co-localize (Freestone *et al.*, 2002; Markel *et al.*, 2007). Conditions that enhance catecholamine hormone levels, such as stress, neuron damage, or therapeutic treatment (Eisenhofer *et al.*, 1996; Freestone *et al.*, 2002, 2008), could further enhance the risk of staphylococcal infection.

Acquisition of iron from haem/haemoproteins

Despite possessing numerous siderophore-mediated systems for the acquisition of iron from serum transferrin, a preferred iron source for *S. aureus* is haem (Skaar *et al.*, 2004a). Haem-iron represents the major iron reservoir within the host, accounting for approximately 75% of the total iron pool (Stojiljkovic and Perkins-Balding, 2002). The vast majority of haem-iron is bound within haemoglobin in circulating erythrocytes (Griffiths, 1999), and is released upon lysis and subsequent haemoglobin degradation. Owing to the potential for haem and haemoglobin derivatives to generate ROS through Fenton chemistry (Everse and Hsia, 1997), both are rapidly sequestered in the host by haemopexin and haptoglobin, respectively (Evans *et al.*, 1999; Everse and Hsia, 1997; Gutteridge, 1987; Gutteridge and Smith, 1988). To access the lucrative host haem-iron pool, the staphylococci express not only haemolysins to facilitate erythrocyte lysis (Wiseman, 1975), but also at least one high-affinity acquisition system to capture free haem-iron.

Isd: a haem/haemoprotein acquisition system

The iron-regulated surface determinant (Isd) pathway represents the principal means by which haem-iron is acquired by *S. aureus* (Fig. 9.1). In this species, the Isd system is composed of nine proteins directly involved in haem-iron acquisition, along with sortase B (SrtB) (Mazmanian *et al.*, 2002, 2003). In general, the pathway involves haem-binding proteins IsdA, IsdB, and IsdH (also known as HarA), which are C-terminally anchored to peptidoglycan by sortase A (SrtA) (Marraffini *et al.*, 2004). IsdC, a fourth haem-binding protein, is secured to the cell wall by Sortase B (SrtB) (Marraffini and Schneewind, 2005; Mazmanian *et al.*, 2002, 2003; Zong *et al.*, 2004). It is believed that the architectural arrangement of these four proteins effectively serves as a funnel to shuttle haem, extracted from haemoglobin at the cell surface, through the thick cell wall structure. IsdE and IsdF are the substrate-binding lipoprotein and permease components of an ABC transporter (Grigg *et al.*, 2007a; Pluym *et al.*, 2007), respectively, which facilitate the translocation of haem across the membrane and into the cell, energized by an as yet unidentified ATPase. Once inside the cell, haem is degraded by haem monooxygenases, IsdG and IsdI, which serve to liberate iron from the porphyrin ring (Haley *et al.*, 2011; Skaar *et al.*, 2004b).

The recognition of haem and haemoproteins by IsdA, IsdB, IsdC and IsdH is facilitated by near iron-transport (NEAT) domains. So termed for their presence in proteins whose genes are found in close proximity to genes encoding iron transporters, these ~125 amino acid domains are widely distributed amongst Gram-positive bacteria (Andrade *et al.*, 2002). The NEAT domain is an eight-stranded immunoglobulin-like β-sandwich fold that is responsible for coordination of haem/haemoproteins, and are found in variable number in Isd proteins (e.g. IsdA has one, IsdB has two, IsdC has one, and IsdH has three) (Gaudin *et al.*, 2011; Grigg *et al.*, 2007b; Pilpa *et al.*, 2006; Sharp *et al.*, 2007; Villareal *et al.*, 2008; Watanabe *et al.*, 2008).

The initial recognition and binding of haemoglobin is mediated by IsdB and IsdH (Dryla *et al.*, 2007; Pishchany *et al.*, 2013; Torres *et al.*, 2006). The N-terminal NEAT domains of these proteins (IsdB-N1, and IsdH-N1 and IsdH-N2) are specific to haemoglobin (or haemoglobin-haptoglobin) (Krishna Kumar *et al.*, 2011; Pilpa *et al.*, 2006), whereas the C-terminal NEAT domains (IsdB-N2 and IsdH-N3) bind haem (Gaudin *et al.*, 2011; Pilpa *et al.*, 2009). A conserved α-helical linker unit between the N-terminal haemoglobin-binding NEAT domain (IsdB-N1 or IsdH-N2) and the C-terminal haem-binding NEAT domain (IsdB-N2 or IsdH-N3) has been shown to be essential for both the binding of haemoglobin and haem extraction; the linker likely induces steric strain on haemoglobin, forcing its dissociation and subsequent release of haem (Bowden *et al.*, 2014; Dickson *et al.*, 2014; Fonner *et al.*, 2014; Spirig *et al.*, 2013).

Following extraction, haem is transferred rapidly from IsdB-N2 or IsdH-N3 (or bidirectionally between the two) to IsdA or IsdC (Muryoi *et al.*, 2008). Haem bound to IsdA is transferred unidirectionally to IsdC (Liu *et al.*, 2008), which effectively serves as a central conduit for haem travelling from the cell surface to the cell membrane. Each Isd protein possesses a higher affinity for haem than the preceding protein in the pathway, driving the energy-independent transfer of haem from IsdB/IsdH to IsdA to IsdC (Moriwaki *et al.*, 2013; Tiedemann *et al.*, 2012). From IsdC, haem is transferred to IsdE, a bi-lobate SBP in the same family as the aforementioned HtsA and SirA proteins. IsdE coordinates haem in a deep groove formed at the interface between the two domains. Extensive protein-substrate contacts between IsdE and haem, coupled with minimal substrate solvent exposure (~19% in IsdE vs. ~35% in NEAT domains) likely affords IsdE with greater specificity and tighter binding to haem than IsdC (Grigg *et al.*, 2007a; Pluym *et al.*, 2007). From IsdE, haem is purportedly transported into the cytoplasm by the IsdF permease (Grigg *et al.*, 2010c; Mazmanian *et al.*, 2003). While no ATPase has been identified to drive this transport, it has been proposed that FhuC, as described, functions in this capacity as well (Beasley *et al.*, 2009; Speziali *et al.*, 2006). Notably, disruption of *isdE* (which presumably should abolish haem transport through the cell membrane), reduces but does not eliminate growth on haem as a sole iron source (Grigg *et al.*, 2007a). More recent studies have shown that the Isd system appears to have maximal utility at low nanomolar concentrations of haem/haemoglobin, suggesting it serves as a high-affinity acquisition system; indeed, when haem is applied in excess, growth-defect phenotypes associated with *isd* mutations are effectively lost (Pishchany *et al.*, 2013). Together these results suggest that an alternate, and as of yet unidentified, lower affinity mechanism of haem-iron acquisition likely exists in the staphylococci.

The extensive characterization of the Isd system, as described above has, by and large, been performed in *S. aureus*. However, recently its utility in haem-iron acquisition was also recognized in *S. lugdunensis*, a member of the CoNS (Brozyna *et al.*, 2014; Haley *et al.*, 2011; Zapotoczna *et al.*, 2012). Additionally, the relevance of Isd in haem-iron acquisition within the host has been intimated by the upregulation of *isd* genes in human blood and serum (Malachowa *et al.*, 2011), as well as specificity of the Isd proteins for human haemoglobin (Pishchany and Skaar, 2012; Pishchany *et al.*, 2010, 2013). Indeed, several *isd* genes have been implicated in staphylococcal pathogenesis in various murine models of infection where deletion of *isdA*, *isdB*, *isdC*, *isdG*, *isdH*, or *isdI* alone is sufficient to attenuate bacterial burden and/or overall virulence (Cheng *et al.*, 2009; Pishchany *et al.*, 2013; Reniere and Skaar, 2008; Torres *et al.*, 2006; Visai *et al.*, 2009). Given their cell surface exposure, role in virulence, and the detection of anti-IsdH and IsdA antibodies in individuals suffering from staphylococcal infection (Clarke *et al.*, 2006), it is not surprising that IsdA and IsdB, in particular, have been targeted as potential vaccine antigens (Kim *et al.*, 2010; Kuklin *et al.*, 2006; Stranger-Jones *et al.*, 2006).

FepABC: a putative haem acquisition system

Given that disruption to the Isd system does not abolish haem-iron acquisition (Hurd *et al.*, 2012; Wright and Nair, 2012), it is assumed that there exists another mechanism to incorporate haem-iron into the cell. Preliminary investigations identified a Fur-regulated locus bearing homology to the *efeUOB* haem-iron acquisition system of *E. coli* (Letoffe *et al.*, 2006, 2009), as potentially being involved in the acquisition of inorganic and/or haem iron. Termed *fepABC* (for Fe-dependent peroxidase), this operon, which is distributed sparsely amongst both coagulase-positive and negative staphylococci, was initially characterized in an effort to identify substrates of the twin-arginine translocation (Tat) pathway and is transcribed contiguous to *tatAC* (Biswas *et al.*, 2009). FepB, a homologue of EfeB, and putative DyP family peroxidase, was isolated as the sole substrate of Tat in *S. aureus* and *S. carnosus* (Biswas *et al.*, 2009; Meissner *et al.*, 2007). Dye-decolorizing (DyP)-type peroxidases have diverse functions, however, insofar as iron is concerned, and *E. coli* DyP peroxidases YfeX and EfeB function in the release of iron from haem through deferrochelatase activity (Letoffe *et al.*, 2009). The ability of YfeX and EfeB to remove iron from haem, while leaving the porphyrin ring intact, is contentious and a competing hypothesis has been suggested, where these proteins act in the oxidation of porphyrinogens to porphyrins (Dailey *et al.*, 2011). Should FepB possess deferrochelatase activity, it suggests a means by which the staphylococci may acquire iron from haem in the absence of a dedicated haem transporter.

In addition to FepB, *fepABC* also encodes a putative iron-binding lipoprotein, FepA, which bears homology to EfeO and other members of the imelysin-like protein superfamily. Members of this family possess a conserved sequence (GxHxxE), which has been implicated in the binding of divalent metal cations, specifically ferrous iron (Rajasekaran *et al.*, 2010a, b; Xu *et al.*, 2011), and are almost always located in close proximity to a DyP-type peroxidase. Additionally, the *fep* operon encodes for FepC, an integral transmembrane protein and EfeU homologue, both of which share ~25% sequence similarity to the ferric iron permease, Ftrp1, from *Saccharomyces cerevisiae* (Cao *et al.*, 2007; Kwok *et al.*, 2006; Turlin *et al.*, 2013). In *E. coli*, EfeU is required for ferrous iron uptake at low pH (Cao *et al.*, 2007; Grosse *et al.*, 2006). Together the similarities between Fep proteins and other iron transporters, coupled with the finding that *fep-tat* mutants are debilitated for radioactive iron uptake (Biswas *et al.*, 2009), support a role for *fepABC* in staphylococcal iron acquisition. Further, FepB was recently demonstrated to bind both haem and the haem precursor protoporphyrin IX (PPIX), and both FepA and FepB together were able to restore haem iron utilization when heterologously expressed in an *E. coli* mutant deficient for this ability (Turlin *et al.*, 2013). A model has been proposed, analogous to the activity of *E. coli* EfeUOB, whereby iron is extracted from haem through FepB deferrochelatase activity, captured by FepA, and transported into the cell via FepC (Turlin *et al.*, 2013). Alternatively, it has been shown in *B. subtilis* that EfeUOB is responsible for both ferric and ferrous iron acquisition, where EfeB utilizes its peroxidase activity to oxidize Fe^{2+} to Fe^{3+} prior to binding to EfeO (Miethke *et al.*, 2013). Although not yet proven experimentally, it is likely *fepABC* plays a role in the acquisition of iron in the staphylococci, and may potentially be involved in both haem-iron extraction and inorganic iron uptake, however the preferred iron substrate for this system has yet to be determined.

Acquisition of inorganic iron

The acquisition of sequestered/complexed iron, as detailed extensively above, has attracted markedly greater research interest than the means by which the staphylococci acquire free inorganic iron. In addition to the potential role of FepABC in ferric and/or ferrous iron uptake, two additional, yet functionally uncharacterized, transporters have been implicated in the uptake of Fe^{2+}: the preemptively termed staphylococcal iron transporter (SitABC) of *S. epidermidis* (Cockayne *et al.*, 1998), and the ubiquitous ferrous iron transporter, FeoABC. Although SitABC is transcriptionally repressed in the presence of both Fe^{2+} and Mn^{2+} by staphylococcal iron regulator repressor (SirR) (Hill *et al.*, 1998), the actual substrate of the transporter is unknown and its homologue in *S. aureus* is dedicated to manganese uptake. Accordingly, further discussion of SitABC and SirR can be found below in a discussion of manganese uptake and its regulation.

First described in *E. coli*, FeoABC is the archetype of ferrous iron uptake systems in bacteria (Hantke, 1987, 2003; Kammler *et al.*, 1993). The Fur-regulated *feo* operon includes a cytoplasmic protein of unknown function, FeoA, and the transmembrane transporter of Fe^{2+}, FeoB. FeoA is proposed to interact with the cytoplasmic domain of FeoB, and indeed is required for its maximal function in Fe^{2+} uptake (Kim *et al.*, 2012; Lau *et al.*, 2013). In γ-proteobacteria, this operon also typically encodes FeoC, a putative transcriptional repressor bearing a winged helix–turn–helix motif, characteristic of DNA-binding proteins, as well as a proposed iron–sulfur cluster [Fe–S] binding site (Cartron *et al.*, 2006; Hung *et al.*, 2012). Like FeoA, the function of FeoC has remained elusive, but it is proposed to function as a [Fe–S]-dependent transcriptional regulator of *feoABC*. In *E. coli*, this system is also regulated by Fnr, an anaerobically induced transcriptional activator (Cartron *et al.*, 2006). Notably, while Fe^{3+} is insoluble under physiological conditions, low pH and/or anaerobiosis facilitate reduction of Fe^{3+} to Fe^{2+} and, indeed, such conditions have been shown to induce expression of *feoABC* in pathogens such as *Shigella flexneri* and *E. coli* O157:H7 (Boulette and Payne, 2007; Cao *et al.*, 2007). The *feoAB* operon has been identified in the staphylococci, and similarities in domain architecture between *E. coli* FeoB and the staphylococcal homologue suggest it likely functions in Fe^{2+} acquisition. Further, while *feoC* homologues have not been conclusively identified in the staphylococci, small proteins of similar structure do exist within the staphylococcal *feo* gene cluster (Cartron *et al.*, 2006). These commonalities, coupled with the known upregulation of *feoB* during iron starvation (Ledala *et al.*, 2014; Lin *et al.*, 2012), suggest *feoAB* (and *feoC*) may function in Fe^{2+} acquisition in the staphylococci. Indeed, further investigation is also required to determine if *feoAB* plays a role in staphylococcal virulence, particularly in anoxic and/or acidic niches within the host where Fe^{2+} may preside.

The fate of intracellular iron

Mechanisms of intracellular iron release

Iron in complex with siderophores or haem must be liberated in order to participate in biological processes within the cell. While the mechanisms of siderophore-iron dissociation have not been well characterized, they are usually contingent upon the reduction of iron from Fe^{3+} to Fe^{2+}, given the reduced affinity of siderophores for the latter. Two general models have been proposed thus far for iron release from siderophores: (i) reduction of iron by ferrisiderophore reductases with no modification to the siderophore, as shown for pyoverdine (Hallé and Meyer, 1992; Schalk *et al.*, 2002), and (ii) reduction of iron concurrent with modification or degradation of the siderophore, as with ferrichrome, enterobactin, and bacillibactin (Brickman and McIntosh, 1992; Hannauer *et al.*, 2010; Hartmann and Braun, 1980; Matzanke *et al.*, 2004; Miethke *et al.*, 2006, 2011). Dissociation without siderophore destruction is presumed to be the more prevalent mechanism, given reductive iron release does not necessitate highly specialized enzymes, and the siderophore is available for reutilization, thus presumably reducing carbon and energy expenditures (Miethke and Marahiel, 2007). The means by which the staphylococci dissociate iron from siderophores have not yet been identified, however the staphylococcal genome encodes for upwards of 80 putative reductases, six of which are upregulated during growth in blood and serum versus growth in a nutrient-rich laboratory medium (Malachowa *et al.*, 2011).

Upon uptake, haem faces two potential intracellular fates: (i) utilization as a nutrient iron source which, in a manner conceptually similar to siderophores, necessitates iron release from the haem–iron complex, or (ii) direct incorporation into haemoproteins. IsdG, and its paralogue, IsdI, represent the haem-degrading constituents of the Isd pathway, and are founding members of a novel family of haem oxygenases (HOs) (Haley *et al.*, 2011; Lee *et al.*, 2008; Skaar *et al.*, 2004b; Wu *et al.*, 2004). While canonical HOs typically degrade haem to biliverdin with liberation of iron and carbon monoxide, the reaction of IsdG and IsdI with haem yields alternative tetrapyrroles called the staphylobilins, as well as formaldehyde and iron (Lee *et al.*, 2008; Matsui *et al.*, 2013; Reniere *et al.*, 2010; Wilks and Ikeda-Saito, 2014). The formation of staphylobilin and formaldehyde is mediated by a unique means of haem binding to these enzymes, which results in substantial distortion to the planarity of the haem molecule, an effect referred to as 'haem ruffling' (Lee *et al.*, 2008; Matsui *et al.*, 2013). Ruffling is predicted to enhance the reactivity of IsdG and IsdI with haem and alter their regioselectivity relative to traditional HOs, thus preventing the formation of carbon monoxide, a potentially toxic metabolite to staphylococci (Nobre *et al.*, 2007, 2012). It should be noted that sustained activity of IsdG and IsdI requires a dedicated source of electrons and molecular oxygen. Recently, the iron utilization oxidoreductase, IruO, was identified as the terminal electron donor for IsdG and IsdI (Loutet *et al.*, 2013). IruO promoted faster haem degradation by IsdI than IsdG, suggesting that another reductase may be involved in haem degradation by the latter. The *in vivo* role of IruO in haem-iron utilization has yet to be determined. Also to be determined is the intracellular fate of the haem degradation product, staphylobilin, as well as the fate of intact exogenously acquired haem versus endogenously synthesized haem (see haem biosynthesis section below).

Mechanisms of intracellular iron storage and utilization

Once iron is internalized, bacteria require means to safely house or compartmentalize potentially toxic Fe^{2+}, yet keep it bioavailable within the cell. Bacteria possess three types of intracellular iron storage molecules: (i) The ubiquitous ferritins (Ftns), found in eukaryotic and prokaryotic cells alike, (ii) the haem-containing bacterioferritins (Bfrs), found only in eubacteria, and (iii) small Dps proteins (DNA protection during starvation), found in the prokaryotes. Although these proteins are evolutionarily distinct, they bear similar structural and functional characteristics. Both Bfrs and Ftns are composed of 24 identical (or highly similar) protein subunits, whereas Dps proteins are comprised of 12. In both cases, the protein subunits form a nanocage, a spherical protein shell about a hollow central cavity which acts as an iron storage reservoir (Andrews *et al.*, 2003). Bacterial Ftns and Bfrs can accommodate at least 1800 iron atoms (Smith, 2004), whereas Dps can house approximately 500 (Hudson *et al.*, 1993). All three iron storage proteins take up soluble ferrous ion pairs, which are oxidized to dinuclear ferric iron by a ferroxidase-like domain and subsequently stored within the core in mineralized form. When iron concentrations decrease, Fe^{3+} is reduced back to Fe^{2+}, assisted in part by a reductant (e.g. a ferredoxin or flavoprotein), and liberated iron can then be utilized to fulfil metabolic demands (Yasmin *et al.*, 2011). Notably, the Bfrs contain 12 haem molecules, each located between adjacent pairs of subunits, predicted to be involved in the reduction and subsequent release of iron from within the nanocage (Andrews *et al.*, 1989).

While bacterioferritins have yet to be identified in the staphylococci, ferritins have been characterized in both *S. aureus* and *S. epidermidis* (FtnA and SefA, respectively) and both encode for a Dps-like protein (MrgA), which protects DNA from oxidative damage (Morrissey *et al.*, 2004) (Fig. 9.1). The protection Dps proteins afford is twofold: firstly, their function in iron sequestration helps to reduce further generation of ROS, and, secondly, and more importantly, Dps proteins bind DNA non-specifically, forming highly stable crystals in which DNA is sequestered and shielded from various assaults (Wolf *et al.*, 1999). Unsurprisingly, both the staphylococcal ferritins and MrgA are regulated by PerR and Fur (Morrissey *et al.*, 2004), indicative of their overlapping roles in iron homeostasis and the oxidative stress response.

Unlike the ferritins and Dps proteins, which function in iron storage and protection from oxidative stress, iron–sulfur-containing proteins represent the major functional iron-containing proteins within the cell. Characterized by the presence of iron–sulfur clusters ([Fe-S]), which are thought to be amongst the most ancient and ubiquitous of inorganic cofactors, [Fe-S]-proteins fulfil diverse roles, functioning as catalysts or redox sensors in an array of metabolic processes including DNA replication, gene regulation, respiration, central metabolism, and various adaptive stress responses (Johnson *et al.*, 2005; Roche *et al.*, 2013). Two main types of [Fe–S] exist, [2Fe–2S] consisting of two iron atoms bridged by two sulfide ions and adopting a rhombic conformation, and [4Fe–4S] consisting of four iron atoms and four sulfide ions at alternating vertices of a cubane-type conformation. Both [Fe–S] types are structurally plastic and can contain either Fe^{2+} or Fe^{3+}, contributing to their versatility as enzymatic cofactors.

Three pathways exist for the synthesis of [Fe–S] in the prokaryotes: ISC (iron–sulfur cluster), SUF (sulfur assimilation), and NIF (nitrogen fixation), the latter of which is responsible solely for the maturation of nitrogenase. Generally, and as is true of the staphylococci, Gram-positive bacteria possess only the SUF biosynthetic pathway. Both ISC and SUF follow a similar model for [Fe–S] assembly, although a detailed description is outside the scope of this chapter (for a review see Roche *et al.*, 2013). In brief, [Fe–S] formation consists of three steps, (i) liberation of elemental sulfur from a cysteine desulfurase, (ii) assembly of the Fe–S cluster on a protein scaffold, and (iii) incorporation of the formed [Fe–S] into the target apo-protein. In the *Firmicutes*, [Fe–S] assembly is performed by gene products of the well-conserved *sufCDSUB* operon, although mechanistic data on [Fe–S] assembly in these bacteria is incomplete. In a proposed model, SufS serves as the cysteine desulfurase and is activated by SufU to acquire sulfur from cysteine. SufS transfers sulfur to SufU, which also serves as the scaffold protein upon which iron and sulfur are combined, using energy derived from the SufC ATPase. SufB and SufD are proposed to contribute to the protein scaffold, where SufD may also assist in the acquisition and incorporation of ferrous iron; however, these activities have yet to be proven *in vitro* or *in vivo*. Lastly, SufA, encoded elsewhere on the chromosome, delivers the assembled [Fe-S] to the apo-protein target (Riboldi *et al.*, 2013; Roche *et al.*, 2013).

The regulation of [Fe–S] assembly provides compelling evidence that the SUF biosynthetic pathway is necessary for ensuring iron homeostasis and resistance to oxidative stress. Indeed, in *S. aureus*, oxygen species (e.g. O_2, H_2O_2, NO) are sensed through a [2Fe–2S]-containing two-component regulatory system referred to as AirRS (anaerobic iron–sulfur cluster-containing redox sensor regulator) that regulates approximately 355 genes under anoxic conditions, including a number involved in response to oxidative stress (Sun *et al.*, 2012). Notably, [2Fe–2S] in AirS can be reconstituted *in vitro* by a cysteine disulfurase bearing homology to IscS (SufS). In *E. coli* SUF transcription is regulated by Fur, whereby SUF is expressed under low iron conditions, by the global transcriptional regulator OxyR (a PerR analogue) in response to H_2O_2, and through auto-regulation by sensing the [Fe–S] status of the cell through the transcriptional regulator IscR (Riboldi *et al.*, 2013; Roche *et al.*, 2013). While the regulation of [Fe–S] synthesis in Gram-positive bacteria has not yet been investigated, expression of *sufU* is known to be essential to maintaining viability of *B. subtilis* in minimal media and is required for proper activity of [Fe–S]-containing enzymes (Albrecht *et al.*, 2010), suggesting that this pathway is essential to providing resistance to adverse environmental conditions.

Regulating iron homeostasis

Fur: global transcriptional regulator of iron homeostasis

The essentiality of iron as a micronutrient, and its propensity to facilitate Fenton chemistry, necessitates strict control over intracellular iron concentrations. Iron deprivation is a key sensory trigger in many bacteria, where the response is mediated by the ferric uptake regulator, Fur (Escolar *et al.*, 1999). Functioning largely as a transcriptional repressor, this homodimeric metalloprotein directly controls expression of iron-responsive genes bearing the 19-bp inverted repeat consensus sequence 'GATAATGATAATCATTATC', known as the Fur box, within their promoter/operator regions (Baichoo and Helmann, 2002; Escolar *et al.*, 1999; Horsburgh *et al.*, 2001a; de Lorenzo *et al.*, 1987). Under iron-replete conditions, association of Fe^{2+} with each subunit of the Fur dimer yields a competent DNA-binding complex, which then engages Fur box sequences. The Fur–Fe^{2+} complex blocks accessibility of the promoter to RNA polymerase, thereby inhibiting transcription. Under iron-deplete conditions, Fe^{2+} is no longer available to associate with Fur, resulting in its dissociation from Fur boxes and thus alleviating transcriptional repression (Fig. 9.1).

In the staphylococci, Fur is known or hypothesized to regulate the transcription of all dedicated iron acquisition systems discussed above (Torres *et al.*, 2010; Xiong *et al.*, 2000). Additionally, Fur acts indirectly on the expression of select virulence factors and immunomodulatory factors, through control of other regulatory proteins such as CodY, RsbU, and Sae, or small RNAs (described below) (Johnson *et al.*, 2011; Torres *et al.*, 2010). While an in-depth discussion of these complex regulatory networks is outside the scope of this chapter, it is evident that Fur plays an important role in the pathophysiology of the staphylococci. Reciprocal expression of iron uptake genes and immune evasion factors suggest the staphylococci prioritize iron acquisition over escape from immune cells under iron restriction and, indeed, *fur* mutants are more susceptible to neutrophil-mediated killing. Fur does, however, coordinate up-regulation of secreted toxins under iron restriction, specifically cytolysins and haemolysins, allowing for access to host-sequestered iron, and therefore is an important facet of iron homeostasis *in vivo* (Torres *et al.*, 2010).

Iron homeostasis is also closely linked to the oxidative stress response, whereby *fur* itself is controlled, in part, by the peroxide stress regulator, PerR (Horsburgh *et al.*, 2001a,b, 2002a). Briefly, under conditions of oxidative stress, PerR helps to mitigate production of iron-induced ROS by derepressing *fur*, allowing Fur to repress iron acquisition. Fur is also a positive regulator of catalase expression (*katA*), which further helps to allay the generation of ROS by catalysing the decomposition of H_2O_2 to H_2O and O_2 when iron is in excess (Horsburgh *et al.*, 2001a). The means by which Fur positively regulates genes in the staphylococci is unknown but, as with *E. coli* and *B. subtilis*, is likely due to the action of small non-coding RNAs (sRNA), that are under negative regulation by Fur (Gaballa *et al.*, 2008; Massé and Gottesman, 2002). While not yet demonstrated for the staphylococci, it is likely that sRNAs are involved both in the regulation of *katA*, as well as the iron-sparing response (see below). The multidimensional nature of the regulation of iron uptake is necessary both to ensure adequate acquisition for metabolic purposes, while at the same time, eschewing the production of toxic free radicals.

The 'iron-sparing response': metabolic rearrangement during iron-starvation

Fur fulfils two essential roles in iron homeostasis; in addition to derepressing iron acquisition systems to facilitate uptake during starvation, Fur also coordinates a metabolic rearrangement that helps to reduce cellular demands for iron, referred to as the 'iron-sparing response'. Characterized in both *S. aureus* and *B. subtilis*, the iron-sparing response suppresses expression of 'low priority' iron-containing enzymes, thus reserving iron for indispensable metabolic pathways (Friedman *et al.*, 2006; Gaballa *et al.*, 2008; Smaldone *et al.*, 2012a). At a mechanistic level, sRNA coordinates the iron-sparing response. In *B. subtilis*, the sRNA FsrA, aided by three small Fur basic proteins (FbpA-C, proposed to act as RNA chaperones), post-transcriptionally represses expression of some non-essential iron-containing enzymes (Gaballa *et al.*, 2008; Smaldone *et al.*, 2012a,b). Most notably, TCA cycle activity is significantly reduced under iron restriction (e.g. aconitase and succinate dehydrogenase are [Fe–S]-containing enzymes and major targets of FsrA in *B. subtilis*), whereas the expression of glycolytic enzymes is increased. The accumulation and subsequent fermentation of glycolysis-derived pyruvate is proposed to facilitate two additional activities beneficial to growth under iron starvation. Firstly, it allows for the production of SB precursors (OAA and acetyl-CoA) in the absence of TCA cycle activity and, secondly, the accumulation of acidic end-products within the vicinity of invading staphylococci may help liberate iron from transferrin (Friedman *et al.*, 2006; Sheldon *et al.*, 2014). In summary, the actions of the iron-sparing response allow bacteria to prioritize iron utilization within the cell, and may further promote iron acquisition during starvation.

Haem synthesis and small colony variants

In addition to possessing a plethora of iron uptake systems, the staphylococci are also capable of synthesizing haem endogenously through the coordinated expression of *hemAXCDBL*, *hemEHY*, and *hemN* (Johansson and Hederstedt, 1999; Kafala and Sasarman, 1997). While these genes are largely uncharacterized in the staphylococci, early steps of haem synthesis from generation of the porphyrin precursor δ-aminolaevulinic acid (ALA) to the cyclic haem precursor, uroporphyrinogen III, have been attributed to expression of *hemCDBL* in *S. aureus* (Kafala and Sasarman, 1994, 1997). Notably, *S. aureus* mutants unable to synthesize haem (e.g. *hemB* mutants) are known to display atypical morphology, referred to as small colony variants (SCV), enhanced resistance to antibiotics, and unusual pervasiveness within the host (Proctor *et al.*, 2006). The phenotype of SCV mutants has been attributed to their inability to respire, as haem is an essential cofactor in cytochromes of the electron transport chain.

The last step of haem synthesis is insertion of iron into the tetrapyrrole ring, and thus it is likely that haem synthesis is activated when iron is not limiting. As previously described, the transport of exogenous haem is favoured when iron concentrations are low, and is likely incorporated directly into haem-containing proteins, as well as degraded and utilized as a source of iron.

Sensing and avoiding haem toxicity

High concentrations of haem are toxic to the bacterial cell and, as such, mechanisms of haem acquisition, synthesis, and catabolism must be tightly balanced. The exact mechanism of haem toxicity is unknown, and likely to be multifaceted. The most plausible explanation, to date, is that haem itself is toxic due to oxidative damage inflicted on DNA through the generation of ROS (Anzaldi and Skaar, 2010; Wakeman *et al.*, 2012, 2014); however, this hypothesis has yet to be fully borne out experimentally. Three main mechanisms exist to subvert haem toxicity in the staphylococci. First, Fur mediates repression of haem uptake systems when iron is present in excess, as previously discussed, although rapid uptake of haem by Isd and other haem transporters may still induce toxicity. Second, IsdG and IsdI promote concurrent haem degradation and iron release, where Ftn or Bfr may then sequester free iron. Notably, IsdG and IsdI are differentially regulated by iron and haem, where IsdG protein stability is enhanced in the presence of haem, suggesting an ability of these enzymes to fine-tune haem homeostasis in different environments (Reniere and Skaar, 2008). Third, the staphylococci express a two-component haem sensing system (HssRS), capable of detecting and exporting excess haem through a haem-regulated transporter (HrtAB) (Fig. 9.1) (Juárez-Verdayes *et al.*, 2012; Stauff *et al.*, 2007; Torres *et al.*, 2007).

Although the exact mechanism of haem recognition by HssS is unknown, exposure of this membrane-localized sensor histidine kinase to haem stimulates its autophosphorylation. Through a phosphorylation cascade, HssS activates the response regulator, HssR, which in its phosphorylated form is capable of binding to a direct repeat within the promoter of *hrtAB*, inducing expression of the encoded haem efflux pump (HrtAB).

Through the coordinated efforts of HssRS-HrtAB, excess haem is then exported into the extracellular environment, thereby reducing potential toxicity. As with free iron, the maintenance of haem homeostasis is essential to maintaining staphylococcal viability; hence, it is not surprising that the *hssRS-hrtAB* system is found widely distributed amongst both pathogenic and commensal *Staphylococcus* spp., in addition to other Gram-positive genera, including *Listeria, Corynebacterium, Bacillus* and *Enterococcus*.

Manganese

Biological function of manganese in the staphylococci

Manganese is an essential micronutrient for eukaryotes and prokaryotes alike, and is increasingly being recognized as centrally important in host–pathogen interactions. In bacteria, Mn metalloenzymes fulfil diverse roles in sugar and amino acid metabolism, signal transduction and the stringent response, photosynthesis and, perhaps most importantly, resistance to oxidative stress (Christianson, 1997; Horsburgh *et al.*, 2002b; Jakubovics and Jenkinson, 2001; Yocum and Pecoraro, 1999) (Table 9.1). As discussed previously with respect to Fenton chemistry, normal growth under aerobiosis, and the threat of oxidative killing by immune cells within the host, necessitates means by which invading pathogens can detoxify ROS generated through these processes.

S. aureus counters the deleterious effects of superoxide radicals through the expression of two manganese-dependent superoxide dismutases (Mn-SOD), SodA and SodM (Clements *et al.*, 1999; Valderas *et al.*, 2002). Both Mn-SODs catalyse the conversion of $\cdot O_2^-$ to H_2O_2, which is subsequently degraded further to hydrogen and water by catalase. Notably, the CoNS encode only one Mn-SOD (SodA), suggesting that SodM may have a unique role in *S. aureus* pathogenesis (Ballal and Manna, 2009; Valderas *et al.*, 2002). Additionally, there is evidence to suggest that a manganese-dependent, non-SOD mechanism of protection against ROS exists in a number of bacterial species (Daly *et al.*, 2010; Inaoka *et al.*, 1999; Al-Maghrebi *et al.*, 2002; Tseng *et al.*, 2001). First characterized in the SOD-lacking *Lactobacillus plantarum*, low molecular weight, intracellular Mn^{2+}–metabolite complexes and Mn^{2+}-salts have been shown to scavenge both superoxide and hydrogen peroxide, through mechanisms that are not yet fully understood (Archibald and Fridovich, 1981, 1982a,b; Coassin *et al.*, 1992). A similar non-enzymatic role for manganese-conferred protection against ROS has also been implicated in *S. aureus* (Horsburgh *et al.*, 2002a; Kehl-Fie *et al.*, 2011). In an effort to sensitize invading pathogens to oxidative stress, the host counters these bacterial defences with manganese sequestration strategies (discussed below). Like with iron, the staphylococci require means to acquire manganese within the host in order to maintain viability and establish an infection.

MntABC and MntH: manganese acquisition systems

Unlike siderophores for iron, there exists no known low molecular weight, high-affinity, microbial chelator of manganese. Instead, cellular Mn^{2+} requirements are fulfilled by direct import systems, two classes of which have been described in bacteria. MntABC, so named for Mn transporter, was first identified in the cyanobacterium *Synechocystis* sp. PCC 6803, and belongs to the ABC transporter superfamily (Bartsevich and Pakrasi, 1995). Accordingly, *mntABC* encodes an ATPase (MntA), a membrane permease (MntB), and a high-affinity, membrane-anchored SBP (MntC) (Davidson *et al.*, 2008; Hutchings *et al.*, 2009). A staphylococcal homologue of *mntABC* was initially characterized in *S. epidermidis*, as the Mn^{2+}- and Fe^{2+}-repressible operon, *sitABC* (staphylococcal iron transporter) (Cockayne *et al.*, 1998; Hill *et al.*, 1998). Whilst the closest structural homologues of SitC are known divalent metal binding proteins (Sheldon and Heinrichs, 2012), the actual substrate of SitABC in *S. epidermidis* has not been experimentally proven. Speculation that SitABC transports divalent metals, such as Mn^{2+} and Fe^{2+}, is supported by its known regulation by these metals (Cockayne *et al.*, 1998; Hill *et al.*, 1998), *in vivo* expression under iron-restriction (Modun *et al.*, 1992; Smith *et al.*, 1991; Wilcox *et al.*, 1991; Williams *et al.*, 1988), and similarity to known staphylococcal Mn^{2+} transporters (Horsburgh *et al.*, 2002a). Notably, anti-MntC monoclonal antibodies raised against *S. aureus* MntC (described below) elicit cross-reactivity against *S. epidermidis*, suggesting, at the very least, that these cell surface proteins are antigenically similar, and indeed share 77% pairwise protein identity (Anderson *et al.*, 2012a). High conservation of MntC in the staphylococci, coupled with its early expression during infection and role in protection against ROS, has prompted efforts towards its incorporation in a multivalent vaccine against staphylococcal infection (Anderson *et al.*, 2012a,b; Gribenko *et al.*, 2013; Handke *et al.*, 2013).

Unlike *sitABC* in *S. epidermidis*, its orthologue in *S. aureus* has been well-characterized as a Mn^{2+} transporter and, accordingly, retains the name *mntABC* (Horsburgh *et al.*, 2002a). Expression of *mntABC* is required for maximal growth in Mn^{2+}-depleted media (Corbin *et al.*, 2008), where supplementation with excess manganese restores growth of the mutant. Some of this characterization was performed in a laboratory isolate of *S. aureus* that contained a truncated version of *mntH* (see below) (Horsburgh *et al.*, 2002a; O'Neill, 2010). MntABC also confers manganese-dependent resistance to oxidative stress (Horsburgh *et al.*, 2002a; Kehl-Fie *et al.*, 2013), where disruption of *mntC* alone was shown to be sufficient to increase sensitivity to intracellular superoxide radicals (Handke *et al.*, 2013).

Structural characterization reveals that, like other cluster A-I metal ion SBPs (Berntsson *et al.*, 2010), MntC possesses a bi-lobate structure consisting of two independently folded domains each comprised of central β-strands surrounded by α-helices. The domains are linked by a long α-helical backbone, with the metal-binding site located at the interdomain interface. The affinity of MntC for Mn^{2+} is in the low nanomolar range and, due to limited hinge movement along the α-helical spine, binding results in little conformational change to the protein (Gribenko *et al.*, 2013). Interestingly, and consistent with the Irving-Williams series, referring to the general stability of metal ion complexes ($Mn^{2+} < Fe^{2+} < Co^{2+} < Ni^{2+} < Cu^{2+} > Zn^{2+}$), MntC may possess two binding modes, one reversible and the other irreversible, depending on the metal. Zn^{2+} and Cd^{2+} bind MntC irreversibly, which may prevent the uptake of toxic metals by sequestering them at the cell surface or, alternatively, may poison the transporter against Mn^{2+} uptake thereby increasing sensitivity to ROS (Gribenko *et al.*, 2013; McDevitt *et al.*, 2011).

A second class of bacterial manganese transporters was discovered during characterization of the metalloregulatory protein, MntR (see below), in *B. subtilis* (Que and Helmann, 2000). In performing a transposon mutagenesis screen for insertions that alleviated the Mn^{2+} sensitivity of an MntR mutant, such integrants were identified in a gene encoding a member of the natural resistance-associated macrophage protein (Nramp) family of proton-coupled metal ion transporters. Termed *mntH*, for (H^+)-dependent Mn transport, homologues of this dedicated manganese transporter have been identified in both Gram-negative and Gram-positive bacteria, including ubiquitous distribution amongst the staphylococci (Hao *et al.*, 1999; Horsburgh *et al.*, 2002a; Kehres *et al.*, 2000; Papp-Wallace and Maguire, 2006; Que and Helmann, 2000; Reeve *et al.*, 2002).

MntH does not appear to be the dominant Mn^{2+} transporter in *S. aureus*. Deletion of *mntH* alone does not significantly impact growth or response to ROS under Mn^{2+} limitation, or hinder Mn^{2+} uptake, at least under the *in vitro* conditions assessed. The mutation of *mntH* does, however, exacerbate the Mn^{2+}-dependent phenotypes of an *mntABC* mutant when these mutations are combined (Horsburgh *et al.*, 2002a; Kehl-Fie *et al.*, 2013). Similarly, while singular mutations in genes encoding MntABC or MntH have not been shown to significantly impact staphylococcal pathogenesis, strains possessing disruptions in both transporters demonstrate impaired survival in human endothelial cells, and decreased bacterial burden in both murine skin abscess and bacteraemia models (Horsburgh *et al.*, 2002a; Kehl-Fie *et al.*, 2013). The latter defect was shown to be mitigated by the absence of calprotectin within the host, suggesting that loss of *mntABC* and *mntH* impairs the ability of the bacterium to acquire manganese in staphylococcal abscesses (Corbin *et al.*, 2008; Kehl-Fie *et al.*, 2011, 2013). Together, these findings suggest that both *mntABC* and *mntH* have an overlapping role in the survival and pathogenesis of *S. aureus*, through conferring manganese-dependent resistance to host-derived ROS.

MntR/SirR and PerR; regulators of manganese homeostasis and resistance to oxidative stress

Control of manganese acquisition in the staphylococci represents a coordinated regulation of both metal ion homeostasis and oxidative stress resistance. MntR, a metalloregulatory protein related to the *Corynebacterium diphtheriae* diphtheria toxin repressor (DtxR), was extensively characterized by Helmann and colleagues in *B. subtilis* as a manganese-dependent regulator of Mn^{2+} uptake and homeostasis (Glasfeld *et al.*, 2003; Lee and Helmann, 2007; Que and Helmann, 2000). As in *B. subtilis*, MntR controls transcription of *mntABC* and *mntH* in *S. aureus* (Ando *et al.*, 2003; Horsburgh *et al.*, 2002a), and its orthologue in *S. epidermidis*, SirR, regulates transcription of *sitABC* (Hill *et al.*, 1998). Both MntR and SirR repress transcription of *mntABC* and *sitABC* in the presence of excess Mn^{2+}, however it has yet to be conclusively determined if Fe^{2+} also directly facilitates this repression (Ando *et al.*, 2003; Hill *et al.*, 1998; Horsburgh *et al.*, 2002a). Similar to Fur, divalent metal cations facilitate binding of MntR/SirR to a conserved MntR/SirR box, a region of dyad symmetry overlapping the transcriptional start of *mntABC/sitABC* and bearing homology to the DtxR operator consensus sequence (Ando *et al.*, 2003; Hill *et al.*, 1998). Binding of these repressors to their respective operons inhibits transcription, which is relieved when the metal ion concentration is depleted. Physiological concentrations of Mn^{2+}, reported to range from 7.3 nM to 273 nM in serum and blood (Agency for Toxic Substances and Disease Registry Website), have been shown to be sufficiently low to allow for *mntABC* expression (Handke *et al.*, 2013). MntR appears to be both a negative and positive regulator of manganese acquisition (Que and Helmann, 2000), where *mntH* is not repressed by high Mn^{2+} concentrations, and is overexpressed in an *mntR* mutant (Horsburgh *et al.*, 2002a). The reasons for bifunctional regulation by MntR of manganese acquisition are unclear, but may allow for tight control of manganese homeostasis and resistance to oxidative stress under both Mn^{2+}-deplete and -replete conditions.

In addition to direct regulation by MntR, Mn^{2+} acquisition, and indeed metal ion acquisition in general, is integrated with responses to ROS (Horsburgh *et al.*, 2001b, 2002a). Owing to the propensity of Fe^{2+} to generate free radicals through Fenton chemistry, and the ability of Mn^{2+} to scavenge ROS through both enzymatic and non-enzymatic means, it is not surprising that both iron and manganese acquisition are regulated, in part, by PerR. In response to elevated Mn^{2+}, this Mn^{2+}-dependent, peroxide-sensitive transcriptional repressor constrains transcription of genes encoding antioxidant proteins (i.e. KatA, AhpCF, Bcp, and TrxB), proteins responsible for the safe storage of iron within the cell (i.e. FtnA and MrgA), and the metalloregulatory proteins Fur (notably, a positive regulator of catalase), and PerR itself (Horsburgh *et al.*, 2001a, b, 2002b). Under Mn^{2+}-depleted conditions, PerR repression is alleviated, promoting resistance to oxidative stress, and Fur-dependent control of iron acquisition (Horsburgh *et al.*, 2001b). Included in genes derepressed by PerR in response to Mn^{2+} limitation or peroxide stress is *mntABC*, which possesses a putative PerR box in its promoter region. The interaction of Mn^{2+} with PerR and, in turn, the regulation of Mn^{2+} uptake directly by PerR, further highlights the essentiality of Mn^{2+} in the intricate response to oxidative stress.

Zinc

Biological function of zinc in the staphylococci

Zinc is unique amongst the biologically relevant transition metals in that it is redox-inert, with a single oxidation state of 2+ in solution. Therefore, instead of functioning in electron transfer, zinc fulfils roles both as a Lewis acid catalyst and as a structural component of proteins, such as zinc fingers (Andreini and Bertini, 2012; Waldron and Robinson, 2009). All six classes within the enzyme commission (EC) system have zinc-associated representatives, with zinc functioning in catalysis in the majority of these enzymes, frequently through facilitating substrate binding within the active site of the enzyme (Andreini and Bertini, 2012). In bacteria, zinc is found in a wide range of hydrolytic enzymes (Andreini *et al.*, 2006), and to a lesser extent in proteins controlling gene expression, such as DNA and RNA polymerases (Barnes *et al.*, 1998; Scrutton *et al.*, 1971), ribosomal proteins (Hensley *et al.*, 2011), and transcriptional regulators, including the zinc-sensing regulator, CzrA (ZntR), in *S. aureus* (Kuroda *et al.*, 1999; Singh *et al.*, 1999) (Table 9.1). Together, approximately 5% of the bacterial proteome is represented by zinc enzymes (Andreini *et al.*, 2006), which constitute the largest fraction of metalloproteins within the cell.

The ability of Zn^{2+} to form stable bonds with oxygen, nitrogen, and sulfur contributes to its utility as an enzyme cofactor. However, as described by the Irving–Williams series (Irving and Williams, 1948), Zn^{2+} is highly competitive and has the potential to displace all other transition metals from the active site of enzymes (Waldron and Robinson, 2009). The intracellular pool of free zinc, and hence its uptake and efflux, must therefore be tightly regulated to avoid mismetallation of other metalloproteins. Indeed, while the total concentration of Zn^{2+} within the bacterial cell is approximately 0.2 mM, only 0.0001% of this (20 pM) is considered to be readily exchangeable (Wang *et al.*, 2011). Despite the requirement for tight restrictions on zinc ion homeostasis, a paucity of information exists, however, on these mechanisms in the staphylococci.

MreA and MreB (ZnuABC): putative zinc acquisition systems

While little is known about zinc acquisition in the staphylococci, other bacteria have been shown to possess both high and low-affinity transporters for Zn^{2+} acquisition under deplete and replete conditions, respectively. High-affinity zinc uptake systems were first described in *Streptococcus pneumoniae* (AdcCAB, a Zn^{2+}-ABC transporter) and *Haemophilus influenzae* (PZP1, a key periplasmic Zn^{2+}-binding protein), but are most extensively characterized in *E. coli* as the Zn^{2+}-uptake system, ZnuABC (Dintilhac *et al.*, 1997; Lu *et al.*, 1997; Patzer and Hantke, 1998). ZnuABC is a homologue of AdcCAB and, like other metal ABC importers in Gram-negative bacteria, consists of a soluble periplasmic substrate binding protein (ZnuA), permease (ZnuB), and ATPase (ZnuC). Expression of *znuABC* is modulated by the zinc uptake regulator, and Fur homologue, Zur (see below) (Patzer and Hantke, 1998). Homologues of *znuABC* have been functionally characterized in several pathogenic bacteria (Ammendola *et al.*, 2007; Chen and Morse, 2001; Dalet *et al.*, 1999; Davis *et al.*, 2009; Desrosiers *et al.*, 2010; Dintilhac *et al.*, 1997; Gunasekera *et al.*, 2009; Janulczyk *et al.*, 1999; Lee *et al.*, 1999; Lewis *et al.*, 1999), and are identified in *S. aureus* as the metal responsive elements, *mreA* and *mreB*, located upstream of Zur. Given their genomic location and homology to known Zn^{2+} transporters, *mreA* and *mreB* were assessed for their role in Zn^{2+} uptake. While Zur was able to repress transcription of the *mreA mreB zur* operon in a Zn^{2+}-dependent manner, no appreciable growth defect was noted for strains inactivated for these genes under Zn^{2+} limitation (Lindsay and Foster, 2001). Further efforts are required to determine the role of *mreA* and *mreB* (*znuABC*) in zinc homeostasis in the staphylococci.

Like the high-affinity Zn^{2+} transporters described above, low-affinity transporters of Zn^{2+} have also been robustly characterized in *E. coli*. Deletion of *znuABC* significantly reduced, but did not abolish *E. coli* growth under Zn^{2+} restriction (Grass *et al.*, 2002; Patzer and Hantke, 1998), suggesting the existence of an alternate means of uptake. The zinc uptake transporter, ZupT, was identified as a homologue of eukaryotic zinc transporters of the ZIP family (zinc regulated transporter, iron regulated transporter-like protein) (Gaither and Eide, 2001; Grass *et al.*, 2002; Guerinot, 2000). While demonstrating a slight preference for Zn^{2+}, ZupT possesses broad specificity for divalent metal cations; the means by which ZupT facilitates metal ion uptake are unclear, although transport may be energized by the proton motive force (Grass *et al.*, 2002, 2005a; Karlinsey *et al.*, 2010; Taudte and Grass, 2010). Furthermore, *zupT* is not demonstrably metal-regulated and exhibits low level, constitutive expression (Grass *et al.*, 2005a). Characterized prokaryotic homologues of ZupT are sparse (Cerasi *et al.*, 2014; Herzberg *et al.*, 2014; Karlinsey *et al.*, 2010), and while several staphylococcal genomes have annotations for ZIP family proteins, none have been experimentally solved.

Zur and CzrA: regulators of zinc uptake and metal ion efflux

As described above, stringent control over the intracellular concentration of Zn^{2+} is required to maintain a sufficient bioavailable supply for diverse metabolic processes, while at the same time avoiding the toxicity associated with excess. Tight regulation of zinc homeostasis is achieved through the complementary actions of Zn^{2+} sensor pairs, whose binding to the operator/promoter region of their cognate genes is mediated by the presence of Zn^{2+}. Sensor pairs consist of uptake repressors, which facilitate inhibition of zinc uptake systems, and efflux regulators, which promote derepression/activation of zinc efflux proteins (Braymer and

Giedroc, 2014; Guerra and Giedroc, 2012). In *S. aureus*, these Zn^{2+} uptake and efflux regulators are represented by Zur and CzrA (ZntR).

As stated, Zur, is a homologue of Fur, and in many bacterial species, serves specifically in the control of Zn^{2+} uptake through high-affinity transporters, such as *znuABC*, along with Zn^{2+}-containing ribosomal proteins (Cerasi *et al.*, 2013; Graham *et al.*, 2009; Li *et al.*, 2009; Lim *et al.*, 2013; Lindsay and Foster, 2001; Panina *et al.*, 2003). Phylogenetic analysis of Zur proteins reveals that they actually comprise three unrelated subgroups encompassing the Enterobacteriaceae, and high and low G+C content Gram-positive bacteria (Hantke, 2005). Depending on the genus and species, one or two Zn^{2+} atoms are bound by each Zur monomer, for a maximum of four bound Zn^{2+} atoms upon dimerization (Lucarelli *et al.*, 2007; Ma *et al.*, 2011; Outten *et al.*, 2001; Shin *et al.*, 2011). When two Zn^{2+}-binding sites are present, one typically fulfils a structural role, while the other aids in the repressor function of the protein by facilitating tight interaction between the DNA-binding domain of Zur and its cognate consensus sequence (Cerasi *et al.*, 2013). As modelled in *B. subtilis*, sequential Zn^{2+} binding to the two aforementioned sites is proposed to allow for stepwise activation of Zur, thereby allowing for differential responses to a broad range of intracellular Zn^{2+} concentrations through gradual repression of the Zur-regulated genes (Ma *et al.*, 2011). Although Fur family transcriptional regulators are widely distributed amongst the staphylococci, and Zur has been identified as a putative regulator of *mreA* (*znuC*) and *mreB* (*znuB*) in *S. aureus* (Lindsay and Foster, 2001), the Zur regulon in the staphylococci has yet to be described.

In contrast to Zur, which largely controls the high-affinity uptake of Zn^{2+}, CzrA (ZntR) is a well-characterized staphylococcal metalloregulatory protein that functions in controlling resistance to extraphysiological concentrations of Zn^{2+}, as well as Co^{2+}, through regulating efflux (Kuroda *et al.*, 1999; Singh *et al.*, 1999; Xiong and Jayaswal, 1998). CrzA is a Zn^{2+}/Co^{2+}-responsive transcriptional regulator of the ubiquitous arsenic repressor (ArsR) family of homodimeric metal sensing proteins. Co-transcribed with *czrA* is *czrB* (*zntA*), which encodes a member of the cation diffusion facilitator (CDF) family of metal exporters (Guilhen *et al.*, 2013). Binding of Zn^{2+} or Co^{2+} to the active site of CzrA results in negative allosteric regulation, whereby transcription of *czrAB* is derepressed and efflux of Zn^{2+} and/or Co^{2+} is allowed to proceed through expressed CzrB (Busenlehner *et al.*, 2003; Campanello *et al.*, 2013; Chakravorty *et al.*, 2012). The exact mechanism of Zn^{2+} and Co^{2+} efflux by CzrB has not been defined, but it likely functions similarly to other prokaryotic members of this family, where metal ion export is coupled to H^{+} and/or K^{+} antiport (Grass *et al.*, 2005b; Guffanti *et al.*, 2002). The specificity of *czrAB* regulation by CzrA, and efflux of Zn^{2+} and Co^{2+} by CzrB, has been defined, and is largely controlled by the coordination geometry of the metal cofactor within the repressor (CzrA) binding site (Pennella and Giedroc, 2005; Pennella *et al.*, 2006; Xiong and Jayaswal, 1998). As such, CzrA has no specificity for, and hence does not regulate resistance to, metals that cannot adopt a coordination geometry comparable to Zn^{2+}/Co^{2+} including Ni^{2+}, Mn^{2+}, $Fe^{2+/3+}$, or $Cu^{1+/2+}$, or other toxic metals such as arsenic, cadmium or mercury (Pennella *et al.*, 2006). Mechanisms of resistance to such toxic metals are discussed in the section 'Avoiding excess: mechanisms of defence against toxic metals'.

Nickel and cobalt

Biological function of nickel and cobalt in the staphylococci

Unlike iron, manganese, and zinc, the biological functions of nickel and cobalt are relatively limited and both are considered trace elements. To date, only nine prokaryotic Ni-containing enzymes have been characterized: urease, [NiFe]-hydrogenase, carbon monoxide dehydrogenase, certain Ni-dependent SODs, glyoxalase I, acidreductone dioxygenase, acetyl-coenzyme A synthase/decarbonylase, lactate racemase, and methyl-coenzyme M reductase (Boer *et al.*, 2014) (Table 9.1). In the staphylococci, the only known Ni-containing protein is the multimeric enzyme, urease (Grüter *et al.*, 1992; Jose *et al.*, 1991). Urease catalyses the hydrolysis of urea, the major nitrogenous waste product of mammals, leading to the production of carbon dioxide and ammonia (Burne and Chen,

2000). In aqueous environments, the latter molecule becomes protonated, yielding ammonium and hydroxide ions, and potentiating an increase in environmental pH. Expression of urease in highly acidic environments, such as the stomach and urinary tract, has been shown to enhance both the survival and virulence of ureolytic bacteria such as *Helicobacter pylori* (Mobley *et al.*, 1991; Tsuda *et al.*, 1994), and *S. saprophyticus* (Gatermann and Crossley, 2009; Gatermann and Marre, 1989; Gatermann *et al.*, 1989). Enhanced pathogenicity conferred by urease may not only be due to the pH-neutralizing effects of ammonium, but also its use as a preferred nitrogen source and as a cytotoxic agent against host epithelial cells (Collins and D'Orazio, 1993). The multifaceted role of urease as a virulence factor, especially in acidic environments, highlights the importance of Ni acquisition by bacteria employing this enzyme as part of their arsenal.

Like Ni, Co is found less frequently as a component of metalloproteins, compared with other biologically relevant transition metals. Literature on the role of Co in staphylococcal metabolism is limited. Co is redox active, with biologically relevant oxidation states, like Fe, of 2+ and 3+. Co^{2+} exists mainly as a cofactor in vitamin B_{12} (cobalamin)-containing enzymes, which catalyse transmethylation, rearrangement, and electron transport reactions (Banerjee and Ragsdale, 2003; Okamoto and Eltis, 2011). Only eight non-cobalamin Co^{2+}-containing enzymes are known (Kobayashi and Shimizu, 1999; Okamoto and Eltis, 2011), and two of these enzymes have been characterized in *S. aureus*: methionyl aminopeptidase (MetAP-1), responsible for removing the N-terminal methionine from proteins (Oefner *et al.*, 2003), and aminopeptidase S (AmpS) (Odintsov *et al.*, 2005). Both enzymes can coordinate either Co^{2+} or Zn^{2+}, however the former displays highest specific activity and efficiency in the presence of Co^{2+}, suggesting a requirement for Co^{2+} in Zn^{2+}-deprived environments (Oefner *et al.*, 2003). The limited role of both Ni and Co in metabolism likely reflects their scarcity. Further, given Co bears a similarly sized radius and comparable redox state to Fe, limiting Co involvement in the cell may serve as a means to prevent its misincorporation, potentially irreversibly and toxically, into Fe-metalloproteins (Okamoto and Eltis, 2011).

NikA and NikBCDE, and NixA: nickel acquisition systems

High-affinity Ni^{2+} and Co^{2+} transporters are necessitated by the fact that in most natural environments, these metal ions are typically available only in trace amounts. Three major types of microbial Ni/Co have been defined: NikABCDE ABC-type transporters which are specific for Ni^{2+}, a second group of Ni^{2+} and Co^{2+} transporters bearing homology to ABC-type systems known as Cbi/NikMNQO and characterized by an apparent lack of substrate binding protein, and secondary permeases of the NiCoT (nickel/cobalt transporter) family (Eitinger *et al.*, 2005; Mulrooney and Hausinger, 2003; Rodionov *et al.*, 2006; Zhang *et al.*, 2009).

The archetypal representative of the first group of Ni^{2+} uptake systems is NikABCDE of *E. coli*, a member of the nickel/opine/peptide transporter family, PepT (peptide transport) (Rodionov *et al.*, 2006). In addition to Ni^{2+}, members of this family have also been shown to transport di- and oligopeptides (Abouhamad *et al.*, 1991; Borezee-Durant *et al.*, 2009; Hiron *et al.*, 2007; Levdikov *et al.*, 2005). Nik consists of a periplasmic protein for substrate binding (NikA), two hydrophobic transmembrane domains for translocation (NikB and NikC), and two ATPases for coupling energy to transport of the substrate (NikD and NikE) (Navarro *et al.*, 1993). In *E. coli*, NikABCDE is specific for Ni^{2+}, where metal ions such as Co^{2+}, Fe^{2+} and Cu^{2+} bind NikA with 10-fold lower affinity than Ni^{2+} (de Pina *et al.*, 1995). In investigating four putative oligopeptide permeases of the PepT family in *S. aureus* (Opp1 to Opp4), only Opp3 was demonstrated to transport peptides for nitrogen metabolism (Borezee-Durant *et al.*, 2009; Hiron *et al.*, 2007). Opp2BCDF, consisting of two integral transmembrane domains (Opp2B and Opp2C), two ATPases (Opp2D and Opp2F), and a genetically unlinked, orphan substrate binding protein, Opp5A, were identified instead as contributing to the uptake of Ni^{2+} and were accordingly renamed NikBCDE and NikA (Hiron *et al.*, 2010). However, when intracellular Ni^{2+} accumulation of a *nikA-nikBCDE* mutant was not found to be substantially reduced relative to wild-type *S. aureus*, further genome analysis revealed another putative Ni^{2+} transporter, bearing homology to the *H. pylori* NiCoT family member, NixA (Hiron *et al.*, 2010; Mobley *et al.*, 1995).

Functional characterization of both NikA/NikBCDE and NixA in *S. aureus* indicates that both contribute to Ni^{2+} acquisition, as well as Ni^{2+}-dependent urease activity *in vitro*. For reasons that are yet unexplained, only NikA/NikBCDE was shown to contribute to alkalinization of urine, and was required for colonization of murine kidneys and bladders in a urinary tract infection (UTI) model. NixA was shown to be partially functionally redundant with NikA *in vivo*, where colonization was further impaired in a *nikA-nixA* mutant, relative to *nikA* alone (Hiron *et al.*, 2010). The genes encoding NikA/NikBCDE are highly conserved within the staphylococci, and with the exception of *S. capitis*, which lacks the gene encoding NikA, have been identified in all sequenced staphylococcal strains. Notably, in *S. saprophyticus*, a common causative agent of UTIs (Kuroda *et al.*, 2005), the orthologous genes follow the prototypical, single-operon organization of NikABCDE and are located adjacent to putative urea uptake and urease encoding genes (Hiron *et al.*, 2010). Given the requirement of Ni^{2+} for urease activity and the contribution of this enzyme to the survival and virulence of uropathogens, the genetic organization of these genes in *S. saprophyticus* may contribute to its penchant for the urinary tract.

While a role for NikA/NikBCDE in Ni^{2+} acquisition has been elucidated, it has not been shown to be involved in the transport of other divalent metals, such as Co^{2+} (Hiron *et al.*, 2010). Heterologous expression of *S. aureus* NixA in *E. coli* revealed that it is a class III member of the NiCoT family, meaning it transports both Ni^{2+} and Co^{2+}, but possesses a significantly greater capacity to transport Ni^{2+} (Hebbeln and Eitinger, 2004). An additional putative staphylococcal oligopeptide permease, Opp1ABCDE was implicated in the uptake of both Ni^{2+} and Co^{2+} under conditions of Zn^{2+}-limitation and was renamed CntABCDF (Co and Ni transporter) (Remy *et al.*, 2012). Consistent with the identification of *opp1* (*cnt*) as a genetic locus impacting *in vivo* growth and survival of *S. aureus* (Coulter *et al.*, 1998; Mei *et al.*, 1997), *cnt* mutants displayed reduced overall mortality, and reduced bacterial burden, in murine models of systemic and urinary tract infection, respectively (Remy *et al.*, 2012). While it is evident that *cnt* influences staphylococcal pathogenesis, the significance of its expression under metal-deplete conditions, as well of the binding affinity of CntA for Co^{2+} and Ni^{2+} have not been elucidated. No known homologues of the Cbi/NikMNQO family of Ni^{2+}/Co^{2+} transporters have been identified, suggesting that NikA/NikBCDE, NixA, and CntABCDF may be solely responsible for the provision of Ni^{2+} for urease activity in the staphylococci. More research is required to conclusively establish the substrate specificity of these transporters, the means by which they are regulated and, correspondingly, the host specific niche(s) in which they are expressed.

Avoiding excess: mechanisms of defence against toxic metals

As defined by the Irving–Williams series, Cu^{2+} is capable of outcompeting all other divalent transition metals for the binding of enzyme active sites (Irving and Williams, 1948), in addition to binding non-specifically to thiol groups and inhibiting protein folding (Baker *et al.*, 2010; Teitzel *et al.*, 2006). Like iron, Cu also possesses two readily available oxidation states, Cu^{1+} and Cu^{2+}, thus is redox-active and capable of generating ROS (Baker *et al.*, 2010). The aforementioned properties necessitate stringent control over intracellular Cu^{1+} concentrations, which are reportedly in the attomolar range (10^{-18}) (Braymer and Giedroc, 2014). A role for Cu, if any, in staphylococcal metabolism has not been identified, and indeed debate exists over whether Cu is required to enter the bacterial cytosol (Tottey *et al.*, 2005). To date, no dedicated transporter for Cu has been identified, and it is likely that Cu, along with other heavy metals, enter the cytosol non-specifically through the activity of broad substrate range ion transporters (Nies, 1992). As such, mechanisms of Cu homeostasis, along with control of other toxic metals, are largely focused on sequestration and efflux.

All staphylococcal genomes sequenced, to date, possess chromosomally-encoded genes for a copper-translocating P1-type ATPase (CopA) and a copper metallochaperone (CopZ) (Huffman and O'Halloran, 2001; Sitthisak *et al.*, 2007). Transcription of *copA* and *copZ* are induced by Cu^{1+} (the redox state of Cu in the cytoplasm) and, to a lesser extent, Fe^{3+} and Pb^{2+}. Mutants of *copA* exhibit increased intracellular concentrations of, and enhanced sensitivity to, these toxic metals, suggesting that CopA functions as an exporter (Sitthisak *et al.*, 2007). Copper chaperones, like CopZ, function to bind Cu^{1+} in the cytosol, and deliver it safely to intracellular targets such as Cu^{1+} sensors and efflux proteins. A limited number of *S. aureus* strains encode a second P1-type ATPase (CopB), and a multicopper oxidase (Mco), which help to confer resistance to copper and oxidative stress (Sitthisak *et al.*, 2005).

In addition to chromosomally encoded copper resistance genes, some *S. aureus* strains also express a plasmid-borne copy of the *copB mco* operon, which confers a transmissible hyper-copper-resistant phenotype (Baker *et al.*, 2011). Both chromosomal and plasmid-encoded copper resistance genes are regulated by the copper-sensitive operon repressor, CsoR (Baker *et al.*, 2011; Liu *et al.*, 2007; Smaldone and Helmann, 2007). CsoR is co-transcribed on the chromosome with two additional putative copper chaperones (Baker *et al.*, 2010), and represses transcription of the aforementioned copper resistance mechanisms under low copper conditions. When copper is present, CsoR derepresses these genes and copper is effluxed from the cytoplasm. Given the lack of biological function of copper in the staphylococci, and its toxicity to bacterial cells, it is unsurprising that host nutritional immunity does not focus on the withholding of copper from invading microbes, but rather on its potential use as an intoxicant. The means by which the host exploits concentrations of biologically relevant transition metals to prevent infection are discussed in the subsequent section.

The importance of metal ion homeostasis in the host

Metal ion homeostasis is essential to human health and perturbation of the processes that control metal transport and utilization contribute to the development of disease. Transition metals such as iron, copper, and zinc play essential and unique roles within the human body, as they are integral to the function of countless enzymes that participate in diverse intra- and extracellular processes (Table 9.1). As in bacteria, the utility of these metals stems from their ability to accept and donate electrons and thus participate in a profusion of redox reactions, a trait that also renders them toxic to the mammalian host when present in excess. Indeed, accumulation of iron in the body due to accelerated intestinal absorption, referred to as haemochromatosis, can promote severe oxidative damage to vital organs like the liver (Pietrangelo, 2010). Conversely, devastating pathological conditions can also result from metal depletion, as evidenced by Menkes disease, a lethal neurodegenerative disorder attributed to a defective copper-transporting ATPase that results in low copper levels in the serum, liver, and brain (Kaler, 2011). The abundance of individual trace metals in specific organs and bodily fluids varies, and is a reflection of the physiological function of the organ, as well as the overall total body metal content. For instance, the liver and spleen have elevated Fe concentrations due to their role in Fe storage and recycling. In the spleen, macrophages of the reticuloendothelial system ingest effete red blood cells (RBCs) and recycle Fe that is liberated from haem and haemoglobin upon degradation (Knutson *et al.*, 2005; White *et al.*, 2013). In the liver, resident macrophages also contribute to the clearance of senescent RBCs and Fe recycling, and hepatocytes comprise an important cellular Fe repository, where the metal is stored intracellularly in complex with ferritin (Harrison and Arosio, 1996).

Infection profoundly affects the distribution of transition metals such as Fe, Mn and Zn within the body, and these metals are actively sequestered from invading microbes through a process termed 'nutritional immunity'. In contrast to other innate immune mechanisms that attack and directly kill invading microorganisms, the nutritional immune response is designed to deprive microbes of essential metals that they require for growth (e.g. Fe and Zn), or that help provide resistance to other facets of the host innate immune system (e.g. Mn). Evidence that metal ion homeostasis is an essential factor that underpins immunity emerges from pathological conditions (e.g. thalassemia or haemochromatosis) that perturb Fe concentrations within the body, resulting in enhanced susceptibility to, and severity of, bacterial infection (Khan *et al.*, 2007; Vento *et al.*, 2006).

Evidence of a metal-based nutritional immune response during staphylococcal infection

As discussed in other chapters, members of the staphylococci can colonize diverse niches on and within the human body, producing a spectrum of pathologies ranging from mild skin infections and soft tissue abscesses, to more severe diseases like sepsis, endocarditis, and osteomyelitis. Ostensibly, the ability of *Staphylococcus* spp., such as *S. aureus*, to inhabit such varied environments is evidence of their ability to acquire essential trace metals, like Fe, by employing the multitude of metal ion acquisition mechanisms described earlier in this chapter. Indeed, cultivation of *S. aureus* in human blood or serum results in a dramatic increase in transcription of genes involved in iron acquisition and copper homeostasis, as compared to growth in routine laboratory media (Malachowa *et al.*, 2011; Oogai *et al.*, 2011). Moreover, cultivation in human blood or serum results in an array of cytolytic and cytotoxic factors being up-regulated, a phenomenon that is negated by the addition of inorganic Fe (Oogai *et al.*, 2011). Interestingly, similar patterns of gene expression are also observed in bacteria recovered from murine kidneys or human cutaneous abscesses (Date *et al.*, 2014). Taken together, these observations indicate that trace metals are not readily available to colonizing staphylococci and indirectly demonstrate the existence of host factors that exert nutritional immunity. In this section, the salient aspects of how host trace metals, relevant to the nutritional immune response, are metabolized are described, along with the specific extracellular and intracellular mechanisms by which nutritional immunity is implemented. Additionally, the countermeasures bacteria employ within the host to subvert nutritional immunity are discussed, highlighting mechanisms that apply directly to the staphylococci.

Host iron metabolism

The human body is an abundant source of Fe and there exists sophisticated regulatory networks that govern Fe homeostasis in mammals (for a recent review see Winter *et al.*, 2014). On average the adult body contains ~ 3.5–5 g of iron; however, free iron is virtually non-existent because it is bound by proteins such as transferrin, in complex with haem in haemoproteins, or sequestered by secreted proteins and mammalian siderophores such as siderocalin and 2,5-dihydroxybenzoic acid (DHBA) (Gomme *et al.*, 2005; Ajioka *et al.*, 2006; Flo *et al.*, 2004; Devireddy *et al.*, 2010). The sequestration of Fe by these host molecules serves multiple functions: iron binding facilitates metal transport and storage, reduces the potential for Fe to participate in toxic radical-producing reactions, and sequestration of Fe by these molecules constitutes an element of the nutritional immune response by restricting its availability to microbial invaders.

Regulation of extracellular iron availability

Transferrin and lactoferrin

The transferrins, a family of host glycoproteins comprised of transferrin (Tf) (present in the serum) and lactoferrin (Ltf) (present in body secretions such as milk and mucus) function as high-affinity Fe^{3+} chelators. Transport of elemental Fe throughout the body is mediated by serum Tf. With the exception of mature anucleated RBCs, most cells throughout the body express high-affinity surface receptors for Tf, TfR1 and TfR2 that, upon binding of Tf-Fe^{3+}, are internalized via endocytosis. The affinity of Tf for Fe^{3+} at near neutral pH (outside the cell) is extremely high ($\sim 10^{23}\ M^{-1}$); however, at acidic pH its affinity decreases dramatically (Aisen *et al.*, 1978). This ensures dissociation of Fe from Tf in the acidic lumen of the endosome. Upon release of Fe^{3+}, TfR–Tf complexes are recycled back to the cell membrane, where Tf is released to the extracellular milieu (Aisen, 1994; Dautry-Varsat *et al.*, 1983). In order for the cell to utilize Fe^{3+} delivered through Tf, it is extruded from the endosome to the cytosol, which is catalysed by divalent metal transporter 1 (DMT1) a limiting endosomal membrane protein (Touret *et al.*, 2003). Transport through DMT1 requires reduction of Fe^{3+} to Fe^{2+}, a reaction catalysed by proteins of the six transmembrane epithelial antigen of the prostate (STEAP) family. STEAPs have emerged as important oxidoreductases that also reside in the endosomal membrane and their significance to Fe homeostasis is evident in mice lacking STEAP3 that are anaemic due to diminished haemoglobin biosynthesis (Sendamarai *et al.*, 2008). Ltf, like Tf also chelates Fe^{3+}, however unlike Tf, Ltf does not serve Fe-transport functions. As mentioned, Ltf is enriched at mucosal surfaces and it is also enriched within the secondary granules of neutrophils where it is released upon immune activation and neutrophil degranulation (Saito *et al.*, 1993). Unlike Tf, Ltf does not release Fe^{3+} under acidic conditions, making it an effective chelator of Fe in diverse physiological niches (Baker and Baker, 2004). In fact the primary function of Ltf is to sequester Fe at sites (e.g. mucosal surfaces of the respiratory and gastrointestinal tract) that are inundated with microbes. In the context of nutritional immunity, both Tf and Ltf constitute an effective element of the innate immune response that can deprive pathogens of much needed Fe (Fig. 9.2). This is evident for *Staphylococcus* spp. that lack the ability to produce iron-chelating siderophores, such as *S. lugdunensis*, where Tf and Ltf are sufficient to curtail efficient microbial growth *in vitro* (Brozyna *et al.*, 2014). To counter the nutrient limiting effects of Ltf and Tf, the majority of the staphylococci secrete the siderophore SA, if not also SB (see above), which through *in vitro* studies enhance staphylococcal growth in the presence of serum Tf (Fig. 9.2) (Beasley *et al.*, 2011; Haag *et al.*, 1994; Lindsay *et al.*, 1994; Meiwes *et al.*, 1990; Szarapińska-Kwaszewska and Farkas, 2003).

Hepcidin-dependent regulation the Fe export protein ferroportin (SLC40A1)

To maintain homeostasis, serum transferrin is ~30% Fe saturated, which ensures an adequate abundance of Fe to sustain physiological processes like erythropoiesis, yet at the same time allowing

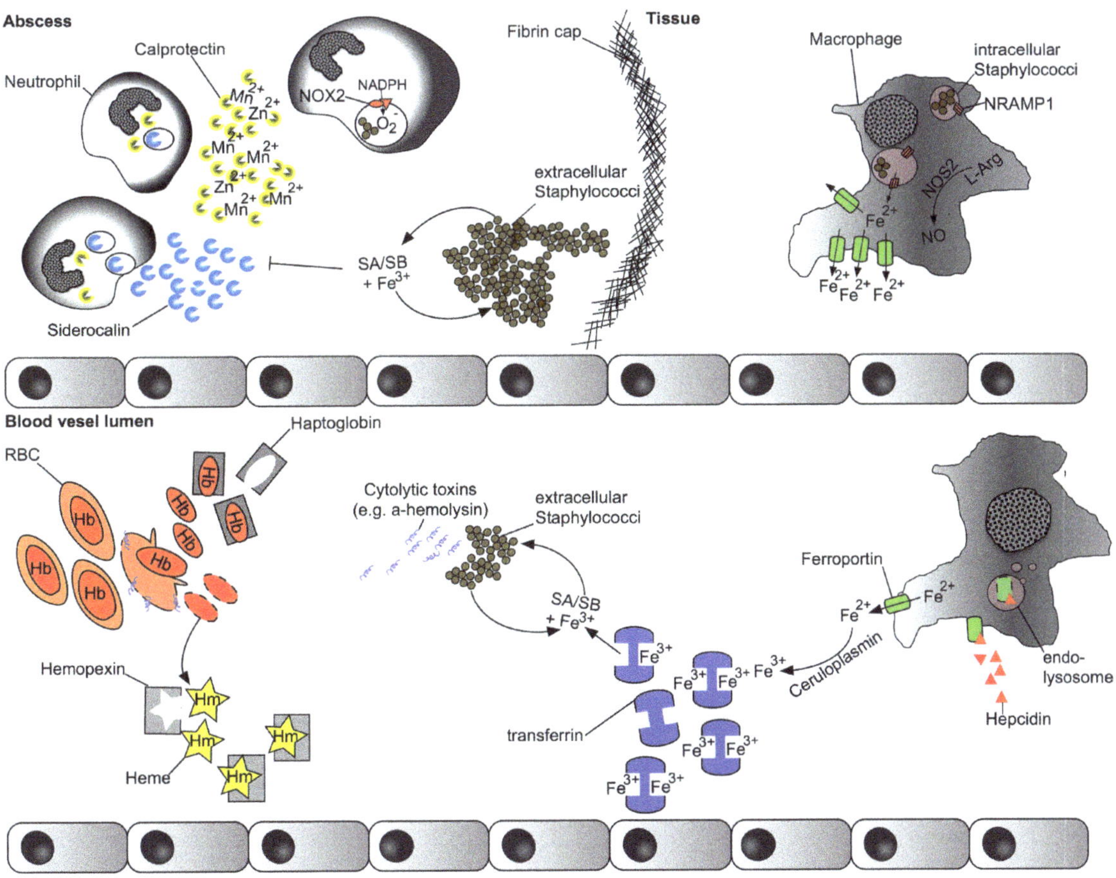

Figure 9.2 Host effectors of metal-associated nutritional immunity. During infection host innate immune effectors are mobilized that function to curtail bacterial growth by limiting the accessibility of essential nutrients to invading bacteria (e.g. *S. aureus*). This nutritional immune response, which is not in itself microbicidal, limits the formation of a significant bacterial burden, better enabling microbicidal effectors of innate immunity to eradicate the infection. Not surprisingly bacteria like the staphylococci with a propensity to infect have evolved strategies that circumvent the nutritional immune response to enable successful colonization of the host. Established host factors that restrict staphylococcal nutrient acquisition in the blood, abscesses and tissue macrophages are depicted, and described in the text. Abbreviations: Hb, haemoglobin; Hm, haem; RBC, red blood cell; L-Arg, L-arginine; NOS2, nitric oxide synthase 2; NO, nitric oxide; NOX2, NADPH oxidase; SA/SB; staphyloferrin A and B.

sufficient binding capacity to quickly scavenge trace Fe released into the blood through desquamation of endothelial cells, for example. Maintenance of this relatively constant Tf saturation level requires systemic coordination of Fe absorption and efflux from the cells at the epicentre of Fe metabolism such as intestinal enterocytes, macrophages and hepatocytes (reviewed in Winter *et al.*, 2014). Systemic co-ordination of Fe absorption and efflux is accomplished through the action of hepcidin, a member of the defensin family of antimicrobial peptides, which is produced largely by the liver (Nicolas *et al.*, 2001). Initially secreted as an 84 amino acid peptide, hepcidin is processed into a 25 amino acid bioactive peptide by the Ca^{2+}-dependent serine protease, furin. The molecular target of hepcidin is the Fe-export protein ferroportin (Fpn), which effects Fe egress from the cell (Gagliardo *et al.*, 2009; Valore and Ganz, 2008). Fpn is expressed in the plasmalemma of macrophages and at the basolateral membrane of intestinal enterocytes where it functions to extrude intracellular Fe^{2+} to the extracellular milieu (Donovan *et al.*, 2005) (Fig. 9.2). Hepcidin-dependent regulation of Fpn function occurs at the protein level, where direct binding of hepcidin to Fpn induces its endocytosis and subsequent proteolysis (Nemeth *et al.*, 2004a). Hepcidin may also repress the absorption of inorganic Fe from dietary sources by impairing the aforementioned DMT1 in the apical membrane of intestinal enterocytes, where it would otherwise import Fe from the intestinal lumen (Yamaji *et al.*, 2004). Fe^{2+} exported into circulation through Fpn is rapidly oxidized to the ferric state through the concerted activity of the copper-dependent ferroxidases hephaestin and ceruloplasmin for Tf binding (Kuo *et al.*, 2004; Griffiths *et al.*, 2005; Fuqua *et al.*, 2014; Eid *et al.*, 2014) (see Table 9.1 and Fig. 9.2).

Hepcidin, ferroportin and hypoferraemia of infection

In response to infection and inflammation, a dramatic systemic decrease of extracellular iron occurs in humans called hypoferraemia (Kemna *et al.*, 2005; Nemeth *et al.*, 2004b). As a consequence of this response, the availability of extracellular Fe to invading microbes becomes greatly reduced and represents a powerful systemic strategy to limit Fe availability for microbial growth. This acute state of extracellular Fe depletion is largely attributed to decreased expression of Fpn in macrophages, intestinal enterocytes and hepatocytes (Nemeth *et al.*, 2004a; Kim *et al.*, 2014). As described above, removal of Fpn from the plasmalemma of cells is a consequence of hepcidin-induced endocytosis and Fpn degradation in lysosomes (Fig. 9.2). During inflammatory responses, hepatocyte-dependent production of hepcidin is amplified in response to pro-inflammatory cytokines including IL-6 (Nemeth *et al.*, 2004b; Kemna *et al.*, 2005; Kim *et al.*, 2014). A recent study has provided evidence that IL-6 evokes enhanced hepcidin production through activation of the transcription factors signal transducer and activation of transcription 3 (STAT3) and oestrogen-related receptor γ (ERRγ), which interact with the hepcidin promoter element (Kim *et al.*, 2014). Thus, through IL-6 signalling, enhanced hepcidin gene transcription and hormone production occurs in the liver (Kim *et al.*, 2014). Interestingly, hepcidin can also be produced from myeloid sources like macrophages due to the pathogen-associated molecular pattern stimulation of toll-like receptors, indicating that macrophages possess the ability to modulate Fe egress in response to infection (Armitage *et al.*, 2011; Peyssonnaux *et al.*, 2006). Indeed, myeloid derived hepcidin can act in an autocrine and paracrine manner to suppress Fpn function at the protein level, as described above, and can suppress Fpn at the level of transcription (Peyssonnaux *et al.*, 2006). The occurrence of these phenomena indicates that myeloid-derived hepcidin exerts multiple mechanisms of regulation on ferroportin function in macrophages.

At present, the induction of hepcidin and its physiological effects in response to staphylococcal infections remain to be determined. It is, however, interesting to note that *S. aureus* infection evokes IL-6 and acute phase hypoferraemia so it is likely that hepcidin is active in this context. Furthermore, the importance of the host response in the context of staphylococcal infection is equivocal, as are the iron acquisition systems in these bacteria, presumably to allow bacteria to circumvent this element of the host's iron starving response.

2,5-Dihydroxybenzoic acid: a mammalian siderophore

While staphylococci, like many microbes, synthesize siderophores, it is now recognized that mammals also produce a siderophore-like compound, 2,5-dihydroxybenzoic acid (DHBA) (Devireddy *et al.*, 2010). 2,5-DHBA is a catechol molecule that is closely related to 2, 3-DHBA, a component of the siderophore enterobactin that is synthesized by *E. coli*. Given its structure, it is possible that the mammalian 2,5-DHBA could also support the growth of *S. aureus* via the Sst transport system (see above), although this has not been demonstrated. 2,5-DHBA functions as a siderophore and is important in maintaining Fe homeostasis in mice. Mammalian cells unable to synthesize 2,5-DHBA present with increased cytosolic Fe, increased ROS damage and have a decrease in mitochondrial function because of decreased Fe content (Devireddy *et al.*, 2010). Mice, however, genetically engineered to be deficient in 2,5-DHBA synthesis demonstrate an increased resistance to *E. coli* infection but not *S. aureus* (Liu *et al.*, 2014). The increased resistance to infection in this model can be attributed to the inability of *E. coli* to utilize host derived 2,5-DHBA as a siderophore for growth. Presumably, ablation of 2,5-DHBA synthesis is without affect in the context of *S. aureus* infection because of the enhanced ability of *S. aureus* to acquire host Fe through multiple pathways. Consistent with this notion, a role for catechol mediated Fe acquisition *in vitro* and *in vivo* is only evident when staphyloferrin A and B production have genetically been inactivated (Beasley *et al.*, 2011). Interestingly, during infection the host gene encoding 3-hydroxybutyrate dehydrogenase-2 (Bdh2), which is critical for 2,5-DHBA production, is down-regulated as a result of toll-like receptor signalling (Liu *et al.*, 2014). In response to both *E. coli* and *S. aureus*, Bdh2 mRNA is down-regulated in a murine infection model, indicating that suppression of 2,5-DHBA synthesis is part of a more generalized response to Gram-positive and Gram-negative infections. These observations suggest it is advantageous to the host to suppress the production of a molecule that can be usurped by pathogenic bacteria as an iron source for growth.

Siderocalin: a siderophore-binding protein

Siderophore-mediated Fe acquisition is an important aspect of microbial physiology that enables growth when iron is scarce. To counter the activity of siderophores, the mammalian host is endowed with the acute phase protein siderocalin (also known as neutrophil gelatinase (NGAL), lipocalin 2, and lipocalin 24p3) (Fig. 9.2) (Flo *et al.*, 2004; Goetz *et al.*, 2002). Siderocalin, a constitutively expressed protein enriched in the secondary granules of neutrophils, is also induced through TLR-dependent signalling in hepatocytes, macrophages and endothelia allowing for copious amounts of siderocalin to be produced in response to infection (Flo *et al.*, 2004). The bacteriostatic effect of siderocalin derives from its capacity to sequester bacterial catechol-type siderophores such as enterobactin and mycobactin from *E. coli* and *Mycobacterium tuberculosis*, respectively (Flo *et al.*, 2004; Goetz *et al.*, 2002; Holmes *et al.*, 2005). Interestingly, siderocalin expression is also induced under aseptic conditions and has been implicated to function as a Fe transport protein, although it lacks the intrinsic capacity to bind Fe directly (Bao *et al.*, 2010). An explanation for this observation, emerges from the realization that mammalian systems produce Fe^{3+}-binding cathechol type compounds such as 2,5-DHBA that are also bound by siderocalin (Bao *et al.*, 2010). Although the citrate-based staphylococcal siderophores SA and SB presumably evade sequestration by siderocalin, expression of siderocalin during staphylococcal infection would ostensibly restrict the accessibility of catechol-related Fe binding molecules such as 2,5-DHBA to the bacteria. In an early seminal study, mice deficient for siderocalin production were shown to be equally susceptible to *S. aureus* infection as compared to siderocalin-competent animals and *in vitro* growth of *S. aureus* was also uninhibited by siderocalin (Flo *et al.*, 2004). This is entirely consistent with the notion that the staphylococcal siderophores are unperturbed by siderocalin. More recently evidence has suggested that neutrophils derived from siderocalin-deficient mice demonstrate enhanced susceptibility to *S. aureus* infection because of inherent defects in neutrophil innate immune function (Liu *et al.*, 2013). The mechanism underlying this neutrophil dysfunction and impaired *S. aureus* killing is unknown; however, it is presumably independent of siderocalin modulation of iron acquisition by *S. aureus*.

Regulation of intracellular iron availability: ferroportin function in infected macrophages

Whether due to steady state delivery of Fe through Tf or as a consequence of the hypoferraemic response, the interior of host cells can represent an iron-replete environment that could support microbial growth. Consequently, under circumstances of intracellular infection hepcidin-dependent suppression of Fpn function could prove deleterious, as the pool of intracellular Fe could be increased and utilized by the intracellular pathogen for growth. As stated, macrophages display multiple mechanisms of Fpn regulation at both the protein and transcriptional level. One transcription factor known to regulate Fpn expression in macrophages is the redox-sensitive nuclear factor erythroid 2 (Nrf2) (Harada *et al.*, 2011; Nairz *et al.*, 2013). Interestingly, Nrf2 activation can override inflammatory-dependent suppression of ferroportin 1 transcription, which during an intracellular infection, could be beneficial as sustained Fpn expression and Fe egress could deplete the intracellular Fe pool in macrophages harbouring bacteria (Harada *et al.*, 2011; Nairz *et al.*, 2013). Consistent with this notion, nitric oxide (NO), an important antimicrobial effector, is also implicated in the modulation of Fe homeostasis in macrophages through regulation of ferroportin transcription (Nairz *et al.*, 2013). In this scenario inducible nitric oxide synthase 2 (NOS2)-dependent production of NO in response to *Salmonella enterica* serovar Typhimurium infection induces increased Fpn expression in human and murine macrophages with a concomitant decrease in intracellular iron (Fig. 9.2). The NO effect on Fpn is abrogated in NOS2-deficient macrophages as a consequence of decreased activity of the transcription factor Nrf2, which as previously mentioned, regulates the Fpn1 gene *slc40a1*. Interestingly, despite increased Fpn1 expression in macrophages Nairz *et al.* (2013) report the occurrence of hypoferraemia in mice, presumably due to hepcidin effector function on cells other than the infected macrophages. This would indicate that localized induction of Fpn expression through NO could enhance the ability of infected macrophages to combat intracellular infections by depleting cellular stores of iron that, if acquired by the bacteria, could promote growth. Like *Salmonella*, *S. aureus* is viewed as a facultative intracellular pathogen and it would be of interest to establish if similar phenomena occur during *S. aureus* infection, since staphylococcal infection induces iNOS expression and NO production (Choi *et al.*, 1998; Richardson *et al.*, 2008; Shay *et al.*, 2003).

NRAMP1: an exporter of divalent metal cations

As described in the preceding section, retention of intracellular iron in the context of intracellular infection could be disastrous, as it would enhance the intracellular replication of bacteria. Indeed, pathogens like *Legionella pneumophila*, *Chlamydia* spp. and *S. enterica* demonstrate increased intracellular replication when iron is abundant (Paradkar *et al.*, 2008; Chlosta *et al.*, 2006). While sustained Fpn expression would ostensibly deplete the cytosol of infected macrophages of Fe, many intracellular bacteria that replicate within macrophages reside with membrane-bound vacuoles or phagosomes that are created during bacterial uptake. In macrophages, the natural resistance-associated macrophage protein 1 (NRAMP1), Slc11A1, is a proton-dependent transporter of divalent cations that extrudes metal cations, like iron and magnesium, from the phagosome lumen into the cytosol (see Fig. 9.2) (Gruenheid *et al.*, 1997; Jabado *et al.*, 2000; Searle *et al.*, 1998). Phagocytosis and phagosome formation are beyond the scope of this review, however, the antimicrobial properties of the macrophage phagosome are essential to macrophage innate immune function (Flannagan *et al.*, 2012). The contribution of NRAMP1 to restrict microbial growth through metal depletion is evident in humans with genetic polymorphisms in the NRAMP1 gene as these individuals have an increased risk of mycobacterial infection (van Crevel *et al.*, 2009; Malik *et al.*, 2005). Investigation of the effect of NRAMP1 mutations on intra-macrophage proliferation of *Staphylococcus* spp. has not been performed, but presumably NRAMP1 operates to combat intraphagosomal Fe acquisition mediated by staphylococci. Export of Fe into the cytosol requires that the extruded Fe be rapidly detoxified through sequestration by cytosolic chaperone proteins. The family of chaperone proteins that fulfil this function are the poly(rC)-binding proteins (PCBPs), of which mammalian genomes encode four. These ubiquitously expressed proteins sequester cytosolic Fe transported through DMT1 and presumably NRAMP1 and deliver bound Fe to ferritins for storage or Fpn for export (Leidgens *et al.*, 2013; Shi *et al.*, 2008; Yanatori *et al.*, 2014).

Host haem-iron metabolism

A significant fraction of the total body iron in mammals is consumed through the biogenesis of haem and haemoproteins. As such, haemoproteins represent a significant source of iron that can be seized by pathogenic microbes like *S. aureus* to fulfil their Fe requirements. Haem comprises a porphyrin ring coordinating a single Fe atom and is essential for life, as it serves as an integral prosthetic group for enzymes involved in innate immune responses, mitochondrial electron transport and gas exchange (Ajioka *et al.*, 2006) (Table 9.1). The biosynthesis of haem occurs in virtually every cell of the body with the exception of mature RBCs. Importantly, RBCs do contain an extraordinary amount of haem in the form of haemoglobin, which is synthesized in nucleated RBC precursor cells during erythropoiesis (Vaisman *et al.*, 1997). Haem biosynthesis is initiated and completed in the matrix of the mitochondria where, upon completion, ferrated haem is exported out of the mitochondrion for incorporation into proteins (Ajioka *et al.*, 2006; Hamza and Dailey, 2012). To date the only identified mitochondrial haem export protein is FLVCR1b (feline leukaemia virus subgroup C receptor 1) (Chiabrando *et al.*, 2012). Evidence for its role in haem transport stems from *in vitro* studies demonstrating that FLVCR1b depletion is accompanied by mitochondrial haem retention, which in erythroid progenitors ablates differentiation (Chiabrando *et al.*, 2012).

Host haem factors: Fe sources for staphylococci during infection

Intact RBCs contain in their interior an enormous amount of haem in complex with haemoglobin, which accounts for ~ 96% of the RBC dry weight. The intracellular isolation of haemoglobin represents an effective physical barrier that sequesters Fe away from invading bacteria. Notably, *in vitro* studies of *S. aureus* have revealed that it is endowed with the ability to utilize haem and haem-containing proteins, such as haemoglobin, as iron sources (Mazmanian *et al.*, 2003; Skaar *et al.*, 2004a; Spirig *et al.*, 2013; Villareal *et al.*, 2011). As haemoglobin is confined within intact RBCs, it must first be liberated if it is to be accessible to staphylococcal bacteria. *S. aureus*, in particular, carries multiple genes encoding cytolytic toxins that participate in the disruption of host cell membranes including RBCs, which can release haemoglobin to the extracellular milieu for bacterial consumption (Fig. 9.2). Furthermore, *in vitro* analysis of the secretion of cytolytic toxins such as α-haemolysin has demonstrated that *S. aureus* increases the secretion of cytolytic proteins under iron depleted conditions (Malachowa *et al.*, 2011; Oogai *et al.*, 2011; Torres *et al.*, 2010). Thus, the Fe restriction imposed by the host represents an environmental cue that evokes transcription of genes that aid in the circumvention of the host nutritional immune response.

With the release of haemoglobin and haem from intracellular stores there exists the potential for toxicity to host cells due to the reactivity of Fe. To negate this possibility, there exist scavenger proteins that bind free haemoglobin and haem. Haptoglobin is an abundant blood plasma protein that binds with high-affinity to free haemoglobin limiting its ability to participate in damaging redox reactions. Normally haptoglobin–haemoglobin complexes present in the blood are removed by macrophages of the reticuloendothelial system through recognition of the complex by the macrophage scavenger receptor CD163 (Kristiansen *et al.*, 2001). Additionally, the plasma protein haemopexin binds free haem that may be liberated from haemoproteins in the blood (Vinchi *et al.*, 2008). Surprisingly, despite the versatility of *S. aureus* in terms of iron acquisition, haemopexin in complex with haem represents a nutritional barrier where as haptoglobin–haemoglobin complexes do not (Torres *et al.*, 2006; Dryla *et al.*, 2003).

Host Mn, Cu and Zn metabolism

Manganese, an extremely abundant element on the Earth's crust, is primarily acquired by mammals through ingestion and intestinal absorption. While intoxication as a result of Mn ingestion is rare, Mn absorption can be highly toxic if it is acquired via alternative routes like inhalation. The P5-ATPase ATP13A2, which is heavily expressed in brain tissue and elsewhere, is one protein reported to function in the maintenance of intracellular Mn homeostasis (Chesi *et al.*, 2012; Gitler *et al.*, 2009; Tan *et al.*, 2011). This being said, strong evidence exists that suggest ATP13A2 functions in Zn homeostasis (Kong *et al.*, 2014; Tsunemi and Krainc, 2014). Conceivably ATP13A2 may function as a non-specific transporter of divalent cations across the membranes of late endosome/lysosomes where the protein localizes. Regardless of whether ATP13A2 transports Mn or Zn, it is clear that mutations in this protein profoundly compromise lysosomal function (Tsunemi and Krainc, 2014; Usenovic *et al.*, 2012). The extreme importance of ATP13A2 in metal ion homeostasis and disease emerges from the observation that mutations in this gene are directly linked to early onset Parkinson's and Parkinsonian-like diseases (Di Fonzo *et al.*, 2007; Ramirez *et al.*, 2006).

More recently the gene encoding SLC30A10, a member of the action diffusion facilitator superfamily, has been identified as an Mn transport protein that, when mutated in humans, results in hypermanganesaemia, cirrhosis of the liver and dystonia (Quadri *et al.*, 2012; Tuschl *et al.*, 2012). Direct experimental evidence supporting the role of this protein as a *bona fide* Mn transporter emerges from *in vitro* studies where its expression complements Mn-related growth defects of *Saccharomyces cerevisiae* that are deficient for Mn transport (Tuschl *et al.*, 2012). While this protein is heavily expressed in the brain, liver and to some extent the retina, a complete assessment of its pattern of expression in human tissue has yet to be performed (Quadri *et al.*, 2012). Furthermore, the cellular distribution of SLC30A10, and whether it pumps Mn into the cytosol and/or from the cytosol into the lumen of intracellular organelles, is also not known.

Copper is integral to mammalian development and abnormalities in copper homeostasis underpin devastating human diseases. Because of its redox-potential, copper like iron, is an essential co-factor that is integral to a variety of enzymes like the aforementioned ceruloplasmin, an enzyme that oxidizes Fe in the blood and is necessary for Fe homeostasis (Vashchenko and MacGillivray, 2013) (Table 9.1 and Fig. 9.2). More recently, exciting evidence has emerged indicating copper ions may also participate in the innate microbicidal functions of macrophages (Achard *et al.*, 2012; White *et al.*, 2009), an aspect of Cu function that is discussed below.

A significant fraction of acquired copper comes from dietary sources and the activity of metal transport proteins expressed in the apical membrane of intestinal enterocytes. Indeed, the high-affinity copper transporter 1 (CT1) and possibly the divalent metal ion transporter 1 (DMT1), which also transports Fe, move Cu from the intestinal lumen to the interior of the enterocyte (Arredondo *et al.*, 2003; Zimnicka *et al.*, 2007). Imported copper ions are accepted by intracellular chaperones such as glutathione and Atox1 (also known as HAH1) (Hamza *et al.*, 2001; Maryon *et al.*, 2013), and these metallochaperones prevent metal toxicity within the cell and also facilitate the trafficking of copper to its transport proteins ATP7A and ATP7B (Hamza *et al.*, 2003; Walker *et al.*, 2004). These proteins are P-type ATPases that hydrolyse ATP to power the extrusion of copper across cellular membranes. Although these proteins demonstrate significant functional similarities, they have entirely different patterns of expression in that ATP7B is largely restricted to hepatocytes, while ATP7A is ubiquitously expressed (Lutsenko *et al.*, 2007). The differences in cellular expression of ATP7A and ATP7B have profound effects on the role they play in copper homeostasis. This is evident from the fact that loss of function mutations in ATP7B cause Wilson's disease, which is characterized by copper overload and severe liver pathology in neonates (Tanzi *et al.*, 1993; Thomas *et al.*, 1995). In contrast, mutations impairing ATP7A function cause Menkes Disease, an X-linked disorder characterized by poor distribution of copper throughout the body and severe neurological dysfunction (Vulpe *et al.*, 1993; Kaler, 2011). At steady state both ATP7A and B reside in the trans–golgi network (TGN) where they receive copper from metallochaperones and transport it into the lumen of the golgi apparatus (Braiterman *et al.*, 2011; Petris and Mercer, 1999; Polishchuk *et al.*, 2014). This is entirely consistent with the notion that most of the incorporation of copper into metalloproteins occurs in the golgi. Upon elevation of the intracellular copper concentration, ATP7A and B are redistributed via vesicular transport to the plasma membrane and to the bile canaliculis of the liver (Pase *et al.*, 2004; Petris *et al.*, 1996; Polishchuk *et al.*, 2014), respectively, and consequently, the copper exporter becomes enriched at the plasma membrane where copper extrusion to the extracellular milieu occurs.

Despite the importance of Zn, there is a significant void in our understanding of mechanisms that govern zinc homeostasis. We do know that zinc is transported across membranes through the action of two distinct families of Zn transport proteins. In mammalian cells, the extrusion of cytosolic Zn to the extracellular milieu or to the lumen of intracellular organelles, is catalysed by the SLC30 (also known as ZnT) family of cation diffusion facilitator proteins of which there are 10 (SLC30A1 to 10) (for a recent review see Huang and Tepaamorndech, 2013). Import of Zn from the extracellular environment to the cytosol, or from the lumen of intracellular organelles into the cytosol, is catalysed by the SLC39 family (also known as the ZIP family) of Zinc-iron permease proteins of which there are 14 (SLC39a1 to 14) (Jeong and Eide, 2013). In addition to transport intracellular Zn toxicity is avoided through the action of metal chelating metallothionein (MT) proteins. MTs are low molecular weight, cysteine-rich cytosolic proteins that bind Zn with high affinity and can sequester as many as seven Zn ions per MT protein (reviewed in Davis and Cousins, 2000). MT expression can be induced via activation of the Zinc-sensitive transcription factor MTF-1 in response to increased intracellular concentration of Zn (Heuchel *et al.*, 1994). Additionally, MT expression can be induced in response to cytokine stimulation (i.e. GM-CSF or IL-6) and pattern recognition receptor signalling (i.e. NOD2 and TLR9) indicating infection can modulate MT production and Zn homeostasis (Lahiri and Abraham, 2014; Schroeder and Cousins, 1990; Winters *et al.*, 2010). Interestingly, during acute phase responses, hypozincaemia, the depletion of extracellular Zn, occurs suggesting that the sequestration of extracellular zinc may provide a host advantage to combat infection during early innate immune responses (Liuzzi *et al.*, 2005).

Modulation of Mn, Cu and Zn availability in nutritional immunity

In preceding sections, aspects of the nutritional immune response with respect to Fe sequestration have been discussed; however, additional mechanisms exist to deprive the staphylococci of other important trace metals such as Zn and Mn. Moreover, sequestration may not be the only element of the nutritional immune response, as recent advances indicate that in addition to starving bacteria of trace metals, host immune cells like macrophages may actually employ Cu and Zn to directly intoxicate bacteria that have been phagocytosed.

Calprotectin: a chelator of extracellular Mn and Zn

A common outcome of staphylococcal infection in humans and animal models is the formation of abscesses, fibrin-enclosed lesions containing massive numbers of neutrophils and bacteria (Cheng *et al.*, 2009). Incredibly, within a *S. aureus* abscess, the concentration of Mn and Zn is negligible, a phenomenon attributed to the activity of calprotectin (CP), which is highly expressed by neutrophils and monocytes (Corbin *et al.*, 2008; Dale *et al.*, 1985). Calprotectin is a member of the S100 subfamily of EF-hand Ca^{2+}-binding proteins that exists as a heterotetramer comprised of two S100A8/S100A9 heterodimers (Korndörfer *et al.*, 2007; Damo *et al.*, 2013). In the cytosol of a neutrophil, calprotectin may account for 45% of the total protein content and in abscesses it may reach concentrations as high as 20 mg/ml (Dale *et al.*, 1985). Whether calprotectin is itself directly microbicidal has been a matter of debate; however, it is quite clear that CP-dependent sequestration of divalent metals including Mn and Zn contributes significantly to its antimicrobial functions (Fig. 9.2). Each S100A8/S100A9 dimer possesses two distinct metal binding sites that reside at the dimer interface, with each site being generated through amino acids contributed from each subunit (Damo *et al.*, 2013). As mentioned above CP is a cytosolic protein that is in abundance in abscesses, indicating it is released to the extracellular milieu. The mechanism by which CP is released from the cytoplasm of neutrophils is not entirely clear, however, *in vitro* experiments indicate that, in some instances, the presence of extracellular CP is accompanied by a loss of neutrophil viability (Voganatsi *et al.*, 2001). In contrast, other studies suggest that CP is secreted through actin- and microtubule-dependent mechanisms without a significant decrease in cell viability (Rammes *et al.*, 1997; Urban *et al.*, 2009). Additionally, CP is also a component of neutrophil-derived extracellular traps (NETs) generated through NETosis, the programmed ejection of neutrophil chromatin decorated with its arsenal of antimicrobial proteins (Urban *et al.*, 2009) and reviewed in (Branzk and Papayannopoulos, 2013). Regardless of the mechanisms by which CP is released it is clear that its ability to sequester Mn and Zn inhibits the growth of many staphylococci, including *S. aureus*, *S. lugdunensis* and *S. epidermidis* (Damo *et al.*, 2013).

Physiologically, the consequence of CP-dependent metal sequestration is that staphylococci are more readily killed by ROS generated through cell metabolism because of impaired Mn-dependent SOD function (Kehl-Fie *et al.*, 2011). Furthermore, CP-dependent reduction in SOD function renders bacteria more susceptible to rapid neutrophil mediated killing due to NOX2-derived ROS production and intoxication of the bacteria (Kehl-Fie *et al.*, 2011). Although CP sequesters Mn and Zn, *in vitro* studies investigating the antimicrobial effects of CP on *S. aureus*, *S. epidermidis* and *S. lugdunensis* reveal that its broad-spectrum antimicrobial function can, in part, be attributed to its Mn-binding activity (Damo *et al.*, 2013).

Intracellular mechanisms of Mn and Zn sequestration

NRAMP1: also a transporter of Mn

In addition to CP-mediated Mn and Zn starvation in the extracellular milieu, the aforementioned phagocyte protein NRAMP1 (SlcA11) can catalyse the extrusion of Mn across the membranes of vacuoles harbouring intracellular bacteria within macrophages (Gruenheid *et al.*, 1997; Searle *et al.*, 1998). Like the function of calprotectin extracellularly, depletion of Mn from bacteria confined within a phagosome would render them more sensitive to oxidative killing (Jabado *et al.*, 2000). The consequence of NRAMP Mn starvation may be particularly effective in classically activated M1 macrophages, which are endowed with a robust NOX2-dependent oxidative burst (reviewed in Canton, 2014).

Infection induced intracellular Zn depletion

In addition to NRAMP-dependent depletion of Mn (and Fe) from the macrophage phagosome, recent work indicates that under certain circumstances Zn is also sequestered from intracellular pathogens. Although not demonstrated in the context of a bacterial infection, intracellular infection of macrophages by the fungal pathogen *Histoplasma capsulatum* induces Zn sequestration to deprive this metal from the phagosomes harbouring the yeast (Winters *et al.*, 2010; Subramanian Vignesh *et al.*, 2013). The caveat in this scenario is that Zn depletion is in response to GM-CSF stimulation and requires activation of a transcriptional program regulated by the factors STAT3 and STAT5 (Subramanian Vignesh *et al.*, 2013). Mechanistically G-MCSF cytokine stimulation induces Zn sequestration through a two-pronged approach: First, cytokine induction is reported to induce expression of the Zn transport proteins Slc30A4 and Slc30A7 that localize to the golgi where they extrude Zn from the cytosol into the organelle. Second, in response to G-MCSF stimulation macrophage expression of metallothioneins, specifically MT1, MT2 and MT3 are induced (Subramanian Vignesh *et al.*, 2013). The role of intracellular Zn sequestration by macrophages in the context of staphylococcal infection has yet to be described; however, since *S. aureus* infections can elicit GM-CSF secretion, it is tempting to speculate that a similar mechanism of Zn sequestration could operate on *S. aureus* within the macrophage phagosome. It is difficult to predict how this aspect of macrophage biology would affect intracellular *S. aureus* infection as some studies indicate that the bacteria may persist for a long duration within the macrophage without replicating while others suggest that wild-type bacteria may use cytolytic toxins like the phenol soluble modulins to mediate phagosome escape and replication (Kubica *et al.*, 2008; Grosz *et al.*, 2014). Furthermore, the importance of Zn for colonization and growth of *S. aureus* during infection remain ill defined.

Phagosomal copper and zinc: death metals of the macrophage

The paradigm that has prevailed in the discussion of nutritional immunity thus far is that it is beneficial to deprive microbial invaders of essential trace metals. However, a novel immune function related to Zn and Cu has only recently been appreciated. Just as there exist elaborate mechanisms to prevent metal ion-dependent toxicity, macrophages harness the redox and destructive potential of Cu and Zn to intoxicate phagocytosed bacteria (reviewed in Stafford *et al.*, 2013). Evidence of this phenomenon was revealed nearly a decade ago when analysis of the metal content of phagosomes harbouring *M. tuberculosis* was analysed. Indeed, macrophage-derived phagosomes harbouring *M. tuberculosis* had increased Zn and Cu content as a function of time, a phenomenon that was accompanied by depletion of Fe (Wagner *et al.*, 2005). More recently, transcriptomic analysis of intra-macrophage *M. tuberculosis* revealed that within the phagosome, several genes encoding P1-type ATPases involved in heavy metal export from the bacteria were induced, suggesting the bacteria have a need to detoxify their cytosol (Botella *et al.*, 2011). Furthermore, evidence of metal poisoning within the macrophage emerge from the fact that *M. tuberculosis* infection induces expression of host cell metallothioneins and the zinc exporter Znt1, which extrudes Zn into the phagosome lumen (Botella *et al.*, 2011). Altered Cu transport in macrophages in response to intracellular infection is also believed to affect their innate antimicrobial activity (Achard *et al.*, 2012; White *et al.*, 2009). Indeed, stimulation of macrophages with IFN-γ, a pro-inflammatory cytokine, has been shown to induce the expression of the Cu-transporter Ctr1, which would mediate Cu uptake into the cytosol from the extracellular milieu. The observed increase in Ctr1 expression was accompanied by trafficking of the Cu-transporter ATP7A from the golgi to vesicular compartments overlapping with phagosomal membranes. Strikingly, depletion of ATP7A impaired bacterial killing of *E. coli*, which was exacerbated in a bacterial mutant lacking a high-affinity copper efflux protein (White *et al.*, 2009). In studies investigating the role of copper in the intramacrophage survival of *S. enterica*, bacteria deficient in copper efflux were also less able to survive intracellularly, consistent with the notion that within macrophages, bacteria are intoxicated with copper ions (Osman *et al.*, 2010). Moreover, chelation of copper from macrophages infected with *S. enterica* promoted bacterial survival, indicating these macrophages were compromised in their ability to combat the intracellular *Salmonella* (Achard *et al.*, 2012). Taken together, these observations reveal that, in addition to sequestering essential transition metals, the redox potential of metals and ostensibly their ability to compete for metal binding sites in proteins can be harnessed to inflict cellular damage on microbial prey. The role of macrophage dependent Cu and/or Zn intoxication of staphylococci has not been established, however, *S. aureus* is endowed with genes encoding Zn and Cu export proteins that could contribute to intracellular survival within the macrophage phagosome (Grossoehme *et al.*, 2011; Sitthisak *et al.*, 2007). Lastly, it is of interest that in response to culture in human blood *S. aureus* enhances transcription of copper efflux proteins (Malachowa *et al.*, 2011). It is tempting to speculate that prevention of copper accumulation by the bacteria may represent an important adaptation to the host environment; however, this remains to be established experimentally for staphylococcal species.

Concluding remarks

Transition metals are crucial to life in all kingdoms and this is evident from the ubiquitous incorporation of these metals into the biological function of many essential proteins. The essentiality of these metals and, at the same time, their toxicity, necessitates that organisms maintain homeostatic levels to achieve proper cell function. In this chapter we have highlighted current knowledge on the varied mechanisms that the staphylococci use to acquire these essential metals, their general functions inside the cell, and how homeostatic levels are maintained. For the mammalian host, we have highlighted the mechanisms that are employed to not only sequester these nutrients from invading bacteria (e.g. staphylococci) but also to use their toxic potential to actively poison bacterial invaders. Metal acquisition pathways represent attractive therapeutic targets in our struggle to control the ever-evolving antibiotic-resistance among the staphylococci, and may present viable means of reducing the significant morbidity and mortality that these bacteria inflict on humans and animals.

Acknowledgements

We gratefully acknowledge the funding support from the Canadian Institutes of Health Research and the Natural Sciences and Engineering Research Council of Canada.

References

Abouhamad, W.N., Manson, M., Gibson, M.M., and Higgins, C.F. (1991). Peptide transport and chemotaxis in *Escherichia coli* and *Salmonella typhimurium*: characterization of the dipeptide permease (Dpp) and the dipeptide-binding protein. Mol. Microbiol. *5*, 1035–1047.

Achard, M.E., Stafford, S.L., Bokil, N.J., Chartres, J., Bernhardt, P.V., Schembri, M.A., Sweet, M.J., and McEwan, A.G. (2012). Copper redistribution in murine macrophages in response to Salmonella infection. Biochem. J. *444*, 51–57.

Agency for Toxic Substances and Disease Registry Website Toxicological Profile: Manganese.

Aisen, P. (1994). The transferrin receptor and the release of iron from transferrin. Adv. Exp. Med. Biol. *356*, 31–40.

Aisen, P., and Listowsky, I. (1980). Iron transport and storage proteins. Annu. Rev. Biochem. *49*, 357–393.
Aisen, P., Leibman, A., and Zweier, J. (1978). Stoichiometric and site characteristics of the binding of iron to human transferrin. J. Biol. Chem. *253*, 1930–1937.
Ajioka, R.S., Phillips, J.D., and Kushner, J.P. (2006). Biosynthesis of heme in mammals. Biochim. Biophys. Acta *1763*, 723–736.
Albrecht, A.G., Netz, D.J., Miethke, M., Pierik, A.J., Burghaus, O., Peuckert, F., Lill, R., and Marahiel, M.A. (2010). SufU is an essential iron-sulfur cluster scaffold protein in *Bacillus subtilis*. J. Bacteriol. *192*, 1643–1651.
American Diabetes Association. (2013). Standards of medical care in diabetes–2013. Diabetes Care. *36* (Suppl. 1), S11–S66.
Ammendola, S., Pasquali, P., Pistoia, C., Petrucci, P., Petrarca, P., Rotilio, G., and Battistoni, A. (2007). High-affinity Zn2+ uptake system ZnuABC is required for bacterial zinc homeostasis in intracellular environments and contributes to the virulence of *Salmonella enterica*. Infect. Immun. *75*, 5867–5876.
Anderson, A.S., Scully, I.L., Timofeyeva, Y., Murphy, E., McNeil, L.K., Mininni, T., Nuñez, L., Carriere, M., Singer, C., Dilts, D.A., et al. (2012a). *Staphylococcus aureus* manganese transport protein C is a highly conserved cell surface protein that elicits protective immunity against *S. aureus* and *Staphylococcus epidermidis*. J. Infect. Dis. *205*, 1688–1696.
Anderson, A.S., Miller, A.A., Donald, R.G.K., Scully, I.L., Nanra, J.S., Cooper, D., and Jansen, K.U. (2012b). Development of a multicomponent *Staphylococcus aureus* vaccine designed to counter multiple bacterial virulence factors. Hum. Vaccines Immunother. *8*, 1585–1594.
Ando, M., Manabe, Y.C., Converse, P.J., Miyazaki, E., Harrison, R., Murphy, J.R., and Bishai, W.R. (2003). Characterization of the role of the divalent metal ion-dependent transcriptional repressor MntR in the virulence of *Staphylococcus aureus*. Infect. Immun. *71*, 2584–2590.
Andrade, M.A., Ciccarelli, F.D., Perez-Iratxeta, C., and Bork, P. (2002). NEAT: a domain duplicated in genes near the components of a putative Fe3+ siderophore transporter from Gram-positive pathogenic bacteria. Genome Biol. *3*, research0047.1-research0047.5.
Andreini, C., and Bertini, I. (2012). A bioinformatics view of zinc enzymes. J. Inorg. Biochem. *111*, 150–156.
Andreini, C., Banci, L., Bertini, I., and Rosato, A. (2006). Zinc through the three domains of life. J. Proteome. Res. *5*, 3173–3178.
Andreini, C., Bertini, I., Cavallaro, G., Holliday, G.L., and Thornton, J.M. (2008). Metal ions in biological catalysis: from enzyme databases to general principles. J. Biol. Inorg. Chem. *13*, 1205–1218.
Andrews, S.C. (2011). Making DNA without iron - induction of a manganese-dependent ribonucleotide reductase in response to iron starvation. Mol. Microbiol. *80*, 286–289.
Andrews, S.C., Harrison, P.M., and Guest, J.R. (1989). Cloning, sequencing, and mapping of the bacterioferritin gene (bfr) of *Escherichia coli* K-12. J. Bacteriol. *171*, 3940–3947.
Andrews, S.C., Robinson, A.K., and Rodríguez-Quiñones, F. (2003). Bacterial iron homeostasis. FEMS Microbiol. Rev. *27*, 215–237.
Anguera, I., Del Río, A., Miró, J.M., Matínez-Lacasa, X., Marco, F., Gumá, J.R., Quaglio, G., Claramonte, X., Moreno, A., Mestres, C.A., et al. (2005). *Staphylococcus lugdunensis* infective endocarditis: description of 10 cases and analysis of native valve, prosthetic valve, and pacemaker lead endocarditis clinical profiles. Heart Br. Card. Soc. *91*, e10.
Anzaldi, L.L., and Skaar, E.P. (2010). Overcoming the heme paradox: heme toxicity and tolerance in bacterial pathogens. Infect. Immun. *78*, 4977–4989.
Archibald, F. (1983). *Lactobacillus plantarum*, an organism not requiring iron. FEMS Microbiol. Lett. *19*, 29–32.
Archibald, F.S., and Fridovich, I. (1981). Manganese and defenses against oxygen toxicity in *Lactobacillus plantarum*. J. Bacteriol. *145*, 442–451.
Archibald, F.S., and Fridovich, I. (1982a). The scavenging of superoxide radical by manganous complexes: *in vitro*. Arch. Biochem. Biophys. *214*, 452–463.
Archibald, F.S., and Fridovich, I. (1982b). Investigations of the state of the manganese in *Lactobacillus plantarum*. Arch. Biochem. Biophys. *215*, 589–596.
Armitage, A.E., Eddowes, L.A., Gileadi, U., Cole, S., Spottiswoode, N., Selvakumar, T.A., Ho, L.P., Townsend, A.R., and Drakesmith, H. (2011). Hepcidin regulation by innate immune and infectious stimuli. Blood *118*, 4129–4139.
Arredondo, M., Muñoz, P., Mura, C.V., and Nùñez, M.T. (2003). DMT1, a physiologically relevant apical Cu1+ transporter of intestinal cells. Am. J. Physiol. Cell Physiol. *284*, C1525–C1530.
Baichoo, N., and Helmann, J.D. (2002). Recognition of DNA by Fur: a reinterpretation of the Fur box consensus sequence. J. Bacteriol. *184*, 5826–5832.
Baker, E.N., and Baker, H.M. (2005). Molecular structure, binding properties and dynamics of lactoferrin. Cell. Mol. Life. Sci. *62*, 2531–2539.
Baker, H.M., and Baker, E.N. (2004). Lactoferrin and iron: structural and dynamic aspects of binding and release. Biometals *17*, 209–216.
Baker, J., Sitthisak, S., Sengupta, M., Johnson, M., Jayaswal, R.K., and Morrissey, J.A. (2010). Copper stress induces a global stress response in *Staphylococcus aureus* and represses *sae* and *agr* expression and biofilm formation. Appl. Environ. Microbiol. *76*, 150–160.
Baker, J., Sengupta, M., Jayaswal, R.K., and Morrissey, J.A. (2011). The *Staphylococcus aureus* CsoR regulates both chromosomal and plasmid-encoded copper resistance mechanisms. Environ. Microbiol. *13*, 2495–2507.
Ballal, A., and Manna, A.C. (2009). Regulation of superoxide dismutase (*sod*) genes by SarA in *Staphylococcus aureus*. J. Bacteriol. *191*, 3301–3310.
Banerjee, R., and Ragsdale, S.W. (2003). The many faces of vitamin B12: catalysis by cobalamin-dependent enzymes. Annu. Rev. Biochem. *72*, 209–247.
Bao, G., Clifton, M., Hoette, T.M., Mori, K., Deng, S.X., Qiu, A., Viltard, M., Williams, D., Paragas, N., Leete, T., *et al.* (2010). Iron traffics in circulation bound to a siderocalin (Ngal)-catechol complex. Nat. Chem. Biol. *6*, 602–609.
Barnes, M.H., Leo, C.J., and Brown, N.C. (1998). DNA polymerase III of Gram-positive eubacteria is a zinc

metalloprotein conserving an essential finger-like domain. Biochemistry 37, 15254–15260.

Barra, D., Schininà, M.E., Bannister, W.H., Bannister, J.V., and Bossa, F. (1987). The primary structure of iron-superoxide dismutase from *Photobacterium leiognathi*. J. Biol. Chem. *262*, 1001–1009.

Bartsevich, V.V., and Pakrasi, H.B. (1995). Molecular identification of an ABC transporter complex for manganese: analysis of a cyanobacterial mutant strain impaired in the photosynthetic oxygen evolution process. EMBO J. *14*, 1845–1853.

Beasley, F.C., and Heinrichs, D.E. (2010). Siderophore-mediated iron acquisition in the staphylococci. J. Inorg. Biochem. *104*, 282–288.

Beasley, F.C., Vinés, E.D., Grigg, J.C., Zheng, Q., Liu, S., Lajoie, G.A., Murphy, M.E., and Heinrichs, D.E. (2009). Characterization of staphyloferrin A biosynthetic and transport mutants in *Staphylococcus aureus*. Mol. Microbiol. 72, 947–963.

Beasley, F.C., Marolda, C.L., Cheung, J., Buac, S., and Heinrichs, D.E. (2011). *Staphylococcus aureus* transporters Hts, Sir, and Sst capture iron liberated from human transferrin by Staphyloferrin A, Staphyloferrin B, and catecholamine stress hormones, respectively, and contribute to virulence. Infect. Immun. *79*, 2345–2355.

Berntsson, R.P., Smits, S.H., Schmitt, L., Slotboom, D.J., and Poolman, B. (2010). A structural classification of substrate-binding proteins. FEBS Lett. *584*, 2606–2617.

Biswas, G.D., and Sparling, P.F. (1995). Characterization of lbpA, the structural gene for a lactoferrin receptor in Neisseria gonorrhoeae. Infect. Immun. *63*, 2958–2967.

Biswas, L., Biswas, R., Nerz, C., Ohlsen, K., Schlag, M., Schäfer, T., Lamkemeyer, T., Ziebandt, A.K., Hantke, K., Rosenstein, R., *et al.* (2009). Role of the twin-arginine translocation pathway in Staphylococcus. J. Bacteriol. *191*, 5921–5929.

Bleackley, M.R., and Macgillivray, R.T. (2011). Transition metal homeostasis: from yeast to human disease. Biometals *24*, 785–809.

Böcher, S., Tønning, B., Skov, R.L., and Prag, J. (2009). *Staphylococcus lugdunensis*, a common cause of skin and soft tissue infections in the community. J. Clin. Microbiol. *47*, 946–950.

Boer, J.L., Mulrooney, S.B., and Hausinger, R.P. (2014). Nickel-dependent metalloenzymes. Arch. Biochem. Biophys. *544*, 142–152.

Borezée-Durant, E., Hiron, A., Piard, J.C., and Juillard, V. (2009). Dual role of the oligopeptide permease Opp3 during growth of *Staphylococcus aureus* in milk. Appl. Environ. Microbiol. *75*, 3355–3357.

Botella, H., Peyron, P., Levillain, F., Poincloux, R., Poquet, Y., Brandli, I., Wang, C., Tailleux, L., Tilleul, S., Charrière, G.M., et al. (2011). Mycobacterial p(1)-type ATPases mediate resistance to zinc poisoning in human macrophages. Cell Host Microbe *10*, 248–259.

Boulette, M.L., and Payne, S.M. (2007). Anaerobic regulation of *Shigella flexneri* virulence: ArcA regulates Fur and iron acquisition genes. J. Bacteriol. *189*, 6957–6967.

Bowden, C.F., Verstraete, M.M., Eltis, L.D., and Murphy, M.E. (2014). Hemoglobin binding and catalytic heme extraction by IsdB near iron transporter domains. Biochemistry *53*, 2286–2294.

Braiterman, L., Nyasae, L., Leves, F., and Hubbard, A.L. (2011). Critical roles for the COOH terminus of the Cu-ATPase ATP7B in protein stability, trans-Golgi network retention, copper sensing, and retrograde trafficking. Am. J. Physiol. - Gastrointest. Liver Physiol. *301*, G69–G81.

Branzk, N., and Papayannopoulos, V. (2013). Molecular mechanisms regulating NETosis in infection and disease. Semin. Immunopathol. *35*, 513–530.

Braymer, J.J., and Giedroc, D.P. (2014). Recent developments in copper and zinc homeostasis in bacterial pathogens. Curr. Opin. Chem. Biol. *19*, 59–66.

Brickman, T.J., and McIntosh, M.A. (1992). Overexpression and purification of ferric enterobactin esterase from *Escherichia coli*. Demonstration of enzymatic hydrolysis of enterobactin and its iron complex. J. Biol. Chem. *267*, 12350–12355.

Brock, J.H., and Ng, J. (1983). The effect of desferrioxamine on the growth of *Staphylococcus aureus*, *Yersinia enterocolitica*, and *Streptococcus faecalis* in human serum. Uptake of desferrioxamine-bound iron. FEMS Microbiol. Lett. *20*, 439–442.

Brozyna, J.R., Sheldon, J.R., and Heinrichs, D.E. (2014). Growth promotion of the opportunistic human pathogen, *Staphylococcus lugdunensis*, by heme, hemoglobin, and coculture with *Staphylococcus aureus*. MicrobiologyOpen 3, 182–195.

Le Brun, N.E., Bengtsson, J., and Hederstedt, L. (2000). Genes required for cytochrome c synthesis in *Bacillus subtilis*. Mol. Microbiol. *36*, 638–650.

Bullen, J.J. (1981). The significance of iron in infection. Rev. Infect. Dis. 3, 1127–1138.

Burne, R.A., and Chen, Y.Y. (2000). Bacterial ureases in infectious diseases. Microbes Infect. *2*, 533–542.

Busenlehner, L.S., Pennella, M.A., and Giedroc, D.P. (2003). The SmtB/ArsR family of metalloregulatory transcriptional repressors: Structural insights into prokaryotic metal resistance. FEMS Microbiol. Rev. *27*, 131–143.

Cabiscol, E., Tamarit, J., and Ros, J. (2000). Oxidative stress in bacteria and protein damage by reactive oxygen species. Int. Microbiol. *3*, 3–8.

Campanello, G.C., Ma, Z., Grossoehme, N.E., Guerra, A.J., Ward, B.P., Dimarchi, R.D., Ye, Y., Dann, C.E., and Giedroc, D.P. (2013). Allosteric inhibition of a zinc-sensing transcriptional repressor: insights into the arsenic repressor (ArsR) family. J. Mol. Biol. *425*, 1143–1157.

Canton, J. (2014). Phagosome maturation in polarized macrophages. J. Leucoc. Biol. *96*, 729–738.

Cao, J., Woodhall, M.R., Alvarez, J., Cartron, M.L., and Andrews, S.C. (2007). EfeUOB (YcdNOB) is a tripartite, acid-induced and CpxAR-regulated, low-pH Fe2+ transporter that is cryptic in Escherichia coli K-12 but functional in E. coli O157:H7. Mol. Microbiol. *65*, 857–875.

Carrano, C.J., and Raymond, K.N. (1979). Ferric ion sequestering agents. 2. Kinetics and mechanism of iron removal from transferrin by enterobactin and synthetic tricatechols. J. Am. Chem. Soc. *101*, 5401–5404.

Cartron, M.L., Maddocks, S., Gillingham, P., Craven, C.J., and Andrews, S.C. (2006). Feo–transport of ferrous iron into bacteria. Biometals *19*, 143–157.

Cerasi, M., Ammendola, S., and Battistoni, A. (2013). Competition for zinc binding in the host-pathogen interaction. Front Cell. Infect. Microbiol. *3*, 108.

Cerasi, M., Liu, J.Z., Ammendola, S., Poe, A.J., Petrarca, P., Pesciaroli, M., Pasquali, P., Raffatellu, M., and Battistoni, A. (2014). The ZupT transporter plays an important role in zinc homeostasis and contributes to *Salmonella enterica* virulence. Metallomics *6*, 845–853.

Chakravorty, D.K., Wang, B., Lee, C.W., Giedroc, D.P., and Merz, K.M. (2012). Simulations of allosteric motions in the zinc sensor CzrA. J. Am. Chem. Soc. *134*, 3367–3376.

Chen, C.Y., and Morse, S.A. (2001). Identification and characterization of a high-affinity zinc uptake system in *Neisseria gonorrhoeae*. FEMS Microbiol. Lett. *202*, 67–71.

Cheng, A.G., Kim, H.K., Burts, M.L., Krausz, T., Schneewind, O., and Missiakas, D.M. (2009). Genetic requirements for *Staphylococcus aureus* abscess formation and persistence in host tissues. FASEB J. *23*, 3393–3404.

Chesi, A., Kilaru, A., Fang, X., Cooper, A.A., and Gitler, A.D. (2012). The role of the Parkinson's disease gene PARK9 in essential cellular pathways and the manganese homeostasis network in yeast. PLoS One *7*, e34178.

Cheung, J., Beasley, F.C., Liu, S., Lajoie, G.A., and Heinrichs, D.E. (2009). Molecular characterization of staphyloferrin B biosynthesis in *Staphylococcus aureus*. Mol. Microbiol. *74*, 594–608.

Cheung, J., Murphy, M.E., and Heinrichs, D.E. (2012). Discovery of an iron-regulated citrate synthase in *Staphylococcus aureus*. Chem. Biol. *19*, 1568–1578.

Chiabrando, D., Marro, S., Mercurio, S., Giorgi, C., Petrillo, S., Vinchi, F., Fiorito, V., Fagoonee, S., Camporeale, A., Turco, E., *et al.* (2012). The mitochondrial heme exporter FLVCR1b mediates erythroid differentiation. J. Clin. Invest. *122*, 4569–4579.

Chlosta, S., Fishman, D.S., Harrington, L., Johnson, E.E., Knutson, M.D., Wessling-Resnick, M., and Cherayil, B.J. (2006). The iron efflux protein ferroportin regulates the intracellular growth of *Salmonella enterica*. Infect. Immun. *74*, 3065–3067.

Choi, K.C., Jeong, T.K., Lee, S.C., Kim, S.W., Kim, N.H., and Lee, K.Y. (1998). Nitric oxide is a marker of peritonitis in patients on continuous ambulatory peritoneal dialysis. Adv. Perit. Dial. *14*, 173–179.

Christianson, D.W. (1997). Structural chemistry and biology of manganese metalloenzymes. Prog. Biophys. Mol. Biol. *67*, 217–252.

Clarke, S.R., Brummell, K.J., Horsburgh, M.J., McDowell, P.W., Mohamad, S.A., Stapleton, M.R., Acevedo, J., Read, R.C., Day, N.P., Peacock, S.J., *et al.* (2006). Identification of *in vivo*-expressed antigens of *Staphylococcus aureus* and their use in vaccinations for protection against nasal carriage. J. Infect. Dis. *193*, 1098–1108.

Clements, M.O., Watson, S.P., and Foster, S.J. (1999). Characterization of the major superoxide dismutase of *Staphylococcus aureus* and its role in starvation survival, stress resistance, and pathogenicity. J. Bacteriol. *181*, 3898–3903.

Coassin, M., Ursini, F., and Bindoli, A. (1992). Antioxidant effect of manganese. Arch. Biochem. Biophys. *299*, 330–333.

Cockayne, A., Hill, P.J., Powell, N.B., Bishop, K., Sims, C., and Williams, P. (1998). Molecular cloning of a 32-kilodalton lipoprotein component of a novel iron-regulated *Staphylococcus epidermidis* ABC transporter. Infect. Immun. *66*, 3767–3774.

Collins, C.M., and D'Orazio, S.E. (1993). Bacterial ureases: structure, regulation of expression and role in pathogenesis. Mol. Microbiol. *9*, 907–913.

Collins, F.M., and Lascelles, J. (1962). The effect of growth conditions on oxidative and dehydrogenase activity in *Staphylococcus aureus*. J. Gen. Microbiol. *29*, 531–535.

Cooper, J.D., Hannauer, M., Marolda, C.L., Briere, L.A., and Heinrichs, D.E. (2014). Identification of a positively charged platform in *Staphylococcus aureus* HtsA that is essential for ferric staphyloferrin A transport. Biochemistry *53*, 5060–5069.

Corbin, B.D., Seeley, E.H., Raab, A., Feldmann, J., Miller, M.R., Torres, V.J., Anderson, K.L., Dattilo, B.M., Dunman, P.M., Gerads, R., *et al.* (2008). Metal chelation and inhibition of bacterial growth in tissue abscesses. Science *319*, 962–965.

Cornelissen, C.N., Biswas, G.D., Tsai, J., Paruchuri, D.K., Thompson, S.A., and Sparling, P.F. (1992). Gonococcal transferrin-binding protein 1 is required for transferrin utilization and is homologous to TonB-dependent outer membrane receptors. J. Bacteriol. *174*, 5788–5797.

Cotton, J.L., Tao, J., and Balibar, C.J. (2009). Identification and characterization of the *Staphylococcus aureus* gene cluster coding for staphyloferrin A. Biochemistry *48*, 1025–1035.

Coulter, S.N., Schwan, W.R., Ng, E.Y., Langhorne, M.H., Ritchie, H.D., Westbrock-Wadman, S., Hufnagle, W.O., Folger, K.R., Bayer, A.S., and Stover, C.K. (1998). *Staphylococcus aureus* genetic loci impacting growth and survival in multiple infection environments. Mol. Microbiol. *30*, 393–404.

Courcol, R.J., Trivier, D., Bissinger, M.C., Martin, G.R., and Brown, M.R. (1997). Siderophore production by *Staphylococcus aureus* and identification of iron-regulated proteins. Infect. Immun. *65*, 1944–1948.

Cox, G., Thompson, G.S., Jenkins, H.T., Peske, F., Savelsbergh, A., Rodnina, M.V., Wintermeyer, W., Homans, S.W., Edwards, T.A., and O'Neill, A.J. (2012). Ribosome clearance by FusB-type proteins mediates resistance to the antibiotic fusidic acid. Proc. Natl. Acad. Sci. U.S.A. *109*, 2102–2107.

van Crevel, R., Parwati, I., Sahiratmadja, E., Marzuki, S., Ottenhoff, T.H., Netea, M.G., van der Ven, A., Nelwan, R.H., van der Meer, J.W., Alisjahbana, B., *et al.* (2009). Infection with *Mycobacterium tuberculosis* Beijing genotype strains is associated with polymorphisms in SLC11A1/NRAMP1 in Indonesian patients with tuberculosis. J. Infect. Dis. *200*, 1671–1674.

Dailey, H.A., Septer, A.N., Daugherty, L., Thames, D., Gerdes, S., Stabb, E.V., Dunn, A.K., Dailey, T.A., and Phillips, J.D. (2011). The *Escherichia coli* protein YfeX functions as a porphyrinogen oxidase, not a heme dechelatase. MBio *2*, e00248–11.

Dale, I., Brandtzaeg, P., Fagerhol, M.K., and Scott, H. (1985). Distribution of a new myelomonocytic antigen (L1) in human peripheral blood leucocytes. Immunofluorescence and immunoperoxidase staining features in comparison with lysozyme and lactoferrin. Am. J. Clin. Pathol. *84*, 24–34.

Dale, S.E., Doherty-Kirby, A., Lajoie, G., and Heinrichs, D.E. (2004a). Role of siderophore biosynthesis in virulence of *Staphylococcus aureus*: identification and characterization of genes involved in production of a siderophore. Infect. Immun. *72*, 29–37.

Dale, S.E., Sebulsky, M.T., and Heinrichs, D.E. (2004b). Involvement of SirABC in iron-siderophore import in *Staphylococcus aureus*. J. Bacteriol. *186*, 8356–8362.

Dalet, K., Gouin, E., Cenatiempo, Y., Cossart, P., and Héchard, Y. (1999). Characterisation of a new operon encoding a Zur-like protein and an associated ABC zinc permease in Listeria monocytogenes. FEMS Microbiol. Lett. *174*, 111–116.

Daly, M.J., Gaidamakova, E.K., Matrosova, V.Y., Kiang, J.G., Fukumoto, R., Lee, D.Y., Wehr, N.B., Viteri, G.A., Berlett, B.S., and Levine, R.L. (2010). Small-molecule antioxidant proteome-shields in *Deinococcus radiodurans*. PLoS One 5, e12570.

Damo, S.M., Kehl-Fie, T.E., Sugitani, N., Holt, M.E., Rathi, S., Murphy, W.J., Zhang, Y., Betz, C., Hench, L., Fritz, G., *et al.* (2013). Molecular basis for manganese sequestration by calprotectin and roles in the innate immune response to invading bacterial pathogens. Proc. Natl. Acad. Sci. U.S.A. *110*, 3841–3846.

Date, S.V., Modrusan, Z., Lawrence, M., Morisaki, J.H., Toy, K., Shah, I.M., Kim, J., Park, S., Xu, M., Basuino, L., *et al.* (2014). Global gene expression of methicillin-resistant *Staphylococcus aureus* USA300 during human and mouse infection. J. Infect. Dis. *209*, 1542–1550.

Dautry-Varsat, A., Ciechanover, A., and Lodish, H.F. (1983). pH and the recycling of transferrin during receptor-mediated endocytosis. Proc. Natl. Acad. Sci. U.S.A. *80*, 2258–2262.

Davidson, A.L., Dassa, E., Orelle, C., and Chen, J. (2008). Structure, function, and evolution of bacterial ATP-binding cassette systems. Microbiol. Mol. Biol. Rev. *72*, 317–364.

Davis, S.R., and Cousins, R.J. (2000). Metallothionein expression in animals: a physiological perspective on function. J. Nutr. *130*, 1085–1088.

Davis, L.M., Kakuda, T., and DiRita, V.J. (2009). A *Campylobacter jejuni znuA* orthologue is essential for growth in low-zinc environments and chick colonization. J. Bacteriol. *191*, 1631–1640.

Desrosiers, D.C., Bearden, S.W., Mier, I., Abney, J., Paulley, J.T., Fetherston, J.D., Salazar, J.C., Radolf, J.D., and Perry, R.D. (2010). Znu is the predominant zinc importer in *Yersinia pestis* during *in vitro* growth but is not essential for virulence. Infect. Immun. *78*, 5163–5177.

Devireddy, L.R., Hart, D.O., Goetz, D.H., and Green, M.R. (2010). A mammalian siderophore synthesized by an enzyme with a bacterial homolog involved in enterobactin production. Cell *141*, 1006–1017.

Dickson, C.F., Kumar, K.K., Jacques, D.A., Malmirchegini, G.R., Spirig, T., Mackay, J.P., Clubb, R.T., Guss, J.M., and Gell, D.A. (2014). Structure of the hemoglobin-IsdH complex reveals the molecular basis of iron capture by *Staphylococcus aureus*. J. Biol. Chem. *289*, 6728–6738.

Dintilhac, A., Alloing, G., Granadel, C., and Claverys, J.P. (1997). Competence and virulence of *Streptococcus pneumoniae*: Adc and PsaA mutants exhibit a requirement for Zn and Mn resulting from inactivation of putative ABC metal permeases. Mol. Microbiol. *25*, 727–739.

Donovan, A., Lima, C.A., Pinkus, J.L., Pinkus, G.S., Zon, L.I., Robine, S., and Andrews, N.C. (2005). The iron exporter ferroportin/Slc40a1 is essential for iron homeostasis. Cell Metab. *1*, 191–200.

Drechsel, H., Freund, S., Nicholson, G., Haag, H., Jung, O., Zähner, H., and Jung, G. (1993). Purification and chemical characterization of staphyloferrin B, a hydrophilic siderophore from staphylococci. Biometals *6*, 185–192.

Dryla, A., Gelbmann, D., von Gabain, A., and Nagy, E. (2003). Identification of a novel iron regulated staphylococcal surface protein with haptoglobin-haemoglobin binding activity. Mol. Microbiol. *49*, 37–53.

Dryla, A., Hoffmann, B., Gelbmann, D., Giefing, C., Hanner, M., Meinke, A., Anderson, A.S., Koppensteiner, W., Konrat, R., von Gabain, A., *et al.* (2007). High-affinity binding of the staphylococcal HarA protein to haptoglobin and hemoglobin involves a domain with an antiparallel eight-stranded beta-barrel fold. J. Bacteriol. *189*, 254–264.

Duan, X., Yang, J., Ren, B., Tan, G., and Ding, H. (2009). Reactivity of nitric oxide with the [4Fe-4S] cluster of dihydroxyacid dehydratase from *Escherichia coli*. Biochem. J. *417*, 783–789.

Duncan, T., Trewick, S.C., Koivisto, P., Bates, P.A., Lindahl, T., and Sedgwick, B. (2002). Reversal of DNA alkylation damage by two human dioxygenases. Proc. Natl. Acad. Sci. U.S.A. *99*, 16660–16665.

Eid, C., Hémadi, M., Ha-Duong, N.T., and El Hage Chahine, J.M. (2014). Iron uptake and transfer from ceruloplasmin to transferrin. Biochim. Biophys. Acta *1840*, 1771–1781.

Eisenhofer, G., Aneman, A., Hooper, D., Rundqvist, B., and Friberg, P. (1996). Mesenteric organ production, hepatic metabolism, and renal elimination of norepinephrine and its metabolites in humans. J. Neurochem. *66*, 1565–1573.

Eitinger, T., Suhr, J., Moore, L., and Smith, J.A. (2005). Secondary transporters for nickel and cobalt ions: theme and variations. Biometals *18*, 399–405.

Escolar, L., Pérez-Martín, J., and de Lorenzo, V. (1999). Opening the iron box: transcriptional metalloregulation by the Fur protein. J. Bacteriol. *181*, 6223–6229.

Evans, R.W., Crawley, J.B., Joannou, C.L., and Sharma, N.D. (1999). Iron proteins. In Iron and Infection, J.J. Bullen, and E. Griffiths, eds. (New York, New York: John Wiley and Sons Ltd.), pp. 27–86.

Everse, J., and Hsia, N. (1997). The toxicities of native and modified hemoglobins. Free. Radic. Biol. Med. *22*, 1075–1099.

Fenton, H.J.H. (1894). Oxidation of tartaric acid in the presence of iron. J. Chem. Soc. *65*, 899–910.

Fisher, W.R., Taniuchi, H., and Anfinsen, C.B. (1973). On the role of heme in the formation of the structure of cytochrome c. J. Biol. Chem. *248*, 3188–3195.

Flannagan, R.S., Jaumouillé, V., and Grinstein, S. (2012). The cell biology of phagocytosis. Annu. Rev. Pathol. *7*, 61–98.

Flo, T.H., Smith, K.D., Sato, S., Rodriguez, D.J., Holmes, M.A., Strong, R.K., Akira, S., and Aderem, A. (2004). Lipocalin 2 mediates an innate immune response to

bacterial infection by sequestrating iron. Nature *432*, 917–921.
Fonner, B.A., Tripet, B.P., Eilers, B., Stanisich, J., Sullivan-Springhetti, R.K., Moore, R., Liu, M., Lei, B., and Copie, V. (2014). Solution structure and molecular determinants of hemoglobin binding of the first NEAT domain of IsdB in *Staphylococcus aureus*. Biochemistry *53*, 3922–3933.
Di Fonzo, A., Chien, H.F., Socal, M., Giraudo, S., Tassorelli, C., Iliceto, G., Fabbrini, G., Marconi, R., Fincati, E., Abbruzzese, G., *et al.* (2007). ATP13A2 missense mutations in juvenile parkinsonism and young onset Parkinson disease. Neurology *68*, 1557–1562.
Freedman, L.P., Luisi, B.F., Korszun, Z.R., Basavappa, R., Sigler, P.B., and Yamamoto, K.R. (1988). The function and structure of the metal coordination sites within the glucocorticoid receptor DNA binding domain. Nature *334*, 543–546.
Freestone, P.P., Lyte, M., Neal, C.P., Maggs, A.F., Haigh, R.D., and Williams, P.H. (2000). The mammalian neuroendocrine hormone norepinephrine supplies iron for bacterial growth in the presence of transferrin or lactoferrin. J. Bacteriol. *182*, 6091–6098.
Freestone, P.P., Williams, P.H., Haigh, R.D., Maggs, A.F., Neal, C.P., and Lyte, M. (2002). Growth stimulation of intestinal commensal *Escherichia coli* by catecholamines: a possible contributory factor in trauma-induced sepsis. Shock *18*, 465–470.
Freestone, P.P., Sandrini, S.M., Haigh, R.D., and Lyte, M. (2008). Microbial endocrinology: how stress influences susceptibility to infection. Trends Microbiol. *16*, 55–64.
Friedman, D.B., Stauff, D.L., Pishchany, G., Whitwell, C.W., Torres, V.J., and Skaar, E.P. (2006). *Staphylococcus aureus* redirects central metabolism to increase iron availability. PLOS Pathog. *2*, e87.
Fuqua, B.K., Lu, Y., Darshan, D., Frazer, D.M., Wilkins, S.J., Wolkow, N., Bell, A.G., Hsu, J., Yu, C.C., Chen, H., *et al.* (2014). The multicopper ferroxidase hephaestin enhances intestinal iron absorption in mice. PLoS One *9*, e98792.
Gaballa, A., Antelmann, H., Aguilar, C., Khakh, S.K., Song, K.B., Smaldone, G.T., and Helmann, J.D. (2008). The *Bacillus subtilis* iron-sparing response is mediated by a Fur-regulated small RNA and three small, basic proteins. Proc. Natl. Acad. Sci. U.S.A. *105*, 11927–11932.
Gagliardo, B., Kubat, N., Faye, A., Jaouen, M., Durel, B., Deschemin, J.C., Canonne-Hergaux, F., Sari, M.A., and Vaulont, S. (2009). Pro-hepcidin is unable to degrade the iron exporter ferroportin unless maturated by a furin-dependent process. J. Hepatol. *50*, 394–401.
Gaither, L.A., and Eide, D.J. (2001). Eukaryotic zinc transporters and their regulation. Biometals *14*, 251–270.
Ganz, T. (2009). Iron in innate immunity: starve the invaders. Curr. Opin. Immunol. *21*, 63–67.
Gatermann, S., and Crossley, K.B. (2009). Urinary tract infections. In Staphylococci in Human Disease, K.B. Crossley, K.K. Jefferson, G.L. Archer, and V.G. Fowler, eds. (Wiley-Blackwell), pp. 455–469.
Gatermann, S., and Marre, R. (1989). Cloning and expression of *Staphylococcus saprophyticus* urease gene sequences in *Staphylococcus carnosus* and contribution of the enzyme to virulence. Infect. Immun. *57*, 2998–3002.
Gatermann, S., John, J., and Marre, R. (1989). *Staphylococcus saprophyticus* urease: characterization and contribution to uropathogenicity in unobstructed urinary tract infection of rats. Infect. Immun. *57*, 110–116.
Gaudin, C.F., Grigg, J.C., Arrieta, A.L., and Murphy, M.E. (2011). Unique heme-iron coordination by the hemoglobin receptor IsdB of *Staphylococcus aureus*. Biochemistry *50*, 5443–5452.
Gaupp, R., Ledala, N., and Somerville, G.A. (2012). Staphylococcal response to oxidative stress. Front. Cell. Infect. Microbiol. *2*, 33.
Gitler, A.D., Chesi, A., Geddie, M.L., Strathearn, K.E., Hamamichi, S., Hill, K.J., Caldwell, K.A., Caldwell, G.A., Cooper, A.A., Rochet, J.C., *et al.* (2009). Alpha-synuclein is part of a diverse and highly conserved interaction network that includes PARK9 and manganese toxicity. Nat. Genet. *41*, 308–315.
Gitlin, D., and Janeway, C.A. (1960). Turnover of the copper and protein moieties of ceruloplasmin. Nature *185*, 693.
Glasfeld, A., Guedon, E., Helmann, J.D., and Brennan, R.G. (2003). Structure of the manganese-bound manganese transport regulator of *Bacillus subtilis*. Nat. Struct. Biol. *10*, 652–657.
Goetz, D.H., Holmes, M.A., Borregaard, N., Bluhm, M.E., Raymond, K.N., and Strong, R.K. (2002). The neutrophil lipocalin NGAL is a bacteriostatic agent that interferes with siderophore-mediated iron acquisition. Mol. Cell *10*, 1033–1043.
Gomme, P.T., McCann, K.B., and Bertolini, J. (2005). Transferrin: structure, function and potential therapeutic actions. Drug Discov. Today *10*, 267–273.
Graham, A.I., Hunt, S., Stokes, S.L., Bramall, N., Bunch, J., Cox, A.G., McLeod, C.W., and Poole, R.K. (2009). Severe zinc depletion of *Escherichia coli*: roles for high affinity zinc binding by ZinT, zinc transport and zinc-independent proteins. J. Biol. Chem. *284*, 18377–18389.
Grass, G., Wong, M.D., Rosen, B.P., Smith, R.L., and Rensing, C. (2002). ZupT is a Zn(II) uptake system in *Escherichia coli*. J. Bacteriol. *184*, 864–866.
Grass, G., Franke, S., Taudte, N., Nies, D.H., Kucharski, L.M., Maguire, M.E., and Rensing, C. (2005a). The metal permease ZupT from *Escherichia coli* is a transporter with a broad substrate spectrum. J. Bacteriol. *187*, 1604–1611.
Grass, G., Otto, M., Fricke, B., Haney, C.J., Rensing, C., Nies, D.H., and Munkelt, D. (2005b). FieF (YiiP) from *Escherichia coli* mediates decreased cellular accumulation of iron and relieves iron stress. Arch. Microbiol. *183*, 9–18.
Gribenko, A., Mosyak, L., Ghosh, S., Parris, K., Svenson, K., Moran, J., Chu, L., Li, S., Liu, T., Woods, V.L., et al. (2013). Three-dimensional structure and biophysical characterization of *Staphylococcus aureus* cell surface antigen-manganese transporter MntC. J. Mol. Biol. *425*, 3429–3445.
Griffiths, E. (1999). Iron in biological systems. In Iron and Infection, J.J. Bullen, and E. Griffiths, eds. (New York: John Wiley & Sons, Ltd.), pp. 1–26.
Griffiths, T.A., Mauk, A.G., and MacGillivray, R.T. (2005). Recombinant expression and functional characterization

of human hephaestin: a multicopper oxidase with ferroxidase activity. Biochemistry *44*, 14725–14731.

Grigg, J.C., Vermeiren, C.L., Heinrichs, D.E., and Murphy, M.E. (2007a). Heme coordination by *Staphylococcus aureus* IsdE. J. Biol. Chem. *282*, 28815–28822.

Grigg, J.C., Vermeiren, C.L., Heinrichs, D.E., and Murphy, M.E. (2007b). Haem recognition by a *Staphylococcus aureus* NEAT domain. Mol. Microbiol. *63*, 139–149.

Grigg, J.C., Cooper, J.D., Cheung, J., Heinrichs, D.E., and Murphy, M.E. (2010a). The *Staphylococcus aureus* siderophore receptor HtsA undergoes localized conformational changes to enclose staphyloferrin A in an arginine-rich binding pocket. J. Biol. Chem. *285*, 11162–11171.

Grigg, J.C., Cheung, J., Heinrichs, D.E., and Murphy, M.E. (2010b). Specificity of Staphyloferrin B recognition by the SirA receptor from *Staphylococcus aureus*. J. Biol. Chem. *285*, 34579–34588.

Grigg, J.C., Ukpabi, G., Gaudin, C.F., and Murphy, M.E. (2010c). Structural biology of heme binding in the *Staphylococcus aureus* Isd system. J. Inorg. Biochem. *104*, 341–348.

Grosse, C., Scherer, J., Koch, D., Otto, M., Taudte, N., and Grass, G. (2006). A new ferrous iron-uptake transporter, EfeU (YcdN), from *Escherichia coli*. Mol. Microbiol. *62*, 120–131.

Grossoehme, N., Kehl-Fie, T.E., Ma, Z., Adams, K.W., Cowart, D.M., Scott, R.A., Skaar, E.P., and Giedroc, D.P. (2011). Control of copper resistance and inorganic sulfur metabolism by paralogous regulators in *Staphylococcus aureus*. J. Biol. Chem. *286*, 13522–13531.

Grosz, M., Kolter, J., Paprotka, K., Winkler, A.-C., Schäfer, D., Chatterjee, S.S., Geiger, T., Wolz, C., Ohlsen, K., Otto, M., et al. (2014). Cytoplasmic replication of *Staphylococcus aureus* upon phagosomal escape triggered by phenol-soluble modulin α. Cell. Microbiol. *16*, 451–465.

Gruenheid, S., Pinner, E., Desjardins, M., and Gros, P. (1997). Natural resistance to infection with intracellular pathogens: the Nramp1 protein is recruited to the membrane of the phagosome. J. Exp. Med. *185*, 717–730.

Grüter, L., Endres, M., and Gatermann, S. (1992). Cloning and expression of *Staphylococcus epidermidis* urease gene sequences in *Staphylococcus carnosus*. FEMS Microbiol. Lett. 72, 33–35.

Guerinot, M.L. (2000). The ZIP family of metal transporters. Biochim. Biophys. Acta *1465*, 190–198.

Guerra, A.J., and Giedroc, D.P. (2012). Metal site occupancy and allosteric switching in bacterial metal sensor proteins. Arch. Biochem. Biophys. *519*, 210–222.

Guffanti, A.A., Wei, Y., Rood, S.V., and Krulwich, T.A. (2002). An antiport mechanism for a member of the cation diffusion facilitator family: divalent cations efflux in exchange for K^+ and H^+. Mol. Microbiol. *45*, 145–153.

Guilhen, C., Taha, M.K., and Veyrier, F.J. (2013). Role of transition metal exporters in virulence: the example of *Neisseria meningitidis*. Front. Cell. Infect. Microbiol. 3, 102.

Gunasekera, T.S., Herre, A.H., and Crowder, M.W. (2009). Absence of ZnuABC-mediated zinc uptake affects virulence-associated phenotypes of uropathogenic *Escherichia coli* CFT073 under Zn(II)-depleted conditions. FEMS Microbiol. Lett. *300*, 36–41.

Gutteridge, J.M. (1987). The antioxidant activity of haptoglobin towards haemoglobin-stimulated lipid peroxidation. Biochim. Biophys. Acta *917*, 219–223.

Gutteridge, J.M., and Smith, A. (1988). Antioxidant protection by haemopexin of haem-stimulated lipid peroxidation. Biochem. J. *256*, 861–865.

Gutteridge, J.M., Rowley, D.A., and Halliwell, B. (1982). Superoxide-dependent formation of hydroxyl radicals and lipid peroxidation in the presence of iron salts. Detection of 'catalytic' iron and anti-oxidant activity in extracellular fluids. Biochem. J. *206*, 605–609.

Haag, H., Fiedler, H.P., Meiwes, J., Drechsel, H., Jung, G., and Zähner, H. (1994). Isolation and biological characterization of staphyloferrin B, a compound with siderophore activity from staphylococci. FEMS Microbiol. Lett. *115*, 125–130.

Haber, F., and Weiss, J. (1932). Über die Katalyse des Hydroperoxydes (On the catalysis of hydroperoxide). Naturwissenschaften *20*, 948–950.

Haley, K.P., Janson, E.M., Heilbronner, S., Foster, T.J., and Skaar, E.P. (2011). *Staphylococcus lugdunensis* IsdG liberates iron from host heme. J. Bacteriol. *193*, 4749–4757.

Hallé, F., and Meyer, J.M. (1992). Ferrisiderophore reductases of *Pseudomonas*. Purification, properties and cellular location of the *Pseudomonas aeruginosa* ferripyoverdine reductase. Eur. J. Biochem. FEBS *209*, 613–620.

Hamza, I., and Dailey, H.A. (2012). One ring to rule them all: trafficking of heme and heme synthesis intermediates in the metazoans. Biochim. Biophys. Acta *1823*, 1617–1632.

Hamza, I., Faisst, A., Prohaska, J., Chen, J., Gruss, P., and Gitlin, J.D. (2001). The metallochaperone Atox1 plays a critical role in perinatal copper homeostasis. Proc. Natl. Acad. Sci. U.S.A. *98*, 6848–6852.

Hamza, I., Prohaska, J., and Gitlin, J.D. (2003). Essential role for Atox1 in the copper-mediated intracellular trafficking of the Menkes ATPase. Proc. Natl. Acad. Sci. U.S.A. *100*, 1215–1220.

Handke, L.D., Hawkins, J.C., Miller, A.A., Jansen, K.U., and Anderson, A.S. (2013). Regulation of *Staphylococcus aureus* MntC expression and its role in response to oxidative stress. PLoS One *8*, e77874.

Hannauer, M., Barda, Y., Mislin, G.L., Shanzer, A., and Schalk, I.J. (2010). The ferrichrome uptake pathway in *Pseudomonas aeruginosa* involves an iron release mechanism with acylation of the siderophore and recycling of the modified desferrichrome. J. Bacteriol. *192*, 1212–1220.

Hantke, K. (1987). Ferrous iron transport mutants in *Escherichia coli* K12. FEMS Microbiol. Lett. *44*, 53–57.

Hantke, K. (2003). Is the bacterial ferrous iron transporter FeoB a living fossil? Trends Microbiol. *11*, 192–195.

Hantke, K. (2005). Bacterial zinc uptake and regulators. Curr. Opin. Microbiol. *8*, 196–202.

Hao, Z., Chen, S., and Wilson, D.B. (1999). Cloning, expression, and characterization of cadmium and manganese uptake genes from *Lactobacillus plantarum*. Appl. Environ. Microbiol. *65*, 4746–4752.

Harada, N., Kanayama, M., Maruyama, A., Yoshida, A., Tazumi, K., Hosoya, T., Mimura, J., Toki, T., Maher, J.M., Yamamoto, M., and Itoh, K. (2011). Nrf2 regulates ferroportin 1-mediated iron efflux and counteracts lipopolysaccharide-induced ferroportin 1 mRNA suppression in macrophages. Arch. Biochem. Biophys. *508*, 101–109.

Harrison, P.M., and Arosio, P. (1996). The ferritins: molecular properties, iron storage function and cellular regulation. Biochim. Biophys. Acta *1275*, 161–203.

Hartford, T., O'Brien, S., Andrew, P.W., Jones, D., and Roberts, I.S. (1993). Utilization of transferrin-bound iron by *Listeria monocytogenes*. FEMS Microbiol. Lett. *108*, 311–318.

Hartmann, A., and Braun, V. (1980). Iron transport in *Escherichia coli*: uptake and modification of ferrichrome. J. Bacteriol. *143*, 246–255.

Hebbeln, P., and Eitinger, T. (2004). Heterologous production and characterization of bacterial nickel/cobalt permeases. FEMS Microbiol. Lett. *230*, 129–135.

Heinrichs, J.H., Gatlin, L.E., Kunsch, C., Choi, G.H., and Hanson, M.S. (1999). Identification and characterization of SirA, an iron-regulated protein from *Staphylococcus aureus*. J. Bacteriol. *181*, 1436–1443.

Hensley, M.P., Tierney, D.L., and Crowder, M.W. (2011). Zn(II) binding to *Escherichia coli* 70S ribosomes. Biochemistry *50*, 9937–9939.

Herzberg, M., Bauer, L., and Nies, D.H. (2014). Deletion of the *zupT* gene for a zinc importer influences zinc pools in Cupriavidus metallidurans CH34. Metallomics *6*, 421–436.

Heuchel, R., Radtke, F., Georgiev, O., Stark, G., Aguet, M., and Schaffner, W. (1994). The transcription factor MTF-1 is essential for basal and heavy metal-induced metallothionein gene expression. EMBO J. *13*, 2870–2875.

Hill, P.J., Cockayne, A., Landers, P., Morrissey, J.A., Sims, C.M., and Williams, P. (1998). SirR, a novel iron-dependent repressor in *Staphylococcus epidermidis*. Infect. Immun. *66*, 4123–4129.

Hiron, A., Borezée-Durant, E., Piard, J.C., and Juillard, V. (2007). Only one of four oligopeptide transport systems mediates nitrogen nutrition in *Staphylococcus aureus*. J. Bacteriol. *189*, 5119–5129.

Hiron, A., Posteraro, B., Carrière, M., Remy, L., Delporte, C., La Sorda, M., Sanguinetti, M., Juillard, V., and Borezée-Durant, E. (2010). A nickel ABC-transporter of *Staphylococcus aureus* is involved in urinary tract infection. Mol. Microbiol. *77*, 1246–1260.

Holmes, M.A., Paulsene, W., Jide, X., Ratledge, C., and Strong, R.K. (2005). Siderocalin (Lcn 2) also binds carboxymycobactins, potentially defending against mycobacterial infections through iron sequestration. Structure *13*, 29–41.

Hood, M.I., and Skaar, E.P. (2012). Nutritional immunity: transition metals at the pathogen-host interface. Nat. Rev. Microbiol. *10*, 525–537.

Horsburgh, M.J., Ingham, E., and Foster, S.J. (2001a). In *Staphylococcus aureus*, *fur* is an interactive regulator with PerR, contributes to virulence, and Is necessary for oxidative stress resistance through positive regulation of catalase and iron homeostasis. J. Bacteriol. *183*, 468–475.

Horsburgh, M.J., Clements, M.O., Crossley, H., Ingham, E., and Foster, S.J. (2001b). PerR controls oxidative stress resistance and iron storage proteins and is required for virulence in *Staphylococcus aureus*. Infect. Immun. *69*, 3744–3754.

Horsburgh, M.J., Wharton, S.J., Cox, A.G., Ingham, E., Peacock, S., and Foster, S.J. (2002a). MntR modulates expression of the PerR regulon and superoxide resistance in *Staphylococcus aureus* through control of manganese uptake. Mol. Microbiol. *44*, 1269–1286.

Horsburgh, M.J., Wharton, S.J., Karavolos, M., and Foster, S.J. (2002b). Manganese: elemental defence for a life with oxygen. Trends Microbiol. *10*, 496–501.

Huang, L., and Tepaamorndech, S. (2013). The SLC30 family of zinc transporters - a review of current understanding of their biological and pathophysiological roles. Mol. Aspects Med. *34*, 548–560.

Hudson, A.J., Andrews, S.C., Hawkins, C., Williams, J.M., Izuhara, M., Meldrum, F.C., Mann, S., Harrison, P.M., and Guest, J.R. (1993). Overproduction, purification and characterization of the *Escherichia coli* ferritin. Eur. J. Biochem. *218*, 985–995.

Huffman, D.L., and O'Halloran, T.V. (2001). Function, structure, and mechanism of intracellular copper trafficking proteins. Annu. Rev. Biochem. *70*, 677–701.

Hung, K.W., Juan, T.H., Hsu, Y.L., and Huang, T.H. (2012). NMR structure note: the ferrous iron transport protein C (FeoC) from *Klebsiella pneumoniae*. J. Biomol. NMR *53*, 161–165.

Hurd, A.F., Garcia-Lara, J., Rauter, Y., Cartron, M., Mohamed, R., and Foster, S.J. (2012). The iron-regulated surface proteins IsdA, IsdB, and IsdH are not required for heme iron utilization in *Staphylococcus aureus*. FEMS Microbiol. Lett. *329*, 93–100.

Hutchings, M.I., Palmer, T., Harrington, D.J., and Sutcliffe, I.C. (2009). Lipoprotein biogenesis in Gram-positive bacteria: knowing when to hold 'em, knowing when to fold 'em. Trends Microbiol. *17*, 13–21.

Imlay, J.A. (2002). How oxygen damages microbes: oxygen tolerance and obligate anaerobiosis. Adv. Microb. Physiol. *46*, 111–153.

Imlay, J.A., and Linn, S. (1988). DNA damage and oxygen radical toxicity. Science *240*, 1302–1309.

Inaoka, T., Matsumura, Y., and Tsuchido, T. (1999). SodA and manganese are essential for resistance to oxidative stress in growing and sporulating cells of *Bacillus subtilis*. J. Bacteriol. *181*, 1939–1943.

Irving, H., and Williams, R.J.P. (1948). Order of stability of metal complexes. Nature *162*, 746–747.

Jabado, N., Jankowski, A., Dougaparsad, S., Picard, V., Grinstein, S., and Gros, P. (2000). Natural resistance to intracellular infections: natural resistance-associated macrophage protein 1 (Nramp1) functions as a pH-dependent manganese transporter at the phagosomal membrane. J. Exp. Med. *192*, 1237–1248.

Jakubovics, N.S., and Jenkinson, H.F. (2001). Out of the iron age: new insights into the critical role of manganese homeostasis in bacteria. Microbiology *147*, 1709–1718.

Janulczyk, R., Pallon, J., and Björck, L. (1999). Identification and characterization of a *Streptococcus pyogenes* ABC transporter with multiple specificity for metal cations. Mol. Microbiol. *34*, 596–606.

Jeong, J., and Eide, D.J. (2013). The SLC39 family of zinc transporters. Mol. Aspects Med. *34*, 612–619.

Johansson, P., and Hederstedt, L. (1999). Organization of genes for tetrapyrrole biosynthesis in gram-positive bacteria. Microbiology *145*, 529–538.

Johnson, D.C., Dean, D.R., Smith, A.D., and Johnson, M.K. (2005). Structure, function, and formation of biological iron-sulfur clusters. Annu. Rev. Biochem. *74*, 247–281.

Johnson, M., Sengupta, M., Purves, J., Tarrant, E., Williams, P.H., Cockayne, A., Muthaiyan, A., Stephenson, R., Ledala, N., Wilkinson, B.J., *et al.* (2011). Fur is required for the activation of virulence gene expression through the induction of the sae regulatory system in *Staphylococcus aureus*. Int. J. Med. Microbiol. *301*, 44–52.

Jose, J., Christians, S., Rosenstein, R., Götz, F., and Kaltwasser, H. (1991). Cloning and expression of various staphylococcal genes encoding urease in *Staphylococcus carnosus*. FEMS Microbiol. Lett. *64*, 277–281.

Juárez-Verdayes, M.A., González-Uribe, P.M., Peralta, H., Rodríguez-Martínez, S., Jan-Roblero, J., Escamilla-Hernández, R., Cancino-Diaz, M.E., and Cancino-Diaz, J.C. (2012). Detection of *hssS*, *hssR*, *hrtA*, and *hrtB* genes and their expression by hemin in *Staphylococcus epidermidis*. Can. J. Microbiol. *58*, 1063–1072.

Kafala, B., and Sasarman, A. (1994). Cloning and sequence analysis of the hemB gene of *Staphylococcus aureus*. Can. J. Microbiol. *40*, 651–657.

Kafala, B., and Sasarman, A. (1997). Isolation of the *Staphylococcus aureus hemCDBL* gene cluster coding for early steps in heme biosynthesis. Gene *199*, 231–239.

Kaler, S.G. (2011). ATP7A-related copper transport diseases-emerging concepts and future trends. Nat. Rev. Neurol. *7*, 15–29.

Kammler, M., Schön, C., and Hantke, K. (1993). Characterization of the ferrous iron uptake system of *Escherichia coli*. J. Bacteriol. *175*, 6212–6219.

Kanyo, Z.F., Scolnick, L.R., Ash, D.E., and Christianson, D.W. (1996). Structure of a unique binuclear manganese cluster in arginase. Nature *383*, 554–557.

Karlinsey, J.E., Maguire, M.E., Becker, L.A., Crouch, M.L., and Fang, F.C. (2010). The phage shock protein PspA facilitates divalent metal transport and is required for virulence of *Salmonella enterica* sv. Typhimurium. Mol. Microbiol. *78*, 669–685.

Kehl-Fie, T.E., Chitayat, S., Hood, M.I., Damo, S., Restrepo, N., Garcia, C., Munro, K.A., Chazin, W.J., and Skaar, E.P. (2011). Nutrient metal sequestration by calprotectin inhibits bacterial superoxide defense, enhancing neutrophil killing of *Staphylococcus aureus*. Cell Host Microbe *10*, 158–164.

Kehl-Fie, T.E., Zhang, Y., Moore, J.L., Farrand, A.J., Hood, M.I., Rathi, S., Chazin, W.J., Caprioli, R.M., and Skaar, E.P. (2013). MntABC and MntH contribute to systemic *Staphylococcus aureus* infection by competing with calprotectin for nutrient manganese. Infect. Immun. *81*, 3395–3405.

Kehres, D.G., Zaharik, M.L., Finlay, B.B., and Maguire, M.E. (2000). The NRAMP proteins of *Salmonella typhimurium* and *Escherichia coli* are selective manganese transporters involved in the response to reactive oxygen. Mol. Microbiol. *36*, 1085–1100.

Kemna, E., Pickkers, P., Nemeth, E., van der Hoeven, H., and Swinkels, D. (2005). Time-course analysis of hepcidin, serum iron, and plasma cytokine levels in humans injected with LPS. Blood *106*, 1864–1866.

Khan, F.A., Fisher, M.A., and Khakoo, R.A. (2007). Association of hemochromatosis with infectious diseases: expanding spectrum. Int. J. Infect. Dis. *11*, 482–487.

Kim, D.K., Jeong, J.H., Lee, J.M., Kim, K.S., Park, S.H., Kim, Y.D., Koh, M., Shin, M., Jung, Y.S., Kim, H.S., *et al.* (2014). Inverse agonist of estrogen-related receptor γ controls *Salmonella typhimurium* infection by modulating host iron homeostasis. Nat. Med. *20*, 419–424.

Kim, H., Lee, H., and Shin, D. (2012). The FeoA protein is necessary for the FeoB transporter to import ferrous iron. Biochem. Biophys. Res. Commun. *423*, 733–738.

Kim, H.K., DeDent, A., Cheng, A.G., McAdow, M., Bagnoli, F., Missiakas, D.M., and Schneewind, O. (2010). IsdA and IsdB antibodies protect mice against *Staphylococcus aureus* abscess formation and lethal challenge. Vaccine *28*, 6382–6392.

Kim, Y.S., Kang, K.R., Wolff, E.C., Bell, J.K., McPhie, P., and Park, M.H. (2006). Deoxyhypusine hydroxylase is an Fe(II)-dependent, heat-repeat enzyme. Identification of amino acid residues critical for Fe(II) binding and catalysis. J. Biol. Chem. *281*, 13217–13225.

Knutson, M.D., Oukka, M., Koss, L.M., Aydemir, F., and Wessling-Resnick, M. (2005). Iron release from macrophages after erythrophagocytosis is up-regulated by ferroportin 1 overexpression and down-regulated by hepcidin. Proc. Natl. Acad. Sci. U.S.A. *102*, 1324–1328.

Kobayashi, M., and Shimizu, S. (1999). Cobalt proteins. Eur. J. Biochem. *261*, 1–9.

Kobylarz, M.J., Grigg, J.C., Takayama, S.J., Rai, D.K., Heinrichs, D.E., and Murphy, M.E. (2014). Synthesis of L-2,3-diaminopropionic acid, a siderophore and antibiotic precursor. Chem. Biol. *21*, 379–388.

Kochan, I. (1973). The role of iron in bacterial infections, with special consideration of host-tubercle bacillus interaction. Curr. Top. Microbiol. Immunol. *60*, 1–30.

Konetschny-Rapp, S., Jung, G., Meiwes, J., and Zähner, H. (1990). Staphyloferrin A: a structurally new siderophore from staphylococci. Eur. J. Biochem. *191*, 65–74.

Kong, S.M., Chan, B.K., Park, J.S., Hill, K.J., Aitken, J.B., Cottle, L., Farghaian, H., Cole, A.R., Lay, P.A., Sue, C.M., *et al.* (2014). Parkinson's disease-linked human PARK9/ATP13A2 maintains zinc homeostasis and promotes α-Synuclein externalization via exosomes. Hum. Mol. Genet. *23*, 2816–2833.

Korndörfer, I.P., Brueckner, F., and Skerra, A. (2007). The crystal structure of the human (S100A8/S100A9)2 heterotetramer, calprotectin, illustrates how conformational changes of interacting alpha-helices can determine specific association of two EF-hand proteins. J. Mol. Biol. *370*, 887–898.

Krishna Kumar, K., Jacques, D.A., Pishchany, G., Caradoc-Davies, T., Spirig, T., Malmirchegini, G.R., Langley, D.B., Dickson, C.F., Mackay, J.P., Clubb, R.T., *et al.* (2011). Structural basis for hemoglobin capture by *Staphylococcus aureus* cell-surface protein, IsdH. J. Biol. Chem. *286*, 38439–38447.

Kristiansen, M., Graversen, J.H., Jacobsen, C., Sonne, O., Hoffman, H.J., Law, S.K., and Moestrup, S.K. (2001). Identification of the haemoglobin scavenger receptor. Nature *409*, 198–201.

Kubica, M., Guzik, K., Koziel, J., Zarebski, M., Richter, W., Gajkowska, B., Golda, A., Maciag-Gudowska, A., Brix,

K., Shaw, L., et al. (2008). A potential new pathway for *Staphylococcus aureus* dissemination: the silent survival of *S. aureus* phagocytosed by human monocyte-derived macrophages. PloS One *3*, e1409.

Kuklin, N.A., Clark, D.J., Secore, S., Cook, J., Cope, L.D., McNeely, T., Noble, L., Brown, M.J., Zorman, J.K., Wang, X.M., et al. (2006). A novel *Staphylococcus aureus* vaccine: iron surface determinant B induces rapid antibody responses in rhesus macaques and specific increased survival in a murine *S. aureus* sepsis model. Infect. Immun. *74*, 2215–2223.

Kuo, Y.M., Su, T., Chen, H., Attieh, Z., Syed, B.A., McKie, A.T., Anderson, G.J., Gitschier, J., and Vulpe, C.D. (2004). Mislocalisation of hephaestin, a multicopper ferroxidase involved in basolateral intestinal iron transport, in the sex linked anaemia mouse. Gut *53*, 201–206.

Kuroda, M., Hayashi, H., and Ohta, T. (1999). Chromosome-determined zinc-responsible operon czr in *Staphylococcus aureus* strain 912. Microbiol. Immunol. *43*, 115–125.

Kuroda, M., Yamashita, A., Hirakawa, H., Kumano, M., Morikawa, K., Higashide, M., Maruyama, A., Inose, Y., Matoba, K., Toh, H., *et al.* (2005). Whole genome sequence of *Staphylococcus saprophyticus* reveals the pathogenesis of uncomplicated urinary tract infection. Proc. Natl. Acad. Sci. U.S.A. *102*, 13272–13277.

Kwok, E.Y., Severance, S., and Kosman, D.J. (2006). Evidence for iron channeling in the Fet3p-Ftr1p high-affinity iron uptake complex in the yeast plasma membrane. Biochemistry *45*, 6317–6327.

Lahiri, A., and Abraham, C. (2014). Activation of pattern recognition receptors upregulates metallothioneins, thereby increasing intracellular accumulation of zinc, autophagy, and bacterial clearance by macrophages. Gastroenterology. *147*, 835–846.

Lau, C.K., Ishida, H., Liu, Z., and Vogel, H.J. (2013). Solution structure of *Escherichia coli* FeoA and its potential role in bacterial ferrous iron transport. J. Bacteriol. *195*, 46–55.

Ledala, N., Zhang, B., Seravalli, J., Powers, R., and Somerville, G.A. (2014). Influence of iron and aeration on *Staphylococcus aureus* growth, metabolism, and transcription. J. Bacteriol. *196*, 2178–2189.

Lee, J.W., and Helmann, J.D. (2007). Functional specialization within the Fur family of metalloregulators. Biometals *20*, 485–499.

Lee, W.C., Reniere, M.L., Skaar, E.P., and Murphy, M.E. (2008). Ruffling of metalloporphyrins bound to IsdG and IsdI, two heme-degrading enzymes in *Staphylococcus aureus*. J. Biol. Chem. *283*, 30957–30963.

Lee, Y.H., Deka, R.K., Norgard, M.V., Radolf, J.D., and Hasemann, C.A. (1999). Treponema pallidum TroA is a periplasmic zinc-binding protein with a helical backbone. Nat. Struct. Biol. *6*, 628–633.

Leidgens, S., Bullough, K.Z., Shi, H., Li, F., Shakoury-Elizeh, M., Yabe, T., Subramanian, P., Hsu, E., Natarajan, N., Nandal, A., *et al.* (2013). Each member of the poly-r(C)-binding protein 1 (PCBP) family exhibits iron chaperone activity toward ferritin. J. Biol. Chem. *288*, 17791–17802.

Létoffé, S., Delepelaire, P., and Wandersman, C. (2006). The housekeeping dipeptide permease is the *Escherichia coli* heme transporter and functions with two optional peptide binding proteins. Proc. Natl. Acad. Sci. U.S.A. *103*, 12891–12896.

Létoffé, S., Heuck, G., Delepelaire, P., Lange, N., and Wandersman, C. (2009). Bacteria capture iron from heme by keeping tetrapyrrol skeleton intact. Proc. Natl. Acad. Sci. U.S.A. *106*, 11719–11724.

Levdikov, V.M., Blagova, E.V., Brannigan, J.A., Wright, L., Vagin, A.A., and Wilkinson, A.J. (2005). The structure of the oligopeptide-binding protein, AppA, from *Bacillus subtilis* in complex with a nonapeptide. J. Mol. Biol. *345*, 879–892.

Lewis, D.A., Klesney-Tait, J., Lumbley, S.R., Ward, C.K., Latimer, J.L., Ison, C.A., and Hansen, E.J. (1999). Identification of the znuA-encoded periplasmic zinc transport protein of *Haemophilus ducreyi*. Infect. Immun. *67*, 5060–5068.

Li, Y., Qiu, Y., Gao, H., Guo, Z., Han, Y., Song, Y., Du, Z., Wang, X., Zhou, D., and Yang, R. (2009). Characterization of Zur-dependent genes and direct Zur targets in *Yersinia pestis*. BMC Microbiol. *9*, 128.

Lim, C.K., Hassan, K.A., Penesyan, A., Loper, J.E., and Paulsen, I.T. (2013). The effect of zinc limitation on the transcriptome of *Pseudomonas protegens* Pf-5. Environ. Microbiol. *15*, 702–715.

Lin, M.H., Shu, J.C., Huang, H.Y., and Cheng, Y.C. (2012). Involvement of iron in biofilm formation by *Staphylococcus aureus*. PLoS One *7*, e34388.

Lindsay, J.A., and Foster, S.J. (2001). *zur*: a Zn(2+)-responsive regulatory element of *Staphylococcus aureus*. Microbiology *147* 3, 1259–1266.

Lindsay, J.A., Riley, T.V., and Mee, B.J. (1994). Production of siderophore by coagulase-negative staphylococci and its relation to virulence. Eur. J. Clin. Microbiol. Infect. Dis. *13*, 1063–1066.

Lisher, J.P., and Giedroc, D.P. (2013). Manganese acquisition and homeostasis at the host-pathogen interface. Front. Cell. Infect. Microbiol. *3*, 91.

Liu, M., Tanaka, W.N., Zhu, H., Xie, G., Dooley, D.M., and Lei, B. (2008). Direct hemin transfer from IsdA to IsdC in the iron-regulated surface determinant (Isd) heme acquisition system of *Staphylococcus aureus*. J. Biol. Chem. *283*, 6668–6676.

Liu, T., Ramesh, A., Ma, Z., Ward, S.K., Zhang, L., George, G.N., Talaat, A.M., Sacchettini, J.C., and Giedroc, D.P. (2007). CsoR is a novel *Mycobacterium tuberculosis* copper-sensing transcriptional regulator. Nat. Chem. Biol. *3*, 60–68.

Liu, Z., Petersen, R., and Devireddy, L. (2013). Impaired neutrophil function in 24p3 null mice contributes to enhanced susceptibility to bacterial infections. J. Immunol. *190*, 4692–4706.

Liu, Z., Reba, S., Chen, W.D., Porwal, S.K., Boom, W.H., Petersen, R.B., Rojas, R., Viswanathan, R., and Devireddy, L. (2014). Regulation of mammalian siderophore 2, 5-DHBA in the innate immune response to infection. J. Exp. Med. *211*, 1197–1213.

Liuzzi, J.P., Lichten, L.A., Rivera, S., Blanchard, R.K., Aydemir, T.B., Knutson, M.D., Ganz, T., and Cousins, R.J. (2005). Interleukin-6 regulates the zinc transporter Zip14 in liver and contributes to the hypozincemia of the acute-phase response. Proc. Natl. Acad. Sci. U.S.A. *102*, 6843–6848.

de Lorenzo, V., Wee, S., Herrero, M., and Neilands, J.B. (1987). Operator sequences of the aerobactin operon of plasmid ColV-K30 binding the ferric uptake regulation (*fur*) repressor. J. Bacteriol. *169*, 2624–2630.

Loutet, S.A., Kobylarz, M.J., Chau, C.H., and Murphy, M.E. (2013). IruO is a reductase for heme degradation by IsdI and IsdG proteins in *Staphylococcus aureus*. J. Biol. Chem. *288*, 25749–25759.

Lu, D., Boyd, B., and Lingwood, C.A. (1997). Identification of the key protein for zinc uptake in *Hemophilus influenzae*. J. Biol. Chem. *272*, 29033–29038.

Lucarelli, D., Russo, S., Garman, E., Milano, A., Meyer-Klaucke, W., and Pohl, E. (2007). Crystal structure and function of the zinc uptake regulator FurB from *Mycobacterium tuberculosis*. J. Biol. Chem. *282*, 9914–9922.

Lutsenko, S., Barnes, N.L., Bartee, M.Y., and Dmitriev, O.Y. (2007). Function and regulation of human copper-transporting ATPases. Physiol. Rev. *87*, 1011–1046.

Lyte, M., Freestone, P.P., Neal, C.P., Olson, B.A., Haigh, R.D., Bayston, R., and Williams, P.H. (2003). Stimulation of *Staphylococcus epidermidis* growth and biofilm formation by catecholamine inotropes. Lancet *361*, 130–135.

Ma, Z., Gabriel, S.E., and Helmann, J.D. (2011). Sequential binding and sensing of Zn(II) by *Bacillus subtilis* Zur. Nucleic Acids Res. *39*, 9130–9138.

Al-Maghrebi, M., Fridovich, I., and Benov, L. (2002). Manganese supplementation relieves the phenotypic deficits seen in superoxide-dismutase-null *Escherichia coli*. Arch. Biochem. Biophys. *402*, 104–109.

Malachowa, N., Whitney, A.R., Kobayashi, S.D., Sturdevant, D.E., Kennedy, A.D., Braughton, K.R., Shabb, D.W., Diep, B.A., Chambers, H.F., Otto, M., *et al.* (2011). Global changes in *Staphylococcus aureus* gene expression in human blood. PLoS One *6*, e18617.

Malik, S., Abel, L., Tooker, H., Poon, A., Simkin, L., Girard, M., Adams, G.J., Starke, J.R., Smith, K.C., Graviss, E.A., *et al.* (2005). Alleles of the NRAMP1 gene are risk factors for pediatric tuberculosis disease. Proc. Natl. Acad. Sci. U.S.A. *102*, 12183–12188.

Maret, W. (2010). Metalloproteomics, metalloproteomes, and the annotation of metalloproteins. Metallomics *2*, 117–125.

Mariotti, P., Malito, E., Biancucci, M., Lo Surdo, P., Mishra, R.P., Nardi-Dei, V., Savino, S., Nissum, M., Spraggon, G., Grandi, G., *et al.* (2013). Structural and functional characterization of the *Staphylococcus aureus* virulence factor and vaccine candidate FhuD2. Biochem. J. *449*, 683–693.

Markel, T.A., Crisostomo, P.R., Wang, M., Herring, C.M., Meldrum, K.K., Lillemoe, K.D., and Meldrum, D.R. (2007). The struggle for iron: gastrointestinal microbes modulate the host immune response during infection. J. Leucoc. Biol. *81*, 393–400.

Marraffini, L.A., and Schneewind, O. (2005). Anchor structure of staphylococcal surface proteins. V. Anchor structure of the sortase B substrate IsdC. J. Biol. Chem. *280*, 16263–16271.

Marraffini, L.A., Ton-That, H., Zong, Y., Narayana, S.V., and Schneewind, O. (2004). Anchoring of surface proteins to the cell wall of *Staphylococcus aureus*. A conserved arginine residue is required for efficient catalysis of sortase A. J. Biol. Chem. *279*, 37763–37770.

Martin, J.E., and Imlay, J.A. (2011). The alternative aerobic ribonucleotide reductase of *Escherichia coli*, NrdEF, is a manganese-dependent enzyme that enables cell replication during periods of iron starvation. Mol. Microbiol. *80*, 319–334.

Maryon, E.B., Molloy, S.A., and Kaplan, J.H. (2013). Cellular glutathione plays a key role in copper uptake mediated by human copper transporter 1. Am. J. Physiol. Cell Physiol. *304*, C768–C779.

Massé, E., and Gottesman, S. (2002). A small RNA regulates the expression of genes involved in iron metabolism in *Escherichia coli*. Proc. Natl. Acad. Sci. U.S.A. *99*, 4620–4625.

Matsui, T., Nambu, S., Ono, Y., Goulding, C.W., Tsumoto, K., and Ikeda-Saito, M. (2013). Heme degradation by *Staphylococcus aureus* IsdG and IsdI liberates formaldehyde rather than carbon monoxide. Biochemistry *52*, 3025–3027.

Matzanke, B.F., Anemüller, S., Schünemann, V., Trautwein, A.X., and Hantke, K. (2004). FhuF, part of a siderophore-reductase system. Biochemistry *43*, 1386–1392.

Mazmanian, S.K., Ton-That, H., Su, K., and Schneewind, O. (2002). An iron-regulated sortase anchors a class of surface protein during *Staphylococcus aureus* pathogenesis. Proc. Natl. Acad. Sci. U.S.A. *99*, 2293–2298.

Mazmanian, S.K., Skaar, E.P., Gaspar, A.H., Humayun, M., Gornicki, P., Jelenska, J., Joachmiak, A., Missiakas, D.M., and Schneewind, O. (2003). Passage of heme-iron across the envelope of *Staphylococcus aureus*. Science *299*, 906–909.

McDevitt, C.A., Ogunniyi, A.D., Valkov, E., Lawrence, M.C., Kobe, B., McEwan, A.G., and Paton, J.C. (2011). A molecular mechanism for bacterial susceptibility to zinc. PLOS Pathog. *7*, e1002357.

McManus, D.C., and Josephy, P.D. (1995). Superoxide dismutase protects *Escherichia coli* against killing by human serum. Arch. Biochem. Biophys. *317*, 57–61.

Mei, J.M., Nourbakhsh, F., Ford, C.W., and Holden, D.W. (1997). Identification of *Staphylococcus aureus* virulence genes in a murine model of bacteraemia using signature-tagged mutagenesis. Mol. Microbiol. *26*, 399–407.

Meissner, D., Vollstedt, A., van Dijl, J.M., and Freudl, R. (2007). Comparative analysis of twin-arginine (Tat)-dependent protein secretion of a heterologous model protein (GFP) in three different Gram-positive bacteria. Appl. Microbiol. Biotechnol. *76*, 633–642.

Meiwes, J., Fiedler, H.P., Haag, H., Zähner, H., Konetschny-Rapp, S., and Jung, G. (1990). Isolation and characterization of staphyloferrin A, a compound with siderophore activity from *Staphylococcus hyicus* DSM 20459. FEMS Microbiol. Lett. *55*, 201–205.

Miethke, M., and Marahiel, M.A. (2007). Siderophore-based iron acquisition and pathogen control. Microbiol. Mol. Biol. Rev. *71*, 413–451.

Miethke, M., Klotz, O., Linne, U., May, J.J., Beckering, C.L., and Marahiel, M.A. (2006). Ferri-bacillibactin uptake and hydrolysis in *Bacillus subtilis*. Mol. Microbiol. *61*, 1413–1427.

Miethke, M., Pierik, A.J., Peuckert, F., Seubert, A., and Marahiel, M.A. (2011). Identification and characterization of a novel-type ferric siderophore

reductase from a gram-positive extremophile. J. Biol. Chem. *286*, 2245–2260.
Miethke, M., Monteferrante, C.G., Marahiel, M.A., van Dijl, J.M., and van Dijl, J.M. (2013). The *Bacillus subtilis* EfeUOB transporter is essential for high-affinity acquisition of ferrous and ferric iron. Biochim. Biophys. Acta *1833*, 2267–2278.
Mishra, A., Eathiraj, S., Corvera, S., and Lambright, D.G. (2010). Structural basis for Rab GTPase recognition and endosome tethering by the C2H2 zinc finger of Early Endosomal Autoantigen 1 (EEA1). Proc. Natl. Acad. Sci. U.S.A. *107*, 10866–10871.
Mishra, R.P.N., Mariotti, P., Fiaschi, L., Nosari, S., Maccari, S., Liberatori, S., Fontana, M.R., Pezzicoli, A., De Falco, M.G., Falugi, F., et al. (2012). *Staphylococcus aureus* FhuD2 is involved in the early phase of staphylococcal dissemination and generates protective immunity in mice. J. Infect. Dis. *206*, 1041–1049.
Mobley, H.L., Hu, L.T., and Foxal, P.A. (1991). *Helicobacter pylori* urease: properties and role in pathogenesis. Scand. J. Gastroenterol. Suppl. *187*, 39–46.
Mobley, H.L., Garner, R.M., and Bauerfeind, P. (1995). *Helicobacter pylori* nickel-transport gene *nixA*: synthesis of catalytically active urease in *Escherichia coli* independent of growth conditions. Mol. Microbiol. *16*, 97–109.
Modun, B., Williams, P., Pike, W.J., Cockayne, A., Arbuthnott, J.P., Finch, R., and Denyer, S.P. (1992). Cell envelope proteins of *Staphylococcus epidermidis* grown *in vivo* in a peritoneal chamber implant. Infect. Immun. *60*, 2551–2553.
Modun, B., Kendall, D., and Williams, P. (1994). Staphylococci express a receptor for human transferrin: identification of a 42-kilodalton cell wall transferrin-binding protein. Infect. Immun. *62*, 3850–3858.
Modun, B., Evans, R.W., Joannou, C.L., and Williams, P. (1998). Receptor-mediated recognition and uptake of iron from human transferrin by *Staphylococcus aureus* and *Staphylococcus epidermidis*. Infect. Immun. *66*, 3591–3596.
Modun, B., Morrissey, J., and Williams, P. (2000). The staphylococcal transferrin receptor: a glycolytic enzyme with novel functions. Trends Microbiol. *8*, 231–237.
Moriwaki, Y., Terada, T., Caaveiro, J.M.M., Takaoka, Y., Hamachi, I., Tsumoto, K., and Shimizu, K. (2013). Heme binding mechanism of structurally similar iron-regulated surface determinant near transporter domains of *Staphylococcus aureus* exhibiting different affinities for heme. Biochemistry *52*, 8866–8877.
Morrissey, J.A., Cockayne, A., Hill, P.J., and Williams, P. (2000). Molecular cloning and analysis of a putative siderophore ABC transporter from *Staphylococcus aureus*. Infect. Immun. *68*, 6281–6288.
Morrissey, J.A., Cockayne, A., Brummell, K., and Williams, P. (2004). The staphylococcal ferritins are differentially regulated in response to iron and manganese and via PerR and Fur. Infect. Immun. *72*, 972–979.
Mulrooney, S.B., and Hausinger, R.P. (2003). Nickel uptake and utilization by microorganisms. FEMS Microbiol. Rev. *27*, 239–261.
Muryoi, N., Tiedemann, M.T., Pluym, M., Cheung, J., Heinrichs, D.E., and Stillman, M.J. (2008). Demonstration of the iron-regulated surface determinant (Isd) heme transfer pathway in *Staphylococcus aureus*. J. Biol. Chem. *283*, 28125–28136.
Nagata, M., Kaito, C., and Sekimizu, K. (2008). Phosphodiesterase activity of CvfA is required for virulence in *Staphylococcus aureus*. J. Biol. Chem. *283*, 2176–2184.
Nairz, M., Schleicher, U., Schroll, A., Sonnweber, T., Theurl, I., Ludwiczek, S., Talasz, H., Brandacher, G., Moser, P.L., Muckenthaler, M.U., *et al.* (2013). Nitric oxide-mediated regulation of ferroportin-1 controls macrophage iron homeostasis and immune function in Salmonella infection. J. Exp. Med. *210*, 855–873.
Navarro, C., Wu, L.F., and Mandrand-Berthelot, M.A. (1993). The nik operon of *Escherichia coli* encodes a periplasmic binding-protein-dependent transport system for nickel. Mol. Microbiol. *9*, 1181–1191.
Neal, C.P., Freestone, P.P., Maggs, A.F., Haigh, R.D., Williams, P.H., and Lyte, M. (2001). Catecholamine inotropes as growth factors for *Staphylococcus epidermidis* and other coagulase-negative staphylococci. FEMS Microbiol. Lett. *194*, 163–169.
Nemeth, E., Tuttle, M.S., Powelson, J., Vaughn, M.B., Donovan, A., Ward, D.M., Ganz, T., and Kaplan, J. (2004a). Hepcidin regulates cellular iron efflux by binding to ferroportin and inducing its internalization. Science *306*, 2090–2093.
Nemeth, E., Rivera, S., Gabayan, V., Keller, C., Taudorf, S., Pedersen, B.K., and Ganz, T. (2004b). IL-6 mediates hypoferraemia of inflammation by inducing the synthesis of the iron regulatory hormone hepcidin. J. Clin. Invest. *113*, 1271–1276.
Netz, D.J., Stith, C.M., Stümpfig, M., Köpf, G., Vogel, D., Genau, H.M., Stodola, J.L., Lill, R., Burgers, P.M., and Pierik, A.J. (2012). Eukaryotic DNA polymerases require an iron-sulfur cluster for the formation of active complexes. Nat. Chem. Biol. *8*, 125–132.
Nicolas, G., Bennoun, M., Devaux, I., Beaumont, C., Grandchamp, B., Kahn, A., and Vaulont, S. (2001). Lack of hepcidin gene expression and severe tissue iron overload in upstream stimulatory factor 2 (USF2) knockout mice. Proc. Natl. Acad. Sci. U.S.A. *98*, 8780–8785.
Nies, D.H. (1992). Resistance to cadmium, cobalt, zinc, and nickel in microbes. Plasmid *27*, 17–28.
Nobre, L.S., Seixas, J.D., Romão, C.C., and Saraiva, L.M. (2007). Antimicrobial action of carbon monoxide-releasing compounds. Antimicrob. Agents Chemother. *51*, 4303–4307.
O'Neill, A.J. (2010). *Staphylococcus aureus* SH1000 and 8325-4: comparative genome sequences of key laboratory strains in staphylococcal research. Lett. Appl. Microbiol. *51*, 358–361.
Odintsov, S.G., Sabała, I., Bourenkov, G., Rybin, V., and Bochtler, M. (2005). *Staphylococcus aureus* aminopeptidase S is a founding member of a new peptidase clan. J. Biol. Chem. *280*, 27792–27799.
Oefner, C., Douangamath, A., D'Arcy, A., Häfeli, S., Mareque, D., Mac Sweeney, A., Padilla, J., Pierau, S., Schulz, H., Thormann, M., *et al.* (2003). The 1.15A crystal structure of the *Staphylococcus aureus* methionyl-aminopeptidase and complexes with triazole based inhibitors. J. Mol. Biol. *332*, 13–21.

Ogunnariwo, J.A., and Schryvers, A.B. (1990). Iron acquisition in *Pasteurella haemolytica*: expression and identification of a bovine-specific transferrin receptor. Infect. Immun. *58*, 2091–2097.

Ogunnariwo, J.A., Alcantara, J., and Schryvers, A.B. (1991). Evidence for non-siderophore-mediated acquisition of transferrin-bound iron by *Pasteurella multocida*. Microb. Pathog. *11*, 47–56.

Okamoto, S., and Eltis, L.D. (2011). The biological occurrence and trafficking of cobalt. Metallomics *3*, 963–970.

Oogai, Y., Matsuo, M., Hashimoto, M., Kato, F., Sugai, M., and Komatsuzawa, H. (2011). Expression of virulence factors by *Staphylococcus aureus* grown in serum. Appl. Environ. Microbiol. *77*, 8097–8105.

Osman, D., Waldron, K.J., Denton, H., Taylor, C.M., Grant, A.J., Mastroeni, P., Robinson, N.J., and Cavet, J.S. (2010). Copper homeostasis in Salmonella is atypical and copper-CueP is a major periplasmic metal complex. J. Biol. Chem. *285*, 25259–25268.

Outten, C.E., Tobin, D.A., Penner-Hahn, J.E., and O'Halloran, T.V. (2001). Characterization of the metal receptor sites in *Escherichia coli* Zur, an ultrasensitive zinc(II) metalloregulatory protein. Biochemistry *40*, 10417–10423.

Overton, T.W., Justino, M.C., Li, Y., Baptista, J.M., Melo, A.M., Cole, J.A., and Saraiva, L.M. (2008). Widespread distribution in pathogenic bacteria of di-iron proteins that repair oxidative and nitrosative damage to iron-sulfur centers. J. Bacteriol. *190*, 2004–2013.

Panina, E.M., Mironov, A.A., and Gelfand, M.S. (2003). Comparative genomics of bacterial zinc regulons: enhanced ion transport, pathogenesis, and rearrangement of ribosomal proteins. Proc. Natl. Acad. Sci. U.S.A. *100*, 9912–9917.

Papp-Wallace, K.M., and Maguire, M.E. (2006). Manganese transport and the role of manganese in virulence. Annu. Rev. Microbiol. *60*, 187–209.

Paradkar, P.N., De Domenico, I., Durchfort, N., Zohn, I., Kaplan, J., and Ward, D.M. (2008). Iron depletion limits intracellular bacterial growth in macrophages. Blood *112*, 866–874.

Park, R.Y., Sun, H.Y., Choi, M.H., Bai, Y.H., and Shin, S.H. (2005). *Staphylococcus aureus* siderophore-mediated iron-acquisition system plays a dominant and essential role in the utilization of transferrin-bound iron. J. Microbiol. *43*, 183–190.

Pase, L., Voskoboinik, I., Greenough, M., and Camakaris, J. (2004). Copper stimulates trafficking of a distinct pool of the Menkes copper ATPase (ATP7A) to the plasma membrane and diverts it into a rapid recycling pool. Biochem. J. *378* 2, 1031–1037.

Patzer, S.I., and Hantke, K. (1998). The ZnuABC high-affinity zinc uptake system and its regulator Zur in *Escherichia coli*. Mol. Microbiol. *28*, 1199–1210.

Pennella, M.A., and Giedroc, D.P. (2005). Structural determinants of metal selectivity in prokaryotic metal-responsive transcriptional regulators. Biometals *18*, 413–428.

Pennella, M.A., Arunkumar, A.I., and Giedroc, D.P. (2006). Individual metal ligands play distinct functional roles in the zinc sensor *Staphylococcus aureus* CzrA. J. Mol. Biol. 356, 1124–1136.

Petris, M.J., and Mercer, J.F. (1999). The Menkes protein (ATP7A; MNK) cycles via the plasma membrane both in basal and elevated extracellular copper using a C-terminal di-leucine endocytic signal. Hum. Mol. Genet. *8*, 2107–2115.

Petris, M.J., Mercer, J.F., Culvenor, J.G., Lockhart, P., Gleeson, P.A., and Camakaris, J. (1996). Ligand-regulated transport of the Menkes copper P-type ATPase efflux pump from the Golgi apparatus to the plasma membrane: a novel mechanism of regulated trafficking. EMBO J. *15*, 6084–6095.

Peyssonnaux, C., Zinkernagel, A.S., Datta, V., Lauth, X., Johnson, R.S., and Nizet, V. (2006). TLR4-dependent hepcidin expression by myeloid cells in response to bacterial pathogens. Blood *107*, 3727–3732.

Pietrangelo, A. (2010). Hereditary hemochromatosis: pathogenesis, diagnosis, and treatment. Gastroenterology *139*, 393–408.

Pilpa, R.M., Fadeev, E.A., Villareal, V.A., Wong, M.L., Phillips, M., and Clubb, R.T. (2006). Solution structure of the NEAT (NEAr Transporter) domain from IsdH/HarA: the human hemoglobin receptor in *Staphylococcus aureus*. J. Mol. Biol. *360*, 435–447.

Pilpa, R.M., Robson, S.A., Villareal, V.A., Wong, M.L., Phillips, M., and Clubb, R.T. (2009). Functionally distinct NEAT (NEAr Transporter) domains within the *Staphylococcus aureus* IsdH/HarA protein extract heme from methemoglobin. J. Biol. Chem. *284*, 1166–1176.

de Pina, K., Navarro, C., McWalter, L., Boxer, D.H., Price, N.C., Kelly, S.M., Mandrand-Berthelot, M.A., and Wu, L.F. (1995). Purification and characterization of the periplasmic nickel-binding protein NikA of *Escherichia coli* K12. Eur. J. Biochem. *227*, 857–865.

Pishchany, G., and Skaar, E.P. (2012). Taste for blood: hemoglobin as a nutrient source for pathogens. PLOS Pathog. *8*, e1002535.

Pishchany, G., McCoy, A.L., Torres, V.J., Krause, J.C., Crowe, J.E., Fabry, M.E., and Skaar, E.P. (2010). Specificity for human hemoglobin enhances *Staphylococcus aureus* infection. Cell Host Microbe *8*, 544–550.

Pishchany, G., Sheldon, J.R., Dickson, C.F., Alam, M.T., Read, T.D., Gell, D.A., Heinrichs, D.E., and Skaar, E.P. (2014). IsdB-dependent hemoglobin binding is required for acquisition of heme by *Staphylococcus aureus*. J. Infect. Dis. *209*, 1764–1772.

Pluym, M., Vermeiren, C.L., Mack, J., Heinrichs, D.E., and Stillman, M.J. (2007). Heme binding properties of *Staphylococcus aureus* IsdE. Biochemistry *46*, 12777–12787.

Podkowa, K.J., Briere, L.A., Heinrichs, D.E., and Shilton, B.H. (2014). Crystal and solution structure analysis of FhuD2 from *Staphylococcus aureus* in multiple unliganded conformations and bound to ferrioxamine-B. Biochemistry 53, 2017–2031.

Polishchuk, E.V., Concilli, M., Iacobacci, S., Chesi, G., Pastore, N., Piccolo, P., Paladino, S., Baldantoni, D., van IJzendoorn, S.C., Chan, J., *et al.* (2014). Wilson disease protein ATP7B utilizes lysosomal exocytosis to maintain copper homeostasis. Dev. Cell 29, 686–700.

Posey, J.E., and Gherardini, F.C. (2000). Lack of a role for iron in the Lyme disease pathogen. Science *288*, 1651–1653.

Proctor, R.A., von Eiff, C., Kahl, B.C., Becker, K., McNamara, P., Herrmann, M., and Peters, G. (2006). Small colony variants: a pathogenic form of bacteria that facilitates persistent and recurrent infections. Nat. Rev. Microbiol. *4*, 295–305.

Psychogios, N., Hau, D.D., Peng, J., Guo, A.C., Mandal, R., Bouatra, S., Sinelnikov, I., Krishnamurthy, R., Eisner, R., Gautam, B., *et al.* (2011). The human serum metabolome. PLoS One *6*, e16957.

Quadri, M., Federico, A., Zhao, T., Breedveld, G.J., Battisti, C., Delnooz, C., Severijnen, L.A., Di Toro Mammarella, L., Mignarri, A., Monti, L., *et al.* (2012). Mutations in SLC30A10 cause parkinsonism and dystonia with hypermanganesemia, polycythemia, and chronic liver disease. Am. J. Hum. Genet. *90*, 467–477.

Que, Q., and Helmann, J.D. (2000). Manganese homeostasis in *Bacillus subtilis* is regulated by MntR, a bifunctional regulator related to the diphtheria toxin repressor family of proteins. Mol. Microbiol. *35*, 1454–1468.

Rajasekaran, M.B., Nilapwar, S., Andrews, S.C., and Watson, K.A. (2010a). EfeO-cupredoxins: major new members of the cupredoxin superfamily with roles in bacterial iron transport. Biometals *23*, 1–17.

Rajasekaran, M.B., Mitchell, S.A., Gibson, T.M., Hussain, R., Siligardi, G., Andrews, S.C., and Watson, K.A. (2010b). Isolation and characterisation of EfeM, a periplasmic component of the putative EfeUOBM iron transporter of *Pseudomonas syringae* pv. syringae. Biochem. Biophys. Res. Commun. *398*, 366–371.

Ramirez, A., Heimbach, A., Gründemann, J., Stiller, B., Hampshire, D., Cid, L.P., Goebel, I., Mubaidin, A.F., Wriekat, A.L., Roeper, J., *et al.* (2006). Hereditary parkinsonism with dementia is caused by mutations in ATP13A2, encoding a lysosomal type 5 P-type ATPase. Nat. Genet. *38*, 1184–1191.

Rammes, A., Roth, J., Goebeler, M., Klempt, M., Hartmann, M., and Sorg, C. (1997). Myeloid-related protein (MRP) 8 and MRP14, calcium-binding proteins of the S100 family, are secreted by activated monocytes via a novel, tubulin-dependent pathway. J. Biol. Chem. *272*, 9496–9502.

Ratledge, C. (2007). Iron metabolism and infection. Food Nutr. Bull. *28* (Suppl. 4), S515–S523.

Ratledge, C., and Dover, L.G. (2000). Iron metabolism in pathogenic bacteria. Annu. Rev. Microbiol. *54*, 881–941.

Reeve, I., Hummel, D., Nelson, N., Voss, J., and Hummell, D. (2002). Overexpression, purification, and site-directed spin labeling of the Nramp metal transporter from *Mycobacterium leprae*. Proc. Natl. Acad. Sci. U.S.A. *99*, 8608–8613.

Remy, L., Carrière, M., Derré-Bobillot, A., Martini, C., Sanguinetti, M., and Borezée-Durant, E. (2013). The *Staphylococcus aureus* Opp1 ABC transporter imports nickel and cobalt in zinc-depleted conditions and contributes to virulence. Mol. Microbiol. *87*, 730–743.

Reniere, M.L., and Skaar, E.P. (2008). *Staphylococcus aureus* haem oxygenases are differentially regulated by iron and haem. Mol. Microbiol. *69*, 1304–1315.

Reniere, M.L., Ukpabi, G.N., Harry, S.R., Stec, D.F., Krull, R., Wright, D.W., Bachmann, B.O., Murphy, M.E., and Skaar, E.P. (2010). The IsdG-family of haem oxygenases degrades haem to a novel chromophore. Mol. Microbiol. *75*, 1529–1538.

Riboldi, G.P., de Mattos, E.P., and Frazzon, J. (2013). Biogenesis of [Fe-S] cluster in Firmicutes: an unexploited field of investigation. Antonie Van Leeuwenhoek *104*, 283–300.

Richardson, A.R., Libby, S.J., and Fang, F.C. (2008). A nitric oxide-inducible lactate dehydrogenase enables *Staphylococcus aureus* to resist innate immunity. Science *319*, 1672–1676.

Roche, B., Aussel, L., Ezraty, B., Mandin, P., Py, B., and Barras, F. (2013). Iron/sulfur proteins biogenesis in prokaryotes: Formation, regulation and diversity. Biochim. Biophys. Acta *1827*, 455–469.

Rodionov, D.A., Hebbeln, P., Gelfand, M.S., and Eitinger, T. (2006). Comparative and functional genomic analysis of prokaryotic nickel and cobalt uptake transporters: evidence for a novel group of ATP-binding cassette transporters. J. Bacteriol. *188*, 317–327.

Saito, N., Takemori, N., Hirai, K., Onodera, R., Watanabe, S., and Namiki, M. (1993). Ultrastructural localization of lactoferrin in the granules other than typical secondary granules of human neutrophils. Hum. Cell *6*, 42–48.

Sandrini, S.M., Shergill, R., Woodward, J., Muralikuttan, R., Haigh, R.D., Lyte, M., and Freestone, P.P. (2010). Elucidation of the mechanism by which catecholamine stress hormones liberate iron from the innate immune defense proteins transferrin and lactoferrin. J. Bacteriol. *192*, 587–594.

Schalk, I.J., Abdallah, M.A., and Pattus, F. (2002). Recycling of pyoverdin on the FpvA receptor after ferric pyoverdin uptake and dissociation in *Pseudomonas aeruginosa*. Biochemistry *41*, 1663–1671.

Schneider, R., and Hantke, K. (1993). Iron-hydroxamate uptake systems in *Bacillus subtilis*: identification of a lipoprotein as part of a binding protein-dependent transport system. Mol. Microbiol. *8*, 111–121.

Schroeder, J.J., and Cousins, R.J. (1990). Interleukin 6 regulates metallothionein gene expression and zinc metabolism in hepatocyte monolayer cultures. Proc. Natl. Acad. Sci. U.S.A. *87*, 3137–3141.

Scrutton, M.C., Wu, C.W., and Goldthwait, D.A. (1971). The presence and possible role of zinc in RNA polymerase obtained from *Escherichia coli*. Proc. Natl. Acad. Sci. U.S.A. *68*, 2497–2501.

Searle, S., Bright, N.A., Roach, T.I., Atkinson, P.G., Barton, C.H., Meloen, R.H., and Blackwell, J.M. (1998). Localisation of Nramp1 in macrophages: modulation with activation and infection. J. Cell Sci. *111*, 2855–2866.

Sebulsky, M.T., and Heinrichs, D.E. (2001). Identification and characterization of *fhuD1* and *fhuD2*, two genes involved in iron-hydroxamate uptake in *Staphylococcus aureus*. J. Bacteriol. *183*, 4994–5000.

Sebulsky, M.T., Hohnstein, D., Hunter, M.D., and Heinrichs, D.E. (2000). Identification and characterization of a membrane permease involved in iron-hydroxamate transport in *Staphylococcus aureus*. J. Bacteriol. *182*, 4394–4400.

Sebulsky, M.T., Shilton, B.H., Speziali, C.D., and Heinrichs, D.E. (2003). The role of FhuD2 in iron(III)-hydroxamate transport in *Staphylococcus aureus*. Demonstration that FhuD2 binds iron(III)-hydroxamates but with minimal conformational change and implication of mutations on transport. J. Biol. Chem. *278*, 49890–49900.

Sebulsky, M.T., Speziali, C.D., Shilton, B.H., Edgell, D.R., and Heinrichs, D.E. (2004). FhuD1, a ferric hydroxamate-binding lipoprotein in *Staphylococcus aureus*: a case of gene duplication and lateral transfer. J. Biol. Chem. *279*, 53152–53159.

Selbach, B.P., Chung, A.H., Scott, A.D., George, S.J., Cramer, S.P., and Dos Santos, P.C. (2014). Fe-S cluster biogenesis in Gram-positive bacteria: SufU is a zinc-dependent sulfur transfer protein. Biochemistry *53*, 152–160.

Sendamarai, A.K., Ohgami, R.S., Fleming, M.D., and Lawrence, C.M. (2008). Structure of the membrane proximal oxidoreductase domain of human Steap3, the dominant ferrireductase of the erythroid transferrin cycle. Proc. Natl. Acad. Sci. *105*, 7410–7415.

Sharp, K.H., Schneider, S., Cockayne, A., and Paoli, M. (2007). Crystal structure of the heme-IsdC complex, the central conduit of the Isd iron/heme uptake system in *Staphylococcus aureus*. J. Biol. Chem. *282*, 10625–10631.

Shay, A.H., Choi, R., Whittaker, K., Salehi, K., Kitchen, C.M., Tashkin, D.P., Roth, M.D., and Baldwin, G.C. (2003). Impairment of antimicrobial activity and nitric oxide production in alveolar macrophages from smokers of marijuana and cocaine. J. Infect. Dis. *187*, 700–704.

Sheldon, J.R., and Heinrichs, D.E. (2012). The iron-regulated staphylococcal lipoproteins. Front. Cell. Infect. Microbiol. *2*, 41.

Sheldon, J.R., Marolda, C.L., and Heinrichs, D.E. (2014). TCA cycle activity in *Staphylococcus aureus* is essential for iron-regulated synthesis of staphyloferrin A, but not staphyloferrin B: the benefit of a second citrate synthase. Mol. Microbiol. *92*, 824–839.

Shen, L., Song, C.X., He, C., and Zhang, Y. (2014). Mechanism and function of oxidative reversal of DNA and RNA methylation. Annu. Rev. Biochem. *83*, 585–614.

Shi, H., Bencze, K.Z., Stemmler, T.L., and Philpott, C.C. (2008). A cytosolic iron chaperone that delivers iron to ferritin. Science *320*, 1207–1210.

Shin, J.H., Jung, H.J., An, Y.J., Cho, Y.B., Cha, S.S., and Roe, J.H. (2011). Graded expression of zinc-responsive genes through two regulatory zinc-binding sites in Zur. Proc. Natl. Acad. Sci. U.S.A. *108*, 5045–5050.

Shouldice, S.R., Dougan, D.R., Williams, P.A., Skene, R.J., Snell, G., Scheibe, D., Kirby, S., Hosfield, D.J., McRee, D.E., Schryvers, A.B., *et al.* (2003). Crystal structure of Pasteurella haemolytica ferric ion-binding protein A reveals a novel class of bacterial iron-binding proteins. J. Biol. Chem. *278*, 41093–41098.

Singh, A.K., Singh, N., Sharma, S., Singh, S.B., Kaur, P., Bhushan, A., Srinivasan, A., and Singh, T.P. (2008). Crystal structure of lactoperoxidase at 2.4 A resolution. J. Mol. Biol. *376*, 1060–1075.

Singh, V.K., Xiong, A., Usgaard, T.R., Chakrabarti, S., Deora, R., Misra, T.K., and Jayaswal, R.K. (1999). ZntR is an autoregulatory protein and negatively regulates the chromosomal zinc resistance operon *znt* of *Staphylococcus aureus*. Mol. Microbiol. *33*, 200–207.

Sitthisak, S., Howieson, K., Amezola, C., and Jayaswal, R.K. (2005). Characterization of a multicopper oxidase gene from *Staphylococcus aureus*. Appl. Environ. Microbiol. *71*, 5650–5653.

Sitthisak, S., Knutsson, L., Webb, J.W., and Jayaswal, R.K. (2007). Molecular characterization of the copper transport system in *Staphylococcus aureus*. Microbiology *153* 9, 4274–4283.

Skaar, E.P., Humayun, M., Bae, T., DeBord, K.L., and Schneewind, O. (2004a). Iron-source preference of *Staphylococcus aureus* infections. Science *305*, 1626–1628.

Skaar, E.P., Gaspar, A.H., and Schneewind, O. (2004b). IsdG and IsdI, heme-degrading enzymes in the cytoplasm of *Staphylococcus aureus*. J. Biol. Chem. *279*, 436–443.

Smaldone, G.T., and Helmann, J.D. (2007). CsoR regulates the copper efflux operon *copZA* in *Bacillus subtilis*. Microbiology *153* 1, 4123–4128.

Smaldone, G.T., Revelles, O., Gaballa, A., Sauer, U., Antelmann, H., and Helmann, J.D. (2012a). A global investigation of the *Bacillus subtilis* iron-sparing response identifies major changes in metabolism. J. Bacteriol. *194*, 2594–2605.

Smaldone, G.T., Antelmann, H., Gaballa, A., and Helmann, J.D. (2012b). The FsrA sRNA and FbpB protein mediate the iron-dependent induction of the *Bacillus subtilis lutABC* iron-sulfur-containing oxidases. J. Bacteriol. *194*, 2586–2593.

Smith, J.L. (2004). The physiological role of ferritin-like compounds in bacteria. Crit. Rev. Microbiol. *30*, 173–185.

Smith, D.G., Wilcox, M.H., Williams, P., Finch, R.G., and Denyer, S.P. (1991). Characterization of cell envelope proteins of *Staphylococcus epidermidis* cultured in human peritoneal dialysate. Infect. Immun. *59*, 617–624.

Solioz, M., Abicht, H.K., Mermod, M., and Mancini, S. (2010). Response of gram-positive bacteria to copper stress. J. Biol. Inorg. Chem. *15*, 3–14.

Speziali, C.D., Dale, S.E., Henderson, J.A., Vinés, E.D., and Heinrichs, D.E. (2006). Requirement of *Staphylococcus aureus* ATP-binding cassette-ATPase FhuC for iron-restricted growth and evidence that it functions with more than one iron transporter. J. Bacteriol. *188*, 2048–2055.

Spirig, T., Malmirchegini, G.R., Zhang, J., Robson, S.A., Sjodt, M., Liu, M., Krishna Kumar, K., Dickson, C.F., Gell, D.A., Lei, B., *et al.* (2013). *Staphylococcus aureus* uses a novel multidomain receptor to break apart human hemoglobin and steal its heme. J. Biol. Chem. *288*, 1065–1078.

Stafford, S.L., Bokil, N.J., Achard, M.E., Kapetanovic, R., Schembri, M.A., McEwan, A.G., and Sweet, M.J. (2013). Metal ions in macrophage antimicrobial pathways: emerging roles for zinc and copper. Biosci. Rep. 33, e00049.

Stauff, D.L., Torres, V.J., and Skaar, E.P. (2007). Signaling and DNA-binding activities of the *Staphylococcus aureus* HssR-HssS two-component system required for heme sensing. J. Biol. Chem. *282*, 26111–26121.

Stojiljkovic, I., and Perkins-Balding, D. (2002). Processing of heme and heme-containing proteins by bacteria. DNA. Cell. Biol. *21*, 281–295.

Stranger-Jones, Y.K., Bae, T., and Schneewind, O. (2006). Vaccine assembly from surface proteins of *Staphylococcus aureus*. Proc. Natl. Acad. Sci. U.S.A. *103*, 16942–16947.

Strasters, K.C., and Winkler, K.C. (1963). Carbohydrate metabolism of *Staphylococcus aureus*. J. Gen. Microbiol. *33*, 213–229.

Subramanian Vignesh, K., Landero Figueroa, J.A., Porollo, A., Caruso, J.A., and Deepe, G.S. (2013). Granulocyte macrophage-colony stimulating factor induced Zn sequestration enhances macrophage superoxide and limits intracellular pathogen survival. Immunity *39*, 697–710.

Sun, F., Ji, Q., Jones, M.B., Deng, X., Liang, H., Frank, B., Telser, J., Peterson, S.N., Bae, T., and He, C. (2012). AirSR, a [2Fe-2S] cluster-containing two-component system, mediates global oxygen sensing and redox signaling in *Staphylococcus aureus*. J. Am. Chem. Soc. *134*, 305–314.

Szarapińska-Kwaszewska, J., and Farkas, L.I. (2003). Synthesis of siderophores by strains of *Staphylococcus cohnii* isolated from various environments. Acta Microbiol. Pol. *52*, 261–269.

Tan, J., Zhang, T., Jiang, L., Chi, J., Hu, D., Pan, Q., Wang, D., and Zhang, Z. (2011). Regulation of intracellular manganese homeostasis by Kufor-Rakeb syndrome-associated ATP13A2 protein. J. Biol. Chem. *286*, 29654–29662.

Tanzi, R.E., Petrukhin, K., Chernov, I., Pellequer, J.L., Wasco, W., Ross, B., Romano, D.M., Parano, E., Pavone, L., and Brzustowicz, L.M. (1993). The Wilson disease gene is a copper transporting ATPase with homology to the Menkes disease gene. Nat. Genet. *5*, 344–350.

Taudte, N., and Grass, G. (2010). Point mutations change specificity and kinetics of metal uptake by ZupT from *Escherichia coli*. Biometals *23*, 643–656.

Tavares, A.F., Nobre, L.S., and Saraiva, L.M. (2012). A role for reactive oxygen species in the antibacterial properties of carbon monoxide-releasing molecules. FEMS Microbiol. Lett. *336*, 1–10.

Teitzel, G.M., Geddie, A., De Long, S.K., Kirisits, M.J., Whiteley, M., and Parsek, M.R. (2006). Survival and growth in the presence of elevated copper: transcriptional profiling of copper-stressed *Pseudomonas aeruginosa*. J. Bacteriol. *188*, 7242–7256.

Thomas, G.R., Forbes, J.R., Roberts, E.A., Walshe, J.M., and Cox, D.W. (1995). The Wilson disease gene: spectrum of mutations and their consequences. Nat. Genet. *9*, 210–217.

Tiedemann, M.T., Heinrichs, D.E., and Stillman, M.J. (2012). Multiprotein heme shuttle pathway in *Staphylococcus aureus*: iron-regulated surface determinant cog-wheel kinetics. J. Am. Chem. Soc. *134*, 16578–16585.

Torres, V.J., Pishchany, G., Humayun, M., Schneewind, O., and Skaar, E.P. (2006). *Staphylococcus aureus* IsdB is a hemoglobin receptor required for heme iron utilization. J. Bacteriol. *188*, 8421–8429.

Torres, V.J., Stauff, D.L., Pishchany, G., Bezbradica, J.S., Gordy, L.E., Iturregui, J., Anderson, K.L., Dunman, P.M., Joyce, S., and Skaar, E.P. (2007). A *Staphylococcus aureus* regulatory system that responds to host heme and modulates virulence. Cell Host Microbe *1*, 109–119.

Torres, V.J., Attia, A.S., Mason, W.J., Hood, M.I., Corbin, B.D., Beasley, F.C., Anderson, K.L., Stauff, D.L., McDonald, W.H., Zimmerman, L.J., *et al.* (2010). *Staphylococcus aureus* fur regulates the expression of virulence factors that contribute to the pathogenesis of pneumonia. Infect. Immun. *78*, 1618–1628.

Tottey, S., Harvie, D.R., and Robinson, N.J. (2005). Understanding how cells allocate metals using metal sensors and metallochaperones. Acc. Chem. Res. *38*, 775–783.

Touret, N., Furuya, W., Forbes, J., Gros, P., and Grinstein, S. (2003). Dynamic traffic through the recycling compartment couples the metal transporter Nramp2 (DMT1) with the transferrin receptor. J. Biol. Chem. *278*, 25548–25557.

Tseng, H.J., Srikhanta, Y., McEwan, A.G., and Jennings, M.P. (2001). Accumulation of manganese in Neisseria gonorrhoeae correlates with resistance to oxidative killing by superoxide anion and is independent of superoxide dismutase activity. Mol. Microbiol. *40*, 1175–1186.

Tsuda, M., Karita, M., Morshed, M.G., Okita, K., and Nakazawa, T. (1994). A urease-negative mutant of *Helicobacter pylori* constructed by allelic exchange mutagenesis lacks the ability to colonize the nude mouse stomach. Infect. Immun. *62*, 3586–3589.

Tsunemi, T., and Krainc, D. (2014). Zn^{2+} dyshomeostasis caused by loss of ATP13A2/PARK9 leads to lysosomal dysfunction and alpha-synuclein accumulation. Hum. Mol. Genet. *23*, 2791–2801.

Turlin, E., Débarbouillé, M., Augustyniak, K., Gilles, A.M., and Wandersman, C. (2013). *Staphylococcus aureus* FepA and FepB proteins drive heme iron utilization in *Escherichia coli*. PLoS One *8*, e56529.

Tuschl, K., Clayton, P.T., Gospe, S.M., Gulab, S., Ibrahim, S., Singhi, P., Aulakh, R., Ribeiro, R.T., Barsottini, O.G., Zaki, M.S., *et al.* (2012). Syndrome of hepatic cirrhosis, dystonia, polycythemia, and hypermanganesemia caused by mutations in SLC30A10, a manganese transporter in man. Am. J. Hum. Genet. *90*, 457–466.

Urban, C.F., Ermert, D., Schmid, M., Abu-Abed, U., Goosmann, C., Nacken, W., Brinkmann, V., Jungblut, P.R., and Zychlinsky, A. (2009). Neutrophil extracellular traps contain calprotectin, a cytosolic protein complex involved in host defense against *Candida albicans*. PLOS Pathog. *5*, e1000639.

Usenovic, M., Tresse, E., Mazzulli, J.R., Taylor, J.P., and Krainc, D. (2012). Deficiency of ATP13A2 leads to lysosomal dysfunction, α-synuclein accumulation, and neurotoxicity. J. Neurosci. *32*, 4240–4246.

Vaisman, B., Fibach, E., and Konijn, A.M. (1997). Utilization of intracellular ferritin iron for hemoglobin synthesis in developing human erythroid precursors. Blood *90*, 831–838.

Valderas, M.W., Gatson, J.W., Wreyford, N., and Hart, M.E. (2002). The superoxide dismutase gene *sodM* is unique to *Staphylococcus aureus*: absence of sodM in coagulase-negative staphylococci. J. Bacteriol. *184*, 2465–2472.

Valore, E.V., and Ganz, T. (2008). Posttranslational processing of hepcidin in human hepatocytes is mediated by the prohormone convertase furin. Blood Cells Mol. Dis. *40*, 132–138.

Vashchenko, G., and MacGillivray, R.T. (2013). Multi-copper oxidases and human iron metabolism. Nutrients *5*, 2289–2313.

Vento, S., Cainelli, F., and Cesario, F. (2006). Infections and thalassaemia. Lancet. Infect. Dis. *6*, 226–233.

Villareal, V.A., Pilpa, R.M., Robson, S.A., Fadeev, E.A., and Clubb, R.T. (2008). The IsdC protein from *Staphylococcus aureus* uses a flexible binding pocket to capture heme. J. Biol. Chem. *283*, 31591–31600.

Villareal, V.A., Spirig, T., Robson, S.A., Liu, M., Lei, B., and Clubb, R.T. (2011). Transient weak protein-protein complexes transfer heme across the cell wall of *Staphylococcus aureus*. J. Am. Chem. Soc. *133*, 14176–14179.

Vinchi, F., Gastaldi, S., Silengo, L., Altruda, F., and Tolosano, E. (2008). Hemopexin prevents endothelial damage and liver congestion in a mouse model of heme overload. Am. J. Pathol. *173*, 289–299.

Visai, L., Yanagisawa, N., Josefsson, E., Tarkowski, A., Pezzali, I., Rooijakkers, S.H., Foster, T.J., and Speziale, P. (2009). Immune evasion by *Staphylococcus aureus* conferred by iron-regulated surface determinant protein IsdH. Microbiology *155* 1, 667–679.

Voganatsi, A., Panyutich, A., Miyasaki, K.T., and Murthy, R.K. (2001). Mechanism of extracellular release of human neutrophil calprotectin complex. J. Leucoc. Biol. *70*, 130–134.

Vulpe, C., Levinson, B., Whitney, S., Packman, S., and Gitschier, J. (1993). Isolation of a candidate gene for Menkes disease and evidence that it encodes a copper-transporting ATPase. Nat. Genet. 3, 7–13.

Wagner, D., Maser, J., Lai, B., Cai, Z., Barry, C.E., Höner Zu Bentrup, K., Russell, D.G., and Bermudez, L.E. (2005). Elemental analysis of *Mycobacterium avium-*, *Mycobacterium tuberculosis-*, and *Mycobacterium smegmatis*-containing phagosomes indicates pathogen-induced microenvironments within the host cell's endosomal system. J. Immunol. *174*, 1491–1500.

Wakeman, C.A., Hammer, N.D., Stauff, D.L., Attia, A.S., Anzaldi, L.L., Dikalov, S.I., Calcutt, M.W., and Skaar, E.P. (2012). Menaquinone biosynthesis potentiates haem toxicity in *Staphylococcus aureus*. Mol. Microbiol. *86*, 1376–1392.

Wakeman, C.A., Stauff, D.L., Zhang, Y., and Skaar, E.P. (2014). Differential activation of *Staphylococcus aureus* heme detoxification machinery by heme analogues. J. Bacteriol. *196*, 1335–1342.

Walden, W.E. (2002). From bacteria to mitochondria: aconitase yields surprises. Proc. Natl. Acad. Sci. U.S.A. *99*, 4138–4140.

Waldron, K.J., and Robinson, N.J. (2009). How do bacterial cells ensure that metalloproteins get the correct metal? Nat. Rev. Microbiol. 7, 25–35.

Waldron, K.J., Rutherford, J.C., Ford, D., and Robinson, N.J. (2009). Metalloproteins and metal sensing. Nature *460*, 823–830.

Walker, J.M., Huster, D., Ralle, M., Morgan, C.T., Blackburn, N.J., and Lutsenko, S. (2004). The N-terminal metal-binding site 2 of the Wilson's Disease Protein plays a key role in the transfer of copper from Atox1. J. Biol. Chem. *279*, 15376–15384.

Wang, D., Hurst, T.K., Thompson, R.B., and Fierke, C.A. (2011). Genetically encoded ratiometric biosensors to measure intracellular exchangeable zinc in *Escherichia coli*. J. Biomed. Opt. *16*, 087011.

Watanabe, M., Tanaka, Y., Suenaga, A., Kuroda, M., Yao, M., Watanabe, N., Arisaka, F., Ohta, T., Tanaka, I., and Tsumoto, K. (2008). Structural basis for multimeric heme complexation through a specific protein-heme interaction: the case of the third neat domain of IsdH from *Staphylococcus aureus*. J. Biol. Chem. *283*, 28649–28659.

Weinberg, E.D. (1974). Iron and susceptibility to infectious disease. Science *184*, 952–956.

Weinberg, E.D. (1975). Nutritional immunity. Host's attempt to withold iron from microbial invaders. JAMA *231*, 39–41.

Weinberg, E.D. (1997). The Lactobacillus anomaly: total iron abstinence. Perspect. Biol. Med. *40*, 578–583.

Vander Wende, C., and Wainio, W.W. (1960). The state of the copper in cytochrome c oxidase. J. Biol. Chem. *235*, PC11–PC12.

White, C., Lee, J., Kambe, T., Fritsche, K., and Petris, M.J. (2009). A role for the ATP7A copper-transporting ATPase in macrophage bactericidal activity. J. Biol. Chem. *284*, 33949–33956.

White, C., Yuan, X., Schmidt, P.J., Bresciani, E., Samuel, T.K., Campagna, D., Hall, C., Bishop, K., Calicchio, M.L., Lapierre, A., *et al.* (2013). HRG1 is essential for heme transport from the phagolysosome of macrophages during erythrophagocytosis. Cell Metab. *17*, 261–270.

Wilcox, M.H., Williams, P., Smith, D.G., Modun, B., Finch, R.G., and Denyer, S.P. (1991). Variation in the expression of cell envelope proteins of coagulase-negative staphylococci cultured under iron-restricted conditions in human peritoneal dialysate. J. Gen. Microbiol. *137*, 2561–2570.

Wilks, A., and Ikeda-Saito, M. (2014). Heme utilization by pathogenic bacteria: not all pathways lead to biliverdin. Acc. Chem. Res. *47*, 2291–2298.

Williams, P., Denyer, S.P., and Finch, R.G. (1988). Protein antigens of *Staphylococcus epidermidis* grown under iron-restricted conditions in human peritoneal dialysate. FEMS Microbiol. Lett. *50*, 29–33.

Winter, W.E., Bazydlo, L.A., and Harris, N.S. (2014). The molecular biology of human iron metabolism. Lab. Med. *45*, 92–102.

Winters, M.S., Chan, Q., Caruso, J.A., and Deepe, G.S. (2010). Metallomic analysis of macrophages infected with *Histoplasma capsulatum* reveals a fundamental role for zinc in host defenses. J. Infect. Dis. *202*, 1136–1145.

Wiseman, G.M. (1975). The hemolysins of *Staphylococcus aureus*. Bacteriol. Rev. *39*, 317–344.

Wolf, S.G., Frenkiel, D., Arad, T., Finkel, S.E., Kolter, R., and Minsky, A. (1999). DNA protection by stress-induced biocrystallization. Nature *400*, 83–85.

Wright, J.A., and Nair, S.P. (2012). The lipoprotein components of the Isd and Hts transport systems are dispensable for acquisition of heme by *Staphylococcus aureus*. FEMS Microbiol. Lett. *329*, 177–185.

Wu, R., Skaar, E.P., Zhang, R., Joachimiak, G., Gornicki, P., Schneewind, O., and Joachimiak, A. (2005). *Staphylococcus aureus* IsdG and IsdI, heme-degrading enzymes with structural similarity to monooxygenases. J. Biol. Chem. *280*, 2840–2846.

Xiong, A., and Jayaswal, R.K. (1998). Molecular characterization of a chromosomal determinant conferring resistance to zinc and cobalt ions in *Staphylococcus aureus*. J. Bacteriol. *180*, 4024–4029.

Xiong, A., Singh, V.K., Cabrera, G., and Jayaswal, R.K. (2000). Molecular characterization of the ferric-uptake regulator, fur, from *Staphylococcus aureus*. Microbiology *146*, 659–668.

Xu, Q., Rawlings, N.D., Farr, C.L., Chiu, H.J., Grant, J.C., Jaroszewski, L., Klock, H.E., Knuth, M.W., Miller, M.D., Weekes, D., *et al.* (2011). Structural and sequence analysis of imelysin-like proteins implicated in bacterial iron uptake. PLoS One *6*, e21875.

Yamaji, S., Sharp, P., Ramesh, B., and Srai, S.K. (2004). Inhibition of iron transport across human intestinal epithelial cells by hepcidin. Blood *104*, 2178–2180.

Yanatori, I., Yasui, Y., Tabuchi, M., and Kishi, F. (2014). Chaperone protein involved in transmembrane transport of iron. Biochem. J. *462*, 25–37.

Yasmin, S., Andrews, S.C., Moore, G.R., and Le Brun, N.E. (2011). A new role for heme, facilitating release of iron from the bacterioferritin iron biomineral. J. Biol. Chem. *286*, 3473–3483.

Yocum, C.F., and Pecoraro, V.L. (1999). Recent advances in the understanding of the biological chemistry of manganese. Curr. Opin. Chem. Biol. *3*, 182–187.

Zapotoczna, M., Heilbronner, S., Speziale, P., and Foster, T.J. (2012). Iron-regulated surface determinant (Isd) proteins of *Staphylococcus lugdunensis*. J. Bacteriol. *194*, 6453–6467.

Zhang, Y., Rodionov, D.A., Gelfand, M.S., and Gladyshev, V.N. (2009). Comparative genomic analyses of nickel, cobalt and vitamin B12 utilization. BMC Genomics *10*, 78.

Zimnicka, A.M., Maryon, E.B., and Kaplan, J.H. (2007). Human copper transporter hCTR1 mediates basolateral uptake of copper into enterocytes: implications for copper homeostasis. J. Biol. Chem. *282*, 26471–26480.

Zong, Y., Mazmanian, S.K., Schneewind, O., and Narayana, S.V. (2004). The structure of sortase B, a cysteine transpeptidase that tethers surface protein to the *Staphylococcus aureus* cell wall. Structure *12*, 105–112.

Stress Responses in *Staphylococcus aureus* 10

Dorte Frees and Hanne Ingmer

Abstract

Staphylococcus aureus is a prominent member of the normal flora of humans and animals, but is also a major cause of mild and severe infections. To persist and disseminate in the human host, and to survive in environmental settings, such as hospitals, *S. aureus* has developed a plethora of cellular stress responses allowing it to sense and adapt to its very different niches. The stress responses often involve dramatic cellular reprogramming, and the technological advances provided by the access to whole genome sequences have let to an unprecedented insight into the global reorganization of gene and protein expression following stress exposure. Characterization of global gene responses has been very helpful both in identifying regulators sensing specific environmental stress signals and overlaps between different stress responses. In this chapter we review the recent progress in our understanding of the specific and general *S. aureus* stress responses, with a special emphasis on how stress responses contribute to virulence and antibiotic resistance in this important human pathogen.

Introduction

In order to ensure growth and survival in an ever-changing environment, bacteria have evolved adaptive responses to almost all environmental stresses. The bacterial stress responses have been a preferred research subject for more than three decades and our knowledge has increased tremendously. The basic paradigms of bacterial stress-responses have been established in the Gram-negative model organism *Escherichia coli* and in its Gram-positive counterpart, the soil bacterium *Bacillus subtilis*. While the central features of stress adaptation appear to be conserved among bacterial species, it is reasonable to believe that the stress-responses of a given bacterial species are shaped by the environmental niche that it occupies. In the case of *Staphylococcus aureus,* its commensal lifestyle probably involves nutritional hardship, competition with other microbes, and very importantly a constant battle with the immune defence mechanisms of the host. Furthermore, *S. aureus* is capable of persisting outside a host for extended times in dry environments, like dust, which contributes to its transmissibility in the hospital environment. *S. aureus* stress responses conceivably have developed to deal with these very different niches. Understanding *S. aureus*'s global stress responses has been greatly aided by the introduction of genome-based transcriptomic and proteomic tools, both in identifying regulators that sense and respond to specific environmental stress signals, and in identifying overlaps between different stress responses. In this chapter, we summarize the current state of knowledge on *S. aureus* stress responses, and the specific defence mechanisms for each type of stress, the global changes in gene and protein expression elicited by the stress, and the global regulators responsible for sensing and transmitting the stress signals.

The heat shock response: a general response to protein damage

Chaperones and proteases

A sudden increase in growth temperature induces unfolding and misfolding of proteins that exposes their hydrophobic regions and makes them susceptible to aggregation (Goff and Goldberg, 1985; Somero, 1995). The accumulation of aggregates of misfolded proteins can impede important cellular processes and is associated with cellular malfunctions and, in eukaryotic cells, with pathophysiological cellular states (Buchberger *et al.*, 2010). To restore protein homeostasis, all living cells induce a set of chaperones and proteases that are conserved from bacteria to man. These proteins were initially designated 'heat shock proteins' (Hsps), but it soon became clear that heat-induced chaperones and proteases are important for protein folding, repair, and degradation also in the non-stressed cells. In fact, studies of Hsps contributed substantially to the realization that folding of newly synthesized proteins, in the crowded environment of the cell, is not a spontaneous process but often requires the assistance of molecular chaperones. Unfolding and aggregating of proteins is a hallmark of heat-stress, however, as most conditions of stress exacerbate misfolding of proteins, chaperones and proteases are cornerstones in the cell's general defence mechanisms against stress. During infection, bacterial pathogens are likely to encounter multiple stresses; therefore, it is not surprising that chaperones and proteases contribute to virulence.

The best studied heat shock inducible bacterial chaperones are the DnaK-DnaJ-GrpE, the GroEL-GroES functional complexes and the Clp chaperones that all belong to highly conserved, ubiquitous family of chaperones – respectively the Hsp70, Hsp60 and Hsp100 chaperone families (Bukau *et al.*, 2006). Chaperones favour the native folding of proteins as either 'holdases', sequestering hydrophobic regions in misfolded polypeptides, or by active, energy-dependent dis-aggregation and refolding of misfolded proteins (Bukau *et al.*, 2006). In addition to chaperones, all cells contain a repertoire of highly conserved energy-dependent, intracellular proteases that shape the cellular proteome by ridding the cell of damaged proteins and by regulating abundance of functional proteins such as regulatory proteins. In *S. aureus*, only the DnaK chaperone, the Clp chaperones, and the ClpP protease have been studied in some detail, and these studies have clearly demonstrated that chaperones and proteases are required for coping with stress both under laboratory conditions and in infection models. In addition, the chaperones and proteases contribute to the virulence potential of *S. aureus* by controlling virulence regulation and antibiotic resistance. These results are summarized below.

The DnaK chaperone complex

The DnaK chaperone belongs to the widely conserved Hsp70 family of classical chaperones that assist a large variety of protein folding processes in the cell by transient association with short peptide segments of target proteins (Bukau *et al.*, 2006). The substrate binding and release cycle is driven by switching from the low affinity ATP bound state to the high affinity ADP bound state of DnaK. This cycling is assisted by the nucleotide exchange factor GrpE and the DnaJ co-chaperone encoded within the *hrcA-grpE-dnaK-dnaJ-prmA* operon. Deletion of *dnaK* in *S. aureus* resulted in poor growth at 37°C and completely inhibited growth at 43°C emphasizing the importance of the DnaK chaperone system for protein folding at optimal growth temperatures and, even more so, under conditions generating misfolded proteins (Singh *et al.*, 2007). The *dnaK* mutant additionally had a growth defect under conditions of oxidative stress presumably because misfolded proteins are more susceptible to irreversible oxidation like carbonylation (Dukan *et al.*, 2000). In contrast, DnaK was not required for growth under acidic, alkaline, or high-salt conditions, indicating that these stresses do not significantly interfere with protein homeostasis (Singh *et al.*, 2007). Furthermore, the *dnaK* mutant exhibited a range of pleiotropic phenotypes, including altered pigmentation, autolytic capacity, and tolerance to beta-lactams antibiotics (Singh *et al.*, 2007, 2012). It remains to be established if these phenotypes are linked to a specific requirement for DnaK chaperone activity in some of the processes linked to cell wall synthesis.

The Clp chaperones

The Clp ATPases (Hsp100) constitute a family of closely related proteins that are divided into subfamilies on the basis of the presence of specific signature sequences and the number and spacing of the nucleotide binding sites (Frees *et al.*, 2007). The specific number and types of Clp ATPases vary, even between closely related genera; however, all Clp ATPases are believed to function as molecular chaperones in protein folding, activation, or disaggregation (Frees *et al.*, 2007). Importantly, not all Clp ATPases take part in the heat stress response. As an example, *S. aureus* encodes two Clp ATPases, ClpX and ClpY that are neither induced by heat shock nor required for growth and survival under heat stress conditions (Frees *et al.*, 2003, 2005; Fleury *et al.*, 2009). Additionally, *S. aureus* encodes three heat-inducible Clp ATPases, namely ClpB, ClpC, and ClpL that are not required under normal growth conditions but increase survival many thousand-fold when *S. aureus* cells are exposed to heat shock (Frees *et al.*, 2004). ClpC homologues are generally found in Gram-positive bacteria, where they are believed to function as classical chaperones that reduce protein damage following heat shock by either refolding damaged proteins or in collaboration with the ClpP proteolytic component, degrade irreversibly damaged proteins – see below. The widely conserved family of ClpB chaperones, on the other hand, do not promote the 'forward' folding of proteins, but instead have the unique ability to rescue proteins from an aggregated state by mediating solubilization of protein aggregates (Goloubinoff *et al.*, 1999; Mogk *et al.*, 1999; Haslberger *et al.*, 2008). Full recovery of disaggregated proteins to active proteins requires the assistance of the DnaK chaperone system (Mogk *et al.*, 1999; Haslberger *et al.*, 2008). During conditions of severe heat stress, aggregation of heat-damaged protein through non-specific hydrophobic interaction presents a major challenge for the cell. Accordingly, we found that ClpB was essential for growth of *S. aureus* at very high temperatures and for inducing thermotolerance (Frees *et al.*, 2003, 2004, 2012). Induced thermotolerance describes the phenomenon that a short pre-adaptation at 46°C enables *S. aureus* to survive for several minutes at 60°C, a treatment that is lethal without preadaptation (Frees *et al.*, 2004; Singh *et al.*, 2007). This induced thermotolerance is the result of increased protein folding capacity due to the increased synthesis of ClpB and other heat-shock chaperones during pre-adaptation (Weibezahn *et al.*, 2004). The finding that both ClpB and DnaK contribute to induction of thermotolerance in *S. aureus* strongly suggests that ClpB and DnaK function as a bi-chaperone dis-aggregase in *S. aureus* as in other bacteria. ClpL homologues are only found in Gram-positive bacteria, including *S. aureus*. We found that ClpL and ClpB are functionally overlapping in *S. aureus*, suggesting that *S. aureus* has two chaperones with dis-aggregation capacity (Frees *et al.*, 2004). Interestingly, transcription of *clpL* and *clpB* is subject to regulation by two different transcriptional regulators, σ^B and CtsR, respectively, which may ensure protein-disaggregating capacity under different conditions of severe stress – see below.

The ClpXP and ClpCP proteases

ClpP proteases are compartmentalized proteases composed of separately encoded proteolytic subunits and ATPase subunits: The ClpP subunits contain the conserved catalytic triad (Ser-His-Asp) that is typical of serine-proteases (Frees *et al.*, 2007). Two heptameric rings of ClpP subunits are stacked to form a secluded proteolytic chamber that can only be accessed through pores too narrow to allow access of folded proteins (Wang *et al.*, 1997). In order to degrade protein substrates, ClpP must associate with the hexameric rings of Clp ATPases. The Clp ATPses are responsible for substrate recognition and the subsequent unfolding and translocation of the substrate into the ClpP proteolytic chamber for degradation (Fig. 10.1). In *S. aureus* two different Clp ATPases, ClpC and ClpX can associate with ClpP to form the ClpCP and ClpXP protease, respectively (Frees *et al.*, 2004; Feng *et al.*, 2013; Graham *et al.*, 2013). The ClpCP and ClpXP proteases occupy strikingly different roles in *S. aureus* stress responses: ClpC is the specificity factor that allows ClpP to degrade misfolded proteins, and inactivation of *clpC* or *clpP* confers a strong heat sensitive phenotype to *S. aureus* (Frees *et al.*, 2003, 2004, 20012). In contrast, a *clpX* mutant grew better than the wild-type at high temperatures, signifying that neither the ClpX chaperone nor the ClpXP protease contribute to protein homeostasis during conditions generating misfolded proteins (Frees *et al.*, 2003). In accordance with this finding, *clpC* and *clpP* are regulated by the heat-shock regulator, CtsR, while *clpX* is not. Conversely, the ClpXP protease is essential for growth and survival in the presence of osmotic stress and some types of oxidative stress (Frees *et al.*, 2003). These data suggest that ClpXP protease controls responses to other types of stress by degrading transcriptional regulators in response to intra- or extracellular stress signals. One example is the thiol-stress regulator, Spx that accumulates to high levels in mutants lacking the ClpXP protease (Pamp *et al.*, 2006; Feng *et al.*, 2013).

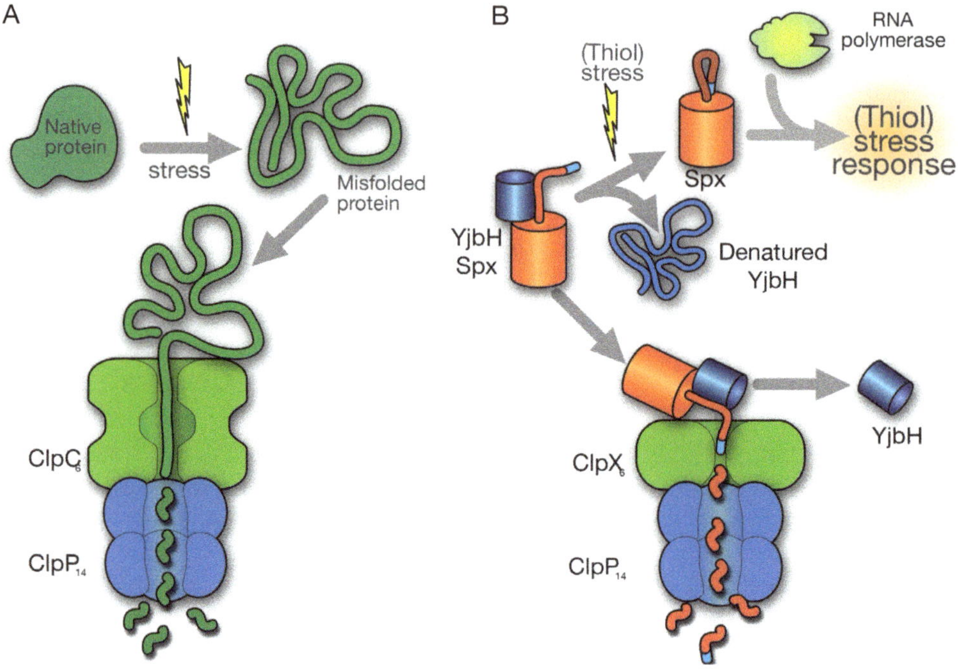

Figure 10.1 Model for the conditional degradation of the Spx transcriptional stress regulator. (A) In the unstressed cell the YjbH adaptor protein interacts with Spx such that the C-terminal region of Spx becomes accessible to the ClpXP protease resulting in rapid degradation of Spx. (B) In response to thiol stress and other stresses interfering with protein folding YjbH un-folds and loses its ability to target Spx for degradation by the ClpXP complex leading to transcriptional induction of an Spx controlled global stress response. The model is based on studies from *S. aureus* and *B. subtilis* as detailed in the text.

Under non-stress conditions, Spx is kept at a low concentration through continuous degradation by the ClpXP protease in a process stimulated by the adaptor protein YjbH, but in response to protein-stress Spx degradation is inhibited by the stress induced aggregation of YjbH (Fig. 10.1) (Engman *et al.*, 2012; Engman and von Wachenfeldt, 2015). Inactivation of *spx* in *S. aureus* rendered the cells hypersensitive to a wide range of stress conditions including high and low temperature, high osmolarity, and hydrogen peroxide, supporting that the Spx regulons contribute to growth and survival during a wide range of stresses (Pamp *et al.*, 2006).

Chaperones and proteases contribute to virulence of *S. aureus*

The ability to maintain protein homeostasis in the stressful environment of the host is predicted to be important for bacterial virulence. This prediction was supported by the observations that inactivation of *dnaK, clpX, clpP,* or *clpC* decreased virulence of *S. aureus* in murine models of systemic infection (Singh *et al.*, 2007; Farrand *et al.*, 2013). Similar, intracellular replication in bovine MAC- T cells was completely inhibited in the absence of ClpB, ClpX, or ClpP, and significantly reduced in mutants lacking ClpC or ClpL (Frees *et al.*, 2004). The finding that heat shock chaperones ClpC and ClpL, and ClpB are required for intracellular replication supports the idea that heat-induced Clp chaperones are required to counteract accumulation of denatured polypeptides and aggregates *in vivo*. In contrast, mutants lacking ClpC and ClpB were as virulent as the parental strain in a murine abscess model, while mutants lacking the components of the ClpXP protease were severely attenuated in the same model (Frees *et al.*, 2003; D. Frees, unpublished). Taken together, these data emphasize that in some types of infections, stress-imposed damage of proteins is not a problem, and that ClpP mediated degradation of stress-damaged proteins does not contribute to virulence via the ClpC specificity factor. Rather, the indispensable role of ClpP in abscess models of virulence seems to be mediated via the ClpXP protease, which controls osmotic and oxidative stress responses, facilitates the acquisition of iron, and alters expression of numerous virulence genes and virulence regulators (Frees *et al.*, 2012, 2013; Farrand *et al.*, 2013).

Transcriptional regulators sensing protein stress

Although the chaperones and proteases induced by heat stress are extremely well conserved, heat shock regulatory mechanisms exhibit great diversity among bacteria. In *E. coli* transcription of most heat shock genes is under the control of two alternative sigma factors (σ^{32} and σ^{40}), while heat shock genes in *B. subtilis* have been grouped into 4 classes according to their heat shock regulation by the HrcA repressor, the σ^{B} stress sigma factor, the CtsR repressor, or by unknown mechanisms (Schumann, 2003). The heat-shock regulators HrcA and CtsR are widely conserved among Gram-positive bacteria, including *S. aureus*. In most organisms, the two regulons do not overlap, and HrcA controls transcription of the *dnaK* and *groES-EL* operons, while CtsR controls transcription of *clpP* and most genes encoding *clp* ATPases. In *S. aureus*, CtsR controls transcription of the monocistronic *clpP* and *clpB* genes and the *ctsr-mcsA-mcsB-ctsR*, *hrcA-grpE-dnaK-dnaJ-prmA*, and *groES-groEL* operons (Frees *et al.*, 2004; Chastanet *et al.*, 2003). In other words, the entire HrcA regulon (*dnaK* and *groESL* operons) is embedded within the CtsR regulon in staphylococci, where HrcA and CtsR seem to act synergistically to control expression of the *dnaK* and *groESL* operons (Chastanet *et al.*, 2003). As the DnaK and GroE chaperones are highly expressed during non-stress conditions (more than 10,000 molecules per cell in *B. subtilis*), the dual repression by HrcA and CtsR does not shut down transcription of these operons (Maaß *et al.*, 2014). Instead, the dual regulation may ensure maximal induction by a wide range of stress conditions.

Pioneering studies in *Bacillus* have clarified the mechanisms that allow HrcA and CtsR activity to be modulated in response to the presence of misfolded proteins. In the case of the HrcA repressor, its DNA-binding activity is dependent on the chaperone activity of the GroES-GroEL chaperonin. In the presence of misfolded protein, the GroES-GroEL chaperonin will be titrated away from HrcA in a dose dependent manner leading to dissociation of HrcA from its operator sequences and induction of the HrcA regulon (Mogk *et al.*, 1997). In contrast, CtsR itself functions as a protein thermometer and reacts intrinsically to the temperature of the surroundings (Elsholz *et al.*, 2010). After dissociation from the DNA, CtsR is degraded by the *B. subtilis* ClpEP protease (Miethke *et al.*, 2006) and later by the ClpCP protease (Elsholz *et al.*, 2010, 2011). Interestingly, CtsR can be inactivated by additional mechanisms making the CtsR-regulon responsive to a wider range of stresses including thiol stress. Briefly, thiol-mediated inactivation of CtsR is controlled by McsA and McsB that are encoded within the same operon as CtsR and ClpC. Thiol-specific stress is sensed by the zinc coordinating cysteines present in McsA (Elsholz *et al.*, 2011). These thiols act as a molecular redox switch to regulate McsB activity by dissociating it from McsA. When McsB is not associated with McsA, it binds and inactivates DNA-bound CtsR, leading to de-repression of the CtsR-regulated genes (Elsholz *et al.*, 2011). Although little is known of the role of these proteins in *S. aureus*, inactivation of *mcsB* leads to oxidative stress sensitivity, indicating that homologues of McsB and McsA may perform similar functions in this organism (Wozniak *et al.*, 2012; Frees *et al.*, 2013).

Global responses to heat stress

The global transcriptomic response to heat stress in *S. aureus* has been described in two different studies (Andersson *et al.*, 2006; Fleury *et al.*, 2009). In one study, the transcriptomic response was analysed 30 min after a temperature up-shift from 37°C to 42°C that did not greatly affect the growth rate (Andersson *et al.*, 2006). In the second study, transcription was recorded 10 min after the temperature was increased to either 43°C (only marginal effect on growth) or 48°C (severe heat shock leading to growth arrest) (Fleury *et al.*, 2009). The studies confirmed the classical heat shock genes controlled by CtsR and HrcA were strongly induced 10 min following a heat shock. Interestingly, 30 min following a temperature upshift no or very modest induction of *clpP* and the HrcA-regulon was observed, confirming that strong induction of classical heat shock genes is a transient phenomenon that is only required immediately following a temperature upshift (Andersson *et al.*, 2006; Frees *et al.*, 2003). Interestingly, the studies revealed a lack of heat shock induction of the FtsH, HtrA and ClpYQ proteases, supporting that ClpP is the only heat-induced protease in *S. aureus*. The urease operon encoding the enzymes responsible for breakdown of urea was strongly induced by heat stress in both studies. Since urea has protein denaturing properties, continuous removal of urea may help the cell to maintain protein homeostasis during heat stress. A severe heat shock treatment also induced transcription of genes encoding enzymes coping with oxidative stress and DNA damage (Fleury *et al.*, 2009).

The SigB regulon

Is SigB the master regulator of a general stress response in *S. aureus*?

In the model organisms *E. coli* and *B. subtilis*, exposure to growth-limiting environmental or metabolic stress activate alternative sigma factors (RpoS and σ^B, respectively) that function as master regulators of large regulons encoding general stress and starvation protective mechanisms (reviewed by Hecker *et al.*, 2007 and Klauck *et al.*, 2007). The contribution of these regulons to survival during stress is illustrated by the increased sensitivity of *sigB* and *sigS* mutants to multiple stresses including acid, heat, osmotic, and oxidative stress (Hecker *et al.*, 2007; Hengge-Aronis, 1993). *S. aureus* encodes a σ^B homologue that is closely related to *Bacillus* σ^B at the sequence level (54% identical residues) indicating a conserved role of σ^B in the closely related bacterial species. This assumption was supported by a similar genetic organization of the staphylococcal *sigB* operon and the distal part of the homologous *B. subtilis* operon (*rsbU-rsbV-rsbW-sigB*), which in addition to σ^B, encodes proteins that control activity of σ^B in response to stress stimuli by a partner-switching mechanism – see below (Senn *et al.*, 2005, Hecker *et al.*, 2007). As detailed in the review by Hecker *et al.* (2007), the function of σ^B as a general stress regulator in *S. aureus* was questioned by the lack of experimental data supporting a critical role for σ^B in growth or survival following diverse stress treatments such as starvation and UV-, hydrogen-peroxide-, acid-, and high osmolarity-stress (Kullik *et al.*, 1998; Nicholas *et al.*, 1999; Giachino *et al.*, 2001; Pané-Farré *et al.*, 2006; Frees *et al.*, 2012). A functional σ^B did, however, promote growth and survival under conditions of severe heat stress or when heat stress was combined with high pH (Cebrián *et al.*, 2009a,b; Frees *et al.*, 2012). Furthermore, the overlap between the *S. aureus* and the *B. subtilis* σ^B regulons turned out to be surprisingly small with only around 10% of the genes under the control of σ^B in *S. aureus* having an orthologue in the *B. subtilis* σ^B regulon and vice versa (Bischoff *et al.*, 2004; Pané-Farré et a., 2006). In addition, the transcriptomic analysis revealed that genes encoding proteins with a verified role in stress tolerance were generally not σ^B regulated in *S. aureus*. As an example, the general stress genes *clpC* and *clpP* are controlled by σ^B in *B. subtilis* but not in *S. aureus*. Finally, σ^B activation in *S. aureus* deviates on several points from the paradigm established in *B. subtilis* – see below (Hecker *et al.*, 2007). Taken together, these data do not support that σ^B in *S. aureus* controls a general, non-specific, and preventive stress response similar to the *B. subtilis* σ^B response. Rather, it seems that σ^B more generally impact on cellular processes related to cell envelope composition, membrane transport processes, and virulence in *S. aureus*. That is to say, σ^B is an example of a highly conserved transcriptional regulator that controls different regulons in closely related bacterial species, signifying that its regulatory features have been shaped by evolution to meet the physiological requirements and lifestyle of the individual organisms.

Activation of σ^B in *S. aureus*

σ^B activity is regulated at the protein level by a partner-switching mechanism in *S. aureus* (Pané-Farré *et al.*, 2006). The proteins regulating the transcriptional activity of σ^B are encoded within the hexa-cistronic operon (*mazEF, rsbU-rsbV-rsbW-sigB*) that also encodes a MazF like toxin–antitoxin system (Senn *et al.*, 2005). The paradigm for this partner-switching mechanism was established in *B. subtilis*: Under non-inducing conditions, σ^B is inactivated by binding tightly to the anti-sigma factor, RsbW. RsbV functions as an anti-anti-sigma factor that in its non-phosphorylated state can bind competitively to RsbW resulting in the release of active σ^B. Central to this mechanism is the phosphorylation state of RsbV. In *B. subtilis*, phosphorylation of RsbV is regulated by the kinase activity of RsbW and by the phosphatase activity of RsbP and RsbU (Hecker *et al.*, 2007). The RsbP phosphatase is required to activate σ^B in response to perturbations of the cellular energy level (energy stress pathway), while RsbU regulates activation of σ^B in response to environmental stress. Interestingly, *S. aureus* does not encode a homologue of RsbP, and consistent with this finding σ^B was not activated by a reduction in the cellular ATP-pool (Pané-Farré *et al.*, 2006). In *S. aureus*, activation of σ^B has been observed following imposition of alkaline shock, temperature upshift, or salt stress and during entry into the stationary growth phase (Kullik and Giachino, 1997, Pane-Farre *et al.*, 2006, 2009). Activation was dependent on RsbU, suggesting that the function of RsbU in σ^B stress activation is conserved between *S. aureus* and *B. subtilis* (Pané-Farré *et al.*, 2006). That being said, additional studies have shown subtle differences in RsbU-mediated activation of σ^B between the *S. aureus* and *B. subtilis*; specifically, the *S. aureus* RsbU appears to be constitutively active compared to its more strictly regulated *B. subtilis* counterpart (Pané-Farré *et al.*, 2009). Consistent with these observations, *S. aureus* do not encode homologues of RsbR, RsbS and RsbT, the proteins responsible for stress activation of RsbU in *B. subtilis*. The high constitutive activity of RsbU in *S. aureus*, probably explains why σ^B activation only resulted in a modest increase in target genes transcription, compared to the dramatic stress induction of target genes in *B. subtilis* (reviewed by Hecker *et al.*, 2007). Interestingly, a recent study revealed fast and transient activation of σ^B following internalization by human host cells (Pförtner *et al.*, 2014). Additionally, subinhibitory concentrations of aminoglycoside antibiotics such as gentamycin and tobramycin were shown to induce σ^B -activity (Mitchell *et al.*, 2010) and a similar effect was observed for subinhibitory concentrations of cell wall-targeting antibiotics such as vancomycin and ampicillin (Chen *et al.*, 2011). Taken together, these data suggest that cell wall stress might be a signal mediating σ^B activation.

Oxidative stress

Reactive oxygen species and defence mechanisms

Oxygen is the final electron receptor in aerobic respiration. Complete reduction results in the formation of H_2O, while incomplete reduction results in the generation of reactive oxygen species (ROS), such as hydrogen peroxide (H_2O_2), superoxide anions (O_2^-), and hydroxyl radicals (HO·). ROS are extremely reactive molecules that have the ability to oxidize and damage any oxidizable moiety in a biological molecule (Imlay, 2003). In addition to being an evitable by-product of aerobic energy generation, ROS are also among the major weapons used by phagocytes to kill engulfed bacteria in the phagolysosome (Rigby and deLeo, 2012). In order to avoid damage to major macromolecules such as DNA, lipids, and proteins, *S. aureus* must be able to eliminate these potent oxidizing agents and to repair oxidatively damaged macromolecules. The major defence mechanisms used by *S. aureus* to repair damaged DNA- and proteins are described elsewhere in this chapter. Here we will focus on the highly conserved set of enzymes including catalases and superoxide dismutases that are responsible for breaking down ROS, and are key players in the bacterial oxidative stress responses elicited by ROS.

Superoxide dismutases (SODs) are metalloenzymes that catalyse the dismutation of O_2^- to oxygen and H_2O_2. SODs are classified by their cellular localization and by the type of metal cofactor utilized (Touati, 1997). In *S. aureus* two SODs (SodA and SodM), sharing 75% sequence identity and both requiring manganese for activity, have been identified (Clements *et al.*, 1999; Valderas and Hart, 2001). SodA is responsible for the majority of SOD activity, but both SodA and SodM activity can be detected throughout the growth cycle (Clements *et al.*, 1999; Valderas and Hart, 2001). While SOD activity is present during the entire growth cycle, SOD activity is increased significantly upon entry into the post-exponential growth phase, and remains high during the stationary phase signifying an increased need to eliminate ROS due to the transition from using substrate-level phosphorylation to using oxidative phosphorylation (Clements *et al.*, 1999; Valderas and Hart, 2001). Furthermore, *sodA* transcription was highly inducible by the addition of methyl viologen that is a potent generator of internal O_2^-, or by a pH downshift from 7 to 4 (Clements *et al.*, 1999). None the less, *sodA* or *sodM* single mutations conferred only minor reductions in growth and survival during oxidative stress, while a *sodA*, *sodM* double mutant was more severely affected, demonstrating overlapping activity of the two SODs (Clements *et al.*, 1999; Valderas and Hart, 2001).

The detoxification of hydrogen peroxide to water and oxygen is accomplished by catalases or by the peroxiredoxin family of enzymes, which additionally have activity against organic peroxides, and peroxynitrite (Poole, 2005). *S. aureus* possesses a single monofunctional haem-containing catalase encoded by the *katA* gene that seems to be the major scavenger of hydrogen peroxide in this organism (Horsburgh *et al.*, 2001; Cosgrove *et al.*, 2007). The alkyl hydroperoxide reductase AhpC on the other hand was important for detoxification of organic-peroxides (Cosgrove *et al.*, 2007). Mutation of both *katA* and *ahpC* led to a severe growth defect under aerobic conditions in defined media (Cosgrove *et al.*, 2007). The potential toxicity of H_2O_2 is exacerbated in the presence of Fe(II) due to the Fe(II)-mediated reduction of H_2O_2 and the subsequent formation of hydroxyl radicals (Fenton chemistry). For this reason, bacteria need to control the level of Fe(II) and other metal ions very tightly. For a more thorough discussion of general oxidative stress resistance mechanisms, we refer to the excellent recent review by Gaupp *et al.* (2012).

Oxidative defence mechanisms in virulence

Phagocytes like neutrophils and macrophage represent the first barrier of defence against invading bacteria. Phagocytosis is accompanied by an oxidative burst that generates the bactericidal superoxide anion, via the NADPH oxidase, which can dismutate into H_2O_2 (Rigby and deLeo, 2012). For this reason, oxidative stress responses were predicted to contribute to the virulence of bacterial pathogens like *S. aureus*. That being said, it was surprising to find that a mutant lacking both of the major enzymes for elimination of hydrogen peroxide (i.e. SodA and SodM), catalase (KatA) and alkylhydroperoxide reductase (AhpC), was fully virulent in two independent models of infection (i.e. murine models of septic arthritis and atopic dermatitis) (Cosgrove *et al.*, 2007). In contrast, both AhpC and KatA were required for nasal colonization (Cosgrove *et al.*, 2007); hence, the contribution of superoxide dismutase to *S. aureus* virulence could not be unambiguously established (Gaupp *et al.*, 2012), indicating that the primary function of these highly conserved detoxifying enzymes is not linked to virulence in *S. aureus*. Instead *S. aureus* seems to have developed more specific tools to combat oxidative stress during infection. One example is the yellowish-orange pigment staphyloxanthin produced by most *S. aureus* isolates from human infections. Staphyloxanthin is a triterpenoid that like other carotenoid-containing pigments functions as a potent antioxidant due to the numerous conjugated double bonds (Marshall and Wilmoth, 1981). The membrane-bound staphyloxanthin is synthesized from enzymes encoded by the *crtOPQMN* operon, which is transcribed from a σ^B-dependent promoter (Kullik *et al.*, 1998; Pelz *et al.*, 2005). Interestingly, defined deletion mutants in the *crt* operon rendered *S. aureus* non-pigmented but also significantly reduced virulence in two different murine models of infection (Liu *et al.*, 2005, 2008). Consistent with carotenoids functioning as antioxidants, the non-pigmented *crt* mutants also exhibited diminished resistance to phagocytic killing (Liu *et al.*, 2005). This is also consistent with the observation that non-pigmented *crt* mutants were as virulent as the wild-type if the oxidative burst was inhibited, or, if transgenic mice not producing the Nox NADPH oxidase were used in an infection model (Liu *et al.*, 2005). Due to these convincing results, inhibition of carotenoid biosynthesis has been proposed as a potential therapeutic target for treating *S. aureus* infections (Liu and Nizet, 2009). However, such a strategy will be challenged by the variable amount of staphyloxanthine produced by clinical *S. aureus* strains as inferred from the white to orange colour of colonies displayed by clinical isolates on standard solid culture media. Furthermore, examination of gene expression in the mouse airway surprisingly revealed that expression of the staphyloxanthine encoding operon *crtOPQMN*, as well as other oxidative stress response genes, was lower in this model of infection than in cells grown under laboratory conditions, suggesting that oxidative stress defences may be of minor importance in some types of infections (Chaffin *et al.*, 2012.)

Global responses to ROS *in vivo* and *in vitro*

The global transcriptomic response to ROS has been determined both *in vitro* and *in vivo*: *in vivo*, the transcriptomic response to phagocytosis was assessed immediately following phagocytosis, a period of time that corresponds with maximum generation of neutrophil ROS (Voyich *et al.*,

2005). Interestingly, this study confirmed strong induction of genes encoding catalase (*katA*), thioredoxin (*trxA*), thioredoxin reductase (*trxB*), superoxide dis- mutase (*sodA, sodM*), alkyl hydroperoxide reductase (*ahpC, ahpF*), and glutathione peroxidase (*gpxA*) in clinical isolates derived both from the community and hospital, supporting the idea that antioxidant defence enzymes do have a conserved function in the initial response to phagocytosis (Voyich *et al.*, 2005). *In vitro*, the transcriptomic response to exogenous hydrogen peroxide was characterized in the *S. aureus* model strain NCTC8325 (Chang *et al.*, 2006). In this study, exponential growth phase cultures of NCTC8325 were exposed to 10 mM hydrogen peroxide for 10 and 20 min. Addition of hydrogen peroxide halted bacterial growth for 10 min, whereafter the cells resumed growth at the same rate as untreated cells, presumably due to the break down of hydrogen peroxide. Strikingly, the addition of H_2O_2 did not induce genes encoding proteins involved in elimination of oxygen radicals. Instead, the most strongly induced genes encode proteins with an established role in DNA-damage repair (*lexA, uvrA, uvrB*) and to a lesser extent in protein homeostasis (*clpP*), emphasizing the potency of oxygen radicals in conferring damage to the cellular macromolecules. The strong induction of DNA-damage repair systems by exposure to hydrogen-peroxide is explained by the close proximity of Fe(II) with DNA, as Fe(II) catalyses reduction of hydrogen peroxide which results in the formation of the extremely reactive hydroxyl radicals via Fenton chemistry (Imlay, 2003). Support for this observation, can be seen in a proteomic study that analysed the response of *S. aureus* to three different stimuli of oxidative stress (methyl viologen, diamide, and H_2O_2) (Wolf *et al.*, 2008). The most striking result in this study was that the three responses had very limited overlap, indicating that the stress imposed by hydrogen-peroxide, methyl viologen, and diamide induce different signalling pathways leading to very specific responses (Wolf *et al.*, 2008). As an example, while exposure to hydrogen peroxide significantly increased DNA repair systems, treatment with the thiol specific oxidizer diamide primarily induced proteins involved in protein repair and degradation emphasizing that the most sensitive macromolecules to thiol-stress are proteins. Synthesis of only two proteins, AhpC and SACOL0815, was induced by all three stressors (Wolf *et al.*, 2008). The hypothetical protein SACOL0815 was recently shown to function as a ribosome dimerization factor in *S. aureus*, as it catalysed the dimerization of 70S ribosomes to 100S ribosomes *in vitro* (Ueta *et al.*, 2010). 100S ribosomes have mainly been studied in *E. coli*, where they were found to be translationally inactive, but believed to play an important role in preserving ribosomes under conditions that do not support growth (Yoshida *et al.*, 2004). For this reason, SACOL0815 homologues have been called *S. aureus* ribosome hibernation factor, SaHpf (Ueta *et al.*, 2010). Interestingly, SaHpf was also induced by heat stress; hence, ribosome dimerization may be one of the conserved features of the *S. aureus* stress responses (Andersson *et al.*, 2006).

Transcriptional regulators sensing oxidative stress

Different oxidative stress conditions induce differing types of damage to cellular macromolecules that are reflected in very specific changes in global gene transcription. A number of regulators responding to oxidative stress have been identified in *S. aureus*, CtsR was described above, and here we will briefly summarize the oxidative sensing regulators PerR, Spx, MgrA, and SarZ.

PerR

PerR is a conserved transcriptional repressor that controls expression of target genes by binding to a consensus DNA sequence termed the PerR box. In *Bacillus*, PerR was identified as a prototype for a novel type of oxidative regulators that senses peroxides by metal-catalysed oxidation (Lee and Helmann, 2006): The DNA-binding activity of PerR is enhanced when complexed with the divalent ions of Fe or Mn. Fe indirectly functions as the sensor of hydrogen peroxide because in the presence of hydrogen-peroxide Fe(II) generate hydoxyl-radicals via Fenton chemistry, which will oxidize the iron-coordinating His-residues, leading to release of iron and loss of DNA-binding activity (Lee and Helmann, 2006). In *S. aureus*, PerR boxes are localized upstream of genes encoding enzymes with a function in detoxification of ROS and in repair mechanisms including *katA, ahpCF, trxB, mrgA*, and *ftn* (iron storage protein ferritin) (Horsburgh *et al.*, 2001). Accordingly, most of these genes are induced in a *perR* mutant, but as most of these genes are controlled by additional regulators expression patterns are often complex (Horsburgh *et al.*, 2001). Notably, inactivation of *perR* reduced virulence in a murine abscess model of infection (Horsburgh *et al.*, 2001).

Spx

Spx is a novel type of regulator whose activity is controlled in response to the redox status of the cell. Spx has been extensively characterized in *B. subtilis* where its primary function is in the global control of the thiol stress response (reviewed by Zuber, 2009). Spx exerts both positive and negative control on gene transcription through direct interaction with the C-terminal domain of the α-subunit of RNA polymerase (Zuber, 2009). Interestingly, Spx-dependent transcriptional activation is under redox control at several levels. First, the Spx protein contains a CXXC disulfide centre at the N-terminus that must be oxidized in order to activate transcription of genes involved in thiol homeostasis, as exemplified by *trxA* and *trxB* (Nakano *et al.*, 2005). Secondly, the cellular level of Spx is controlled by regulated proteolysis responding to the stress status of the cell: under non-stress conditions, Spx is kept at a low concentration through continuous degradation by the ClpXP protease in a process stimulated by the adaptor protein YjbH (Fig. 10.1) (Zuber, 2009; Garg *et al.*, 2009). Under disulfide stress, Spx turnover is decreased due to the inactivation of YjbH-mediated proteolysis by the release of Zn from both the ClpX ATPase subunit and the YjbH adaptor protein (Zuber, 2009; Garg *et al.*, 2009). The basic features in Spx function and regulation of activity appear to be conserved among *S. aureus* and *B. Subtilis* (Pamp *et al.*, 2006; Engman *et al.*, 2012). As an example, Spx is required for transcription of *trxB* in un-stressed *S. aureus* and for transcriptional induction of *trxB* in response to disulfide or heat stress (Pamp *et al.*, 2006). Notably, inactivation of *spx* confers much more severe growth defects to *S. aureus* than to *B. subtilis*, and recent research suggest that the *S. aureus spx* mutant is only viable because it has accumulated additional suppressor mutations (Pamp *et al.*, 2006; W.L. Kelley, unpublished). Interestingly, Spx activity also seems to function in biofilm formation and in resistance to cell wall targeting antibiotics (Pamp *et al.*, 2006; Jousselin *et al.*, 2013).

MgrA/SarZ

S. aureus encodes two transcriptional regulators of the MarR (multiple antibiotic resistance) family, MgrA and SarZ, whose activity is controlled by a redox-sensitive cysteine, which can be oxidized by ROS such as hydrogen peroxide and organic hydroperoxide (Chen *et al.*, 2006, 2009). In that sense, MgrA and SarZ are similar to the *B. subtilis* peroxide sensing transcriptional regulator OhrR, whose DNA binding activity is modulated in response to cysteine oxidation (Lee *et al.*, 2007). Interestingly, there is only a very limited overlap between the MgrA- and SarZ-controlled regulons (Luong *et al.*, 2006; Chen *et al.*, 2009): MgrA controls a large number of genes encoding proteins with a presumed role in virulence including capsule biosynthesis enzymes, autolytic enzymes, surface bound proteins, like protein A, and several multidrug efflux pumps (Luong *et al.*, 2006), while SarZ regulates expression of genes encoding proteins with a role in intermediary metabolism of nucleotides, fatty acid, amino acid, and sugar metabolism (Chen *et al.*, 2009). Thus, MgrA and SarZ represent examples of transcriptional regulators that, despite a conserved signalling mechanism, control different subset of genes allowing the cell to fine-tune adaptation to different physiological conditions. In the case of MgrA, the oxidation-sensing mechanism seems to be used as a signal for turning on a number of global defensive pathways that help *S. aureus* to establish an infection. In support of the importance of MgrA for virulence of *S. aureus*, an *mgrA* mutant was substantially attenuated in two different murine infection models (Chen *et al.*, 2006). It should be noted that in the same virulence model the *sarZ* mutant displayed only a statistically insignificant decrease in virulence, emphasizing the non-overlapping nature of the two regulons.

DNA damage response

In response to extensive DNA damage, bacteria elicit a response collectively known as the SOS response (Schlacher and Goodman, 2007). Although mostly characterized in *E. coli* a similar response has been recognized in many bacterial cells including *S. aureus*, but also with some prominent exceptions like *Campylobacter jejuni* and *Helicobacter pylori*. Two key elements of the SOS response are the transcription repressor, LexA and the recombination protein, RecA (Gudas and Pardee, 1975). When a bacterial chromosome is damaged, regions of single-stranded DNA are formed and allow assembly of activated RecA nucleoprotein filaments. Activation of RecA in turn stimulates autocleavage of the LexA repressor that allows transcription of LexA repressed genes (Little *et al.*, 1980). LexA binds to operator sites composed of 20 bp-consensus sequences with differential affinities, so that genes carrying a low affinity site are induced early (Fernández De Henestrosa *et al.*, 2000). In *E. coli*, these genes include *lexA* itself as well as those encoding proteins involved in DNA repair such as nucleotide excision repair (*uvr*) and recombination repair (*recA*). Upon continued DNA damage, a later response is initiated comprising genes with high-affinity LexA binding sites. The primary function of these genes is to allow replication of the DNA across damaged regions in a process termed translesion synthesis, a replicative process that is inherently mutagenic. Translesion synthesis in *E. coli* is accomplished by three polymerases encoded by PolIV (*dinB*), PolII (*dinA*) and PolV (*umuDC*). The PolV is responsible for the vast majority of mutations occurring in rapidly growing cells, PolIV for adaptive mutations in non-growing cells (Foster, 2000), and all three polymerases for contributing to cellular fitness during stationary phase (Yeiser *et al.*, 2002).

The SOS regulon in *S. aureus*

In *S. aureus*, the LexA repressor was originally identified as part of a study in which the regulation of fibronectin binding proteins was examined (Bisognano *et al.*, 2004). It was demonstrated that sub-inhibitory concentrations of ciprofloxacin induced transcription of *fnbB* and that this response was abolished in a *recA* mutant strain. (Ciprofloxacin is a fluoroquinolone that inhibits DNA gyrase and topoisomerase activity and consequently stalls DNA replication.) Upon purification of the *S. aureus* LexA homologue, electrophoretic mobility shift assays revealed specific binding to both *recA* and *fnbB* promoters. DNase I footprint analysis demonstrated LexA binding overlaps the core promoter elements in *fnbB* and the binding site resembles that identified in *B. subtilis* (Bisognano *et al.*, 2004).

An overview of the SOS response in *S. aureus* was determined using DNA microarrays of cells from exponentially growing cells of strain UAMS1 that were exposed to the DNA cross-linking agent, mitomycin C (Anderson *et al.*, 2006). A total of 73 genes were found to be induced by mitomycin C, including *lexA*, components of the nucleotide excision repair genes (i.e. *uvrA* and *uvrB* and SACOL1400), and a gene with homology to *umuC*, encoding one of the translesion polymerases. Also included in the SOS response are several phage replication and packaging genes and the recombination repair proteins, SbcC and SbcD, which upon SOS induction lead to repression of capsule synthesis (Chen *et al.*, 2007). To define the LexA regulon more precisely, a mutant was constructed that expresses a LexA variant unable to be cleaved, thus, remains bound to the target sites even in the presence of SOS inducing conditions (Maiques *et al.*, 2006, Cirz *et al.*, 2007). Upon exposure to ciprofloxacin 18 genes were induced in wild-type bacteria but not in bacteria carrying the LexA variant (Cirz *et al.*, 2007). Potential LexA binding sites were found in front of 16 out of the 18 genes. Among the genes identified were *lexA* and *recA*, as well as the *uvrA*, *uvrB* and SACOL1400, and the *parC* and *parE* encoding subunits of the topoisomerase IV (Cirz *et al.*, 2007).

LexA activity

During investigations of the Clp proteolytic complex, it was observed that several of the LexA regulated genes are up-regulated in a *clpP* mutant (Michel *et al.*, 2006). As described above, the Clp complex is composed of subunits of ClpP forming the proteolytic core flanked by Clp ATPases responsible for selective access to the proteolytic chamber (Frees *et al.*, 2014). Targets of the Clp complex include both unfolded proteins and central, short-lived regulatory proteins that are inherently unstable (Frees *et al.*, 2014). Among the genes up-regulated in a *clpP* deletion mutant were 12 out of 20 putative LexA-dependent genes including *recA*, *fnbB*, *umuC*, *uvrA*, and *lexA* itself. Importantly, the LexA recognition motif was identified upstream of two genes belonging to the oxidative stress regulon controlled by PerR (*ahpC* and the ferric ABC transporter SA0891 gene). Based on these data, de-repression of the LexA regulon in the *clpP* mutant was suggested to be a consequence of increased DNA damage due to the reduced capability of the Δ*clpP* strain to cope with oxidative stress and remove unfolded proteins (Michel *et al.*, 2006).

During SOS inducing conditions, the LexA repressor undergoes autocleavage resulting in a C-terminal dimerization domain (CTD) and an N-terminal DNA-binding domain (NTD) known to retain some DNA-binding activity (Fig. 10.2) (Neher *et al.*, 2003). Using an antibody specific for *S. aureus* LexA, we could show that LexA undergoes autocleavage and that both fragments are highly unstable. In the absence of the proteolytic subunit, ClpP, or one or both of the Clp ATPases, ClpX or ClpC, the LexA domains are stabilized after autocleavage. When we ectopically overexpressed a peptide corresponding to the N-terminal half of LexA, we observed that *sosA*, a gene divergently transcribed from *lexA*, remained repressed following mitomycin treatment, while transcription of *lexA* was induced. Based on our results, we proposed that sequential proteolysis of LexA is conserved in *S. aureus* and that a subset of the LexA controlled genes remains repressed in the presence of the N-terminal domain, as illustrated in Fig. 10.2 (Cohn *et al.*, 2011). If so, then we expect conditions to exist where ClpP mediated proteolysis is halted, allowing repression of the genes in the LexA regulon that are repressed by the N-terminal fragment. The complex nature of this regulation is further complicated by our findings that RecA itself appears a target for ClpP and that degradation of RecA by ClpP is dependent on the presence of SOS inducing conditions (Feng *et al.*, 2013). The two-step degradation process of LexA also involves the Clp proteolytic complex, which allows integration of signals other than those associated with DNA damage to be integrated in modulation of the SOS response.

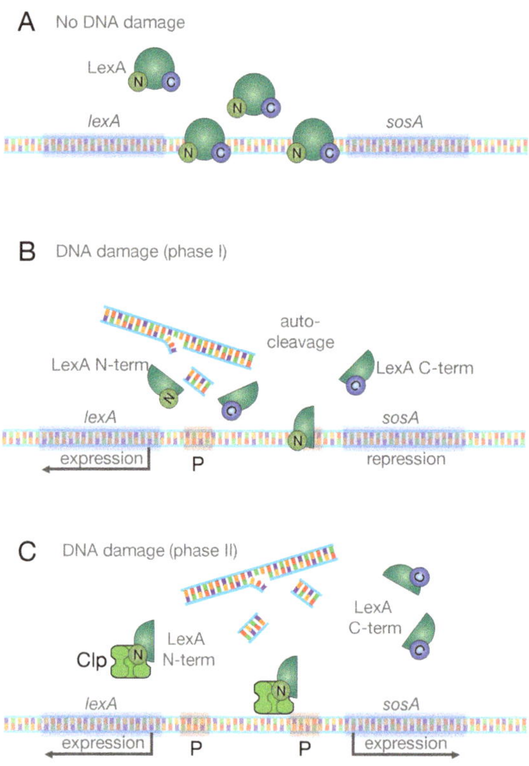

Figure 10.2 Model for derepression of the LexA regulon. In the absence of DNA damage the LexA controlled genes are repressed by the LexA repressor (A) that in response to initial DNA damage (phase I, B) undergoes RecA mediated auto-cleavage upon which the N-terminal fragment of LexA retains DNA binding capacity. Upon prolonged or severe DNA damage stress (phase II, C) the LexA N-terminal fragment is degraded by the Clp protease and full induction of the LexA regulon is observed.

SOS response in prophage induction and virulence

Induction of lysogenic prophages has been one of the hallmarks of an active SOS response. In *S. aureus*, mobile genetic elements are strong contributors to diversification, with phages and phage-related genomic islands playing a major role in this diversification. Most phages and phage-related genomic islands encode accessory virulence determinants such as staphylokinase, enterotoxins, exfoilative toxin, or the Panton–Valentine leucocidin, PVL (Xia and Wolz, 2014, McCarthy *et al.*, 2012). In strains of human origin, phages are commonly found integrated in the chromosomal gene encoding the beta-haemolysin (*hlb*). Excision of such lysogens can be accomplished by exposure to subinhibitory concentrations of ciprofloxacin, which enhances *recA* transcription and in a dose-dependent manner promotes replication of phages in the bacterial host (Goerke *et al.*, 2006). In addition to excision of prophages, induction of the SOS response may also influence expression of virulence factors residing on phages. One such example is expression of the Panton–Valentine leucocidin (PVL), which is a pore-forming, bi-component toxin secreted by *S. aureus* strains associated with diseases such as necrotizing pneumonia and skin and soft-tissue infections. Transcription of the phage-encoded PVL from the *luk-PV* operon could be induced by mitomycin C as a result of read-through transcription from latent phage promoters and from an increase in phage copy numbers. In contrast in strains, where prophage excision failed, a constant expression of *luk-PV* was observed, demonstrating the link between excision and expression of virulence factors (Wirtz *et al.*, 2009).

In addition to traditional phages, *S. aureus* also harbours phage-related pathogenicity islands, the SaPI elements, which are extremely common and found in almost all strains. These SaPI usually carry superantigens and are responsible for most superantigen-related human diseases, especially staphylococcal toxic shock syndrome (Novick and Subedi, 2007). Interestingly, the SaPI elements are induced to excise and replicate after SOS induction of different temperate phages, and when packaged into phage-like particles are transferred at high frequency (Ubeda *et al.*, 2005, 2007). As mentioned, fluoroquinolones can induce the SOS response, but beta-lactams also induce a bona fide SOS response in *S. aureus*, with the activation of the RecA and LexA proteins (Maiques *et al.*, 2006). Moreover, the beta-lactams trigger prophage induction in *S. aureus* lysogens as well as replication and high-frequency transfer of the SaPI (Maiques *et al.*, 2006); thus, the SOS response plays a central role in dissemination of, and expression from, mobile genetic elements. An unintended consequence of antibiotic treatment may be increased SaPI replication and transcription, which could exacerbate a superantigen based illness.

Response to antimicrobial peptides and antibiotics

In addition to the classical induction of the SOS response by fluoroquinolones, there are other links between antibiotics and the SOS response system. One well-studied area has been the role of the SOS response in methicillin resistance. Methicillin-resistant *S. aureus* (MRSA) are commonly multidrug-resistant strains that give rise to difficult-to-treat infections in humans. Curiously, MRSA strains can be either hetero-resistant or homo-resistant, with the former being mostly associated with persistent infections, and composed of mixed cell populations that contain organisms with low levels of resistance and the latter being those that display high levels of drug resistance (Finan *et al.*, 2002). Importantly, the homo-resistant phenotype can be induced in a hetero-resistant population by exposure to β-lactam antibiotics, a process that depends on the SOS response (Cuirolo *et al.*, 2009). Induction of the SOS response by β-lactam antibiotics recently was shown to be mediated by tricarboxylic acid (TCA) cycle activity through generation of reactive oxygen species, which affects DNA integrity, triggers activation of the SOS response and results in enhanced mutagenesis (Rosato *et al.*, 2014). The latter finding provides a mechanistic explanation for the previous observation that β-lactams induce a bona fide SOS response (Maiques *et al.*, 2006).

The increase in bacterial antibiotic resistance has led to renewed interest in development of alternative antimicrobial treatment strategies such as antimicrobial peptides (AMPs) that are either naturally occurring or synthetically derived. Although little is known of the mode of action of such peptides they have, in a few cases, been shown to bind directly to chromosomal DNA with the possibility of inducing the SOS response. In one case, the lysine–peptoid hybrid, LP5, at concentrations above the minimal inhibitory concentration (MIC), led to ATP leakage; however, at lower concentrations, it bound DNA and inhibited DNA gyrase and topoisomerase IV activity and caused induction of the SOS response (Gottschalk *et al.*, 2013). In another study, a number of cyclic peptides were found to interfere with dimerization of the β-sliding clamp of the chromosomal replication complex in *S. aureus* (Kjelstrup *et al.*, 2013). Of these cyclic peptides, two 8-mer peptides were analysed in greater detail. Both inhibited DNA replication, induction of the SOS response, and altered cell morphology, ultimately resulting in cell death. The target of the peptides were determined to be DnaN, the β-clamp of the replication machinery termed the replisome (Kjelstrup *et al.*, 2013). Although little is known of how *S. aureus* responds to human antimicrobial peptides, a critical component of the innate immune defence, it is likely the SOS DNA damage response will play a critical role.

Osmotic stress

S. aureus is capable of growing in food products with exceptionally low water activity (aw), a phenotypic trait that has been exploited in isolation and enrichment procedures for *S. aureus* (Lancette *et al.*, 1986). Presumably, the high osmotolerance of *S. aureus* supports growth of the bacterium in its preferred niches (i.e. skin and mucous membranes of mammals) and in blood or other high-osmolarity body sites during infection. Moreover, salt tolerance is important for the ability of *S. aureus* to cause food-borne intoxication, as it enables *S. aureus* to grow in food with high salt concentrations, for example on ham (Portocarrero *et al.*, 2002). For these reasons, *S. aureus* represents an interesting model organism for studying osmotolerance, and recent publications in this area have shed new light on osmotolerance. Generally, bacteria adapt to growth in environments of high osmolarity by cytoplasmic accumulation of solutes that restore the turgor pressure and help the cell to retain water and avoid plasmolysis (reviewed by Sleator and Hill, 2002). The accumulated solutes are compatible with macromolecular and cellular functions and are often small organic zwitterions such as glutamine, proline, and glycine betaine (Sleator and Hill, 2002). In *S. aureus*, choline, glycine betaine, L-proline, and taurine act as osmoprotectants, with glycine betaine being most efficient in restoring growth in medium containing high concentrations of NaCl (Miller *et al.*, 1991; Graham and Wilkinson, 1992). *S. aureus* cultivated in the presence of 2.5 M NaCl exhibited significantly increased cell size and abnormal septum formation; however, the addition of glycine betaine abolished both of these phenotypes (Vijaranakul *et al.*, 1997). Pre-genetic studies identified two transport systems that mediate uptake of glycine betaine, however, the responsible genes and proteins were not identified (Pourkomailian and Booth, 1992, 1994). More recently, two transport systems responsible for uptake of proline have been characterized genetically: OpuD is a low-affinity proline transporter that contribute to proline uptake only when proline is abundant (Wetzel *et al.*, 2011), while PutP is a high-affinity proline uptake system of importance to virulence (Wengender and Miller, 1995; Schwan *et al.*, 1998). The contribution of these transporters to osmotolerance has yet to be elucidated.

Recently, research on osmotolerance has established the importance of K^+ in *S. aureus*. Typically, K^+ is the most abundant cation in bacteria, and the initial response to high osmotic stress generally involves the rapid accumulation of K^+ (Sleator and Hill, 2002). The transcriptional response of *S. aureus* to growth in the presence of high NaCl was recently characterized (Price-Whelan *et al.*, 2013); specifically, the transcriptional profiles of the CA-MRSA strain USA300 LAC grown in complex medium with and without 2 M NaCl were determined. Interestingly, this study identified the *kdpFABC* operon encoding an ATP-driven, high-affinity K^+ uptake system as the most highly up-regulated genetic locus in cells grown in the presence of high salt. These results indicate that the K^+-status is important for continued growth of *S. aureus* in the presence of high NaCl, and not only for the initial adaptation following an up-shift in osmolarity. *S. aureus* additionally encodes a low-affinity K^+ uptake system belonging to the Ktr transporter family, which consists of two ion-conducting proteins (KtrB and KtrD) and one regulator (KtrA) controlling activity of the ion-conducting pores (Gries *et al.*, 2013; Price-Whelan *et al.*, 2013). In support of the important role of K^+ in providing *S. aureus* osmotolerance, inactivation of the Ktr system severely reduced growth in medium with high NaCl or high sorbitol (Gries *et al.*, 2013; Price-Whelan *et al.*, 2013). Despite the increased transcription of the *kdp* locus during osmotic stress, a mutant lacking the Kdp transporter grew as well as the parental strain, probably due to increased transcription of the genes encoding the Ktr system (Price-Whelan *et al.*, 2013). Remarkably, up-regulation of the *kdpFABC*-operon was dependent on the KdpDE two-component system that also controls transcription of the *S. aureus* capsule. Consistent with this finding, transcription of the *cap5*-operon encoding the enzymes responsible for capsule biosynthesis was up-regulated more than 10-fold in cells grown in the presence of high salt (Price-Whelan *et al.*, 2013). Transcription of the genes encoding the Ktr uptake system was not induced by addition of NaCl, consistent with the finding that Ktr systems are constitutively expressed in other bacteria and that activity is regulated at the protein level (Matsuda *et al.*, 2004; Tholema *et al.*, 2005). Interestingly, the newly discovered nucleotide signalling molecule c-di-AMP seems to be involved in controlling K^+-uptake, as both the KtrA gating component of the low-affinity Ktr uptake system, and the histidine kinase protein KdpD controlling expression of the Kdp high-affinity K^+ uptake were identified as c-di-AMP receptor proteins (Corrigan *et al.*, 2013).

Cold stress

Prolonged incubation at low temperatures induce formation of small colony variants

The ability of *S. aureus* to survive at suboptimal temperatures is likely an important parameter for the nosocomial spread of *S. aureus* in hospitals. Furthermore, refrigeration is the most commonly used method for extending the shelf life of foods, therefore, the cold shock adaptation of *S. aureus* is crucial for the ability of enterotoxin-producing isolates to cause food-poisoning. Recently, it was investigated how *S. aureus*, *S. epidermidis* and *S. lugdunensis* responded to incubations at 4°C for up to 8 weeks. The study revealed that prolonged incubation at low temperature resulted in the formation of increasing proportions of small colony variant (SCV) phenotypes (i.e. < 1 mm in colony size, reduced pigmentation, decreased haemolysis) (Onyango *et al.*, 2012). SCV formation in *S. aureus* has attracted attention due to its proposed role in chronic infections (von Eiff and Becker, 2007). While *S. aureus* produced SCVs during cultivation at 4°C, it was surprising to find that two other staphylococcal species generated SCVs more readily than did *S. aureus*, which did not produce SCVs until after 5 weeks of incubation (Onyango *et al.*, 2012). The late appearance and time-dependent increase in the proportion of SCV's indicate that the SCV-phenotype is selected in order to improve survival only under conditions of long-term stress that inhibit growth for extended times. Notably, cells from SCV colonies were characterized by thicker and more diffuse appearing cell walls and species-specific alterations in the amino acid composition of cell wall associated proteins (Onyango *et al.*, 2012). The SCV-phenotype was not due to genetic changes, as evidenced by the swift reversion to WT-phenotypes upon subculturing (Onyango *et al.*, 2012).

Inactivation of cold shock proteins changes susceptibility of *S. aureus* to antimicrobials

The major challenges associated with growth and survival at low temperature are a decrease in membrane fluidity and the stabilization of secondary structures in RNA and DNA, which interfere with the efficiency of translation, transcription, and replication. In response to a temperature downshift, bacteria induce a set of low-molecular-weight, cationic cold shock proteins (Csps) (reviewed by Phadtare and Severinov, 2010). In *E. coli*, the major Csp, CspA, acts as an RNA chaperone that prevents the formation of undesired secondary structures that would otherwise impede the transcriptional and translational machinery (Jiang *et al.*, 1997; Phadtare, 2011). Csps have been divided into two classes based on expression patterns (Yamanaka, 1999): Class I includes Csps that are synthesized at low levels at 37°C, but are highly induced by downshift in temperature. Conversely, class II Csps are relatively abundant at 37°C and are only moderately increased during cold shock. In *S. aureus*, the transcriptomic response to cold-shock was studied following a temperature downshift from 37°C to 10°C (Andersson *et al.*, 2006). The *cspB* gene was found to be the major cold shock-inducible gene in *S. aureus* (Andersson *et al.*, 2006). This is consistent with another study demonstrating that the *cspA* mRNA was more abundant than either the *cspB* or the *cspC* mRNAs in cultures maintained at 37°C, but the *cspB* transcript predominated in a culture grown at 15°C (Duval *et al.*, 2010). Interestingly, transcription of *cspB* and *cspC* was also induced in a murine pneumonia model demonstrating that transcription of these genes responds to signals other than cold (Chaffin *et al.*, 2012). The molecular functions of CspA, CspB, and CspC in *S. aureus* remain unidentified, but mutants lacking either CspA and CspB had some interesting phenotypes: Genetic inactivation of *cspB* conferred a growth defect both at low temperature and at 37°C, emphasizing that CspB activity, despite its relatively low expression at 37°C, is important for growth also at optimal temperature (Duval *et al.*, 2010). The *cspB* mutant exhibited reduced pigmentation, and altered levels of susceptibility to many antimicrobials (Duval *et al.*, 2010). In particular, a *cspB* null mutant displayed increased resistance to aminoglycosides like gentamycin that can be used to select for stable SCV-variants (Duval *et al.*, 2010). Additionally, the *cspB* null mutant exhibited increased susceptibility to daptomycin and teicoplanin targeting the cell membrane and the cell wall, respectively (Duval *et al.*, 2010). A *cspA* mutant had similar phenotypes, as it adapted less readily to growth after a downshift in temperature, was less pigmented, and had increased tolerance to some antimicrobial peptides (Katzif *et al.*, 2003). The enzymes responsible for the first steps of the carotenoid synthesis are encoded within a σ^B-dependent transcript, and based on the observation that σ^B activity was decreased in the *cspA* mutant, it was suggested that CspA regulates pigment production in a σ^B-dependent manner (Katzif *et al.*, 2005). In conclusion, we still have very limited knowledge on the functions of *S. aureus* Csps, but we know that Cps's contribute not only to growth at low temperatures but also to the general tolerance of different types of antimicrobials at optimal temperatures. Hopefully, future experiments will reveal the underlying mechanisms of Csps in facilitating the staphylococcal response to cold stress.

The stringent response

The stringent response is a highly conserved physiological response to stress associated with limitations in nutrients and other adverse conditions. In *E. coli*, the stringent response is mediated via the alarmones (p)ppGpp. Guanosine 3′-diphosphate, 5′-triphosphate (pppGpp) is synthesized from ATP and GTP through the combined actions of two enzymes, RelA and SpoT, which is converted to ppGpp (guanosine 3′,5′-bis(diphosphate)) via the enzyme GppA. The RelA synthase is associated with the ribosome and responds to amino acid starvation, whereas, the bi-functional SpoT is a cytoplasmic enzyme that is responsible for maintaining the basal level of (p)ppGpp during normal growth through its (p)ppGpp synthesis and degradation activities. SpoT also responds to changes in external, environmental conditions such as heat shock, fatty acid starvation, iron and phosphate limitations, hyper osmotic shock, and oxidative stress (Boutte and Crosson, 2013). Similar to most firmicutes, *S. aureus* has a bifunctional RelA/SpoT enzyme (i.e. RelA/SpoT homologue, RSH) that combines the synthase and hydrolase activities in one protein (Geiger and Wolz, 2014). Although the *rsh* gene is essential, the synthase activity of the enzymes can be eliminated without compromising viability. Deletion of RSH synthase activity was shown to be responsible for the synthesis of (p)ppGpp in response to Leu/Val amino acid starvation and mupirocin treatment that inhibits the bacterial isoleucyl-tRNA synthetase (IleRS), leading to the accumulation of uncharged isoleucyl-tRNA and induction of the stringent response (Fig. 10.3) (Geiger *et al.*, 2010). The essentiality of RSH hydrolase activity is due to the toxicity associated with (p)ppGpp accumulation, which accumulates due to the activity of two small synthases designated RelP and RelQ that lack hydrolase domains (Geiger *et al.*, 2014). Both of these small synthases have a preference for GDP as a pyrophosphate recipient, leading to the synthesis of ppGpp and in common with RSH, overexpression of either RelP or RelQ leads to induction of the stringent response. The role of RelP/Q appears to be in connecting cell wall distress to the stringent response, as cell wall active antibiotics induce synthase production, via the VraRS wall stress signal transduction system (Fig. 10.3) (Geiger *et al.*, 2014). Although substantial insight has been shed on the synthesis of the (p)ppGpp signal molecules, little is known of how they modulate changes in gene expression. In *E. coli*, (p)ppGpp binds directly to RNA polymerase and alters transcription of many genes including the rRNA operons, whereas, in other bacteria, this appears not to be the case (Krasny and Gourse, 2004). In *B. subtilis*, (p)ppGpp binds and inhibits directly the activity of several GTP biosynthesis enzymes (Kriel *et al.*, 2012), leading to a reduced GTP pool, which is important in at least two ways: first, expression of transcripts that are initiated with a GTP is reduce including the rRNAs (Krasny and Gourse, 2004); second, the reduced GTP pool influences the activity of enzymes for which GTP is a co-factor, importantly including the CodY transcriptional regulator (Handke *et al.*, 2008). Similar studies still remain to be conducted for *S. aureus*.

The stringent response regulon and CodY

Upon induction of (p)ppGpp synthesis, bacterial cells respond by inducing or repressing gene expression. As has been observed for other Gram-positive bacteria, the starvation response in *S. aureus* involves a coordinated action between the stringent response and the transcriptional regulator CodY (Fig. 10.3) (Majerczyk *et al.*, 2008). During exponential growth, CodY is an active transcriptional repressor that binds to DNA. In the transition into the stationary phase, CodY is released from DNA, and genes responsible for adaptation to nutrient limitation can be transcribed. The effector molecules of CodY are GTP and/or branched chain

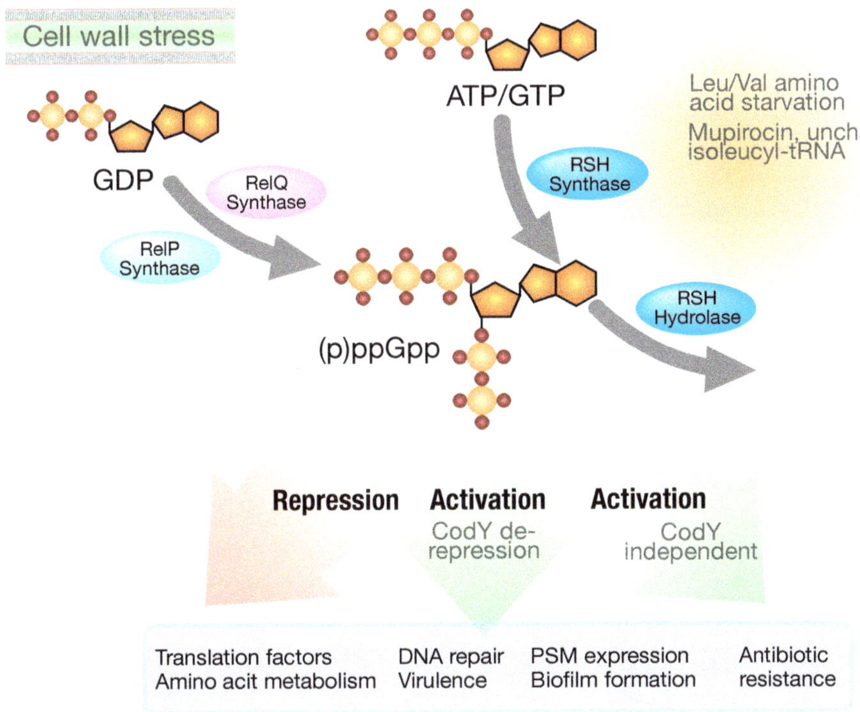

Figure 10.3 The stringent response in *S. aureus.* Synthesis of (p)pGpp occurs via the synthase activity of the bifunctional RSH or the small RelP and RelQ in response to either amino acid starvation or cell wall stress, respectively. Toxic levels of (p)ppGpp are prevented through the hydrolytic activity of RSH. Accumulation of (p)ppGpp influences major physiological functions mediated in part via the transcriptional regulator, CodY.

amino acids (BCAA); hence, when intracellular levels of GTP and BCAA decrease during entry into the stationary phase, CodY decreases its affinity for DNA, leading to the induction of its regulon (Geiger and Wolz, 2014).

In *S. aureus,* the stringent response was investigated where the induction was mediated by leucine and valine starvation in defined medium (Geiger *et al.*, 2012). Under these conditions, a mutant that expresses a RSH variant lacking the synthase domain (RSH_{syn}) showed no (p)ppGpp accumulation, indicating that the contribution of the small syntheses, RelP and RepQ was negligible (Geiger *et al.*, 2010, 2012). Analysis of wild type and mutants expressing either the RSH_{syn} or lacking CodY revealed multiple response pathways upon amino acid starvation. Independent of CodY, a large group of genes are transcriptionally repressed upon induction of the stringent response, including those encoding ribosomal proteins, translation factors and rRNA processing factors (*rps, rpl, infB, tsf, rnpA, rimM*), DNA helicases (*dnaB, dnaC, dnaD*), and DNA repair genes (*ruvAB, dinP*) (Geiger *et al.*, 2012). Another large group of genes are those up-regulated during stringent response, and the majority of those depend on CodY (Fig. 10.3). These genes possess a canonical CodY binding motif and are part of the CodY regulon, primarily involved in amino acid metabolism and transport (Pohl *et al.*, 2009; Majerczyk *et al.*, 2010). A small, third set of genes that are positively regulated by the stringent response are those that are activated independently of CodY and transcribed significantly less in the RSH_{syn} mutant (Fig. 10.3). Interestingly, this small set of genes was revealed to encompass the phenol-soluble modulins (PSMs). PSMs are peptide toxins that cause lysis of many human cell types including leucocytes and erythrocytes, stimulate inflammatory responses, and contribute to biofilm development. Interestingly, the PSMs are directly controlled by the AgrA response regulator of the accessory gene regulator (Agr) quorum sensing system, but stringent response mediated activation of transcription of the PSM genes appears to occur independently of AgrA (Geiger *et al.*, 2012).

Similar to the transcriptional profiling study of Geiger *et al.*, a proteomic approach was taken to study the stringent response (Reiss *et al.*, 2012). In this study, proteins involved in nucleotide biosynthesis, DNA metabolism, energy metabolism, and translation were found to be significantly down-regulated during a stringent response; whereas, expression of isoleucyl-tRNA synthetase, the branched-chain amino acid biosynthetic pathway, and proteins with functions in oxidative-stress resistance (*ahpC* and *katA*) and stress protection in general (the *yvyD* homologue *SACOL0815* and *SACOL1759* and *SACOL2131*), and transport processes were increased. In addition to regulation by CodY, this study also pointed to the general sigma factor, σ^B as shaping the stringent response (Reiss *et al.*, 2012).

Virulence

Several studies have investigated the link between the stringent response and virulence in *S. aureus* in different infection models. Since multiple enzymes contribute to (p)ppGpp synthesis, elimination of (p)ppGpp from *S. aureus* is not trivial. To determine the effect of (p)ppGpp synthesis on virulence, the RSH_{syn} mutant (described above), which lacks the (p)ppGpp synthase domain but does have small RelP and RepQ synthases was used to reduce the intracellular concentration of (p)ppGpp, and virulence was assessed in a haematogenic kidney abscess model (Geiger *et al.*, 2010). In this model, the RSH_{syn} mutant displayed significantly lower virulence compared to wild type cells and importantly, this reduction could be abrogated by an additional mutation in *codY*, indicating that one link between the stringent response and virulence is through repression of the CodY regulon (Geiger *et al.*, 2010). In contrast, hyper-accumulation of (p)ppGpp was observed in a clinical MRSA isolate that displayed a small colony variant (SCV) phenotype, which gave rise to a persistent infection and resistance to several different antibiotics (Gao *et al.*, 2010). The constitutive accumulation of (p)ppGpp was the result of a mutation in *rsh* that inactivated the hydrolase domain and which arose during treatment involving intravenous vancomycin combined with clindamycin and rifampicin and followed by linezolide. This finding is somewhat surprising given the proposed lethality of (p)ppGpp accumulation, but it may reflect that the mutation only reduced hydrolase activity and did not eliminate it. In a later study using this *rsh* mutant strain, the *rsh* mutation was shown to slightly increase survival in a murine abscess model, while significantly increasing resistance to human antimicrobial peptides (Gao *et al.*, 2013).

Interestingly, among the few genes up-regulated during the stringent response in an RSH-dependent, but CodY independent, manner are the cytotoxic phenol-soluble modulins (PSMs) (Fig. 10.3). The PSMs are important for survival of *S. aureus* when encountering human polymorphonuclear neutrophils (PMNs); hence, survival following PMN phagocytosis was examined. Importantly, survival of rsh_{syn} and rsh_{syn} *codY* mutants in neutrophils was reduced comparable to that observed for *psm* mutant strains, and hallmark genes of the stringent response were up-regulated in phagocytosed *S. aureus* cells, demonstrating the importance of the stringent response for intracellular survival (Geiger *et al.*, 2012).

An important element of many staphylococcal infections is the formation of biofilms. Biofilms are multicellular communities that are particularly associated with medical device-related infections and are contributing to chronic infections when colonizing body parts such as heart valves and soft tissue (de la Fuente-Núñez *et al.*, 2013). To deal with these types of infections, compounds have been sought that are able to dissolve and eradicate biofilm formation. In a search for such compounds, a previously characterized immune modulatory peptide was discovered to specifically target and kill both Gram-positive and Gram-negative cells present in a biofilm (de la Fuente-Núñez *et al.*, 2014). Previously, the stringent response was demonstrated to be actively supporting survival of bacteria in biofilms (Nguyen *et al.*, 2011); therefore, the authors speculated that it may be targeted by this immune modulatory peptide. Remarkably, this peptide displayed a broad ability to disrupt biofilms by direct binding of the peptide to (p)ppGpp, leading to repression of the stringent response and reduced survival in biofilms (de la Fuente-Núñez *et al.*, 2014).

Antibiotic resistance

While resistance to antibiotics was initially considered as limited to the traditional resistance mechanisms involving changes in cell wall, protein and nucleotide biosynthesis, recently our understanding has increased and broadened the view of the cellular factors contributing to antimicrobial resistance. For the stringent response, a link between resistance and (p)ppGpp synthesis was indicated by the recent finding that transcription of both *relP* and *relQ* is induced by cell wall active antibiotics (Fig. 10.3) (Geiger *et al.*, 2014). Importantly, a mutant lacking both *relP* and *relQ* was killed significantly faster than wild type bacteria by either ampicillin or vancomycin antibiotics (Geiger *et al.*, 2014). Surprisingly, activity of just one of these small synthases was sufficient to restore survival in the presence of antibiotics to that of wild type cells, suggesting that (p)ppGpp levels are fine-tuned with respect to resisting antibiotics. In an earlier study, (p)ppGpp was linked to β-lactam resistance in MRSA (Mwangi *et al.*, 2013). MRSA express the penicillin binding protein PBP2a encoded by *mecA* and clinical strains are known to differ widely in their susceptibility to β-lactam antibiotics (Tomasz *et al.*, 1991). Interestingly, this variation in susceptibility to β-lactams is due in part to the phenomenon of hetero-resistance, where the majority of a population has low or intermediate levels of resistance to oxacillin, but a subpopulation of cells display very high resistance levels (de Lencastre *et al.*, 1993). Importantly, it was shown that the subpopulation of highly resistant cells carried a mutation in *relA* (another designation for *rsh*), leading to overproduction of (p)ppGpp and high level resistance, which could be reversed by a mutation in *relQ* presumably reducing (p)ppGpp levels (Mwangi *et al.*, 2013). The mechanism behind this increased resistance was found to be through the synthesis of the PBP2a protein, as induction of the stringent response by mupirocin enhanced expression of *mecA* and reverted heterogeneous resistant clones to high-level resistance (Kim *et al.*, 2013). These findings were supported by additional mutations in highly resistant clones also indicating a link between the stringent response and expression of PBP2a (Dordel *et al.*, 2014).

Persister cells formation

The phenomenon of persister cells has recently attracted substantial interest. Persister cells form a small sub-population of a bacterial culture that are able to survive bactericidal concentrations of antibiotics without being genetically distinguishable from the majority of the bacteria in the population that are killed (Maisonneuve and Gerdes, 2014). Persister cell formation has been reported for many different bacterial species, including *S. aureus* (Lechner *et al.*, 2012). Recently, it was demonstrated that in *E. coli*, the concentration of (p)ppGpp varies stochastically in bacterial populations and that the persisters are pre-existing cells that express high concentrations of (p)ppGpp, leading to induction of the toxin–antitoxin (TA) elements (Maisonneuve *et al.*, 2013). The prokaryotic TA loci encode two components, a stable toxin that inhibits cell growth and a labile antitoxin that counteracts toxin activity. Importantly, toxin overproduction is known to inhibit cell growth and induced a non-growing state from which the bacterial cells can be resuscitated by the induction of the cognate antitoxin (Maisonneuve and Gerdes, 2014). In a complicated process involving inorganic poly-phosphate and activation of the Lon protease, which is responsible for anti-toxin instability in *E. coli*, toxin expression was demonstrated in cells specifically entering an antibiotic resistant persister state (Maisonneuve *et al.*, 2013). In *S. aureus*, the role of the stringent response in formation of persister cells has not been extensively addressed. That being said, when *relPQ* were inactivated survival in the presence of some antibiotics was generally reduced indicating that persister formation may also be affected in this mutant strain background (Geiger *et al.*, 2014). In contrast, the reduced level of (p)ppGpp in the RSH_{syn} mutant was not sufficient to induce persister cell formation in strain Newman (M. Bojer, unpublished data), possibly reflecting the difficulties in manipulating (p)ppGpp levels in *S. aureus*.

The TA loci involved in persister cell formation in *E. coli* are also found in *S. aureus*. Similar to *E. coli*, proteolysis is responsible for the lability of the antitoxin; however, where Lon does this in *E. coli*, it is the ClpCP protease that performs this in *S. aureus* (Donegan *et al.*, 2010). Despite several studies the Clp protease has not been implicated in persister cell formation. In contrast, it proved to be a valuable target when eradicating persister cells in *S. aureus*. In a study by Conlon *et al.* (2013), an acyldepsipeptide antibiotic was shown to kill persister cells by hyperactivating the Clp protease independently of the ATPase, thus being active in cells with low metabolic activity. Importantly, this compound was also able to eradicate a persistent infection in a neutropenic mouse model, indicating there is a close linkage between persister formation and persistent infections.

Cell envelope stress

The bacterial cell envelope is the first line of defence against threats from the environment and it is also a major target for antimicrobial agents. The cell envelope stress responses comprise factors that are involved in sensing and responding to perturbations in the cell envelope. In general, cell envelope stress is sensed, transmitted, and coupled to transcriptional regulation via two-component systems (TCSs) or extracytoplasmic sigma factors (ECF) (reviewed by Jordan *et al.*, 2008). These two systems comprise a membrane-anchored sensory component that, upon stimulation, activates a cytoplasmic transcriptional regulator that is kept inactive in the absence of the inducing signal. In *S. aureus*, cell envelope stress has almost entirely been investigated in the context of the response to antibiotics targeting the cell wall. These studies have identified a number of TCSs implicated in controlling processes related to the cell envelope, including the essential WalKR TCS, the GraRS TCS, and the VraSR TCS. The importance of these TCS for resistance to antibiotics targeting the cell wall is emphasized by the frequent identification of point-mutations in *walKR*, *graRS*, *vraSR* in *S. aureus* isolates that have developed tolerance to the last-line antibiotics vancomycin and daptomycin *in vivo* by a step-wise acquisition of point mutations (Howden *et al.*, 2011; Cui *et al.*, 2009; Neoh *et al.*, 2008). Interestingly, no ECF sigma factors have been identified in *S. aureus*, leading to speculation that *S. aureus* does not have a cell envelope stress signalling pathway responding to general stresses like pH, heat, and high salinity which would be analogous to the *E. coli* σ^E (Sig40) regulon or the *B. subtilis* ECF regulons (Jordan *et al.*, 2008). To avoid extensive overlap with the chapter dealing with antibiotic resistance (see Chapter 11), we will not describe the general responses to cell wall damaging antibiotics in detail here. Instead, we will focus on links between cell envelope stress responses and the more general cellular stress responses.

Interference with cell wall synthesis induces the VraSR controlled cell wall stimulon

Treatment of *S. aureus* with cell wall-acting antibiotics from distinct, clinically relevant chemical classes (e.g. oxacillin, vancomycin, and daptomycin) induces a conserved set of genes referred to as the 'cell wall stimulon' (Utaida *et al.*, 2003; Kuroda *et al.*, 2003; Muthaiyan *et al.*, 2008). This same set of genes was induced by reducing expression of enzymes involved in cell wall biosynthesis, emphasizing that the cell wall stimulon is a general response to cell wall perturbations rather than representing a specific response to antibiotics inhibiting cell wall synthesis (Sobral *et al.*, 2007). The TCS VraSR (*vra* = vancomycin resistance associated) composed of the VraS, histidine protein kinase, and the VraR response regulator protein, was linked to the coordinated transcription of the cell wall stimulon (Kuroda *et al.*, 2003). The *vraSR* locus was first identified as a locus that was up-regulated in a vancomycin-intermediate *S. aureus* (VISA) strain compared to its vancomycin sensitive strain counterpart (Kuroda *et al.*, 2000). Subsequent analysis confirmed, that VraSR was induced in response to blockage of cell wall synthesis in general (Kuroda *et al.*, 2003; Utaida *et al.*, 2003; Sorbral *et al.*, 2007; Muthaiyan *et al.*, 2008). Importantly, the VraSR TCS was not induced by stresses such as heat, osmotic shock, or pH shifts that would be predicted to cause protein-folding stress both in the cytoplasm and in the cell envelope (Kuroda *et al.*, 2003). Thus, the signal responsible for stimulating the intra-membrane sensor VraS to activate the response regulator VraR by phosphorylation must be very specific for constricts in cell wall synthesis (Belcheva *et al.*, 2008, 2009). In this sense, it may be important that VraSR belongs to a special branch of TCSs termed intramembrane-sensing histidine kinases that are characterized by the absence of extracellular sensing domains (Mascher, 2014). Instead, intramembrane-sensing histidine kinases often interact with accessory membrane proteins that function as the true sensors (Mascher, 2014). The VraSR induced cell wall stimulon primarily encompasses genes encoding proteins with a role in cell wall synthesis (Kuroda *et al.*, 2003). However, the VraSR TCS also controls transcription of genes encoding proteins with a predicted role in protein homeostasis, suggesting that an important task of the cell wall stimulon is to prevent misfolding of proteins in the extracytoplasmic compartment, when cell wall synthesis is impeded (Kuroda *et al.*, 2003; Jousselin *et al.*, 2012; Pang *et al.*, 2014). This group of proteins include the HtrA protease, the PrsA chaperone, and the methionine sulfoxide reductases, MrsA and MrsB. In bacteria, the highly conserved serine protease HtrA (high temperature requirement) is localized to the cell envelope where it contributes to protein homeostasis by providing both protease and chaperone activity (reviewed by Hansen and Hilgenfeld, 2013). In Gram-positive bacteria, HtrA proteases are localized in the extracellular space between the cytoplasmic membrane and the cell wall. Two HtrA homologues (i.e. HtrA1 and HtrA2) have been identified in *S. aureus*, and preliminary data support that they both function in combating protein-folding stress in the cell envelope (Rigoulay *et al.*, 2005). In comparison, *E. coli* expresses three periplasmic proteases (i.e. DegP, DegQ, and DegS) belonging to the HtrA family. Interestingly, while the Deg proteases contribute directly to protein homeostasis by re-folding or degrading misfolded proteins, the Deg proteases also have a key role in controlling induction of two major cell envelope stress responses controlled by the Cpx two-component system and σ^E (Hansen and Hilgenfeld, 2013). It could be speculated that HtrA proteases have a similar role in *S. aureus*; however, this has yet to be proven. The PrsA chaperone is located on the outer face of the cytoplasmic membrane and is found ubiquitously in Gram-positive bacteria (Sarvas *et al.*, 2004; Jousselin *et al.*, 2012). PrsA catalyses cis–trans isomerization of peptide bonds preceding proline residues, and evidence from *B. subtilis*, supports that PrsA assists both secretion and folding of some extracellular proteins. Thus, *S. aureus* PrsA is predicted to function as a cellular chaperone in the unusual folding environment at the wall–membrane interface, where for instance the terminal steps of *de novo* peptidoglycan biosynthesis take place (Jousselin *et al.*, 2012) (see Chapter 8). Accordingly, inactivation of *prsA* substantially decreased tolerance to beta-lactams and to a lesser extent to glycopeptides (Jousselin *et al.*, 2012). Finally, an interesting study, demonstrated that up-regulation of the genes endcoding methionine sulfoxide reductases MsrA and MsrB, in *S. aureus* ingested by human neutrophils, was partly mediated by the VraSR TCS. This finding demonstrated that transcription of the methionine sulfoxide reductases improved survival of *S. aureus* in neutrophils, revealing an interesting link between the VraSR TCS and the oxidative stress response (Pang *et al.*, 2014).

Cell wall-acting antibiotics in some cases revealed significant induction of cellular stress responses, demonstrating that blocking of cell wall synthesis may lead to intracellular stress. As an example, genes belonging to the protein-stress induced CtsR regulon were induced 3- to 4-fold 15 min following daptomycin treatment, and were also slightly induced by vancomycin and oxacillin (Muthaiyan *et al.*, 2008). In the same study, transcription of *trfA* (teicoplanin resistance factor) was found to be induced by several cell wall acting antibiotics (Muthaiyan *et al.*, 2008). Interestingly, transcription of *trfA* was recently shown to be controlled by the thiol/oxidative stress global regulator Spx, signifying a novel link between a regulator sensing oxidative stress and the cell wall stress response (Jousselin *et al.*, 2013). The *B. subtilis* homologue of TrfA (designated MecA) functions as an adaptor and assembly factor for the heat-stress chaperone ClpC (Kirstein *et al.*, 2006). If TrfA has a similar function in *S. aureus*, then it would be interesting to see if TrfA contributes to glycopeptide resistance via its function in the cellular stress response.

The GraSR two component system contributes to survival during stress

The GraSR TCS was first discovered in *S. aureus* as a locus whose overexpression led to increased vancomycin resistance (GraSR = glycopeptide resistance associated). Since its discovery, GraSR has been shown to be involved in mediating resistance to a range of cationic antimicrobial peptides including colistin and lysozyme (Cui *et al.*, 2005; Herbert *et al.*, 2007; Falord *et al.*, 2011). Interestingly, a mutant lacking the GraRS system grew poorly in the presence of paraquat and growth was completely inhibited at 44°C, suggesting the GraSR regulon directly or indirectly improves tolerance to stresses such as heat-stress and oxidative stress (Falord *et al.*, 2011). The combined sensitivity of the *graRS* mutant to oxidative stress and antimicrobial peptides could explain the important role of the GraSR system for staphylococcal survival in human neutrophils (Falord *et al.*, 2011). Interestingly, this finding also indicates that the GraSR system responds to signals other than the presence of antimicrobial peptides in *S. aureus*. As discussed above, heat stress and oxidative stress trigger unfolding and oxidation of proteins, and in *E. coli* and *B. subtilis*, the ECF sigma factors control regulons involved in maintaining protein homeostasis in the cell envelope (Jordan *et al.*, 2008). The stress-sensitivity of the *graSR* deletion mutant indicates that GraSR-controlled genes may have a similar function in *S. aureus*. The GraR transcriptional regulator binds to a highly conserved palindromic sequence (5′ ACAAA TTTGT 3′) that was identified upstream of 29 genes or operons (Falord *et al.*, 2011). The genes directly regulated by GraSR encompass genes encoding enzymes that contribute to the resistance to cationic antimicrobial peptides by increasing the positive charge of the cell surface (i.e. MprF, DltABCD) and enzymes involved in cell envelope modification (i.e. the TarM operon involved in teichoic acid glycosylation, SpsB signal peptidase, and several autolysins) (Falord *et al.*, 2011). Interestingly, mutants not synthesizing wall teichoic acids similar to the *graSR* mutant had a severe growth defect at high temperature (Vergara-Irigaray *et al.*, 2008). The heat sensitivity of the *graSR* mutant may therefore be linked to its impact on wall teichoic acid modification. Similarly, the effect of GraSR on oxidative stress survival may be through its regulation of the *mntABC* manganese transport operon, which functions in *S. aureus* resistance to superoxide radicals (Falord *et al.*, 2011).

Concluding remarks

The study of bacterial stress responses has provided insight into the general mechanisms that regulate the homeostasis of bacterial cells under both stress and non-stress conditions. As illustrated by studies of *S. aureus*, stress responses are closely integrated with basic bacterial physiology and for pathogens are central for interactions with the human host. In the last decade, stress responses have been investigated extensively by genome-wide transcriptomic and proteomic investigations that have greatly expanded our knowledge on how stress signals are transformed into transcriptional changes, altered protein expression and adaptation at the population level. Advances are within responses to heat, cold, oxidative and osmotic stress, DNA damage, stringent response, cell envelope stress as well as response to sub-lethal concentrations of antibiotics. At the sub-population level there is increasing acceptance that genetically identical bacteria growing under homogenous conditions may respond differently to stressful conditions at the single cell level. Existence of such heterogeneity within *S. aureus* cell populations is already documented from for example the existence of sub-populations of cells that tolerate much higher concentrations of an antibiotic than the majority of the population. We predict that future stress research will be dedicated to assess heterogeneity of *S. aureus* stress responses and the mechanisms underlying this phenomenon as well as how such heterogeneity contribute to pathogenesis and resistance to antimicrobial treatment.

References

Anderson, K.L., Roberts, C., Disz, T., Vonstein, V., Hwang, K., Overbeek, R., Olson, P.D., Projan, S.J., and Dunman, P.M. (2006). Characterization of the *Staphylococcus aureus* heat shock, cold shock, stringent, and SOS responses and their effects on log-phase mRNA turnover. J. Bacteriol. *188*, 6739–6756.

Belcheva, A., and Golemi-Kotra, D. (2008). A close-up view of the VraSR two-component system. A mediator of *Staphylococcus aureus* response to cell wall damage. J. Biol. Chem. *283*, 12354–12364.

Belcheva, A., Verma, V., and Golemi-Kotra, D. (2009). DNA-binding activity of the vancomycin resistance associated regulator protein VraR and the role of phosphorylation in transcriptional regulation of the *vraSR* operon. Biochemistry, *48*, 5592–5601.

Bischoff, M., Dunman, P., Kormanec, J., Macapagal, D., Murphy, E., Mounts, W., Berger-Bächi, B., and Projan, S. (2004). Microarray-based analysis of the *Staphylococcus aureus* sigmaB regulon. J. Bacteriol. *186*, 4085–4099.

Bisognano, C., Kelley, W.L., Estoppey, T., Francois, P., Schrenzel, J., Li, D., Lew, D.P., Hooper, D.C., Cheung, A.L., and Vaudaux, P. (2004). A recA-LexA-dependent pathway mediates ciprofloxacin-induced fibronectin binding in *Staphylococcus aureus*. J. Biol. Chem. *279*, 9064–9071.

Boutte, C.C., and Crosson, S. (2013). Bacterial lifestyle shapes stringent response activation. Trends Microbiol. *21*, 174–180.

Buchberger, A., Bukau, B., and Sommer, T. (2010). Protein quality control in the cytosol and the endoplasmic reticulum: brothers in arms. Mol. Cell *40*, 238–252.

Bukau, B., Weissman, J., and Horwich, A. (2006). Molecular chaperones and protein quality control. Cell *125*, 443–451.

Cebrián, G., Sagarzazu, N., Aertsen, A., Pagán, R., Condón, S., and Mañas, P. (2009a). Role of the alternative sigma factor sigma on *Staphylococcus aureus* resistance to stresses of relevance to food preservation. J. Appl. Microbiol. *107*, 187–196.

Cebrián, G., Condón, S., and Mañas, P. (2009b). Heat-adaptation induced thermotolerance in *Staphylococcus aureus*: Influence of the alternative factor sigmaB. Int. J. Food Microbiol. *135*, 274–280.

Chastanet, A., Fert, J., and Msadek, T. (2003). Comparative genomics reveal novel heat shock regulatory mechanisms in *Staphylococcus aureus* and other Gram-positive bacteria. Mol. Microbiol. *47*, 1061–1073.

Chaffin, D.O., Taylor, D., Skerrett, S.J., and Rubens, C.E. (2012). Changes in the *Staphylococcus aureus* transcriptome during early adaptation to the lung. PLoS One *7*, e41329.

Chang, W., Small, D.A., Toghrol, F., and Bentley, W.E. (2006). Global transcriptome analysis of *Staphylococcus aureus* response to hydrogen peroxide. J. Bacteriol. *188*, 1648–1659.

Chen, P.R., Bae, T., Williams, W.A., Duguid, E.M., Rice, P.A., Schneewind, O., and He, C. (2006). An oxidation-sensing mechanism is used by the global regulator MgrA in *Staphylococcus aureus*. Nat. Chem. Biol. *2*, 591–595.

Chen, P.R., Nishida, S., Poor, C.B., Cheng, A., Bae, T., Kuechenmeister, L., Dunman, P.M., Missiakas, D., and He, C. (2009). A new oxidative sensing and regulation pathway mediated by the MgrA homologue SarZ in *Staphylococcus aureus*. Mol. Microbiol. *71*, 198–211.

Chen, H.Y., Chen, C.C., Fang, C.S., Hsieh, Y.T., Lin, M.H., and Shu, J.C. (2011). Vancomycin activates σ(B) in vancomycin-resistant *Staphylococcus aureus* resulting in the enhancement of cytotoxicity. PLoS One *6*, e24472.

Chen, Z., Luong, T.T., and Lee, C.Y. (2007). The sbcDC locus mediates repression of type 5 capsule production as part of the SOS response in *Staphylococcus aureus*. J. Bacteriol. *189*, 7343–7350.

Cheung, G.Y., Joo, H.S., Chatterjee, S.S., and Otto, M. (2014). Phenol-soluble modulins–critical determinants of staphylococcal virulence. FEMS Microbiol. Rev. *38*, 698–719.

Cirz, R.T., Jones, M.B., Gingles, N.A., Minogue, T.D., Jarrahi, B., Peterson, S.N., and Romesberg, F.E. (2007). Complete and SOS-mediated response of *Staphylococcus aureus* to the antibiotic ciprofloxacin. J. Bacteriol. *189*, 531–539.

Clements, M.O., Watson, S.P., and Foster, S.J. (1999). Characterization of the major superoxide dismutase of *Staphylococcus aureus* and its role in starvation survival, stress resistance, and pathogenicity. J. Bacteriol. *181*, 3898–3903.

Cohn, M.T., Kjelgaard, P., Frees, D., Penadés, J.R., and Ingmer, H. (2011). Clp-dependent proteolysis of the LexA N-terminal domain in *Staphylococcus aureus*. Microbiology *157*, 677–684.

Conlon, B.P., Nakayasu, E.S., Fleck, L.E., LaFleur, M.D., Isabella, V.M., Coleman, K., Leonard, S.N., Smith, R.D., Adkins, J.N., and Lewis, K. (2013). Activated ClpP kills persisters and eradicates a chronic biofilm infection. Nature *503*, 365–370.

Corrigan, R.M., Campeotto, I., Jeganathan, T., Roelofs, K.G., Lee, V.T., and Gründling, A. (2013). Systematic identification of conserved bacterial c-di-AMP receptor proteins. Proc. Natl. Acad. Sci. U.S.A. *110*, 9084–9089.

Cosgrove, K., Coutts, G., Jonsson, I.M., Tarkowski, A., Kokai-Kun, J.F., Mond, J.J., and Foster, S.J. (2007). Catalase (KatA) and alkyl hydroperoxide reductase (AhpC) have compensatory roles in peroxide stress resistance and are required for survival, persistence, and nasal colonization in *Staphylococcus aureus*. J. Bacteriol. *189*, 1025–1035.

Cui, L., Lian, J.Q., Neoh, H.M., Reyes, E., and Hiramatsu, K. (2005). DNA microarray-based identification of genes associated with glycopeptide resistance in *Staphylococcus aureus*. Antimicrob. Agents. Chemother. *49*, 3404–3413.

Cui, L., Neoh, H.M., Shoji, M., and Hiramatsu, K. (2009). Contribution of *vraSR* and *graSR* point mutations to vancomycin resistance in vancomycin-intermediate *Staphylococcus aureus*. Antimicrob. Agents Chemother. *53*, 1231–1234.

Cuirolo, A., Plata, K., and Rosato, A.E. (2009). Development of homogeneous expression of resistance in methicillin-resistant *Staphylococcus aureus* clinical strains is functionally associated with a beta-lactam-mediated SOS response. J. Antimicrob. Chemother. *64*, 37–45.

Donegan, N.P., Thompson, E.T., Fu, Z., and Cheung, A.L. (2010). Proteolytic regulation of toxin-antitoxin systems by ClpPC in *Staphylococcus aureus*. J. Bacteriol. *192*, 1416–1422.

Dordel, J., Kim, C., Chung, M., Pardos de la Gándara, M., Holden, M.T., Parkhill, J., de Lencastre, H., Bentley, S.D., and Tomasz, A. (2014). Novel determinants of antibiotic resistance: identification of mutated loci in highly methicillin-resistant subpopulations of methicillin-resistant *Staphylococcus aureus*. MBio. *5*, e01000.

Dukan, S., Farewell, A., Ballesteros, M., Taddei, F., Radman, M., and Nyström, T. (2000). Protein oxidation in response to increased transcriptional or translational errors. Proc. Natl. Acad. Sci. U.S.A. *97*, 5746–5749.

Duval, B.D., Mathew, A., Satola, S.W., and Shafer, W.M. (2010). Altered growth, pigmentation, and antimicrobial susceptibility properties of *Staphylococcus aureus* due to loss of the major cold shock gene *cspB*. Antimicrob. Agents. Chemother. *54*, 2283–2290.

von Eiff, C., and Becker, K. (2007). Small-colony variants (SCVs) of staphylococci: a role in foreign body-associated infections. Int. J. Artif. Organs *30*, 778–785.

Elsholz, A.K., Michalik, S., Zühlke, D., Hecker, M., and Gerth, U. (2010). CtsR, the Gram-positive master regulator of protein quality control, feels the heat. EMBO J. *29*, 3621–3629.

Elsholz, A.K., Hempel, K., Pöther, D.C., Becher, D., Hecker, M., and Gerth, U. (2011). CtsR inactivation during thiol-specific stress in low GC, Gram+ bacteria. Mol. Microbiol. *79*, 772–785.

Engman, J., and von Wachenfeldt, C. (2015). Regulated protein aggregation: a mechanism to control the activity of the ClpXP adaptor protein YjbH. Mol. Microbiol. *95*, 51–63.

Engman, J., Rogstam, A., Frees, D., Ingmer, H., and von Wachenfeldt, C. (2012). The YjbH adaptor protein enhances proteolysis of the transcriptional regulator Spx in *Staphylococcus aureus*. J. Bacteriol. *194*, 1186–1194.

Falord, M., Mader, U., Hiron, A., Debarbouille, M., and Msadek, T. (2011). Investigation of the *Staphylococcus aureus* GraSR regulon reveals novel links to virulence, stress response and cell wall signal transduction pathways. PLoS One *6*, e21323.

Farrand, A.J., Reniere, M.L., Ingmer, H., Frees, D., and Skaar, E.P. (2013). Regulation of host hemoglobin binding by the *Staphylococcus aureus* Clp proteolytic system. J. Bacteriol. *195*, 5041–5050.

Feng, J., Michalik, S., Varming, A.N., Andersen, J.H., Albrecht, D., Jelsbak, L., Krieger, S., Ohlsen, K., Hecker, M., Gerth, U., *et al.* (2013). Trapping and proteomic identification of cellular substrates of the ClpP protease in *Staphylococcus aureus*. J. Proteome Res. *12*, 547–558.

Fernández De Henestrosa, A.R., Ogi, T., Aoyagi, S., Chafin, D., Hayes, J.J., Ohmori, H., and Woodgate, R. (2000). Identification of additional genes belonging to the LexA regulon in *Escherichia coli*. Mol. Microbiol. *35*, 1560–1572.

Finan, J.E., Rosato, A.E., Dickinson, T.M., Ko, D., and Archer, G.L. (2002). Conversion of oxacillin-resistant staphylococci from heterotypic to homotypic resistance expression. Antimicrob. Agents Chemother. *46*, 24–30.

Fleury, B., Kelley, W.L., Lew, D., Götz, F., Proctor, R.A., and Vaudaux, P. (2009). Transcriptomic and metabolic responses of *Staphylococcus aureus* exposed to supra-physiological temperatures. BMC Microbiol. *9*, 76.

Foster, P.L. (2000). Adaptive mutation in *Escherichia coli*. Cold Spring Harbor Symp. Quant. Biol. *65*, 21–29.

Frees, D., Qazi, S.N., Hill, P.J., and Ingmer, H. (2003). Alternative roles of ClpX and ClpP in *Staphylococcus aureus* stress tolerance and virulence. Mol. Microbiol. *48*, 1565–1578.

Frees, D., Chastanet, A., Qazi, S., Sørensen, K., Hill, P., Msadek, T., and Ingmer, H. (2004). Clp ATPases are required for stress tolerance, intracellular replication and biofilm formation in *Staphylococcus aureus*. Mol. Microbiol. *54*, 1445–1462.

Frees, D., Thomsen, L.E., and Ingmer, H. (2005). *Staphylococcus aureus* ClpYQ plays a minor role in stress survival. Arch. Microbiol. *183*, 286–291.

Frees, D., Savijoki, K., Varmanen, P., and Ingmer, H. (2007). Clp ATPases and ClpP proteolytic complexes regulate vital biological processes in low GC, Gram-positive bacteria. Mol. Microbiol. *63*, 1285–1295.

Frees, D., Andersen, J.H., Hemmingsen, L., Koskenniemi, K., Bæk, K.T., Muhammed, M.K., Gudeta, D.D., Nyman, T.A., Sukura, A., Varmanen, P., *et al.* (2012). New insights into *Staphylococcus aureus* stress tolerance and virulence regulation from an analysis of the role of the ClpP protease in the strains Newman, COL, and SA564. J. Proteome Res. *11*, 95–108.

Frees, D., Brøndsted, L., and Ingmer, H. (2013). Bacterial proteases and virulence. Subcell. Biochem. *66*, 161–192.

Frees, D., Gerth, U., and Ingmer, H. (2014). Clp chaperones and proteases are central in stress survival, virulence and antibiotic resistance of *Staphylococcus aureus*. Int. J. Med. Microbiol. *304*, 142–149.

de la Fuente-Núñez, C., Reffuveille, F., Fernández, L., and Hancock. R.E. (2013). Bacterial biofilm development as a multicellular adaptation: antibiotic resistance and new therapeutic strategies. Curr. Opin. Microbiol. *16*, 580–589.

de la Fuente-Núñez, C., Reffuveille, F., Haney, E.F., Straus, S.K., and Hancock, R.E. (2014). Broad-spectrum anti-biofilm peptide that targets a cellular stress response. PLOS Pathog. *10*, e1004152.

Gao, W., Chua, K., Davies, J.K., Newton, H.J., Seemann, T., Harrison, P.F., Holmes, N.E., Rhee, H.W., Hong, J.I., Hartland, E.L., Stinear, T.P., and Howden, B.P. (2010). Two novel point mutations in clinical *Staphylococcus aureus* reduce linezolid susceptibility and switch on the stringent response to promote persistent infection. PLOS Pathog. *6*, e1000944.

Gao, W., Cameron, D.R., Davies, J.K., Kostoulias, X., Stepnell, J., Tuck, K.L., Yeaman, M.R., Peleg, A.Y., Stinear, T.P., and Howden, B.P. (2013). The RpoB $H_{481}Y$ rifampicin resistance mutation and an active stringent response reduce virulence and increase resistance to innate immune responses in *Staphylococcus aureus*. J. Infect. Dis. *207*, 929–939.

Garg, S.K., Kommineni, S., Henslee, L., Zhang, Y., and Zuber, P. (2009). The YjbH protein of *Bacillus subtilis* enhances ClpXP-catalyzed proteolysis of Spx. J. Bacteriol. *191*, 1268–1277.

Gaupp, R., Ledala, N., and Somerville, G.A. (2012). Staphylococcal response to oxidative stress. Front. Cell. Infect. Microbiol. *2*, 33.

Geiger, T., Goerke, C., Fritz, M., Schäfer, T., Ohlsen, K., Liebeke, M., Lalk, M., and Wolz, C. (2010). Role of the (p)ppGpp synthase RSH, a RelA/SpoT homolog, in stringent response and virulence of *Staphylococcus aureus*. Infect. Immun. *78*, 1873–1883.

Geiger, T., Francois, P., Liebeke, M., Fraunholz, M., Goerke, C., Krismer, B., Schrenzel, J., Lalk, M., and Wolz, C. (2012). The stringent response of *Staphylococcus aureus* and its impact on survival after phagocytosis through the induction of intracellular PSMs expression. PLOS Pathog. *8*, e1003016.

Geiger, T., Kästle, B., Gratani, F.L., Goerke, C., and Wolz, C. (2014). Two small (p)ppGpp synthases in *Staphylococcus aureus* mediate tolerance against cell envelope stress conditions. J. Bacteriol. *196*, 894–902.

Geiger, T., and Wolz, C. (2014). Intersection of the stringent response and the CodY regulon in low GC Gram-positive bacteria. Int. J. Med. Microbiol. *304*, 150–155.

Giachino, P., Engelmann, S., and Bischoff, M. (2001). Sigma(B) activity depends on RsbU in *Staphylococcus aureus*. J. Bacteriol. *183*, 1843–1852.

Goerke, C., Köller, J., and Wolz, C. (2006). Ciprofloxacin and trimethoprim cause phage induction and virulence modulation in *Staphylococcus aureus*. Antimicrob. Agents Chemother. *50*, 171–177.

Goff, S.A., and Goldberg, A.L. (1985). Production of abnormal proteins in E. coli stimulates transcription of *lon* and other heat shock genes. Cell *41*, 587–595.

Goloubinoff, P., Mogk, A., Zvi, A.P., Tomoyasu, T., and Bukau, B. (1999). Sequential mechanism of solubilization and refolding of stable protein aggregates by a bichaperone network. Proc. Natl. Acad. Sci. U.S.A. *96*, 13732–13737.

Gottschalk, S., Ifrah, D., Lerche, S., Gottlieb, C.T., Cohn, M.T., Hiasa, H., Hansen, P.R., Gram, L., Ingmer, H., and Thomsen, L.E. (2013). The antimicrobial lysine-peptoid hybrid LPS inhibits DNA replication and induces the SOS response in *Staphylococcus aureus*. BMC Microbiol. *13*, 192.

Graham, J.W., Lei, M.G., and Lee, C.Y. (2013). Trapping and identification of cellular substrates of the *Staphylococcus aureus* ClpC chaperone. J. Bacteriol. *195*, 4506–4516.

Graham, J.E., and Wilkinson, B.J. (1992). *Staphylococcus aureus* osmoregulation: roles for choline, glycine betaine, proline, and taurine. J. Bacteriol. *174*, 2711–2716.

Gries, C.M., Bose, J.L., Nuxoll, A.S., Fey, P.D., and Bayles, K.W. (2013). The Ktr potassium transport system in *Staphylococcus aureus* and its role in cell physiology, antimicrobial resistance and pathogenesis. Mol. Microbiol. *89*, 760–773.

Gudas, L.J., and Pardee, A.B. (1975). Model for regulation of *Escherichia coli* DNA repair functions. Proc. Natl. Acad. Sci. U.S.A. *72*, 2330–2334.

Handke, L.D., Shivers, R.P., and Sonenshein, A.L. (2008). Interaction of *Bacillus subtilis* CodY with GTP. J. Bacteriol. *190*, 798–806.

Hansen, G., and Hilgenfeld, R. (2013). Architecture and regulation of HtrA-family proteins involved in protein quality control and stress response. Cell. Mol. Life Sci. *70*, 761–775.

Haslberger, T., Zdanowicz, A., Brand, I., Kirstein, J., Turgay, K., Mogk, A., and Bukau, B. (2008). Protein disaggregation by the AAA+ chaperone ClpB involves partial threading of looped polypeptide segments. Nat. Struct. Mol. Biol. *15*, 641–650.

Hecker, M., Pané-Farré, J., and Völker, U. (2007). SigB-dependent general stress response in *Bacillus subtilis* and related gram-positive bacteria. Annu. Rev. Microbiol. *61*, 215–236.

Hengge-Aronis, R. (1993). Survival of hunger and stress: the role of *rpoS* in early stationary phase gene regulation in *E. coli*. Cell 72, 165–168.

Herbert, S., Bera, A., Nerz, C., Kraus, D., Peschel, A., Goerke, C., Meehl, M., Cheung, A., and Götz, F. (2007). Molecular basis of resistance to muramidase and cationic antimicrobial peptide activity of lysozyme in staphylococci. PLOS Pathog. 3, e102.

Horsburgh, M.J., Clements, M.O., Crossley, H., Ingham, E., and Foster, S.J. (2001). PerR controls oxidative stress resistance and iron storage proteins and is required for virulence in *Staphylococcus aureus*. Infect. Immun. *69*, 3744–3754.

Howden, B.P., McEvoy, C.R., Allen, D.L., Chua, K., Gao, W., Harrison, P.F., Bell, J., Coombs, G., Bennett-Wood, V., Porter, J.L., Robins-Browne, R., Davies, J.K., Seemann, T., and Stinear, T.P. (2011). Evolution of multidrug resistance during *Staphylococcus aureus* infection involves mutation of the essential two component regulator WalKR. PLOS Pathog. *7*, e1002359.

Imlay, J.A. (2003). Pathways of oxidative damage. Annu. Rev. Microbiol. *57*, 395–418.

Jiang, W., Hou, Y., and Inouye, M. (1997). CspA, the major cold-shock protein of *Escherichia coli*, is an RNA chaperone. J. Biol. Chem. *272*, 196–202.

Jordan, S., Hutchings, M.I., and Mascher, T. (2008). Cell envelope stress response in Gram-positive bacteria. FEMS Microbiol. Rev. *32*, 107–146.

Jousselin, A., Renzoni, A., Andrey, D.O., Monod, A., Lew, D.P., and Kelley, W.L. (2012). The posttranslocational chaperone lipoprotein PrsA is involved in both glycopeptide and oxacillin resistance in *Staphylococcus aureus*. Antimicrob. Agents Chemother. *56*, 3629–3640.

Jousselin, A., Kelley, W.L., Barras, C., Lew, D.P., and Renzoni, A. (2013). The *Staphylococcus aureus* thiol/oxidative stress global regulator Spx controls *trfA*, a gene implicated in cell wall antibiotic resistance. Antimicrob. Agents. Chemother. *57*, 3283–3292.

Katzif, S., Danavall, D., Bowers, S., Balthazar, J.T., and Shafer, W.M. (2003). The major cold shock gene, *cspA*, is involved in the susceptibility of *Staphylococcus aureus* to an antimicrobial peptide of human cathepsin G. Infect. Immun. *71*, 4304–4312.

Katzif, S., Lee, E.H., Law, A.B., Tzeng, Y.L., and Shafer, W.M. (2005). CspA regulates pigment production in *Staphylococcus aureus* through a SigB-dependent mechanism. J. Bacteriol. *187*, 8181–8184.

Kim, C., Mwangi, M., Chung, M., Milheiriço, C., de Lencastre, H., and Tomasz, A. (2013). The mechanism of heterogeneous beta-lactam resistance in MRSA: key role of the stringent stress response. PLoS One *8*, e82814.

Kirstein, J., Schlothauer, T., Dougan, D.A., Lilie, H., Tischendorf, G., Mogk, A., Bukau, B., and Turgay, K. (2006). Adaptor protein controlled oligomerization activates the AAA+ protein ClpC. EMBO J. *25*, 1481–1491.

Kjelstrup, S., Hansen, P.M., Thomsen, L.E., Hansen, P.R., and Løbner-Olesen, A. (2013). Cyclic peptide inhibitors of the β-sliding clamp in *Staphylococcus aureus*. PLoS One *8*, e72273.

Klauck, E., Typas, A., and Hengge, R. (2007). The sigmaS subunit of RNA polymerase as a signal integrator and network master regulator in the general stress response in Escherichia coli. Sci. Prog. *90* 9, 103–127.

Krásný, L., and Gourse, R.L. (2004). An alternative strategy for bacterial ribosome synthesis: *Bacillus subtilis* rRNA transcription regulation. EMBO J. *23*, 4473–4483.

Kriel, A., Bittner, A.N., Kim, S.H., Liu, K., Tehranchi, A.K., Zou, W.Y., Rendon, S., Chen, R., Tu, B.P., and Wang, J.D. (2012). Direct regulation of GTP homeostasis by (p)ppGpp: a critical component of viability and stress resistance. Mol. Cell *48*, 231–241.

Kullik, I., and Giachino, P. (1997). The alternative sigma factor sigmaB in *Staphylococcus aureus*: regulation of the *sigB* operon in response to growth phase and heat shock Arch. Microbiol. *167*, 151–159.

Kullik, I., Giachino, P., and Fuchs, T. (1998). Deletion of the alternative sigma factor sigmaB in *Staphylococcus aureus* reveals its function as a global regulator of virulence genes. J. Bacteriol. *180*, 4814–4820.

Kuroda, M., Kuwahara-Arai, K., and Hiramatsu, K. (2000). Identification of the up- and down-regulated genes in vancomycin-resistant *Staphylococcus aureus* strains Mu3 and Mu50 by cDNA differential hybridization method. Biochem. Biophys. Res. Commun. *269*, 485–490.

Kuroda, M., Kuroda, H., Oshima, T., Takeuchi, F., Mori, H., and Hiramatsu, K. (2003). Two-component system VraSR positively modulates the regulation of cell wall biosynthesis pathway in *Staphylococcus aureus*. Mol. Microbiol. *49*, 807–821.

Lancette, G.A., Peeler, J.T., and Lanier, J.M. (1986). Evaluation of an improved MPN medium for recovery of stressed and nonstressed *Staphylococcus aureus*. J. Assoc. Off. Anal. Chem. *69*, 44–46.

de Lencastre, H., Figueiredo, A.M., and Tomasz, A. (1993). Genetic control of population structure in heterogeneous strains of methicillin resistant *Staphylococcus aureus*. Eur. J. Clin. Microbiol. Infect. Dis. *12*, S13–18.

Lechner, S., Lewis, K., and Bertram, R. (2012). *Staphylococcus aureus* persisters tolerant to bactericidal antibiotics. J. Mol. Microbiol. Biotechnol. *22*, 235–244.

Lee, J.W., and Helmann, J.D. (2006). The PerR transcription factor senses H2O2 by metal-catalysed histidine oxidation. Nature *440*, 363–367.

Lee, J.W., Soonsanga, S., and Helmann, J.D. (2007). A complex thiolate switch regulates the *Bacillus subtilis* organic peroxide sensor OhrR. Proc. Natl. Acad. Sci. U.S.A. *104*, 8743–8748.

Little, J.W., Edmiston, S.H., Pacelli, L.Z., and Mount, D.W. (1980). Cleavage of the *Escherichia coli* lexA protein by the recA protease. Proc. Natl. Acad. Sci. U.S.A. *77*, 3225–3229.

Liu, G.Y., Essex, A., Buchanan, J.T., Datta, V., Hoffman, H.M., Bastian, J.F., Fierer, J., and Nizet, V. (2005). *Staphylococcus aureus* golden pigment impairs neutrophil killing and promotes virulence through its antioxidant activity. J. Exp. Med. *202*, 209–215.

Liu, C.I., Liu, G.Y., Song, Y., Yin, F., Hensler, M.E., Jeng, W.Y., Nizet, V., Wang, A.H., and Oldfield, E. (2008). A cholesterol biosynthesis inhibitor blocks *Staphylococcus aureus* virulence. Science *319*, 1391–1394.

Liu, G.Y., and Nizet, V. (2009). Color me bad: microbial pigments as virulence factors. Trends Microbiol. *17*, 406–413.

Luong, T.T., Dunman, P.M., Murphy, E., Projan, S.J., and Lee, C.Y. (2006). Transcription Profiling of the mgr*A* Regulon in *Staphylococcus aureus*. J. Bacteriol. *188*, 1899–1910.

Maaß, S., Wachlin, G., Bernhardt, J., Eymann, C., Fromion, V., Riedel, K., Becher, D., and Hecker, M. (2014). Highly precise quantification of protein molecules per cell during stress and starvation responses in *Bacillus subtilis*. Mol. Cell. Proteomics. *13*, 2260–2276.

Maiques, E., Ubeda, C., Campoy, S., Salvador, N., Lasa, I., Novick, R.P., Barbé, J., and Penadés, J.R. (2006). beta-lactam antibiotics induce the SOS response and horizontal transfer of virulence factors in *Staphylococcus aureus*. J. Bacteriol. *188*, 2726–2729.

Maisonneuve, E., and Gerdes, K. (2014). Molecular mechanisms underlying bacterial persisters. Cell *157*, 539–548.

Maisonneuve, E., Castro-Camargo, M., and Gerdes, K. (2013). (p)ppGpp controls bacterial persistence by stochastic induction of toxin-antitoxin activity. Cell *154*, 1140–1150.

Majerczyk, C.D., Sadykov, M.R., Luong, T.T., Lee, C., Somerville, G.A., and Sonenshein, A.L. (2008). *Staphylococcus aureus* CodY negatively regulates virulence gene expression. J. Bacteriol. *190*, 2257–2265.

Majerczyk, C.D., Dunman, P.M., Luong, T.T., Lee, C.Y., Sadykov, M.R., Somerville, G.A., Bodi, K., and Sonenshein, A.L. (2010). Direct targets of CodY in *Staphylococcus aureus*. J. Bacteriol. *192*, 2861–2877.

Marshall, J.H., and Wilmoth, G.J. (1981). Pigments of *Staphylococcus aureus*, a series of triterpenoid carotenoids. J. Bacteriol. *147*, 900–913.

Mascher, T. (2014). Bacterial (intramembrane-sensing) histidine kinases: signal transfer rather than stimulus perception. Trends Microbiol. *22*, 559–565.

Matsuda, N., Kobayashi, H., Katoh, H., Ogawa, T., Futatsugi, L., Nakamura, T., Bakker, E.P., and Uozumi, N. (2004). Na^+-dependent K^+ uptake Ktr system from the cyanobacterium *Synechocystis* sp. PCC 6803 and its role in the early phases of cell adaptation to hyperosmotic shock. J. Biol. Chem. 279, 54952–54962.

McCarthy, A.J., Witney, A.A., and Lindsay, J.A. (2012). *Staphylococcus aureus* temperate bacteriophage: carriage and horizontal gene transfer is lineage associated. Front. Cell. Infect. Microbiol. *2*, 6.

Michel, A., Agerer, F., Hauck, C.R., Herrmann, M., Ullrich, J., Hacker, J., and Ohlsen, K. (2006). Global regulatory impact of ClpP protease of *Staphylococcus aureus* on regulons involved in virulence, oxidative stress response, autolysis, and DNA repair. J. Bacteriol. *188*, 5783–5796.

Miethke, M., Hecker, M., and Gerth, U. (2006). Involvement of *Bacillus subtilis* ClpE in CtsR degradation and protein quality control. J. Bacteriol. *188*, 4610–4619.

Miller, K.J., Zelt, S.C., and Bae, J.H. (1991). Glycine betaine and proline are the principal compatible solutes of *Staphylococcus aureus*. Curr. Microbiol. *23*, 131–137.

Mitchell, G., Séguin, D.L., Asselin, A.E., Déziel, E., Cantin, A.M., Frost, E.H., Michaud, S., and Malouin, F. (2010). *Staphylococcus aureus* sigma B-dependent emergence of small-colony variants and biofilm production following exposure to *Pseudomonas aeruginosa* 4-hydroxy-2-heptylquinoline-N-oxide. BMC Microbiol. *10*, 33.

Mogk, A., Homuth, G., Scholz, C., Kim, L., Schmid, F.X., and Schumann, W. (1997). The GroE chaperonin machine is a major modulator of the CIRCE heat shock regulon of *Bacillus subtilis*. EMBO J. *16*, 4579–4590.

Mogk, A., Tomoyasu, T., Goloubinoff, P., Rüdiger, S., Röder, D., Langen, H., and Bukau, B. (1999). Identification of thermolabile *Escherichia coli* proteins: prevention and reversion of aggregation by DnaK and ClpB. EMBO. J. *18*, 6934–6949.

Muthaiyan, A., Silverman, J.A., Jayaswal, R.K., and Wilkinson, B.J. (2008). Transcriptional profiling reveals that daptomycin induces the *Staphylococcus aureus* cell wall stress stimulon and genes responsive to membrane depolarization. Antimicrob. Agents. Chemother. *52*, 980–990.

Mwangi, M.M., Kim, C., Chung, M., Tsai, J., Vijayadamodar, G., Benitez, M., Jarvie, T.P., Du, L., and Tomasz, A. (2013). Whole-genome sequencing reveals a link between β-lactam resistance and synthetases of the alarmone (p)ppGpp in *Staphylococcus aureus*. Microb. Drug Resist. *19*, 153–159.

Nakano, S., Erwin, K.N., Ralle, M., and Zuber, P. (2005). Redox-sensitive transcriptional control by a thiol/disulphide switch in the global regulator, Spx. Mol. Microbiol. *55*, 498–510.

Neher, S.B., Flynn, J.M., Sauer, R.T., and Baker, T.A. (2003). Latent ClpX-recognition signals ensure LexA destruction after DNA damage. Genes Dev. *17*, 1084–1089.

Neoh, H.M., Cui, L., Yuzawa, H., Takeuchi, F., Matsuo, M., and Hiramatsu, K. (2008). Mutated response regulator *graR* is responsible for phenotypic conversion of *Staphylococcus aureus* from heterogeneous vancomycin-intermediate resistance to vancomycin-intermediate resistance. Antimicrob. Agents Chemother. *52*, 45–53.

Nguyen, D., Joshi-Datar, A., Lepine, F., Bauerle, E., Olakanmi, O., Beer, K., McKay, G., Siehnel, R., Schafhauser, J., Wang, Y., Britigan, B.E., and Singh, P.K. (2011). Active starvation responses mediate antibiotic tolerance in biofilms and nutrient-limited bacteria. Science. *334*, 982–986.

Nicholas, R.O., Li, T., McDevitt, D., Marra, A., Sucoloski, S., Demarsh, P.L., and Gentry, D.R. (1999). Isolation and characterization of a *sigB* deletion mutant of *Staphylococcus aureus*. Infect. Immun. *67*, 3667–3669.

Novick, R.P., and Subedi, A. (2007). The SaPIs: mobile pathogenicity islands of *Staphylococcus*. Chem. Immunol. Allergy *93*, 42–57.

Onyango, L.A., Dunstan, R.H., Gottfries, J., von Eiff, C., and Roberts, T.K. (2012). Effect of low temperature on growth and ultra-structure of *Staphylococcus* spp. PLoS One *7*, e29031.

Onyango, L.A., Hugh Dunstan, R., Roberts, T.K., Macdonald, M.M., and Gottfries, J. (2013). Phenotypic variants of staphylococci and their underlying population distributions following exposure to stress. PLoS One *8*, e77614.

Pamp, S.J., Frees, D., Engelmann, S., Hecker, M., and Ingmer, H. (2006). Spx is a global effector impacting stress tolerance and biofilm formation in *Staphylococcus aureus*. J. Bacteriol. *188*, 4861–4870.

Pané-Farré, J., Jonas, B., Förstner, K., Engelmann, S., and Hecker, M. (2006). The sigmaB regulon in *Staphylococcus aureus* and its regulation. Int. J. Med. Microbiol. *296*, 237–258.

Pané-Farré, J., Jonas, B., Hardwick, S.W., Gronau, K., Lewis, R.J., Hecker, M., and Engelmann, S. (2009). Role of RsbU in controlling SigB activity in *Staphylococcus aureus* following alkaline stress. J. Bacteriol. *191*, 2561–2573.

Pang, Y.Y., Schwartz, J., Bloomberg, S., Boyd, J.M., Horswill, A.R., and Nauseef, W.M. (2014). Methionine sulfoxide reductases protect against oxidative stress in *Staphylococcus aureus* encountering exogenous oxidants and human neutrophils. J. Innate Immun. *6*, 353–364.

Pelz, A., Wieland, K.P., Putzbach, K., Hentschel, P., Albert, K., and Götz, F. (2005). Structure and biosynthesis of staphyloxanthin from *Staphylococcus aureus*. J. Biol. Chem. *280*, 32493–32498.

Pförtner, H., Burian, M.S., Michalik, S., Depke, M., Hildebrandt, P., Dhople, V.M., Pané-Farré, J., Hecker, M., Schmidt, F., and Völker, U. (2014). Activation of the alternative sigma factor SigB of *Staphylococcus aureus* following internalization by epithelial cells - an *in vivo* proteomics perspective. Int. J. Med. Microbiol. *304*, 177–187.

Phadtare, S. (2011). Unwinding activity of cold shock proteins and RNA metabolism. RNA Biol. *8*, 394–397.

Phadtare, S., and Severinov, K. (2010). RNA remodeling and gene regulation by cold shock proteins. RNA Biol. *7*, 788–795.

Pohl, K., Francois, P., Stenz, L., Schlink, F., Geiger, T., Herbert, S., Goerke, C., Schrenzel, J., and Wolz, C. (2009). CodY in *Staphylococcus aureus*: a regulatory link between metabolism and virulence gene expression. J. Bacteriol. *191*, 2953–2963.

Poole, L.B. (2005). Bacterial defenses against oxidants: mechanistic features of cysteine-based peroxidases and their flavoprotein reductases. Arch. Biochem. Biophys. *433*, 240–254.

Portocarrero, S.M., Newman, M., and Mikel, B. (2002). *Staphylococcus aureus* survival, staphylococcal enterotoxin production and shelf stability of country-cured hams manufactured under different processing procedures. Meat Sci. *62*, 267–273.

Pourkomailian, B., and Booth, I.R. (1992). Glycine betaine transport by *Staphylococcus aureus*: evidence for two transport systems and for their possible roles in osmoregulation. J. Gen. Microbiol. *138*, 2515–2518.

Pourkomailian, B., and Booth, I.R. (1994). Glycine betaine transport by *Staphylococcus aureus*: evidence for feedback regulation of the activity of the two transport systems. Microbiology *140*, 3131–3138.

Price-Whelan, A., Poon, C.K., Benson, M.A., Eidem, T.T., Roux, C.M., Boyd, J.M., Dunman, P.M., Torres, V.J., and Krulwich, T.A. (2013). Transcriptional profiling of *Staphylococcus aureus* during growth in 2 M NaCl leads to clarification of physiological roles for Kdp and Ktr K+ uptake systems. MBio *4*, e00407–e00413.

Reiss, S., Pané-Farré, J., Fuchs, S., François, P., Liebeke, M., Schrenzel, J., Lindequist, U., Lalk, M., Wolz, C., Hecker, M., *et al.* (2012). Global analysis of the *Staphylococcus aureus* response to mupirocin. Antimicrob. Agents Chemother. *56*, 787–804.

Rigby, K.M., and DeLeo, F.R. (2012). Neutrophils in innate host defense against *Staphylococcus aureus* infections. Semin. Immunopathol. *34*, 237–259.

Rigoulay, C., Entenza, J.M., Halpern, P., Widmer, E., Moreillon, P., Poquet, I., and Gruss, A. (2005). Comparative analysis of the roles of HtrA-like surface proteases in two virulent *Staphylococcus aureus* strains. Infect. Immun. *73*, 563–572.

Rosato, R.R., Fernandez, R., Paz, L.I., Singh, C.R., and Rosato, A.E. (2014). TCA cycle-mediated generation of ROS is a key mediator for HeR-MRSA survival under β-lactam antibiotic exposure. PLoS One *9*, e99605.

Sarvas, M., Harwood, C.R., Bron, S., and van Dijl, J.M. (2004). Post-translocational folding of secretory proteins in Gram-positive bacteria. Biochim. Biophys. Acta *1694*, 311–327.

Schlacher, K., and Goodman, M.F. (2007). Lessons from 50 years of SOS DNA-damage-induced mutagenesis. Nat. Rev. Mol. Cell Biol. *8*, 587–594.

Schumann, W. (2003). The *Bacillus subtilis* heat shock stimulon. Cell Stress Chaperones *8*, 207–217.

Schwan, W.R., Coulter, S.N., Ng, E.Y., Langhorne, M.H., Ritchie, H.D., Brody, L.L., Westbrock-Wadman, S., Bayer, A.S., Folger, K.R., and Stover, C.K. (1998). Identification and characterization of the PutP proline permease that contributes to *in vivo* survival of *Staphylococcus aureus* in animal models. Infect. Immun. *66*, 567–572.

Senn, M.M., Giachino, P., Homerova, D., Steinhuber, A., Strassner, J., Kormanec, J., Flückiger, U., Berger-Bächi, B., and Bischoff, M. (2005). Molecular analysis and organization of the sigmaB operon in *Staphylococcus aureus*. J. Bacteriol. *187*, 8006–8019.

Singh, V.K., Utaida, S., Jackson, L.S., Jayaswal, R.K., Wilkinson, B.J., and Chamberlain, N.R. (2007). Role for dnaK locus in tolerance of multiple stresses in *Staphylococcus aureus*. Microbiology *153*, 3162–3173.

Singh, V.K., Syring, M., Singh, A., Singhal, K., Dalecki, A., and Johansson, T. (2012). An insight into the significance of the DnaK heat shock system in *Staphylococcus aureus*. Int. J. Med. Microbiol. *302*, 242–252.

Sleator, R.D., and Hill, C. (2002). Bacterial osmoadaptation: the role of osmolytes in bacterial stress and virulence. FEMS Microbiol. Rev. *26*, 49–71.

Sobral, R.G., Jones, A.E., Des Etages, S.G., Dougherty, T.J., Peitzsch, R.M., Gaasterland, T., Ludovice, A.M., de Lencastre, H., and Tomasz, A. (2007). Extensive and genome-wide changes in the transcription profile of *Staphylococcus aureus* induced by modulating the transcription of the cell wall synthesis gene *murF*. J. Bacteriol. *189*, 2376–2391.

Somero, G.N. (1995). Proteins and temperature. Annu. Rev. Physiol. *57*, 43–68.

Tholema, N., Vor der Brüggen, M., Mäser, P., Nakamura, T., Schroeder, J.I., Kobayashi, H., Uozumi, N., and Bakker, E.P. (2005). All four putative selectivity filter glycine residues in KtrB are essential for high affinity and selective K^+ uptake by the KtrAB system from *Vibrio alginolyticus*. J. Biol. Chem. *280*, 41146–41154.

Tomasz, A., Nachman, S., and Leaf, H. (1991). Stable classes of phenotypic expression in methicillin-resistant clinical isolates of staphylococci. Antimicrob. Agents Chemother. *35*, 124–129.

Touati, D. (1997). Superoxide dismutases in bacteria and pathogen protists. In Oxidative Stress and the Molecular Biology of Antioxidant Defenses, J.G. Scadalios, ed. (Cold Spring Harbor, NY: Cold Spring Harbor Laboratory), pp. 447–493.

Ueta, M., Wada, C., and Wada, A. (2010). Formation of 100S ribosomes in *Staphylococcus aureus* by the hibernation promoting factor homolog SaHPF. Genes Cells *15*, 43–58.

Ubeda, C., Maiques, E., Tormo, M.A., Campoy, S., Lasa, I., Barbé, J., Novick, R.P., and Penadés, J.R. (2007). SaPI operon I is required for SaPI packaging and is controlled by LexA. Mol. Microbiol. *65*, 41–50.

Ubeda, C., Maiques, E., Knecht, E., Lasa, I., Novick, R.P., and Penadés, J.R. (2005). Antibiotic-induced SOS response promotes horizontal dissemination of pathogenicity island-encoded virulence factors in staphylococci. Mol. Microbiol. *56*, 836–844.

Utaida, S., Dunman, P.M., Macapagal, D., Murphy, E., Projan, S.J., Singh, V.K., Jayaswal, R.K., and Wilkinson, B.J. (2003). Genome-wide transcriptional profiling of the response of *Staphylococcus aureus* to cell wall -active antibiotics reveals a cell wall -stress stimulon. Microbiology *149* 1, 2719–2732.

Valderas, M.W., and Hart, M.E. (2001). Identification and characterization of a second superoxide dismutase gene (sodM) from *Staphylococcus aureus*. J. Bacteriol. *183*, 3399–3407.

Vergara-Irigaray, M., Maira-Litrán, T., Merino, N., Pier, G.B., Penadés, J.R., and Lasa, I. (2008). Wall teichoic acids are dispensable for anchoring the PNAG exopolysaccharide to the *Staphylococcus aureus* cell surface. Microbiology *154* 11, 865–877.

Vijaranakul, U., Nadakavukaren, M.J., Bayles, D.O., Wilkinson, B.J., and Jayaswal, R.K. (1997). Characterization of an NaCl-sensitive *Staphylococcus aureus* mutant and rescue of the NaCl-sensitive phenotype by glycine betaine but not by other compatible solutes. Appl. Environ. Microbiol. *63*, 1889–1897.

Voyich, J.M., Braughton, K.R., Sturdevant, D.E., Whitney, A.R., Saïd-Salim, B., Porcella, S.F., Long, R.D., Dorward, D.W., Gardner, D.J., Kreiswirth, B.N., *et al.* (2005). Insights into mechanisms used by *Staphylococcus aureus* to avoid destruction by human neutrophils. J. Immunol. *175*, 3907–3919.

Wang, J., Hartling, J.A., and Flanagan, J.M. (1997). The structure of ClpP at 2.3 A resolution suggests a model for ATP-dependent proteolysis. Cell *91*, 447–456.

Weibezahn, J., Tessarz, P., Schlieker, C., Zahn, R., Maglica, Z., Lee, S., Zentgraf, H., Weber-Ban, E.U., Dougan, D.A., Tsai, F.T., *et al.* (2004). Thermotolerance requires refolding of aggregated proteins by substrate translocation through the central pore of ClpB. Cell *119*, 653–665.

Wengender, P.A., and Miller, K.J. (1995). Identification of a PutP proline permease gene homolog from *Staphylococcus aureus* by expression cloning of the high-affinity proline transport system in *Escherichia coli*. Appl. Environ. Microbiol. *61*, 252–259.

Wetzel, K.J., Bjorge, D., and Schwan, W.R. (2011). Mutational and transcriptional analyses of the *Staphylococcus aureus* low-affinity proline transporter OpuD during *in vitro* growth and infection of murine tissues. FEMS Immunol. Med. Microbiol. *61*, 346–355.

Wirtz, C., Witte, W., Wolz, C., and Goerke, C. (2009). Transcription of the phage-encoded Panton–Valentine leucocidin of *Staphylococcus aureus* is dependent on the phage life cycle and on the host background. Microbiology *155* 3, 3491–3499.

Wolf, C., Hochgräfe, F., Kusch, H., Albrecht, D., Hecker, M., and Engelmann, S. (2008). Proteomic analysis of antioxidant strategies of *Staphylococcus aureus*: diverse responses to different oxidants. Proteomics *8*, 3139–3153.

Wozniak, D.J., Tiwari, K.B., Soufan, R., and Jayaswal, R.K. (2012). The *mcsB* gene of the *clpC* operon is required for stress tolerance and virulence in *Staphylococcus aureus*. Microbiology *158* 15, 2568–2576.

Xia, G., and Wolz, C. (2014). Phages of *Staphylococcus aureus* and their impact on host evolution. Infect. Genet. Evol. *21*, 593–601.

Yeiser, B., Pepper, E.D., Goodman, M.F., and Finkel, S.E. (2002). SOS-induced DNA polymerases enhance long-term survival and evolutionary fitness. Proc. Natl. Acad. Sci. U.S.A. *99*, 8737–8741.

Yamanaka, K. (1999). Cold shock response in *Escherichia coli*. J. Mol. Microbiol. Biotechnol. *1*, 193–202.

Yoshida, H., Yamamoto, H., Uchiumi, T., and Wada, A. (2004). RMF inactivates ribosomes by covering the peptidyl transferase centre and entrance of peptide exit tunnel. Genes Cells *9*, 271–278.

Zuber, P. (2009). Management of oxidative stress in *Bacillus*. Annu. Rev. Microbiol. *63*, 575–597.

Molecular Strategies of *Staphylococcus aureus* for Resisting Antibiotics

11

Susan Boyle-Vavra and Robert S. Daum

Abstract

Since there is no vaccine to prevent *Staphylococcus aureus infection*, clinicians must heavily rely on antibiotics to treat *S. aureus infections.* However, *S. aureus* has been able to elude every antibiotic that has been widely introduced into clinical practice. This chapter summarizes some of the most important antibiotics that have been used in the therapy of *S. aureus* infection since the golden age of antibiotic discovery, starting with the introduction of penicillin in 1940s and methicillin in the 1960s. The emphasis of the chapter is placed on the variety of resistance mechanisms that *S. aureus* has evolved to elude antibiotics and the mobile genetic elements that allow horizontal transfer of resistance genes between strains and between species.

S. aureus strains have become resistant to antibiotics by a wide variety of mechanisms. At the genetic level, resistance has occurred through acquiring point mutations in the chromosomal gene or genes encoding the antibiotic target or by horizontal gene exchange resulting in the acquisition of new genes that confer resistance.

Examples of antibiotic resistance strategies that have evolved in *S. aureus* include (a) acquisition of spontaneous point mutations in the native target that decreases the binding of the antibiotic, (b) acquisition of a gene encoding an enzyme that inactivates the antibiotic, (c) acquisition of an antibiotic-insensitive target that can bypass the antibiotic's effect by replacing the function of the native target, (d) acquisition of an enzyme that modifies the target thereby blocking access of the antibiotic to the target, (e) expressing a protein or proteins that transport(s) a drug out of the cytoplasm (for instance via an efflux pump), thereby reducing its cytoplasmic concentration, (f) acquiring mutations that alter the bacterial surface properties that, for instance, decrease the interaction of the antibiotic with the bacteria, (g) ribosomal protection without target modification.

Antibiotic strategies that target an essential process in bacteria place strong selective pressure that ultimately results in resistance. Strategies that potentiate reliable antibiotics or that delay resistance development could increase the lifespan and usefulness of old and new antimicrobial agents. Also, by better understanding how bacteria cause disease and death, it will be possible to design treatment strategies that ameliorate the toxic effects of the bacteria on the host rather than by directly killing the bacteria.

Introduction

Staphylococcus aureus is an important human and animal commensal bacterium, but it is also a bacterial pathogen that can cause a variety of infectious disease syndromes. It was identified as Gram-positive cocci in clusters that were recognized in the pus of people suffering from wound infections. Prior to the discovery and widespread use of penicillin G in the 1950s, nearly 90% of *S. aureus* invasive cases were fatal (Smith and Vickers, 1960). This situation drastically changed after Alexander Fleming made the serendipitous observation, in 1928, that a contaminating mould called *Penicillium notatum* inhibited growth of *S. aureus* on a culture plate (Fleming, 1942). This observation led to the discovery of penicillin, which belongs to the most widely used class of antibiotics still in use today, the β-lactams. The enormous impact of this discovery on public health resulted in a Nobel Prize in Physiology and Medicine awarded in 1945 to Fleming and two chemists (Ernst B Chain and Sir Howard Florey) who made it possible to produce penicillin in large quantities for the masses. The widespread distribution and clinical use of penicillin heralded a drastic reduction in mortality caused by *S. aureus* infections.

While the introduction of penicillin was a tremendous advance in treating infectious diseases, it was not long before penicillin-resistant strains were recognized (Fleming, 1942). Even at the earliest stage of its discovery, Fleming had the foresight to predict the ease with which *S. aureus* could develop resistance to antibiotics when he noticed a few colonies that were able to break through the zone of inhibition surrounding the mould on the culture plate. In his Nobel Prize acceptance speech, Fleming warned of the need for careful stewardship of this powerful and precious resource to prevent resistant mutants from arising that would make future infections untreatable (Fleming, 1945). This message still rings true today. A movement has begun in earnest since a 2012 position paper that was supported by several professional organizations such as the Infectious Disease Society of America (IDSA), the Pediatric Infectious Disease Society (PIDS), and the Society of Healthcare Epidemiology of America (SHEA), that urged steps be taken to preserve antibiotics as one would approach a precious resource (2012). This movement has fostered the awareness of healthcare professionals, has inspired action to promote development of new drugs and has promoted antimicrobial stewardship programs to eliminate inappropriate use of antibiotics in humans. Inappropriate use of antibiotics also occurs in the production of food animals, which accounted for 80% of antibiotic sales in the USA in 2009 and where sub therapeutic doses are used for non-therapeutic purposes such as growth promotion in healthy animals. Evidence that these practices contribute to resistance development in bacterial pathogens that can cause infections in humans is growing (Mole, 2013). The emergence of livestock associated methicillin-resistant *S. aureus* (MRSA) in European countries that have disseminated among humans is but one example (Deiters *et al.*, 2015; Larsen *et al.*, 2015).

It is often taken for granted in the modern era that a common *S. aureus* infection should be easily treatable with antibiotics. Over a decade ago, it was thought that clinical failure of antimicrobial chemotherapy of a staphylococcal infection was rare (Projan, 2000). This situation has changed with the surge of infections caused by MRSA strains in the last decade.

Antibiotic treatment failure of *S. aureus* infection can no longer be considered rare. According to recent data, about a quarter of infections caused by MRSA fail to respond to antibiotic therapy (Eells *et al.*, 2013) and about 19,000 people were estimated to have died from MRSA infection in the USA during a recent year of intense study in many geographic locations by the Centers for Disease Control and Prevention (Klevens *et al.*, 2007). MRSA infections have been reported with increasing frequency over the past decade, even among previously healthy individuals in the community, and cause more deaths than any single infectious agent. At the present time, MRSA isolates cause about 50% of *S. aureus* infections in the USA. Moreover, MRSA strains are often resistant to antibiotic classes in addition to the β-lactams, which further limits treatment options.

In addition to the threat of MRSA, new strains have emerged that are resistant to the mainstay of therapy, glycopeptides, and the recently licensed antibiotics daptomycin, linezolid, dalfopristin/quinupristin (synercid) and tigecycline. It is fortunate that in recent years several new antibiotics that can be used to treat MRSA infections have been licensed, especially for skin and soft tissue infections (SSTIs). These include ceftaroline, telavancin, dalbavancin, oritivancin and tedizolid.

What is an antibiotic?

For the working definition for this chapter, an antibiotic or antibacterial agent is a chemotherapeutic agent that inhibits the growth of, or kills, bacteria. An antimicrobial refers more broadly to a chemical agent that can be used in the therapy of a microbe, including bacteria, fungi, viruses and parasites. We will focus on those antibiotics that target *S. aureus* and the accompanying resistance mechanisms that this stealthy pathogen has acquired to avoid them.

Antibiotics are classified functionally into either bactericidal or bacteriostatic based on their mechanism of destructive capacity. Bactericidal agents will kill a majority of cells in the target bacterial population, whereas bacteriostatic agents inhibit growth of the organism without killing. For both agent types, the host's defences contribute to the elimination of the bacteria from the host's tissues.

How to measure susceptibility or resistance

The standard unit of measure of susceptibility to a drug is the minimal inhibitory concentration (MIC), defined as the lowest concentration required to inhibit growth of the organism. Growth inhibition is assessed by a loss in turbidity. The MIC breakpoints that determine whether an organism is resistant or susceptible are determined empirically for each antibiotic. An organism is resistant when its MIC to the agent is above the susceptibility breakpoint as determined by an accepted standard set by agencies such as the Clinical Laboratory Standards Institute in the USA (CLSI) or EUCAST in Europe.

Whether an antibiotic is bacteriostatic or bactericidal for a given organism is determined by the minimal bactericidal concentration (MBC). This can be determined by assessing the antibiotic concentration at which no viable colonies are recovered from a broth culture. The MBC to MIC ratio of an antibiotic for the majority of clinical isolates is one means to define whether a new agent is bactericidal. An agent is considered bactericidal if it produces an MBC to MIC ratio of < 32 in the majority of isolates tested and an MBC to MIC ratio of ≥ 32 defines a bacteriostatic drug.

Tolerance to an agent is a special category of resistance phenotype that applies to bactericidal drugs. A strain that becomes tolerant to a bactericidal antibiotic is no longer killed by the agent at the usual bactericidal concentration, although its growth is inhibited. Tolerance to a bactericidal antibiotic has often been defined by a strain exhibiting an MBC to MIC ratio ≥ 32 for a drug that usually produces a ratio close to 1.

How do antibiotics work?

Antibiotics act by interfering with a target that can be either a physiological process or a protein that is essential or very important to the life of the bacteria. The ideal antibiotic has minimal toxicity to mammalian hosts. Usually this is accomplished at the mechanistic level by specifically inhibiting a bacterial target that has minimal similarity to proteins or macromolecules that are essential to humans (or other mammalian hosts).

Where do antibiotics come from?

There are approximately 12,000 antibiotics, but only about 160 are used commonly in human therapy (Walsh, 2003d). Although penicillin was obtained from a mould, over 55% of antibiotics in common use are natural products derived from the *Streptomycetes* family of filamentous bacteria that typically inhabit the soil and are members of the Order *Actinomycetales*. This group of bacteria have co-evolved chemical warfare strategies to outcompete the abundant species that inhabit a shared niche. This fact was exploited in pioneering the earliest antibiotic discovery efforts in the twentieth century that involved systematic screening of soil organisms for substances that inhibited bacteria. Using this approach, Selman Waksman identified streptomycin, the first naturally produced antibiotic that gained widespread usage. As it turns out, streptomycin was actually obtained from a bacteria isolated from a chicken's throat, although most of Waksman's efforts focused on soil organisms. Most ensuing discoveries of natural antibiotics have employed a similar screening strategy.

Antibiotics can also be made semisynthetically to improve on the pharmacokinetic or pharmacodynamic properties of naturally occurring scaffolds. Alternatively, structure-based drug design can be used to strategically target the active site of an essential target.

Where do antibiotic resistance genes come from?

Since antibiotic-producing bacteria must be inherently resistant to the toxic effects of the chemicals they produce, they have often co-evolved a defence mechanism for their own protection. It is thought that horizontally transferred antibiotic resistance genes found in resistant pathogenic microorganisms originated from antibiotic producers existing in nature (Walsh, 2003b). Based on this logic it was hoped that, for synthetic antibiotics, there would not be any previously existing resistance genes available for transfer into *S. aureus*. However, for the synthetic antibiotic linezolid, this hope was soon dashed as will be discussed in more detail below.

Antibiotics commonly used in the therapy of *S. aureus* infection

Recent antimicrobial treatment guidelines were published in 2011 on the therapy of methicillin-resistant *Staphylococcus aureus* (MRSA) infection (Liu *et al.*, 2011). The antibiotic choice and dosage can vary widely depending on syndrome, patient age, and other underlying illnesses or patient-dependent factors. The reader is referred to these guidelines for a more detailed understanding of the antibiotic strategies that are in common use to treat MRSA infection.

How *S. aureus* strains develop antibiotic resistance

S. aureus strains have been readily able to develop resistance to every antibiotic used in clinical practice. There are a few newly licensed antibiotics for which resistance has not yet been described; however, if history repeats itself, resistant strains will be identified after increasing reliance on each agent.

S. aureus strains have become resistant to antibiotics by a wide variety of mechanisms. At the genetic level, resistance has occurred through acquiring point mutations in the gene or genes encoding the antibiotic target or by horizontal gene exchange resulting in the acquisition of new genes that confer resistance. The latter mechanism usually results in a high level of resistance.

Examples of antibiotic resistance strategies that have evolved in *S. aureus* include: (a) acquisition of spontaneous point mutations in the native target that decreases the binding of the antibiotic, (b) acquisition of a gene encoding an enzyme that inactivates the antibiotic, (c) acquisition of an antibiotic-insensitive target that can bypass the antibiotic's effect by replacing the function of the native target, (d) acquisition of an enzyme that modifies the target thereby blocking access of the antibiotic to the target, (e) expressing a protein or proteins that transport(s) a drug out of the cytoplasm (for instance via an efflux pump), thereby reducing its cytoplasmic concentration, (f) acquiring mutations that alter the bacterial surface properties that, for instance, decrease the interaction of the antibiotic with the bacteria, and (g) ribosomal protection without modification.

The mechanics of resistance gene acquisition

Many of the genes that are horizontally acquired by *S. aureus* are maintained on transposons (Tns) that are carried on plasmids. Such plasmids can be transferred between cells through conjugation or by transducing phages. Each plasmid can carry more than one resistance gene to confer transmissible multidrug resistance. Of special note is the family of site-specifically integrated elements that harbour the methicillin resistance gene, *mecA*. These elements, called staphylococcal chromosome cassette mec (SCC*mec*), have only been found in the chromosome at a specific site. They are highly variable which has led to the assignment of 12 SCC*mec* types to date. But identical cassettes can be found among diverse staphylococcal species leaving evidence of the horizontal exchange of the cassette. The means by which SCC*mec* elements have transferred between strains has not yet been elucidated, although it has been shown that transducing phage (Scharn *et al.*, 2013) and conjugation (Ray *et al.*, 2016) can accomplish this *in vitro*.

Resistance to antibiotics used in the therapy of *S. aureus* infection

This section will focus on the mechanism of action of those antibiotics that have been used clinically in the therapy of *S. aureus* infections. It will also describe the molecular mechanisms by which *S. aureus* strains have evolved resistance including a description of mobile genetic elements on which many resistance genes are transported into clinical *S. aureus* isolates. The antibiotics to be discussed will be organized by their bacterial targets, such as peptidoglycan synthesis (see 'Antibiotics that target peptidoglycan synthesis'), cell membrane fluidity and homeostasis (see 'Antibiotics that interfere with cell membrane homeostasis'), protein synthesis (see 'Antibiotics that target the ribosome and interfere with protein biosynthesis'), folic acid synthesis (see 'Antibiotics that interfere with folate metabolism'), DNA synthesis (see 'Antibiotics that interfere with DNA replication'), and RNA polymerase (see 'Inhibition of RNA polymerase').

The section on peptidoglycan or cell wall synthesis inhibitors is preceded by a section about cell wall physiology (see 'The *S. aureus* cell wall'). Cell wall synthesis inhibitors encompass the β-lactams and glycopeptides, the two most widely used classes of agents in the therapy of *S. aureus* infection. Peptidoglycan biosynthesis is also the target of several recently FDA licensed antibiotics including ceftaroline, telavancin, dalbavancin and oritivancin.

The *S. aureus* cell wall

Because of the central role of the cell wall as a target of antibiotics in the therapy of *S. aureus*, this section is intended to explain the physiology and structure of this important macromolecule. We also refer the reader to Chapter 8 of this book, in which the staphylococcal cell wall is also described.

The cell wall of *S. aureus* is a complex assembly of molecules that surrounds the bacterium. Since peptidoglycan (synonym = murein) is an essential component of the cell wall, it is an outstanding antibiotic target. Peptidoglycan is an essential multilayered polymer that envelopes the *S. aureus* cell. It is essential for maintaining the shape of the bacterium and preventing osmotic lysis (Shockman *et al.*, 1996). It consists of repeating subunits of a precursor molecule that is first assembled in the cytoplasm and then transported across the cell membrane where it is incorporated into a growing glycan chain on the cell surface. Peptidoglycan has long glycan chains that have stem peptides that are cross-linked via peptide cross-bridges (reviewed in Hao *et al.*, 2012, and Chapter 8). The cell wall peptidoglycan and the enzymes involved in its biosynthesis are outstanding targets for antibiotics since they fulfil two basic criteria of a successful antibiotic target: (1) they are essential to forming an intact cell wall, an essential component to the survival of *S. aureus* and (2) mammalian cells are completely devoid of any target that resembles peptidoglycan or the enzymes used in their synthesis. Thus, an antibiotic that disables peptidoglycan synthesis should have little or no cross-reactivity with essential host proteins, thereby avoiding toxicity in a mammalian host.

In *S. aureus*, peptidoglycan also serves as a scaffold on which a variety of surface proteins are tethered (Navarre and Schneewind, 1999). Such proteins assist *S. aureus* in colonizing tissues or evading the host immune response (e.g. the immunoglobulin binding protein A). Another important component of the cell wall is teichoic acid, membrane associated (LTA) and peptidoglycan-anchored (WTA), that affects the charge of the membrane, acts as an attachment site for bacteriophage and interacts with host cells. The cell wall-acting antibiotics discussed in this chapter will be those that target peptidoglycan biosynthesis.

Peptidoglycan structure

In order to understand how cell wall-acting antibiotics kill or inhibit growth of *S. aureus*, it is important to understand the structure of peptidoglycan and how it is synthesized. As the name implies, peptidoglycan is composed of both peptide and glycan substituents. With few exceptions, bacteria produce peptidoglycan but there are species-specific differences in structure. Much of our knowledge about peptidoglycan is known from studies of *S. aureus*. *S. aureus* peptidoglycan is a three dimensional lattice-like polymer that surrounds the bacterium. It also comprises the cross walls that separate dividing cells. Peptidoglycan is assembled from a precursor that is synthesized in the cytoplasm (with both cytosolic stages and lipid linked stages) and is assembled into about 15 layers of highly cross-linked glycan chains consisting of repeating subunits of the same precursor monomer. Each glycan chain consists of repeating β-(1→4)-linked disaccharide consisting of *N*-acetylglucosamine (GlcNac) and *N*-acetylmuramic acid (MurNac) at the reducing end. The carboxyl group of each MurNAc residue is amidated by the stem pentapeptide L-Ala-D-iso-Gln-L-Lys-D-Ala-D-Ala, and the ε-amino group of the lysine residue is substituted with a pentaglycine cross-bridge that is unique to *S. aureus*. This cross-bridge is highly susceptible to the endopeptidase lysostaphin (Schindler and Schuhardt, 1964), which is produced by another staphylococcal species, *Staphylococcus simulans* (Recsei *et al.*, 1987), and is highly specific for *S. aureus* peptidoglycan. Some minor cross-bridges can also be found consisting of one, three or four glycines; alanine; and alanyl-tetraglycine (de Jonge *et al.*, 1992). These cross linked peptides connect neighbouring 'stem' pentapeptides that extend from each MurNac residue and provide both flexibility and strength to the murein sacculus that envelopes the bacteria. A recent analysis describes the stem pentapeptide linked to the pentaglycine as a branched decapeptide (Dmitriev *et al.*, 2004). However, most models depict the glycan chains arranged parallel to the cell membrane.

Based on studies using high-performance liquid chromatography (HPLC) coupled with mass spectrometry (MS) analysis of products produced following digestion with muramidase (cleaves glycosidic binds) or muramidase and lysostaphin (cleaves the pentaglycine peptide), DeJonge *et al.* (1992) determined that 40% of the peptidoglycan strands in the *S. aureus* cell wall consist of higher oligomers (three to nine per glycan strand) and 15–25% of the peptidoglycan consist of even longer chains. This same approach has provided insight into the effect of β-lactam antibiotics on the cell wall of *S. aureus* and the effect of expression of the *mecA* gene product, PBP2a, on the structure of the cell wall. HPLC-MS has also been employed to elucidate the role of enzymes involved in cell wall biosynthesis and alterations in peptidoglycan structure associated with vancomycin resistance.

Peptidoglycan synthesis provides multiple targets for antibiotics

The assembly of peptidoglycan is a complex process that can be described in three stages involving distinct subcellular localizations in the bacterial cell (Bugg *et al.*, 2011): (a) the cytoplasmic stage (early), (b) the membrane linked cytoplasmic stage (intermediate), and (c) the extracellular, membrane associated stage (late). Each step in the synthesis of peptidoglycan is catalysed by a distinct enzyme that has the potential to be, or is, the target of an antibiotic.

The basic building block of the cell wall is the peptidoglycan precursor. It is synthesized in the cytoplasm, becomes linked to a lipid carrier at the cell membrane, where additional peptide is assembled into a pentaglycine bridge. The entire lipid linked precursor, often called lipid II, is then translocated across the bacterial cell membrane to the exterior face of the bacterial cell where the non-lipid portion of the precursor (the disaccharide-decapeptide) is incorporated into the growing peptidoglycan chain with the use of transglycosylases and transpeptidases. These enzymes, involved in the later steps, include the penicillin-binding proteins (PBPs) so named because they are the direct targets of penicillin and related antibiotics in the β-lactam class. In addition to the PBPs, *S. aureus* produces non-essential monofunctional transglycosylases (i.e. SgtB and SgtA) that are involved in peptidoglycan chain elongation; however, these enzymes do not bind penicillin.

Peptidoglycan precursor assembly

The assembly of the peptidoglycan precursor begins in the cytoplasm and involves thirteen enzymes (MurA/Z, B, C, D, E, F, *ddl and* alanine racemase, MraY, MurG, FemX, FemA and FemB). These enzymes are involved in the early steps in peptidoglycan synthesis and can be further stratified into the enzymes that act entirely in the cytosol (eight enzymes, MurA-E, Ddl and Ala racemase and MurF) or at the cytoplasmic face of the membrane in the cytosol (five enzymes) (Bugg *et al.*, 2011).

The first step in peptidoglycan precursor assembly involves the biosynthesis of the first building block (UDP-N-acetylmuramic acid), which requires a pair of enzymes acting in tandem, MurA (or MurZ) and MurB. MurA and MurZ are homologues of the enzyme UDP-N-acetylglucosamine enolpyruvyl transferase that transfers the enoyl moiety from phosphoenol pyruvate (PEP) to N-acetyl glucosamine, forming UDP-N-acetylenolpyruvylglucosamine (UDP-GlcNAcEP). MurB, a UDP-N-acetylenolpyruvylglucosamine reductase, catalyses the NADPH-dependent reduction of UDP-GlcNAcEP to form UDP-N-acetylmuramic acid (UDP-MurNAc), the amino sugar on which to build the stem peptide of the precursor.

MurC, D and E sequentially add L-alanine, D-glutamic acid, and L-lysine, to the lactyl side-chain of UDP-MurNAc. The terminal two D-Ala residues are added as a dipeptide, D-Ala-D-Ala, which is produced by first forming the D-Ala building block from L-Ala using alanine racemase followed by formation of the dipeptide catalysed by D-Ala-D-Ala ligase (Ddl). The dipeptide is then added to the N-terminus of the tripeptide peptidoglycan precursor using MurF to form the UDP-MurNac-L-Ala-D-iso-Gln-L-Lys-D-Ala-D-Ala pentapeptide.

The beginning of the lipid-linked stage occurs with the addition of the MurNac-L-Ala-D-iso-Gln-L-Lys-D-Ala-D-Ala and *N*-acetylglucosamine (GlcNAc) to a membrane-bound acceptor (bactoprenol, e.g. C55-P) with the use of MraY, forming the lipid linked intermediate called 'Lipid I'. Lipid II is formed by the addition of UDP-GlcNAc to lipid I with the use of MurG. A pentaglycine cross-bridge is then synthesized by substitution of the ε-amino group of the L-lysine of lipid II with a glycine using the non-ribosomal enzyme, FemX. Two more glycines are added using FemA and the last 2 glycines are added by using FemB (Schneider *et al.*, 2004). FemA and FemB were identified as so-called factors essential for methicillin resistance (fem) by using a Tn*551* mutagenesis screen of insertions that reversed or attenuated the methicillin resistance phenotype (Berger-Bachi and Tschierske, 1998). Isolation and sequencing of the chromosomal DNA locus containing the Tn revealed that the insertion site was in the promoter of a two-gene operon (*femAB*). Peptidoglycan analysis of the cell wall using HPLC-MS revealed the mutants had a shortened pentaglycine bridge consisting of a single glycine. The enzyme responsible for addition of the first glycine (FemX) was later identified as the enzyme encoded by the second gene (*fmhB*) in an operon encoding three genes (*fmhABC*) that were homologous to *femAB* (Rohrer *et al.*, 1999). Interestingly, only *fmhB* was essential for growth. Therefore, it was deduced that FmhB was responsible for addition of the first glycine to the peptidoglycan precursor which was strongly supported by the finding that mutants with induced repression of *fmhB* produced peptidoglycan consisting mainly of highly uncross-linked monomers of N-acetylglucosamine-(beta-1,4)-N-acetylmuramic acid(L-Ala-D-iGln-L-Lys-D-Ala-D-Ala) (Rohrer *et al.*, 1999). The function of all three enzymes (FmhA, B, and C) in the assembly of the pentaglycine bridge was finally confirmed biochemically using purified constituents from the pathway (Schneider *et al.*, 2004). The essentiality of FemXAB and the alteration in peptidoglycan precursors produced by the Tn mutants with an attenuated methicillin resistance phenotype suggested that the pentaglycine bridge is critical for recognition by the horizontally acquired low-affinity penicillin-binding transpeptidase, PBP2a, that is encoded by *mecA* that is essential for the methicillin resistance phenotype in virtually all MRSA.

MurA and MurZ provide redundancy in the first step of peptidoglycan precursor synthesis

S. aureus encodes two genes (*murA* and *murZ*) that encode homologues of the enzyme UDP-*N*-acetylglucosamine enolpyruvyl transferase that catalyses the step that leads to production of the first building block (UDP-N-acetylmuramic acid) of the peptidoglycan precursor. Both enzymes have been purified and perform the expected catalytic reaction. However, the presence of two genes in the *S. aureus* chromosome encoding this enzyme is not explained by a simple need for redundancy. Each gene has its own preferred condition for optimal expression. *murA* is expressed at five times the rate of *murZ* under steady state conditions whereas *murZ* expression is induced at least six-fold under antibiotic stress imparted by cell wall acting antibiotics (Blake *et al.*, 2009). Among the cell wall acting antibiotics that induce *murZ* but not *murA* are the β-lactams and glycopeptides. In contrast, the antibiotic fosfomycin, which specifically targets both MurA and MurZ, induces expression of both genes. Thus, the staphylococcal cell has devised a backup plan for times when increased cell wall synthesis is advantageous. Activation of *murZ* expression by cell wall antibiotics is mediated by the three-component regulatory system, VraTSR, that is activated by cell wall antibiotics, is autoregulated and controls a regulon that facilitates resistance, presumably by accelerating cell wall biosynthesis (Boyle-Vavra *et al.*, 2013; Kuroda *et al.*, 2003). In contrast, VraTSR does not activate *murA*.

Late stages of peptidoglycan synthesis

Once the lipid-linked precursor (also commonly referred to as lipid II) is assembled on the bactoprenol carrier, an unknown enzyme (sometimes called 'flippase') translocates it to the exterior side of the membrane where it is incorporated into the glycan chain with the use of peptidoglycan transglycosylases, PBP2, SgtA and SgtB. Cross-linking of the peptidoglycan peptide chains is accomplished by PBP4 along with a transpeptidase domain from PBP2. These activities comprise the terminal stages of peptidoglycan synthesis. In MRSA, an additional high molecular weight PBP (called PBP2a or PBP′) with low affinity for β-lactams supplies the transpeptidase required when the native PBPs are bound and inhibited by a β-lactam.

Antibiotics that target peptidoglycan synthesis

Although the early stages of peptidoglycan synthesis provide thirteen potential enzyme targets for antibiotics, only two of these are targets of clinically relevant antibiotics (Bugg *et al.*, 2011) and none are widely used in the therapy of *S. aureus*. Most of the cell wall-acting antibiotics that are used clinically act at a late stage in peptidoglycan biosynthesis and include β-lactams, glycopeptides (e.g. vancomycin), dalbavancin, telavancin and oritivancin. We will describe these antibiotics and resistance mechanisms according to the stage of peptidoglycan synthesis they affect.

Bacitracin

Bacitracin is an antibiotic that inhibits an intermediate stage of cell wall synthesis, linking the early and late stages. Bacitracin is not used systemically do to toxicity; however, it is used as a topical agent to prevent wound infections and to treat and prevent ophthalmic infections. It is a cyclic dodecylpeptide, assembled independently of the ribosome that inhibits peptidoglycan synthesis at a late stage in precursor assembly in the cytoplasm. It has a unique mechanism of action. It binds to and inhibits dephosphorylation and recycling of the membrane carrier (C_{55} undecaprenol pyrophosphate) of peptidoglycan precursors by forming a cation-dependent complex with the lipid portion of the carrier (Walsh, 2003c,d).

Since bacitracin inserts into the cytoplasmic membrane, it is toxic to eukaryotic cells. This property limits its use to topical ointments for skin and ophthalmic infections. Like penicillin, bacitracin is bactericidal and induces lysis (Smith and Weinberg, 1962), but only if the bacteria are exposed to the antibiotic before stationary phase. Like other cell wall inhibitors, bacitracin induces a cell wall stress response that includes about 100 genes that are also induced by oxacillin and D-cycloserine (Utaida *et al.*, 2003). However, there are 46 genes specifically upregulated by bacitracin but not by other cell wall agents.

Bacitracin resistance in *S. aureus* is mediated by the ABC transporter VraDE (Pietiainen *et al.*, 2009), which likely acts by transporting bacitracin out of the cell to effectively decrease its intracellular concentration to a non-toxic level. VraD encodes an ATPase that uses ATP hydrolysis to derive energy to transport molecules across the membrane through the permease, VraE. A second ABC transporter operon, *braDE*, that also encodes an ATPase and a permease, is also required for bacitracin resistance. Transcription of the operons encoding VraDE and BraDE is induced by bacitracin through the two-component regulatory system, BraSR (Hiron *et al.*, 2011). BraDE does not mediate resistance by elimination of bacitracin as VraDE does; instead, the BraD ATPase has evolved as a sensor of the antibiotic and acts in concert with BraSR. Apparently, BraE is not involved (Hiron *et al.*, 2011).

β-Lactams

The β-lactams are defined chemically by a four-membered β-lactam ring fused to a five membered sulfur ring system, with penicillin being the first prototype derived from a fungus secondary metabolite. Numerous semisynthetic derivatives have been made from this basic β-lactam ring structure by addition of side chains that enhance their activity. Such side chain modifications can protect the β-lactam against the hydrolysis and destruction by β-lactamase or allow penetration of the drug through Gram-negative outer membranes (Walsh, 2003d). The cephalosporins, the most widely prescribed and largest selling class of β-lactams, have a sulfur-containing ring expanded system (Walsh, 2003d).

All β-lactams act by inhibiting late-stage peptidoglycan synthesis by binding to and inhibiting PBPs. By acylation of the transpeptidase domain of PBPs, β-lactams block access to the peptidoglycan precursor, thereby preventing cross-linking of peptidoglycan strands during cell wall synthesis. In *S. aureus*, there are four native PBPs: 1, 2, 3 and 4.

Although penicillin initially was said to provide a miracle cure for *S. aureus* infection, it was widely overused with such practices as dusting of penicillin powder on infected wounds and using penicillin-containing snuff for treating respiratory infections (Livermore, 2000). Penicillin-resistant (PenR) strains of *S. aureus* emerged in hospitalized patients not long after penicillin was initiated into clinical practice in the 1940s, and their incidence rapidly increased in the healthcare setting (Chambers and Deleo, 2009). In one hospital in London, England, PenR *S. aureus* (PRSA) infections increased from 14% to 59% of *S. aureus* infection in just 3 years in 1946–1948 (Barber and Rozwadowska-Dowzenko, 1948). Within a decade the problem of PRSA permeated into the community and currently nearly 99% of all *S. aureus* strains isolated from infections are PRSA.

The major mechanism of penicillin resistance in *S. aureus* is the inactivation of penicillin by elaboration of the β-lactamase enzyme. This mechanism efficiently and irreversibly hydrolyses the β-lactam ring and thus confers very high level penicillin resistance (Dyke, 1969). β-Lactamase-producing PRSA strains emerged early after introduction of penicillin and disseminated rapidly such that they reached pandemic proportions by the 1950s (Chambers and Deleo, 2009). The gene encoding β-lactamase (*blaZ*) was carried on plasmids that were readily transferred by conjugation, thereby easily disseminating *blaZ* among different genetic backgrounds of *S. aureus* (Chambers and Deleo, 2009). Eventually, *blaZ* integrated into the chromosome of some strains, apparently mediated by homologous recombination between insertion sequence elements contained in the plasmid and chromosome.

Combination antibiotics that include a β-lactamase inhibitor with penicillin can counteract the effect of β-lactamase and decrease the MIC of penicillin sufficiently to treat an infection. An example of this combination is Augmentin, an oral antibiotic that combines amoxicillin and clavulanate (the latter being the β-lactamase inhibitor). Clavulanate is structurally related to β-lactams and acts by blocking the active site of the β-lactamase enzyme thereby restoring the activity of amoxicillin.

Another means to deal with the problem of penicillin resistance was to develop β-lactamase-resistant β-lactams such as methicillin and nafcillin, the cephalosporins and oxacillin. These compounds have a substitution of the 6′ phenyl acetyl group of benzylpenicillin with bulky acyl side chains that provide steric hindrance against an attack on the β-lactam ring. However, isolates resistant to these agents, MRSA, were isolated soon after their introduction into clinical practice in the 1960s (Jevons *et al.*, 1961).

Mechanism of methicillin resistance

It took 24 years after the first MRSA isolate was identified before the first observation was made to begin to explain the molecular basis of methicillin resistance. As summarized in Table 11.1, landmark discoveries occurring over the ensuing 15 years led to a clear understanding of the molecular determinant of methicillin resistance and the mobile genetic element responsible for horizontal dissemination of the methicillin resistance gene between strains. It was evident very early that the mechanism of methicillin resistance was independent of β-lactamase production. Although most MRSA isolates produced β-lactamase, curing the β-lactamase plasmid from most MRSA strains did not reverse the methicillin resistance phenotype. Also, there were naturally occurring MRSA isolates that did not produce β-lactamase, which pointed to a separate mechanism (Dyke, 1969). These pieces of evidence strongly suggested that MRSA isolates had evolved a new mechanism for overcoming β-lactamase-insensitive β-lactams and the mechanism was revealed in 1984 by Hartman and Tomasz (Hartman and Tomasz, 1984). By reacting membrane proteins isolated from MRSA isolates with radioactively labelled penicillin followed by SDS-polyacrylamide electrophoresis and autoradiography, an additional high molecular weight PBP (called PBP2a) was identified in a MRSA transductant that was absent from the isogenic methicillin-susceptible strain. The name PBP2a derives from the fact that it migrates close to PBP2 on an SDS PAGE gel. PBP2a was also named PBP2′ in an independent investigation that confirmed the presence of a low affinity PBP in a β-lactamase-negative MRSA strain (Utsui and Yokota, 1985). PBP2a/PBP2′ was found to have a substantially lower affinity for β-lactams compared with the four native β-lactam-sensitive PBPs, as determined in competition experiments with non-radioactive penicillin. Moreover, acylation of PBP2a in the active site serine residue by β-lactams occurs at a decreased rate (two to three orders of magnitude) compared with PBPs 1 through 4; deacylation occurs very slowly with a half-life of over 77 h (Peacock and Paterson, 2015).

Matsuhashi *et al.* (1986) cloned a large DNA fragment from an MRSA strain where it expressed a ca. 70 KDa protein in *E. coli* that could bind penicillin with slow reaction kinetics. The 2.5 kb ORF encoding the PBP2a gene (later called *mecA*) was subsequently identified by DNA sequencing (Song *et al.*, 1987) which predicted a 76,462 Da protein. The sequences of the promoter region and upstream ORFs were largely similar to those of the staphylococcal *blaZ* gene responsible for inducible expression whereas the *mecA* ORF more closely resembled that of PBPs from *E coli*. This led to the proposal that a recombination event between the regulatory region of *blaZ* and a low affinity PBP created this unique resistance gene that could be

Table 11.1 Landmark discoveries in methicillin resistance

Date	Discovery	Reference
1960	Licensure of methicillin/celbenin	Jevons *et al.* (1961)
1960[1]	Identification of first MRSA isolate	Jevons *et al.* (1961)
1984	Demonstration of a new PBP in MRSA, termed PBP2a (and PBP2'), with reduced affinity for β-lactams	Hartman and Tomasz (1984)
1986	Cloning and expression of the low affinity penicillin binding protein PBP2a that confers methicillin resistance	Matsuhashi *et al.* (1986)
1987	Identification of the sequence of the *mecA* gene that encodes PBP2a, conferring resistance to methicillin	Song *et al.* (1987)
1990	Insertional inactivation of *mecA* proves its role as necessary for methicillin resistance	Matthews and Tomasz (1990)
1999	The entire SCC*mec* sequence was published, the mobile genetic element that carries *mecA* in the staphylococcal chromosome	Ito *et al.* (2001)
2000	*ccrA* and *ccrB* recombinases were shown to mediate precise excision and integration of SCC*mec*	Katayama *et al.* (2000)
2001	The first whole genome sequence of an MRSA strain was published (in the first VISA strain)	Kuroda *et al.* (2001)
2006	Whole genome sequence of community associated MRSA strain USA300 was published	Diep *et al.* (2006)

[1]Although this report was published in 1961, the isolates described by Jevons were isolated in 1960 in the months preceding release of methicillin for clinical practice.

induced by the β-lactams to which it encoded resistance.

The crystal structure of PBP2a has been solved in the presence and absence of β-lactam ligands. PBP2a encodes four characteristic structural domains: an N-terminal transmembrane domain (aa 1–23), an N-terminal extension (aa 27–138), a transpeptidase domain that contains the ligand binding site and catalyses peptide cross-links in peptidoglycan (aa 327–668), and a non-penicillin-binding domain (aa 27–326) which nevertheless has an allosteric site that when occupied by the newly licensed cephalosporin ceftaroline opens up the active site, making it more vulnerable to β-lactams (Otero *et al.*, 2013). The active-site serine within the transpeptidase domain is buried in a hydrophobic cleft, thus making the serine residue less accessible to β-lactams.

Although PBP2a is able to remain unbound by β-lactams when the native PBPs are inactivated, PBP2a is not sufficient to carry out cell wall biosynthesis alone. Cell wall biosynthesis requires both transpeptidase and transglycosylase activities and since PBP2a only has transpeptidase activity the transglycosylase must be supplied by another PBP. So how is this accomplished if the other PBPs are bound by penicillin? This dilemma is solved by the native PBP2 which has both a transglycosylase and a transpeptidase domain. Since β-lactams only bind to the transpeptidase domain of PBP2, the PBP2 transglycosylase domain remains unimpeded and available to cooperate with the transpeptidase of PBP2a in the biosynthesis of peptidoglycan (Pinho *et al.*, 2001). The recent licensure of cephalosporins that bind PBP2a with high affinity (ceftaroline and ceftobiprole) provide a β-lactam therapy for methicillin-resistant strains, at least until strains that have resistance to those agents disseminate.

Genetics of methicillin resistance

PBP2a is encoded by the 2.0 kb *mecA* gene that is carried into the staphylococcal chromosome on a family of mobile genetic elements called SCC*mec* (Daum *et al.*, 2002; Ito *et al.*, 2001; Katayama *et al.*, 2000). Most MRSA strains carry a *mecA* homologue with > 98% identity. *mecA* homologues with > 98% identity to *mecA* have also been found in several other staphylococcal species (Ito *et al.*, 2012). In other staphylococcal species there are also *mecA* allotypes (mecA1, mecA2, etc.) that are ≥ 70% but

<95% identical to *mecA* (Ito *et al.*, 2012); these have been proposed to be evolutionary progenitors of *mecA* gene (Wu *et al.*, 1996). Recently, two distantly related homologues of *mecA*, termed *mecB* and *mecC*, were identified. Whereas *mecC* (70% identical to *mecA*) was identified in *S. aureus* strains from dairy cattle and humans (Garcia-Alvarez *et al.*, 2011; Shore *et al.*, 2011), *mecB* (63% identical to *mecA*) has been found in *Macrococcus caseolyticus* on either a plasmid or the chromosome and has not been identified in any staphylococcal species to date. Importantly, since *M. caseolyticus* is considered to be an ancestral species relative to staphylococcal species, the presence of a *mec* homologue present on a plasmid is significant (Baba *et al.*, 2009). To date, no other *mecA* homologue has been identified on a naturally occurring plasmid.

At least twelve SCC*mec* types have been described to date that range in size from 24 to 60 kb (Ito *et al.*, 2012; Wu *et al.*, 2015) but there are four common features present in all SCC*mec* elements: (a) the *mecA* gene, (b) the *mecA* regulatory complex *mecR1* and *mecI* (unless they have been deleted), (c) the *ccr* recombinase gene(s) responsible for integration and excision in to and out of the chromosome and (d) terminal inverted and direct repeats that flank the element act as cis elements for the cognate recombinase genes(s). The variations in SCC*mec* elements have been useful markers for epidemiologic investigations and some have even been used to delineate community associated (CA) from healthcare associated (HA) MRSA.

The basis for the size differences between SCC*mec* elements can be attributed to deletions or insertions of transposons and/or plasmids that carry genes encoding antibiotic resistance to classes of agents besides β-lactams. SCC*mec* elements integrate site specifically into an attachment site located at the 3′ end of an rRNA methyltransferase (*rlmH*), previously called *orfX* (Boundy *et al.*, 2013). Although *rlmH* is not essential *in vitro* (Boundy *et al.*, 2013), integration of SCC*mec* elements into this ORF does not disrupt the coding sequence in clinical isolates. There are also non-*mecA* containing elements (SCC), presumably the progenitors of SCC*mec*, that share similar sequence features of SCC*mec* and insert into the same attB site.

One of the most interesting facets about SCC*mec* elements is that they vary in both their core genes (i.e. the *ccr* and *mec* gene complexes) and in regions outside the core SCC DNA. The 'extra' SCC DNA has included integrated plasmids, insertion sequences, transposons, antibiotic resistance genes, a *pbp4* gene, a capsule 1 gene cluster, heavy metal resistance gene clusters, and an arginine deiminase operon (IWG-SCC, 2009; Diep *et al.*, 2006; Ito *et al.*, 2001; Luong *et al.*, 2002; Mongkolrattanothai *et al.*, 2004; Shore *et al.*, 2011). In other words, SCC elements, including SCC*mec* elements, have served as a mobile source of extra genomic material that might provide a survival benefit in different niches. Indeed, the multi drug resistance phenotype of healthcare-associated MRSA is due in part to the ease with which plasmid DNA can be inserted into SCC*mec* DNA using insertion sequences as sites for homologous recombination (Ito *et al.*, 2001).

The SCC*mec* cassette that harbours *mecC*, called SCC*mec* XI, has a very different *mecI-mecR1* and recombinase complex compared with SCC*mec* cassettes I through X, pointing to a divergent evolutionary origin for this cassette (Garcia-Alvarez *et al.*, 2011; Shore *et al.*, 2011). Moreover, the oxacillin MICs of SCC*mec* XI-*mecC*-containing strains are low compared with other strains containing SCC*mec*. Although strains carrying *mecC* have been reported to cause MRSA infection in humans in Europe, the overwhelming majority of human MRSA cases in the world to date are caused by *mecA*-containing strains (Becker *et al.*, 2014).

The existence of nearly identical SCC*mec* elements (for example SCC*mec* IV) in different genetic backgrounds of *S. aureus* and different staphylococcal species is indirect evidence that SCC*mec* is transferred between strains. The role of the Ccr (cassette chromosome recombinases) encoded in the SCC*mec* element is clearly important to site specific integration of the element into the staphylococcal chromosome. It is primarily maintained as an integrated element until conditions stimulate excision, which results in formation of episomal circles. Excision occurs at a low rate but overexpression of the *ccr* genes on a multicopy plasmid can hasten the excision, allowing the screening of revertant MSSA isolates. Integration of electroporated constructs has been observed as well (Katayama *et al.*, 2000). However, the means by which the element is actually transferred in nature between cells and whether it involves replication of an excised circularized element remain to be elucidated.

Bioinformatic, biochemical and crystallographic

data suggest that SCC*mec* elements also encode machinery that could direct their own replication after excision. This could help stabilize the element by increasing the chances of re-insertion after excision, and/or increase the efficiency of horizontal gene transfer (Phoebe Rice and Ignacio Mir-Sanchis, personal communication).

Some SCC*mec* cassettes carry a pair of recombinase alleles, *ccrA* and *ccrB*, on a two-gene operon, whereas other cassette types carry only a single ccr, *ccrC*. Several studies have shown that, for elements that encode *ccrA* and *ccrB*, both gene products are required for integration and excision of the SCC elements into and out of the staphylococcal chromosome (Katayama *et al.*, 2000; Noto and Archer, 2006; Noto *et al.*, 2008). Misiura *et al.* (2013) recently demonstrated in an *E. coli* host the Ccrs are sufficient for excision and integration in the absence of any additional staphylococcal or SCC*mec* encoded proteins. The Ccrs belong to the 'large serine' family of site-specific DNA recombinases, which usually bind as dimers to inverted repeats. Interestingly, the SCC insertion site in the *S. aureus* chromosome is asymmetric. CcrA binds the left half of this asymmetrical site, while CcrB binds the other half and to the pair of half-sites found in the SCC element before recombination (Misiura *et al.*, 2013). Thus, Misiura *et al.* proposed the role of CcrA is to serve as a host specificity factor. It is unknown how CcrC as the sole recombinase in SCC*mec* V handles this asymmetry, although the CcrC dimer may simply bind more cooperatively, obviating the need for a heterodimer. Unlike many recombinases from the large serine family, which will only catalyse reactions between a host insertion site and a circularized element site (for integration) or a left and right end (for excision), the Ccrs will recombine a circularized element site with the right end of an inserted element, which may therefore afford the flexibility to form tandem integrations of SCC elements that are so commonly found in staphylococcal chromosomes.

Heterotypic and homotypic methicillin resistance

Perhaps the greatest unsolved enigma in our understanding of phenotypic methicillin resistance is why the level of oxacillin resistance varies among clinical MRSA isolates from 1 μg/ml to >1000 μg/ml. Although *mecA* gene expression is repressed in drug free conditions and induced in the presence of β-lactams, there are other factors encoded in the genetic background that influence the level of resistance. When exposed to β-lactams, strains that express low-level resistance can spawn highly resistant subclones when plated on nutrient agar. When *mecA* is deleted from stable highly resistant subclones, the methicillin resistance phenotype is lost. However an imprint remains in the genetic background that encodes the potential for expression of high-level resistance when the *mecA* gene derived from a heterotypic strain is reconstituted back through transduction. The highly resistant phenotype of the homotypic clone rather than the low level phenotype is expressed. The mutation that led to this stable high level resistance phenotype was referred to as chr*, referring to an unknown chromosomal mutation. Despite more than two and a half decades of investigation undertaken to understand this phenomenon, the means by which the genetic background influences the level of phenotypic resistance is still not well understood (Ender *et al.*, 2008; Murakami and Tomasz, 1989; Ryffel *et al.*, 1994). Whereas colonies from the well characterized, homotypic strain COL have a homogeneous high level of resistance, most other strains have heterogeneous subpopulations with varying resistance and are referred to as heterotypic.

Heterotypic versus homotypic resistance cannot be simply explained by the differential regulation of *mecA* gene expression. Varying the level of *mecA* transcription or PBP2a production in a given strain does not necessarily correlate with the level of resistance (Ender *et al.*, 2008; Murakami and Tomasz, 1989; Rosato *et al.*, 2003a; Ryffel *et al.*, 1994); however this situation is strain dependent (Niemeyer *et al.*, 1996).

mecA gene expression is usually repressed at the transcription level in most MRSA strains in the absence of β-lactams. Expression is induced, or de-repressed, in the presence of β-lactams, when *mecA* expression is needed. This strategy conserves energy by maintaining a low level of *mecA* expression in the bacteria when it is not needed. Repression of *mecA* transcription is mediated by either a dedicated cytoplasmic repressor protein (MecI) (Kuwahara-Arai *et al.*, 1996) or by BlaI, the homologous repressor of the β-lactamase gene, *blaZ* (Gregory *et al.*, 1997; Sharma *et al.*, 1998). MecI and BlaI repressors form homodimers that bind to a 30 bp palindrome in the

−10 region of the *mecA* promoter, thereby blocking transcription initiation.

Reversal of MecI-mediated *mecA* repression involves a sentinel membrane protein (MecR1) that binds to, and detects, β-lactams in the environment. The genes encoding MecI and MecR1 are located together in an operon that is adjacent and divergently transcribed relative to *mecA* (Hiramatsu *et al.*, 1992). Many clinical MRSA isolates have deletions or mutations in *mecI* and *mecR1* genes (Archer *et al.*, 1994; Rosato *et al.*, 2003a; Suzuki *et al.*, 1993). Nevertheless, *mecA* repression in most clinical isolates is mediated by the BlaI-BlaR1 homologues that are encoded from an ORF adjacent to the β-lactamase gene, *blaZ*, (Hackbarth and Chambers, 1993) now present in nearly all *S. aureus* strains. Conversely, MecI-MecR1 can repress and induce *blaZ* transcription (Lewis and Dyke, 2000). However, there is no cross-talk between MecI and BlaR1 or between BlaI and MecR1 (Arede *et al.*, 2012; Arede and Oliveira, 2013; McKinney *et al.*, 2001).

BlaR1 and MecR1 use a novel proteolytic mechanism to induce transcription of *mecA* or *blaZ* resistance genes (Zhang *et al.*, 2001; Arede and Oliveira, 2013). BlaR1 binds to a β-lactam via a penicillin-binding domain which in turn triggers autoproteolysis that subsequently releases a peptide that proteolyses BlaI. Upon proteolysis, BlaI no longer binds to the cognate promoter and no longer represses *blaZ* or *mecA* transcription. A similar model has been demonstrated for MecR1-MecI mediated repression of *mecA* (Arede and Oliveira, 2013). Some strains of MRSA encode an antirepressor of MecI, MecR2, which enhances MecR1-mediated proteolysis and prevents binding of MecI to the *mecA* promoter (Arede *et al.*, 2012; Arede and Oliveira, 2013). The *mecR2* gene, as well as *mecI*, is missing from the SCC*mec* type IV element in which repression and induction of *mecA* expression is likely mediated by BlaI.

Why does BlaR1/BlaI cross-repress *mecA* expression? *In vitro* analysis of the binding affinities of MecI and BlaI to the *mecA* promoter revealed that MecI-mediated repression of *mecA* is several orders of magnitude more efficient than BlaI–mediated repression of *mecA* (Ryffel *et al.*, 1992). Thus, BlaI allows for a more rapid response to a β-lactam and a nimbler means of inducing *mecA* expression than MecI-MecR1. Moreover, most clinical MRSA isolates have mutations that attenuate MecI-mediated repression of *mecA* expression (Kuwahara-Arai *et al.*, 1996). These mutations include deletions in *mecI-mecR1* genes, substitutions that decrease the affinity of MecI for the *mecA* promoter, or cis-acting mutations in the *mecA* promoter that decrease binding to MecI. It was hypothesized that selection of mutations that attenuate the repression of *mecA* by MecI was an important evolutionary step that allowed MRSA strains to succeed in the face of increasing use of β-lactams. However, most MRSA isolates contain BlaI, which allows the maintenance of *mecA* repression, albeit with more rapid induction kinetics than realized with MecI. An opposing hypothesis was presented by Rosato *et al.* who provided evidence that BlaI is actually a more effective repressor of *mecA* compared with MecI (Rosato *et al.*, 2003b), suggesting that defects in MecI-mediated control of *mecA* expression were necessary to enable tighter control of *mecA* by BlaI.

This brings to mind the question of why *S. aureus* does not eliminate both repression systems of *mecA* expression? Would not a strain be better off constitutively producing PBP2a to maintain a constant supply in the event that a β-lactam is encountered? Apparently not. Most MRSA isolates maintain repression of *mecA* by one of these two regulatory systems (Rosato *et al.*, 2003b). This suggests the cell–wall that is constructed with the use of PBP2a, could be deleterious to a *S. aureus* cell in the absence of an antibiotic challenge. In support of this idea is the finding that *mecA* is unstable in MRSA isolates that have been cured of the β-lactamase plasmid (Hiramatsu *et al.*, 1990) and that the *blaZ* plasmid is maintained in clinical MRSA isolates. Indeed, HPLC analysis of peptidoglycan isolated from an MRSA strain exposed to methicillin revealed that the structure of the cell wall under inducing conditions has a lower oligopeptide character with more monomeric and dimeric muropeptides compared with the uninduced strain (de Jonge *et al.*, 1992). Such a PBP2a-sculpted cell wall composition in the absence of drug selection might be less beneficial than the normal cell wall.

Ceftaroline (Cfn), a new PBP2a inhibitor belonging to the cephalosporin class

Ceftaroline fosamil (Cfn) is a cephalosporin antibiotic that is unique in its activity against MRSA strains due to its high affinity for binding to PBP2a as well as the four native essential PBPs of *S. aureus*. It has been approved by the US Food and Drug Administration (FDA) for the treatment of acute bacterial SSTIs and community-acquired bacterial pneumonia caused by *S. aureus* as well as other Gram-positive and negative bacteria. Using CLSI and EUCAST clinical guidelines, sensitivity to Cfn is defined as ≤1 µg/ml (Livermore *et al.*, 2015). CLSI and EUCAST standards are at odds with the resistant and intermediate breakpoints. Whereas an MIC > 1 µg/ml is considered resistant by EUCAST guidelines this value is considered intermediate by CLSI guidelines. And according to CLSI and the FDA guidelines, an MIC of ceftaroline > 2 µg/ml is resistant. The intermediate category is meant as a buffer zone especially in consideration of a recent study showing high variability in the interpretations of resistance among clinical isolates (Livermore *et al.*, 2015). To gain further insight into the contribution of bacterial factors affecting the MIC of Cfn, a genetic analysis of PBP2a substitutions associated with MIC of Cfn was performed on MRSA isolates from the global surveillance program referred to as AWARE (Assessing Worldwide Antimicrobial Resistance Evaluation), which tracks the global susceptibility of isolates to Cfn. In a collection of 4324 global MRSA isolates from 2012, the MIC of Cfn of 90% of strains (MIC_{90}) was 1–2 µg/ml (Lahiri *et al.*, 2015), similar to studies in the United States (Richter *et al.*, 2011). However, there were 10 MRSA isolates that had an MIC of Cfn of 8 µg/ml (Lahiri *et al.*, 2015).

A whole genome sequence analysis of 180 representative MRSA isolates from this collection, including those with a Cfn MIC of 8 µg/ml, provided insight, but no clear answers, into the role of mutations in PBP2a in Cfn resistance. Although there were 58 MRSA isolates with MICs of Cfn ≤2 µg/ml that had a wildtype *mecA*, nucleotide variations were found in isolates with MICs of Cfn that range from susceptible to resistant (0.5–4 µg/ml) outside the transpeptidase and Cfn-binding domains of PBP2a (the so called nPBD). Although the nPBD substitutions were enriched in isolates with an MIC of Cfn of 2 µg/ml, there was no direct association of nPBD mutations with the level of susceptibility. In 10 of 12 isolates that had an MIC of 8 µg/ml, a glutamic acid to lysine substitution (E447K) in the PBP2a transpeptidase binding pocket was present that might have a direct effect on cell wall biosynthesis. However, these 10 isolates also had an E239K substitution in the nPBD making it difficult to assess if the E447K substitution was sufficient to increase the MIC to 8 (Lahiri *et al.*, 2015). Notably the E447K substitution was not found in isolates with MICs of ceftaroline between 0.5 to 4 µg/ml.

The most common nPBD substitutions were a glutamic acid to lysine substitution (E150K, E239K) and N146K. The E to K substitutions effectively reverse the local charge effect. These substitutions were found among multiple genetic lineages, suggesting they were independently acquired and selected rather than represent dissemination of a single clone. Among a small fraction (9%) of the isolates with Cfn MIC values of ≥2 µg/ml, there were an additional 11 unique nPBD variations.

The fact that most mutations among Cfn NS isolates are in the nPBD suggests this domain plays a role in the function of PBP2a activity in cell wall biosynthesis or Cfn binding. In support of this hypothesis, the nPBD was shown to be a dimerization domain that can interact with other PBPs and co-crystallize with cell wall fragments that allosterically open the active site of PBP2a (Otero *et al.*, 2013).

It may not be a coincidence that a majority of the nPBD mutations are located in a groove in close proximity to each other. It was proposed that the charge distribution of the outer surface would be influenced by combinations of these substitutions (Lahiri *et al.*, 2015).

Lahiri *et al.* (2015) suggested that, based on these genetic data, the breakpoint for Cfn susceptibility should be raised above 2 µg/ml. This is supported by three pieces of genetic data: (a) isolates with MICs of Cfn of 2 µg/ml often had no substitutions in PBP2a, (b) substitutions in the nPBD of PBP2a were present in strains with MICs of Cfn of 2 µg/ml as well as more susceptible strains and (c) only nPBD substitutions (outside the active site) were associated with MICs of Cfn < 4 µg/ml. The suggestion that the susceptibility breakpoint should be based on these genetic changes should be considered with caution since substitutions in nPBD could play an important role in the allosteric conformation of the PBP2a protein that could promote second step mutations that lead to higher MICs.

Vancomycin

Vancomycin is the only glycopeptide antibiotic licensed in the United States. It is naturally produced by *Amycolatopsis orientalis* and since its discovery in 1956, it has been mainly used to treat Gram-positive infections. The toxic effects caused by contaminants in early versions of the antibiotic delayed its widespread use (Barna and Williams, 1984). Once this problem was solved it became an invaluable antibiotic for treating Gram-positive bacterial infections, including MRSA. Although it took very little time for *S. aureus* to develop resistance to penicillin and methicillin, it took 40 years between the discovery of vancomycin and emergence of vancomycin-resistance in clinical *S. aureus* isolates (Hanaki *et al.*, 1998; Hiramatsu *et al.*, 1997b). However, the increase in the number of infections caused by MRSA isolates led to increased use of vancomycin so that by 2011 it was the second most commonly used antibiotic in hospitals (Jafari Saraf and Wilson, 2011). In turn, the emergence of vancomycin-intermediate resistant *S. aureus* (VISA) isolates, and later vancomycin-resistant *S. aureus* (VRSA) isolates with higher MICs (Weigel *et al.*, 2003), were reported.

Despite the emergence of vancomycin resistance in *S. aureus*, vancomycin remains one of the most widely recommended antibiotic therapies for MRSA infection (Liu *et al.*, 2011). Although it is referred to as a last resort agent, vancomycin continues to be heavily relied upon as a first line of therapy for treating infections caused by MRSA isolates. Indeed, new agents are often compared to vancomycin in clinical trials as the gold standard.

Vancomycin resistance phenotypes in *S. aureus*

To understand the mechanism of resistance to vancomycin, it is helpful to first consider the mechanism of antibacterial action for glycopeptides. Vancomycin inhibits cell wall synthesis by binding to the terminal D-Ala-D-Ala dipeptide of the peptidoglycan lipid II precursor (Barna and Williams, 1984). Although β-lactams and glycopeptides inhibit the same late stage of peptidoglycan synthesis, β-lactams inhibit the enzymes that catalyse peptidoglycan synthesis, whereas vancomycin binds to the D-Ala-D-Ala termini of lipid II, thereby blockading transglycosylation. Although vancomycin acts exterior to the Gram-positive cell, it must pass through the many layered peptidoglycan meshwork in order to access its target, lipid II, at the cytoplasmic membrane. Because there are terminal D-Ala-D-Ala residues in mature peptidoglycan, a portion of the vancomycin molecules become absorbed on its passage through the cell wall thereby diluting them on their way to the target at the cell membrane (Barna and Williams, 1984). This phenomenon can account for the inoculum effect in MIC testing, making it critical to use a standard inoculum in the MIC test of vancomycin susceptibility.

Vancomycin resistance mechanisms in *S. aureus*

Two basic forms of vancomycin resistance in *S. aureus* have been described: vancomycin-intermediate resistant *S. aureus* (VISA) and high level vancomycin-resistant *S. aureus* (VRSA).

The first (VISA or GISA) identified from a human infection was reported in Japan in 1996 (Hanaki *et al.*, 1998; Hiramatsu *et al.*, 1997b), about 40 years after vancomycin was discovered. VISA isolates have an MIC of vancomycin in the intermediate range; however, the MIC cut-off for vancomycin intermediate resistance has changed since 1996. When the first VISA isolates were identified, intermediate vancomycin resistance was defined as an MIC $\geq 8\,\mu g/ml$ but $\leq 16\,\mu g/ml$. (Isolates that express high level resistance with an MIC of vancomycin $\geq 32\,\mu g/ml$ did not emerge until 2002). Since 2006, when the cut-off for vancomycin susceptibility was lowered, isolates with MICs of $\geq 4\,\mu g/ml$ to $\leq 8\,\mu g/ml$ have been considered to be intermediate (Tenover and Moellering, 2007). The basis for this change was the observation that isolates that had an MIC of $4\,\mu g/ml$, which under the previous guidelines would have been considered susceptible, often had hidden resistant subpopulations capable of growing on higher amounts of vancomycin (hetro VISA or hVISA), despite being considered susceptible according to their MIC value (Hiramatsu *et al.*, 1997a). In fact, some isolates with MICs of vancomycin as low as $0.5\,\mu g/ml$ could have resistant subpopulations. Because detection of resistant subpopulations is labour intensive for a typical clinical microbiology laboratory, requiring extensive plating on agar media, it became more convenient to lower the susceptibility cut-off score for vancomycin.

S. aureus clinical isolates with an MIC of vancomycin ≥ 4 μg/ml remain infrequently encountered but a phenomenon called 'vancomycin creep' has been described. This phenomenon posits that the number of clinical *S. aureus* isolates with undetected vancomycin-resistant subpopulations is increasing but might be undetected by conventional automated detection modalities. This finding is geographically specific and may reflect local vancomycin usage patterns (Fridkin *et al.*, 2003). Nevertheless, it is generally agreed that most MRSA isolates still test susceptible to vancomycin. Therefore, vancomycin remains a drug of choice in empiric therapy of bloodstream infections in the absence of culture or susceptibility data. Nevertheless, vancomycin treatment failure occurs in the therapy of infections caused by vancomycin-susceptible MRSA isolates often enough to think that the current methods of detecting vancomycin resistance are still inadequate at predicting treatment failure.

VISA and VRSA

VRSA isolates have high level resistance and are defined as isolates with MICs of vancomycin ≥ 32 μg/ml. The mechanism of resistance in VISA isolates is distinct from that of VRSA. Whereas VRSA isolates carry the *vanA* gene cluster acquired from *Enterococcus faecalis*, VISA isolates develop resistance in a step wise fashion when under selective pressure from vancomycin. From extensive studies of clinical and *in vitro* derived VISA and hVISA isolates it is apparent that *S. aureus* can become intermediately resistant to this drug through multiple mechanisms (Boyle-Vavra *et al.*, 2001a). That being said, there are common characteristics among VISA isolates. Of note, increased cell wall thickness, a trait that was observed among *in vitro* derived VISA prior to the isolation of clinical isolates (Daum *et al.*, 1992) has remained the most constant characteristic among VISA strains. The *in vitro* derived VISA isolates also had many other features in common with the clinical VISA isolates that were identified later. These common features included increased cell wall thickness, increased lysostaphin resistance, decreased phage susceptibility, decreased susceptibility to transformation by electroporation, and decreased expression of the polysaccharide capsule. Altogether, these characteristics pointed towards a change in the structure of the cell wall.

Cui *et al.* (2000, 2003) provided experimental insight into a molecular basis for the cell wall alteration in the first clinical VISA isolate, Mu50. Several features of the clinical VISA strain Mu50 were observed that could decrease the ability of vancomycin to access its vital target at the cell membrane: (1) a decreased amount of cross-linked peptidoglycan, (2) an increased abundance of glutamate-containing muropeptides that could bind to vancomycin more avidly than amidated D-isoglutamine-containing muropeptides, and (3) a thicker cell wall. Since the terminal D-Ala is cleaved upon cross-linking during peptidoglycan biosynthesis, decreased cross-linking leads to an increased number of decoy D-Ala–D-Ala sites in the cell wall available for non-productive vancomycin binding. Combined with the observed thickening of the Mu50 cell wall it was proposed that these off-target binding sites sufficiently divert the amount of vancomycin available for binding to the lethal target, lipid II. However, after a number of VISA isolates were identified in the USA and available for comparative analysis, we collaborated with the late Harald Labischinski to use HPLC analysis of the peptidoglycan hydrolysates from several clinical VISA isolates. We observed that decreased cross-linking was a strain specific characteristic rather than a universal feature of vancomycin intermediate resistance (Boyle-Vavra *et al.*, 2001b). Indeed, we found that one VISA isolate from Illinois had a greater than normal proportion of highly cross-linked species in its peptidoglycan and had a normal abundance of the non-amidated Gln precursor. Nevertheless, we observed that the non-amidated Gln precursor abundance diminished in two of the four vancomycin-susceptible revertants of VISA isolates, an observation that supports the idea they could play a role in resistance in some isolates. From the collective characterization of subsequently identified VISA isolates, it became clear that a VISA phenotype could develop through a number of different pathways through step-wise selection, with the most consistent feature being a thickened cell wall (Cui *et al.*, 2000, 2003; Daum *et al.*, 1992; McAleese *et al.*, 2006).

The role of autolysins in susceptibility to vancomycin

Autolysins are enzymes involved in the cleavage of covalent linkages in peptidoglycan and include amidases (which target peptides) and glycosidases (which target the glycan chains). These enzymes play essential roles in peptidoglycan turnover and in regulating cell wall thickness by sloughing off the outermost layer of peptidoglycan as new wall material is added to the innermost layer at the cytoplasmic membrane. They also play roles in splitting the cross-wall between two daughter cells allowing for cell division. Autolytic activity is amplified in the presence of β-lactams and is responsible for the lysis of bacteria following β-lactam exposure. Although a *S. aureus* cell dies prior to β-lactam-induced lysis, it is clear that a lysed cell will ensure that there is no recovery from the experience.

Since vancomycin and β-lactams inhibit the same stage of peptidoglycan synthesis, and since autolytic activity is stimulated by β-lactams that leads to lysis, the role of autolysis in the mechanism of vancomycin intermediate resistance has been intensively studied. Autolysis assays of whole cells can be performed by exposure to a detergent such as Triton-X 100, since this liberates teichoic acids that shield peptidoglycan from endogenous autolysins. Using triton to induce autolysis in VISA strains, Sieradzki *et al.*, demonstrated that autolysis could be inhibited in VISA strains upon exposure to vancomycin (Sieradzki and Tomasz, 1997). However, VISA strains have decreased autolysis even in the absence of vancomycin exposure (Boyle-Vavra *et al.*, 2001a, 2003). It is important to note that decreased autolysis was not a feature of the clinical VISA strain Mu50, illustrating that decreased autolysis is not a requisite feature for the VISA phenotype.

Several mechanisms of decreased autolysis have been described among VISA strains. Some VISA strains have cell walls that are relatively resistant to autolysis despite having active autolytic enzymes (Boyle-Vavra *et al.*, 2001a). When the cell walls were isolated from clinical VISA isolates and used as the substrate in zymogram assays, the VISA cell wall was resistant to autolysis by extracts from autolysis-susceptible strains. In contrast, the autolytic activity of cell extracts from the same VISA strain was highly active on the autolytically susceptible cell walls of other *S. aureus* strains (Boyle-Vavra *et al.*, 2001a). These same isolates were shown to have increased cross-linking; however, as mentioned, other VISA isolates have decreased cross-linking in their cell walls. Since PBP4 is a carboxypeptidase that removes terminal D-Ala residues from peptidoglycan stem peptides during cross-linking, it was proposed that low or absent PBP4 activity could explain a high number of false D-Ala-D-Ala targets that are available to absorb vancomcyin. Indeed, the Tomasz and Archer laboratories have shown that such VISA isolates with low cross-linking have low PBP4 enzyme activity despite the presence of a wildtype *pbp4* gene (Finan *et al.*, 2001; Sieradzki and Tomasz, 2003). All VISA isolates have increased cell wall thickness (Cui *et al.*, 2003; Reipert *et al.*, 2003), which could also explain slow autolysis and non-specific absorption of vancomycin to the cell wall.

Genetic correlates of vancomycin-intermediate resistance in *S. aureus*

Since each VISA isolate acquires resistance by one or more point mutations, the path to resistance is different for each isolate. Nevertheless a few common themes have emerged in the analysis of gene sequences or gene expression in VISA isolates compared with susceptible strains (Alam *et al.*, 2014; Cui *et al.*, 2009, 2010; Gardete *et al.*, 2012; Mwangi *et al.*, 2007; Watanabe *et al.*, 2011). Indeed mutations in regulatory genes *vraTSR*, *graSR*, *walKR* (aka, *yycGF*, *vicKR*) that control cell wall metabolism have been associated with VISA strains. These will be discussed in more detail below. Surprisingly, mutations in *rpoB* encoding the β-subunit of RNA polymerase, which are normally associated with rifampin resistance, have also been frequently found in association with VISA and hVISA, sometimes without affecting rifampin susceptibility (Alam *et al.*, 2014; Cui *et al.*, 2010). The RpoB (A621E) mutation that is outside the rifampin resistance region affects the shape of the resistance population curve with respect to vancomycin and daptomycin (Cui *et al.*, 2010). This is intriguing since RNA polymerase is not a target of vancomycin or daptomycin. The RpoB A621E polymorphism was also associated with altered expression of the *dlt* operon that could increase D-alanylation of teichoic acid, thereby increasing the net positive surface charge. Although it was thought that *rpoB* mutations might create a permanent state of tolerance by slowing cell growth, this was disproven (Cui *et al.*, 2010).

VRSA strains harbouring Tn*1546* acquired from vancomycin-resistant enterococci

It took 30 years after introduction of vancomycin into clinical practice, before the first vancomycin-resistant *Enterococcus* (VRE) strain was recognized in 1988 (Perichon and Courvalin, 2009) and another 14 years before the first high-level vancomycin-resistant *S. aureus* clinical strain was identified in 2002 (Weigel *et al.*, 2003). *S. aureus* (VRSA) isolates that express high level vancomcyin resistance have acquired a transposon Tn*1546* from enterococci, which carries the *vanA* gene cluster (Kos *et al.*, 2012). The *S. aureus* version of Tn*1546* contains a *vanA* gene identical to that of enterococci and most of the genes carried by Tn*1546* found in VRE. However, there are mutations or deletions occurring in a few VRSA isolates at the extremities of the transposon (Kos *et al.*, 2012; Weigel *et al.*, 2003). The second *vanA*-containing isolate (VRS2) was identified in Pennsylvania in the same year as the first in 2002 (Tenover *et al.*, 2004). Unlike in the first VRSA isolate (VRS1), vancomycin resistance could not be detected in VRS2 using automated susceptibility panels. Moreover, the broth MIC of strain VRS2 was 16–64 μg/ml considerably lower than the MIC of vancomycin of the majority of VRSA strains identified (MIC 1024 μg/ml). Since 2002, about 13 additional *vanA* containing VRSA strains have been identified and confirmed in the USA (Limbago *et al.*, 2014; Rossi *et al.*, 2014). Fortunately these have not spread widely. In addition to vancomycin and β-lactams, all *vanA*-containing VRSA strains have been resistant to methicillin (and oxacillin) and multiple non-β-lactam antibiotic classes (including macrolides, aminoglycosides, fluoroquinolones, and tetracycline). Nevertheless, VRSA isolates have been susceptible to some agents such as linezolid, quinupristin–dalfopristin, trimethoprim–sulfamethoxazole, daptomycin and tigecycline.

The mechanism of *vanA* mediated vancomycin resistance, in both VRE and *S. aureus*, involves the synthesis of an alternative peptidoglycan precursor that terminates in D-Ala-D-lactate rather than D-Ala-D-Ala as reviewed by Courvalin (Courvalin, 2006). This substitution of D-lactate for D-Ala decreases the affinity of vancomycin for its target by 1000 fold, thus drastically increasing the MIC of vancomycin. In VanA mediated resistance, the biosynthesis of the D-Ala-D-lactate containing precursors involves three enzymes (i.e. VanH, A, and X) that are encoded on the Tn*1546* transposon. With the use of a D-specific pyruvate dehydrogenase (VanH), pyruvate is converted to D-lactate. The D-Ala-D-lactate ligase (VanA) is required for synthesis of D-Ala-D-lactate, which is subsequently added to the peptidoglycan precursor terminus. VanX is a peptidase required to search and destroy any native D-Ala-D-Ala dipeptides that would compete with D-Ala-D-lactate for incorporation into the precursor. VanY has a similar role. If not for these last two accessory enzymes, the abundance of D-Ala-D-Ala peptides produced by the native Ddl ligase would compete with D-Ala-D-lactate, wildtype precursors would be synthesized and the VRSA isolate would likely remain susceptible to vancomycin. Homologues of *vanA* (*vanB*, *vanD*, *vanE*, *vanF*, and *vanG*) have also been reported in vancomycin-resistant enterococci. However, only *vanA* gene clusters have been acquired by *S. aureus* to date.

As in enterococci, the expression of *vanHAXY* cluster is induced by vancomycin, which is facilitated by the VanR and VanS two-component regulatory system (Courvalin, 2006). VanRS is encoded on Tn*1546* immediately upstream of, and in the same orientation as, *vanHAXY*. The model for glycopeptide induction of the *van* operons in *S. aureus* parallels that proposed for enterococci. Briefly, the VanS gene product resides in the membrane and is presumed to bind extracellular glycopeptides. This induces dimerization and autophosphorylation of VanS with rapid subsequent transfer of a phosphorlyl group to VanR. Phosphorylated VanR in turn binds to the promoters in the *vanHAXY* and *vanRS* operons resulting in their induction (Perichon and Courvalin, 2009). In the absence of glycopeptides, VanS phosphatase activity removes the phosphate from VanR, thereby reversing binding to the promoter. In *S. aureus*, Qureshi *et al.*, demonstrated the native staphylococcal cell wall stress sensing system, mediated by VraTSR, is able to cross talk with the enterococcal VanRS two-component system (Qureshi *et al.*, 2014). Like VanRS, VraTSR is induced by glycopeptides and modulates induction of the acquired *vanHAXY* genes as shown in a VraTSR deletion strain of VRS1. Accordingly, VraTSR affects the vancomycin MIC as shown by the decreased MIC of vancomycin in a *vraTSR* mutant strain of VRS1.

One remaining question is what causes the

difference in vancomycin resistance phenotypes between low-level resistant (LLR) VRSA strains (MICs of 16–64 μg/ml) and high-level resistant (HLR)-VRSA strains (MIC 1024 μg/ml), despite having identical *vanHAXYZ* genes. Strains VRS2 and VRS3 are considered to be LLR-VRSA compared with most VRSA isolates, which are considered to be high-level resistant (HLR)-VRSA. Despite different MICs of vancomycin, HLR- and LLR-VRSA strains have the same percentage of alternative depsipeptide (D-Alal-D-lactate) containing precursors (Perichon and Courvalin, 2004). Perichon and Courvalin proposed two reasons to explain the difference between HLR and LLR phenotypes. First, there is a slower induction of the *vanA* operon (evident as a long lag phase in the presence of vancomycin) in the LLR-VRSA strains compared with HLR-VRSA strains. Second, the enterococcal plasmids in the LLR-VRSA strains are less stable in the absence of vancomycin selection (Perichon and Courvalin, 2004, 2009). However, these explanations seem insufficient since the MIC measurement is evaluated after 24 h, long after the lag phase is over. Moreover, plasmid DNA maintenance is stable in VRS2 during an MIC experiment, which requires overnight growth in vancomycin (S. Boyle-Vavra, unpublished data). One possibility for the differences between HLR and LLR phenotypes are the genetic differences between the strains in the Tn*1546* genetic clusters, which includes an insertion sequence in the LLR-VRSA strains in the intergenic region between *vanS* and *vanH* and in *orf1* (Perichon and Courvalin, 2009). Another explanation could be the effect of native *S. aureus* genes from different genetic backgrounds on the resistance phenotype. This possibility is supported by experiments that have shown the influence of the native VraTSR cell wall stress response on the MIC of vancomycin in HLR-VRSA strain VRS1. When the *vraTSR* operon was deleted from VRS1, the MIC of vancomycin decreased to that of a LLR-VRSA strain such as VRS2 and it lowered expression the *vanA* operon as well as the *vraSR* regulon (Qureshi *et al.*, 2014). Whether there is differential expression of *vraTSR* expression among LLR- and HLR-VRSA strains has not been studied. Additional studies are needed to more fully explain this phenomenon.

As determined by whole genome sequence comparisons, it is clear that each VRSA isolate arose independently by acquiring Tn*1546* in a separate MRSA isolate rather than by dissemination of a single VRSA clone (Kos *et al.*, 2012). It is interesting to note that although *Enterococcus faecium* is more prevalent among clinical VRE, it is *E. faecalis* rather than *E. faecium* that was associated with acquisition of *vanA* by VRSA. Many of the patients who had a VRSA infection harboured a colonizing vancomycin-resistant *E. faecalis* isolate at the time their VRSA strain was isolated. Occasionally, VRE and *S. aureus* strains were isolated from the same body site in the same person, suggesting the *vanA* gene cluster transferred from the VRE isolate in *situ* during co-habitation in a human microenvironment (Chang *et al.*, 2003). Genetic evidence supports this idea in some, but not all, cases. The best evidence to date is that in several VRSA case patients, the plasmid carrying the Tn*1546* transposon in the VRSA isolate was the same as that in a co-resident vancomycin-resistant *E. faecalis* isolate.

The genetic analysis of plasmid DNA in VRSA strains reveals there have been several routes to acquisition and maintenance of a VRSA plasmid. In one route, the Tn*1546*-containing plasmid that was transferred from the co-resident *E. faecalis* isolate to *S. aureus* adapted itself for replication in *S. aureus*. In another route, the plasmid that harboured Tn*1546* in strains VRS1, VRS8 and VRS9 was a staphylococcal conjugal plasmid (i.e. pSK41). This suggests that the original enterococcal plasmid that carried the *vanA* gene cluster into *S. aureus* served as a suicide vector prior to transposition of Tn*1546* onto a resident *S. aureus* plasmid. In some VRSA strains, there were truncated versions of pSK41 present in VRSA (Kos *et al.*, 2012). In addition, a cointegrate was found in a VRSA strain consisting of several enterococcal plasmids (Kos *et al.*, 2012). In summary, it is clear that *S. aureus* is able to acquire and maintain expression of the *vanA* gene cluster from VRE by using different strategies that involve maintenance of Tn*1546* on a plasmid. To date, none of the VRSA isolates have maintained the *vanA* gene cluster on the chromosome. Moreover, no *vanB* containing VRSA isolates have been identified.

Despite the variety of plasmid constructs that harbour a *vanA* gene cluster in VRSA, there are consistent genetic correlates among the VRSA. First, the Inc18 broad host range conjugal plasmid was present in several VRSA isolates or in VRE that co-resided with VRSA isolates. Second, the conjugal staphylococcal plasmid pSK41 was present in a number of VRSA isolates. Third, a majority of *S. aureus* had the CC5 genetic background.

Filter mating experiments have succeeded in transferring the *vanA* containing plasmids from VRSA to vancomycin-susceptible *S. aureus* (VSSA) *in vitro*, demonstrating the potential of the plasmid to disseminate among isolates of *S. aureus* that could infect and colonize humans (Severin *et al.*, 2004). Importantly, VRSA isolates arising *in vivo* have been uncommon. This is consistent with our studies where it has been difficult to transfer the *vanA*-containing plasmids from VRSA into an MSSA strain that is defective in *S. aureus* restriction systems and otherwise readily capable of accepting foreign DNA (S. Boyle-Vavra, unpublished data). Thus, the difficulty we found in accomplishing such transfer *in vitro* is probably not due to restriction alone.

If production of VRSA isolates is difficult *in vitro*, then what led to the emergence of the 13 VRSA isolates in the USA? Curiously, there has been a geographic bias to emergence of VRSA in the USA. Eight of the 13 VRSA clinical isolates from the USA have been identified from patients in Michigan (Zhu *et al.*, 2008). Moreover, Inc18 plasmids, a family of broad host range plasmids that often carry multiple antibiotic resistance determinants between enterococci, staphylococci, lactobacilli and streptococci, were shown to be five times more common in Michigan VRE compared with VRE from other states (Zhu *et al.*, 2013). Additionally, although Inc18 plasmids are rarely found in *S. aureus* natural isolates, 8 of 13 VRSA isolates have been found to carry Inc18 plasmid DNA. Thus, it is thought that Inc18 plasmids in the enterococcal donor may be a factor that can facilitate transfer of the Tn*1546* transposon to *S. aureus* in a polymicrobial infection. Moreover, Zhu *et al.* have shown *in vitro* that *S. aureus* isolates harbouring the staphylococcal plasmid pSK41 are much more likely to be competent recipients of conjugation by an Inc18 enterococcal plasmid, pAM378 (Zhu *et al.*, 2013). This may explain why pSK41 and Inc18 plasmids or remnants were present in many VRSA isolates (five of the VRSA isolates carry pSK41 staphylococcal plasmids). Altogether, these data suggest that the Inc18 conjugal plasmids in donor VRE and pSK41 plasmids in *S. aureus* can facilitate transfer of vancomycin resistance to *S. aureus*. Hence, the preponderance of VRSA isolated in Michigan patients may be explained by a higher prevalence of VRE with Inc18 plasmids. Maintaining a low prevalence of Inc18 containing VRE is therefore desirable.

It is also possible that strain-specific characteristics restrict the acquisition of Tn*1546*. In support of this, an overwhelming majority of clinical VRSA isolates belong to clonal cluster 5, a healthcare-associated genetic background prevalent in the USA. Although a few VRSA isolates have been described from community-associated backgrounds, ST8 (USA1100) and clonal complex 30 (Limbago *et al.*, 2014). Since most VRSA isolates were obtained following a recent course of vancomycin therapy, the infections are considered healthcare-associated according to CDC criteria, thus have not been easily transmitted to healthy hosts. Accordingly, none of the family contacts of VRSA infected patients have been shown to carry a VRSA isolate.

The rarity of VRSA isolates could be due to enforcement of strict contact precautions in healthcare settings when a VRSA-infected patient was recognized. Alternatively, the acquisition of the heterogeneous genetic transposable element may confer a fitness cost that limits dissemination. Competition experiments for growth performed *in vitro* suggest the latter explanation contributes to limited dissemination (Perichon and Courvalin, 2009). Nevertheless, extreme caution should be exercised to prevent transmission of these highly resistant isolates.

Role of a three component regulatory system VraTSR in resistance to cell wall antibiotics, vancomycin and methicillin

Bacteria often respond to environmental stimuli using two-component regulatory (TCR) systems that are comprised of a histidine kinase (HK) and a response regulator (RR). TCR systems allow the bacteria to conserve energy by turning on genes only when they are required.

S. aureus responds to antibiotics from a variety of chemical classes that damage the cell wall by using a TCR system consisting of VraS (a HK) and VraR (a RR). This system is unique among TCR systems since it is required for MRSA to express full resistance to methicillin (Boyle-Vavra *et al.*, 2006; Kuroda *et al.*, 2003; Zhu *et al.*, 2013). A cartoon to illustrate how VraTSR senses and responds to β-lactams

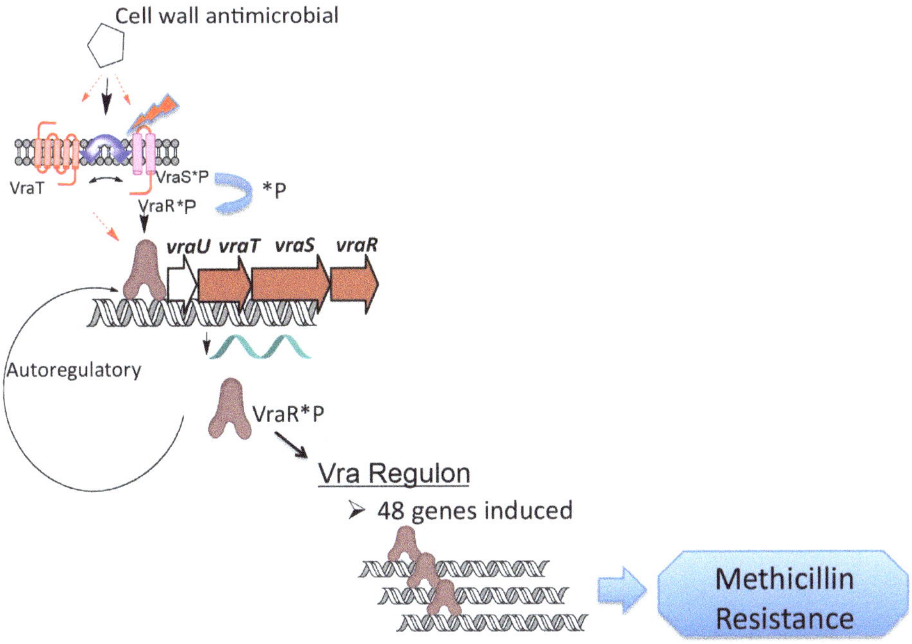

Figure 11.1 Model for signal transduction between VraTSR and antibiotic resistance. The vra operon consists of 4 genes: *vraU*, *vraT*, *vraS* and *vraR*. The three genes in the vra operon that are required for methicillin resistance are indicated by brown arrows (*vraS*, *vraR*, and *vraT*) above the DNA graphic. The *vraU* ORF is white to indicate its non-involvement in the *vra* regulon and methicillin resistance. Cell wall damage elicited by β-lactams or glycopeptides imparts a signal that leads to phosphorylation of VraS (VraS*P) which in turn phosphorylates VraR (VraR*P). VraR*P can transcriptionally activate a cell wall stress response regulon of about 40–50 genes. Altogether the increased expression of these genes increases resistance to methicillin and other cell wall antimicrobials. The exact mechanism by which VraT acts is unknown. Since VraT is a probable membrane protein we speculate that an interaction between VraT and VraS modulates the cell wall stress response, as depicted by intervening arcs between VraT and VraS in the membrane and cytoplasm. The signal for the stimulus is unknown but it could be a by-product of cell wall stress (lightning bolt) elicited by antimicrobials. The blue wavy line beneath the *vra* operon indicates mRNA expression leading to production of VraT, S and R (and U). The wide circular arc indicates that VraR autoinduces its own operon. This figure was adapted from Boyle-Vavra *et al*. (2013).

to facilitate methicillin resistance is shown in Fig. 11.1. *vraS* and *vraR* are encoded within a four gene operon that is present in all *S. aureus* isolates (Yin *et al.*, 2006) and consists of *vraU* (previously *orf1*), *vraT* (previously *yvqF*), *vraS* and *vraR*. We recently proposed to change the name of *orf1* to *vraU* and *yvqF* to *vraT*, to reflect the fact they are all colocalized on the *vra* operon.

VraS belongs to the intramembrane HK family, since it lacks an extracellular sensing domain (Mascher, 2014). This topology suggests the signal that activates VraS does not derive from the external environment. Instead the signal likely originates from either within the membrane, another membrane protein, or within the cytoplasm. It is also possible that VraS receives a signal through VraT, which, like VraS, is a putative membrane protein with four to five predicted transmembrane loops that are available for receiving a cell wall damage signal and interacting with VraS. VraU is the first product encoded in the *vra* operon. Unlike VraT, VraS and VraR, VraU is not required for methicillin and vancomycin resistance, nor is it required for induction of the *vra* regulon in response to oxacillin and vancomycin (Boyle-Vavra *et al.*, 2013; McCallum *et al.*, 2011).

Loss of function mutations in either VraR or VraS have been shown to lead to decreased resistance to vancomycin or β-lactams (Boyle-Vavra *et al.*, 2006, 2013; Cui *et al.*, 2009; Gardete *et al.*, 2012; Kuroda *et al.*, 2003; McCallum *et al.*, 2011). Conversely, polymorphisms in *vraS*, *vra*R or *vraT* that could confer an increase in vancomycin resistance have been reported in clinical and *in vitro* derived

VISA isolates (Hafer *et al.*, 2012; Katayama *et al.*, 2009; Kato *et al.*, 2010; Mwangi *et al.*, 2007; Shoji *et al.*, 2011; Yoo *et al.*, 2013). However, multiple polymorphisms in *vra* operon genes among VISA isolates compared with related susceptible isolates, and administration of complex antibiotic regimens in case patients, has made it difficult to tease out the exact role of a single VraSR polymorphism in resistance. Several studies that have examined serial clinical isolates, or pairs of susceptible and resistant engineered strains, suggest that a *vraS* mutation or polymorphism can be an important initial step towards a VISA phenotype (Cui *et al.*, 2009; Katayama *et al.*, 2009), but that a second step mutation facilitates the transition to intermediate resistance. For example, in a study of serial isolates from the same patient during a complicated treatment course with several antibiotics, a premature translational stop mutation in *vraS* was present in the second isolate. Also, a *vraS* mutation combined with a mutation in another TCR system, *graSR*, was sufficient to produce a VISA phenotype by increasing the vancomycin MIC from 1 to 4 μg/ml (Mwangi *et al.*, 2007). Moreover, when the VISA isolate from that series reverted to a vancomycin-susceptible phenotype, the sole genetic event detected in the revertant was the restoration of the *vraS* ORF to wildtype. In another study, Berscheid *et al.* (2014) demonstrated that two amino acid changes in VraS, L114S and D242G, led to a VISA phenotype in the laboratory strain 8325, although whether one mutation alone was sufficient for the VISA phenotype was not addressed. Accordingly, the recombinant VISA isolate had increased transcription of the cell wall stress stimulon, a thickened cell wall, a decreased growth rate, decreased autolytic activity and increased resistance to lysostaphin-induced lysis, all phenotypes associated with VISA. Thus, it is clear that the VISA phenotype can be influenced by VraS polymorphisms.

The role of VraT in the regulation of the *vraSR* operon has not been determined. VISA strains containing polymorphisms in *vraT* have been described, suggesting that VraT may be important in cells expressing the VISA phenotype (Gardete *et al.*, 2012; Hafer *et al.*, 2012; Kato *et al.*, 2010; Mwangi *et al.*, 2007; Yoo *et al.*, 2013). Two independent studies have demonstrated that VraT is a positive activator of the VraSR operon and its regulon in response to cell wall antibiotics (Boyle-Vavra *et al.*, 2013; McCallum *et al.*, 2011). In contrast, Gardete *et al.* (2012) provided evidence in support of VraT being a negative regulator of the VraSR mediated cell wall stress response. This latter idea is consistent with the fact that the VraT orthologue in other Firmicute bacterial species (LiaF, YvqF) is a negative regulator of the cognate two-component system LiaSR (Boyle-Vavra *et al.*, 2013; Jordan *et al.*, 2006; Mascher, 2014). However, there are many differences in sequence between the *Staphylococcus* and *Bacillus* orthologues and additional genes present in the *Bacillus lia* operon are not present in the *S. aureus vraSR* operon. Such differences allow for the possibility of divergent regulatory strategies between the two genera.

Boyle-Vavra *et al.* and McCallum *et al.* have shown that VraT has a positive role in the *vra* operon/regulon (Boyle-Vavra *et al.*, 2013; McCallum *et al.*, 2011). The approaches used in these studies were to produce a mutation in the chromosomal copy of *vraT* that eliminated VraT production but did not decrease expression of the adjacent downstream genes *vraS* and *vraR*. Whereas McCallum *et al.* inserted a premature signal for translational termination in *vraT*, Boyle-Vavra *et al.* replaced the *vraT* ORF with an antibiotic resistance cassette containing a promoter for maintaining transcription of the downstream genes encoding VraS and VraR. The effect of the *vraT* mutation in both studies was to decrease oxacillin and vancomycin resistance and transcription of genes in the *vra* regulon. This was remarkable since the *vraS* and *vraR* genes were both transcribed and in one case a VraR protein was shown to be produced using a Western blot probed with a VraR antibody (Boyle-Vavra *et al.*, 2013).

In the contrasting study, a *vraT* polymorphism (cysteine to tyrosine in residue 220) in the putative cytoplasmic domain, in a USA300 clinical VISA strain, was associated with increased *vraSR* operon expression, an observation suggesting the wildtype *vraT* gene had a negative influence on *vraSR* expression (Gardete *et al.*, 2012). Furthermore, when a wildtype *vraT* gene was overexpressed from a multicopy plasmid in the *vraT* mutant VISA strain, the expression of *vraSR* and the genes regulated by VraSR decreased; this decrease was accompanied by attenuation of the vancomycin resistance phenotype (Gardete *et al.*, 2012). Conversely, the strain expressing both the wildtype *vraT* chromosomal allele (in single copy) and a plasmid-encoded polymorphic allele from the VISA strain was vancomycin-susceptible and expressed a baseline level of *vraS*. Although these data support the idea that VraT is a negative regulator of *vraSR*, an alternative interpretation is that the polymorphism found in the single *vraT* gene encoded in the VISA strain actually conferred a gain of function that was recessive when the wildtype protein was co-expressed in the same cell. To rule this out, reciprocal gene replacement studies should be carried out so that only a single allele is expressed. Thus, for now, it is still unresolved whether VraT is an activator or repressor of VraTSR function. The differences between these studies involved different strains and different approaches to study the problem. Given the conflicting data surrounding the role of VraT in the VraSR cell wall stress response, additional studies are needed to better inform our understanding of the molecular underpinnings of this three component regulatory system.

WalKR

WalKR encodes another TCR system that has been associated with vancomycin intermediate resistance in *S. aureus* (Alam *et al*., 2014; Gardete *et al*., 2012; Hafer *et al*., 2012; Howden *et al*., 2011; Jansen *et al*., 2007; Mwangi *et al*., 2007; Shoji *et al*., 2011). WalKR (aka, YycFG or VicKR) is unique among TCR systems in that it is essential for viability. Dubrac *et al.* provided a convincing series of experiments that strongly suggested that WalKR is a master gatekeeper of cell wall metabolism. In particular, depletion of WalKR decreased expression of major lytic enzymes and produced a more autolytically resistant *S. aureus*, which in turn leads to increased cross-linking and longer glycan chains in peptidoglycan (Dubrac *et al*., 2007). To date WalKR has been shown to influence 13 genes involved in cell wall biosynthesis; however, it also controls many other genes in the bacterium including virulence genes in the SaeSR regulon (Delaune *et al*., 2012). Since vancomycin targets the cell wall, it might be expected that WalKR polymorphism would be associated with vancomycin tolerance or resistance.

In one survey, WalKR polymorphisms occurred in 61% of clinical VISA isolates (Shoji *et al*., 2011) and in a genome wide association study of 26 VISA isolates *walKR* genes were among those in which a rare specific mutation could predict the VISA phenotype (Alam *et al*., 2014). The variety of polymorphisms in *walKR* ORFs that are associated with the VISA phenotype was recently reviewed by Howden *et al.* (2014). It is remarkable that the VISA or hVISA phenotype is associated with a wide variety of polymorphisms that occur in all functional domains of WalK or WalR. Although this diversity makes the role of these polymorphisms in a cell expressing the VISA phenotype less clear, an association between polymorphisms in WalKR and autolysis-resistance, increased cell wall thickness and increased D-Ala-D-Ala non-specific targets in VISA provides a believable rationale for their role (McEvoy *et al*., 2013; Howden *et al*., 2011). It is possible that the WalKR mutations explain the common (but not universal) autolysis-resistance phenotype and cell wall thickening among VISA isolates (Boyle-Vavra *et al*., 2003; Cui *et al*., 2003; Sieradzki and Tomasz, 1997).

A few VISA-associated polymorphisms in WalKR are worth highlighting since they have been confirmed to influence vancomycin resistance. By using reciprocal allelic exchange studies in clinically derived VISA and VSSA strain pairs, the WalR K208R polymorphism was sufficient to convert the VSSA isolate to the VISA phenotype identical to that of the donor, as judged by population analysis (Howden *et al*., 2014). The 208th residue occurs in close proximity to the α3–β5 DNA binding loop and therefore may affect the function of WalR binding to DNA.

With respect to WalK, allelic replacement of the G with a D at residue 223 converted the parental VSSA strain to a VISA phenotype (MIC of 4 μg/ml as determined by population analysis). The G223D polymorphism is in a highly conserved amino acid required for a sharp reverse turn between the α1 helix and the connector region of the HAMP domain and may affect the functional ability of WalK (Hulko *et al*., 2006). However, a higher, more homogeneous VISA phenotype was documented when the G223D WalK polymorphism was transferred into a GraS T136I hVISA variant (Howden *et al*., 2011).

To complicate matters, there are inconsistent findings between laboratories whether increased or decreased WalKR expression can produce the VISA phenotype. Jansen *et al.* found increased expression of *walKR* in a VISA isolate due to insertion of IS256 in the region upstream of the *walKR* promoter (Jansen *et al.*, 2007), whereas McEvoy *et al.* found that an IS256 insertion in a VISA strain in the *walKR* promoter region decreased *walKR* expression (McEvoy *et al.*, 2013). However, the strains used by these two groups were derived *in vitro* and perhaps other co-residing mutations influence the effect of the expression.

In support of the idea that decreased *walR* expression is not associated with increased vancomycin resistance, constitutive activation of WalR was achieved by replacement of the phosphorylated aspartate (amino acid 55) with glutamate within the receiver domain (Dubrac *et al.*, 2007). This WalR constitutive mutant was not vancomycin resistant (Delaune *et al.*, 2012). In contrast, many of the characteristics of a WalR-depleted strain were similar to VISA strains, such as decreased autolysis phenotype (Dubrac *et al.*, 2007).

Telavancin

Telavancin was the first of several new lipoglycopeptide antibiotics approved for treatment of complicated SSTI caused by susceptible *S. aureus* (and certain other Gram-positive species). It is also approved by the FDA for hospital-acquired and ventilator-associated bacterial pneumonia (Mendes *et al.*, 2015). The FDA licensed clinical formulation is called VIBATIV™. (Two other newly licensed lipoglycopeptides, oritivancin and dalbavancin, will also be discussed below.) One advantage of telavancin over vancomycin is a long half-life (7–9 h) and a 4–6 h post-antibiotic effect allowing for once-daily intravenous dosing. Also, telavancin has greater activity than vancomycin against MSSA MRSA, hVISA, VISA and VRSA (Draghi *et al.*, 2008; Hegde *et al.*, 2009; Hegde and Janc, 2014; Hegde *et al.*, 2008; Hegde *et al.*, 2010; Leuthner *et al.*, 2006). Nevertheless, compared with VISA or VSSA, the MIC of telavancin is higher in VRSA strains containing the *vanA* gene cluster that express alternate peptidoglycan precursors terminating in D-Ala-D-lactate.

It should be noted that, despite its desirable pharmacokinetic properties, a black box warning on the VIBATIV™ label cautions against its use during pregnancy due to adverse developmental outcomes documented for three animal species. Also of concern is the increased mortality observed in patients with pre-existing renal impairment compared to patients who used vancomycin. Also, new onset or worsening renal impairment has occurred. For these reasons, physicians use this antibiotic after vancomycin treatment has failed.

Telavancin, as a lipoglycopeptide, is semisynthetically derived from vancomycin and contains the vancosamine sugar with an attached lipophilic decylaminoethyl side chain. This structure affords telavancin a dual faceted mechanism of action that renders it more effective than vancomycin. Telavancin behaves like glycopeptides by binding of the vancosamine sugar to the terminal D-Ala-D-Ala moiety of membrane anchored peptidoglycan lipid II precursors (Lunde *et al.*, 2009), thereby interfering with transglycosylation and biosynthesis of nascent peptidoglycan. Additionally, telavancin rapidly depolarizes the bacterial membrane potential in a time and concentration dependent fashion and increases membrane permeability, leading to cell death. Depolarization is dependent on the presence of lipid II and the lipophilic side chain of telavancin has been hypothesized to enhance the affinity for lipid II by increasing the membrane anchoring properties (Lunde *et al.*, 2009). The evidence for dependence on lipid II for depolarization came from studies that depleted lipid II in live bacteria. This was accomplished by either pretreatment with antibiotics that inhibit lipid II formation, by shifting the growth temperature or by growth in chemically defined medium to retard growth. All approaches to depleting lipid II antagonized telavancin's effect on depolarization of membrane potential. Further evidence is from the knowledge that pretreatment with the tripeptide, dKAA, that mimics cell wall precursors, antagonizes telavancin-dependent depolarization and antimicrobial susceptibility (Lunde *et al.*, 2009).

A transcriptional analysis of the response to telavancin by *S. aureus* provided further support and insight into telavancin's dual mechanism of action. Telavancin induces genes in the cell wall stress stimulon, including *vraS* and *vraR*, and genes characteristic of membrane depolarization (*lrgA*, *lrgB*, lysine operon genes (*dap*), and *vraD* and *vraE*) (Song *et al.*, 2012). The depolarization activity is achieved either at the higher concentrations in the clinically achievable level (Jafari Saraf and Wilson, 2011) or at later growth phases as detected *in vitro* by using transcriptional microarray analysis (Song *et al.*, 2012).

To date there have been no *S. aureus* clinical isolates reported with decreased susceptibility or resistance to telavancin in the literature. However, several *S. aureus* clinical isolates have been identified by the Public Health Laboratory at the Minnesota Department of Health as being non-susceptible to telavancin (MIC of 2 μg/ml by broth microdilution) (personal communication, Anita Glennen).

Few studies have been documented that evaluate telavancin non-susceptibility derived by *in vitro* passage. In one study, a stable telavancin non-susceptible (TelNS) isolate was generated *in vitro* after 43 stepwise passages of a telavancin susceptible (TelS) wildtype strain, SA248, in telavancin (Kosowska-Shick *et al.*, 2009). The MIC of the TelNS derivative of SA248 was 2 μg/ml, four times higher than the wildtype parent whose MIC was 0.5 μg/ml (Kosowska-Shick *et al.*, 2009). Such decreased susceptibility to telavancin was not observed in nine additional *S. aureus* isolates subjected to the identical passaging protocol. In a subsequent study, Song *et al.* (2013) characterized the non-susceptible derivative of SA248 (called Tlv (DS) MED1952) and found there was cross-non-susceptibility to vancomycin and daptomycin, noting an increase from 1.0 to 2.0 μg/ml in the MIC of daptomycin and an increase in the MIC of vancomycin from 0.5 to 2.0 μg/ml compared with the parent isolate. Also, the MIC of oxacillin increased from 12 μg/ml in the wildtype parent to 32 μg/ml in the TelNS isolate, demonstrating that an inverse relationship (see-saw effect) between oxacillin and telavancin susceptibility is not absolute. Compared with the TelS parent isolate, the TelNS derivative had thickened cell walls, decreased autolytic activity, and membrane fatty acid changes associated with increased membrane fluidity. A similar number of genes were up-regulated ($n = 330$) as down-regulated ($n = 311$) in the TelNS isolate. Many cell surface and virulence genes were down-regulated, a phenomenon that was also noted in VISA isolates (Gardete *et al.*, 2012). Notably, in the TelNS isolate there were several down-regulated genes that were previously considered signature genes induced in response to membrane damaging agents (Song *et al.*, 2012). These include an anti-holin protein encoded by *lrgB* and the lysine biosynthetic operon *dap*. Although it is a paradox that these genes were down-regulated in a TelNS isolate, the transcriptional response of the TelNS isolate was not studied under telavancin inducing conditions. Therefore, further studies are needed to understand transcriptional differences between TelS and TelNS isolates.

Oritivancin

Oritivancin is a semi-synthetic lipoglycopeptide and one of the newer therapies approved by the FDA for treatment of acute SSTIs (August 2014) caused by Gram-positive isolates. Its effectiveness covers a broad range of resistant *S. aureus* isolates such MRSA, VISA, VRSA, daptomycin NS isolates and VRE. In this respect, oritivancin offers an advantage over two other lipoglycopeptides, telavancin and dalbavancin, which are not effective against VRSA (Klinker and Borgert, 2015). As a result of such desirable properties as a long serum half-life (393 h), concentration-dependent bactericidal activity, and a three-pronged mechanism of action, it takes just a single i.v. dose of oritivancin to treat an acute SSTI in adults (Mitra *et al.*, 2015; Saravolatz and Stein, 2015). Moreover, due to slow tissue elimination, no dosing adjustments are needed in patients with renal or hepatic insufficiency (Saravolatz and Stein, 2015). In clinical trials of SSTI therapy, oritivancin has proven to be non-inferior to vancomycin but offers the advantage of requiring just a single dose for the entire treatment regimen.

Oritivancin is highly bactericidal and the MIC of oritivancin for 90% of strains tested is extremely low (0.06 μg/ml). Like vancomycin and other glycopeptides, oritivancin binds to D-Ala–D-Ala carboxyl termini in membrane anchored lipid II peptidoglycan precursors, thereby sterically interfering with the transglycosylation reaction required for incorporation of the precursors into newly synthesized peptidoglycan. In this regard, oritivancin is more effective than vancomycin against VRSA and VRE due to its enhanced ability to bind to D-Ala–D-lactate termini in lipid II precursors expressed in these strains. Such enhanced interaction with lipid II is facilitated by dimerization of oritivancin, made possible by the disaccharide at ring 4 and interaction with unique structural features of oritivancin compared with vancomycin: the 4-*epi*-vancosamine attached to the amino acid on ring 6 and a lipophilic 4-chlorobiphenyl methyl side chain on the disaccharide (Mitra *et al.*, 2015).

Oritivancin is distinguished from vancomycin by having multiple mechanisms of action. In addition to the steric inhibition of transglycosylation, oritivancin can bind pentaglycyl cross-bridges, which inhibits peptidoglycan transpeptidation and cross-linking (Mitra *et al.*, 2015). Oritivancin is also distinguishable from vancomycin by its ability to interact with the bacterial cell membrane, leading to depolarization, increased membrane permeability and cell death of both stationary and exponential growth phase bacteria (Mitra *et al.*, 2015). Membrane depolarization also provides effectiveness against daptomycinNS bacteria.

To date, no oritivancin resistance has been detected in clinical *S. aureus* isolates. This may well be because it has only been recently licensed and has not yet been used on a large enough scale to detect resistant isolates. The effectiveness of oritivancin has been shown to decrease in enterococci with an increasing proportion of D-Ala-D-lactate containing precursors produced by induction of *vanA* operon transcription (Mitra *et al.*, 2015). Since *vanA* expression is also induced by vancomycin in VRSA isolates, it is likely that oritivancin will be less effective against VRSA when used in combination with vancomycin. This hypothesis, however, remains to be examined.

In summary, oritivancin is an alternative to vancomycin in treating acute SSTI infection in adults and is effective even against vancomycin-resistant isolates. It is able to inhibit peptidoglycan synthesis by using a multipronged mechanism of action: by inhibiting transglycosylation and transpeptidation, and by inducing membrane depolarization. Unlike vancomycin, oritivancin requires only a single dose. Despite the convenience offered by a single dose, oritivancin requires intravenous administration and would not be the antibiotic of choice in ambulatory settings. Moreover, since oritivancin costs about 30x more than vancomycin, it is not likely to replace vancomycin until widespread vancomycin resistance among *S. aureus* clinical isolates is realized (Mitra *et al.*, 2015).

Dalbavancin

Dalbavancin is a semisynthetic lipoglycopeptide that was derived from teicoplanin and approved in the USA and Europe (2014 and 2015) for treatment of complicated SSTIs in adults caused by MSSA, MRSA and other Gram-positive bacteria (Klinker and Borgert, 2015). Dalbavancin has not been evaluated in children younger than 12 years of age but the kinetics in children 12–17 years of age appear to be similar to those found in adults.

Dalbavancin is bactericidal, has high potency and a long half-life leading to the desirable property of being dosed just twice for a full therapeutic course. However, dalbavancin is very expensive and must be given parenterally.

Notably, it is more potent *in vitro* than vancomycin against strains with a vancomycin MIC of ≤ 2 μg/ml. Although it has activity against VISA and hVISA, dalbavancin does not overcome the strong vancomycin resistance mechanism provided by the *vanA* gene cluster (Klinker and Borgert, 2015).

The mechanism of action is similar to glycopeptides by binding to the D-Ala-D-Ala terminus of lipid II, thereby interfering with cell wall polymerization leading to cell death. Like other lipopeptides, dalbavancin has a long lipophilic side chain which, in part, is responsible for improved activity by facilitating drug dimer formation, anchoring of the drug to the bacterial cell membrane, and stabilizing interactions with lipid II. It is bactericidal but unlike telavancin and oritivancin, dalbavancin does not destabilize the cell membrane (Klinker and Borgert, 2015).

Susceptibility of clinical *S. aureus* isolates to dalbavancin remains high. Among 2861 clinical isolates (both MSSA and MRSA) obtained from SSTIs from the US and Europe, the MIC_{90} value of dalbavancin was 0.06 μg/ml and only one had an MIC above the susceptibility breakpoint of 0.12 μg/ml (Mendes *et al.*, 2016). Moreover, dalbavancin was found to have MICs 4- to 8-fold greater than daptomycin (a lipopeptide) and 16-fold greater than vancomycin.

As a final technical note about the lipoglycopeptides, the CLSI recommends including polysorbate 80 into the testing media when evaluating susceptibility to this drug class to improve solubility and reduce adherence of the drugs to plastic (Mendes *et al.*, 2015). This should be considered when reviewing earlier data since this recommendation was not always in place.

Antibiotics that interfere with cell membrane homeostasis

Daptomycin

Daptomycin is a bactericidal cyclic lipopeptide produced by *Streptomyces roseosporus.* Daptomycin was approved by the FDA in 2003 for treatment of complicated SSTIs, and in 2006 for bacteraemia and right sided endocarditis (Boucher and Sakoulas, 2007). It has antimicrobial activity against Gram-positive bacteria including MRSA (Boucher and Sakoulas, 2007), VRSA, VISA and vancomycin-resistant *E. faecalis and E. faecium*. Now that daptomycin has been in widespread use for over a decade, failure to effectively treat *S. aureus* infections concomitant with the development of daptomycin non-susceptible (DapNS) isolates has been frequently documented (Enoch *et al.*, 2007; Fowler *et al.*, 2006; Hayden *et al.*, 2005; Jacobson *et al.*, 2009; Julian *et al.*, 2007; Kirby *et al.*, 2009; Murthy *et al.*, 2008; Skiest, 2006; Tran *et al.*, 2013; Yang *et al.*, 2009), especially in association with endocarditis. In addition, DapNS isolates have been obtained from patients with osteomyelitis and bacteraemia, especially when it occurs secondary to an indwelling catheter. By definition, DapNS isolates have an MIC of daptomycin > 2 µg/ml. Perhaps because they have been isolated from patients during daptomycin treatment failures, such DapNS isolates have been frequently referred to as 'resistant', despite the fact that they have MICs of daptomycin below the resistance threshold (Tran *et al.*, 2013).

To understand how strains can become resistant (or NS) to daptomycin, it is worthwhile considering how it acts to kill *S. aureus*. Daptomycin is a complex anionic lipopeptide that is bactericidal for Gram-positive bacteria without inducing lysis (Silverman *et al.*, 2003). To elicit bactericidal activity, daptomycin requires the presence of Ca^{2+}, although the function of Ca^{2+} in the activity of daptomycin has been debated (Straus and Hancock, 2006). According to recent biochemical data, Ca^{2+} interacts with daptomycin causing it to change conformation and oligomerize (Muraih *et al.*, 2011). This forms a cationic micellar structure that interacts with the membrane, followed by the likely dissociation of the micellar structure. Daptomycin, facilitated by interaction with Ca^{2+}, is then competent to interact with anionic phospholipid head groups of phosphatidylglycerol, thereby eliciting a curvature in the cell membrane (Straus and Hancock, 2006). This curvature is thought to be responsible for the observed gradual depolarization, permeabilization and leaking of small ions (such as K^+) that ultimately result in cell death (Silverman *et al.*, 2003). Also, the interaction of daptomycin with cell membranes results in relocalization of proteins involved in cell division and cell wall synthesis (Pogliano *et al.*, 2012). With this in mind, the prominent hypotheses to explain daptomycin resistance strategies evolved by *S. aureus* isolates encompass: the overall cell membrane charge, composition, fluidity and curvature of the membrane and the genetic correlates associated with these phenotypes.

Molecular characteristics of DapNS *S. aureus* isolates

Molecular analysis of DapNS isolates compared with isogenic parental daptomycin-susceptible (DapS) isolates have provided the basis for models to explain the mechanism of daptomycin non-susceptibility in *S. aureus.* Before reviewing the prevailing hypotheses, it is important to note that there does not seem to be a specific route to non-susceptibility for each DapS isolate; cultivation with daptomycin in separate instances can lead to resistance by a different route each time. Like VISA isolates, the prevailing data suggest that DapNS isolates usually (Bertsche *et al.*, 2011), but not always (Boyle-Vavra *et al.*, 2011; Yang *et al.*, 2010), have thicker cell walls than DapS counterparts that would decrease access of daptomycin to its target, the cell membrane. In addition, DapNS isolates usually have more positively charged cell membranes than DapS counterparts, which are thought to effectively repel daptomycin-Ca^{2+} micelles from binding. A more recent hypothesis suggests that negatively charged phosphatidylglycerol abundance in the membrane is key to the docking of daptomycin to the cell membrane, therefore decreased amounts phosphatidylglycerol are associated with non-susceptibility (Tran *et al.*, 2013). In contrast, an increased amount of lysinylated phosphatidylglycerol is associated with daptomycin-non-susceptibility since the positively charged lysine repels daptomycin.

Genetic correlates of daptomycin non-susceptibility

An early study that was performed to identify the daptomycin resistance mechanism examined a series of isolates obtained following *in vitro*, stepwise incubation of a DapS MRSA isolate in daptomycin (Friedman *et al.*, 2006). The isolate series became NS to daptomycin in a stepwise fashion. The isolates were then studied using comparative genomic hybridization. Stepwise polymorphisms were identified in four genes [*mprF*, *yycG* (also known as *walK* and *vicK*), *rpoB* and *rpoC*] in association with the development of daptomycin non-susceptibility.

We had the opportunity to compare the whole genome sequences of a clinical DapNS isolate (considered resistant) obtained after therapy with daptomycin and a preceding DapS isolate obtained from the same patient prior to daptomycin therapy (Boyle-Vavra *et al.*, 2011). The DapS and DapNS isolates were isogenic by pulsed field gel electrophoresis (USA800), belonged to multilocus sequence type 5 (ST5) and carried SCC*mec* type IV. By analysis of the whole genome sequences, the DapNS isolate contained two polymorphisms compared with the DapS parent isolate. One polymorphism was in *mprF* and the other was in *sph* (encoding β-haemolysin). *mprF* encodes lysyl phosphatidyl glycerol transferase, a bifunctional membrane protein, with 14 transmembrane segments (Ernst *et al.*, 2009), that influences the net positive charge on the staphylococcal cell surface by transferring positively charged lysines onto phosphatidyl glycerol. The MprF polymorphism (S337L) was mapped to the tenth transmembrane segment where several other DapNS polymorphisms have also been mapped (Boyle-Vavra *et al.*, 2011). The second change in the genome was the excision of the β-haemolysin-converting phage, which restored the haemolysin ORF, *sph*. The role of this phage excision in the daptomycin resistance phenotype is unclear. Notably, no mutations were found in *rpoB*, *rpoC* or *yycG* in this clinical isolate. Thus, the whole genome sequence analysis of this isogenic pair of isolates from the same patient pointed to an MprF polymorphism as the relevant genetic change in a clinical DapNS strain and was likely the sole factor that conferred daptomycin-non-susceptibility. This idea was supported by other studies that selectively analysed *mprF*, *rpoC*, *rpoB* and *yycG* sequences in clinical DapNS isolates, although other mutations in the genomes were not ruled out by whole genome sequence analysis (Chambers *et al.*, 2009; Julian *et al.*, 2007; Murthy *et al.*, 2008).

In another study, seven of twenty-nine ST5, SCC*mec* type IV-containing clinical isolates were obtained from a single patient during therapy for an endocardial abscess (secondary to a pacemaker infection). Julian *et al.* found an MprF polymorphism (I420N) in several of the DapNS isolates. The *rpoB*, *rpoC*, and *yycG* sequences in the DapNS isolates were all identical to those of the DapS parent (Julian *et al.*, 2007). Curiously, the MprF polymorphism was not universally present in all DapNS isolates from this patient. In another study, Murthy *et al.* selectively sequenced these same four genes in clinical USA300 MRSA DapNS isolate and found a T345A polymorphism in MprF (Murthy *et al.*, 2008). It is possible that the *yycFG*, *rpoC* and *rpoB* mutations seen in Friedlander's *in vitro* derived isolates are less relevant to development of daptomycin-non-susceptibility during daptomycin therapy during infection where host selective pressures and serum binding proteins are present.

Numerous genetic comparisons of DapS and DapNS *S. aureus* isolates have now been performed using genetically related pairs of DapS and DapNS isolates obtained from the same patient. Most DapNS *S. aureus* isolates have either *mprF* polymorphisms and/or increased expression of *mprF*. These data strongly suggest that *mprF* has a role in resistance to daptomycin. However, a few DapNS isolates have been shown to have enhanced wall teichoic acid synthesis concordant with significant upregulation of *tag* genes, responsible for teichoic acid synthesis, and *dlt* operon genes responsible for addition of positively charged D-Ala to wall teichoic acid (Bertsche *et al.*, 2011).

The role of MprF and cell surface charge in the DapNS phenotype

As mentioned, MprF is a membrane protein with 14 transmembrane segments that influences the net positive charge on the staphylococcal cell surface. The carboxyl terminus contains a lysynyl transferase (also referred to as a synthase) that catalyses the addition of positively charged lysine (using lysyl tRNA) to the head groups of phosphatidylglycerol at the cytoplasmic face of the membrane (Ernst *et al.*, 2009). The MprF amino terminus has a domain responsible for translocating the lysinylated phosphatidylglycerol across the cytoplasmic membrane ('flippase' activity) ultimately decorating the outer membrane with the positively charged lysinylated moieties (Ernst *et al.*, 2009; Tran *et al.*, 2013). Thus, by the concerted action of lysinylation and translocation of the lysine-charged moiety across the cytoplasmic membrane, MprF increases the cationic character of the bacterial envelope (Tran *et al.*, 2013). For this reason it was proposed that the

MprF polymorphisms in DapNS isolates enhance the enzyme's ability to increase the positive cell surface charge, which would in turn repel daptomycin (which is cationic when complexed with Ca^{++} ions) and prevent cell death. The prevailing evidence supports this hypothesis, but other hypotheses that involve cell membrane fluidity have emerged that will also be discussed.

DapNS isolates that have polymorphisms in *mprF*, or that overexpress the gene, have an enhanced cationic surface charge and exhibit reduced daptomycin binding compared with DapS counterparts. The best evidence that MprF polymorphisms play a role in daptomycin resistance derives from genetic cross-complementation with 'resistant' *mprF* variants in *mprF* knockout or deletion strains that can elevate the MICs of daptomycin to NS (Mehta *et al.*, 2012; Yang *et al.*, 2013). Interestingly, most of the 'hotspot' polymorphisms associated with DapNS isolates have occurred in the ninth and tenth transmembrane helix of this membrane-bound protein and they have rarely been found in the first eight transmembrane helices (Boyle-Vavra *et al.*, 2011; Tran *et al.*, 2013). A commonly encountered DapNS-associated polymorphism occurs in the eighth transmembrane segment. Since the flippase and lysyl transferase domains overlap in the ninth and tenth membrane helices, it is not known with certainty the mechanism by which the polymorphism in these locations affects the activity of the enzyme at the molecular level. Crystal structure information with wildtype and variant proteins would advance this understanding and possibly facilitate design of inhibitors of MprF that could potentiate daptomycin activity.

Also shown to be associated with daptomycin-non-susceptibility in *S. aureus* was overexpression of the *dlt* operon, which encodes an enzyme that transfers D-Ala to wall teichoic acid. This in turn increases the cell membrane cationic charge and decreases binding to daptomycin. This phenomenon was associated with *in vitro*-derived DapNS by passage in daptomycin.

It is worth noting that although *mprF* homologues exist in other Gram-positive Firmicute species, MprF does not play a major role in daptomycin resistance in enterococci. In contrast to DapNS *S. aureus*, only a few DapNS *Enterococcus* sp. isolates had polymorphisms in *mprF* (Diaz *et al.*, 2014). In addition, the genetic changes in *E. feacalis* and *E. faecium* isolates with daptomycin resistance phenotypes have not been observed in *S. aureus.* These include genes encoding enzymes involved in phospholipid metabolism: cardiolipin synthase (*cls*) and glycerophosphoryl diester phosphodiesterase (*gdpD*).

Vancomycin and daptomycin cross-resistance

Development of vancomycin-intermediate resistance resulting from therapy with vancomycin can sometimes confer cross-non-susceptibility to daptomycin, which complicates the ability to discern mechanisms of daptomycin resistance (Cui *et al.*, 2006; Patel *et al.*, 2006; Sakoulas *et al.*, 2006). Conversely, step-wise incubation in increasing concentrations of daptomycin can increase the MICs of both daptomycin and vancomycin (Camargo *et al.*, 2008; Mishra *et al.*, 2009). Since daptomycin is often used in the therapy of MRSA infection following treatment failure with vancomycin, a better understanding of the mechanism of cross-resistance between daptomycin and vancomycin is needed.

Role of cationic peptides and cross-resistance to daptomycin

Many of the host defence peptides present in platelets and neutrophils are cationic peptides with antibacterial activity. To investigate the possibility that in the absence of daptomycin therapy, persistent exposure of *S. aureus* isolates to host defences could lead to the initial steps of daptomycin non-susceptibility, Mishra *et al.* recently examined 47 unique MRSA isolates obtained from bacteraemic patients in clinical trials that were testing thrombin-induced platelet microbicidal protein (tPMP) and human neutrophil defensin-1 (hNP-1). Isolates from patients were tested for susceptibility to host peptides and daptomycin. Despite the fact that patients had not received daptomycin, slightly higher MICs of daptomycin occurred in all isolates that resisted killing by platelet-derived host defence peptides, although the MIC of daptomycin was not above the susceptibility breakpoint. These data are concerning since they suggest that persistent infection and exposure to host defence peptides during infection can initiate the first step towards daptomycin-non-susceptibility.

Role of metabolism in daptomycin non-susceptibility

Recent studies have shown an effect of daptomycin resistance on the metabolism of MRSA isolates. Gaupp *et al.* recently demonstrated that six DapNS isolates had an altered metabolic state compared with isogenic susceptible parent isolates (Gaupp *et al.*, 2015). Specifically, daptomycin non-susceptibility coincides with a slight delay into the post-exponential phase with an accompanying delay in alkalization of the medium and acetate catabolism. These indicators were associated with decreased carbon flow through the tricarboxylic acid (TCA) cycle (Gaupp *et al.*, 2015). The altered metabolic state was substantiated by the finding that the TCA cycle enzyme, aconitase, had significantly lower activity in six DapNS isolates compared with their isogenic parent isolates. Moreover, the metabalomes of three DapS/DapNS pairs examined by NMR metabolomics demonstrated differences between the DapS and DapNS isolates in both the exponential and post-exponential growth phases, with a greater difference seen in the latter. Metabolic alterations were documented in central, amino acid, pyrimidine, and purine metabolism. Moreover, in DapNS strains, an increase was noted in carbon flow to pathways that provide biosynthetic intermediates involved in cell wall precursors (wall teichoic acid and peptidoglycan), thereby diverting carbon flow from other cellular pathways.

Bacteria that were challenged with daptomycin had decreased levels of the TCA cycle intermediate 2-oxoglutarate and the TCA derived amino acids Asp and Glu; however, significantly lower concentrations were found in the treated DapS parent strain compared with the DapNS strain. In the presence of daptomycin, TCA metabolism was minimally perturbed in the DapNS isolates compared with the DapS isolates. These data suggest that DapNS strains have pre-adapted their metabolism to minimize the metabolic perturbations induced upon exposure to daptomycin.

Overall lower metabolite concentrations were observed in the daptomycin treated, DapS strain compared with the DapNS strain (associated with glycolysis, pentose phosphate pathway, peptidoglycan pathway, and WTA synthesis). The exception was N-acetyl glucosamine, which was higher in the treated DapS strain compared with the DapNS derivative.

The finding that metabolism was altered in DapNS isolates led to an idea for potentiating daptomycin therapy by targeting aconitase since it was in lower abundance in DapNS isolates. Three of the six DapNS isolates became resensitized to daptomycin in the presence of the aconitase specific inhibitor, fluorocitrate. However, this result was not consistent in the three other DapNS isolates tested. This experiment demonstrated that it was not the TCA alterations that conferred daptomycin non-susceptibility; however, it does provide some insight into possible strategies for treatment of DapNS isolates by targeting metabolism.

Antibiotics that target the ribosome and interfere with protein biosynthesis

Macrolides, lincosamides, and streptogramins$_B$ (MLS$_B$)

The macrolides, lincoasamides and streptogramin antibiotics belong to three different chemical classes; however, each inhibits protein synthesis by targeting a similar region in the 50S ribosome. Erythromycin is the most commonly used macrolide in the treatment of *S. aureus* infection in humans, whereas clindamycin is the common lincosamide used in human therapy. Resistance is mediated by target site mutations or by modification of the target, and these can confer cross-resistance to all three classes in the MLS$_B$ group since they share overlapping target sites. In contrast, the *msrA* efflux pump is specific for macrolides and does not confer cross resistance to members of the lincosamide and streptogramin$_B$ classes.

Cross-resistance to all three MLS$_B$ classes in *S. aureus* can be mediated by one of several ribosomal methylases (encoded by an *erm* allele) that modify a nucleotide in the rRNA target by methylation. Resistance to macrolides can also be conferred by an efflux pump (encoded by *msrA*) that specifically extrudes macrolides from the cytoplasm. *S. aureus* isolates that are resistant to macrolides due to MsrA are susceptible to commonly used lincosamides, such as clindamycin, and the streptogramin$_B$ class of agents. In contrast, isolates in which erythromycin resistance is mediated by an *erm* gene have the potential to become cross-resistant to clindamycin and streptogramins$_B$ since the nucleotides modified by the Erm methylases sterically block the overlapping target sites of macrolides, lincosamides and streptogramin$_B$ agents. However, the erm resistance phenotype is controlled by the ability of the *erm* gene to be induced by the antibiotic class. Wildtype *erm* genes are constitutively silent but are induced in the presence of 14 and 15 membered macrolides such as erythromycin. In contrast, lincosamides and streptogramin$_B$ agents do not induce the wildtype *erm* genes; therefore, isolates that harbour an *erm* gene will test susceptible to clindamycin simply because *erm* gene expression is not induced by that drug class. However, cross-resistance to clindamycin and streptogramin$_B$ agents develops readily through point mutations in the 5′ non-coding region of the *erm* gene that disrupt the inducibility region. Such mutations permit constitutive *erm* gene expression, even in the absence of erythromycin, thereby conferring cross-resistance to clindamycin and streptogramin$_B$ as well as macrolides. Isolates that have an inducible *erm* gene are said to exhibit the inducible MLS$_B$ (iMLS$_B$) phenotype whereas isolates that express an *erm* gene constitutively exhibit the constitutive MLS$_B$ (cMLS$_B$) phenotype.

There are few choices for oral therapy of MRSA infection in ambulatory patients. One reason that erythromycin is a poor choice for treating MRSA infection is that most MRSA isolates are resistant to that antibiotic. In contrast, the lincosamide, clindamycin, is a frequently prescribed oral antibiotic for treating MRSA infection caused by isolates that are susceptible to the drug. However, among erythromycin-resistant MRSA that test susceptible to clindamycin, it is important to distinguish between isolates that have an inducible *erm* gene from those that do not, due to the potential for cross-resistance to clindamycin when mutations convert the inducible *erm* gene to one that is constitutively expressed. To this end, clinical microbiology laboratories often use a phenotypic assay called the D-zone test (hereafter called the D-test) on MRSA isolates that are erythromycin resistant and clindamycin susceptible by single agent testing. The D-test is a double disc diffusion assay that uses a paper disc containing erythromycin and a disc containing clindamycin. Each disc is placed at a fixed distance from each other on a Mueller Hinton agar plate that has been freshly inoculated over the entire area with the test strain. After an overnight incubation, growth of the bacteria is observed relative to the placement of the discs. A positive D-test is indicated when the clindamycin zone of inhibition is flattened on the side of the erythromycin disc, thereby producing a D shaped zone of inhibition. This is caused by induction of the *erm* gene by erythromycin that diffuses into the agar, allowing the cells in the clindamycin zone to resist clindamycin and grow. In contrast, the side of the clindamycin disc opposite to that of the erythromycin disc remains clear and round reflecting the susceptibility of the strain when exposed to clindamycin alone. The D-test + phenotype provides evidence that the test isolate contains an *erm* gene. Since use of clindamycin can select for isolates with constitutive *erm* gene expression during therapy, clindamycin would not be the treatment of choice for a D-test positive isolate if that isolate is susceptible to antibiotics from other classes. If however, the isolate maintains a clear round zone around the entire clindamycin disc, the isolate is D-test negative and clindamycin would be a suitable antibiotic for therapy. The D-zone test has been replaced by an alternative automated broth based assay for inducible resistance. Isolates are reported as positive for inducible MLS$_B$ resistance so that healthcare providers can use discretion in using clindamycin for MRSA therapy.

A genetic assay to test for the presence of the *erm* gene would also be a suitable, if not a better, strategy to identify clindamycin-susceptible isolates that harbour an *erm* gene. However, due to the ease of use and low cost, the D-test or a comparable broth double agent assay have gained traction as an accepted standard that is used widely in clinical microbiology laboratories. Whether clindamycin would be used for a clindamycin-susceptible isolate with an inducible *erm* gene is at the discretion of the healthcare provider. However, other antibiotics to which the isolate is susceptible would be a better choice to avoid treatment failure following mutation to constitutive *erm* expression.

The D-test became popular in the USA after the emergence of epidemic SSTIs caused by community-associated (CA)-MRSA strains. The CA-MRSA strains differed from healthcare-associated strains in being resistant to fewer agents. With respect to the MLS$_B$ phenotype, erythromycin resistant CA-MRSA isolates were less likely to carry an *erm* gene compared with HA-MRSA counterparts. Instead, CA-MRSA isolates were more often resistant to erythromycin due to the *msrA* efflux gene that encodes a macrolide specific efflux pump. Therefore, the *msrA*$^+$ strains are D-test negative and are susceptible to clindamycin, unless an *erm* determinant is also present in the strain.

Clindamycin therapy of CA-MRSA infection

Clindamycin is often recommended for empiric therapy of CA skin infections in the USA because the USA300 genetic background that causes the majority of CA skin infections in the USA is usually susceptible to clindamycin, despite the fact that erythromycin resistance is prevalent.

For example, in a surveillance study performed in 2001, Fridkin *et al.* (2005) found that only 12% of CA-MRSA isolates from several US cities were susceptible to erythromycin but 87% were susceptible to clindamycin. Remarkably, this trend has remained consistent over time. We recently evaluated the susceptibilities of MRSA isolates obtained from patients enrolled in a study (between 2009 and 2011) that compared clindamycin and trimethoprim–sulfamethoxazole (TMP-SMX) effectiveness in the therapy of CA-SSTIs (Miller *et al.*, 2015). The prevalence of clindamycin resistance was also low (12%) among *S. aureus* strains isolated from SSTIs. The trial demonstrated that effectiveness of clindamycin and TMP-SMX were not statistically different.

Tetracycline and tigecycline

Tetracyclines are bacteriostatic antibiotics that inhibit protein synthesis. The first tetracycline was identified in 1948 in the form of chlotetracycline. The related drug, tetracycline, was subsequently identified from *Streptomyces aureofaciens* (Walsh, 2003a). Related compounds such as doxycycline and minocycline were introduced in 1967 and 1972, respectively. Tigecycline, the newest tetracycline to be FDA-licensed (in 2005), is a glycylcycline derived from minocycline that can be used to parenterally treat tetracycline-resistant *S. aureus* infections.

Drugs in this class are composed of a linear tetracyclic framework to which various functional groups are attached. Tetracycline and related compounds inhibit protein synthesis by binding reversibly to a high affinity site on the 30S ribosomal subunit and disrupting association of aminoacyl-tRNA with its acceptor site (Strahilevitz and Hooper, 2009). Lower affinity sites have also been reported on 30S and 50S subunits (Strahilevitz and Hooper, 2009). The S7 ribosomal protein plays a role in binding to the drug with other ribosomal proteins contributing to optimal binding (Strahilevitz and Hooper, 2009).

The majority of tetracycline-resistant *S. aureus* isolates use an efflux pump to expel the drug from the cytoplasm. Resistance can also be mediated by a novel ribosomal protection mechanism that does not involve target modification.

S. aureus genetic determinants encoding tetracycline resistance functions have been found on both plasmids and the chromosome and consist of native chromosomal determinants with point mutations as well as horizontally acquired genes harboured on plasmids and transposons. Tet38 is a native chromosomally encoded efflux pump that transports tetracycline and confers resistance when point mutations lead to its overexpression (Jensen and Lyon, 2009). Acquired tetracycline resistance genes that have been found in *S. aureus* on plasmids include those that encode efflux proteins belonging to the major facilitator superfamily *tetA(K)* and *tetA(l)*. Acquired tetracycline resistance genes that encode ribosomal protection proteins are *tetA(M)* (Nesin *et al.*, 1990) and *tetA(O)* (Jensen and Lyon, 2009). *tetA(M)* has been found on a Tn*916*-like transposon, Tn*5801*, that is inserted in the chromosome of the VISA strain Mu50, just upstream of GMP synthase. *tetA(K)* is commonly carried by rolling circle plasmids, such as pT181, but it has also been found in the chromosome as a cointegrated copy of pT181 (Jensen and Lyon, 2009). For example, *tetA(K)* is resident on the prototype SCC*mec* type III cassette as part of a cointegrated multiresistance plasmid (Ito *et al.*, 2001).

Tetracycliine-resistant *S. aureus* strains carrying only *tetA(K)* are minocycline susceptible. In contrast, as expected for a target directed mechanism, the TetA(M) ribosomal protection protein confers broader resistance within the tetracycline group, including tetracycline and minocycline (Schmitz *et al.*, 2001). However, TetA(M) produces higher MICs of tetracycline compared with minocycline. A combination of efflux and ribosomal protection seems to have an additive effect on resistance since strains harbouring both *tetA(K)* and *tetA(M)* have higher MICs than those carrying either one alone (Schmitz *et al.*, 2001).

TetA(M)-mediated resistance is induced by low levels of drug. This is explained by exposure to subinhibitory levels of the drug that induce a higher level of transcription compared to baseline, thereby allowing the bacteria to resist higher drug concentrations. Induction of *tetA(M)* gene transcription occurs by transcriptional and translational attenuation (Strahilevitz and Hooper, 2009).

Mechanism of ribosomal protection by TetA(M)

The mechanism of ribosomal protection elicited by staphylococcal TetA(M) is likely to be similar to that of the closely related (92% amino acid similarity) streptococcal TetM protein which has been well characterized (Strahilevitz and Hooper, 2009). Ribosomes that have been purified from *tetM*+ bacteria are resistant to tetracycline when studied using *in vitro* systems and purified TetM interacts with the ribosome and promotes the release of bound tetracycline in a GTP dependent fashion (Jensen and Lyon, 2009; Strahilevitz and Hooper, 2009). The prevailing model is that ribosomal protection occurs by an interaction between the TetA(M) protein and ribosomes that are actively engaged in protein synthesis (Strahilevitz and Hooper, 2009).

Mechanism of tetracycline efflux

The tetracycline pump proteins are members of the major facilitator family and have 12 or 14 transmembrane segments. A proton or sodium gradient across the membrane is used to drive the extrusion of tetracycline complexed with cations (Strahilevitz and Hooper, 2009). The tetracycline resistance pumps have narrow substrate specificity for drugs in the tetracycline class and each pump has a preference for a specific tetracycline derivative. This is exemplified by TetK which has a preference for tetracycline over minocycline. It remains to be determined if the substrate profiles are dictated by the ability of a given tetracycline derivative to induce expression of the pump or by the specificity of the pump itself for a given derivative.

The selectivity of the tetracycline pumps is potentially beneficial to the development of newer generation tetracycline derivatives that are poor substrates of known resistance pumps. For example, tetracycline-resistant strains that contain *tetA(K)* or *tetA(M)* are susceptible to tigecycline. The basis for such susceptibility is the t-butylglycylamido modification at the 9 position, which increases the affinity for the ribosome about 20-fold (Olson *et al.*, 2006).

Point mutations in *rpsJ* encoding the S10 ribosomal protein have been shown to decrease the susceptibility to tigecycline-passaged *S. aureus* (Beabout *et al.*, 2015). Also, point mutations that increased the expression of the *mepA* gene encoding the MATE efflux pump protein were found in tigecycline-passaged *S. aureus* strains (McAleese *et al.*, 2005).

Tigecycline

Although tigecycline resistance remains very low among *S. aureus* clinical isolates and tigecycline has been shown to be non-inferior to vancomycin in clinical trials (Puzniak *et al.*, 2014), it was not included among the most recent recommendations for treatment of MRSA infection (Liu *et al.*, 2011). This was mainly due to the finding of increased risk of death reported of tigecycline-treated patients versus those treated with other antibiotics for similar syndromes (adjusted risk difference for death was 0.6% with corresponding 95% confidence interval (0.0%, 1.2%) (McGovern *et al.*, 2013). The basis for increased risk of death could be co-morbidities, increased baseline bacteraemia, persistent bacteraemia, or higher MICs of tigecycline (2 μg/ml in isolates obtained following deaths compared with an average MIC <1). More tigecycline-associated deaths occurred in the ventilator-assisted pneumonia patients, for which tigecycline is not indicated. [Tigecycline was approved for the treatment of complicated SSTIs, complicated intra-abdominal infections and community-acquired pneumonia (CAP), but not healthcare-associated pneumonia, ventilator assisted pneumonia, or diabetic foot ulcers.] The finding prompted the FDA in 2013 to require a black box warning to be placed on the drug label, the most serious type of warning for a drug. Moreover the FDA states that 'health care professionals should reserve Tygacil for use in situations when alternative treatments are not suitable' (Administration, 2013). The 2011 IDSA guidelines did recommend tetracycline for certain syndromes, except for children in who tooth discoloration can occur.

For a comprehensive review of tetracycline

resistance mechanisms in Gram-positive bacteria the reader is referred to McMurray and Levy (2000).

Aminoglycosides

The first aminoglycoside antibiotic discovered was Streptomycin in 1944. Aminoglycosides are hydrophilic sugars with multiple amino groups that are protonated at physiological pH thereby rendering them as polycations. They bind the 30S ribosome of 16S rRNA in the acceptor site for aminoacyl tRNA binding which stops the translocation steps of protein translation (Walsh, 2003a) resulting in mistranslation of the genetic code (Strahilevitz and Hooper, 2009). Aminoglycosides are not often used as single agents in front-line therapy of *S. aureus* but they are used in combination with other agents to treat endocarditis (Strahilevitz and Hooper, 2009).

In clinical *S. aureus* isolates, aminoglycoside resistance is mainly due to chemical modification of the binding targets in the ribosome through enzymes that are encoded by horizontally acquired genes. These modifications include (a) ATP dependent O-phosphorylation by phosphotransferases (Aph), (b) ATP-dependent-O-adenylation by nucleotidyl transferases (Ant), and (c) acetyl CoA-dependent N-acetylation by acetyltransferases (Aac). A common aminoglycoside resistance mechanism in *S. aureus* is conferred by a bifunctional enzyme that performs acetylation and phosphorylation encoded by *aacA-aphD* (Strahilevitz and Hooper, 2009). Aminoglycoside resistance genes are often carried on the composite transposon Tn*4001* that can be present on plasmids such as pSK1 and pSK41 or the chromosome.

Alterations in electrochemical gradients can lead to decreased uptake of aminoglycosides across the cell membrane. This mechanism is common among small colony variants. Efflux pumps and methylation of the ribosome have not been found in staphylococci as they have in Gram-negative bacteria. For further details on aminoglycoside resistance mechanisms the reader is referred to reviews by Jensen and Lyon (2009) and Strahilevitz and Hooper (2009).

Quinupristin/dalfopristin

Quinupristin/dalfopristin, often referred to by its trade name Synercid, is a combination antibiotic comprised of a 30:70 ratio (by weight) of two naturally occurring compounds derived from *Streptomyces pristinaspiralis*. Both compounds are water soluble and belong to the streptogramin class of antibiotics. The antibacterial mechanism of action is through inhibition of protein synthesis by sequential binding to distinct sites on the 50S ribosome, first by quinupristin, then by dalfopristin. This results in a stable complex of drug and ribosome that interferes with targets within 23S rRNA, resulting in the inability of nascent peptide chains to be released from the ribosome (Nailor and Sobel, 2011). Due to the extraordinarily high expense of Synercid and its relatively high level of toxicity, it has not been widely used in clinical practice.

The development of resistance to Synercid has been shown to be due to several mechanisms: decreased binding to the target accomplished by either efflux of the drug, mutations that alter the binding site, or modification of the target.

Oxazolidinones: linezolid and tedizolid

Linezolid

Linezolid is an oxazolidinone, a synthetic class of antibiotic in existence since 1987 that was approved in 2000 by the FDA under the brand name Zyvox. It is available in both oral and iv formulations. For staphylococci, linezolid is bacteriostatic and used for the treatment of nosocomial pneumonia and complicated SSTIs caused by *S. aureus*, including MRSA. It is often an option reserved for treatment of vancomycin-resistant organisms, difficult to treat infections, and vancomycin-intolerant patients (Liu *et al.*, 2011).

The means by which linezolid inhibits protein synthesis is by inhibition of the initiation complex of the 50S ribosome which is unique compared with other ribosome synthesis inhibitors. More specifically, it inhibits binding or proper placement of aminoacyl-tRNA in the peptidyl transferase site (Mendes *et al.*, 2014).

Linezolid has favourable pharmacokinetic characteristics compared with vancomycin, the mainstay of MRSA therapy. Linezolid is orally bioavailable and penetrates into respiratory secretions, bone, fat and muscle where concentrations above the susceptibility breakpoint can be attained. In patients with ventilator-associated pneumonia, linezolid has been shown to be statistically superior versus vancomycin in the therapy of MRSA nosocomial pneumonia in a phase IV, randomized, controlled study (Chastre *et al.*, 2014). Importantly, virulence factor gene expression can be inhibited by linezolid even at subinhibitory levels, a desirable property that might complement the antibacterial properties of the drug.

Based on linezolid's unique mechanism of action and a number of other reasons outlined here, it was thought that the likelihood of resistance would be low (Meka and Gold, 2004). First, the 23S rRNA is present in multiple copies (five or six in *S. aureus*); therefore multiple mutations would be required to confer full resistance. Second, since linezolid is a synthetic drug, there are no natural producers that would supply a portable transferable resistance mechanism. Third, the Erm determinants commonly present in MRSA strains that confer MLS_B type resistance by dimethylation of N6 of nucleotide A2058 in 23S rRNA did not confer cross-resistance to linezolid. In fact, none of the genes known to confer resistance to other classes of antibiotics that target the ribosome (i.e. chloramphenicol, macrolides, lincosamides, streptogramins, aminoglycosides and tetracyclines) conferred cross-resistance to linezolid (Fines and Leclercq, 2000). These included modifying enzymes LinA, LinA′, LinB, Vgb, Vat, SatA, ANT(4′) (4″)-I, AAC(6′)-APH(2″), APHA-3 and Cat, the efflux pumps MsrA, MefE, MefA, MreA, Vga, TetK and TeL, the methylases or TetM and TetO (Fines and Leclercq, 2000). Finally selection for resistant mutants occurred at an extremely low frequency *in vitro* (Mendes *et al.*, 2014).

Altogether, these rationales portrayed linezolid as an ideal choice for treating MRSA isolates, particularly due to their nearly universal resistance to erythromycin with cross resistance to clindamycin. Despite such high hopes, linezolid resistance has been reported among multiple genetic lineages in *S. aureus* and *S. epidermidis*, conferred through a variety of mechanisms (Kehrenberg *et al.*, 2009; Locke *et al.*, 2009b, 2014a; Morales *et al.*, 2010; Toh *et al.*, 2007; Wong *et al.*, 2010). The most common mechanism of resistance involves an alteration in domain V within the 23S rRNA genes that alters oxazolidinone binding in the 50S peptidyl transferase centre. This mutation can be found in multiple copies of 23S rRNA, with increasing resistance reported for each additional mutation (Tsakris *et al.*, 2007). Also, mutations in ribosomal proteins L3 and L4 have been found in linezolid resistant *S. aureus* (Locke *et al.*, 2009a; Mendes *et al.*, 2014). A third resistance mechanism involves the horizontally mobile chloramphenicol-florfenicol resistance gene (*cfr*) that encodes a 23S rRNA methyltransferase that modifies the peptidyl transferase centre of the 50S ribosome subunit. This mechanism confers cross-resistance to the phenicols (including chloramphenicol and florfenicol), lincosamides (clindamycin), pleuromutilins, and $streptogramin_A$ (so-called $PhLOPS_A$) (Long *et al.*, 2006). Fortunately, susceptibility to the second generation oxyzolidinone, tedizolid, is retained by *cfr* expressing strains (see below).

The *cfr* gene was first identified in isolates of animal origin from the late 1990s in Europe (Mendes *et al.*, 2014), while the first report of *cfr* in *S. aureus* from a human clinical linezolid-resistant MRSA isolate was obtained in 2005 in a Colombian patient (Toh *et al.*, 2007). Subsequent reports of *cfr* carriage in linezolid-resistant *S. aureus* have occurred, but have been infrequent (Kehrenberg *et al.*, 2009; Locke *et al.*, 2014a; Mendes *et al.*, 2013; Morales *et al.*, 2010; Shore *et al.*, 2010). In many cases, the *cfr* determinant was found in several genotypes on a similar plasmid with the backbone of a single plasmid, pSCFS3 (Mendes *et al.*, 2013), even though pSCFS3 is not a conjugal plasmid. The *cfr* gene has also been found in *S. aureus* on a 37 kb conjugal plasmid that was successfully transferred *in vitro* to another *S. aureus* strain resulting in the full linezolid resistance phenotype in the recipient (along with co-resistance to trimethoprim and aminoglycosides). This element is promiscuous as evidenced by a 7-kb *cfr*-containing region that was identical to that in a *Proteus vulgaris* isolate that is found in both Gram-positive and Gram-negative species (Locke *et al.*, 2014a).

As further evidence of the mobility of *cfr*, there have been *cfr*-expressing *S. aureus* isolates from multiple genotypes and lineages such as ST5, ST239, or ST8 (USA300). Cfr has also been found in porcine MRSA isolates from ST398 and ST90 genetic backgrounds (Kehrenberg *et al.*, 2009), suggesting that overuse of linezolid in agricultural settings and $PhLOPS_A$ antibiotics could promote dissemination of this determinant.

Cfr confers multidrug resistance to phenicols, lincosamides, oxazolidinones, pleuromutilins, and $streptogramin_A$ antibiotics ($PhLOPS_A$), which are antibiotics that share nearby target sites (Kehrenberg *et al.*, 2005; Long *et al.*, 2006). By using drug binding footprinting studies, it was shown that Cfr decreases target binding of chloramphenicol, florfenicol, and clindamycin (Kehrenberg *et al.*, 2005). It was further shown by MALDI-TOF mass spectrometry in combination with tandem mass spectrometry analysis of RNA fragments isolated from cells harbouring the plasmid-borne *cfr* gene that the adenine at position 2503 (A2503) in 23S rRNA is the modification site methylated by Cfr. This finding correlates with an enhanced reverse transcriptase stop signal at nucleotide position 2504 observed in drug binding assays. Thus, Cfr methylates A2503 in the peptidyl transferase centre of 23S rRNA, resulting in decreased drug binding by multiple classes of ribosomal inhibitors, thereby causing cross resistance. Notably, Cfr is the only methyltransferase known to confer antibiotic resistance by acting directly in the peptidyl transferase centre (Kehrenberg *et al.*, 2005).

Tedizolid

The oxyzolidinone tedizolid phosphate (trade name Sivextro) is a prodrug that is converted by plasma phosphatases to microbiologically active tedizolid *in vivo*. It was approved by the FDA in 2014 as a second generation oxazolidinone with 4- to 16-fold increased potency against MRSA when compared with linezolid. For example, the MIC of tedizolid was found to be 0.5 µg/ml among strains that had MICs of linezolid ranging from 8 to 16 µg/ml (Wong and Rab, 2014). Tedizolid phosphate is currently indicated for the treatment of acute SSTIs and is under investigation for the treatment of nosocomial pneumonia (Rybak and Roberts, 2015). Importantly, tedizolid is active against linezolid resistant strains carrying the transferable *cfr* methyltransferase gene, which can be explained by structural differences in A-ring C5 substituents between linezolid and tedizolid (Locke *et al.*, 2014b). Tedizolid is also more potent than linezolid against linezolid-resistant *S. aureus* strains that have mutations in chromosomal genes encoding 23S rRNA or ribosomal proteins L3 or L4. From a safety perspective, tedizolid has a better safety profile and pharmacokinetic properties than linezolid and is associated with a lower incidence of thrombocytopenia. Tedizolid is therefore indicated for a single daily dose in both oral and intravenous forms.

Tedizolid resistance

Among 6884 Gram-positive isolates obtained from the USA and Europe as part of the Surveillance of Tedizolid Activity and Resistance Program in 2011 and 2012, 16 isolates had elevated tedizolid or linezolid MIC values (intermediate or resistant isolates), 10 had mutations in the genes encoding 23S rRNA (primarily G2576T), five had mutations in the genes encoding ribosomal proteins L3 or L4, and five carried the *cfr* multidrug resistance gene (Sahm *et al.*, 2015).

A novel oxazolidinone resistance gene, *optrA*, that encodes an ABC transporter, was found on a conjugal plasmid in an *E. faecalis* isolate (Wang *et al.*, 2015). This plasmid conferred combined resistance or elevated MICs to both linezolid and tedizolid and phenicols (chloramphenicol and florfenicol) (likely due to the transporter gene (*fexA*) that was co-localized on the same plasmid). Although not found in a clinical *S. aureus* isolate to date, *optrA* was transferred to and functionally expressed in *S. aureus* and *E. faecalis* isolates *in vitro*. As a note of caution, *optrA* has been detected more often in *E. faecalis* and *E. faecium* isolated from food-producing animals compared with humans suggesting that antibiotics in animal feed could contribute to the emergence of tedizolid resistance in human pathogens such as *S. aureus* (Wang *et al.*, 2015).

Antibiotics that interfere with folate metabolism

Trimethoprim-sulfamethoxazole

Trimethprim-sulfamethoxazole (TMP-SMX) is a bactericidal combination drug, sometimes referred to as co-trimoxazole. TMP-SMX includes two antibiotics that each inhibits a distinct enzyme in the bacterial folic acid pathway. By inhibiting this pathway, the synthesis of thymidine is blocked. By combining two drugs that target distinct enzymes that are required in the same pathway, the risk of developing resistance mutations in both enzymes is diminished and bactericidal activity is enhanced (Hitchings, 1973).

SMX is a structural analogue of para-amino benzoic acid (PABA), the substrate of dihydropteroate synthase, the penultimate enzyme in the pathway. Due to the structural similarity between SMX and PABA, SMX competes with PABA for binding to the enzyme, which in turn disrupts the synthesis of dihydrofolic acid. The resulting reaction product is a sulfa-containing pteroate that is unable to be used downstream (Strahilevitz and Hooper, 2009). TMP acts by binding to and inhibiting the last enzyme in the tetrahydrofolate pathway, dihydrofolate reductase (DHFR), which deprives the bacteria of tetrahydrofolate that is required for biosynthesis of several essential metabolites, glycine, methionine, purines and pyrimidines. The toxic effect is thought to be due to the lack of thymine (Projan, 2000).

SMX was introduced as a single agent in 1935 but resistance readily developed (Nakhla, 1972). Resistance to SMX was attributed to the *sulA* gene and overproduction of para-amino benzoate (Projan, 2000). After TMP was produced in 1956 the combination drug TMP-SMX was introduced. Although resistance to TMP was reported in London in the early 1970s, the combination of TMP and SMX maintains usefulness against isolates that maintain sensitivity to at least one of the drugs. Nevertheless, TMP-SMX resistance has been reported with varying frequencies, depending on the locale.

Until recently, TMP-SMX was not considered a first-line therapy due to inferiority in clinical trials and to resistance (Strahilevitz and Hooper, 2009). However, this situation changed in the USA with the dawn of the CA-MRSA epidemic. CA-MRSA strains have usually been susceptible to agents other than β-lactams which caused TMP-SMX and clindamycin to become common treatment options for skin and soft tissue infection in the USA. It is therefore noteworthy that TMP-SMX and clindamycin produced similar cure rates in SSTI in a recent clinical trial performed in three major US cities (Miller *et al.*, 2015). In this trial, *S. aureus* was the most common organism isolated from skin lesions (41%) and MRSA consisted of a vast majority (77%) among *S. aureus* isolates. The cure rate with TMP-SMX was 77% in the intention to treat population and 80% in the evaluable population. Accordingly resistance to TMP-SMX among *S. aureus* isolates was very low (1 in 217 isolates = 0.5%).

Mechanism of TMP resistance

Resistance to TMP in *S. aureus* strains is associated with one of two pairs of point mutations in the chromosomally encoded DHFR gene (*dfrB*) resulting in H30N/F98Y or H149R/F98Y exchanges in DHFR. Since the F98Y substitution decreases the effectiveness of the enzyme by 36-fold and is common in both pairs of TMP resistance mutations, it is now considered to be largely responsible for chromosomal TMP resistance. TMP associated point mutations usually cause intermediate resistance (MIC < 256 μg/ml) (Dale *et al.*, 1997).

In addition to chromosomally encoded resistance due to point mutations in the target gene, several acquired resistant *dfr* gene variants harboured by *S. aureus* have been described. *S. aureus* isolates harbouring *dfrA* (also called *dfrS1*), *dfrG* or *dfrK* have MICs of TMP > 256 μg/ml (Archer *et al.*, 1986; Aspiroz *et al.*, 2010; Kadlec and Schwarz, 2009; Rouch *et al.*, 1989). Such acquired TMP resistant *dfr* alleles have been found on plasmids and/or Tns that provided them with a horizontal mode of transfer between strains of *S. aureus*. As evidence that these genes are TMP resistance determinants, when cloned on a plasmid they can even confer TMP resistance in *E. coli*. Interestingly, these transferable alleles have the same point mutation as the native chromosomal TMP-resistant *dfr* gene that produces the F98Y exchange. However, unlike the native *dfr* gene, the acquired genes are usually present in multiple copies resulting in a higher level of resistance compared with the single copy chromosomal mutation.

While the *dfrA* determinant was once considered to be the most common transferable TMP resistance determinant, *dfrG* has become the dominant TMP resistance determinant in some locales. For example, *dfrG* was the only TMP resistance determinant found among TMP-resistant *S. aureus* isolates obtained from people in Sub-Saharan Africa where TMP-resistant *S. aureus* is prevalent (Nurjadi *et al.*, 2014). Also, *dfrG* was present in 93% of TMP-resistant *S. aureus* isolates identified in Europe among people who recently returned from sub-Saharan Africa (Nurjadi *et al.*, 2014). *dfrG* was also found among *S. aureus* isolates in Thailand (Sekiguchi *et al.*, 2005) although it was present on the chromosome rather than a plasmid. A *dfrK* variant, that is more closely related to *dfrG* than *dfrA*, was found in a porcine MRSA isolate belonging to ST398, which has been associated with recent livestock and human infections in Europe (Kadlec and

Schwarz, 2009). *dfrK* has also been identified in Spain among human MRSA isolates belonging to ST398 and ST1 that were isolated from people associated with pig farming (Aspiroz *et al.*, 2010). Interestingly, resistance to TMP-SMX may develop independently of TMP and SMX (Grim *et al.*, 2005).

A special consideration for TMP-SMX susceptibility testing is worth noting. When performing susceptibility testing for TMP-SMX the media should be free of thymidine since its presence will allow the bacteria to bypass the folate pathway thereby giving the appearance of TMP-SMX resistance.

Further details on TMP-SMX pharmacokinetics, dosing, drug interactions and resistance, which are outside the scope of this chapter, can be found in a review article by Grim *et al.* (2005).

Antibiotics that interfere with DNA replication

Quinolones and fluoroquinolones

The quinolones have bactericidal activity for *S. aureus* and block DNA replication, an essential function required for bacterial replication and propagation of daughter cells. However, as will be described in this section, staphylococci readily develop quinolone resistance. Even among CA-MRSA isolates, which tend to be susceptible to non-β-lactam antibiotics, the prevalence of ciprofloxacin-resistance has been high (60% in some studies) (Moran *et al.*, 2006) compared with clindamycin and tetracycline resistance. Thus, since fluoroquinolones have a low barrier to resistance they are not recommended as the sole agent in the therapy of MRSA (Liu *et al.*, 2011). However, in certain situations, fluoroquinolones are recommended in combination with rifampin, an RNA polymerase inhibitor (Liu *et al.*, 2011) in susceptible isolates.

Quinolones are synthetic molecules with a structural modification of the '4-quinolone skeleton (Blondeau, 2004). The first quinolone, nalidixic acid, was useful for treatment of uncomplicated urinary tract infections caused by Gram-negative bacteria. Additional modification of the quinolone (fluorination at the R_6 position) produced fluoroquinolones. A cycloproprYl ring added to the R_1 position led to the development of ciprofloxacin. The methoxy side-chain added to the R_8 position led to the development of fourth-generation quinolone compounds including gatifloxacin and moxifloxacin (Blondeau, 2004).

The two known targets of fluoroquinolones are topoisomerase IV and DNA gyrase (aka, topoisomerase II). In *S. aureus* the primary target of early versions of quinolones and fluoroquinolones is topoisomerase IV and the secondary target is DNA gyrase (Hooper, 2000). This contrasts with *E. coli* in which the primary and secondary targets are DNA gyrase and topoisomerase IV, respectively.

DNA gyrase catalyses the early stages of DNA replication by producing negative superhelical twists in the DNA. DNA gyrase also untangles positive superhelical twists that accumulate ahead of the DNA replication fork or as a result of transcription. Topoisomerase IV is involved in a later stage of DNA replication by allowing intertwined daughter chromosomes to separate, a prerequisite for segregation of DNA into daughter cells. By binding to and stabilizing the DNA-topoisomerase IV complex and the DNA–DNA gyrase complex, quinolones block both late and early stages of DNA replication. This blockage leads to cell death in a process that involves irreversible stabilization of DNA–protein complexes that triggers DNA damage and the generation of DNA-strand breaks that eventually kills the cells (Hooper, 2000).

Resistance to quinolones and fluoroquinolones in *S. aureus* is due to mutations in primary and secondary targets, or elimination of internal concentrations of the drug due to enhanced production of the native NorA efflux pump. To date, horizontal acquisition of resistance determinants to this class of antibiotics has not been described in *S. aureus*. Presumably, resistance to fluoroquinolones could be influenced the ability of this class of drugs to induce damage of bacterial DNA and trigger the SOS DNA repair system that leads to errors during DNA replication, although this has not been proven. For additional details on resistance mechanisms against fluoroquinolones among other Gram-positive bacteria the reader is referred to review articles by Hooper (2000, 2002). The sections below highlight the data specific to *S. aureus*.

Fluoroquinolone resistance in *S. aureus* by alteration of primary and secondary targets.

Fluoroquinolone resistance in *S. aureus* involves stepwise mutations in the primary and secondary targets of fluoroquinolones, one occurring in topoisomerase IV followed by DNA gyrase (Hooper, 2000). In *S. aureus*, DNA gyrase is a tetramer composed of two GyrA and two GyrB proteins, while topoisomerase IV is composed of two GrlA and two GrlB subunits. (GrlA and GrlB homologues in *E. coli* are referred to as ParC and ParE, respectively). In *S. aureus*, a mutation occurring in genes encoding the subunits of topoisomerase IV (*grlA* or *grlB*) can produce a slight fluoroquinolone resistant phenotype, but a mutation occurring solely in a DNA gyrase subunit gene, *gyrA* or *gyrB*, does not. Moreover, mutations occurring in subunits of both topoisomerase IV and DNA gyrase produce a much higher level of resistance than if a mutation occurs solely in topoisomerase IV (Ince and Hooper, 2001). The basis for this phenomenon can be understood since the sensitivity of *S. aureus* topoisomerase IV to fluoroquinolones is much greater than that of DNA gyrase so the increased resistance conferred by an amino acid substitution in DNA gyrase does not overcome the susceptibility of the primary target, topoisomerase IV. For instance, the IC_{50} (the concentration required to inhibit 50% of the enzyme activity *in vitro*) of topoisomerase IV for ciprofloxacin is about 6-fold higher than that of DNA gyrase (Hooper, 2000). In other words, although a mutation can decrease the sensitivity of the secondary target for the fluoroquinolone, the primary target remains the dominant determinant of a strain's MIC.

The newer fluoroquinolones, gatifloxacin and moxifloxacin, are more potent than the first generation members of this class such as ciprofloxacin and ofloxacin. In particular, gatifloxacin is 2- to 16-fold more active than ciprofloxacin due to an 8-methoxyquinolone substitution on the quinolone ring of gatifloxacin that increases potency and bactericidal activity. Moreover, gatifloxacin has greater activity than ciprofloxacin in *grlA* or *grlB* single mutant strains. Even greater differential activity is realized in *grlA/gyrA* dual resistant strains (Ince and Hooper, 2001). For example, in a *grlA/gyrB* double mutant in which the MIC of ciprofloxacin increased up to 256-fold (to 32 μg/ml), the MIC of gatifloxacin only increased up to 64-fold (to just 4 μg/ml) (Hooper, 2000). Gatifloxacin also seems to protect better than ciprofloxacin against development of high level resistance. Ince *et al.* found that selection of resistant mutants with gatifloxacin at twice the MIC was 100- to 1000-fold less frequent than with ciprofloxacin (Ince and Hooper, 2001). Moreover, mutants were rarely selected using four times the MIC of gatifloxacin.

The quinolone resistance-determining region

Fluoroquinolone resistance mutations in target genes have clustered in the 5′ region of *grlA/parC*, encoding the GrlA subunit of topoisomerase IV, or in *gyrA* encoding the GyrA subunit. To some extent, polymorphisms also cluster in the central region of *GrlB/parE* and *gyrB*. The most common polymorphisms in the ciprofloxacin resistant GrlA subunit that cause quinolone resistance include amino acid changes at positions 80 and 84, which are in the quinolone resistance determining region (QRDR), and 116 (discussed in (Ince and Hooper, 2001)). The most common changes found among ciprofloxacin resistant *S. aureus* isolates from Japan were Ser80Phe or Tyr in the GrlA topoisomerase IV subunit and Ser84Leu in the GyrA subunit (Tanaka *et al.*, 2000). Overall, Tanaka *et al.* found 15 types of concomitant mutations in GrlA and GyrA associated with ciprofloxacin resistance. These were also the predominant substitutions found among a more recent collection of ciprofloxacin-resistant *S. aureus* isolates from patients in Turkey in 2003–2006 (Coskun-Ari and Bosgelmez-Tinaz, 2008). In contrast, Ince and Hooper (2001) reported other amino acid changes in the subunits of GyrB and GrlB as well as GyrA and GrlA that conferred increased fluoroquinolone resistance. Moreover, GyrB polymorphisms in residue 437 and 458 contribute to ciprofloxacin resistance only when they occur in combination with a subunit of topoisomerase IV. Notably, the identified resistance-associated mutations occur both inside and outside the QRDR.

Like ciprofloxacin, the primary target of gatifloxacin in *S. aureus* is likely topoisomerase IV. As in ciprofloxacin, resistance to topoisomerase IV precedes that of DNA gyrase and single *gyr* mutant strains are not resistant unless combined with an amino acid change in topoisomerase

IV. Interestingly, novel mutations selected with gatifloxacin suggest that its interactions with topoisomerase IV may differ from those of earlier quinolones. The increased activity of gatifloxacin may be due to the fact that gatifloxacin may simultaneously bind two targets instead of one (Blondeau, 2004).

Role of efflux pumps in fluoroquinolone resistance

Since the targets of quinolones and fluoroquinolones are located in the cytoplasm, the antibiotic must pass through the thick peptidoglycan layer and the cytoplasmic membrane of *S. aureus* and be retained in the cytoplasm long enough and in a sufficient concentration to bind and inhibit its targets. Due to the small size of fluoroquinolones, their translocation into the *S. aureus* cell occurs by simple diffusion. A strategy that can contribute, in part, to fluoroquinolone resistance in *S. aureus* is through mutations that allow the overexpression of native efflux pumps that can extrude fluoroquinolones from the cytoplasm (Poole, 2000). NorA, a member of the major facilitator superfamily of transporters, integrates in the cytoplasmic membrane through 12 membrane-spanning helices and is able to extrude quinolones from the cytoplasm, as determined from accumulation studies in everted vesicles derived from *E. coli* cells expressing *norA* (Hooper, 1999; Ng *et al.*, 1994). Moreover, enhanced expression of cloned *norA* in *S. aureus* produces resistance to hydrophilic fluoroquinolones such as norfloxacin, ciprofloxacin, and ofloxacin (Ng *et al.*, 1994), as well as several other classes of compounds including chloramphenicol, ethidium bromide, rhodamine 6G, and tetraphenylphosphonium (Hooper, 2002). Also, heterologous expression of *norA* on a plasmid in *Bacillus subtilus* confers resistance to norfloxacin (Neyfakh *et al.*, 1993). The hydrophobic fluoroquinolones sparfloxacin and trovafloxacin are poor substrates for efflux by NorA, thus resistance to these compounds is not affected by *norA* overexpression (Hooper, 1999; Poole, 2000). Notably, NorA is not essential for *S. aureus* viability and its role in normal physiology is not yet known (Hooper, 1999).

NorA expression is low in bacteria susceptible to quinolones but in a resistant strain it is induced by quinolones and other substrates. Alternately, *norA* transcription can be upregulated (reviewed in (Poole, 2000)) in resistant strains. The most common mutation leading to *norA* overexpression is a point mutation in the 5′ promoter region that results in an increased steady state expression of the *norA* transcript (Ng *et al.*, 1994). The increased expression was shown to be due enhanced stability of the mRNA transcript due to formation of a predicted additional stem–loop structure in the mutant sequence that alters a putative RNase III cleavage site (Fournier *et al.*, 2001).

In addition to NorA, other efflux pumps are implicated in the efflux of fluoroquinolones in *S. aureus* (Kaatz *et al.*, 2002; Kaatz *et al.*, 2000). The newer fluoroquinolones are not substrates of NorA, but other efflux proteins have been identified that can accommodate the new drugs as substrates. For example, although moxifloxacin is not recognized by NorA, it is a substrate of NorB, another member of the major facilitator superfamily with 30% similarity to NorA (Truong-Bolduc *et al.*, 2005). Another efflux protein, NorC, contributes to low-level resistance to moxifloxacin and sparfloxacin (Truong-Bolduc *et al.*, 2006).

It is not yet clear to what extent overexpression of NorA or other efflux pumps in *S. aureus* mediates clinically important resistance since concomitant mutations in *parC/E* and *gyrA/B* genes usually occur in clinical isolates along with overexpressed efflux systems. Nevertheless, because overexpression of efflux pumps has been linked to clinical fluoroquinolone resistant *S. aureus* strains (Kwak *et al.*, 2013; Poole, 2000), it has been proposed that expression of an efflux mechanism is the first step leading to higher level mutations due to target site mutations (Kaatz *et al.*, 2002).

Regulation of expression of NorA, NorB and NorC efflux pumps

By understanding how the expression of the fluoroquinolone efflux genes are regulated it might be possible to target prevention therapies towards them. In fluorquinolone-susceptible strains of *S. aureus*, *norA* expression is influenced by the two-component regulatory system, ArlSR, and by a small 18 KDa protein that binds to the *norA* promoter (Fournier *et al.*, 2000). Transcription of both *norA* and *norB* is negatively regulated by

MgrC. A second regulator, MgrA (previously called NorR), has differential effects on *norA* and *norB* expression. Whereas *norA* transcription is positively regulated by MgrA (Truong-Bolduc *et al.*, 2006) *norB* is negatively regulated. MgrA also influences *norC* gene expression but it acts indirectly (Truong-Bolduc *et al.*, 2005).

Interestingly, the specificity of MgrA for *norA* and *norB* promoters is related to the phosphorylation state of MgrA (Truong-Bolduc and Hooper, 2010). Phosphorylated MgrA binds to the *norB* promoter but not to that of *norA*. Conversely, the unphosphorylated form of MgrA binds preferentially to *norA*'s promoter. MgrA is also known as a global regulator of the expression of virulence factors (Luong *et al.*, 2006), AbcA (Truong-Bolduc and Hooper, 2007) and the Tet38 tetracycline resistance efflux protein (Truong-Bolduc *et al.*, 2005).

NorG is yet another regulator of the genes encoding the Nor A, B and C efflux pumps; it belongs to the GntR-like family of transcriptional regulators (Truong-Bolduc and Hooper, 2007). Overexpression of *norG* increases *norB* transcript levels and resistance to quinolones. While MgrA acts as a repressor of *norB* expression, NorG directly activates *norB*. In addition, transcriptional profiling revealed that NorG positively activates mgr*A*, *arlS* (shown previously to regulate *norA*), and *sarZ global* regulatory genes and negatively influences *norC* expression (Truong-Bolduc *et al.*, 2011).

In summary, NorA, B and C are chromosomally encoded, multidrug efflux pumps belonging to the major facilitator superfamily with overlapping and distinct specificities for the various fluoroquinolones and other agents that are toxic to *S. aureus*. By overexpression of these efflux proteins, *S. aureus* is able to reduce the intracellular concentration of quinolones or fluoroquinolones to elicit low-level resistance. And this may precede development of point mutations in the target genes leading to higher level resistance. There are multiple regulatory proteins that can cross-modulate the expression of *norA*, *B* and *C* efflux pump genes. However, the role of the identified regulators in development of resistance to fluoroquinolones in clinical isolates is not known.

Inhibition of RNA polymerase

Rifampin

Rifampin (or rifampicin) is a member of the rifampin class of antibiotics that act by targeting the β-subunit of RNA polymerase (RNP). Eukaryotic polymerases are about 100-fold less sensitive than prokaryotic ribosomes (Walsh, 2003c).

Although rifampin is an orally administered antibiotic that has been used to treat *S. aureus* infection, it is not recommended for single agent therapy due to the low threshold to resistance development (Aubry-Damon *et al.*, 1998; Schmitz *et al.*, 2000). However, rifampin is recommended for MRSA therapy in certain circumstances, but only in combination with another antibiotic. Also, rifampin is not recommended for MRSA bacteraemia until bacteria are no longer detected in the blood (Liu *et al.*, 2011). The rationale being that selection of resistant isolates due to spontaneous mutation (which occurs at a rate of 10^{-6} to 10^{-8} in planktonic culture and 10^{-4} to 10^{-7} in biofilms; Schmitz *et al.*, 2000; Tang *et al.*, 2015) is minimized with a low population of bacteria.

The mechanism of resistance to rifampin involves point mutations in the native chromosomal *rpoB* gene encoding the β-subunit of RNP (Aubry-Damon *et al.*, 1998). (RNP has three core subunits and alternate sigma factors for directing RNP to specific promoters). These target alterations lead to decreased affinity of the antibiotic for RNP which blocks transcription initiation. In *S. aureus*, rifampin resistance is associated with amino acid exchanges in RpoB that align with two rifampin resistance clusters described in RpoB in *E. coli* (Aubry-Damon *et al.*, 1998; Tang *et al.*, 2015).

Notably, rifampin resistance can be selected *in vitro* in a single step with a spontaneous mutation rate of 10^{-6} to 10^{-8} (Aubry-Damon *et al.*, 1998; Tang *et al.*, 2015). Even more disturbing is the finding that rifampin resistance mutations leading to high level resistance can occur spontaneously in biofilms, even without exposure to rifampin (Tang *et al.*, 2015).

The point has often been raised that resistance to antibiotics has a fitness cost on the bacterium and that removal of the antibiotic selective pressure could easily counterselect for bacteria that lack the resistance phenotype. Wichelhaus *et al.* found that the His481Asn substitution in RpoB, found in 91% of rifampin-resistant clinical isolates from around the world, conferred low level resistance and had no fitness cost (Wichelhaus *et al.*, 2002). Since their sample size was small (32 isolates), this study bears repeating with a larger sample size.

Since *rpoB* mutations were found to be present among VISA isolates, Watanabe *et al.* investigated whether selection for *rpoB* mutations could be sufficient to produce the VISA phenotype. Surprisingly, rifampin alone was able to select for *rpoB* mutations *in vitro* that created a VISA or hVISA phenotype (Watanabe *et al.*, 2011). These experiments support the clinical guidelines that recommend rifampin be used sparingly to treat *S. aureus* infection.

Discussion

There is currently no vaccine for prevention of *S. aureus* infection. Therefore, antibiotics are heavily relied upon to treat illness caused by *S. aureus*. However, as described in this chapter, *S. aureus* has devised a myriad of strategies to elude the inhibitory and killing effects of antibiotics. For some isolates that have accumulated multiple resistance mechanisms, few options remain for therapy. Compounding this problem is the fact that the pipeline of new antibiotic development has slowed in recent years. This is due to the financial disincentive of large pharmaceutical companies to risk the high cost and time investment needed to get an antibiotic through the scientific and regulatory hurdles required to advance the antibiotic to market. Also, the short time a patient actually needs the antibiotic and the short-lived usefulness of antibiotics, driven by the rapid rate with which the bacteria evolve resistance, limit the post licensing profit margins. Paradoxically, profits are also negatively impacted by stewardship programs that limit use and demand for antibiotics. Nevertheless, a recent surge in licensed antibiotics for treatment of SSTI will provide some relief, at least for awhile, until staphylococci devise strategies to elude these agents.

It is time for developers of antibiotics to think out of the box. Antibiotic strategies that target an essential process in bacteria place strong selective pressure that ultimately results in the survival of the most resistant strains. Strategies that potentiate reliable antibiotics or that delay resistance development could increase the lifespan and usefulness of old and new antimicrobial agents. Also, by better understanding how bacteria cause disease and death, it will be possible to target strategies that ameliorate the toxic effects of the bacteria on the host rather than by directly killing the bacteria.

There has been a recent increase in the attention given to antibiotic resistance as a public health threat. In 2015 the World Health Assembly (WHA) endorsed a global action plan to combat antibiotic resistance and called for the first World Antibiotic Awareness Week. In the USA, President Barack Obama signed an Executive Order in 2014 to increase federal efforts to recognize and discourage the rising trend of antibiotic resistance. The President's Council of Advisors on Science and Technology also made recommendations to address the antibiotic-resistance crisis. It is hoped that such awareness focused on this issue will bring much needed resources and innovation to develop new antimicrobials and treatment strategies.

References

Alam, M.T., Petit, R.A., Crispell, E.K., Thornton, T.A., Conneely, K.N., Jiang, Y., Satola, S.W., and Read, T.D. (2014). Dissecting vancomycin-intermediate resistance in *Staphylococcus aureus* using genome-wide association. Genome Biol. Evol. *6*, 1174–1185.

Archer, G.L., Coughter, J.P., and Johnston, J.L. (1986). Plasmid-encoded trimethoprim resistance in staphylococci. Antimicrob. Agents Chemother. *29*, 733–740.

Archer, G.L., Niemeyer, D.M., Thanassi, J.A., and Pucci, M.J. (1994). Dissemination among staphylococci of DNA sequences associated with methicillin resistance. Antimicrob. Agents. Chemother. *38*, 447–454.

Arêde, P., Milheiriço, C., de Lencastre, H., and Oliveira, D.C. (2012). The anti-repressor MecR2 promotes the proteolysis of the *mecA* repressor and enables optimal expression of β-lactam resistance in MRSA. PLOS Pathog. *8*, e1002816.

Arêde, P., and Oliveira, D.C. (2013). Proteolysis of *mecA* repressor is essential for expression of methicillin resistance by *Staphylococcus aureus*. Antimicrob. Agents Chemother. *57*, 2001–2002.

Aspiroz, C., Lozano, C., Vindel, A., Lasarte, J.J., Zarazaga, M., and Torres, C. (2010). Skin lesion caused by ST398 and ST1 MRSA, Spain Emerg. Infect. Dis. *16*, 157–159.

Aubry-Damon, H., Soussy, C.J., and Courvalin, P. (1998). Characterization of mutations in the rpoB gene that confer rifampin resistance in *Staphylococcus aureus*. Antimicrob. Agents Chemother. *42*, 2590–2594.

Baba, T., Kuwahara-Arai, K., Uchiyama, I., Takeuchi, F., Ito, T., and Hiramatsu, K. (2009). Complete genome sequence of *Macrococcus caseolyticus* strain JCSCS5402, [corrected] reflecting the ancestral genome of the human-pathogenic staphylococci. J. Bacteriol. *191*, 1180–1190.

Barber, M., and Rozwadowska-Dowzenko, M. (1948). Infection by penicillin-resistant staphylococci. Lancet *2*, 641–644.

Barna, J.C., and Williams, D.H. (1984). The structure and mode of action of glycopeptide antibiotics of the vancomycin group. Annu. Rev. Microbiol. *38*, 339–357.

Beabout, K., Hammerstrom, T.G., Perez, A.M., Magalhaes, B.F., Prater, A.G., Clements, T.P., Arias, C.A., Saxer, G., and Shamoo, Y. (2015). The ribosomal S10 protein is a general target for decreased tigecycline susceptibility. Antimicrob. Agents. Chemother. *59*, 5561–5566.

Becker, K., Ballhausen, B., Köck, R., and Kriegeskorte, A. (2014). Methicillin resistance in *Staphylococcus* isolates: the "mec alphabet" with specific consideration of *mecC*, a mec homolog associated with zoonotic *S. aureus* lineages. Int. J. Med. Microbiol. *304*, 794–804.

Berger-Bächi, B., and Tschierske, M. (1998). Role of fem factors in methicillin resistance. Drug Resist. Updat. *1*, 325–335.

Berscheid, A., François, P., Strittmatter, A., Gottschalk, G., Schrenzel, J., Sass, P., and Bierbaum, G. (2014). Generation of a vancomycin-intermediate *Staphylococcus aureus* (VISA) strain by two amino acid exchanges in VraS. J. Antimicrob. Chemother. *69*, 3190–3198.

Bertsche, U., Weidenmaier, C., Kuehner, D., Yang, S.J., Baur, S., Wanner, S., Francois, P., Schrenzel, J., Yeaman, M.R., and Bayer, A.S. (2011). Correlation of daptomycin resistance in a clinical *Staphylococcus aureus* strain with increased cell wall teichoic acid production and D-alanylation. Antimicrob. Agents. Chemother. *55*, 3922–3928.

Blake, K.L., O'Neill, A.J., Mengin-Lecreulx, D., Henderson, P.J., Bostock, J.M., Dunsmore, C.J., Simmons, K.J., Fishwick, C.W., Leeds, J.A., and Chopra, I. (2009). The nature of *Staphylococcus aureus* MurA and MurZ and approaches for detection of peptidoglycan biosynthesis inhibitors. Mol. Microbiol. *72*, 335–343.

Blondeau, J.M. (2004). Fluoroquinolones: mechanism of action, classification, and development of resistance. Surv. Ophthalmol. *49* (Suppl. 2), S73–8.

Boucher, H.W., and Sakoulas, G. (2007). Perspectives on Daptomycin resistance, with emphasis on resistance in *Staphylococcus aureus*. Clin. Infect. Dis. *45*, 601–608.

Boundy, S., Safo, M.K., Wang, L., Musayev, F.N., O'Farrell, H.C., Rife, J.P., and Archer, G.L. (2013). Characterization of the *Staphylococcus aureus* rRNA methyltransferase encoded by *orfX*, the gene containing the staphylococcal chromosome Cassette mec (SCCmec) insertion site. J. Biol. Chem. *288*, 132–140.

Boyle-Vavra, S., Carey, R.B., and Daum, R.S. (2001a). Development of vancomycin and lysostaphin resistance in a methicillin-resistant *Staphylococcus aureus* isolate. J. Antimicrob. Chemother. *48*, 617–625.

Boyle-Vavra, S., Challapalli, M., and Daum, R.S. (2003). Resistance to autolysis in vancomycin-selected *Staphylococcus aureus* isolates precedes vancomycin-intermediate resistance. Antimicrob. Agents. Chemother. *47*, 2036–2039.

Boyle-Vavra, S., Jones, M., Gourley, B.L., Holmes, M., Ruf, R., Balsam, A.R., Boulware, D.R., Kline, S., Jawahir, S., Devries, A., *et al.* (2011). Comparative genome sequencing of an isogenic pair of USA800 clinical methicillin-resistant *Staphylococcus aureus* isolates obtained before and after daptomycin treatment failure. Antimicrob. Agents. Chemother. *55*, 2018–2025.

Boyle-Vavra, S., Labischinski, H., Ebert, C.C., Ehlert, K., and Daum, R.S. (2001b). A spectrum of changes occurs in peptidoglycan composition of glycopeptide-intermediate clinical *Staphylococcus aureus* isolates. Antimicrob. Agents. Chemother. *45*, 280–287.

Boyle-Vavra, S., Yin, S., and Daum, R.S. (2006). The VraS/VraR two-component regulatory system required for oxacillin resistance in community-acquired methicillin-resistant *Staphylococcus aureus*. FEMS Microbiol. Lett. *262*, 163–171.

Boyle-Vavra, S., Yin, S., Jo, D.S., Montgomery, C.P., and Daum, R.S. (2013). VraT/YvqF is required for methicillin resistance and activation of the VraSR regulon in *Staphylococcus aureus*. Antimicrob. Agents Chemother. *57*, 83–95.

Bugg, T.D., Braddick, D., Dowson, C.G., and Roper, D.I. (2011). Bacterial cell wall assembly: still an attractive antibacterial target. Trends Biotechnol. *29*, 167–173.

Camargo, I.L., Neoh, H.M., Cui, L., and Hiramatsu, K. (2008). Serial daptomycin selection generates daptomycin-nonsusceptible *Staphylococcus aureus* strains with a heterogeneous vancomycin-intermediate phenotype. Antimicrob. Agents. Chemother. *52*, 4289–4299.

Chambers, H.F., Basuino, L., Diep, B.A., Steenbergen, J., Zhang, S., Tattevin, P., and Alder, J. (2009). Relationship between susceptibility to daptomycin *in vitro* and activity *in vivo* in a rabbit model of aortic valve endocarditis. Antimicrob. Agents. Chemother. *53*, 1463–1467.

Chambers, H.F., and Deleo, F.R. (2009). Waves of resistance: *Staphylococcus aureus* in the antibiotic era. Nat. Rev. Microbiol. *7*, 629–641.

Chang, S., Sievert, D.M., Hageman, J.C., Boulton, M.L., Tenover, F.C., Downes, F.P., Shah, S., Rudrik, J.T., Pupp, G.R., Brown, W.J., *et al.* (2003). Infection with vancomycin-resistant *Staphylococcus aureus* containing the vanA resistance gene. N. Engl. J. Med. *348*, 1342–1347.

Chastre, J., Blasi, F., Masterton, R.G., Rello, J., Torres, A., and Welte, T. (2014). European perspective and update on the management of nosocomial pneumonia due to methicillin-resistant *Staphylococcus aureus* after more than 10 years of experience with linezolid. Clin. Microbiol. Infect. *20 Suppl 4*, 19–36.

Coskun-Ari, F.F., and Bosgelmez-Tinaz, G. (2008). *grlA* and *gyrA* mutations and antimicrobial susceptibility in clinical isolates of ciprofloxacin- methicillin-resistant *Staphylococcus aureus*. Eur. J. Med. Res. *13*, 366–370.

Courvalin, P. (2006). Vancomycin resistance in gram-positive cocci. Clin. Infect. Dis. *42* (Suppl. 1), S25–34.

Cui, L., Isii, T., Fukuda, M., Ochiai, T., Neoh, H.M., Camargo, I.L., Watanabe, Y., Shoji, M., Hishinuma, T., and Hiramatsu, K. (2010). An RpoB mutation confers dual heteroresistance to daptomycin and vancomycin in *Staphylococcus aureus*. Antimicrob. Agents Chemother. *54*, 5222–5233.

Cui, L., Ma, X., Sato, K., Okuma, K., Tenover, F.C., Mamizuka, E.M., Gemmell, C.G., Kim, M.N., Ploy, M.C., El-Solh, N., *et al.* (2003). Cell wall thickening is a common feature of vancomycin resistance in *Staphylococcus aureus*. J. Clin. Microbiol. *41*, 5–14.

Cui, L., Murakami, H., Kuwahara-Arai, K., Hanaki, H., and Hiramatsu, K. (2000). Contribution of a thickened cell wall and its glutamine nonamidated component to the vancomycin resistance expressed by *Staphylococcus aureus* Mu50. Antimicrob. Agents. Chemother. *44*, 2276–2285.

Cui, L., Neoh, H.M., Shoji, M., and Hiramatsu, K. (2009). Contribution of *vraSR* and *graSR* point mutations to vancomycin resistance in vancomycin-intermediate *Staphylococcus aureus*. Antimicrob. Agents. Chemother. *53*, 1231–1234.

Cui, L., Tominaga, E., Neoh, H.M., and Hiramatsu, K. (2006). Correlation between reduced daptomycin susceptibility and vancomycin Resistance in Vancomycin-Intermediate *Staphylococcus aureus*. Antimicrob. Agents. Chemother. *50*, 1079–1082.

Dale, G.E., Broger, C., D'Arcy, A., Hartman, P.G., DeHoogt, R., Jolidon, S., Kompis, I., Labhardt, A.M., Langen, H., Locher, H., *et al.* (1997). A single amino acid substitution in *Staphylococcus aureus* dihydrofolate reductase determines trimethoprim resistance. J. Mol. Biol. *266*, 23–30.

Daum, R.S., Gupta, S., Sabbagh, R., and Milewski, W.M. (1992). Characterization of *Staphylococcus aureus* isolates with decreased susceptibility to vancomycin and teicoplanin: isolation and purification of a constitutively produced protein associated with decreased susceptibility. J. Infect. Dis. *166*, 1066–1072.

Daum, R.S., Ito, T., Hiramatsu, K., Hussain, F., Mongkolrattanothai, K., Jamklang, M., and Boyle-Vavra, S. (2002). A novel methicillin-resistance cassette in community-acquired methicillin-resistant *Staphylococcus aureus* isolates of diverse genetic backgrounds. J. Infect. Dis. *186*, 1344–1347.

de Jonge, B.L., Chang, Y.S., Gage, D., and Tomasz, A. (1992). Peptidoglycan composition of a highly methicillin-resistant *Staphylococcus aureus* strain. The role of penicillin binding protein 2A. J. Biol. Chem. *267*, 11248–11254.

Deiters, C., Gunnewig, V., Friedrich, A.W., Mellmann, A., and Kock, R. (2015). Are cases of Methicillin-resistant *Staphylococcus aureus* clonal complex (CC) 398 among humans still livestock-associated? Int. J. Med. Microbiol. *305*, 110–113.

Delaune, A., Dubrac, S., Blanchet, C., Poupel, O., Mader, U., Hiron, A., Leduc, A., Fitting, C., Nicolas, P., Cavaillon, J.M., *et al.* (2012). The WalKR system controls major staphylococcal virulence genes and is involved in triggering the host inflammatory response. Infect. Immun. *80*, 3438–3453.

Diaz, L., Tran, T.T., Munita, J.M., Miller, W.R., Rincon, S., Carvajal, L.P., Wollam, A., Reyes, J., Panesso, D., Rojas, N.L., *et al.* (2014). Whole-genome analyses of *Enterococcus faecium* isolates with diverse daptomycin MICs. Antimicrob. Agents Chemother. *58*, 4527–4534.

Diep, B.A., Gill, S.R., Chang, R.F., Phan, T.H., Chen, J.H., Davidson, M.G., Lin, F., Lin, J., Carleton, H.A., Mongodin, E.F., *et al.* (2006). Complete genome sequence of USA300, an epidemic clone of community-acquired meticillin-resistant *Staphylococcus aureus*. Lancet *367*, 731–739.

Dmitriev, B.A., Toukach, F.V., Holst, O., Rietschel, E.T., and Ehlers, S. (2004). Tertiary structure of *Staphylococcus aureus* cell wall murein. J. Bacteriol. *186*, 7141–7148.

Draghi, D.C., Benton, B.M., Krause, K.M., Thornsberry, C., Pillar, C., and Sahm, D.F. (2008). Comparative surveillance study of telavancin activity against recently collected gram-positive clinical isolates from across the United States. Antimicrob. Agents. Chemother. *52*, 2383–2388.

Dubrac, S., Boneca, I.G., Poupel, O., and Msadek, T. (2007). New insights into the WalK/WalR (YycG/YycF) essential signal transduction pathway reveal a major role in controlling cell wall metabolism and biofilm formation in *Staphylococcus aureus*. J. Bacteriol. *189*, 8257–8269.

Dyke, K.G. (1969). Penicillinase production and intrinsic resistance to penicillins in methicillin-resistant cultures of *Staphylococcus aureus*. J. Med. Microbiol. *2*, 261–278.

Eells, S.J., McKinnell, J.A., Wang, A.A., Green, N.L., Whang, D., O'Hara, P., Brown, M.L., and Miller, L.G. (2013). A comparison of clinical outcomes between healthcare-associated infections due to community-associated methicillin-resistant *Staphylococcus aureus* strains and healthcare-associated methicillin-resistant *S. aureus* strains. Epidemiol. Infect. *141*, 2140–2148.

Ender, M., McCallum, N., and Berger-Bächi, B. (2008). Impact of *mecA* promoter mutations on *mecA* expression and β-lactam resistance levels. Int. J. Med. Microbiol. *298*, 607–617.

Enoch, D.A., Bygott, J.M., Daly, M.L., and Karas, J.A. (2007). Daptomycin. J. Infect. *55*, 205–213.

Ernst, C.M., Staubitz, P., Mishra, N.N., Yang, S.J., Hornig, G., Kalbacher, H., Bayer, A.S., Kraus, D., and Peschel, A. (2009). The bacterial defensin resistance protein MprF consists of separable domains for lipid lysinylation and antimicrobial peptide repulsion. PLOS Pathog. *5*, e1000660.

FDA Drug Safety Communication: Increased risk of death with Tygacil (tigecycline) compared to other antibiotics used to treat similar infections. (2013).

Finan, J.E., Archer, G.L., Pucci, M.J., and Climo, M.W. (2001). Role of penicillin-binding protein 4 in expression of vancomycin resistance among clinical isolates of oxacillin-resistant *Staphylococcus aureus*. Antimicrob. Agents. Chemother. *45*, 3070–3075.

Fines, M., and Leclercq, R. (2000). Activity of linezolid against Gram-positive cocci possessing genes conferring resistance to protein synthesis inhibitors. Journal of Antimicrob. Chemother. *45*, 797–802.

Fleming, A. (1942). *In vitro* tests of penicillin potency. Lancet, *i*, 732.

Fleming, A. (1945). Penicillin. "Sir Alexander Fleming - Nobel Lecture: Penicillin". *Nobelprize.org.* Nobel Media AB 2014. Web. 3 Mar 2016. <http://www.nobelprize.org/nobel_prizes/medicine/laureates/1945/fleming-lecture.html

Fournier, B., Aras, R., and Hooper, D.C. (2000). Expression of the multidrug resistance transporter NorA from *Staphylococcus aureus* is modified by a two-component regulatory system. J. Bacteriol. *182*, 664–671.

Fournier, B., Truong-Bolduc, Q.C., Zhang, X., and Hooper, D.C. (2001). A mutation in the 5' untranslated region increases stability of *norA* mRNA, encoding a multidrug resistance transporter of *Staphylococcus aureus*. J. Bacteriol. *183*, 2367–2371.

Fowler, V.G., Boucher, H.W., Corey, G.R., Abrutyn, E., Karchmer, A.W., Rupp, M.E., Levine, D.P., Chambers, H.F., Tally, F.P., Vigliani, G.A., *et al.* (2006). Daptomycin versus standard therapy for bacteremia and endocarditis caused by *Staphylococcus aureus*. N. Engl. J. Med. *355*, 653–665.

Fridkin, S.K., Hageman, J., McDougal, L.K., Mohammed, J., Jarvis, W.R., Perl, T.M., and Tenover, F.C. (2003). Epidemiological and microbiological characterization of infections caused by *Staphylococcus aureus* with reduced susceptibility to vancomycin, United States, 1997–2001. Clin. Infect. Dis. *36*, 429–439.

Fridkin, S.K., Hageman, J.C., Morrison, M., Sanza, L.T., Como-Sabetti, K., Jernigan, J.A., Harriman, K., Harrison, L.H., Lynfield, R., Farley, M.M., and Active Bacterial Core Surveillance Program of the Emerging Infections Program Network. (2005). Methicillin-resistant *Staphylococcus aureus* disease in three communities. N. Engl. J. Med. *352*, 1436–1444.

Friedman, L., Alder, J.D., and Silverman, J.A. (2006). Genetic changes that correlate with reduced susceptibility to daptomycin in *Staphylococcus aureus*. Antimicrob. Agents Chemother. *50*, 2137–2145.

Garcia-Alvarez, L., Holden, M.T., Lindsay, H., Webb, C.R., Brown, D.F., Curran, M.D., Walpole, E., Brooks, K., Pickard, D.J., Teale, C., *et al.* (2011). Meticillin-resistant *Staphylococcus aureus* with a novel *mecA* homologue in human and bovine populations in the UK and Denmark: a descriptive study. Lancet Infect. Dis. *11*, 595–603.

Gardete, S., Kim, C., Hartmann, B.M., Mwangi, M., Roux, C.M., Dunman, P.M., Chambers, H.F., and Tomasz, A. (2012). Genetic pathway in acquisition and loss of vancomycin resistance in a methicillin resistant *Staphylococcus aureus* (MRSA) strain of clonal type USA300. PLOS Pathog. *8*, e1002505.

Gaupp, R., Lei, S., Reed, J.M., Peisker, H., Boyle-Vavra, S., Bayer, A.S., Bischoff, M., Herrmann, M., Daum, R.S., Powers, R., *et al.* (2015). *Staphylococcus aureus* metabolic adaptations during the transition from a daptomycin susceptibility phenotype to a daptomycin nonsusceptibility phenotype. Antimicrob. Agents. Chemother. *59*, 4226–4238.

Gregory, P.D., Lewis, R.A., Curnock, S.P., and Dyke, K.G. (1997). Studies of the repressor (BlaI) of β-lactamase synthesis in *Staphylococcus aureus*. Mol. Microbiol. *24*, 1025–1037.

Grim, S.A., Rapp, R.P., Martin, C.A., and Evans, M.E. (2005). Trimethoprim-sulfamethoxazole as a viable treatment option for infections caused by methicillin-resistant *Staphylococcus aureus*. Pharmacotherapy *25*, 253–264.

Hackbarth, C.J., and Chambers, H.F. (1993). *blaI* and *blaR1* regulate β-lactamase and PBP 2a production in methicillin-resistant *Staphylococcus aureus*. Antimicrob. Agents Chemother. *37*, 1144–1149.

Hafer, C., Lin, Y., Kornblum, J., Lowy, F.D., and Uhlemann, A.C. (2012). Contribution of selected gene mutations to resistance in clinical isolates of vancomycin-intermediate *Staphylococcus aureus*. Antimicrob. Agents. Chemother. *56*, 5845–5851.

Hanaki, H., Kuwahara-Arai, K., Boyle-Vavra, S., Daum, R.S., Labischinski, H., and Hiramatsu, K. (1998). Activated cell wall synthesis is associated with vancomycin resistance in methicillin-resistant *Staphylococcus aureus* clinical strains Mu3 and Mu50. J. Antimicrob. Chemother. *42*, 199–209.

Hao, H., Cheng, G., Dai, M., Wu, Q., and Yuan, Z. (2012). Inhibitors targeting on cell wall biosynthesis pathway of MRSA. Mol. Biosyst. *8*, 2828–2838.

Hartman, B.J., and Tomasz, A. (1984). Low-affinity penicillin-binding protein associated with β-lactam resistance in *Staphylococcus aureus*. J. Bacteriol. *158*, 513–516.

Hayden, M.K., Rezai, K., Hayes, R.A., Lolans, K., Quinn, J.P., and Weinstein, R.A. (2005). Development of Daptomycin resistance *in vivo* in methicillin-resistant *Staphylococcus aureus*. J. Clin. Microbiol. *43*, 5285–5287.

Hegde, S.S., Difuntorum, S., Skinner, R., Trumbull, J., and Krause, K.M. (2009). Efficacy of telavancin against glycopeptide-intermediate *Staphylococcus aureus* in the neutropenic mouse bacteraemia model. J. Antimicrob. Chemother. *63*, 763–766.

Hegde, S.S., and Janc, J.W. (2014). Efficacy of telavancin, a lipoglycopeptide antibiotic, in experimental models of Gram-positive infection. Expert. Rev. Anti. Infect. Ther. *12*, 1463–1475.

Hegde, S.S., Reyes, N., Skinner, R., and Difuntorum, S. (2008). Efficacy of telavancin in a murine model of pneumonia induced by methicillin-susceptible *Staphylococcus aureus*. J. Antimicrob. Chemother. *61*, 169–172.

Hegde, S.S., Skinner, R., Lewis, S.R., Krause, K.M., Blais, J., and Benton, B.M. (2010). Activity of telavancin against heterogeneous vancomycin-intermediate *Staphylococcus aureus* (hVISA) *in vitro* and in an *in vivo* mouse model of bacteraemia. J. Antimicrob. Chemother. *65*, 725–728.

Hiramatsu, K., Aritaka, N., Hanaki, H., Kawasaki, S., Hosoda, Y., Hori, S., Fukuchi, Y., and Kobayashi, I. (1997a). Dissemination in Japanese hospitals of strains of *Staphylococcus aureus* heterogeneously resistant to vancomycin. Lancet *350*, 1670–1673.

Hiramatsu, K., Asada, K., Suzuki, E., Okonogi, K., and Yokota, T. (1992). Molecular cloning and nucleotide sequence determination of the regulator region of *mecA* gene in methicillin-resistant *Staphylococcus aureus* (MRSA). FEBS Lett *298*, 133–136.

Hiramatsu, K., Hanaki, H., Ino, T., Yabuta, K., Oguri, T., and Tenover, F.C. (1997b). Methicillin-resistant *Staphylococcus aureus* clinical strain with reduced vancomycin susceptibility. J. Antimicrob. Chemother. *40*, 135–136.

Hiramatsu, K., Suzuki, E., Takayama, H., Katayama, Y., and Yokota, T. (1990). Role of penicillinase plasmids in the stability of the *mecA* gene in methicillin-resistant *Staphylococcus aureus*. Antimicrob. Agents Chemother. *34*, 600–604.

Hiron, A., Falord, M., Valle, J., Débarbouillé, M., and Msadek, T. (2011). Bacitracin and nisin resistance in *Staphylococcus aureus*: a novel pathway involving the BraS/BraR two-component system (SA2417/SA2418) and both the BraD/BraE and VraD/VraE ABC transporters. Mol. Microbiol. *81*, 602–622.

Hitchings, G.H. (1973). Mechanism of action of trimethoprim-sulfamethoxazole. J. Infect. Dis. *128* (Suppl.), 433–436.

Hooper, D.C. (1999). Mechanisms of fluoroquinolone resistance. Drug. Resist. Updat. *2*, 38–55.

Hooper, D.C. (2000). Mechanisms of action and resistance of older and newer fluoroquinolones. Clin. Infect. Dis. *31* (Suppl. 2), S24–S28.

Hooper, D.C. (2002). Fluoroquinolone resistance among Gram-positive cocci. Lancet. Infect. Dis. *2*, 530–538.

Howden, B.P., McEvoy, C.R., Allen, D.L., Chua, K., Gao, W., Harrison, P.F., Bell, J., Coombs, G., Bennett-Wood, V., Porter, J.L., *et al.* (2011). Evolution of multidrug resistance during *Staphylococcus aureus* infection involves mutation of the essential two component regulator WalKR. PLOS Pathog. *7*, e1002359.

Howden, B.P., Peleg, A.Y., and Stinear, T.P. (2014). The evolution of vancomycin intermediate *Staphylococcus aureus* (VISA) and heterogenous-VISA. Infect. Genet. Evol. *21*, 575–582.

Hulko, M., Berndt, F., Gruber, M., Linder, J.U., Truffault, V., Schultz, A., Martin, J., Schultz, J.E., Lupas, A.N., and Coles, M. (2006). The HAMP domain structure implies helix rotation in transmembrane signaling. Cell *126*, 929–940.

Ince, D., and Hooper, D.C. (2001). Mechanisms and frequency of resistance to gatifloxacin in comparison to AM-1121 and ciprofloxacin in *Staphylococcus aureus.* Antimicrob. Agents Chemother. *45*, 2755–2764.

Ito, T., Hiramatsu, K., Tomasz, A., de Lencastre, H., Perreten, V., Holden, M.T., Coleman, D.C., Goering, R., Giffard, P.M., Skov, R.L., *et al.* (2012). Guidelines for reporting novel *mecA* gene homologues. Antimicrob. Agents Chemother. *56*, 4997–4999.

Ito, T., Hiramatsu, K., Tomasz, A., de Lencastre, H., Perreten, V., Holden, M.T., Coleman, D.C., Goering, R., Giffard, P.M., Skov, R.L., *et al.* (2009). Classification of Staphylococcal Cassette Chromosome mec (SCCmec): Guidelines for Reporting Novel SCCmec Elements. Antimicrob. Agents. Chemother. *53*, 4961–4967.

Ito, T., Katayama, Y., Asada, K., Mori, N., Tsutsumimoto, K., Tiensasitorn, C., and Hiramatsu, K. (2001). Structural comparison of three types of staphylococcal cassette chromosome *mec* integrated in the chromosome in methicillin-resistant *Staphylococcus aureus.* Antimicrob. Agents. Chemother. *45*, 1323–1336.

Jacobson, L.M., Milstone, A.M., Zenilman, J., Carroll, K.C., and Arav-Boger, R. (2009). Daptomycin therapy failure in an adolescent with methicillin-resistant *Staphylococcus aureus* bacteremia. Pediatr. Infect. Dis. J. *28*, 445–447.

Jafari Saraf, L., and Wilson, S.E. (2011). Telavancin, a new lipoglycopeptide antimicrobial, in complicated skin and soft tissue infections. Infect. Drug Resist. *4*, 87–95.

Jansen, A., Türck, M., Szekat, C., Nagel, M., Clever, I., and Bierbaum, G. (2007). Role of insertion elements and *yycFG* in the development of decreased susceptibility to vancomycin in *Staphylococcus aureus.* Int. J. Med. Microbiol. *297*, 205–215.

Jensen, S.O., and Lyon, B.R. (2009). Genetics of antimicrobial resistance in *Staphylococcus aureus.* Future Microbiol. *4*, 565–582.

Jevons, M., Rolinson, G.N., and Knox, R. (1961). "Celbenin"-Resistant Staphylococci. The British Medical Journal *1*, 124–126.

Jordan, S., Junker, A., Helmann, J.D., and Mascher, T. (2006). Regulation of LiaRS-dependent gene expression in bacillus subtilis: identification of inhibitor proteins, regulator binding sites, and target genes of a conserved cell envelope stress-sensing two-component system. J. Bacteriol. *188*, 5153–5166.

Julian, K., Kosowska-Shick, K., Whitener, C., Roos, M., Labischinski, H., Rubio, A., Parent, L., Ednie, L., Koeth, L., Bogdanovich, T., *et al.* (2007). Characterization of a daptomycin-nonsusceptible vancomycin-intermediate *Staphylococcus aureus* strain in a patient with endocarditis. Antimicrob. Agents. Chemother. *51*, 3445–3448.

Kaatz, G.W., Moudgal, V.V., and Seo, S.M. (2002). Identification and characterization of a novel efflux-related multidrug resistance phenotype in *Staphylococcus aureus.* J. Antimicrob. Chemother. *50*, 833–838.

Kaatz, G.W., Seo, S.M., O'Brien, L., Wahiduzzaman, M., and Foster, T.J. (2000). Evidence for the existence of a multidrug efflux transporter distinct from NorA in *Staphylococcus aureus.* Antimicrob. Agents Chemother. *44*, 1404–1406.

Kadlec, K., and Schwarz, S. (2009). Identification of a novel trimethoprim resistance gene, *dfrK*, in a methicillin-resistant *Staphylococcus aureus* ST398 strain and its physical linkage to the tetracycline resistance gene *tet*(L). Antimicrob. Agents. Chemother. *53*, 776–778.

Katayama, Y., Ito, T., and Hiramatsu, K. (2000). A new class of genetic element, staphylococcus cassette chromosome mec, encodes methicillin resistance in *Staphylococcus aureus.* Antimicrob. Agents. Chemother. *44*, 1549–1555.

Katayama, Y., Murakami-Kuroda, H., Cui, L., and Hiramatsu, K. (2009). Selection of heterogeneous vancomycin-intermediate *Staphylococcus aureus* by imipenem. Antimicrob. Agents. Chemother. *53*, 3190–3196.

Kato, Y., Suzuki, T., Ida, T., and Maebashi, K. (2010). Genetic changes associated with glycopeptide resistance in *Staphylococcus aureus*: predominance of amino acid substitutions in YvqF/VraSR. J. Antimicrob. Chemother. *65*, 37–45.

Kehrenberg, C., Cuny, C., Strommenger, B., Schwarz, S., and Witte, W. (2009). Methicillin-resistant and -susceptible *Staphylococcus aureus* strains of clonal lineages ST398 and ST9 from swine carry the multidrug resistance gene cfr. Antimicrob. Agents. Chemother. *53*, 779–781.

Kehrenberg, C., Schwarz, S., Jacobsen, L., Hansen, L.H., and Vester, B. (2005). A new mechanism for chloramphenicol, florfenicol and clindamycin resistance: methylation of 23S ribosomal RNA at A2503. Mol. Microbiol. *57*, 1064–1073.

Kirby, A., Mohandas, K., Broughton, C., Neal, T.J., Smith, G.W., Pai, P., and Nistal de Paz, C. (2009). *In vivo* development of heterogeneous glycopeptide-intermediate *Staphylococcus aureus* (hGISA), GISA and daptomycin resistance in a patient with meticillin-resistant *S. aureus* endocarditis. J. Med. Microbiol. *58*, 376–380.

Klevens, R.M., Morrison, M.A., Nadle, J., Petit, S., Gershman, K., Ray, S., Harrison, L.H., Lynfield, R.,

Dumyati, G., Townes, J.M., *et al.* (2007). Invasive methicillin-resistant *Staphylococcus aureus* infections in the United States. JAMA *298*, 1763–1771.

Klinker, K.P., and Borgert, S.J. (2015). Beyond Vancomycin: The Tail of the Lipoglycopeptides. Clin. Ther. *37*, 2619–2636.

Kos, V.N., Desjardins, C.A., Griggs, A., Cerqueira, G., in Tonder, A., Holden, M.T., Godfrey, P., Palmer, K.L., Bodi, K., Mongodin, E.F., *et al.* (2012). Comparative genomics of vancomycin-resistant *Staphylococcus aureus* strains and their positions within the clade most commonly associated with Methicillin-resistant *S. aureus* hospital-acquired infection in the United States. MBio 3, 00112–00112.

Kosowska-Shick, K., Clark, C., Pankuch, G.A., McGhee, P., Dewasse, B., Beachel, L., and Appelbaum, P.C. (2009). Activity of telavancin against staphylococci and enterococci determined by MIC and resistance selection studies. Antimicrob. Agents. Chemother. *53*, 4217–4224.

Kuroda, M., Ohta, T., Uchiyama, I., Baba, T., Yuzawa, H., Kobayashi, I., Cui, L., Oguchi, A., Aoki, K., Nagai, Y., *et al.* (2001). Whole genome sequencing of meticillin-resistant Staphylococcus aureus. Lancet *357*, 1225–1240

Kuroda, M., Kuroda, H., Oshima, T., Takeuchi, F., Mori, H., and Hiramatsu, K. (2003). Two-component system VraSR positively modulates the regulation of cell wall biosynthesis pathway in *Staphylococcus aureus.* Mol. Microbiol. *49*, 807–821.

Kuwahara-Arai, K., Kondo, N., Hori, S., Tateda-Suzuki, E., and Hiramatsu, K. (1996). Suppression of methicillin resistance in a *mecA*-containing pre-methicillin-resistant *Staphylococcus aureus* strain is caused by the *mecI*-mediated repression of PBP 2' production. Antimicrob. Agents. Chemother. *40*, 2680–2685.

Kwak, Y.G., Truong-Bolduc, Q.C., Bin Kim, H., Song, K.H., Kim, E.S., and Hooper, D.C. (2013). Association of *norB* overexpression and fluoroquinolone resistance in clinical isolates of *Staphylococcus aureus* from Korea. J. Antimicrob. Chemother. *68*, 2766–2772.

Lahiri, S.D., McLaughlin, R.E., Whiteaker, J.D., Ambler, J.E., and Alm, R.A. (2015). Molecular characterization of MRSA isolates bracketing the current EUCAST ceftaroline-susceptible breakpoint for *Staphylococcus aureus*: the role of PBP2a in the activity of ceftaroline. J. Antimicrob. Chemother. *70*, 2488–2498.

Larsen, J., Petersen, A., Sørum, M., Stegger, M., van Alphen, L., Valentiner-Branth, P., Knudsen, L.K., Larsen, L.S., Feingold, B., Price, L.B., *et al.* (2015). Meticillin-resistant *Staphylococcus aureus* CC398 is an increasing cause of disease in people with no livestock contact in Denmark, 1999 to 2011. Euro. Surveill. *20*.

Leuthner, K.D., Cheung, C.M., and Rybak, M.J. (2006). Comparative activity of the new lipoglycopeptide telavancin in the presence and absence of serum against 50 glycopeptide non-susceptible staphylococci and three vancomycin-resistant *Staphylococcus aureus.* J. Antimicrob. Chemother. *58*, 338–343.

Lewis, R.A., and Dyke, K.G. (2000). MecI represses synthesis from the β-lactamase operon of *Staphylococcus aureus.* J. Antimicrob. Chemother. *45*, 139–144.

Limbago, B.M., Kallen, A.J., Zhu, W., Eggers, P., McDougal, L.K., and Albrecht, V.S. (2014). Report of the 13th vancomycin-resistant *Staphylococcus aureus* isolate from the United States. J. Clin. Microbiol. *52*, 998–1002.

Liu, C., Bayer, A., Cosgrove, S.E., Daum, R.S., Fridkin, S.K., Gorwitz, R.J., Kaplan, S.L., Karchmer, A.W., Levine, D.P., Murray, B.E., *et al.* (2011). Clinical practice guidelines by the Infectious Diseases Society of America for the treatment of methicillin-resistant *Staphylococcus aureus* infections in adults and children: executive summary. Clin. Infect. Dis. *52*, 285–292.

Livermore, D.M. (2000). Antibiotic resistance in staphylococci. Int. J. Antimicrob. Agents *16*(Suppl. 1), S3–S10.

Livermore, D.M., Mushtaq, S., Warner, M., James, D., and Woodford, N. (2015). Susceptibility testing challenges with ceftaroline, MRSA and a 1 mg/L breakpoint. J. Antimicrob. Chemother. *70*, 3259–3266.

Locke, J.B., Hilgers, M., and Shaw, K.J. (2009a). Mutations in ribosomal protein L3 are associated with oxazolidinone resistance in staphylococci of clinical origin. Antimicrob. Agents Chemother. *53*, 5275–5278.

Locke, J.B., Hilgers, M., and Shaw, K.J. (2009b). Novel ribosomal mutations in *Staphylococcus aureus* strains identified through selection with the oxazolidinones linezolid and torezolid (TR-700). Antimicrob. Agents. Chemother. *53*, 5265–5274.

Locke, J.B., Zuill, D.E., Scharn, C.R., Deane, J., Sahm, D.F., Denys, G.A., Goering, R.V., and Shaw, K.J. (2014a). Linezolid-resistant *Staphylococcus aureus* strain 1128105, the first known clinical isolate possessing the cfr multidrug resistance gene. Antimicrob. Agents. Chemother. *58*, 6592–6598.

Locke, J.B., Zurenko, G.E., Shaw, K.J., and Bartizal, K. (2014b). Tedizolid for the management of human infections: *in vitro* characteristics. Clin. Infect. Dis. *58* (Suppl. 1), S35–S42.

Long, K.S., Poehlsgaard, J., Kehrenberg, C., Schwarz, S., and Vester, B. (2006). The Cfr rRNA methyltransferase confers resistance to Phenicols, Lincosamides, Oxazolidinones, Pleuromutilins, and Streptogramin A antibiotics. Antimicrob. Agents. Chemother. *50*, 2500–2505.

Lunde, C.S., Hartouni, S.R., Janc, J.W., Mammen, M., Humphrey, P.P., and Benton, B.M. (2009). Telavancin disrupts the functional integrity of the bacterial membrane through targeted interaction with the cell wall precursor lipid II. Antimicrob. Agents. Chemother. *53*, 3375–3383.

Luong, T.T., Dunman, P.M., Murphy, E., Projan, S.J., and Lee, C.Y. (2006). Transcription profiling of the mgrA Regulon in *Staphylococcus aureus.* J. Bacteriol. *188*, 1899–1910.

Luong, T.T., Ouyang, S., Bush, K., and Lee, C.Y. (2002). Type 1 capsule genes of *Staphylococcus aureus* are carried in a staphylococcal cassette chromosome genetic element. J. Bacteriol. *184*, 3623–3629.

Mascher, T. (2014). Bacterial (intramembrane-sensing) histidine kinases: signal transfer rather than stimulus perception. Trends Microbiol. *22*, 559–565.

Matsuhashi, M., Song, M.D., Ishino, F., Wachi, M., Doi, M., Inoue, M., Ubukata, K., Yamashita, N., and Konno, M.

(1986). Molecular cloning of the gene of a penicillin-binding protein supposed to cause high resistance to β-lactam antibiotics in *Staphylococcus aureus*. J. Bacteriol. *167*, 975–980.

McAleese, F., Petersen, P., Ruzin, A., Dunman, P.M., Murphy, E., Projan, S.J., and Bradford, P.A. (2005). A novel MATE family efflux pump contributes to the reduced susceptibility of laboratory-derived *Staphylococcus aureus* mutants to tigecycline. Antimicrob. Agents. Chemother. *49*, 1865–1871.

McAleese, F., Wu, S.W., Sieradzki, K., Dunman, P., Murphy, E., Projan, S., and Tomasz, A. (2006). Overexpression of genes of the cell wall stimulon in clinical isolates of *Staphylococcus aureus* exhibiting vancomycin-intermediate- *S. aureus*-type resistance to vancomycin. J. Bacteriol. *188*, 1120–1133.

McCallum, N., Meier, P.S., Heusser, R., and Berger-Bachi, B. (2011). Mutational analyses of open reading frames within the *vraSR* operon and their roles in the cell wall stress response of *Staphylococcus aureus*. Antimicrob. Agents. Chemother. *55*, 1391–1402.

McEvoy, C.R., Tsuji, B., Gao, W., Seemann, T., Porter, J.L., Doig, K., Ngo, D., Howden, B.P., and Stinear, T.P. (2013). Decreased vancomycin susceptibility in *Staphylococcus aureus* caused by IS*256* tempering of WalKR expression. Antimicrob. Agents. Chemother. *57*, 3240–3249.

McGovern, P.C., Wible, M., El-Tahtawy, A., Biswas, P., and Meyer, R.D. (2013). All-cause mortality imbalance in the tigecycline phase 3 and 4 clinical trials. Int. J. Antimicrob. Agents *41*, 463–467.

McKinney, T.K., Sharma, V.K., Craig, W.A., and Archer, G.L. (2001). Transcription of the gene mediating methicillin resistance in *Staphylococcus aureus* (*mecA*) is corepressed but not coinduced by cognate *mecA* and β-lactamase regulators. J. Bacteriol. *183*, 6862–6868.

McMurray, L.M., and Levy, S.B. (2000). Tetracycline Resistance in Gram-Positive Bacteria. In Gram-Positive Pathogens, V.A. Fischetti, R.P. Novick, J.J. Ferritti, D.A. Portnoy, and J.I. Rood, eds. (Washington, DC: ASM Press), pp. 660–667.

Mehta, S., Cuirolo, A.X., Plata, K.B., Riosa, S., Silverman, J.A., Rubio, A., Rosato, R.R., and Rosato, A.E. (2012). VraSR two-component regulatory system contributes to *mprF*-mediated decreased susceptibility to daptomycin in *in vivo*-selected clinical strains of methicillin-resistant *Staphylococcus aureus*. Antimicrob. Agents. Chemother. *56*, 92–102.

Meka, V.G., and Gold, H.S. (2004). Antimicrobial resistance to linezolid. Clin. Infect. Dis. *39*, 1010–1015.

Mendes, R.E., Castanheira, M., Farrell, D.J., Flamm, R.K., Sader, H.S., and Jones, R.N. (2016). Update on dalbavancin activity tested against Gram-positive clinical isolates responsible for documented skin and skin-structure infections in US and European hospitals (2011–13). J. Antimicrob. Chemother. *71*, 276–278.

Mendes, R.E., Deshpande, L.M., Bonilla, H.F., Schwarz, S., Huband, M.D., Jones, R.N., and Quinn, J.P. (2013). Dissemination of a pSCFS3-like *cfr*-carrying plasmid in *Staphylococcus aureus* and *Staphylococcus epidermidis* clinical isolates recovered from hospitals in Ohio. Antimicrob. Agents. Chemother. *57*, 2923–2928.

Mendes, R.E., Deshpande, L.M., and Jones, R.N. (2014). Linezolid update: stable *in vitro* activity following more than a decade of clinical use and summary of associated resistance mechanisms. Drug Resist. Updat. *17*, 1–12.

Mendes, R.E., Farrell, D.J., Sader, H.S., Streit, J.M., and Jones, R.N. (2015). Update of the telavancin activity *in vitro* tested against a worldwide collection of Gram-positive clinical isolates (2013), when applying the revised susceptibility testing method. Diagn. Microbiol. Infect. Dis. *81*, 275–279.

Miller, L.G., Daum, R.S., Creech, C.B., Young, D., Downing, M.D., Eells, S.J., Pettibone, S., Hoagland, R.J., and Chambers, H.F. (2015). Clindamycin versus trimethoprim-sulfamethoxazole for uncomplicated skin infections. N. Engl. J. Med. *372*, 1093–1103.

Mishra, N.N., Yang, S.J., Sawa, A., Rubio, A., Nast, C.C., Yeaman, M.R., and Bayer, A.S. (2009). Analysis of cell membrane characteristics of *in vitro*-selected daptomycin-resistant strains of methicillin-resistant *Staphylococcus aureus*. Antimicrob. Agents. Chemother. *53*, 2312–2318.

Misiura, A., Pigli, Y.Z., Boyle-Vavra, S., Daum, R.S., Boocock, M.R., and Rice, P.A. (2013). Roles of two large serine recombinases in mobilizing the methicillin-resistance cassette SCCmec. Mol. Microbiol. *88*, 1218–1229.

Mitra, S., Saeed, U., Havlichek, D.H., and Stein, G.E. (2015). Profile of oritavancin and its potential in the treatment of acute bacterial skin structure infections. Infect. Drug Resist. *8*, 189–197.

Mole, B. (2013). MRSA: Farming up trouble. Nature *499*, 398–400.

Mongkolrattanothai, K., Boyle, S., Murphy, T.V., and Daum, R.S. (2004). Novel non-*mecA*-containing staphylococcal chromosomal cassette composite island containing pbp4 and tagF genes in a commensal staphylococcal species: a possible reservoir for antibiotic resistance islands in *Staphylococcus aureus*. Antimicrob. Agents. Chemother. *48*, 1823–1836.

Morales, G., Picazo, J.J., Baos, E., Candel, F.J., Arribi, A., Pelaez, B., Andrade, R., de la Torre, M.A., Fereres, J., and Sanchez-Garcia, M. (2010). Resistance to linezolid is mediated by the *cfr* gene in the first report of an outbreak of linezolid-resistant *Staphylococcus aureus*. Clin. Infect. Dis. *50*, 821–825.

Moran, G.J., Krishnadasan, A., Gorwitz, R.J., Fosheim, G.E., McDougal, L.K., Carey, R.B., Talan, D.A., EMERGEncy ID Net Study Group. (2006). Methicillin-resistant *S. aureus* infections among patients in the emergency department. N. Engl. J. Med. *355*, 666–674.

Muraih, J.K., Pearson, A., Silverman, J., and Palmer, M. (2011). Oligomerization of daptomycin on membranes. Biochim. Biophys. Acta *1808*, 1154–1160.

Murakami, K., and Tomasz, A. (1989). Involvement of multiple genetic determinants in high-level methicillin resistance in *Staphylococcus aureus*. J. Bacteriol. *171*, 874–879.

Murthy, M.H., Olson, M.E., Wickert, R.W., Fey, P.D., and Jalali, Z. (2008). Daptomycin non-susceptible meticillin-resistant *Staphylococcus aureus* USA 300 isolate. J. Med. Microbiol. 57 2, 1036–1038.

Mwangi, M.M., Wu, S.W., Zhou, Y., Sieradzki, K., de Lencastre, H., Richardson, P., Bruce, D., Rubin, E.,

Myers, E., Siggia, E.D., *et al.* (2007). Tracking the *in vivo* evolution of multidrug resistance in *Staphylococcus aureus* by whole-genome sequencing. Proc. Natl. Acad. Sci. U.S.A. *104*, 9451–9456.

Nailor, M.D., and Sobel, J.D. (2011). Antibiotics for gram-positive bacterial infection: vancomycin, teicoplanin, quinupristin/dalfopristin, oxazolidinones, daptomycin, telavancin, and ceftaroline. Med. Clin. North. Am. *95*, 723–742, vii.

Nakhla, L.S. (1972). Resistance of *Staphylococcus aureus* to sulphamethoxazole and trimethoprim. J. Clin. Pathol. *25*, 708–712.

Navarre, W.W., and Schneewind, O. (1999). Surface proteins of gram-positive bacteria and mechanisms of their targeting to the cell wall envelope. Microbiol. Mol. Biol. Rev. *63*, 174–229.

Nesin, M., Svec, P., Lupski, J.R., Godson, G.N., Kreiswirth, B., Kornblum, J., and Projan, S.J. (1990). Cloning and nucleotide sequence of a chromosomally encoded tetracycline resistance determinant, *tetA*(M), from a pathogenic, methicillin-resistant strain of *Staphylococcus aureus*. Antimicrob. Agents. Chemother. *34*, 2273–2276.

Neyfakh, A.A., Borsch, C.M., and Kaatz, G.W. (1993). Fluoroquinolone resistance protein NorA of *Staphylococcus aureus* is a multidrug efflux transporter. Antimicrob. Agents. Chemother. *37*, 128–129.

Ng, E.Y., Trucksis, M., and Hooper, D.C. (1994). Quinolone resistance mediated by *norA*: physiologic characterization and relationship to *flqB*, a quinolone resistance locus on the *Staphylococcus aureus* chromosome. Antimicrob. Agents. Chemother. *38*, 1345–1355.

Niemeyer, D.M., Pucci, M.J., Thanassi, J.A., Sharma, V.K., and Archer, G.L. (1996). Role of *mecA* transcriptional regulation in the phenotypic expression of methicillin resistance in *Staphylococcus aureus*. J. Bacteriol. *178*, 5464–5471.

Noto, M.J., and Archer, G.L. (2006). A subset of *Staphylococcus aureus* strains harboring staphylococcal cassette chromosome mec (SCCmec) type IV is deficient in CcrAB-mediated SCCmec excision. Antimicrob. Agents. Chemother. *50*, 2782–2788.

Noto, M.J., Fox, P.M., and Archer, G.L. (2008). Spontaneous deletion of the methicillin resistance determinant, *mecA*, partially compensates for the fitness cost associated with high-level vancomycin resistance in *Staphylococcus aureus*. Antimicrob. Agents. Chemother. *52*, 1221–1229.

Nurjadi, D., Olalekan, A.O., Layer, F., Shittu, A.O., Alabi, A., Ghebremedhin, B., Schaumburg, F., Hofmann-Eifler, J., Van Genderen, P.J., Caumes, E., *et al.* (2014). Emergence of trimethoprim resistance gene *dfrG* in *Staphylococcus aureus* causing human infection and colonization in sub-Saharan Africa and its import to Europe. J. Antimicrob. Chemother. *69*, 2361–2368.

Olson, M.W., Ruzin, A., Feyfant, E., Rush, T.S., O'Connell, J., and Bradford, P.A. (2006). Functional, biophysical, and structural bases for antibacterial activity of tigecycline. Antimicrob. Agents Chemother. *50*, 2156–2166.

Otero, L.H., Rojas-Altuve, A., Llarrull, L.I., Carrasco-Lopez, C., Kumarasiri, M., Lastochkin, E., Fishovitz, J., Dawley, M., Hesek, D., Lee, M., *et al.* (2013). How allosteric control of *Staphylococcus aureus* penicillin binding protein 2a enables methicillin resistance and physiological function. Proc. Natl. Acad. Sci. U.S.A. *110*, 16808–16813.

Patel, J.B., Jevitt, L.A., Hageman, J., McDonald, L.C., and Tenover, F.C. (2006). An association between reduced susceptibility to daptomycin and reduced susceptibility to vancomycin in *Staphylococcus aureus*. Clin. Infect. Dis. *42*, 1652–1653.

Peacock, S.J., and Paterson, G.K. (2015). Mechanisms of Methicillin Resistance in *Staphylococcus aureus*. Annu. Rev. Biochem. *84*, 577–601.

Périchon, B., and Courvalin, P. (2004). Heterologous expression of the enterococcal *vanA* operon in methicillin-resistant *Staphylococcus aureus*. Antimicrob. Agents Chemother. *48*, 4281–4285.

Périchon, B., and Courvalin, P. (2009). VanA-type vancomycin-resistant *Staphylococcus aureus*. Antimicrob. Agents Chemother. *53*, 4580–4587.

Pietiäinen, M., François, P., Hyyryläinen, H.L., Tangomo, M., Sass, V., Sahl, H.G., Schrenzel, J., and Kontinen, V.P. (2009). Transcriptome analysis of the responses of *Staphylococcus aureus* to antimicrobial peptides and characterization of the roles of *vraDE* and *vraSR* in antimicrobial resistance. BMC Genomics *10*, 429.

Pinho, M.G., Filipe, S.R., de Lencastre, H., and Tomasz, A. (2001). Complementation of the essential peptidoglycan transpeptidase function of penicillin-binding protein 2 (PBP2) by the drug resistance protein PBP2A in *Staphylococcus aureus*. J. Bacteriol. *183*, 6525–6531.

Pogliano, J., Pogliano, N., and Silverman, J.A. (2012). Daptomycin-mediated reorganization of membrane architecture causes mislocalization of essential cell division proteins. J. Bacteriol. *194*, 4494–4504.

Policy statement on antimicrobial stewardship by the Society for Healthcare Epidemiology of America (SHEA), the Infectious Diseases Society of America (IDSA), and the Pediatric Infectious Diseases Society (PIDS). Infect. Control. Hosp. Epidemiol. *33*, 322–327. (2012).

Poole, K. (2000). Efflux-mediated resistance to fluoroquinolones in gram-positive bacteria and the mycobacteria. Antimicrob. Agents Chemother. *44*, 2595–2599.

Projan, S. (2000). Antibiotic Resistance in the Staphylococci. In Gram-Positive Pathogens, V.A. Fischetti, R.P. Novick, J.J. Ferritti, D.A. Portnoy, and J.I. Rood, eds. (Washington, DC: ASM Press), p. 463.

Puzniak, L.A., Quintana, A., Wible, M., Babinchak, T., and McGovern, P.C. (2014). Methicillin-resistant *Staphylococcus aureus* infection epidemiology and clinical response from tigecycline soft tissue infection trials. Diagn. Microbiol. Infect. Dis. *79*, 261–265.

Qureshi, N.K., Yin, S., and Boyle-Vavra, S. (2014). The role of the staphylococcal VraTSR regulatory system on vancomycin resistance and *vanA* operon expression in vancomycin-resistant *Staphylococcus aureus*. PLoS One *9*, e85873.

Ray, M.D., Boundy, S., and Archer, G.L. (2016). Transfer of the methicillin resistance genomic island among staphylococci by conjugation. Mol. Microbiol. *100*, 675–685.

Recsei, P.A., Gruss, A.D., and Novick, R.P. (1987). Cloning, sequence, and expression of the lysostaphin gene from

Staphylococcus simulans. Proc. Natl. Acad. Sci. U.S.A. *84*, 1127–1131.

Reipert, A., Ehlert, K., Kast, T., and Bierbaum, G. (2003). Morphological and genetic differences in two isogenic *Staphylococcus aureus* strains with decreased susceptibilities to vancomycin. Antimicrob. Agents. Chemother. *47*, 568–576.

Richter, S.S., Heilmann, K.P., Dohrn, C.L., Riahi, F., Costello, A.J., Kroeger, J.S., Biek, D., Critchley, I.A., Diekema, D.J., and Doern, G.V. (2011). Activity of ceftaroline and epidemiologic trends in *Staphylococcus aureus* isolates collected from 43 medical centers in the United States in 2009. Antimicrob. Agents Chemother. *55*, 4154–4160.

Rohrer, S., Ehlert, K., Tschierske, M., Labischinski, H., and Berger-Bächi, B. (1999). The essential *Staphylococcus aureus* gene *fmhB* is involved in the first step of peptidoglycan pentaglycine interpeptide formation. Proc. Natl. Acad. Sci. U.S.A. *96*, 9351–9356.

Rosato, A.E., Craig, W.A., and Archer, G.L. (2003a). Quantitation of *mecA* transcription in oxacillin-resistant *Staphylococcus aureus* clinical isolates. J. Bacteriol. *185*, 3446–3452.

Rosato, A.E., Kreiswirth, B.N., Craig, W.A., Eisner, W., Climo, M.W., and Archer, G.L. (2003b). *mecA-blaZ* corepressors in clinical *Staphylococcus aureus* isolates. Antimicrob. Agents Chemother. *47*, 1460–1463.

Rossi, F., Diaz, L., Wollam, A., Panesso, D., Zhou, Y., Rincon, S., Narechania, A., Xing, G., Di Gioia, T.S., Doi, A., *et al.* (2014). Transferable vancomycin resistance in a community-associated MRSA lineage. N. Engl. J. Med. *370*, 1524–1531.

Rouch, D.A., Messerotti, L.J., Loo, L.S., Jackson, C.A., and Skurray, R.A. (1989). Trimethoprim resistance transposon Tn*4003* from *Staphylococcus aureus* encodes genes for a dihydrofolate reductase and thymidylate synthetase flanked by three copies of IS257. Mol. Microbiol. 3, 161–175.

Rybak, J.M., and Roberts, K. (2015). Tedizolid Phosphate: a Next-Generation Oxazolidinone. Infect. Dis. Ther. *4*, 1–14.

Ryffel, C., Kayser, F.H., and Berger-Bächi, B. (1992). Correlation between regulation of *mecA* transcription and expression of methicillin resistance in staphylococci. Antimicrob. Agents Chemother. *36*, 25–31.

Ryffel, C., Strässle, A., Kayser, F.H., and Berger-Bächi, B. (1994). Mechanisms of heteroresistance in methicillin-resistant *Staphylococcus aureus*. Antimicrob. Agents Chemother. *38*, 724–728.

Sahm, D.F., Deane, J., Bien, P.A., Locke, J.B., Zuill, D.E., Shaw, K.J., and Bartizal, K.F. (2015). Results of the surveillance of Tedizolid activity and resistance program: *in vitro* susceptibility of gram-positive pathogens collected in 2011 and 2012 from the United States and Europe. Diagn. Microbiol. Infect. Dis. *81*, 112–118.

Sakoulas, G., Alder, J., Thauvin-Eliopoulos, C., Moellering, R.C., Jr., and Eliopoulos, G.M. (2006). Induction of daptomycin heterogeneous susceptibility in *Staphylococcus aureus* by exposure to vancomycin. Antimicrob. Agents. Chemother. *50*, 1581–1585.

Saravolatz, L.D., and Stein, G.E. (2015). Oritavancin: A Long-Half-Life Lipoglycopeptide. Clin. Infect. Dis. *61*, 627–632.

Scharn, C.R., Tenover, F.C., and Goering, R.V. (2013). Transduction of staphylococcal cassette chromosome mec elements between strains of *Staphylococcus aureus*. Antimicrob. Agents Chemother. *57*, 5233–5238.

Schindler, C.A., and schuhardt, V.T. (1964). Lysostaphin: a new bacteriolytic agent for the Staphylococcus. Proc. Natl. Acad. Sci. U.S.A. *51*, 414–421.

Schmitz, F.-J., Krey, A., Sadurski, R., Verhoef, J., Milatovic, D., Fluit, A.C., and for the European SENTRY participants (2001). Resistance to tetracycline and distribution of tetracycline resistance genes in European *Staphylococcus aureus* isolates. J. Antimicrob. Chemother. *47*, 239–240.

Schmitz, F.J., Fluit, A.C., Hafner, D., Beeck, A., Perdikouli, M., Boos, M., Scheuring, S., Verhoef, J., Kohrer, K., and Von Eiff, C. (2000). Development of resistance to ciprofloxacin, rifampin, and mupirocin in methicillin-susceptible and -resistant *Staphylococcus aureus* isolates. Antimicrob. Agents. Chemother. *44*, 3229–3231.

Schneider, T., Senn, M.M., Berger-Bächi, B., Tossi, A., Sahl, H.G., and Wiedemann, I. (2004). *In vitro* assembly of a complete, pentaglycine interpeptide bridge containing cell wall precursor (lipid II-Gly5) of *Staphylococcus aureus*. Mol. Microbiol. *53*, 675–685.

Sekiguchi, J., Tharavichitkul, P., Miyoshi-Akiyama, T., Chupia, V., Fujino, T., Araake, M., Irie, A., Morita, K., Kuratsuji, T., and Kirikae, T. (2005). Cloning and characterization of a novel trimethoprim-resistant dihydrofolate reductase from a nosocomial isolate of *Staphylococcus aureus* CM.S2 (IMCJ1454). Antimicrob. Agents. Chemother. *49*, 3948–3951.

Severin, A., Tabei, K., Tenover, F., Chung, M., Clarke, N., and Tomasz, A. (2004). High level oxacillin and vancomycin resistance and altered cell wall composition in *Staphylococcus aureus* carrying the staphylococcal *mecA* and the enterococcal *vanA* gene complex. J. Biol. Chem. *279*, 3398–3407.

Sharma, V.K., Hackbarth, C.J., Dickinson, T.M., and Archer, G.L. (1998). Interaction of native and mutant MecI repressors with sequences that regulate *mecA*, the gene encoding penicillin binding protein 2a in methicillin-resistant staphylococci. J. Bacteriol. *180*, 2160–2166.

Shockman, G.D., Daneo-Moore, L., Kariyama, R., and Massidda, O. (1996). Bacterial walls, peptidoglycan hydrolases, autolysins, and autolysis. Microb. Drug Resist. *2*, 95–98.

Shoji, M., Cui, L., Iizuka, R., Komoto, A., Neoh, H.M., Watanabe, Y., Hishinuma, T., and Hiramatsu, K. (2011). *walK* and *clpP* mutations confer reduced vancomycin susceptibility in *Staphylococcus aureus*. Antimicrob. Agents Chemother. *55*, 3870–3881.

Shore, A.C., Brennan, O.M., Ehricht, R., Monecke, S., Schwarz, S., Slickers, P., and Coleman, D.C. (2010). Identification and characterization of the multidrug resistance gene *cfr* in a Panton–Valentine leucocidin-positive sequence type 8 methicillin-resistant *Staphylococcus aureus* IVa (USA300) isolate. Antimicrob. Agents. Chemother. *54*, 4978–4984.

Shore, A.C., Deasy, E.C., Slickers, P., Brennan, G., O'Connell, B., Monecke, S., Ehricht, R., and Coleman, D.C. (2011). Detection of staphylococcal cassette chromosome mec type XI carrying highly divergent *mecA*, *mecI*, *mecR1*, *blaZ*, and *ccr* genes in human clinical isolates of clonal complex 130 methicillin-resistant *Staphylococcus aureus.* Antimicrob. Agents. Chemother. *55*, 3765–3773.

Sieradzki, K., and Tomasz, A. (1997). Inhibition of cell wall turnover and autolysis by vancomycin in a highly vancomycin-resistant mutant of *Staphylococcus aureus.* J. Bacteriol. *179*, 2557–2566.

Sieradzki, K., and Tomasz, A. (2003). Alterations of cell wall structure and metabolism accompany reduced susceptibility to vancomycin in an isogenic series of clinical isolates of *Staphylococcus aureus.* J. Bacteriol. *185*, 7103–7110.

Silverman, J.A., Perlmutter, N.G., and Shapiro, H.M. (2003). Correlation of daptomycin bactericidal activity and membrane depolarization in *Staphylococcus aureus.* Antimicrob. Agents Chemother. *47*, 2538–2544.

Skiest, D.J. (2006). Treatment failure resulting from resistance of *Staphylococcus aureus* to daptomycin. J. Clin. Microbiol. *44*, 655–656.

Smith, I.M., and Vickers, A.B. (1960). Natural history of 338 treated and untreated patients with staphylococcal septicaemia (1936–1955). Lancet *1*, 1318–1322.

Smith, J.L., and Weinberg, E.D. (1962). Mechanisms of antibacterial action of bacitracin. J. Gen. Microbiol. *28*, 559–569.

Song, M.D., Wachi, M., Doi, M., Ishino, F., and Matsuhashi, M. (1987). Evolution of an inducible penicillin-target protein in methicillin-resistant *Staphylococcus aureus* by gene fusion. FEBS Lett. *221*, 167–171.

Song, Y., Lunde, C.S., Benton, B.M., and Wilkinson, B.J. (2012). Further insights into the mode of action of the lipoglycopeptide telavancin through global gene expression studies. Antimicrob. Agents Chemother. *56*, 3157–3164.

Song, Y., Lunde, C.S., Benton, B.M., and Wilkinson, B.J. (2013). Studies on the mechanism of telavancin decreased susceptibility in a laboratory-derived mutant. Microb. Drug Resist. *19*, 247–255.

Strahilevitz, J., and Hooper, D.C. (2009). Resistance to Other Agents. In Staphylococci in Human Disease, K.B. Crossley, K.K. Jefferson, G.L. Archer, and V.G. Fowler, eds. (Blackwell Publishing Ltd.), pp. 210–234.

Straus, S.K., and Hancock, R.E. (2006). Mode of action of the new antibiotic for Gram-positive pathogens daptomycin: comparison with cationic antimicrobial peptides and lipopeptides. Biochim. Biophys. Acta *1758*, 1215–1223.

Suzuki, E., Kuwahara-Arai, K., Richardson, J.F., and Hiramatsu, K. (1993). Distribution of mec regulator genes in methicillin-resistant Staphylococcus clinical strains. Antimicrob. Agents. Chemother. 37, 1219–1226.

Tanaka, M., Wang, T., Onodera, Y., Uchida, Y., and Sato, K. (2000). Mechanism of quinolone resistance in *Staphylococcus aureus.* J. Infect. Chemother. *6*, 131–139.

Tang, H.J., Lai, C.C., Hsueh, P.R., Chen, C.C., Wu, K.Y., Lin, Y.C., Zhang, C.C., Weng, T.C., Chiu, Y.H., Toh, H.S., *et al.* (2015). RNA polymerase B subunit gene mutations in biofilm-embedded methicillin-resistant *Staphylococcus aureus* following rifampin treatment. J. Microbiol. Immunol. Infect. *doi:10.1016/j.jmii.2015.06.006.*

Tenover, F.C., and Moellering, R.C. (2007). The Rationale for Revising the Clinical and Laboratory Standards Institute Vancomycin Minimal Inhibitory Concentration Interpretive Criteria for *Staphylococcus aureus.* Clin. Infect. Dis. *44*, 1208–1215.

Tenover, F.C., Weigel, L.M., Appelbaum, P.C., McDougal, L.K., Chaitram, J., McAllister, S., Clark, N., Killgore, G., O'Hara, C.M., Jevitt, L., *et al.* (2004). Vancomycin-resistant *Staphylococcus aureus* isolate from a patient in Pennsylvania. Antimicrob. Agents Chemother. *48*, 275–280.

Toh, S.M., Xiong, L., Arias, C.A., Villegas, M.V., Lolans, K., Quinn, J., and Mankin, A.S. (2007). Acquisition of a natural resistance gene renders a clinical strain of methicillin-resistant *Staphylococcus aureus* resistant to the synthetic antibiotic linezolid. Mol. Microbiol. *64*, 1506–1514.

Tran, T.T., Panesso, D., Mishra, N.N., Mileykovskaya, E., Guan, Z., Munita, J.M., Reyes, J., Diaz, L., Weinstock, G.M., Murray, B.E., *et al.* (2013). Daptomycin-resistant *Enterococcus faecalis* diverts the antibiotic molecule from the division septum and remodels cell membrane phospholipids. MBio *4*, e00281–13.

Truong-Bolduc, Q.C., Dunman, P.M., Eidem, T., and Hooper, D.C. (2011). Transcriptional profiling analysis of the global regulator NorG, a GntR-like protein of *Staphylococcus aureus.* J. Bacteriol. *193*, 6207–6214.

Truong-Bolduc, Q.C., Dunman, P.M., Strahilevitz, J., Projan, S.J., and Hooper, D.C. (2005). MgrA is a multiple regulator of two new efflux pumps in *Staphylococcus aureus.* J. Bacteriol. *187*, 2395–2405.

Truong-Bolduc, Q.C., and Hooper, D.C. (2007). The transcriptional regulators NorG and MgrA modulate resistance to both quinolones and β-lactams in *Staphylococcus aureus.* J. Bacteriol. *189*, 2996–3005.

Truong-Bolduc, Q.C., and Hooper, D.C. (2010). Phosphorylation of MgrA and its effect on expression of the NorA and NorB efflux pumps of *Staphylococcus aureus.* J. Bacteriol. *192*, 2525–2534.

Truong-Bolduc, Q.C., Strahilevitz, J., and Hooper, D.C. (2006). NorC, a new efflux pump regulated by MgrA of *Staphylococcus aureus.* Antimicrob. Agents Chemother. *50*, 1104–1107.

Tsakris, A., Pillai, S.K., Gold, H.S., Thauvin-Eliopoulos, C., Venkataraman, L., Wennersten, C., Moellering, R.C., and Eliopoulos, G.M. (2007). Persistence of rRNA operon mutated copies and rapid re-emergence of linezolid resistance in *Staphylococcus aureus.* J. Antimicrob. Chemother. *60*, 649–651.

Utaida, S., Dunman, P.M., Macapagal, D., Murphy, E., Projan, S.J., Singh, V.K., Jayaswal, R.K., and Wilkinson, B.J. (2003). Genome-wide transcriptional profiling of the response of *Staphylococcus aureus* to cell wall -active antibiotics reveals a cell wall -stress stimulon. Microbiology *149* 3, 2719–2732.

Utsui, Y., and Yokota, T. (1985). Role of an altered penicillin-binding protein in methicillin- and cephem-resistant *Staphylococcus aureus.* Antimicrob. Agents. Chemother. *28*, 397–403.

Walsh, C. (2003a). Antibiotics That Block Bacterial Protein Biosynthesis. In Antibiotics, C. Walsh, ed. (Washington, D.C.: ASM Press), p. 64.

Walsh, C. (2003b). Antibiotics: Initial Concepts. In Antibiotics, C. Walsh, ed. (Washington, D.C.: ASM Press), pp. 23–49.

Walsh, C. (2003c). Other Targets of Antibacterial Drugs. In Antibiotics (Washington, D.C.: ASM Press), p. 83.

Walsh, C. (2003d). Regulation of Antibiotic Biosynthesis in Producer Organisms In Antibiotics (Washington, D.C.: ASM Press), pp. 23–49.

Wang, Y., Lv, Y., Cai, J., Schwarz, S., Cui, L., Hu, Z., Zhang, R., Li, J., Zhao, Q., He, T., *et al.* (2015). A novel gene, *optrA*, that confers transferable resistance to oxazolidinones and phenicols and its presence in *Enterococcus faecalis* and *Enterococcus faecium* of human and animal origin. J. Antimicrob. Chemother. *70*, 2182–2190.

Watanabe, Y., Cui, L., Katayama, Y., Kozue, K., and Hiramatsu, K. (2011). Impact of *rpoB* mutations on reduced vancomycin susceptibility in *Staphylococcus aureus*. J. Clin. Microbiol. *49*, 2680–2684.

Weigel, L.M., Clewell, D.B., Gill, S.R., Clark, N.C., McDougal, L.K., Flannagan, S.E., Kolonay, J.F., Shetty, J., Killgore, G.E., and Tenover, F.C. (2003). Genetic analysis of a high-level vancomycin-resistant isolate of *Staphylococcus aureus*. Science *302*, 1569–1571.

Wichelhaus, T.A., Böddinghaus, B., Besier, S., Schäfer, V., Brade, V., and Ludwig, A. (2002). Biological cost of rifampin resistance from the perspective of *Staphylococcus aureus*. Antimicrob. Agents Chemother. *46*, 3381–3385.

Wong, A., Reddy, S.P., Smyth, D.S., Aguero-Rosenfeld, M.E., Sakoulas, G., and Robinson, D.A. (2010). Polyphyletic emergence of linezolid-resistant staphylococci in the United States. Antimicrob. Agents Chemother. *54*, 742–748.

Wong, E., and Rab, S. (2014). Tedizolid phosphate (sivextro): a second-generation oxazolidinone to treat acute bacterial skin and skin structure infections. P.T. *39*, 555–579.

Wu, S., Piscitelli, C., de Lencastre, H., and Tomasz, A. (1996). Tracking the evolutionary origin of the methicillin resistance gene: cloning and sequencing of a homologue of *mecA* from a methicillin susceptible strain of *Staphylococcus sciuri*. Microb. Drug Resist. 2, 435–441.

Wu, Z., Li, F., Liu, D., Xue, H., and Zhao, X. (2015). Novel type XII staphylococcal cassette chromosome mec harboring a new cassette chromosome recombinase, CcrC2. Antimicrob. Agents Chemother. *59*, 7597–7601.

Yang, S.J., Kreiswirth, B.N., Sakoulas, G., Yeaman, M.R., Xiong, Y.Q., Sawa, A., and Bayer, A.S. (2009). Enhanced expression of *dltABCD* is associated with the development of daptomycin nonsusceptibility in a clinical endocarditis isolate of *Staphylococcus aureus*. J. Infect. Dis. *200*, 1916–1920.

Yang, S.J., Mishra, N.N., Rubio, A., and Bayer, A.S. (2013). Causal role of single nucleotide polymorphisms within the *mprF* gene of *Staphylococcus aureus* in daptomycin resistance. Antimicrob. Agents. Chemother. *57*, 5658–5664.

Yang, S.J., Nast, C.C., Mishra, N.N., Yeaman, M.R., Fey, P.D., and Bayer, A.S. (2010). Cell wall thickening is not a universal accompaniment of the daptomycin nonsusceptibility phenotype in *Staphylococcus aureus*: evidence for multiple resistance mechanisms. Antimicrob. Agents Chemother. *54*, 3079–3085.

Yin, S., Daum, R.S., and Boyle-Vavra, S. (2006). VraSR two-component regulatory system and its role in induction of *pbp2* and *vraSR* expression by cell wall antimicrobials in *Staphylococcus aureus*. Antimicrob. Agents Chemother. *50*, 336–343.

Yoo, J.I., Kim, J.W., Kang, G.S., Kim, H.S., Yoo, J.S., and Lee, Y.S. (2013). Prevalence of amino acid changes in the *yvqF*, *vraSR*, *graSR*, and *tcaRAB* genes from vancomycin intermediate resistant *Staphylococcus aureus*. J. Microbiol. *51*, 160–165.

Zhang, H.Z., Hackbarth, C.J., Chansky, K.M., and Chambers, H.F. (2001). A proteolytic transmembrane signaling pathway and resistance to β-lactams in staphylococci. Science *291*, 1962–1965.

Zhu, W., Clark, N., and Patel, J.B. (2013). pSK41-like plasmid is necessary for Inc18-like *vanA* plasmid transfer from *Enterococcus faecalis* to *Staphylococcus aureus in vitro*. Antimicrob. Agents Chemother. *57*, 212–219.

Genetic Regulation

Markus Bischoff and Pascale Romby

12

Abstract

The ability of *Staphylococcus aureus* to accurately respond to changing environments is one of the prerequisites for its success as a versatile human pathogen. This bacterium is equipped with an armamentarium of virulence factors that facilitate adaptation to nearly all ecological niches within a host. In addition, this armamentarium allows *S. aureus* to counteract most of the immune response mechanisms used by the host to combat the invading pathogen. Regulation of this armamentarium is carried out by at least 112 transcription factors than can be divided into 36 regulatory families. The largest families of transcription factors found in *S. aureus* are one-component systems such as MarR-type, GntR/DeoR-type, and Xre-type of regulators, and two-component system response regulators. Alternative sigma factors, the third most common class of proteins involved in bacterial signal transduction, are underrepresented in *S. aureus*. Notably, less than half of *S. aureus* transcription factors have been functionally characterized, suggesting that our understanding of the regulatory network utilized by this pathogen to adapt to a changing environment is incomplete.

Introduction

Staphylococcus aureus is a member of the normal microflora of humans, with about 50% of the human population being transiently colonized in the anterior nares, or on the skin (Wertheim *et al.*, 2005). When *S. aureus* enters a host through a wounds or by inhalation, it can become a serious problem capable of causing live-threatening infections such as endocarditis, osteomyelitis, and bacteraemia/sepsis (Gordon and Lowy, 2008). Importantly, methicillin-resistant variants of *S. aureus*, known as MRSA, have become a major health concern, and are currently the predominant nosocomial pathogen in the Western hemisphere (Köck *et al.*, 2010; Sievert *et al.*, 2013). The ability of *S. aureus* to colonize its host and/or to cause a wide array of diseases has been attributed to its large arsenal of virulence factors and global regulators (Gordon and Lowy, 2008). These regulators form a sophisticated network (Fig. 12.1A), allowing *S. aureus* to rapidly respond to changing conditions, and to fine-tune the expression of its virulence determinants (Santos *et al.*, 2013). The necessity of *S. aureus* to tightly control virulence determinant expression is seen by the fact that transcription and translation of almost all of the major virulence determinants is under multiple levels of control. As an example, one of the most important virulence factors of *S. aureus*, α-toxin (encoded by *hla*; see Chapter 5), is regulated by at least 21 regulators (Fig. 12.1B), which modulate transcription and translation of this cytotoxin in response to environmental stimuli such as cell density, nutritional status, oxygen and iron availability, and pH. Inactivation of any one of these regulators has a severe impact on the expression of α-toxin; hence, it can be assumed that each of these regulators has unique functions in the virulence determinant regulatory network, rather than forming a backup system for another regulatory molecule.

While the overall number of transcription factors (TFs) varies from strain to strain, the *S. aureus* core genome encodes 112TFs (Ibarra *et al.*, 2013). The variation is likely due to genomic rearrangements, duplications, and the acquisition of mobile genetic elements (Ibarra *et al.*, 2013; see also Chapter 4). Remarkably, less than half of these regulators have

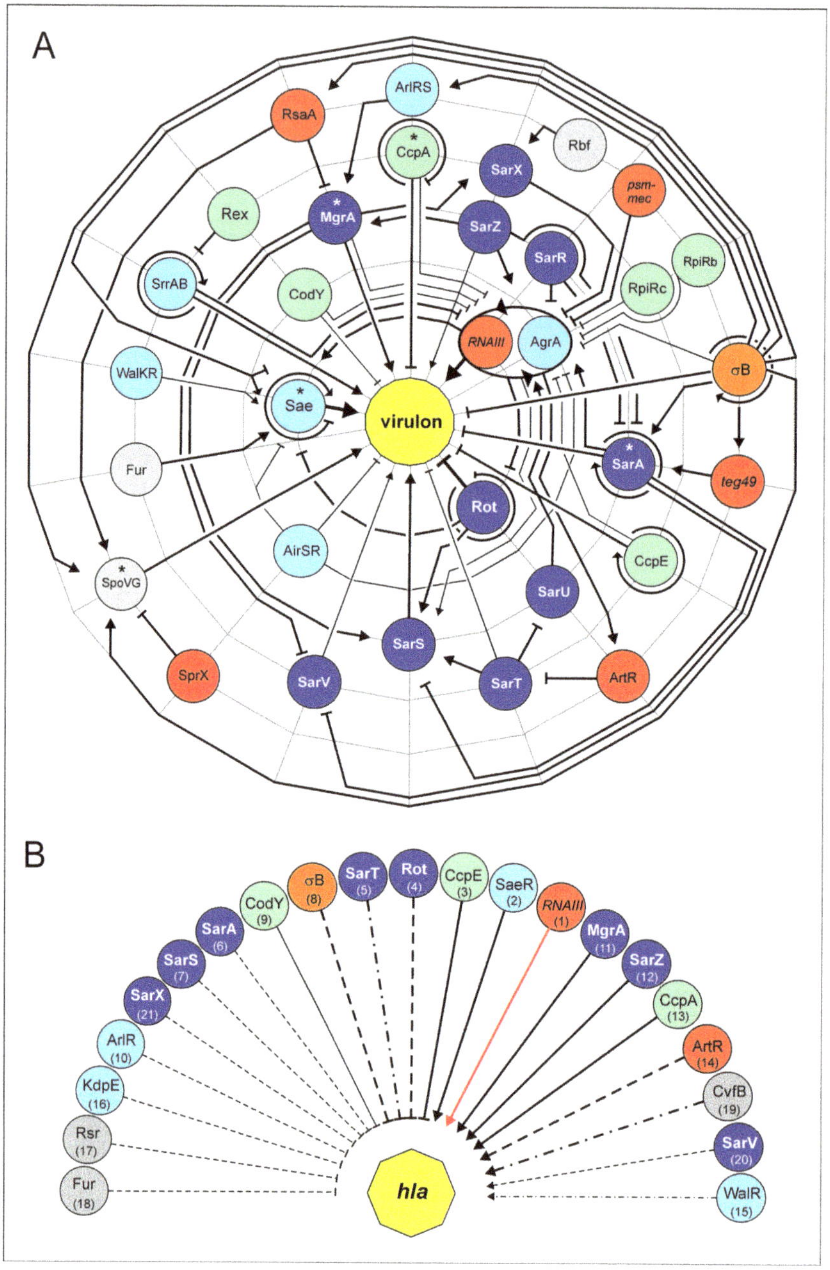

Figure 12.1 Transcriptional control of the virulon in *S. aureus*. (A) Interplay of key regulators affecting virulence factor synthesis in *S. aureus*. TCS components are labelled in light blue, Sar family proteins in dark blue, metabolite responsive elements in light green, regulatory RNAs in red, and other TFs in grey. Arrows indicate a stimulatory effect, and perpendicular lines a repressive effect. Degrees of influence are depicted by coat thickness. Modulation of TF activity by serine-threonine kinase 1 (Stk1)-driven phosphorylation is indicated by asterisks. (B) TFs reported to control the expression of α-toxin. Suggested direct effects are depicted by continuous lines and indirect effects by broken lines. TFs shown to affect *hla* transcription and for which the mode of action has not been determined yet are indicated by broken/dotted lines. Positive/repressive effects and colour-codes are displayed as described in A. Red lines indicate post transcriptional effects on Hla synthesis. Numbers given in parenthesis denote references reporting this effect. (1) Morfeldt *et al.*, 1995; (2) Nygaard *et al.*, 2010; (3) Hartmann *et al.*, 2014; (4) Li and Cheung, 2008; (5) Schmidt *et al.*, 2001; (6) Zielinska *et al.*, 2011; (7) Oscarsson *et al.*, 2006; (8) Cheung *et al.*, 1999; (9) Majerczyk *et al.*, 2010; (10) Fournier *et al.*, 2001; (11) Luong *et al.*, 2006; (12) Kaito *et al.*, 2006; (13) Leiba *et al.*, 2012; (14) Xue *et al.*, 2014; (15) Delaune *et al.*, 2012; (16) Xue *et al.*, 2011; (17) Tamber *et al.*, 2010; (18) Torres *et al.*, 2010; (19) Matsumoto *et al.*, 2007; (20) Manna *et al.*, 2004; (21) Manna and Cheung, 2006a.

been experimentally characterized, indicating that a substantial part of the regulatory network formed by *S. aureus* has yet to be discovered. These TFs can be classified into 36 regulatory families, with the largest families being the MarR type of regulators, two-component systems (TCSs) consisting of a sensor kinase and a response regulator, the GntR family, and the Xre family of transcription factors (Ibarra *et al.*, 2013).

This chapter will focus on regulatory factors that modulate the expression of virulence determinants, especially TCSs, the Sar family of DNA-binding proteins, metabolite responsive elements, alternative sigma (σ) factors, and regulatory RNAs.

Two-component systems

Two-component systems (TCSs) are the major regulatory systems by which bacteria sense and respond to environmental stimuli (Stock *et al.*, 2000). These sophisticated signal transduction systems typically consist of a transmembrane-spanning sensor histidine kinase (SHK), and a cognate response regulator (RR). The SHK usually responds to an environmental stimuli through the autophosphorylation of the intracellular histidine kinase (HK) domain of the SHK, and this phosphoryl group is subsequently transferred to the RR, which activates the effector domain of the RR to elicit a specific response (Stock *et al.*, 2000). Most of the *S. aureus* strains for which fully annotated whole genome sequences are available, contain 16 TCSs (Table 12.1) (Ibarra *et al.*, 2013), and many of these TCSs are part of larger operons, in which the TCS is integrated into complex regulatory circuits.

The *agr* operon

The accessory gene regulator (*agr*) locus is an excellent example of a TCS being integrated into a complex regulatory circuit, and probably one of the most fascinating regulatory systems found in the bacterial kingdom. Given the complexity of this system, its impact on virulence factor production *in vitro* and *in vivo*, and the number of excellent reviews on this topic (George and Muir, 2007; Novick, 2003; Novick and Geisinger, 2008; Thoendel *et al.*, 2011), we will focus here on the genetic organization of this locus, its transcriptional regulation, and the regulatory molecules encoded by this locus.

The *agr* locus is composed of two divergently orientated transcriptional units termed RNAII and RNAIII, whose transcription is driven by the P_2 and P_3 promoters, respectively (Table 12.1). RNAII encodes for one of the quorum-sensing systems of *S. aureus*; namely, AgrBDCA. AgrD is a 46 amino acid long precursor of the autoinducing peptide (AIP), which is processed by AgrB into its active form. The AIP is a seven to nine amino acid long cyclic peptide that is exported by AgrB into the extracellular space after it is processed into the active quorum sensing signalling molecule. Increased levels of the AIP in the environment are sensed by AgrC the SHK of the TCS encoded by this locus. Binding of the AIP to the membrane-embedded sensor AgrC causes the autophosphorylation of its intracellular HK domain. This phosphoryl group is subsequently passed to AgrA, the cognate RR of this system. Upon phosphorylation of AgrA, it becomes activated to bind the P_2-P_3 promoter region of the *agr* locus to activate the transcription of both, RNAII and RNAIII. Activating transcription of the P_2-P_3 promoter region creates a strong positive feedback loop that induces transcription of the major effector molecule, the small regulatory RNA (sRNA) RNAIII (reviewed in Thoendel *et al.*, 2011), and RNAII. Importantly, RNAIII is one of the master regulators controlling exoprotein synthesis in *S. aureus*, and one of the few examples for a regulatory RNA that also encodes a protein, δ-toxin (*hld*), a small cytolytic peptide belonging to the group of phenol soluble modulins (Wang *et al.*, 2007). For a detailed description of RNAIII activity, please see the subchapter 'regulatory RNAs'.

The *agr* quorum sensing system, although universally found among staphylococci, includes a hypervariable region covering the AIP precursor AgrD, the C-terminal part of AgrB, and the N-terminal part of AgrC (containing the sensor region of this SHK). Variation in this region has led to the formation of at least four *agr* specificity groups within *S. aureus* (Jarraud *et al.*, 2000; Ji *et al.*, 1997), and further *agr* specificity groups exist in other staphylococcal species (Dufour *et al.*, 2002). As the AIP and sensor portion of AgrC form a highly specific recognition system within each of the specificity groups, staphylococci use this variation to control their *agr* regulons by ensuring that only the AIP of the same specificity group can activate the HK domain of its cognate SHK. AIPs

Table 12.1 Two-component systems of *S. aureus*

Name	Gene ID[1]	Genetic organization[2]	Function	Regulator family	Regulated by
AgrCA	SA1843SA1844	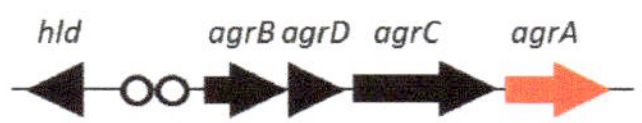	Quorum sensing, regulation of exoprotein synthesis on a transcriptional and post-transcriptional level	LytTR	AgrA, ArlR, CcpA, CcpE, CodY, RpiRC, Rsr, SarA, SigB, SrrA
AirSR (YhcSR)	SA1667SA1666	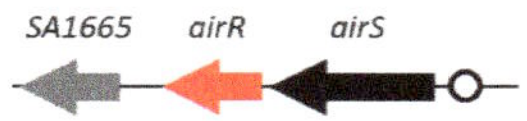	Redox-dependent regulation of virulence and cell wall metabolism, modulation of the nitrate respiratory pathway under anaerobic conditions, transcriptional control of the *lac* and *opuC* operons; essential	LuxR	NsaR
ArlRS	SA1247SA1248	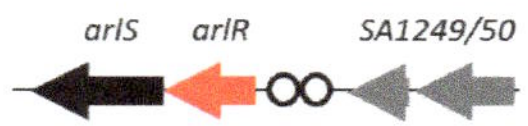	Regulation of virulence factor synthesis and biofilm formation; non-essential	OmpR	Agr, SarA
GraRS	SA0614SA0615	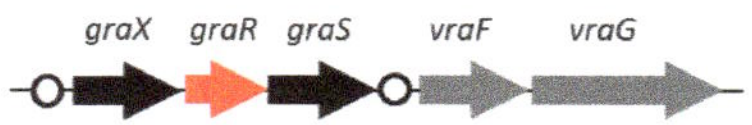	Part of a five-component signal transduction network controlling cationic antimicrobial peptide sensing and resistance in *S. aureus*	OmpR	
HssRS	SA2151SA2152		Haem sensing TCS involved in haem tetoxification; non-essential	OmpR	MgrA
KdpDE (KdpD-PstA)	SA1882SA1883		Quorum sensing-dependent regulation of capsule synthesis, c-di-AMP dependent regulation of ion transport, K(+)-dependent regulation of virulence factors; non-essential	OmpR	Agr, LuxS
LytSR	SA0250SA0251	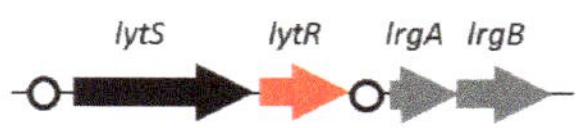	Membrane potential-dependent regulation of holin/antiholin systems and biofilm formation, adaptation to host defence cationic antimicrobial peptides (HDPs); non-essential	LytR	MgrA
NreBC	SA2179SA2180	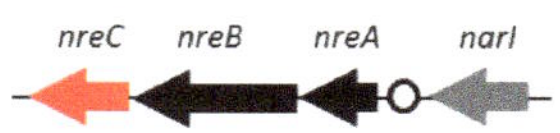	Oxygen-dependent regulation of genes involved in dissimilatory reduction and transport of nitrate and nitrite; non-essential	OmpR	Rex, AirR

NsaRS (BceRS) (BraRS)	SA2417SA2418	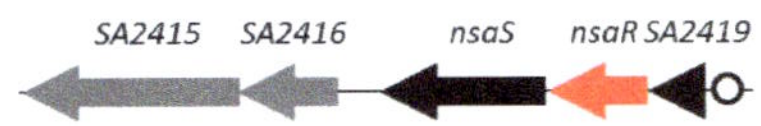	Cell wall and membrane damage sensing TCS affecting biofilm formation, virulence and cell wall biosynthesis. Putative part of a five-component signal transduction network; non-essential	OmpR	Agr
PhoPR	SA1515SA1516	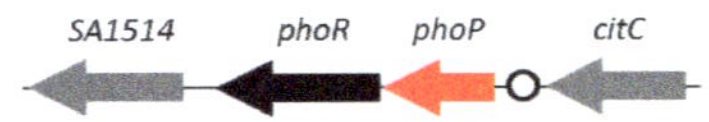	Function unknown; non-essential	OmpR	
SaeRS	SA0660SA0661	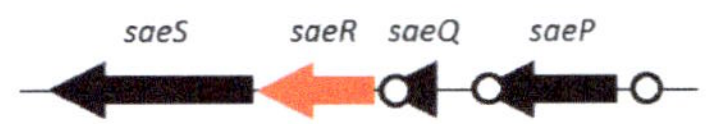	Regulation of exoprotein synthesis, essentially involved in immune evasion; non-essential	OmpR	Agr, Fur, Rot, SaeR, SigB, WalR
SrrAB (SrhSR)	SA1322SA1323	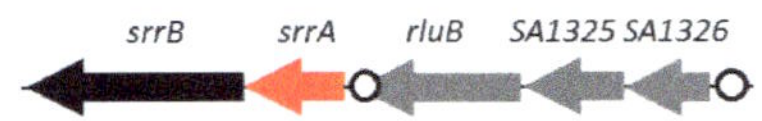	Repression of exoprotein synthesis and activation of polysaccharide intercellular adhesion (PIA) formation under low-oxygen conditions; non-essential	OmpR	Rex, SrrA
VraSR	SA1700SA1701	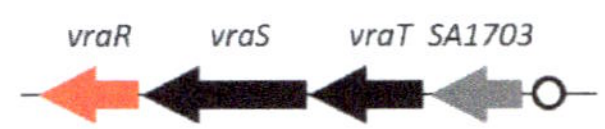	Regulation of the 'cell wall stimulon' in response to cell wall active antimicrobials; non-essential	LuxR	VraR
WalKR (VicRS) (YycGF)	SA0017SA0018	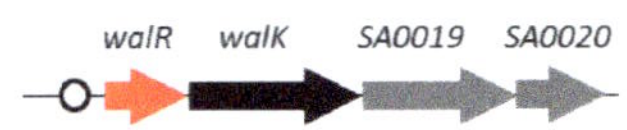	Regulation of virulence and cell wall metabolism, assumed role in nasal colonization; essential	OmpR	Rex, WalR
–	SA0215SA0216		Function unknown, suggested to play a role in the anaerobic response of *S. aureus*; non-essential	AraC	NsaR
–	SA1158SA1159	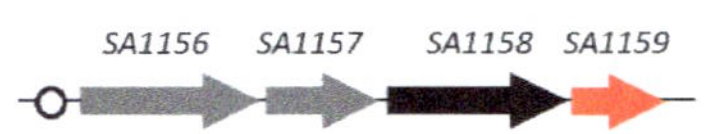	Function unknown; non-essential	LuxR	

[1]Based on the genome sequence of *S. aureus* strain N315 (NCBI Reference Sequence: NC_002745).
[2]Genes encoding RRs are indicated by red arrows, and genes encoding other components of the regulatory systems by black arrows. Surrounding genes and ORFs with yet unknown functions are represented by grey arrows, and promoters by open circles.

of other specificity groups can bind to the sensor domain of AgrC from another specificity group; however, this does not activate the HK domain of the non-cognate SHK and, in fact, it leads to a cross-inhibition of RNAIII transcription in these bacteria (Ji *et al.*, 1997).

Transcription of RNAII and/or RNAIII is affected by a multitude of other regulatory molecules (Table 12.1), making it one of the most regulated loci found in *S. aureus*. While basal transcription from the *agr* P_2-P_3 promoters was reported *in vitro* in the absence of AgrA (Reynolds and Wigneshweraraj, 2011), full transcription of RNAII and RNAIII requires the RR of the *agr* locus (Reyes *et al.*, 2011). As a member of the LytR family of response regulators, AgrA binds to a pair of direct repeats that are similar to the consensus sequence of LytR type of RRs ([TA][AC][CA] GTT[ATGC][AG][TG], separated by a 12- to 13-bp spacer region) (Nikolskaya and Galperin, 2002). Two such elements are present in the *agr* P_2-P_3 intergenic region, and both can be bound by AgrA (Koenig *et al.*, 2004), which facilitates the binding of RNA polymerase to both promoters and fosters the formation of an open promoter complex (Reynolds and Wigneshweraraj, 2011). In addition to its activating effect on RNAII and RNAIII transcription, AgrA controls the expression of a number of additional genes including the phenol-soluble modulin (PSM) clusters (Queck *et al.*, 2008).

Probably the most important regulator affecting *agr* transcription, besides AgrA, is the staphylococcal accessory regulator, SarA (Cheung and Projan, 1994) (see below). The small DNA-binding protein binds to the *agr* P_2-P_3 promoter region to activate transcription of RNAII by AgrA, presumably by enhancing the bending of the *agr* P_2 promoter (Reyes *et al.*, 2011). The *agr* P_2-P_3 promoter region is also directly bound by another member of the Sar family of DNA-binding proteins, SarR, which is thought to repress transcription of RNAII by displacing SarA from the *agr* P_2-P_3 promoter (Reyes *et al.*, 2011), and by the TCS-RR SrrA that acts as a repressor of RNAIII transcription (Pragman *et al.*, 2004; Yarwood *et al.*, 2001). Expression of the *agr* locus is also affected by the carbon catabolite-responsive regulator CodY, which acts as a repressor of RNAII transcription, presumably by a roadblock mechanism via direct binding of CodY to *agrC* (Majerczyk *et al.*, 2008). Numerous other regulatory factors have been reported to affect RNAII and/or RNAIII transcription as well, including the TCS RR ArlR (Fournier *et al.*, 2001), the catabolite responsive elements CcpA (Seidl *et al.*, 2006), CcpE (Hartmann *et al.*, 2014), RpiRb and RpiRc (Zhu *et al.*, 2011), the *rsr* locus (Tamber *et al.*, 2010), and the alternative sigma factor σ^B (Bischoff *et al.*, 2001). At present, it is unclear if these regulators directly bind to the *agr* P_2-P_3 promoter, suggesting they may indirectly affect the transcription of the *agr* locus.

Given its prominent role in regulating virulence factor synthesis in *S. aureus*, the *agr* system was suggested as a potential target for the development of novel antibacterial therapeutics (reviewed inArya and Princy, 2013; Gordon *et al.*, 2013; Grey *et al.*, 2013). Consistent with this suggestion, Sully *et al.* (2014) reported that a small molecule inhibitor named savirin disrupts *agr*-mediated quorum sensing in *S. aureus* by AgrA interference. Savirin reduced infectivity in two murine skin infection models without provoking stress responses usually observed with conventional antibiotics. Unfortunately, the frequent isolation of *agr*-defective *S. aureus* isolates from carriers and hospitals (Chong *et al.*, 2013; Paulander *et al.*, 2013; Shopsin *et al.*, 2008; Smyth *et al.*, 2012; Traber *et al.*, 2008; Tsuji *et al.*, 2011) make targeting of quorum sensing problematic, and may lead *S. aureus* to cause biofilm-associated infections.

The *arl* locus

The autolysis-related locus (*arl*) encodes one of the few TCS systems found in *S. aureus* that is not embedded into larger operons (Table 12.1). The locus consists of the RR ArlR, a member of the OmpR-PhoB family of response regulators, and the SHK ArlS (Fournier and Hooper, 2000). The corresponding genes are transcribed as a bi-cistronic transcript driven by σ^B- and σ^A-dependent promoters in front of *arlR* (Meier *et al.*, 2007). Transcription of *arl* is additionally influenced by other regulatory factors such as *agr* and SarA (Fournier *et al.*, 2001).

Microarray analyses indicate ArlRS controls at least 144 genes/operons involved in different functions, including autolysis, cell division, gene regulation, growth, and pathogenesis (Liang *et al.*, 2005). Despite the large regulon, a consensus binding motif for ArlR has not been determined. ArlRS also affects the expression of a number of regulators including *agr*, MgrA, Rot, and SarS, and it has been suggested that this TCS exerts most of its effects on virulence factor production via its interaction with the *agr*, *mgrA*, and *sarA* regulatory loci (Fournier *et al.*, 2001; Luong and Lee, 2006). Another feature of ArlRS is its influence on *spa* (codes for protein A) transcription under high osmolarity. The transcription of *spa* is regulated by DNA supercoiling in response to high osmolarity, and the Arl TCS seems to facilitate interactions with proteins controlling DNA topology (Fournier and Klier, 2004).

As mentioned, ArlRS is one of the TCSs of *S. aureus* that markedly affects virulence determinant synthesis. ArlRS modulates autolytic, biofilm, and extracellular proteolytic capabilities (Fournier and Hooper, 2000) by repressing the synthesis of secreted proteins such as α-toxin, β-haemolysin, lipase, coagulase, and serine protease (Ssp), and protein A (Spa) (Fournier *et al.*, 2001). In addition, ArlRS promotes capsule formation in *S. aureus* (Luong and Lee, 2006). Interestingly, *arlRS* deletions differentially affect the autolytic phenotypes of methicillin-susceptible *S. aureus* (MSSA) and MRSA (Memmi *et al.*, 2012).

The *sae* operon

The *sae* operon is one of the better studied regulatory loci of *S. aureus*, which is in part due to its major effect on the virulon of this pathogen (Giraudo *et al.*, 1994, 1997; Nygaard *et al.*, 2010; Rogasch *et al.*, 2006; Steinhuber *et al.*, 2003). Sae is primarily known as an activator of exotoxin synthesis (Benson *et al.*, 2012; Giraudo *et al.*, 1994, 1997; Nygaard *et al.*, 2010; Rogasch *et al.*, 2006), and is thought to act downstream of *agr* and SarA (Novick and Jiang, 2003). A special regulation was recently observed for staphylococcal superantigen-like protein (Ssl) encoding genes, which require the cooperation of SaeR and Rot for full transcription (Benson *et al.*, 2012). *In vivo* studies suggest that *sae* is one of the major regulatory loci for α-toxin expression during the course of infection. This observation is in contrast to the functions of *agr*, *sarA*, and *sigB*, which are critical for *hla* transcription *in vitro*, but appear to be less important in *in vivo* conditions (Goerke *et al.*, 2005; Xiong *et al.*, 2006). Consistent with its prominent effect on exoprotein production, *sae* was found to be important for virulence of *S. aureus* in numerous infection models (Cassat *et al.*, 2013; Montgomery *et al.*, 2010; Nygaard *et al.*, 2010; Voyich *et al.*, 2009; Watkins *et al.*, 2011). The *sae* locus also highlights a common theme with *S. aureus*; namely, that regulatory molecules and/or their activity can differ wildly between individual isolates. Indeed, the detailed molecular characterization of the *sae* regulon in two different isolates led to the publication of a number of conflicting observations, until it was determined there was a point mutation in the SHK of strain Newman, which is a widely used strain that is highly pathogenic to mice (Adhikari and Novick, 2008; Schafer *et al.*, 2009). This missense mutation results in an amino acid exchange that renders the TCS constitutively active (Mainiero *et al.*, 2010).

The *sae* locus consists of four open reading frames: *saePQRS* (Novick and Jiang, 2003; Steinhuber *et al.*, 2003), encoding the auxiliary proteins SaeP and SaeQ, which activate the sensor kinase's phosphatase activity (Jeong *et al.*, 2012); the RR SaeR, a member of the OmpR-PhoB family of RRs; and the SHK SaeS (Table 12.1). Transcription of this locus is driven by at least two promoters, promoter P_1 (aka P_C) in front of *saeP*, giving rise to a 3-kb *saePQRS* transcript, and promoter P_3 (aka P_A) which is centred within the ORF of *saeQ* giving rise to a 2-kb transcript encoding *saeRS* (Adhikari and Novick, 2008; Geiger *et al.*, 2008; Steinhuber *et al.*, 2003). Promoter P_1 also drives transcription of a 0.7-kb *saeP* mRNA, probably due to a stem–loop being located in the intergenic region between *saeP* and *saeQ* that might lead to a premature stop of *saePQRS* mRNA synthesis (Novick and Jiang, 2003; Steinhuber *et al.*, 2003). Earlier work suggested the presence of a third promoter, P_2, in front of the stem–loop between *saeP* and *saeQ* that gave rise to a 2.4-kb *saeQRS* mRNA (Steinhuber *et al.*, 2003); however, later studies suggested the 2.4-kb *saeQRS* transcript was more likely an RNA processing product rather than an mRNA arising from an internal promoter (Adhikari and Novick, 2008; Geiger *et al.*, 2008). Both P_1 and P_3 promoters are autoregulated, with P_1 being activated by the *sae* system, while P_3 is autorepressed by the Sae products (Adhikari and Novick, 2008; Geiger *et al.*, 2008). Promoter P_1, which is considered being the primary promoter of this locus, is activated by the *agr* system and repressed by Rot and σ^B activity (Geiger *et al.*, 2008; Li and Cheung, 2008). Activity of P_1 is also modulated by a number of environmental signals, such as low pH, salt stress, and subinhibitory concentrations of clindamycin, which all decrease the transcription of *saePQRS* mRNA. In contrast, phagocytosis related effector molecules such as hydrogen peroxide and α-defensins strongly activate transcription from the P_1 promoter (Geiger *et al.*, 2008; Novick and Jiang, 2003). Transcription of *sae* was also reported to be activated by the cell wall metabolism controlling TCS WalKR (Delaune *et al.*, 2012) (Table 12.1), and by the iron-responsive regulator Fur during growth in low-iron (Johnson *et al.*, 2011).

Activated SaeR binds to the direct DNA repeat (GTTAA[ATCG]$_6$GTTAA) (Sun *et al.*, 2010), and this binding is influenced by SaeS activity. A *saeR* mutagenesis study suggested that a subset of the *sae* regulon, the class II target genes, can be activated by low-level SaeR phosphorylation even in the absence of SaeS. The majority of *sae* regulated genes, the class I target genes, require a high level of SaeR phosphorylation by its cognate SHK to be actively transcribed (Mainiero *et al.*, 2010).

The *srr* locus

The staphylococcal respiratory response (*srr*) locus encodes a TCS, SrrAB (aka, SrhSR), that is involved in the global regulation of virulence factors in *S. aureus* (Pragman and Schlievert, 2004; Yarwood *et al.*, 2001) and in protecting the facultative pathogen against non-oxidative killing by neutrophils in an anaerobic environment (Ulrich *et al.*, 2007). This TCS is thought to link environmental signals, respiratory metabolism, and regulation of virulence factors (Yarwood *et al.*, 2001) under microaerobic/anaerobic conditions. Inactivation of *srrAB* impairs *in vitro* growth under anaerobic conditions, changes the expression of proteins involved in energy metabolism and other metabolic processes, and to attenuate growth in a murine pyelonephritis model (Throup *et al.*, 2001). The attenuation of virulence is due in part to the fact that SrrAB represses the transcription of RNAIII, *tst* (encoding toxic shock syndrome toxin 1), and *spa* (Yarwood *et al.*, 2001), presumably by direct interference of the RR, SrrA, with the respective promoter elements (Pragman *et al.*, 2004). In addition, SrrAB also functions as a major activator of *ica* transcription and PIA production under low-oxygen conditions. [The *ica* operon is required for the synthesis of the polysaccharide intercellular adhesin (PIA), a major component of the biofilm matrix which is found in the majority of biofilm-positive staphylococcal isolates (Pragman *et al.*, 2007; Ulrich *et al.*, 2007).] Recent work indicates that SrrAB also regulates the expression of Hmp, a flavohaemoglobin required for nitric oxide (NO) resistance, and of the conserved sRNA RseA in response to NO and hypoxia (Durand *et al.*, 2015; Kinkel *et al.*, 2013).

Not only is SrrAB important under microaerobic conditions, Pragman and colleagues reported that a functional SrrAB may also act as a transcriptional activator of SrrA-responsive genes such *tst*, *spa*, and *icaR* under aerobic conditions (Pragman *et al.*, 2007).

Transcriptional analyses suggest that *srrAB* are co-transcribed as a bi-cistronic transcript driven by a σ^A-dependent promoter in front of *srrA*, giving rise to a 2.5-knt transcript covering *srrAB*, and a 0.7-knt transcript covering *srrA* only (Pragman *et al.*, 2004; Yarwood *et al.*, 2001). However, recent RNA-seq studies indicate that *srrAB* might also form part of a larger polycistronic transcript including the preceding open reading frames for *rluB*, *scpB*, and *scpA* (Lasa *et al.*, 2011; Ruiz de los Mozos *et al.*, 2013). *srrAB* transcription seems to be autoregulated (Yarwood *et al.*, 2001), and SrrA production appears to be negatively affected by the redox-responsive regulator Rex at the post-transcriptional level, rather than on the transcriptional level. On this latter point, it is interesting that regulation occurs at the post-transcriptional level, as there is a Rex binding site in the promoter region of *srrA* (Pagels *et al.*, 2010). A consensus binding motif for SrrA has not been suggested yet.

The Sar family of DNA-binding proteins

The Sar family of DNA-binding proteins represents a major group of TFs found in *S. aureus*, with many members being exclusively found in the family of *Staphylococcaceae* (Ibarra *et al.*, 2013). Eponymous for this family of regulators is SarA, a 14.7-kDa (124 amino acids) DNA binding protein with a dimeric winged helix structure constituted by a helix–turn–helix motif and a β-hairpin turn wing (Liu *et al.*, 2006). Members of this family can be grouped into three subfamilies: single domain proteins, double domain proteins, and MarR homologues (Cheung *et al.*, 2008). The SarA family consists of at least 11 members (SarA, SarR, SarS, SarT,

SarU, SarV, SarX, SarY, SarZ, MgrA, and Rot), with most of them being found in all *S. aureus* genomes analysed (Ibarra *et al.*, 2013). It is assumed that all SarA-family proteins recognize DNA in a similar fashion, with the helix–turn–helix motif binding to the major groove and the winged region interacting with the minor groove, while the activation domains differ between the Sar homologues, probably reflecting their specific roles in gene regulation (Cheung *et al.*, 2008). Many of the Sar homologues (SarA, SarR, SarS, SarU, SarV, SarX, SarZ, MgrA, and Rot) modulate transcription of the *S. aureus* virulon (reviewed in Cheung *et al.*, 2008).

SarA

The staphylococcal accessory regulator A is the prototypic member of the Sar family proteins and was initially identified as a DNA binding protein (Cheung *et al.*, 1992). Its transcription is growth-phase-dependent, and driven by the three promoters P1, P2 and P3 (Bayer *et al.*, 1996). While the P1 and P2 transcripts are transcribed predominantly during the exponential growth phase, the P3 transcript is produced primarily post-exponentially. Hence, the three overlapping *sarA* transcripts contain large 5′UTRs, appropriate for embedded regulatory elements. Indeed, recent studies revealed the existence of two small non-coding RNAs (teg49, teg48) arising from the 5′UTR of *sarA* mRNAs (Beaume *et al.*, 2010; Kim *et al.*, 2014). Transcription of P3 and the sRNA teg49 is dependent on the alternative σ factor σ^B (see below) and the DEAD-box RNA helicase encoded by *cshA* (Kim *et al.*, 2014). Taken together, these data strongly suggest that teg49 is the result of a processing event of the P3 transcript. The importance of teg49 can be seen in teg49 mutants and in *trans* complemented strains, which suggest that this sRNA might play an important role in modulating the synthesis of SarA by a mechanism not yet solved.

Functional studies have revealed that SarA promotes the synthesis of fibronectin binding proteins and a number of secreted toxins, while repressing the expression of protein A and several proteases (reviewed in Arvidson and Tegmark, 2001). SarA regulates transcription of the virulon through direct binding to target gene promoters at conserved AT-rich DNA motifs (ATTTTAT) (Sterba *et al.*, 2003), and indirectly by regulating the transcription of major global regulatory loci such as *agr* and *arl* [see above]. More surprisingly, recent studies have shown that SarA also acts at the post-transcriptional level by modulating the degradation of mRNAs (Morrison *et al.*, 2012a; Roberts *et al.*, 2006). Several mRNAs encoding virulence factors such as protein A and collagen-binding protein were stabilized in the exponential growth phase in a SarA-dependent manner. This effect was proposed to be the result of a direct interaction of SarA with the mRNAs. Hence, this dual RNA/DNA binding activity of SarA extends the numbers of targets and demonstrates the intricate interactions between transcriptional and post-transcriptional regulatory circuits in this organism.

SarA homologs

Most of the SarA-like TFs contribute to regulation of the virulon, often in cooperation with or by counteracting SarA, presumably to allow *S. aureus* to fine tune the synthesis of its virulence factor arsenal in a highly dynamic manner (Cheung *et al.*, 2008). Some of these homologues involved in virulence determinant production and/or influenced by regulatory RNAs are briefly introduced here. For a detailed description of this family, please refer to (Cheung *et al.*, 2008).

SarR

Probably the most prominent antagonist of SarA is SarR (Manna and Cheung, 2001). This 13.6-kDa, single domain SarA paralogue counteracts the action of SarA in several ways: SarR binds to the *sarA* promoters to inhibit transcription (Manna and Cheung, 2001), and competes with SarA in binding to the *agr* P_2-P_3 promoter region (Manna and Cheung, 2006b; Reyes *et al.*, 2011). DNase footprint assays and electrophoretic mobility shift assays (EMSA) suggest SarR binds to the consensus sequence (TAAATTAA) (Manna and Cheung, 2006b). In addition to being an antagonist of SarA, recent phenotypic analysis revealed that a *sarR* mutant strain is more sensitive to autolytic activities than its parental strain (Lucas and Manna, 2013).

SarS

The 29-kDa, two-domain Sar family protein SarS (aka SarH1) was the first Sar homologue identified in *S. aureus* (Cheung *et al.*, 2001; Tegmark *et al.*, 2000). Unlike the single domain SarA paralogues, which require the formation of a homodimer to be functional, SarS contains two heterologous but similar domains that are connected by a well-ordered loop, with each domain sharing sequence similarity to the single domain SarA paralogues SarA and SarR, respectively (Li *et al.*, 2003).

Transcription of *sarS* is probably controlled by three promoters, giving rise to 950-nt, 1550-nt, and 3000-nt transcripts that all encode SarS (Tegmark *et al.*, 2000). While the origin of the largest transcript is unknown, transcription of 1550-nt transcripts is driven by σ^B, and the smallest transcription unit is preceded by a typical σ^A-dependent promoter. Transcription of *sarS* is repressed by the *agr* system (Cheung *et al.*, 2001), while the ClpXP proteolytic complex (Frees *et al.*, 2005) and the SarA homologues MgrA, Rot and SarT seem to enhance SarS synthesis (Oscarsson *et al.*, 2005; Saïd-Salim *et al.*, 2003; Schmidt *et al.*, 2003).

Like SarR, SarS seems to antagonize SarA activity by stimulating the synthesis of protein A and repressing the production of α-toxin (Cheung *et al.*, 2001; Tegmark *et al.*, 2000). A DNA binding consensus for this TF was not determined yet.

SarT-SarU

The 16-kDa single domain Sar homologue SarT and the 29.3-kDa two-domain Sar family protein SarU are further members of TFs affecting *agr* and *hla* expression (Manna and Cheung, 2003; Schmidt *et al.*, 2001). The corresponding genes are likely to form an operon with two divergently orientated monocistronic transcriptional units that are separated by a 323-bp intergenic sequence (Manna and Cheung, 2003). Notably, the *sarT-sarU* operon is missing in some *S. aureus* strains (Ibarra *et al.*, 2013). The transcription of both genes is driven by overlapping promoters (Manna and Cheung, 2003). The *in vitro* expression of SarT is very low, and strongly repressed by SarA and *agr*, suggesting that *sarT* induction may require specific environmental signals that are not produced under standard laboratory growth conditions (Schmidt *et al.*, 2001). Similarly, transcription of *sarU* is only poorly detectable under *in vitro* growth conditions, and controlled mainly by SarT, which strongly represses *sarU* transcription by direct binding to the *sarU* promoter (Manna and Cheung, 2003).

SarT exerts at least some of its repressive effect on *agr* via the transcriptional control of *sarU*, which acts as an activator of *agr* (Manna and Cheung, 2003). SarT was shown to directly interact with the *sarS* promoter to activate transcription of *sarS* (Schmidt *et al.*, 2003), and to repress *hla* transcription (Schmidt *et al.*, 2001). Recent data indicate that the SarT-dependent effect on α-toxin production is modulated by a regulatory sRNA, ArtR (Xue *et al.*, 2014) (see below). DNase I footprint assays suggest that SarT binds to the DNA consensus sequence (AAATG[A/T]CATTTT) (Schmitt *et al.*, 2012). The DNA binding consensus for SarU is unknown.

MgrA

The multiple gene regulator A (MgrA, aka Rat) is a 19-kDa single domain Sar homologue that affects transcription of at least 355 genes and operons, and plays a key role in regulation of several major virulence factors (Luong *et al.*, 2006; Luong *et al.*, 2003). It activates synthesis of capsular polysaccharides, sortase A, leucotoxins, and α-toxin, while several cell surface associated proteins are down-regulated (Ingavale *et al.*, 2005; Luong *et al.*, 2003). MgrA was also shown to repress the biofilm formation abilities of different *S. aureus* strains (Trotonda *et al.*, 2008). In addition, MgrA controls transcription of efflux pumps and autolysins, which affect the resistance of *S. aureus* to several anti-microbial drugs such as the fluoroquinolones and glycopeptides (Cui *et al.*, 2005; Truong-Bolduc *et al.*, 2005, 2006; Truong-Bolduc and Hooper, 2007). The effect of MgrA on autolysins is most likely indirect through the transcriptional control of TCSs involved in autolysis such as ArlRS and LytSR (Ingavale *et al.*, 2003), and the repression of *sarV* (Manna *et al.*, 2004). MgrA was also reported to enhance *sarS* transcription (Ingavale *et al.*, 2005), and to decrease *agr* expression, presumably via the transcriptional control of *sarX*, another member of the Sar protein family (Manna and Cheung, 2006a). Interestingly, MgrA responds to reactive oxygen species through oxidation of a single cysteine located at the dimer interface, which leads to the dissociation of MgrA from its DNA target sequences and the derepression of antibiotic resistance genes (Chen *et al.*, 2006). MgrA activity was also shown being modulated by the Stk1 (aka PknB)-Stp1 kinase/phosphatase pair (Sun *et al.*, 2012; Truong-Bolduc *et al.*, 2008; Truong-Bolduc and Hooper, 2010). Its relevance for infectivity of *S. aureus* has been documented in a number of infection models (Gupta *et al.*, 2013; Jonsson *et al.*, 2008; Liu *et al.*, 2014).

Transcription of mgr*A* is driven by two promoters, P_1 and P_2, giving rise to 0.56-knt and 0.75-knt transcripts, respectively (Ingavale *et al.*, 2003). Interestingly, transcription of MgrA was found to vary considerably between different laboratory strains (Ballal and Manna, 2009). Expression of MgrA is controlled both at the transcriptional level through the action of another Sar homologue, SarZ (Tamber and Cheung, 2009), and at the translational level by the σ^B-dependent sRNA RsaA (Romilly *et al.*, 2014) and by RNAIII (Gupta *et al.*, 2015)(see below). The DNA binding consensus of MgrA has not been identified.

Rot

The repressor of toxins (Rot) is another master regulator of virulence that acts primarily during the exponential phase of growth to regulate more than 150 genes (Saïd-Salim *et al.*, 2003). The 16-kDa single domain Sar homologue acts as a transcriptional repressor of toxins and lipases and as an activator of cell-surface associated proteins (McNamara *et al.*, 2000; Saïd-Salim *et al.*, 2003), presumably by counteracting the action of the *agr* system (McNamara and Bayer, 2005). Some of the effect of Rot on exoprotein synthesis is exerted via direct interaction with the *sae* system (Benson *et al.*, 2012; Li and Cheung, 2008) (see above).

In contrast to other members of the Sar family, which are highly basic proteins, Rot has an isoelectric point of about 5, indicating that it might function in a different way than the other Sar homologues (Arvidson and Tegmark, 2001). In line with this assumption, a recently solved crystal structure indicates that Rot adopts a novel dimerization model that differs from that of other Sar homologues (Zhu *et al.*, 2014). While site-directed mutagenesis studies have demonstrated that Rot differentially interacts with target promoters (Killikelly *et al.*, 2014), a consensus binding motif for Rot has not been determined.

Regulation of *rot* transcription is incompletely understood. As with many *S. aureus* regulators, *rot* transcription varies between different clinical isolates (Jelsbak *et al.*, 2010). Northern blot analyses and RNA-seq studies indicate that *rot* is transcribed as a monocistronic transcript (Manna and Ray, 2007; Ruiz de los Mozos *et al.*, 2013; Saïd-Salim *et al.*, 2003); however, primer extension identified a number of closely spaced transcriptional start site candidates, but most were not preceded by clear promoter consensus sequences (Hsieh *et al.*, 2008; Manna and Ray, 2007). Interestingly, Rot synthesis is repressed during the post-exponential phase of growth, both at the transcriptional level by SarA (Manna and Ray, 2007), and at the post-transcriptional level through base-pairing interactions with RNAIII (Boisset *et al.*, 2007; Geisinger *et al.*, 2006) (see below). Furthermore, affinity purification studies identified a binding of SarR in the *rot* promoter region, suggesting that this Sar homologue directly affects the transcription of *rot* (Hsieh *et al.*, 2008). The alternative σ factor σ^B was also reported to modulate the transcription of *rot* (Hsieh *et al.*, 2008), probably due to its transcriptional control of *sarA* (see below).

Metabolite responsive elements

Central metabolism and virulence determinant production is intimately linked in all pathogenic bacteria. *S. aureus* is capable of utilizing all central metabolic pathways (for a detailed description please refer to Chapter 6) that supply the organism with the 13 biosynthetic intermediates necessary to synthesize all biomolecules (Somerville and Proctor, 2009). The linkage between central metabolism and virulence factor synthesis in *S. aureus* is mediated by at least six regulatory elements, CcpA, CcpE, CodY, Rex, and the RpiR homologues RpiRb and RpiRc (Hartmann *et al.*, 2014; Richardson *et al.*, 2015). In addition, comparative genome analyses suggest further regulators exist in this bacterium that may link metabolism and virulence (Hartmann *et al.*, 2013).

CcpA

The catabolite control protein A (CcpA) is a member of the LacI/GalR family of bacterial regulators (Ibarra *et al.*, 2013), and the master-regulator of carbon catabolite repression (CCR) in low G+C Gram-positive bacteria (Fujita, 2009). CCR is a global regulatory phenomenon in which the transcription of genes and operons involved in the catabolism of non-preferred carbon sources is repressed when preferred carbon sources are present (reviewed in Görke and Stülke, 2008). In contrast to its well-studied orthologue in *B. subtilis* (reviewed in Deutscher *et al.*, 2006; Fujita, 2009; Sonenshein, 2007), little is known about the circuits that activate this global regulator in *S. aureus*. That being said, it is assumed that parts of the catabolite-responsive pathway for CcpA activation are similar between *S. aureus* and *B. subtilis*. In the latter organism, CcpA is activated in the presence of glucose or other rapidly metabolized carbon sources by complex formation with its co-regulator HPr (histidine-containing phosphocarrier protein). This small phosphocarrier protein is intimately involved in carbohydrate transport and phosphorylation via the phosphotransferase system (PTS) and can be activated by phosphorylation at histidine residue 15 and serine residue 46. In order to bind CcpA, HPr needs to be phosphorylated on Ser-46 by its cognate kinase/phosphorylase HprK/P, which is activated in presence of the glycolytic intermediate fructose bisphosphate. Binding of HPr-Ser(P)–46 to CcpA enhances the affinity of CcpA for specific *cis*-acting sequences, called catabolite-responsive element(s) (*cre*), which promotes or represses transcription depending on the position of the *cre* in relation to the operator sequence. In *B. subtilis*, binding of CcpA to *cre*-sites is also fostered by the HPr homologue Crh (Seidel *et al.*, 2005), however, this CcpA co-effector is missing in *S. aureus*. The DNA binding consensus sequence for the *S. aureus* CcpA has not been experimentally determined yet, but is thought to be similar to the *cre* boxes of *B. subtilis* that are highly degenerate semi-palindromes with the consensus sequence of (TGAAAGCGCTTTCA) and (TGAAA[AG]CG[TC]TTTC[AT]) for strong and weak *cre* boxes, respectively (Marciniak *et al.*, 2012). Recent microarray analyses indicate that *S. aureus* CcpA exerts a substantial part of its regulatory effects in absence of glucose, suggesting that CcpA might bind to some promoters in the absence of its inducer HPr-Ser(P)-46 (Seidl *et al.*, 2009). Another difference in the regulation of CcpA activity between *S. aureus* and *B. subtilis* was recently identified by Leiba and colleagues (Leiba *et al.*, 2012), who demonstrated that *S. aureus* CcpA activity is influenced by the serine/threonine kinase Stk1. Stk1-driven phosphorylation of CcpA inhibits DNA binding activity, while the CcpA of *B. subtilis* is not phosphorylated by the cognate Stk1 orthologue PrkC. Transcriptional studies suggest that *ccpA* is transcribed on a rather constant level throughout growth (Seidl *et al.*, 2006), although CcpA probably represses its own transcription (Leiba *et al.*, 2012).

In addition to its role in CCR, CcpA markedly affects the virulon of *S. aureus* (Seidl *et al.*, 2009). The global regulator was shown to modulate the transcription of toxin encoding genes such as *hla* and *tst* (Seidl *et al.*, 2006, 2008a), to suppress capsule formation in presence of glucose (Seidl *et al.*, 2006), and to be essential for biofilm formation in a glucose-rich environment (Seidl *et al.*, 2008b). Consistent with its profound effect on virulence factor synthesis, CcpA was found to decrease bacterial load in a murine soft tissue infection model of *S. aureus* (Li *et al.*, 2010), in particular under hyperglycaemic conditions (M. Bischoff, unpublished results).

CcpE

The catabolite control protein E (CcpE) was recently identified as a regulator affecting central metabolism and virulence in *S. aureus* (Hartmann *et al.*, 2014). This LysR-type of transcriptional regulator (LTTR) shares a certain degree of similarity to the catabolite control protein C (CcpC) of *B. subtilis*, a citrate-responsive regulator of TCA cycle gene transcription (Mittal *et al.*, 2013). While there is similarity to the *B. subtilis* CcpC on the amino acid level and it is involved in transcription of TCA cycle genes, the *S. aureus* CcpE seems to fulfil a different function than the *B. subtilis* CcpC. CcpE functions as a major positive regulator of TCA cycle activity, presumably due to its direct positive effect on transcription of the TCA cycle gene *citB* (Ding *et al.*, 2014; Hartmann *et al.*, 2013). Inactivation of *ccpE* was shown to enhance transcription of *hla* and RNAIII, while the transcription of *capA* (required for capsule synthesis) and *psmα* (encoding the phenol soluble modulin cluster α), and capsule formation were negatively affected in the *ccpE* mutant (Hartmann *et al.*, 2014). Deletion of *ccpE* also increased the infectivity of the mutant in different murine infection models, suggesting that CcpE is a negative regulator of virulence in *S. aureus* (Ding *et al.*, 2014; Hartmann *et al.*, 2014).

Recent EMSA analyses indicate that CcpE might serve as a citrate-sensing regulator (Ding *et al.*, 2014). Notably, a CcpE/*citB* promoter complex was only observed when a molar ratio of at least 10,000 citrate molecules per 1 molecule of CcpE (Ding *et al.*, 2014). This indicates that citrate is a rather poor inducer of CcpE, suggesting other cofactors to exist that activate this LysR-type of regulator in *S. aureus* at more physiological concentrations.

As a member of the LTTR, CcpE is hypothesized to bind the so-called LTTR box (T[ATCG]$_{11}$A) (Maddocks and Oyston, 2008). Preliminary DNase I footprint assays indicate that CcpE indeed binds to such a DNA motif (Ding *et al.*, 2014; M. Bischoff, unpublished results); however, the exact DNA binding consensus for CcpE awaits further confirmation.

CodY

Another important regulatory link between central metabolism and virulence in *S. aureus* is CodY, a global regulator that can be found in nearly all low G+C Gram-positive bacteria (Majerczyk *et al.*, 2010; Pohl *et al.*, 2009). The 259-amino acid protein is a chimera of two folding units, with a GAF (cGMP-specific and –stimulated phosphodiesterases, *Anabaena* adenylate cyclase and *E. coli* FhlA) like domain at the N-terminal part involved in cofactor binding, and a winged helix–turn–helix DNA binding motif in the C-terminus (Levdikov *et al.*, 2006). As a dimer, CodY primarily acts as a repressor that is activated by two types of ligands, the branched-chain amino acid isoleucine (Pohl *et al.*, 2009) and GTP (Majerczyk *et al.*, 2010), thereby coupling the activity of this regulator to the availability of quickly metabolized carbon and nitrogen sources (Sonenshein, 2005).

In *S. aureus*, CodY controls a regulon of at least 200 genes and operons, whose products are primarily involved in amino acid metabolism, and a number of virulence factors (Majerczyk *et al.*, 2010; Pohl *et al.*, 2009). CodY-mediated repression of virulence genes is due in part to inhibition of the *agr* quorum sensing system, presumably via an indirect mechanism that overrides the cell density signals (Roux *et al.*, 2014). In line with its predominantly repressive effects on the transcription of virulence genes, in particular the exotoxins (Majerczyk *et al.*, 2008, 2010; Pohl *et al.*, 2009; Rivera *et al.*, 2012), inactivation of *codY* was found to increase infectivity (Montgomery *et al.*, 2012), highlighting once again that *S. aureus* adapts the expression of its virulon in response to nutrient availability. CodY activity is also linked to the stringent response in *S. aureus*, and gene activation under stringent conditions is due to relief of CodY-mediated repression (Geiger *et al.*, 2012). CodY-DNA pull-down assays suggest that the activated regulator binds to a 21-bp DNA motif ([TA][AT][AT][TA][CT][TA][GA]AA[TA][AT]TT[CT][TA][GA]A[ATC][AT]A[TA]) (Majerczyk *et al.*, 2010) that resembles the canonical CodY-box found in other species (AATTTTC[AT]GAAAATT) (Sonenshein, 2005). Notably, nearly all of the CodY-affected promoters harbour one and more mismatches to the CodY box, and many have two overlapping copies of the palindromic sequence, each with several mismatches (Majerczyk *et al.*, 2010). In fact, a study carried out in *B. subtilis* in which an imperfect CodY box was changed to a perfect CodY-box in a CodY regulated promoter led to an almost complete repression of transcription of the downstream gene, even under conditions of very low CodY activity (Belitsky and Sonenshein, 2011). This observation suggests that a certain amount of mismatches to the CodY-box are needed to allow for an effector-driven control of gene transcription by CodY.

Rex

The redox sensing repressor Rex, a 211-amino acid DNA-binding protein, is another unique type of transcriptional regulator found in nearly all low G+C Gram-positive bacteria (Richardson *et al.*, 2015). Proteins of the Rex family are typically composed of two structural domains: an N-terminal winged helix–turn–helix domain and a C-terminal NADH binding Rossmann fold (Sickmier *et al.*, 2005). Rex binds as a dimer to its target site, an inverted repeat (TTGTGAA[AT]$_4$TTCACAA) that is highly conserved in *S. aureus* (Pagels *et al.*, 2010). Binding of Rex to its operator sequence is modulated by the intracellular NAD^+/NADH pools, with NAD^+ stimulating binding to the target sequences, while NADH suppresses binding (Pagels *et al.*, 2010), probably by changing the conformation of the dimer in a way that excludes DNA binding (Sickmier *et al.*, 2005). *In vitro*, NADH binds to Rex with a much higher affinity than NAD^+ (Pagels *et al.*, 2010), suggesting that Rex is only active under conditions leading to high intracellular NAD^+ level.

Proteomic experiments using a *rex*-mutant indicate that Rex serves as a central regulator of anaerobic metabolism in *S. aureus* (Pagels *et al.*, 2010). Rex mainly represses genes whose products are involved in pathways leading to anaerobic NAD^+ regeneration, such as ethanol, formate and lactate formation, nitrate respiration, and ATP synthesis (Pagels *et al.*, 2010). TFs affected by Rex are SrrAB (see above) and VicKR (aka WalKR), a TCS involved in regulation of cell wall homeostasis (Table 12.1). Although Rex does not directly control transcription of virulence factors, it is thought to be of importance for infectivity of *S. aureus* by controlling the expression of NO·-inducible L-lactate dehydrogenase (Pagels *et al.*, 2010), an enzyme crucial for *S. aureus* to resist the host innate immune response (Richardson *et al.*, 2008).

RpiR

TFs of the RpiR type constitute a distinct family that can be found in both Gram-negative and Gram-positive bacteria (Richardson *et al.*, 2015). RpiR, like many TFs, typically contain a DNA-binding domain near the N-terminal region, and a C-terminal sugar isomerase (SIS) binding domain, which is found in many phosphosugar isomerases and phosphosugar binding proteins (Bateman, 1999), suggesting that phosphosugars might modulate the activity of these TFs.

Genomic analyses suggest that *S. aureus* produces four RpiR-like proteins that are all part of the core genome of this species (Ibarra *et al.*, 2013). Three of them, RpiRa-c, have been experimentally analysed (Zhu *et al.*, 2011). While no obvious role for RpiRa was found in this study, much clearer effects were observed for RpiRb and RpiRc, which were found to promote the transcription of pentose phosphate pathway genes such as *rpiA* (encoding ribose phosphate isomerase A) and *zwf* (encoding glucose-6-phosphate 1-dehydrogenase). Interestingly, both RpiRb and c were also found to modulate transcription of RNAIII; however, the homologues differed in their effect on RNAIII. RpiRb decreased RNAIII transcript levels primarily during the later growth stages, while RpiRc repressed RNAIII transcription mainly during the early stages of growth. RpiRc was additionally found to be essential for protein A production, to affect biofilm formation by decreasing the binding of *S. aureus* cells to polystyrene surfaces, and to suppress capsule formation (Zhu *et al.*, 2011). Inactivation of *rpiRc* was recently shown to increase the infectivity of the mutant in two independent murine infection models (Gaupp *et al.*, 2016), suggesting that RpiRc, like CcpE and CodY, serves as an attenuator of virulence of *S. aureus* by controlling the synthesis of virulence factors in response to the bacterial metabolic status.

Alternative σ factors

Alternative σ factors constitute an elegant mechanism in eubacteria to allow the simultaneous transcription of genetically unlinked genes and operons, provided these genes and operons share the same promoter specificities. As part of the RNA-polymerase holoenzyme (E-σ), the σ subunit guides the catalytically competent RNA polymerase core enzyme (E) to specific promoter sequences and allows efficient initiation of transcription (reviewed inBurgess and Anthony, 2001). The *S. aureus* genome encodes three alternative σ factors (σ^B, σ^H, and σ^S) that are found in all strains sequenced to date (Ibarra *et al.*, 2013).

σ^B

The best studied alternative σ factor of *S. aureus* is σ^B. It is transcribed as part of a polycistronic operon consisting of *mazEF-rsbUVW-sigB*, and the transcription of this operon is driven by at least two promoters, a σ^A-dependent promoter in front of *mazE*, and a σ^B-dependent promoter preceding *rsbV* (Senn *et al.*, 2005b). The co-transcribed gene products RsbU, RsbV, and RsbW form a post-translational control system for σ^B activity, in which RsbW acts as an anti-σ^B factor that binds to σ^B to prevent binding of this σ subunit to the core enzyme. RsbV acts as an anti-anti-σ^B factor that in an unphosphorylated state can bind to RsbW to prevent binding of RsbW to σ^B. Interestingly, RsbW possesses an RsbV-specific kinase activity to phosphorylate this anti-anti-σ^B factor, which in its phosphorylated state (RsbV-P) cannot bind to RsbW. This phosphorylation process is counteracted by RsbU, an RsbV-specific phosphatase that dephosphorylates RsbV-P to reactivate its binding affinity towards RsbW (reviewed inPane-Farre *et al.*, 2006; van Schaik and Abee, 2005). MazE and MazF constitute a classical bacterial toxin/anti-toxin (TA) system that does not directly affect σ^B activity (Donegan and Cheung, 2009), but transcription of the *sigB* operon (Fu *et al.*, 2009).

The molecular mechanism(s) by which the *S. aureus* σ^B becomes activated are still incompletely understood. It is assumed that under favourable growth conditions, σ^B is kept inactive mainly by RsbW. When the cell encounters nutritional or physical/chemical stresses, σ^B becomes activated by a process that probably involves RsbU (Giachino *et al.*, 2001; Palma and Cheung, 2001; Pane-Farre *et al.*, 2009). Although it is not known yet, which and how the internal and/or external stimuli activate the phosphatase activity of RsbU towards RsbV, it is assumed that activated RsbU shifts the proportion of RsbV/RsbV-P into a direction favouring the interaction of RsbV with RsbW, thereby releasing σ^B from the RsbW/σ^B complex to interact with the core enzyme (Pane-Farre *et al.*, 2009; Senn *et al.*, 2005b). In *S. aureus*, RsbU is expressed at a relatively constant level throughout growth, and overexpression of RsbU is sufficient to activate σ^B. In contrast to RsbU, σ^B activity varies significantly between growth phases, strongly suggesting that RsbU activity is modulated by additional factors (Senn *et al.*, 2005b).

During *in vitro* growth, σ^B is active mainly in the late exponential growth phase and during the stationary phase (Giachino *et al.*, 2001; Senn *et al.*, 2005a). Exposure of *S. aureus* cells to physical/chemical stresses such as heat shock (Gertz *et al.*, 1999, 2000; Giachino *et al.*, 2001; Inose *et al.*, 2006; Pane-Farre *et al.*, 2006; Senn *et al.*, 2005b), pH alterations (Palma and Cheung, 2001; Pane-Farre *et al.*, 2006, 2009), salt stress (Pane-Farre *et al.*, 2006; Senn *et al.*, 2005b), or exposure to antibiotics (Chen *et al.*, 2011; Mitchell *et al.*, 2010; Reiss *et al.*, 2012) induce a fast but often short-lived increase in σ^B activity (Gertz *et al.*, 2000; Giachino *et al.*, 2001; Inose *et al.*, 2006; Pane-Farre *et al.*, 2006; Senn *et al.*, 2005b). *In vivo* activity of σ^B was also demonstrated in different murine infection models (Depke *et al.*, 2012; Goerke *et al.*, 2005), and in a number of cell culture based internalization and persistence assays (Garzoni *et al.*, 2007; Mitchell *et al.*, 2013; Moisan *et al.*, 2006; Pfortner *et al.*, 2014; Surmann *et al.*, 2014), suggesting that this alternative σ factor plays an important role in the pathogens ability to adapt to the intracellular milieu and to survive for prolonged periods of time within host cells.

Transcriptomic and proteomic studies have demonstrated that σ^B controls a rather large regulon (≥229 genes or operons), including numerous virulence factors (Bischoff *et al.*, 2004; Hempel *et al.*, 2010; Pane-Farre *et al.*, 2006; Ziebandt *et al.*, 2001). However, σ^B seems to exert most of its effects on virulence determinant production via transcriptional control of other TFs such as *arlRS*, *sarA*, *sarS*, and *yabJ-spoVG*, and by modulating RNAIII and Rot production (Bischoff *et al.*, 2001, 2004; Hsieh *et al.*, 2008; Kim *et al.*, 2014; Meier *et al.*, 2007). By promoting the expression of numerous adhesins and repressing the production of many exotoxins, σ^B largely antagonizes RNAIII activity, suggesting that this alternative σ factor serves primarily as a modulator of virulence gene expression in order to fine-tune its virulence factor production in response to changing host environments (Bischoff *et al.*, 2004). The proposed DNA binding consensus for σ^B is GttTaa (–35 region) and gGGTAt (–10 region), separated by a spacer region of 12–15 base pairs (Homerova *et al.*, 2004).

Given its profound effect on virulence determinant regulation, one can assume that σ^B has a major impact on the pathogenesis of *S. aureus* as well; however, testing of *sigB* mutants in different infection models has yielded inconclusive results. Specifically, a clear infection enhancing effect of σ^B was observed in an experimental central venous catheter-related infection model (Lorenz *et al.*, 2008) and a septic arthritis model (Jonsson *et al.*, 2004), but no such effect was observed in a wound infection model and haematogenous pyelonephritis model (Nicholas *et al.*, 1999), an experimental endocarditis model (Entenza *et al.*, 2005), and a kidney infection model (Depke *et al.*, 2012). Results from the latter infection model were surprising because transcription of a number of important virulence factors was affected by σ^B (Depke *et al.*, 2012), supporting the idea that σ^B serves primarily as virulence modulator rather than being a virulence determinant by itself (Depke *et al.*, 2012; Entenza *et al.*, 2005).

σ^H

The second alternative σ factor identified in *S. aureus* was σ^H. Although sharing only a low degree of similarity to its counterpart in *B. subtilis* (23% identity at the amino acid level), the gene arrangement around the σ factor is highly conserved between these two species, which facilitated its identification in *S. aureus* (Morikawa *et al.*, 2003). Sequencing of the *sigH* loci of 39 staphylococcal species revealed an unusually high degree of divergence in *sigH* sequences within this genus, suggesting that *sigH* might serve as useful molecular target to differentiate staphylococci (Morikawa *et al.*, 2008).

sigH is transcribed as part of a polycistronic operon consisting of *yazC-yacOP-sigH-rpmG* (Dumoulin, 2005), and translation of *sigH* is post-transcriptionally suppressed by an inverted repeat immediately upstream of the initiation codon (Dumoulin, 2005; Morikawa *et al.*, 2012). The conditions activating σ^H are still largely unknown. Morikawa and colleagues could demonstrate σ^H activity in *S. aureus* grown aerobically in a complete synthetic medium, CS2, but not when cells were grown in standard laboratory media such as brain heart infusion (BHI) or RPMI (Morikawa *et al.*, 2012). In addition, the same authors demonstrated that σ^H-active *S. aureus* cells spontaneously occur at a low frequency due to chromosomal rearrangements caused by short junction duplications (Edlund and Normark, 1981; Kugelberg *et al.*, 2006), leading to the formation of active chimeric SigH proteins (Morikawa *et al.*, 2012).

The first two genetic loci found to be affected by σ^H in *S. aureus* were the *comE* and *comG* operons, suggesting that this σ factor might be involved in genetic competence of this organism (Morikawa *et al.*, 2003). More recently, it was shown that *S. aureus* producing active σ^H become competent for transformation by plasmid or chromosomal DNA, albeit on a very low level, and that this process requires the presence of functional *comE* and *comG* operons (Morikawa *et al.*, 2012). Overexpression of a functional σ^H is insufficient to induce genetic competence in *S. aureus*, although it does activate transcription of the *comE* and *comG* operons (Fagerlund *et al.*, 2014), suggesting the requirement of further σ^H activating factor(s) to become competent for DNA uptake. Recent RNA-sequencing studies indicate that σ^H controls a comparably small regulon, with forty of the genes/operons being positively affected by this alternative σ factor (Fagerlund *et al.*, 2014). The proposed DNA binding consensus for σ^H is AGTAA (–35 region) and ACGTG (–10 region), separated by a spacer region of 14 base pairs (Fagerlund *et al.*, 2014).

σ^H was also reported to modulate prophage integration and excision in *S. aureus* via transcriptional control of the integrase encoding gene *int* (Tao *et al.*, 2010). However, the promoter region of *int* lacks an apparent σ^H consensus sequence, and the gene was not identified as being affected by σ^H in the RNA-seq studies (Fagerlund *et al.*, 2014).

σ^S

The latest σ factor identified in *S. aureus* was σ^S (Shaw *et al.*, 2008). This alternative σ factor shares homology to extracytoplasmic function (ECF) σ factors, the largest and most diverse subfamily of σ factors (Staron *et al.*, 2009). A common feature of ECF s factors is that they contain only two of the four conserved domains, σ_{1-4}, typically found in primary σ factors such as σ^A or σ^{70}. Moreover, most of the ECF σ factors autoregulate their own transcription, and are co-expressed with a cognate anti-σ factor that is usually membrane-anchored and keeps the ECF σ factor inactive in the absence of a specific stimulus (Helmann, 2002). A phylogenetic study suggested that σ^S appertains to one of the minor groups of ECF σ factors, ECF104, with all members belonging to the genus *Staphylococcus* (Staron *et al.*, 2009).

Biochemical analyses confirmed that σ^S of *S. aureus* bind to core enzyme to form a functional holoenzyme, and that this alternative σ factor positively affects its own transcription (Shaw *et al.*, 2008). Efforts to elucidate the function of σ^S have been hampered due to the fact that it lacks a cognate anti-σ factor immediately downstream of *sigS*, and is only weakly transcribed under standard laboratory conditions (Shaw *et al.*, 2008). More recent analyses demonstrated that *sigH* transcription is driven by three partially overlapping promoters, with promoter P_2 being autoregulated, and that *sigS* expression is strongly induced by chemical stressors that cause DNA damage and alter cell wall stability (Miller *et al.*, 2012). High level transcription of *sigS* was also observed in cells grown in serum or after digestion by macrophage-like cells. Inactivation of *sigH* was found to decrease the fitness under elevated temperatures, long term survival, and the infectivity in a murine model of septic arthritis, suggesting a role for σ^S in the stress response and virulence (Shaw *et al.*, 2008). Neither the regulon that is controlled by σ^S nor the DNA binding consensus of this alternative σ factor has been identified.

Regulatory RNAs

The impact of overlapping transcription in *S. aureus* gene regulation

Every step leading to an active gene product can be the target of a regulatory event. In bacteria, TFs are important players in gene regulation as described above. They precisely tune gene expression and determine the molecular machineries for developmental plasticity, homeostasis, stress adaptation, and virulence. Besides, the development of genome-wide RNA methods in *S. aureus* has enabled the analysis of post-transcriptional regulation on a global scale (Lasa *et al.*, 2011), revealing complex and versatile RNA-based regulatory networks (e.g. Guillet *et al.*, 2013; Lasa *et al.*, 2012; Tomasini *et al.*, 2014). Fundamental to RNA-based regulation are changes in mRNA turnover, which are commonly observed in response to growth phase, environmental cues, antibiotic exposure, adherence to epithelial cells, and pathogenesis (Morrison and Dunman, 2011). Linking RNA-based regulation to other cellular networks has the advantage of providing a very rapid reprogramming of gene expression, which is essential when responding to changing physiological conditions. This RNA-based regulation is governed by *trans*-acting factors including RNA-binding proteins, various RNAs, and metabolites that coordinate regulation of functionally related sets of mRNAs through binding to specific structures and/or sequence elements in the mRNA targets. Hence, post-transcriptional regulation may be comparable in terms of diversity and complexity to protein-mediated transcriptional regulatory systems.

One of the more important observations about the *S. aureus* genome was that it had extensive overlapping transcription, which raised the question of the importance of such a phenomenon in gene regulation. Most pervasive transcription led to the synthesis of unexpectedly long and cryptic antisense to protein-coding transcripts, which were rapidly degraded in a manner dependent on the double strand specific endoribonuclease III, RNase III (Lasa *et al.*, 2011). It is unclear presently how many cryptic transcripts are byproducts of functional or spurious transcription events. We cannot exclude that such RNAs may have some impact on gene regulation, on the evolution of the genomes, in maintaining the nucleoid structure, or that they may encode for unexpected ORFs. Conversely, it has been proposed that pervasive transcription could reflect a global surveillance mechanism of the genome integrity based on recent data showing that transcription is coupled to nucleotide excision repair (Kamarthapu and Nudler, 2015).

The existence of overlapping operons or overlapping 5′ and 3′ UTRs of divergent or convergent mRNAs led to the recent observations that antisense transcription does impact *S. aureus* gene regulation (Lasa *et al.*, 2011; Lioliou *et al.*, 2012). The presence of mRNAs with overlaps of either their entire length or over their 5′ or 3′ UTRs seems to be the rule rather than an exception in the *S. aureus* genome (Fig. 12.2A and B) (Lasa *et al.*, 2011; Ruiz de los Mozos *et al.*, 2013). The presence of mRNAs in bacteria having extensive overlaps is not unique to *S. aureus*, a new gene organization in *Listeria monocytogenes*, the so-called excludon, corresponded to overlapping UTRs that regulated the expression of divergent genes encoding proteins with opposing functions (Wurtzel *et al.*, 2012). Although such a gene organization has not been demonstrated in *S. aureus*, positively coordinated regulation of enzyme synthesis through overlapping 5′ UTRs was recently suggested (Lioliou *et al.*, 2012). This post-transcriptional regulation concerned the synthesis of the enzymes TagG and TagH that are required for the synthesis of teichoic acids (see Chapter 8). Teichoic acids are polymers covalently linked to the peptidoglycan or cytoplasmic membrane and contribute to fitness, cell wall maintenance, colonization, infection, and immune evasion (Xia *et al.*, 2010). Regulation of *tagG* and *tagH* requires RNase III-dependent processing of the overlapping 5′ UTRs of *tag*H and *tag*G mRNAs leading to shorter 5′ ends starting a few nucleotides upstream of the Shine and Dalgarno (SD) sequence. This shortening of the 5′ UTRs might facilitate recruitment of ribosomes. Taken together, overlapping UTRs represent a novel mechanism to regulate and coordinate gene expression between neighbouring and functionally related genes.

Several stable RNAs (SSRs), which respond to growth phase and various stress conditions, have been identified by microarray analysis (Anderson *et al.*, 2006; Roberts *et al.*, 2006). Among these SSRs, the stability of the 891-long SSR42 RNA was greatly enhanced during the stationary phase

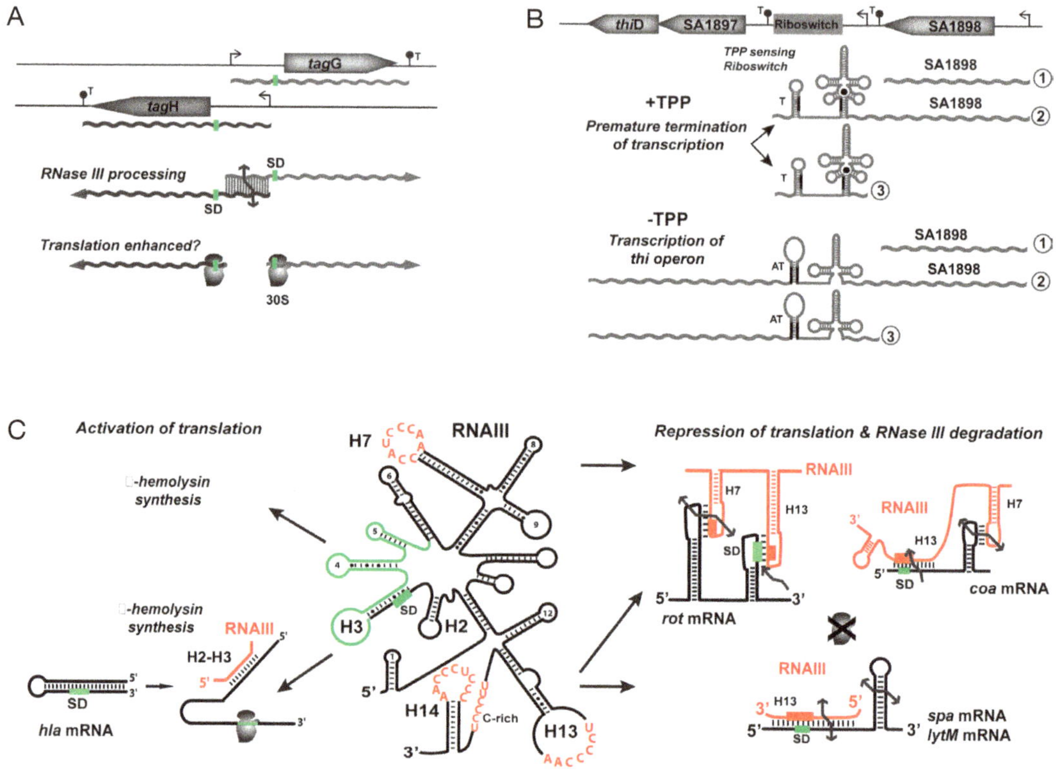

Figure 12.2 Examples of post-transcriptional regulation in *S. aureus*. (A) Schematic representation of the *tagGH* locus. The two divergent genes produced mRNAs with large 5′ UTRs, which partially overlapped. The endoribonuclease III (RNase III) processed the two complementary 5′UTRs (Lioliou *et al*., 2012), leading to mRNAs with shorter leaders. It is proposed that this cleavage facilitates ribosome recruitment at the initiation step of protein synthesis (Lioliou *et al*., 2012). SD is for Shine and Dalgarno sequence. (B) Transcription readthrough to generate long 3′UTRs. RNA transcripts can stop at a Rho-independent transcription terminator that is part of a riboswitch (Ruiz de los Mozos *et al*., 2013). When the metabolite (black circle) binds to the riboswitch, it stabilizes the formation of a Rho-independent terminator (T). As a consequence, the transcription of the downstream gene is arrested while the upstream gene generates a mRNA with a long 3′ UTR containing the riboswitch. At low concentration of the metabolite, an anti-terminator structure (AT) is formed allowing transcription, and a long polycistronic RNA is generated. A typical example is given for the thiamine pyrophosphate (TPP)-sensing riboswitch. (C) Mechanism of action of the quorum sensing induced RNAIII. The schematic and modular secondary structure of RNAIII shows the *hld* gene encoding δ-haemolysin in green and the non-coding parts in black. The various C-rich sequence motifs of RNAIII (in red) represent the seed sequences that bind to the SD sequence of mRNA targets. Various topologies of RNAIII-mRNA are given. These topologies either activated the translation of *hla* (left) or prevented the access of the 30S ribosomal subunit to several mRNAs to repress translation (right). For details and references, see text.

of growth (Anderson *et al.*, 2006; Morrison *et al.*, 2012b). Although the direct targets of SSR42 are not known, SSR42 upregulated genes include those involved in capsule biosynthesis and the down-regulated 80 genes include several cell surface virulence factors. Deletion of SSR42 altered erythrocyte lysis, resistance to opsonization killing, and pathogenesis in a murine skin and soft tissue infection model (Morrison *et al.*, 2012b). Recent high throughput sequencing analysis suggested the 3′ end of SSR42 overlaps with the 3′UTR of the upstream and divergent gene (Lioliou *et al.*, 2012; Ruiz de los Mozos *et al.*, 2013). Of note, these two RNAs were also co-immunoprecipitated with RNase III, suggesting that antisense regulation would regulate either the stability of SSR42 or of the flanking genes in a growth phase dependent manner.

In addition to long antisense RNAs, many small and structured antisense RNAs (asRNA) have been identified (Abu-Qatouseh *et al.*, 2010; Beaume *et al.*, 2010; Bohn *et al.*, 2007; Pinel-Marie *et al.*, 2014). Interestingly, several of these asRNAs are transcribed from the same locus as the transposase encoding genes they regulate (Beaume *et al.*, 2010). Other asRNAs are part of the type I toxin–antitoxin (TA) system that represses the synthesis of cytolytic toxin genes, and most of those are located on pathogenicity islands (Guillet *et al.*, 2013). Recently, an unconventional type I TA system, SprF1/SprG1, was described, in which the sRNA SprG1 was found to encode two cytotoxic peptides produced from two in-frame initiation codons that are endowed with haemolytic and antibacterial activities (Pinel-Marie *et al.*, 2014). Under normal growth conditions, the asRNA SprF1 prevents the mortality of *S. aureus* cells by triggering the degradation of the SprG1 sRNA (Pinel-Marie *et al.*, 2014). However, the active region of a short and structured anti-toxin asRNA is not always the full complementary region to the target sRNA encoding the toxin. Indeed, the interacting region of the anti-toxin asSprA1 is not the expected 3′ end domain, which is fully complementary to the toxin-encoding sRNA SprA1. Instead, it is the 5′ end that is partially complementary to the ribosome binding site of SprA1 that inhibits translation (Sayed *et al.*, 2012). The example of SprA1 sRNA highlights the diversity of RNA regulatory mechanisms and emphasizes the important need for deep mechanistic studies to delineate the active regions of regulatory RNAs and their mechanisms of action. Another example demonstrating the importance of understanding sRNA regulation originates from recent work suggesting that persistence of *Escherichia coli* depends on type II TA system (reviewed inGerdes and Maisonneuve, 2012). Persistence is caused by the presence of slow growing bacteria that are tolerant against antibiotics and other environmental stresses. It is likely that persistence might also be intimately linked to RNA-dependent regulation. Interestingly, transcription of several sRNAs in *S. aureus* were found being deregulated in small colony variants (SCV), which are slow-growing and persistent bacteria (Abu-Qatouseh *et al.*, 2010; Garcia *et al.*, 2013). Some of these sRNAs are antisense or partially overlapping coding genes, with six of them encoded on pathogenicity islands or inserted phage sequences, and one SCV-specific asRNA might have a role in controlling inorganic ion transport (Abu-Qatouseh *et al.*, 2010). Whether the persistence of *S. aureus* bacteria also depends on type I TA loci or of the deregulation of some sRNAs has yet to be determined.

Proteins associated with post-transcriptional regulation

A full perspective on post-transcriptional regulation requires improved understanding of regulatory networks and of the interconnections existing between RNAs and proteins in the cell environment. These interactions are still poorly explored in *S. aureus* and the functions of many RNA-binding proteins remain enigmatic. Purification of ribonucleoprotein particles combined with high throughput sequencing should help identifying RNA-binding proteins and their target RNAs that play key roles in gene regulation.

The role of several ribonucleases has been well documented in *S. aureus* (Morrison and Dunman, 2011). Not only are ribonucleases associated with RNA decay and maturation, they are important in gene regulation and contribute to the stress response and pathogenesis. RNase III was first associated with degradation of translationally repressed mRNAs by the quorum sensing-dependent RNAIII (Boisset *et al.*, 2007; Chevalier *et al.*, 2010; Huntzinger *et al.*, 2005; Romilly *et al.*, 2012). In addition, RNase III was found to be a major partner for antisense regulation that cleaved weakly expressed asRNAs bound to mRNAs (Lasa *et al.*, 2011; Lioliou *et al.*, 2012). Other endonucleases, namely RNase Y, RNase J1 and RNase J2, have also been analysed (Linder *et al.*, 2014; Marincola *et al.*, 2012). The 5′ ends of total RNAs have been precisely analysed in a *S. aureus* wild-type strain and in RNase J1, J2, and Y deletion mutants (Linder *et al.*, 2014). This study revealed that RNase J1, which is a dual enzyme endowed with endo- and 5′ exo-ribonuclease activities, is a major factor in the mRNA decay and maturation processes of several RNAs (mRNA, rRNA, and RNase P), while RNase J2 stimulates the activity of RNase J1. In contrast, RNase Y, which is anchored to the membrane, has a more specific action on the processing and stabilization of *saeRS* mRNA, on the turnover of several regulatory RNAs (Marincola *et al.*, 2012), and on the regulatory regions of mRNAs carrying T-box leaders (Linder *et al.*, 2014). Together, RNase Y, RNases J1/J2, polynucleotide phosphorylase

(PNPase), enolase, phosphofructokinase, RpnA (the main co-factor of the RNase P), and CshA have been proposed to form a degradosome in *S. aureus*, although direct interaction was not clearly demonstrated (Roux *et al.*, 2011). Among these proteins, CshA has been reported to fine-tune the steady-state level and stability of the *agr* locus, which in turn indirectly affects the level of RNAIII and virulence of *S. aureus* (Oun *et al.*, 2013). More surprisingly, RpnA displays an RNase activity *in vitro*, and this activity can be inhibited by a specific drug, RNPA1000 (Olson *et al.*, 2011). This compound attenuated virulence in a mouse infection model and had an anti-microbial activity against biofilm-associated *S. aureus*. This study suggested that anti-microbial therapy could be targeted against RNA metabolism and essential RNases (Olson *et al.*, 2011).

Another RNA-binding protein having a key role in post-transcriptional regulation in many bacteria is the Sm-like protein Hfq, which acts cooperatively with sRNAs to initiate the decay of mRNAs (Lalaouna *et al.*, 2013). The *hfq* gene is present in most of the staphylococcal genomes but the protein is not always expressed. In addition, an *hfq* deletion mutant had no significant phenotype on growth in several genetic backgrounds, despite its postulated function in pathogenesis and stress responses for several clinical strains (Castro *et al.*, 2011; Liu *et al.*, 2010; Romilly *et al.*, 2012). This homohexameric protein carries various RNA binding sites such as proximal and distal sites recognizing unpaired uridine-rich and adenine-rich sequences, respectively (e.g. Brennan and Link, 2007; Horstmann *et al.*, 2012). In contrast to *E. coli*, *S. aureus* Hfq carries a short C-terminal tail and does not contain a conserved arginine rich motif (RRER) that is required for the chaperone activity of *E. coli* Hfq and for annealing of two interacting RNAs (Panja and Woodson, 2012; Zheng *et al.*, 2016). The structural differences between *E. coli* and *S. aureus* Hfq have physiological consequences in that the *S. aureus* Hfq cannot totally complement *S. typhymurium* or *E. coli hfq* mutant strains, demonstrating there are evolutionary differences between Hfq proteins across genera (Rochat *et al.*, 2012; Vecerek *et al.*, 2008). The implication of *S. aureus* Hfq in gene regulation is still a matter of debate, and the gain of function in strains expressing Hfq remains to be studied.

S. aureus CvfB has also been described as an RNA-binding protein that contributes to the expression of several exotoxins via both *agr*-dependent and *agr*-independent pathways (Matsumoto *et al.*, 2007). This protein contains three OB-fold domains and a winged helix (WH) motif, which are all required for cooperative binding to polyU sequences (Matsumoto *et al.*, 2010). Identification of the RNA targets of CvfB would certainly help to identify the mechanism of action of this protein and its role in virulence.

A further multifunctional protein is SarA, which was unexpectedly identified as an RNA binding protein using *in vitro* binding assays (Morrison *et al.*, 2012a) (see above). It is not yet known if SarA uses the same determinants to recognize DNA promoters and RNA regulatory domains. It would also not be surprising if other members of the SarA family of proteins also have this dual function.

mRNAs regulating essential metabolic pathways and biofilm formation

As with many bacteria, *S. aureus* exploits the extraordinary properties of mRNA to fold into different structures in response to various signals that regulate transcription, translation or mRNA stability. Although the UTRs of mRNAs are generally of small size, mRNAs encoding proteins involved in virulence, stress responses and metabolism may carry large 5′ or 3′ UTRs (Lasa *et al.*, 2011; Ruiz de los Mozos *et al.*, 2013). In large 5′ UTRs, regulatory domains are often embedded and many of them function as direct sensors for intracellular metabolite concentrations, the so-called riboswitches (Ramesh and Winkler, 2014). These *cis*-acting regulatory elements exhibit a structured receptor domain that specifically binds a small metabolite, which provokes premature transcription, translation repression/activation, or cleavage (Fig. 12.2B). They regulate downstream mRNAs, which are involved in the uptake or metabolism of the bound metabolite. In *S. aureus*, at least seven operons are under the control of riboswitches that respond to the intracellular concentrations of S-adenosylmethionine (SAM), thiamine pyrophosphate (TPP), flavin mononucleotide (FMN), lysine, glycine, guanine, 7-aminomethyl-7-deazaguanine (preQ1), and glucosamine-6-phosphate (Glc-6P) (Caldelari *et al.*, 2012). Because these elements regulate essential metabolic genes and bind to small molecules, they offer novel opportunities for the design of drugs that could stably bind to the riboswitch and constitutively repress or activate expression of the downstream mRNAs. As an example, a novel pyrimidine derivative, 2,5,6-triaminopyrimidin-4-one (PC1), was designed to bind the purine sensing riboswitch and repress the downstream gene (Mulhbacher *et al.*, 2010). PC1 has a selective bactericidal activity against a sub-group of bacteria, including *S. aureus*, which contain *guaA* (coding for GMP synthetase) under the control of the purine riboswitch. Importantly, treatment of mice with PC1 reduced *S. aureus* infection in a murine mastitis model. In addition to treating an infection, the narrow spectrum of bactericidal activity has the advantage of reducing the selective pressure for resistance, an important consideration in an era when antibiotic resistance is rapidly rising. The modular architecture of riboswitches has also been recently exploited to rewire bacterial regulatory circuits in response to synthetic non-natural effector molecules (Groher and Suess, 2014; Robinson *et al.*, 2014). For several riboswitches, regulating termination of transcription, the Rho-independent terminator can have dual functions. When it is formed as the result of the metabolite binding, it prevents the synthesis of the downstream gene, but in addition it may serve as a terminator for the upstream gene (Fig. 12.2B).

Large 3′ UTRs often contain regulatory signals that can affect mRNA stability. A recent study revealed that *icaR* mRNA, which encodes a transcriptional repressor of biofilm formation, is only weakly translated. IcaR is a transcriptional repressor of the *icaADBC* operon, which produces the exopolysaccharide polymer PIA/PNAG that constitutes the biofilm matrix. The weak translation of IcaR is due to the presence of a C-rich motif in the 390 nt-long 3′ UTR which acts as a negative determinant of *icaR* translation (Ruiz de los Mozos *et al.*, 2013). This sequence forms an unexpectedly long–range interaction with the SD sequence to prevent ribosome binding and to recruit RNase III for rapid mRNA degradation. In this mechanism, the 5′ UTR structure would prevent efficient ribosome recognition during transcription and would undergo a rearrangement mediated by the 3′ UTR binding. Although these observations are intriguing, many questions remain to be addressed: What is the mechanism to relieve the 5′–3′ UTR pairings? Is there a *trans*-acting factor involved in this mechanism? Does the RNA polymerase alter the RNA folding pathway and the formation of the long–distance pairings? Do the pairings occur *in cis* or *in trans*? Given the fact that many mRNAs carry large UTRs, one might expect to reveal novel post-transcriptional mechanisms. Another notable example of mRNAs having large UTRs is RNAIII, however, due to its importance in regulating *S. aureus* virulence it will be considered separately (see below).

With the new development of high-throughput sequencing methodologies such as ribosome profiling (Ingolia, 2014), it is now possible to examine post-transcriptional regulation on a global scale in both *S. aureus* and the infected host. This information will allow unravelling of the importance of such regulation in pathogenesis, persistence, and evolution, and to provide new insights into the identity of proteins and peptides produced in *S. aureus*.

trans-Acting RNAs regulating virulence gene expression in *S. aureus*

Bioinformatic prediction, genomic studies and high throughput approaches have identified more than 200 genes encoding small and large sRNAs that are expressed in *S. aureus* strains (Abu-Qatouseh *et al.*, 2010; Beaume *et al.*, 2010; Bohn *et al.*, 2010; Geissmann *et al.*, 2009; Lioliou *et al.*, 2012; Morrison and Dunman, 2011; Novick *et al.*, 1993; Pichon and Felden, 2005; Roberts *et al.*, 2006; Mäder *et al.*, 2016). We will describe briefly several sRNAs for which a function in pathogenesis has been documented, and illustrate the complexity of networks involved in the regulation of virulence factors. In addition, detailed examples have been described in recent reviews (Felden *et al.*, 2011; Guillet *et al.*, 2013; Tomasini *et al.*, 2014).

RNAIII: a paradigm for RNA regulation

As mentioned earlier in this chapter, the *S. aureus* quorum sensing system involves two main intracellular effectors, the response regulator AgrA and the regulatory RNAIII (reviewed in Novick, 2003). The bi-functional RNAIII acts as an mRNA encoding the PSM δ-haemolysin (Hld), and as a regulator mediating the switch between the expression of surface proteins and adhesins and the synthesis of excreted toxins. It also represses the synthesis of several hydrolytic enzymes that are involved in peptidoglycan turnover, cell separation, protein secretion, biofilm formation, and programmed cell death (Boisset *et al.*, 2007; Chunhua *et al.*, 2012). In other words, RNAIII helps to prevent cell wall damage as the bacteria adapt their metabolism to the entry into the stationary phase of growth and to changes in environmental conditions.

How RNAIII regulates gene expression

The UTRs of *hld* are the important regulatory domains of RNAIII. The 5′ UTR contains the complementary sequences to the leader region of *hla* mRNA and binding of RNAIII would facilitate the ribosome recruitment to initiate the translation of α-toxin (Morfeldt *et al.*, 1995). A recent work showed that two distant domains of RNAIII including its 5′ and 3′ ends bind to the 5′ UTR of *mgrA* mRNA to enhance the synthesis of MgrA (Gupta *et al.*, 2015). In contrast, the large 3′UTR is acting as a repressor domain (Fig. 12.2C). This domain carries several unpaired C-rich sequence motifs, which are the seed sequences for base-pairing interactions with target mRNAs. The topologies of mRNA–RNAIII complexes are diverse, but they all share a similar repression mechanism that is to occlude the formation of the ribosomal initiation complex, and in most cases they also recruit RNase III to initiate RNA degradation (Boisset *et al.*, 2007; Chabelskaya *et al.*, 2014; Chevalier *et al.*, 2010; Huntzinger *et al.*, 2005). These mRNAs encode mainly cell surface-localized virulence factors, and the TF Rot (see above). Through the inhibition of Rot, the transcription of exotoxins is indirectly activated while the transcription of several adhesins is arrested (Boisset *et al.*, 2007; Geisinger *et al.*, 2006). RNAIII is at the heart of complex regulatory networks involving double feedforward loops, which contribute to spreading and dissemination of the pathogen, and to the establishment of an infection following a quorum sensing signal.

Functional interactions between the quorum sensing system and other sRNAs

The above view is certainly too simplistic because the transcription of RNAIII varies considerably in clinical isolates (Song *et al.*, 2012). Indeed, it is well known that the *agr* operon is controlled at the transcriptional level by many global regulatory factors (see above). Controlling the *agr* system independently of the quorum sensing signal can be advantageous for the bacteria to respond to multiple signals or environmental cues and to produce virulence factors under appropriate conditions. Recent studies have shown that other sRNAs can interfere with the functions of RNAIII or synergistically coordinate the regulation of specific virulence factors. One characteristic example is the pathogenicity-island encoded sRNA SprD, which at low cell density represses translation of the *sbi* mRNA encoding a cell-surface localized immune evasion protein (Chabelskaya *et al.*, 2014). At high cell density, the level of SprD decreases, and RNAIII takes over to repress translation of *sbi* mRNA using a similar mechanism. Based on these observations, it was postulated that the action of RNAIII and SprD would restrain the synthesis of Sbi to a short specific window during growth. However, it is not yet known if such precise timing of Sbi expression is essential during an infection.

As mentioned above, the *agr* encoded TCS-RR AgrA is able to regulate gene expression in an RNAIII-independent manner (Queck *et al.*, 2008). A recent study revealed that AgrA repressed at high cell density a long antisense RNA, named AgrA-repressed toxin-regulating sRNA (ArtR) (Xue *et al.*, 2014). ArtR overlapped with the 3′ end of *luxS* mRNA and is expected to reduce the steady state levels of *luxS* mRNA. LuxS (aka autoinducer-2, AI-2) is a signalling pathway that regulates biofilm formation in an IcaR-dependent manner and capsular polysaccharide synthesis via the TCS KdpDE (Yu *et al.*, 2012). In addition, ArtR forms basepairing interactions with the 5′ UTR of *sarT* mRNA leading to a rapid degradation by RNase III (Xue *et al.*, 2014). The Sar homologue SarT is a repressor of the synthesis of *hla* mRNA (Schmidt *et al.*, 2001), suggesting that ArtR would activate the synthesis of α-toxin in an indirect manner. Because *hla* mRNA carries a highly structured ribosome binding site (Morfeldt *et al.*, 1995), and because ArtR and RNAIII are not produced at the same time, one can wonder how translation of *hla* mRNA would be promoted. It may be that ArtR would act as a fully complementary asRNA to regulate *luxS* and as a *trans*-acting sRNA to regulate *sarT* through imperfect interactions under conditions when the *agr* system is not activated.

Impact of sRNAs on the evolution of *S. aureus* strains and pathogenesis

S. aureus is a commensal bacterium that has evolved into a major human pathogen causing a wide range of diseases. It is assumed that its advancements as an opportunistic pathogen are primarily due to the acquisition/loss of functions through horizontal gene transfers (for additional information see Chapter 4). Currently, some strains have acquired resistance to most major antibiotic classes, and others have increased their virulence and transmissibility. Two major modes of evolution have been thought to explain the virulence traits between strains: First, mutations may cause changes in the regulation and expression of genes. Second, genes encoding virulence factors may be acquired through mobile elements by lateral transfer from other organisms. A third strategy was recently described, where the transposition of an IS element in the promoter sequence of the *rot* gene in an MRSA USA500 strain resulted in hypervirulence (Benson *et al.*, 2014). This insertion resulted in the inactivation of *rot*, which led to the derepression of many exotoxin genes, and enhanced virulence. Importantly, this study showed how perturbation of the temporal expression of exotoxin production had a significant impact on virulence, and on the evolution of strains. In addition, this study highlights again that *S. aureus* genomes constantly evolve in response to a changing environment, in part due to the horizontally acquired mobile genetic elements (MGE) including phages, pathogenicity islands, plasmids and transposons.

As mentioned above, several pathogenicity island encoded sRNAs were shown to regulate genes present in the core genome, such as SprD, which represses the translation of *sbi* mRNA (Chabelskaya *et al.*, 2010, 2014). In other words, these sRNAs contribute to the host regulatory networks by providing either additive or independent regulatory effects (Guillet *et al.*, 2013). Another recent study has illustrated how the acquisition of a novel regulatory RNA interferes with the quorum sensing system (Kaito *et al.*, 2013). The repeated acquisition of the staphylococcal cassette chromosome *mec* (SCC-*mec*) by different *S. aureus* strains, which confers β-lactam resistance, has caused major problems in the treatment of infections. This acquisition has given rise to community and hospital acquired-methicillin-resistant *S. aureus* (CA- and HA-MRSA, respectively) that can cause severe infections in healthy patients. Specifically, CA-MRSA strains are usually more susceptible to anti-microbial drugs but are more virulent than HA-MRSA strains. The altered virulence of CA-MRSA strains might be due to the observation that SCC-*mec* elements of some MRSA include an additional PSM cluster, *psm-mec* (Queck *et al.*, 2009). Recently, it was determined that the psm-mec RNA has a dual function in that it also acts as an asRNA that binds to the coding region of *agrA* on RNAII to repress AgrA synthesis (Kaito *et al.*, 2013). Interestingly, a C-rich sequence motif of *agrA* mRNA was predicted to interact with the SD sequence of psm-mec suggesting that the translation of the cytolytic peptide would also be inhibited. For this reason, it is difficult to assess which is the target and which is the regulator. That being said, deletion of the *psm-mec* gene from HA-MRSA increased the level of AgrA and enhanced virulence in a murine infection model. In addition, recently it was found that *psm-mec* RNA had a major effect on the synthesis of protein A (Cheung *et al.*, 2014), although the mechanism was not determined. Taken together, these data illustrate how the bi-functional *psm-mec* reduces virulence in HA-MRSA by modulating the *agr* response.

A non-coding RNA as an attenuator of acute infections

Among the sRNAs that have been functionally characterized in *S. aureus*, several of them regulate mRNAs encoding TFs mediating interactions between regulators. This is also the case for the sRNA RsaA, which inhibits the translation of MgrA by an antisense mechanism (Romilly *et al.*, 2014). MgrA is an activator of capsular polysaccharide synthesis that protects *S. aureus* against the host immune system (see above). The synthesis of capsular polysaccharide was noted predominantly in isolates from patients with acute infections, while the loss of capsule biosynthesis has been associated with persistence. Analysis of an RsaA mutant revealed that RsaA enhances biofilm formation, strongly represses capsule synthesis, and promotes opsonophagocytic killing of *S. aureus* by polymorphonuclear leucocytes (Romilly *et al.*, 2014). *In vivo*, the expression of RsaA attenuated the severity of systemic infections and enhanced chronic infections, suggesting that RsaA functions as a virulence suppressor of acute infections (Romilly *et al.*, 2014).

Because RsaA is expressed in all staphylococci, it is possible that this RNA has been selected for commensal interactions with the host. This possibility is consistent with the observations that RsaA transcription is under the control of σ^B (Geissmann *et al.*, 2009), which responds to stress and regulates the transcription of virulence genes by reducing the levels of RNAIII (Bischoff *et al.*, 2001). In contrast, MgrA was shown to indirectly activate the *agr* system (Ingavale *et al.*, 2005; Ingavale *et al.*, 2003) and conversely RNAIII was proposed to activate the synthesis of MgrA (Gupta *et al.*, 2015). Hence, RsaA functionally links two global and antagonistic regulators, σ^B and MgrA, with possible effects on the temporal expression of the virulon. Interestingly, RsaA and the pathogenicity island sRNA SprX share a similar mRNA target encoding SpoVG (Eyraud *et al.*, 2014; Romilly *et al.*, 2014), which is required for capsule synthesis and bacterial resistance to methicillin and glycopeptides (Meier *et al.*, 2007). Whether the two RNAs act synergistically or in an independent manner remains to be determined.

These data show the complex interplay between sRNAs, RNA-binding proteins, and TFs to modulate the expression of the virulon in response to the global physiology of the cell and the environmental conditions.

Conclusions and future trends

The present review provides a glimpse of the complexity of gene regulation and regulatory networks in this major human opportunistic pathogen. It particularly illustrates the intricate interactions between global TFs, TCSs, RNA-binding proteins, and RNAs to modulate the expression of virulence genes, in response to changes in the global physiology of the cell and the environmental conditions (Figs. 12.1 and 12.2). While our review attempts to provide a comprehensive overview of *S. aureus* regulation, it is difficult to generalize the impact of specific regulators amongst clinical *S. aureus* isolates due to their genetic and phenotypic heterogeneity, which results from genome rearrangements, duplication, and acquisition of mobile genetic elements (MGE). Horizontally acquired virulence factors, TFs, and sRNAs are likely to provide a gain of function that mediates crosstalk with the core genome TFs, and may contribute to gene evolution, genome plasticity and virulence (Malachowa and DeLeo, 2010).

Although a rather good knowledge of the TF repertoire in specific *S. aureus* strains is now available, we are still far away from getting a full appreciation of their primary targets and of the regulatory networks. Particularly, the specific DNA-binding site consensi for most of the *S. aureus* TFs have not been determined yet. The use of ChIP-seq and Rip-ChIP approaches will help to unravel the direct targets of specific transcriptional or post-transcriptional regulators, respectively (e.g. Davis *et al.*, 2011; Lioliou *et al.*, 2012).

The examples given here illuminate how TFs and sRNAs converge to form extensive regulatory networks to control virulence factor expressions (Figs. 12.1 and 12.2). One of the future tasks will be to understand the depths to which these networks are interconnected, and how they influence each other to alter the pattern of virulence gene expression in a dynamic manner. Along this line, an attempt at a more global analysis of quorum-sensing in *S. aureus* was recently performed and stressed the important roles of *sae* and *agr* in fine-tuning biofilm repression and/or dissemination (Audretsch *et al.*, 2013).

Another major challenge will be to understand the contribution of regulatory proteins and sRNAs in the context of colonization and/or infection. Indeed, the host response and the human microbiome are expected to influence the expression pattern of regulatory factors and *vice versa*. Dual RNA-seq approaches performed on the host and *S. aureus* might provide a more comprehensive view of gene expression changes occurring during the course of infection (Westermann *et al.*, 2012).

Importantly, most of the studies carried out to date have only provided a comparably static picture of gene regulation in a population of cells. However, the identical genotype and growth environment are insufficient to ensure that individual cells within a clonal microbial culture will exhibit the same phenotype (Gefen and Balaban, 2009). At present, most of the transcriptional effects measured are the mean transcript rates of several to multiple subpopulations that may largely differ in their transcriptional activities. Cell heterogeneity most likely confers the ability of a bacterial population to persist in response to external changes and to exploit new niches, a postulate noted in population dynamics of antibiotic-resistant strains (Koch *et al.*,

2014; Lee *et al.*, 2010). A particularly interesting feature of MRSA strains is the occurrence of competitive interactions between bacterial cells. It has been observed that isolates of MRSA evolve and diversify spontaneously into two distinct strains that react differently to antimicrobial compounds. This strain diversification illustrates how competition between coevolving bacterial subpopulations may generate antibiotic resistance in the context of an infection (Koch *et al.*, 2014). Understanding the population behaviour and single cell analyses should led to a better characterization of the mechanisms that enhance the survival capacity of the overall population under stressful environmental conditions, and to better understand the cell to cell communication.

Web resources

http://staph.unavarra.es Mapped reads from the deep sequencing of RNAs from *S. aureus* strain 15981 have been loaded at the *S. aureus* transcriptome.

http://srd.genouest.org, a resource for sRNAs identified in staphylococci.

http://www.bioinformatics.org/sammd/SATMD, a *S. aureus* transcriptome meta-database that contains post-analysis data from many transcriptome experiments done in *S. aureus*.

References

Abu-Qatouseh, L.F., Chinni, S.V., Seggewiss, J., Proctor, R.A., Brosius, J., Rozhdestvensky, T.S., Peters, G., von Eiff, C., and Becker, K. (2010). Identification of differentially expressed small non-protein-coding RNAs in *Staphylococcus aureus* displaying both the normal and the small-colony variant phenotype. J. Mol. Med. *88*, 565–575.

Adhikari, R.P., and Novick, R.P. (2008). Regulatory organization of the staphylococcal *sae* locus. Microbiology *154*, 949–959.

Anderson, K.L., Roberts, C., Disz, T., Vonstein, V., Hwang, K., Overbeek, R., Olson, P.D., Projan, S.J., and Dunman, P.M. (2006). Characterization of the *Staphylococcus aureus* heat shock, cold shock, stringent, and SOS responses and their effects on log-phase mRNA turnover. J. Bacteriol. *188*, 6739–6756.

Arvidson, S., and Tegmark, K. (2001). Regulation of virulence determinants in *Staphylococcus aureus*. Int. J. Med. Microbiol. *291*, 159–170.

Arya, R., and Princy, S.A. (2013). An insight into pleiotropic regulators Agr and Sar: molecular probes paving the new way for antivirulent therapy. Future Microbiol. *8*, 1339–1353.

Audretsch, C., Lopez, D., Srivastava, M., Wolz, C., and Dandekar, T. (2013). A semi-quantitative model of Quorum-Sensing in *Staphylococcus aureus*, approved by microarray meta-analyses and tested by mutation studies. Mol. Biosyst. *9*, 2665–2680.

Ballal, A., and Manna, A.C. (2009). Expression of the sarA family of genes in different strains of *Staphylococcus aureus*. Microbiology *155*, 2342–2352.

Bateman, A. (1999). The SIS domain: a phosphosugar-binding domain. Trends Biochem. Sci *24*, 94–95.

Beaume, M., Hernandez, D., Farinelli, L., Deluen, C., Linder, P., Gaspin, C., Romby, P., Schrenzel, J., and Francois, P. (2010). Cartography of methicillin-resistant *Staphylococcus aureus* transcripts: detection, orientation and temporal expression during growth phase and stress conditions. PloS ONE *5*, e10725.

Belitsky, B.R., and Sonenshein, A.L. (2011). Contributions of multiple binding sites and effector-independent binding to CodY-mediated regulation in *Bacillus subtilis*. J. Bacteriol. *193*, 473–484.

Benson, M.A., Lilo, S., Nygaard, T., Voyich, J.M., and Torres, V.J. (2012). Rot and SaeRS cooperate to activate expression of the staphylococcal superantigen-like exoproteins. J. Bacteriol. *194*, 4355–4365.

Benson, M.A., Ohneck, E.A., Ryan, C., Alonzo, F., Smith, H., Narechania, A., Kolokotronis, S.O., Satola, S.W., Uhlemann, A.C., Sebra, R., *et al.* (2014). Evolution of hypervirulence by a MRSA clone through acquisition of a transposable element. Mol. Microbiol. *93*, 664–681.

Bischoff, M., Dunman, P., Kormanec, J., Macapagal, D., Murphy, E., Mounts, W., Berger-Bächi, B., and Projan, S. (2004). Microarray-based analysis of the *Staphylococcus aureus* sigmaB regulon. J. Bacteriol. *186*, 4085–4099.

Bischoff, M., Entenza, J.M., and Giachino, P. (2001). Influence of a functional *sigB* operon on the global regulators *sar* and *agr* in *Staphylococcus aureus*. J. Bacteriol. *183*, 5171–5179.

Bohn, C., Rigoulay, C., and Bouloc, P. (2007). No detectable effect of RNA-binding protein Hfq absence in *Staphylococcus aureus*. BMC Microbiol. *7*, 10.

Bohn, C., Rigoulay, C., Chabelskaya, S., Sharma, C.M., Marchais, A., Skorski, P., Borezée-Durant, E., Barbet, R., Jacquet, E., Jacq, A., *et al.* (2010). Experimental discovery of small RNAs in *Staphylococcus aureus* reveals a riboregulator of central metabolism. Nucleic Acids Res. *38*, 6620–6636.

Boisset, S., Geissmann, T., Huntzinger, E., Fechter, P., Bendridi, N., Possedko, M., Chevalier, C., Helfer, A.C., Benito, Y., Jacquier, A., *et al.* (2007). *Staphylococcus aureus* RNAIII coordinately represses the synthesis of virulence factors and the transcription regulator Rot by an antisense mechanism. Genes Dev. *21*, 1353–1366.

Brennan, R.G., and Link, T.M. (2007). Hfq structure, function and ligand binding. Curr. Opin. Microbiol. *10*, 125–133.

Burgess, R.R., and Anthony, L. (2001). How sigma docks to RNA polymerase and what sigma does. Curr. Opin. Microbiol. *4*, 126–131.

Caldelari, I., Fechter, P., Lioliou, E., Romilly, C., Chevalier, C., Gaspin, C., and Romby, P. (2012). A Current Overview of Regulatory RNAs in *Staphylococcus Aureus*. In Regulatory RNAs in Prokaryotes (Springer Vienna), pp. 51–75.

Cassat, J.E., Hammer, N.D., Campbell, J.P., Benson, M.A., Perrien, D.S., Mrak, L.N., Smeltzer, M.S., Torres, V.J., and Skaar, E.P. (2013). A secreted bacterial protease tailors the *Staphylococcus aureus* virulence repertoire to modulate bone remodeling during osteomyelitis. Cell Host Microbe. *13*, 759–772.

Castro, S.L., Nelman-Gonzalez, M., Nickerson, C.A., and Ott, C.M. (2011). Induction of attachment-independent biofilm formation and repression of Hfq expression by low-fluid-shear culture of *Staphylococcus aureus*. Appl. Environ. Microbiol.77, 6368–6378.

Chabelskaya, S., Bordeau, V., and Felden, B. (2014). Dual RNA regulatory control of a *Staphylococcus aureus* virulence factor. Nucleic Acids Res. *42*, 4847–4858.

Chabelskaya, S., Gaillot, O., and Felden, B. (2010). A *Staphylococcus aureus* small RNA is required for bacterial virulence and regulates the expression of an immune-evasion molecule. PLOS Pathog. *6*, e1000927.

Chen, H.Y., Chen, C.C., Fang, C.S., Hsieh, Y.T., Lin, M.H., and Shu, J.C. (2011). Vancomycin activates σ^B in vancomycin-resistant *Staphylococcus aureus* resulting in the enhancement of cytotoxicity. PLoS One *6*, e24472.

Chen, P.R., Bae, T., Williams, W.A., Duguid, E.M., Rice, P.A., Schneewind, O., and He, C. (2006). An oxidation-sensing mechanism is used by the global regulator MgrA in *Staphylococcus aureus*. Nat. Chem. Biol. *2*, 591–595.

Cheung, A.L., Chien, Y.T., and Bayer, A.S. (1999). Hyperproduction of alpha-hemolysin in a *sigB* mutant is associated with elevated SarA expression in *Staphylococcus aureus*. Infect. Immun. *67*, 1331–1337.

Cheung, A.L., Koomey, J.M., Butler, C.A., Projan, S.J., and Fischetti, V.A. (1992). Regulation of exoprotein expression in *Staphylococcus aureus* by a locus (*sar*) distinct from *agr*. Proc. Natl. Acad. Sci. U.S.A. *89*, 6462–6466.

Cheung, A.L., Nishina, K.A., Trotonda, M.P., and Tamber, S. (2008). The SarA protein family of *Staphylococcus aureus*. Int. J. Biochem. Cell Biol. *40*, 355–361.

Cheung, A.L., and Projan, S.J. (1994). Cloning and sequencing of *sarA* of *Staphylococcus aureus*, a gene required for the expression of *agr*. J. Bacteriol. *176*, 4168–4172.

Cheung, A.L., Schmidt, K., Bateman, B., and Manna, A.C. (2001). SarS, a SarA homolog repressible by *agr*, is an activator of protein A synthesis in *Staphylococcus aureus*. Infect. Immun. *69*, 2448–2455.

Cheung, G.Y., Villaruz, A.E., Joo, H.S., Duong, A.C., Yeh, A.J., Nguyen, T.H., Sturdevant, D.E., Queck, S.Y., and Otto, M. (2014). Genome-wide analysis of the regulatory function mediated by the small regulatory *psm-mec* RNA of methicillin-resistant *Staphylococcus aureus*. Int. J. Med. Microbiol. *304*, 637–644.

Chevalier, C., Boisset, S., Romilly, C., Masquida, B., Fechter, P., Geissmann, T., Vandenesch, F., and Romby, P. (2010). *Staphylococcus aureus* RNAIII binds to two distant regions of *coa* mRNA to arrest translation and promote mRNA degradation. PLOS Pathog. *6*, e1000809.

Chong, Y.P., Kim, E.S., Park, S.J., Park, K.H., Kim, T., Kim, M.N., Kim, S.H., Lee, S.O., Choi, S.H., Woo, J.H., *et al.* (2013). Accessory gene regulator (*agr*) dysfunction in *Staphylococcus aureus* bloodstream isolates from South Korean patients. Antimicrob. Agents Chemother. *57*, 1509–1512.

Chunhua, M., Yu, L., Yaping, G., Jie, D., Qiang, L., Xiaorong, T., and Guang, Y. (2012). The expression of LytM is down-regulated by RNAIII in *Staphylococcus aureus*. J. Basic Microbiol. *52*, 636–641.

Cui, L., Lian, J.Q., Neoh, H.M., Reyes, E., and Hiramatsu, K. (2005). DNA microarray-based identification of genes associated with glycopeptide resistance in *Staphylococcus aureus*. Antimicrob. Agents Chemother. *49*, 3404–3413.

Davis, S.E., Mooney, R.A., Kanin, E.I., Grass, J., Landick, R., and Ansari, A.Z. (2011). Mapping *E. coli* RNA polymerase and associated transcription factors and identifying promoters genome-wide. Methods Enzymol. *498*, 449–471.

Delaune, A., Dubrac, S., Blanchet, C., Poupel, O., Mader, U., Hiron, A., Leduc, A., Fitting, C., Nicolas, P., Cavaillon, J.M., *et al.* (2012). The WalKR system controls major staphylococcal virulence genes and is involved in triggering the host inflammatory response. Inf. Immun. *80*, 3438–3453.

Depke, M., Burian, M., Schafer, T., Broker, B.M., Ohlsen, K., and Volker, U. (2012). The alternative sigma factor B modulates virulence gene expression in a murine *Staphylococcus aureus* infection model but does not influence kidney gene expression pattern of the host. Int. J. Med. Microbiol. *302*, 33–39.

Deutscher, J., Francke, C., and Postma, P.W. (2006). How phosphotransferase system-related protein phosphorylation regulates carbohydrate metabolism in bacteria. Microbiol. Mol. Biol. Rev. *70*, 939–1031.

Ding, Y., Liu, X., Chen, F., Di, H., Xu, B., Zhou, L., Deng, X., Wu, M., Yang, C.G., and Lan, L. (2014). Metabolic sensor governing bacterial virulence in *Staphylococcus aureus*. Proc. Natl. Acad. Sci. U.S.A. *111*, E4981–90.

Donegan, N.P., and Cheung, A.L. (2009). Regulation of the *mazEF* toxin-antitoxin module in *Staphylococcus aureus* and its impact on *sigB* expression. J. Bacteriol. *191*, 2795–2805.

Dufour, P., Jarraud, S., Vandenesch, F., Greenland, T., Novick, R.P., Bes, M., Etienne, J., and Lina, G. (2002). High genetic variability of the agr locus in *Staphylococcus* species. J. Bacteriol. *184*, 1180–1186.

Dumoulin, A.D. (2005). Molecular genetic studies on the thiol-disulfide oxidoreductase lipoprotein DsbA and the alternative sigma factor SigH of *Staphylococcus aureus*. ETH. http://dx.doi.org/10.3929/ethz-a-004959495.

Durand, S., Braun, F., Lioliou, E., Romilly, C., Helfer, A.C., Kuhn, L., Quittot, N., Nicolas, P., Romby, P., and Condon, C. (2015). A nitric oxide regulated small RNA controls expression of genes involved in redox homeostasis in *Bacillus subtilis*. PLOS Genet. *11*, e1004957.

Edlund, T., and Normark, S. (1981). Recombination between short DNA homologies causes tandem duplication. Nature *292*, 269–271.

Entenza, J.M., Moreillon, P., Senn, M.M., Kormanec, J., Dunman, P.M., Berger-Bächi, B., Projan, S., and Bischoff, M. (2005). Role of sigmaB in the expression of *Staphylococcus aureus* cell wall adhesins ClfA and FnbA and contribution to infectivity in a rat model of experimental endocarditis. Infect. Immun. *73*, 990–998.

Eyraud, A., Tattevin, P., Chabelskaya, S., and Felden, B. (2014). A small RNA controls a protein regulator involved in antibiotic resistance in *Staphylococcus aureus*. Nucleic Acids Res. *42*, 4892–4905.

Fagerlund, A., Granum, P.E., and Håvarstein, L.S. (2014). *Staphylococcus aureus* competence genes: mapping of the

SigH, ComK1 and ComK2 regulons by transcriptome sequencing. Mol. Microbiol. *94*, 557–579.

Felden, B., Vandenesch, F., Bouloc, P., and Romby, P. (2011). The *Staphylococcus aureus* RNome and its commitment to virulence. PLOS Pathog. *7*, e1002006.

Fournier, B., and Hooper, D.C. (2000). A new two-component regulatory system involved in adhesion, autolysis, and extracellular proteolytic activity of *Staphylococcus aureus*. J. Bacteriol. *182*, 3955–3964.

Fournier, B., and Klier, A. (2004). Protein A gene expression is regulated by DNA supercoiling which is modified by the ArlS-ArlR two-component system of *Staphylococcus aureus*. Microbiology *150*, 3807–3819.

Fournier, B., Klier, A., and Rapoport, G. (2001). The two-component system ArlS-ArlR is a regulator of virulence gene expression in *Staphylococcus aureus*. Mol. Microbiol. *41*, 247–261.

Frees, D., Sørensen, K., and Ingmer, H. (2005). Global virulence regulation in *Staphylococcus aureus*: pinpointing the roles of ClpP and ClpX in the sar/agr regulatory network. Infect. Immun. *73*, 8100–8108.

Fu, Z., Tamber, S., Memmi, G., Donegan, N.P., and Cheung, A.L. (2009). Overexpression of MazFsa in *Staphylococcus aureus* induces bacteriostasis by selectively targeting mRNAs for cleavage. J. Bacteriol. *191*, 2051–2059.

Fujita, Y. (2009). Carbon catabolite control of the metabolic network in *Bacillus subtilis*. Biosci. Biotechnol. Biochem. *73*, 245–259.

Garcia, L.G., Lemaire, S., Kahl, B.C., Becker, K., Proctor, R.A., Denis, O., Tulkens, P.M., and Van Bambeke, F. (2013). Antibiotic activity against small-colony variants of *Staphylococcus aureus*: review of *in vitro*, animal and clinical data. J. Antimicrob. Chemother. *68*, 1455–1464.

Garzoni, C., Francois, P., Huyghe, A., Couzinet, S., Tapparel, C., Charbonnier, Y., Renzoni, A., Lucchini, S., Lew, D.P., Vaudaux, P., *et al.* (2007). A global view of *Staphylococcus aureus* whole genome expression upon internalization in human epithelial cells. BMC Genomics *8*, 171.

Gaupp, R., Wirf, J., Wonnenberg, B., Biegel, T., Eisenbeis, J., Graham, J., Herrmann, M., Lee, C.Y., Beisswenger, C., Wolz, C., *et al.* (2016). RpiRc is a pleiotropic effector of virulence determinant synthesis and attenuates pathogenicity in *Staphylococcus aureus*. Infect. Immun. Accepted manuscript posted online 25 April 2016, doi:10.1128/IAI.00285-16.

Gefen, O., and Balaban, N.Q. (2009). The importance of being persistent: heterogeneity of bacterial populations under antibiotic stress. FEMS Microbiol. Rev. *33*, 704–717.

Geiger, T., Francois, P., Liebeke, M., Fraunholz, M., Goerke, C., Krismer, B., Schrenzel, J., Lalk, M., and Wolz, C. (2012). The stringent response of *Staphylococcus aureus* and its impact on survival after phagocytosis through the induction of intracellular PSMs expression. PLOS Pathog. *8*, e1003016.

Geiger, T., Goerke, C., Mainiero, M., Kraus, D., and Wolz, C. (2008). The virulence regulator Sae of *Staphylococcus aureus*: promoter activities and response to phagocytosis-related signals. J. Bacteriol. *190*, 3419–3428.

Geisinger, E., Adhikari, R.P., Jin, R., Ross, H.F., and Novick, R.P. (2006). Inhibition of rot translation by RNAIII, a key feature of agr function. Mol. Microbiol. *61*, 1038–1048.

Geissmann, T., Chevalier, C., Cros, M.J., Boisset, S., Fechter, P., Noirot, C., Schrenzel, J., François, P., Vandenesch, F., Gaspin, C., *et al.* (2009). A search for small noncoding RNAs in *Staphylococcus aureus* reveals a conserved sequence motif for regulation. Nucleic Acids Res. *37*, 7239–7257.

George, E.A., and Muir, T.W. (2007). Molecular mechanisms of *agr* quorum sensing in virulent staphylococci. Chembiochem *8*, 847–855.

Gerdes, K., and Maisonneuve, E. (2012). Bacterial persistence and toxin-antitoxin loci. Annu. Rev. Microbiol. *66*, 103–123.

Gertz, S., Engelmann, S., Schmid, R., Ohlsen, K., Hacker, J., and Hecker, M. (1999). Regulation of sigmaB-dependent transcription of *sigB* and *asp23* in two different *Staphylococcus aureus* strains. Mol. Gen. Genet. *261*, 558–566.

Gertz, S., Engelmann, S., Schmid, R., Ziebandt, A.K., Tischer, K., Scharf, C., Hacker, J., and Hecker, M. (2000). Characterization of the sigma(B) regulon in *Staphylococcus aureus*. J. Bacteriol. *182*, 6983–6991.

Giachino, P., Engelmann, S., and Bischoff, M. (2001). Sigma(B) activity depends on RsbU in *Staphylococcus aureus*. J. Bacteriol. *183*, 1843–1852.

Giraudo, A.T., Cheung, A.L., and Nagel, R. (1997). The *sae* locus of *Staphylococcus aureus* controls exoprotein synthesis at the transcriptional level. Arch. Microbiol. *168*, 53–58.

Giraudo, A.T., Raspanti, C.G., Calzolari, A., and Nagel, R. (1994). Characterization of a Tn551-mutant of *Staphylococcus aureus* defective in the production of several exoproteins. Can. J. Microbiol. *40*, 677–681.

Goerke, C., Fluckiger, U., Steinhuber, A., Bisanzio, V., Ulrich, M., Bischoff, M., Patti, J.M., and Wolz, C. (2005). Role of *Staphylococcus aureus* global regulators *sae* and sigmaB in virulence gene expression during device-related infection. Inf. Immun. *73*, 3415–3421.

Gordon, C.P., Williams, P., and Chan, W.C. (2013). Attenuating *Staphylococcus aureus* virulence gene regulation: a medicinal chemistry perspective. J. Med. Chem. *56*, 1389–1404.

Gordon, R.J., and Lowy, F.D. (2008). Pathogenesis of methicillin-resistant *Staphylococcus aureus* infection. Clin. Infect. Dis. *46 Suppl 5*, S350–359.

Görke, B., and Stülke, J. (2008). Carbon catabolite repression in bacteria: many ways to make the most out of nutrients. Nat. Rev. Microbiol. *6*, 613–624.

Gray, B., Hall, P., and Gresham, H. (2013). Targeting *agr*- and *agr*-like quorum sensing systems for development of common therapeutics to treat multiple gram-positive bacterial infections. Sensors *13*, 5130–5166.

Groher, F., and Suess, B. (2014). Synthetic riboswitches – A tool comes of age. Biochim. Biophys. Acta *1839*, 964–973.

Guillet, J., Hallier, M., and Felden, B. (2013). Emerging functions for the *Staphylococcus aureus* RNome. PLOS Pathog. *9*, e1003767.

Gupta, R.K., Alba, J., Xiong, Y.Q., Bayer, A.S., and Lee, C.Y. (2013). MgrA activates expression of capsule genes, but not the alpha-toxin gene in experimental *Staphylococcus aureus* endocarditis. J. Infect. Dis. *208*, 1841–1848.

Gupta, R.K., Luong, T.T., and Lee, C.Y. (2015). RNAIII of the *Staphylococcus aureus agr* system activates global regulator MgrA by stabilizing mRNA. Proc. Natl. Acad. Sci. USA *112*, 14036–14041.

Hartmann, T., Baronian, G., Nippe, N., Voss, M., Schulthess, B., Wolz, C., Eisenbeis, J., Schmidt-Hohagen, K., Gaupp, R., Sunderkotter, C., *et al.* (2014). The catabolite control protein E (CcpE) affects virulence determinant production and pathogenesis of *Staphylococcus aureus*. J. Biol. Chem. *289*, 29701–29711.

Hartmann, T., Zhang, B., Baronian, G., Schulthess, B., Homerova, D., Grubmuller, S., Kutzner, E., Gaupp, R., Bertram, R., Powers, R., *et al.* (2013). Catabolite control protein E (CcpE) is a LysR-type transcriptional regulator of tricarboxylic acid cycle activity in *Staphylococcus aureus*. J. Biol. Chem. *288*, 36116–36128.

Helmann, J.D. (2002). The extracytoplasmic function (ECF) sigma factors. Adv. Microb. Physiol. *46*, 47–110.

Hempel, K., Pane-Farre, J., Otto, A., Sievers, S., Hecker, M., and Becher, D. (2010). Quantitative cell surface proteome profiling for SigB-dependent protein expression in the human pathogen *Staphylococcus aureus* via biotinylation approach. J. Proteome Res. *9*, 1579–1590.

Homerova, D., Bischoff, M., Dumolin, A., and Kormanec, J. (2004). Optimization of a two-plasmid system for the identification of promoters recognized by RNA polymerase containing *Staphylococcus aureus* alternative sigma factor sigmaB. FEMS Microbiol. Lett. *232*, 173–179.

Horstmann, N., Orans, J., Valentin-Hansen, P., Shelburne, S.A., and Brennan, R.G. (2012). Structural mechanism of *Staphylococcus aureus* Hfq binding to an RNA A-tract. Nucleic Acids Res. *40*, 11023–11035.

Hsieh, H.Y., Tseng, C.W., and Stewart, G.C. (2008). Regulation of Rot expression in *Staphylococcus aureus*. J. Bacteriol. *190*, 546–554.

Huntzinger, E., Boisset, S., Saveanu, C., Benito, Y., Geissmann, T., Namane, A., Lina, G., Etienne, J., Ehresmann, B., Ehresmann, C., *et al.* (2005). *Staphylococcus aureus* RNAIII and the endoribonuclease III coordinately regulate spa gene expression. EMBO J. *24*, 824–835.

Ibarra, J.A., Pérez-Rueda, E., Carroll, R.K., and Shaw, L.N. (2013). Global analysis of transcriptional regulators in *Staphylococcus aureus*. BMC Genomics *14*, 126.

Ingavale, S., van Wamel, W., Luong, T.T., Lee, C.Y., and Cheung, A.L. (2005). Rat/MgrA, a regulator of autolysis, is a regulator of virulence genes in *Staphylococcus aureus*. Infect. Immun. *73*, 1423–1431.

Ingavale, S.S., Van Wamel, W., and Cheung, A.L. (2003). Characterization of RAT, an autolysis regulator in *Staphylococcus aureus*. Mol. Microbiol. *48*, 1451–1466.

Ingolia, N.T. (2014). Ribosome profiling: new views of translation, from single codons to genome scale. Nat. Rev. Genet. *15*, 205–213.

Inose, Y., Takeshita, S.L., Hidaka, T., Higashide, M., Maruyama, A., Hayashi, H., Morikawa, K., and Ohta, T. (2006). Genetic characterization of the natural SigB variants found in clinical isolates of *Staphylococcus aureus*. J. Gen. Appl. Microbiol. *52*, 259–271.

Jarraud, S., Lyon, G.J., Figueiredo, A.M., Lina, G., Gérard, L., Vandenesch, F., Etienne, J., Muir, T.W., and Novick, R.P. (2000). Exfoliatin-producing strains define a fourth *agr* specificity group in *Staphylococcus aureus*. J. Bacteriol. *182*, 6517–6522.

Jelsbak, L., Hemmingsen, L., Donat, S., Ohlsen, K., Boye, K., Westh, H., Ingmer, H., and Frees, D. (2010). Growth phase-dependent regulation of the global virulence regulator Rot in clinical isolates of *Staphylococcus aureus*. Int. J. Med. Microbiol. *300*, 229–236.

Jeong, D.W., Cho, H., Jones, M.B., Shatzkes, K., Sun, F., Ji, Q., Liu, Q., Peterson, S.N., He, C., and Bae, T. (2012). The auxiliary protein complex SaePQ activates the phosphatase activity of sensor kinase SaeS in the SaeRS two-component system of *Staphylococcus aureus*. Mol. Microbiol. *86*, 331–348.

Ji, G., Beavis, R., and Novick, R.P. (1997). Bacterial interference caused by autoinducing peptide variants. Science *276*, 2027–2030.

Johnson, M., Sengupta, M., Purves, J., Tarrant, E., Williams, P.H., Cockayne, A., Muthaiyan, A., Stephenson, R., Ledala, N., Wilkinson, B.J., *et al.* (2011). Fur is required for the activation of virulence gene expression through the induction of the sae regulatory system in *Staphylococcus aureus*. Int. J. Med. Microbiol. *301*, 44–52.

Jonsson, I.M., Arvidson, S., Foster, S., and Tarkowski, A. (2004). Sigma factor B and RsbU are required for virulence in *Staphylococcus aureus*-induced arthritis and sepsis. Infect. Immun. *72*, 6106–6111.

Jonsson, I.M., Lindholm, C., Luong, T.T., Lee, C.Y., and Tarkowski, A. (2008). MgrA regulates staphylococcal virulence important for induction and progression of septic arthritis and sepsis. Microbes. Infect. *10*, 1229–1235.

Kaito, C., Morishita, D., Matsumoto, Y., Kurokawa, K., and Sekimizu, K. (2006). Novel DNA binding protein SarZ contributes to virulence in *Staphylococcus aureus*. Mol. Microbiol. *62*, 1601–1617.

Kaito, C., Saito, Y., Ikuo, M., Omae, Y., Mao, H., Nagano, G., Fujiyuki, T., Numata, S., Han, X., Obata, K., *et al.* (2013). Mobile genetic element SCCmec-encoded psm-mec RNA suppresses translation of agrA and attenuates MRSA virulence. PLOS Pathog. *9*, e1003269.

Kamarthapu, V., and Nudler, E. (2015). Rethinking transcription coupled DNA repair. Curr. Opin. Microbiol. *24*, 15–20.

Killikelly, A., Benson, M.A., Ohneck, E.A., Sampson, J.M., Jakoncic, J., Spurrier, B., Torres, V.J., and Kong, X.P. (2015). Structure-based functional characterization of repressor of toxin (Rot), a central regulator of *Staphylococcus aureus* virulence. J. Bacteriol. *197*, 188–200.

Kim, S., Reyes, D., Beaume, M., Francois, P., and Cheung, A. (2014). Contribution of teg49 small RNA in the 5′ upstream transcriptional region of sarA to virulence in *Staphylococcus aureus*. Infect. Immun. *82*, 4369–4379.

Kinkel, T.L., Roux, C.M., Dunman, P.M., and Fang, F.C. (2013). The *Staphylococcus aureus* SrrAB two-component system promotes resistance to nitrosative stress and hypoxia. MBio *4*, e00696–13.

Koch, G., Yepes, A., Förstner, K.U., Wermser, C., Stengel, S.T., Modamio, J., Ohlsen, K., Foster, K.R., and Lopez, D. (2014). Evolution of resistance to a last-resort antibiotic in *Staphylococcus aureus* via bacterial competition. Cell *158*, 1060–1071.

Köck, R., Becker, K., Cookson, B., van Gemert-Pijnen, J.E., Harbarth, S., Kluytmans, J., Mielke, M., Peters, G., Skov, R.L., Struelens, M.J., Tacconelli, E., Navarro Torné, A., Witte, W., and Friedrich, A.W. (2010). Methicillin-resistant *Staphylococcus aureus* (MRSA): burden of disease and control challenges in Europe. Euro. Surveill. *15*, 19688.

Koenig, R.L., Ray, J.L., Maleki, S.J., Smeltzer, M.S., and Hurlburt, B.K. (2004). *Staphylococcus aureus* AgrA binding to the RNAIII-*agr* regulatory region. J. Bacteriol. *186*, 7549–7555.

Kugelberg, E., Kofoid, E., Reams, A.B., Andersson, D.I., and Roth, J.R. (2006). Multiple pathways of selected gene amplification during adaptive mutation. Proc. Natl. Acad. Sci. U.S.A. *103*, 17319–17324.

Lalaouna, D., Simoneau-Roy, M., Lafontaine, D., and Massé, E. (2013). Regulatory RNAs and target mRNA decay in prokaryotes. Biochim. Biophys. Acta *1829*, 742–747.

Lasa, I., Toledo-Arana, A., Dobin, A., Villanueva, M., de los Mozos, I.R., Vergara-Irigaray, M., Segura, V., Fagegaltier, D., Penades, J.R., Valle, J., *et al.* (2011). Genome-wide antisense transcription drives mRNA processing in bacteria. Proc. Natl. Acad. Sci. U.S.A. *108*, 20172–20177.

Lasa, I., Toledo-Arana, A., and Gingeras, T.R. (2012). An effort to make sense of antisense transcription in bacteria. RNA Biol. *9*, 1039–1044.

Lee, H.H., Molla, M.N., Cantor, C.R., and Collins, J.J. (2010). Bacterial charity work leads to population-wide resistance. Nature *467*, 82–85.

Leiba, J., Hartmann, T., Cluzel, M.E., Cohen-Gonsaud, M., Delolme, F., Bischoff, M., and Molle, V. (2012). A novel mode of regulation of the *Staphylococcus aureus* catabolite control protein A (CcpA) mediated by Stk1 protein phosphorylation. J. Biol. Chem. *287*, 43607–43619.

Levdikov, V.M., Blagova, E., Joseph, P., Sonenshein, A.L., and Wilkinson, A.J. (2006). The structure of CodY, a GTP- and isoleucine-responsive regulator of stationary phase and virulence in gram-positive bacteria. J. Biol. Chem. *281*, 11366–11373.

Li, C., Sun, F., Cho, H., Yelavarthi, V., Sohn, C., He, C., Schneewind, O., and Bae, T. (2010). CcpA mediates proline auxotrophy and is required for *Staphylococcus aureus* pathogenesis. J. Bacteriol. *192*, 3883–3892.

Li, D., and Cheung, A. (2008). Repression of *hla* by *rot* is dependent on sae in *Staphylococcus aureus*. Infect. Immun. *76*, 1068–1075.

Li, R., Manna, A.C., Dai, S., Cheung, A.L., and Zhang, G. (2003). Crystal structure of the SarS protein from *Staphylococcus aureus*. J. Bacteriol. *185*, 4219–4225.

Liang, X., Zheng, L., Landwehr, C., Lunsford, D., Holmes, D., and Ji, Y. (2005). Global regulation of gene expression by ArlRS, a two-component signal transduction regulatory system of *Staphylococcus aureus*. J. Bacteriol. *187*, 5486–5492.

Linder, P., Lemeille, S., and Redder, P. (2014). Transcriptome-wide analyses of 5′-ends in RNase J mutants of a gram-positive pathogen reveal a role in RNA maturation, regulation and degradation. PLOS Genet. *10*, e1004207.

Lioliou, E., Sharma, C.M., Caldelari, I., Helfer, A.C., Fechter, P., Vandenesch, F., Vogel, J., and Romby, P. (2012). Global regulatory functions of the *Staphylococcus aureus* endoribonuclease III in gene expression. PLOS Genet. *8*, e1002782.

Liu, C.L., Chen, Z.J., Wang, F., Hou, H.Y., Wang, Y., Zhu, X.H., Jian, C., Tian, L., Yan, S.Z., Xu, L.Q., *et al.* (2014). The impact of mg*rA* on progression of *Staphylococcus aureus* sepsis. Microb. Pathog. *71–72*, 56–61.

Liu, Y., Manna, A.C., Pan, C.H., Kriksunov, I.A., Thiel, D.J., Cheung, A.L., and Zhang, G. (2006). Structural and function analyses of the global regulatory protein SarA from *Staphylococcus aureus*. Proc. Natl. Acad. Sci. U.S.A. *103*, 2392–2397.

Liu, Y., Wu, N., Dong, J., Gao, Y., Zhang, X., Mu, C., Shao, N., and Yang, G. (2010). Hfq is a global regulator that controls the pathogenicity of *Staphylococcus aureus*. PLoS One 5, e13069.

Lorenz, U., Huttinger, C., Schafer, T., Ziebuhr, W., Thiede, A., Hacker, J., Engelmann, S., Hecker, M., and Ohlsen, K. (2008). The alternative sigma factor sigma B of *Staphylococcus aureus* modulates virulence in experimental central venous catheter-related infections. Microbes Infect. *10*, 217–223.

Lucas, A.L., and Manna, A.C. (2013). Phenotypic characterization of *sarR* mutant in *Staphylococcus aureus*. Microb. Pathog. *57*, 52–61.

Luong, T.T., Dunman, P.M., Murphy, E., Projan, S.J., and Lee, C.Y. (2006). Transcription profiling of the mg*rA* regulon in *Staphylococcus aureus*. J. Bacteriol. *188*, 1899–1910.

Luong, T.T., and Lee, C.Y. (2006). The arl locus positively regulates *Staphylococcus aureus* type 5 capsule via an mgrA-dependent pathway. Microbiology *152* 5, 3123–3131.

Luong, T.T., Newell, S.W., and Lee, C.Y. (2003). Mgr, a novel global regulator in *Staphylococcus aureus*. J. Bacteriol. *185*, 3703–3710.

Maddocks, S.E., and Oyston, P.C. (2008). Structure and function of the LysR-type transcriptional regulator (LTTR) family proteins. Microbiology *154*, 3609–3623.

Mäder, U., Nicolas, P., Depke, M., Pané-Farré, J., Debarbouille, M., van der Kooi-Pol, M.M., Guérin, C., Dérozier, S., Hiron, A., Jarmer, H., *et al.* (2016). *Staphylococcus aureus* transcriptome architecture. From laboratory to infection-mimicking conditions. PLoS Genet. *12*, e1005962.

Mainiero, M., Goerke, C., Geiger, T., Gonser, C., Herbert, S., and Wolz, C. (2010). Differential target gene activation by the *Staphylococcus aureus* two-component system saeRS. J. Bacteriol. *192*, 613–623.

Majerczyk, C.D., Dunman, P.M., Luong, T.T., Lee, C.Y., Sadykov, M.R., Somerville, G.A., Bodi, K., and Sonenshein, A.L. (2010). Direct targets of CodY in *Staphylococcus aureus*. J. Bacteriol. *192*, 2861–2877.

Majerczyk, C.D., Sadykov, M.R., Luong, T.T., Lee, C., Somerville, G.A., and Sonenshein, A.L. (2008). *Staphylococcus aureus* CodY negatively regulates virulence gene expression. J. Bacteriol. *190*, 2257–2265.

Malachowa, N., and DeLeo, F.R. (2010). Mobile genetic elements of *Staphylococcus aureus*. Cell. Mol. Life. Sci. *67*, 3057–3071.

Manna, A., and Cheung, A.L. (2001). Characterization of *sarR*, a modulator of sar expression in *Staphylococcus aureus*. Infect. Immun. *69*, 885–896.

Manna, A.C., and Cheung, A.L. (2003). *sarU*, a *sarA* homolog, is repressed by SarT and regulates virulence

genes in *Staphylococcus aureus*. Infect. Immun. *71*, 343–353.
Manna, A.C., and Cheung, A.L. (2006a). Expression of SarX, a negative regulator of *agr* and exoprotein synthesis, is activated by MgrA in *Staphylococcus aureus*. J. Bacteriol. *188*, 4288–4299.
Manna, A.C., and Cheung, A.L. (2006b). Transcriptional regulation of the *agr* locus and the identification of DNA binding residues of the global regulatory protein SarR in *Staphylococcus aureus*. Mol. Microbiol. *60*, 1289–1301.
Manna, A.C., Ingavale, S.S., Maloney, M., van Wamel, W., and Cheung, A.L. (2004). Identification of *sarV* (SA2062), a new transcriptional regulator, is repressed by SarA and MgrA (SA0641) and involved in the regulation of autolysis in *Staphylococcus aureus*. J. Bacteriol. *186*, 5267–5280.
Manna, A.C., and Ray, B. (2007). Regulation and characterization of *rot* transcription in *Staphylococcus aureus*. Microbiology *153*, 1538–1545.
Marciniak, B.C., Pabijaniak, M., de Jong, A., Dűhring, R., Seidel, G., Hillen, W., and Kuipers, O.P. (2012). High- and low-affinity *cre* boxes for CcpA binding in Bacillus subtilis revealed by genome-wide analysis. BMC Genomics *13*, 401.
Marincola, G., Schäfer, T., Behler, J., Bernhardt, J., Ohlsen, K., Goerke, C., and Wolz, C. (2012). RNase Y of *Staphylococcus aureus* and its role in the activation of virulence genes. Mol. Microbiol. *85*, 817–832.
Matsumoto, Y., Kaito, C., Morishita, D., Kurokawa, K., and Sekimizu, K. (2007). Regulation of exoprotein gene expression by the *Staphylococcus aureus cvfB* gene. Infect. Immun. 75, 1964–1972.
Matsumoto, Y., Xu, Q., Miyazaki, S., Kaito, C., Farr, C.L., Axelrod, H.L., Chiu, H.J., Klock, H.E., Knuth, M.W., Miller, M.D., *et al.* (2010). Structure of a virulence regulatory factor CvfB reveals a novel winged helix RNA binding module. Structure *18*, 537–547.
McNamara, P.J., and Bayer, A.S. (2005). A *rot* mutation restores parental virulence to an *agr*-null *Staphylococcus aureus* strain in a rabbit model of endocarditis. Infect. Immun. 73, 3806–3809.
McNamara, P.J., Milligan-Monroe, K.C., Khalili, S., and Proctor, R.A. (2000). Identification, cloning, and initial characterization of *rot*, a locus encoding a regulator of virulence factor expression in *Staphylococcus aureus*. J. Bacteriol. *182*, 3197–3203.
Meier, S., Goerke, C., Wolz, C., Seidl, K., Homerova, D., Schulthess, B., Kormanec, J., Berger-Bächi, B., and Bischoff, M. (2007). sigmaB and the sigmaB-dependent *arlRS* and *yabJ-spoVG* loci affect capsule formation in *Staphylococcus aureus*. Infect. Immun. 75, 4562–4571.
Memmi, G., Nair, D.R., and Cheung, A. (2012). Role of ArlRS in autolysis in methicillin-sensitive and methicillin-resistant *Staphylococcus aureus* strains. J. Bacteriol. *194*, 759–767.
Miller, H.K., Carroll, R.K., Burda, W.N., Krute, C.N., Davenport, J.E., and Shaw, L.N. (2012). The extracytoplasmic function sigma factor σ^S protects against both intracellular and extracytoplasmic stresses in *Staphylococcus aureus*. J. Bacteriol. *194*, 4342–4354.
Mitchell, G., Brouillette, E., Seguin, D.L., Asselin, A.E., Jacob, C.L., and Malouin, F. (2010). A role for sigma factor B in the emergence of *Staphylococcus aureus* small-colony variants and elevated biofilm production resulting from an exposure to aminoglycosides. Microb. Pathog. *48*, 18–27.
Mitchell, G., Fugère, A., Pépin Gaudreau, K., Brouillette, E., Frost, E.H., Cantin, A.M., and Malouin, F. (2013). SigB is a dominant regulator of virulence in *Staphylococcus aureus* small-colony variants. PLoS One *8*, e65018.
Mittal, M., Pechter, K.B., Picossi, S., Kim, H.J., Kerstein, K.O., and Sonenshein, A.L. (2013). Dual role of CcpC protein in regulation of aconitase gene expression in *Listeria monocytogenes* and *Bacillus subtilis*. Microbiology *159*, 68–76.
Moisan, H., Brouillette, E., Jacob, C.L., Langlois-Bégin, P., Michaud, S., and Malouin, F. (2006). Transcription of virulence factors in *Staphylococcus aureus* small-colony variants isolated from cystic fibrosis patients is influenced by SigB. J. Bacteriol. *188*, 64–76.
Montgomery, C.P., Boyle-Vavra, S., and Daum, R.S. (2010). Importance of the global regulators Agr and SaeRS in the pathogenesis of CA-MRSA USA300 infection. PLoS One *5*, e15177.
Montgomery, C.P., Boyle-Vavra, S., Roux, A., Ebine, K., Sonenshein, A.L., and Daum, R.S. (2012). CodY deletion enhances *in vivo* virulence of community-associated methicillin-resistant *Staphylococcus aureus* clone USA300. Inf. Immun. *80*, 2382–2389.
Morfeldt, E., Taylor, D., von Gabain, A., and Arvidson, S. (1995). Activation of alpha-toxin translation in *Staphylococcus aureus* by the trans-encoded antisense RNA, RNAIII. EMBO J. *14*, 4569–4577.
Morikawa, K., Inose, Y., Okamura, H., Maruyama, A., Hayashi, H., Takeyasu, K., and Ohta, T. (2003). A new staphylococcal sigma factor in the conserved gene cassette: functional significance and implication for the evolutionary processes. Genes Cells. *8*, 699–712.
Morikawa, K., Ohniwa, R.L., Kumano, M., Okamura, H., Saito, S., and Ohta, T. (2008). The *sigH* gene sequence can subspeciate staphylococci. Diagn. Microbiol. Infect. Dis. *61*, 373–380.
Morikawa, K., Takemura, A.J., Inose, Y., Tsai, M., Nguyen Thi le, T., Ohta, T., and Msadek, T. (2012). Expression of a cryptic secondary sigma factor gene unveils natural competence for DNA transformation in *Staphylococcus aureus*. PLOS Pathog. *8*, e1003003.
Morrison, J.M., Anderson, K.L., Beenken, K.E., Smeltzer, M.S., and Dunman, P.M. (2012a). The staphylococcal accessory regulator, SarA, is an RNA-binding protein that modulates the mRNA turnover properties of late-exponential and stationary phase *Staphylococcus aureus* cells. Front. Cell Infect. Microbiol. *2*, 26.
Morrison, J.M., and Dunman, P.M. (2011). The modulation of *Staphylococcus aureus* mRNA turnover. Future Microbiol. *6*, 1141–1150.
Morrison, J.M., Miller, E.W., Benson, M.A., Alonzo, F., Yoong, P., Torres, V.J., Hinrichs, S.H., and Dunman, P.M. (2012b). Characterization of SSR42, a novel virulence factor regulatory RNA that contributes to the pathogenesis of a *Staphylococcus aureus* USA300 representative. J. Bacteriol. *194*, 2924–2938.
Mulhbacher, J., Brouillette, E., Allard, M., Fortier, L.C., Malouin, F., and Lafontaine, D.A. (2010). Novel riboswitch ligand analogs as selective inhibitors of

guanine-related metabolic pathways. PLOS Pathog. *6*, e1000865.

Nicholas, R.O., Li, T., McDevitt, D., Marra, A., Sucoloski, S., Demarsh, P.L., and Gentry, D.R. (1999). Isolation and characterization of a *sigB* deletion mutant of *Staphylococcus aureus*. Infect. Immun. *67*, 3667–3669.

Nikolskaya, A.N., and Galperin, M.Y. (2002). A novel type of conserved DNA-binding domain in the transcriptional regulators of the AlgR/AgrA/LytR family. Nucleic Acids Res. *30*, 2453–2459.

Novick, R.P. (2003). Autoinduction and signal transduction in the regulation of staphylococcal virulence. Mol. Microbiol. *48*, 1429–1449.

Novick, R.P., and Geisinger, E. (2008). Quorum sensing in staphylococci. Annu. Rev. Genet. *42*, 541–564.

Novick, R.P., and Jiang, D. (2003). The staphylococcal *saeRS* system coordinates environmental signals with *agr* quorum sensing. Microbiology *149*, 2709–2717.

Novick, R.P., Ross, H.F., Projan, S.J., Kornblum, J., Kreiswirth, B., and Moghazeh, S. (1993). Synthesis of staphylococcal virulence factors is controlled by a regulatory RNA molecule. EMBO J. *12*, 3967–3975.

Nygaard, T.K., Pallister, K.B., Ruzevich, P., Griffith, S., Vuong, C., and Voyich, J.M. (2010). SaeR binds a consensus sequence within virulence gene promoters to advance USA300 pathogenesis. J. Infect. Dis. *201*, 241–254.

Olson, P.D., Kuechenmeister, L.J., Anderson, K.L., Daily, S., Beenken, K.E., Roux, C.M., Reniere, M.L., Lewis, T.L., Weiss, W.J., Pulse, M., *et al.* (2011). Small molecule inhibitors of *Staphylococcus aureus* RnpA alter cellular mRNA turnover, exhibit antimicrobial activity, and attenuate pathogenesis. PLOS Pathog. *7*, e1001287.

Oscarsson, J., Harlos, C., and Arvidson, S. (2005). Regulatory role of proteins binding to the *spa* (protein A) and *sarS* (staphylococcal accessory regulator) promoter regions in *Staphylococcus aureus* NTCC 8325-4. Int. J. Med. Microbiol. *295*, 253–266.

Oscarsson, J., Kanth, A., Tegmark-Wisell, K., and Arvidson, S. (2006). SarA is a repressor of hla (alpha-hemolysin) transcription in *Staphylococcus aureus*: its apparent role as an activator of *hla* in the prototype strain NCTC 8325 depends on reduced expression of *sarS*. J. Bacteriol. *188*, 8526–8533.

Oun, S., Redder, P., Didier, J.P., François, P., Corvaglia, A.R., Buttazzoni, E., Giraud, C., Girard, M., Schrenzel, J., and Linder, P. (2013). The CshA DEAD-box RNA helicase is important for quorum sensing control in *Staphylococcus aureus*. RNA Biol. *10*, 157–165.

Pagels, M., Fuchs, S., Pané-Farré, J., Kohler, C., Menschner, L., Hecker, M., McNamarra, P.J., Bauer, M.C., von Wachenfeldt, C., Liebeke, M., *et al.* (2010). Redox sensing by a Rex-family repressor is involved in the regulation of anaerobic gene expression in *Staphylococcus aureus*. Mol. Microbiol. *76*, 1142–1161.

Palma, M., and Cheung, A.L. (2001). sigma(B) activity in *Staphylococcus aureus* is controlled by RsbU and an additional factor(s) during bacterial growth. Infect. Immun. *69*, 7858–7865.

Pané-Farré, J., Jonas, B., Förstner, K., Engelmann, S., and Hecker, M. (2006). The sigmaB regulon in *Staphylococcus aureus* and its regulation. Int. J. Med. Microbiol. *296*, 237–258.

Pané-Farré, J., Jonas, B., Hardwick, S.W., Gronau, K., Lewis, R.J., Hecker, M., and Engelmann, S. (2009). Role of RsbU in controlling SigB activity in *Staphylococcus aureus* following alkaline stress. J. Bacteriol. *191*, 2561–2573.

Panja, S., and Woodson, S.A. (2012). Hfq proximity and orientation controls RNA annealing. Nucleic Acids Res. *40*, 8690–8697.

Paulander, W., Nissen Varming, A., Bæk, K.T., Haaber, J., Frees, D., and Ingmer, H. (2013). Antibiotic-mediated selection of quorum-sensing-negative *Staphylococcus aureus*. MBio 3, e00459-12.

Pfortner, H., Burian, M.S., Michalik, S., Depke, M., Hildebrandt, P., Dhople, V.M., Pane-Farre, J., Hecker, M., Schmidt, F., and Volker, U. (2014). Activation of the alternative sigma factor SigB of *Staphylococcus aureus* following internalization by epithelial cells - an *in vivo* proteomics perspective. Int. J. Med. Microbiol. *304*, 177–187.

Pichon, C., and Felden, B. (2005). Small RNA genes expressed from *Staphylococcus aureus* genomic and pathogenicity islands with specific expression among pathogenic strains. Proc. Natl. Acad. Sci. U.S.A. *102*, 14249–14254.

Pinel-Marie, M.L., Brielle, R., and Felden, B. (2014). Dual toxic-peptide-coding *Staphylococcus aureus* RNA under antisense regulation targets host cells and bacterial rivals unequally. Cell Rep. *7*, 424–435.

Pohl, K., Francois, P., Stenz, L., Schlink, F., Geiger, T., Herbert, S., Goerke, C., Schrenzel, J., and Wolz, C. (2009). CodY in *Staphylococcus aureus*: a regulatory link between metabolism and virulence gene expression. J. Bacteriol. *191*, 2953–2963.

Pragman, A.A., Ji, Y., and Schlievert, P.M. (2007). Repression of *Staphylococcus aureus* SrrAB using inducible antisense *srrA* alters growth and virulence factor transcript levels. Biochemistry *46*, 314–321.

Pragman, A.A., and Schlievert, P.M. (2004). Virulence regulation in *Staphylococcus aureus*: the need for *in vivo* analysis of virulence factor regulation. FEMS Immunol. Med. Microbiol. 42, 147–154.

Pragman, A.A., Yarwood, J.M., Tripp, T.J., and Schlievert, P.M. (2004). Characterization of virulence factor regulation by SrrAB, a two-component system in *Staphylococcus aureus*. J. Bacteriol. *186*, 2430–2438.

Queck, S.Y., Jameson-Lee, M., Villaruz, A.E., Bach, T.H., Khan, B.A., Sturdevant, D.E., Ricklefs, S.M., Li, M., and Otto, M. (2008). RNAIII-independent target gene control by the *agr* quorum-sensing system: insight into the evolution of virulence regulation in *Staphylococcus aureus*. Mol. Cell 32, 150–158.

Queck, S.Y., Khan, B.A., Wang, R., Bach, T.H., Kretschmer, D., Chen, L., Kreiswirth, B.N., Peschel, A., Deleo, F.R., and Otto, M. (2009). Mobile genetic element-encoded cytolysin connects virulence to methicillin resistance in MRSA. PLOS Pathog. *5*, e1000533.

Ramesh, A., and Winkler, W.C. (2014). Metabolite-binding ribozymes. Biochim. Biophys. Acta *1839*, 989–994.

Reiss, S., Pané-Farré, J., Fuchs, S., François, P., Liebeke, M., Schrenzel, J., Lindequist, U., Lalk, M., Wolz, C., Hecker, M., *et al.* (2012). Global analysis of the *Staphylococcus aureus* response to mupirocin. Antimicrob. Agents Chemother. *56*, 787–804.

Reyes, D., Andrey, D.O., Monod, A., Kelley, W.L., Zhang, G., and Cheung, A.L. (2011). Coordinated regulation by AgrA, SarA, and SarR to control *agr* expression in *Staphylococcus aureus*. J. Bacteriol. *193*, 6020–6031.

Reynolds, J., and Wigneshweraraj, S. (2011). Molecular insights into the control of transcription initiation at the *Staphylococcus aureus agr* operon. J. Mol. Biol. *412*, 862–881.

Richardson, A.R., Libby, S.J., and Fang, F.C. (2008). A nitric oxide-inducible lactate dehydrogenase enables *Staphylococcus aureus* to resist innate immunity. Science *319*, 1672–1676.

Richardson, A.R., Somerville, G.A., and Sonenshein, A.L. (2015). Regulating the Intersection of Metabolism and Pathogenesis in Gram-positive Bacteria. Microbiol. Spectr. *3*.

Rivera, F.E., Miller, H.K., Kolar, S.L., Stevens, S.M., and Shaw, L.N. (2012). The impact of CodY on virulence determinant production in community-associated methicillin-resistant *Staphylococcus aureus*. Proteomics *12*, 263–268.

Roberts, C., Anderson, K.L., Murphy, E., Projan, S.J., Mounts, W., Hurlburt, B., Smeltzer, M., Overbeek, R., Disz, T., and Dunman, P.M. (2006). Characterizing the effect of the *Staphylococcus aureus* virulence factor regulator, SarA, on log-phase mRNA half-lives. J. Bacteriol. *188*, 2593–2603.

Robinson, C.J., Vincent, H.A., Wu, M.C., Lowe, P.T., Dunstan, M.S., Leys, D., and Micklefield, J. (2014). Modular riboswitch toolsets for synthetic genetic control in diverse bacterial species. J. Am. Chem. Soc. *136*, 10615–10624.

Rochat, T., Bouloc, P., Yang, Q., Bossi, L., and Figueroa-Bossi, N. (2012). Lack of interchangeability of Hfq-like proteins. Biochimie *94*, 1554–1559.

Rogasch, K., Rühmling, V., Pané-Farré, J., Höper, D., Weinberg, C., Fuchs, S., Schmudde, M., Bröker, B.M., Wolz, C., Hecker, M., *et al.* (2006). Influence of the two-component system SaeRS on global gene expression in two different *Staphylococcus aureus* strains. J. Bacteriol. *188*, 7742–7758.

Romilly, C., Caldelari, I., Parmentier, D., Lioliou, E., Romby, P., and Fechter, P. (2012). Current knowledge on regulatory RNAs and their machineries in *Staphylococcus aureus*. RNA Biol. *9*, 402–413.

Romilly, C., Lays, C., Tomasini, A., Caldelari, I., Benito, Y., Hammann, P., Geissmann, T., Boisset, S., Romby, P., and Vandenesch, F. (2014). A non-coding RNA promotes bacterial persistence and decreases virulence by regulating a regulator in *Staphylococcus aureus*. PLOS Pathog. *10*, e1003979.

Roux, A., Todd, D.A., Velázquez, J.V., Cech, N.B., and Sonenshein, A.L. (2014). CodY-mediated regulation of the *Staphylococcus aureus* Agr system integrates nutritional and population density signals. J. Bacteriol. *196*, 1184–1196.

Roux, C.M., DeMuth, J.P., and Dunman, P.M. (2011). Characterization of components of the *Staphylococcus aureus* mRNA degradosome holoenzyme-like complex. J. Bacteriol. *193*, 5520–5526.

Ruiz de los Mozos, I., Vergara-Irigaray, M., Segura, V., Villanueva, M., Bitarte, N., Saramago, M., Domingues, S., Arraiano, C.M., Fechter, P., Romby, P., *et al.* (2013). Base pairing interaction between 5'- and 3'-UTRs controls icaR mRNA translation in *Staphylococcus aureus*. PLOS Genet. *9*, e1004001.

Saïd-Salim, B., Dunman, P.M., McAleese, F.M., Macapagal, D., Murphy, E., McNamara, P.J., Arvidson, S., Foster, T.J., Projan, S.J., and Kreiswirth, B.N. (2003). Global regulation of *Staphylococcus aureus* genes by Rot. J. Bacteriol. *185*, 610–619.

Castelhano Santos, N., Pereira, M.O., and Lourenço, A. (2015). Pathogenicity phenomena in three model systems: from network mining to emerging system-level properties. Briefings Bioinf. *16*, 169–182.

Sayed, N., Jousselin, A., and Felden, B. (2012). A cis-antisense RNA acts in trans in *Staphylococcus aureus* to control translation of a human cytolytic peptide. Nat. Struct. Mol. Biol. *19*, 105–112.

Schäfer, D., Lâm, T.T., Geiger, T., Mainiero, M., Engelmann, S., Hussain, M., Bosserhoff, A., Frosch, M., Bischoff, M., Wolz, C., *et al.* (2009). A point mutation in the sensor histidine kinase SaeS of *Staphylococcus aureus* strain Newman alters the response to biocide exposure. J. Bacteriol. *191*, 7306–7314.

Schmidt, K.A., Manna, A.C., and Cheung, A.L. (2003). SarT influences *sarS* expression in *Staphylococcus aureus*. Infect. Immun. *71*, 5139–5148.

Schmidt, K.A., Manna, A.C., Gill, S., and Cheung, A.L. (2001). SarT, a repressor of α-hemolysin in *Staphylococcus aureus*. Infect. Immun. *69*, 4749–4758.

Schmitt, J., Joost, I., Skaar, E.P., Herrmann, M., and Bischoff, M. (2012). Hemin represses the hemolytic activity of *Staphylococcus aureus* in an Sae-dependent manner. Microbiology *158*, 2619–2631.

Seidel, G., Diel, M., Fuchsbauer, N., and Hillen, W. (2005). Quantitative interdependence of coeffectors, CcpA and *cre* in carbon catabolite regulation of *Bacillus subtilis*. FEBS J. *272*, 2566–2577.

Seidl, K., Bischoff, M., and Berger-Bächi, B. (2008a). CcpA mediates the catabolite repression of *tst* in *Staphylococcus aureus*. Infect. Immun. *76*, 5093–5099.

Seidl, K., Goerke, C., Wolz, C., Mack, D., Berger-Bächi, B., and Bischoff, M. (2008b). *Staphylococcus aureus* CcpA affects biofilm formation. Infect. Immun. *76*, 2044–2050.

Seidl, K., Müller, S., François, P., Kriebitzsch, C., Schrenzel, J., Engelmann, S., Bischoff, M., and Berger-Bächi, B. (2009). Effect of a glucose impulse on the CcpA regulon in *Staphylococcus aureus*. BMC Microbiol. *9*, 95.

Seidl, K., Stucki, M., Rügg, M., Goerke, C., Wolz, C., Harris, L., Berger-Bächi, B., and Bischoff, M. (2006). *Staphylococcus aureus* CcpA affects virulence determinant production and antibiotic resistance. Antimicrob. Agents Chemother. *50*, 1183–1194.

Senn, M.M., Bischoff, M., von Eiff, C., and Berger-Bächi, B. (2005a). sigmaB activity in a *Staphylococcus aureus hemB* mutant. J. Bacteriol. *187*, 7397–7406.

Senn, M.M., Giachino, P., Homerova, D., Steinhuber, A., Strassner, J., Kormanec, J., Flückiger, U., Berger-Bächi, B., and Bischoff, M. (2005b). Molecular analysis and organization of the sigmaB operon in *Staphylococcus aureus*. J. Bacteriol. *187*, 8006–8019.

Shaw, L.N., Lindholm, C., Prajsnar, T.K., Miller, H.K., Brown, M.C., Golonka, E., Stewart, G.C., Tarkowski, A., and Potempa, J. (2008). Identification and characterization of sigma, a novel component of the *Staphylococcus aureus* stress and virulence responses. PLoS One 3, e3844.

Shopsin, B., Drlica-Wagner, A., Mathema, B., Adhikari, R.P., Kreiswirth, B.N., and Novick, R.P. (2008). Prevalence of *agr* dysfunction among colonizing *Staphylococcus aureus* strains. J. Infect. Dis. *198*, 1171–1174.

Sickmier, E.A., Brekasis, D., Paranawithana, S., Bonanno, J.B., Paget, M.S., Burley, S.K., and Kielkopf, C.L. (2005). X-ray structure of a Rex-family repressor/NADH complex insights into the mechanism of redox sensing. Structure *13*, 43–54.

Sievert, D.M., Ricks, P., Edwards, J.R., Schneider, A., Patel, J., Srinivasan, A., Kallen, A., Limbago, B., and Fridkin, S. (2013). Antimicrobial-resistant pathogens associated with healthcare-associated infections: summary of data reported to the National Healthcare Safety Network at the Centers for Disease Control and Prevention, 2009–2010. Infect. Control Hosp. Epidemiol. *34*, 1–14.

Smyth, D.S., Kafer, J.M., Wasserman, G.A., Velickovic, L., Mathema, B., Holzman, R.S., Knipe, T.A., Becker, K., von Eiff, C., Peters, G., *et al.* (2012). Nasal carriage as a source of *agr*-defective *Staphylococcus aureus* bacteremia. J. Infect. Dis. *206*, 1168–1177.

Somerville, G.A., and Proctor, R.A. (2009). At the crossroads of bacterial metabolism and virulence factor synthesis in Staphylococci. Microbiol. Mol. Biol. Rev. *73*, 233–248.

Sonenshein, A.L. (2005). CodY, a global regulator of stationary phase and virulence in Gram-positive bacteria. Curr. Opin. Microbiol. *8*, 203–207.

Sonenshein, A.L. (2007). Control of key metabolic intersections in *Bacillus subtilis*. Nat. Rev. Microbiol. *5*, 917–927.

Song, J., Lays, C., Vandenesch, F., Benito, Y., Bes, M., Chu, Y., Lina, G., Romby, P., Geissmann, T., and Boisset, S. (2012). The expression of small regulatory RNAs in clinical samples reflects the different life styles of *Staphylococcus aureus* in colonization vs. infection. PLoS One *7*, e37294.

Staron, A., Sofia, H.J., Dietrich, S., Ulrich, L.E., Liesegang, H., and Mascher, T. (2009). The third pillar of bacterial signal transduction: classification of the extracytoplasmic function (ECF) sigma factor protein family. Mol. Microbiol. *74*, 557–581.

Steinhuber, A., Goerke, C., Bayer, M.G., Döring, G., and Wolz, C. (2003). Molecular architecture of the regulatory Locus sae of *Staphylococcus aureus* and its impact on expression of virulence factors. J. Bacteriol. *185*, 6278–6286.

Sterba, K.M., Mackintosh, S.G., Blevins, J.S., Hurlburt, B.K., and Smeltzer, M.S. (2003). Characterization of *Staphylococcus aureus* SarA binding sites. J. Bacteriol. *185*, 4410–4417.

Stock, A.M., Robinson, V.L., and Goudreau, P.N. (2000). Two-component signal transduction. Annu. Rev. Biochem. *69*, 183–215.

Sully, E.K., Malachowa, N., Elmore, B.O., Alexander, S.M., Femling, J.K., Gray, B.M., DeLeo, F.R., Otto, M., Cheung, A.L., Edwards, B.S., *et al.* (2014). Selective chemical inhibition of *agr* quorum sensing in *Staphylococcus aureus* promotes host defense with minimal impact on resistance. PLOS Pathog. *10*, e1004174.

Sun, F., Ding, Y., Ji, Q., Liang, Z., Deng, X., Wong, C.C., Yi, C., Zhang, L., Xie, S., Alvarez, S., *et al.* (2012). Protein cysteine phosphorylation of SarA/MgrA family transcriptional regulators mediates bacterial virulence and antibiotic resistance. Proc. Natl. Acad. Sci. U.S.A. *109*, 15461–15466.

Sun, F., Li, C., Jeong, D., Sohn, C., He, C., and Bae, T. (2010). In the *Staphylococcus aureus* two-component system *sae*, the response regulator SaeR binds to a direct repeat sequence and DNA binding requires phosphorylation by the sensor kinase SaeS. J. Bacteriol. *192*, 2111–2127.

Surmann, K., Michalik, S., Hildebrandt, P., Gierok, P., Depke, M., Brinkmann, L., Bernhardt, J., Salazar, M.G., Sun, Z., Shteynberg, D., Kusebauch, U., Moritz, R.L., Wollscheid, B., Lalk, M., Völker, U., and Schmidt, F. (2014). Comparative proteome analysis reveals conserved and specific adaptation patterns of *Staphylococcus aureus* after internalization by different types of human non-professional phagocytic host cells. Front. Microbiol. *5*, 392.

Tamber, S., and Cheung, A.L. (2009). SarZ promotes the expression of virulence factors and represses biofilm formation by modulating SarA and *agr* in *Staphylococcus aureus*. Infect. Immun. *77*, 419–428.

Tamber, S., Reyes, D., Donegan, N.P., Schwartzman, J.D., Cheung, A.L., and Memmi, G. (2010). The staphylococcus-specific gene *rsr* represses *agr* and virulence in *Staphylococcus aureus*. Infect. Immun. *78*, 4384–4391.

Tao, L., Wu, X., and Sun, B. (2010). Alternative sigma factor sigmaH modulates prophage integration and excision in *Staphylococcus aureus*. PLOS Pathog. *6*, e1000888.

Tegmark, K., Karlsson, A., and Arvidson, S. (2000). Identification and characterization of SarH1, a new global regulator of virulence gene expression in *Staphylococcus aureus*. Mol. Microbiol. *37*, 398–409.

Thoendel, M., Kavanaugh, J.S., Flack, C.E., and Horswill, A.R. (2011). Peptide signaling in the staphylococci. Chem. Rev. *111*, 117–151.

Throup, J.P., Zappacosta, F., Lunsford, R.D., Annan, R.S., Carr, S.A., Lonsdale, J.T., Bryant, A.P., McDevitt, D., Rosenberg, M., and Burnham, M.K. (2001). The srhSR gene pair from *Staphylococcus aureus*: genomic and proteomic approaches to the identification and characterization of gene function. Biochemistry *40*, 10392–10401.

Tomasini, A., Francois, P., Howden, B.P., Fechter, P., Romby, P., and Caldelari, I. (2014). The importance of regulatory RNAs in *Staphylococcus aureus*. Infect. Genet. Evol. *21*, 616–626.

Torres, V.J., Attia, A.S., Mason, W.J., Hood, M.I., Corbin, B.D., Beasley, F.C., Anderson, K.L., Stauff, D.L., McDonald, W.H., Zimmerman, L.J., Friedman, D.B., Heinrichs, D.E., Dunman, P.M., and Skaar, E.P. (2010). *Staphylococcus aureus* fur regulates the expression of virulence factors that contribute to the pathogenesis of pneumonia. Infect. Immun. *78*, 1618–1628.

Traber, K.E., Lee, E., Benson, S., Corrigan, R., Cantera, M., Shopsin, B., and Novick, R.P. (2008). *agr* function in clinical *Staphylococcus aureus* isolates. Microbiology *154*, 2265–2274.

Trotonda, M.P., Tamber, S., Memmi, G., and Cheung, A.L. (2008). MgrA represses biofilm formation in *Staphylococcus aureus*. Infect. Immun. *76*, 5645–5654.

Truong-Bolduc, Q.C., Ding, Y., and Hooper, D.C. (2008). Posttranslational modification influences the effects of MgrA on *norA* expression in *Staphylococcus aureus*. J. Bacteriol. *190*, 7375–7381.

Truong-Bolduc, Q.C., Dunman, P.M., Strahilevitz, J., Projan, S.J., and Hooper, D.C. (2005). MgrA is a multiple regulator of two new efflux pumps in *Staphylococcus aureus*. J. Bacteriol. *187*, 2395–2405.

Truong-Bolduc, Q.C., and Hooper, D.C. (2007). The transcriptional regulators NorG and MgrA modulate resistance to both quinolones and beta-lactams in *Staphylococcus aureus*. J. Bacteriol. *189*, 2996–3005.

Truong-Bolduc, Q.C., and Hooper, D.C. (2010). Phosphorylation of MgrA and its effect on expression of the NorA and NorB efflux pumps of *Staphylococcus aureus*. J. Bacteriol. *192*, 2525–2534.

Truong-Bolduc, Q.C., Strahilevitz, J., and Hooper, D.C. (2006). NorC, a new efflux pump regulated by MgrA of *Staphylococcus aureus*. Antimicrob. Agents Chemother. *50*, 1104–1107.

Tsuji, B.T., MacLean, R.D., Dresser, L.D., McGavin, M.J., and Simor, A.E. (2011). Impact of accessory gene regulator (agr) dysfunction on vancomycin pharmacodynamics among Canadian community and health-care associated methicillin-resistant *Staphylococcus aureus*. Ann. Clin. Microbiol. Antimicrob. *10*, 20.

Ulrich, M., Bastian, M., Cramton, S.E., Ziegler, K., Pragman, A.A., Bragonzi, A., Memmi, G., Wolz, C., Schlievert, P.M., Cheung, A., and Döring, G. (2007). The staphylococcal respiratory response regulator SrrAB induces *ica* gene transcription and polysaccharide intercellular adhesin expression, protecting *Staphylococcus aureus* from neutrophil killing under anaerobic growth conditions. Mol. Microbiol. 65, 1276–1287.

van Schaik, W., and Abee, T. (2005). The role of sigmaB in the stress response of Gram-positive bacteria – targets for food preservation and safety. Curr. Opin. Biotechnol. *16*, 218–224.

Vecerek, B., Rajkowitsch, L., Sonnleitner, E., Schroeder, R., and Bläsi, U. (2008). The C-terminal domain of *Escherichia coli* Hfq is required for regulation. Nucleic Acids Res. *36*, 133–143.

Voyich, J.M., Vuong, C., DeWald, M., Nygaard, T.K., Kocianova, S., Griffith, S., Jones, J., Iverson, C., Sturdevant, D.E., Braughton, K.R., *et al.* (2009). The SaeR/S gene regulatory system is essential for innate immune evasion by *Staphylococcus aureus*. J. Infect. Dis. *199*, 1698–1706.

Wang, R., Braughton, K.R., Kretschmer, D., Bach, T.H., Queck, S.Y., Li, M., Kennedy, A.D., Dorward, D.W., Klebanoff, S.J., Peschel, A., *et al.* (2007). Identification of novel cytolytic peptides as key virulence determinants for community-associated MRSA. Nat. Med. *13*, 1510–1514.

Watkins, R.L., Pallister, K.B., and Voyich, J.M. (2011). The SaeR/S gene regulatory system induces a pro-inflammatory cytokine response during *Staphylococcus aureus* infection. PLoS One *6*, e19939.

Wertheim, H.F., Melles, D.C., Vos, M.C., van Leeuwen, W., van Belkum, A., Verbrugh, H.A., and Nouwen, J.L. (2005). The role of nasal carriage in *Staphylococcus aureus* infections. Lancet Infect. Dis. *5*, 751–762.

Westermann, A.J., Gorski, S.A., and Vogel, J. (2012). Dual RNA-seq of pathogen and host. Nat. Rev. Microbiol. *10*, 618–630.

Wurtzel, O., Sesto, N., Mellin, J.R., Karunker, I., Edelheit, S., Bécavin, C., Archambaud, C., Cossart, P., and Sorek, R. (2012). Comparative transcriptomics of pathogenic and non-pathogenic Listeria species. Mol. Syst. Biol. *8*, 583.

Xia, G., Kohler, T., and Peschel, A. (2010). The wall teichoic acid and lipoteichoic acid polymers of *Staphylococcus aureus*. Int. J. Med. Microbiol. *300*, 148–154.

Xiong, Y.Q., Willard, J., Yeaman, M.R., Cheung, A.L., and Bayer, A.S. (2006). Regulation of *Staphylococcus aureus* α-toxin gene (*hla*) expression by *agr*, *sarA*, and *sae in vitro* and in experimental infective endocarditis. J. Infect. Dis. *194*, 1267–1275.

Xue, T., You, Y., Hong, D., Sun, H., and Sun, B. (2011). The *Staphylococcus aureus* KdpDE two-component system couples extracellular K+ sensing and Agr signaling to infection programming. Infect. Immun. *79*, 2154–2167.

Xue, T., Zhang, X., Sun, H., and Sun, B. (2014). ArtR, a novel sRNA of *Staphylococcus aureus*, regulates α-toxin expression by targeting the 5' UTR of *sarT* mRNA. Med. Microbiol. Immunol. *203*, 1–12.

Yarwood, J.M., McCormick, J.K., and Schlievert, P.M. (2001). Identification of a novel two-component regulatory system that acts in global regulation of virulence factors of *Staphylococcus aureus*. J. Bacteriol. *183*, 1113–1123.

Yu, D., Zhao, L., Xue, T., and Sun, B. (2012). *Staphylococcus aureus* autoinducer-2 quorum sensing decreases biofilm formation in an *icaR*-dependent manner. BMC Microbiol. *12*, 288.

Zheng, A., Panja, S., and Woodson, S.A. (2016). Arginine patch predicts the RNA annealing activity of Hfq from Gram-negative and Gram-positive bacteria. J. Mol. Biol. *428*, 2259–2264.

Zhu, Y., Fan, X., Zhang, X., Jiang, X., Niu, L., Teng, M., and Li, X. (2014). Structure of Rot, a global regulator of virulence genes in *Staphylococcus aureus*. Acta Crystallogr. D. Biol. Crystallogr. *70*, 2467–2476.

Zhu, Y., Nandakumar, R., Sadykov, M.R., Madayiputhiya, N., Luong, T.T., Gaupp, R., Lee, C.Y., and Somerville, G.A. (2011). RpiR homologues may link *Staphylococcus aureus* RNAIII synthesis and pentose phosphate pathway regulation. J. Bacteriol. *193*, 6187–6196.

Ziebandt, A.K., Weber, H., Rudolph, J., Schmid, R., Höper, D., Engelmann, S., and Hecker, M. (2001). Extracellular proteins of *Staphylococcus aureus* and the role of SarA and sigma B. Proteomics *1*, 480–493.

Zielinska, A.K., Beenken, K.E., Joo, H.S., Mrak, L.N., Griffin, L.M., Luong, T.T., Lee, C.Y., Otto, M., Shaw, L.N., and Smeltzer, M.S. (2011). Defining the strain-dependent impact of the staphylococcal accessory regulator (*sarA*) on the α-toxin phenotype of *Staphylococcus aureus*. J. Bacteriol. *193*, 2948–2958.

Immune Response to *Staphylococcus aureus*

13

Aisling F. Brown and Rachel M. McLoughlin

Abstract

Staphylococcus aureus can act as either a harmless commensal or a potentially lethal pathogen. Disease is just one of a number of possible outcomes of host–microbe interactions. It is the interplay between both host and microbial factors that determines whether or not disease occurs and its severity. Here, we outline the main functions and elements of the host immune system and consider how they interact with *S. aureus* to shape clinical outcomes. We further discuss how the organism seeks to evade host immune mechanisms and review the immunodeficiency states that render certain hosts susceptible to *S. aureus* infection. Finally, we review the status of anti-*S. aureus* vaccine development and examine how we might modulate the immune response in order to improve the outcome of infection.

Introduction

The importance of immunity

Staphylococcus aureus colonizes a substantial proportion of the human population, for either long or short periods, at various stages throughout life. It replicates and evolves in this reservoir as a harmless non-pathogenic commensal (Brown *et al.*, 2014). Approximately 20% of individuals have persistent nasal colonization, and the majority of the remainder intermittently carry lower numbers of bacteria (van Belkum *et al.*, 2009).

In apparent contrast, *S. aureus* is also a leading cause of community- and hospital-acquired bacterial infections. The greatest morbidity burden is from skin and soft tissue infections (SSTIs). Invasive infections, however, are considerably more serious and often life-threatening. *S. aureus* is one of the most common causes of bloodstream infection (BSI), and carries a higher mortality than any other bacteraemia, with a 20–40% mortality within 30 days despite appropriate treatment (Melzer and Welch, 2013). In addition, it is the principal cause of other serious infections including osteomyelitis, septic arthritis, endocarditis and device-related infections, and results in significant morbidity and healthcare costs (de Kraker *et al.*, 2011) (see Chapter 2).

S. aureus can therefore act either as a harmless commensal or a potentially lethal opportunistic pathogen. Interestingly, the incidence of colonization is over 1000 times greater than that of serious invasive infection (Laupland *et al.*, 2008). Disease – as characterized by damage to the host – is therefore just one of a number of possible outcomes of host–microbe interactions (Casadevall and Pirofski, 2003). An entirely pathogen-centred view of infectious diseases ignores the fact that microbial 'virulence' can only be expressed in a relatively 'susceptible' host. While microbial components certainly contribute to pathogenicity, they do not exclusively determine it. Infection may not necessarily lead to disease, and disease is not solely the result of insufficient host response allowing unchecked microbial replication. Disease may also result from excessive, uncontrolled, or misplaced host inflammatory responses against the invading organism. Consequently, the outcome of host–microbe interactions is a complex, multifactorial and dynamic process, shaped by interplay between both microbial and host factors (Fig. 13.1).

Efforts to identify microbial factors that distinguish pathogenic from colonizing strains of *S.*

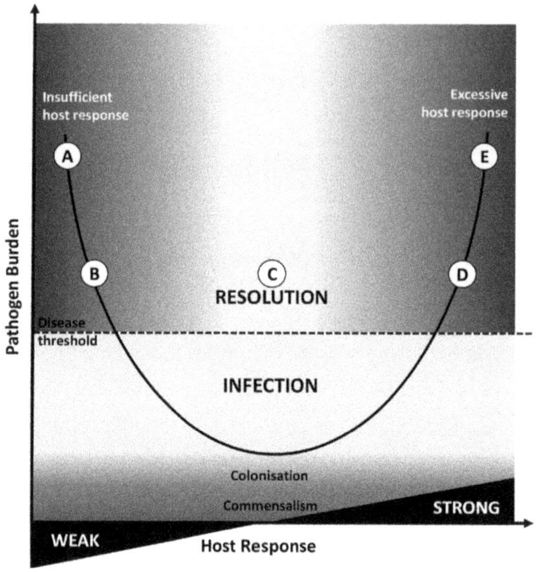

Figure 13.1 Host–microbe interactions determine disease severity. Possible outcomes following the contact of a host with a potential pathogen. A microorganism may contact or enter the host organism and not result in any host damage (commensalism). When colonization occurs there may be variable host–pathogen interaction with the potential to lead to host damage, but this is often avoided by host immune response. Finally, the interaction may result in infection. Infection may evolve differentially according to the host response. If immune response is entirely insufficient, overwhelming infection may lead to death (A). Where host response is somewhat weak, recurrent or chronic infection may occur (B). An ideal host response results in disease being avoided or easily resolved without host damage (C). Persistently dysregulated or excessive host immune response results in chronic inflammation (D). Severely uncontrolled host immune response may result in death due to excessive inflammation.

aureus have largely failed (Feil *et al.*, 2003; van Hal *et al.*, 2012) and – with the exception of diseases mediated by specific staphylococcal toxins (such as food poisoning and toxic shock syndromes) – no single virulence factor is sufficient to recapitulate a *S. aureus* infection. On the other hand, compelling clinical evidence suggests that the outcome of human *S. aureus* infection may depend considerably on host factors (Greenberg *et al.*, 2014; Lesens *et al.*, 2003; van Hal *et al.*, 2012). Human hosts bring different genetic backgrounds and immunological histories to the point of host-microbe contact, and such immunity may impact on the outcome of infection. Thus, the duality of *S. aureus*'s interaction with humans as both a pathogen and colonizer is a fascinating paradox that cannot be answered by focusing on the microbe in isolation.

What is the immune system?

The immune system comprises cells and molecules responsible for protection against disease. It is best characterized by its role in preventing and eliminating infection by micro-organisms. The immune system can be broadly divided into two arms, although these work in synergy. The innate immune system is a front line of defence providing an immediate, non-specific response to neutralize or eliminate a pathogen. The adaptive immune system is an acquired response to specific pathogen elements that emerges days after initial infection and, crucially, has the capacity to form immunological memory. This memory allows the host to mount faster and more precise attacks on subsequent encounters with the same pathogen, and forms the basis of vaccination. Both the innate and adaptive immune systems are comprised of humoral and cellular components. Humoral elements are secreted soluble molecules particularly important in protection against extracellular pathogens. Cellular elements are specialized cells which can directly engulf and kill or otherwise orchestrate attacks on pathogens resulting in their elimination (Table 13.1).

Innate immune response to bacterial infection

A pathogen initially encounters non-specific physical barriers including intact skin or mucous membranes, broad-spectrum antimicrobial peptides, mucus, cilia and resident microflora. Once the invading organism overcomes these general non-specific defences, it is detected by the host, resulting in activation of an initial inflammatory response. This response is characterized by a coordinated release of inflammatory mediators and delivery of blood components to the site of infection. Thus, the functions of the innate immune system during bacterial infection are to recognize the pathogen, actively attempt to clear infection, and facilitate activation of adaptive immunity specific to the bacteria.

Table 13.1 Innate and adaptive immunity

Property	External barriers (non-induced)	Innate immune system (induced)	Adaptive immune system (induced)
Distribution	Host–environment boundaries	Widespread all tissues	Primarily concentrated in lymphoid organs. Smaller populations tissue-resident
Recognition	None	Non-specific: recognize PAMPS (common microbial motifs)	Antigen-specific: receptors recognize particular epitopes or details of a pathogen's molecular structure
Receptors	None	PRRs: fixed in germline genome, rearrangement not necessary	B cell (BCR) and antibodies; T cell receptors (TCR): encoded in gene segments, rearrangement necessary
Response time	Continuous	Immediate activation of effectors	Delay between exposure and maximal activation of effectors
Immunological memory	None	Not established	Memory of prior exposures
Major humoral component(s)		Complement	Antibodies
Major cellular component(s)		Phagocytes, NK cells	Lymphocytes
Major responses	Maintenance of barrier. Antimicrobial peptide and lipid production	Phagocytosis; chemokine and cytokine production (IL-1β, IL-6, IL-8); Facilitate adaptive response (antigen presentation, co-stimulation, polarizing cytokines)	Clonal expansion of lymphocytes. Differentiation to effector B and T cells; effector cytokine production (IFNγ, IL-4 IL-17)

PAMPs, pathogen-associated molecular patterns; PRRs, pattern recognition receptors; NK, natural killer; IL, interleukin; IFN, interferon.

Intact skin and antimicrobial peptides

The initial physical barrier of intact skin is almost certainly the most important host defence element in preventing *S. aureus* infection. Because few organisms have the ability to directly penetrate the skin, they usually gain access by some physical means, such as trauma, surgical incision or i.v. catheters. Compromised skin integrity is consistently associated with *S. aureus* SSTIs (Begier *et al.*, 2004; Huang *et al.*, 2008; Kazakova *et al.*, 2005; Miller and Kaplan, 2009). In addition, breaches in the skin barrier often precede invasive disease, as evidenced by the increased incidence of *S. aureus* BSI in patients with indwelling catheters, repeated injections or burns (Kaasch *et al.*, 2014; Laupland *et al.*, 2003; Oncul *et al.*, 2014).

Antimicrobial peptides (AMPs) and lipids are a collection of host defence products generated by epithelial and immune cells, and are particularly important at the skin surface. These have broad-spectrum antimicrobial activity, can rapidly and directly kill micro-organisms, as well as modulate the innate immune response. Some of these antimicrobial peptides have been shown to interact with *S. aureus*. Human β-defensin 3 (hBD-3) is an epithelial cell peptide with potent anti-*S. aureus* effect *in vitro* and its production is induced by the presence of the bacteria and by skin damage (Menzies and Kenoyer, 2005; Midorikawa *et al.*, 2003). The cutaneous cathelicidin LL-37 is also impressively effective at *in vitro* killing of both extra- and intracellular *S. aureus* (Noore *et al.*, 2013). *S. aureus* has, however, adapted to overcome many of these local immune challenges by producing proteases to degrade AMPs, proteins to bind and inactivate them, and by altering the charge of its cell wall to reduce their attachment affinity (Rieg *et al.*, 2011).

Some defects in AMPs and lipids have been described in the skin of highly colonized populations who are more susceptible to *S. aureus* skin infections. Constitutive and induced levels of protective hBD-3 in skin are significantly lower in persistent carriers, creating a more permissive environment for successful colonization and/or infection (Brown *et al.*, 2014; Nurjadi *et al.*, 2013). The skin of atopic dermatitis (AD) patients – who are almost universally colonized with *S. aureus* – also exhibits impaired antimicrobial peptide activity (Arikawa *et al.*, 2002; Takigawa *et al.*, 2005).

Innate immune recognition of *S. aureus*

Innate immune recognition is accomplished via a diverse series of host molecules known as **p**attern **r**ecognition **r**eceptors (PRRs), which recognize highly conserved pathogen-associated molecular patterns (PAMPs). PAMPs are essential and common to multiple microbes, yet not present in host cells (Akira *et al.*, 2006). PRRs are germline-encoded receptors constitutively expressed on innate immune cells, as well as some non-immune cells such as fibroblasts and epithelial cells. A number of PRR families have been identified – Toll-like receptors (TLRs), NOD-like receptors (NLRs), C-type lectin receptors (CLRs) and RIG-1 like receptors (RLRs). These encompass transmembrane, cytosolic and secreted extracellular receptors. Different PRRs react with subsets of PAMPs, such that distinct responses are generated by different types of pathogen. TLRs 1, 2, 4, 6, 9 and NLRs NOD1 and NOD2 among the best-characterized of these, and are implicated in response to bacterial infections. PRR:PAMP binding triggers downstream signalling cascades, resulting in production of inflammatory and immunomodulatory cytokines and chemokines which contribute to initial defence against infecting micro-organisms, and these pathways have been extensively reviewed elsewhere (Akira *et al.*, 2006).

Several components of *S. aureus*'s cell wall can act as PAMPs and ligate certain PRRs to stimulate innate immunity. PRR ligation triggers a cascade of protein signalling which results in the activation or suppression of inflammatory cytokine and chemokine production. While innate recognition of *S. aureus* is not comprehensively understood, some proposed signalling pathways are presented in Fig. 13.2. Activation of TLR2 is to-date the best-characterized of *S. aureus*–PRR interactions. Staphylococcal lipoproteins, and perhaps to a lesser extent lipoteichoic acid (LTA) and peptidoglycan (PG) fragments are recognized by this receptor (Fournier, 2012). TLR2 is primarily expressed on circulating innate immune cells (particularly monocytes), endothelial cells and in lymphoid tissues (Flo *et al.*, 2001; Wilhelmsen *et al.*, 2012). It interacts with TLR1 or TLR6 to form functional heterodimers. Upon staphylococcal lipoprotein binding, TLRs undergo conformational changes and initiate a cascade of signalling events resulting in the activation of NF-κB, activator protein 1 (AP-1) or interferon regulatory factor (IRF7) – transcription factors mediating the expression of a number of pro- and anti-inflammatory genes.

Alternative PRRs have also been implicated in *S. aureus* recognition, and phagocytosis can amplify immune recognition. Engulfed bacteria are degraded in the phagosome, and staphylococcal cell wall components and DNA released, allowing for activation of PRRs in the cytosol. Cytosolic PRRs include TLR9, NOD2 or endocytosed TLR2 (Stack *et al.*, 2014). Intracellular TLR9 detects unmethylated bacterial cytosine-phosphate-guanosine (CpG) DNA, and while its specific role in staphylococcal infections is not well established, it can induce type I interferon production (Parker and Prince, 2012). NOD2 is mainly expressed in the cytosol of monocytes and macrophages and detects degraded peptidoglycan (Shimada *et al.*, 2010). Both of these pathways result in a final common pathway of NF-κB activation.

The importance of innate immune recognition in initiating host defence against Gram-positive bacteria has been demonstrated using TLR2-deficient, MyD88-deficient, and NOD2-deficient mice, which were found to be highly susceptible to systemic challenge with *S. aureus*. These deficiencies resulted in altered cytokine production, impaired bacterial clearance and significantly increased mortality (Deshmukh *et al.*, 2009; Takeuchi *et al.*, 2000). The adaptor protein MyD88 and IL-1R-associated kinase (IRAK) are central elements of the TLR2 signalling pathway (Fig. 13.2). MyD88 also acts as an adaptor protein for IL-1R signalling. Some of these findings have been correlated with human pathology, while also highlighting some of the key caveats of translation. Humans, in general, have a greater complexity and reserve of redundant ancillary mechanisms of achieving immune responses, such that isolated deletions/mutations are often not sufficient to recapitulate the catastrophic consequences seen in animal models. A number of human autosomal recessive mutations can, however, result in

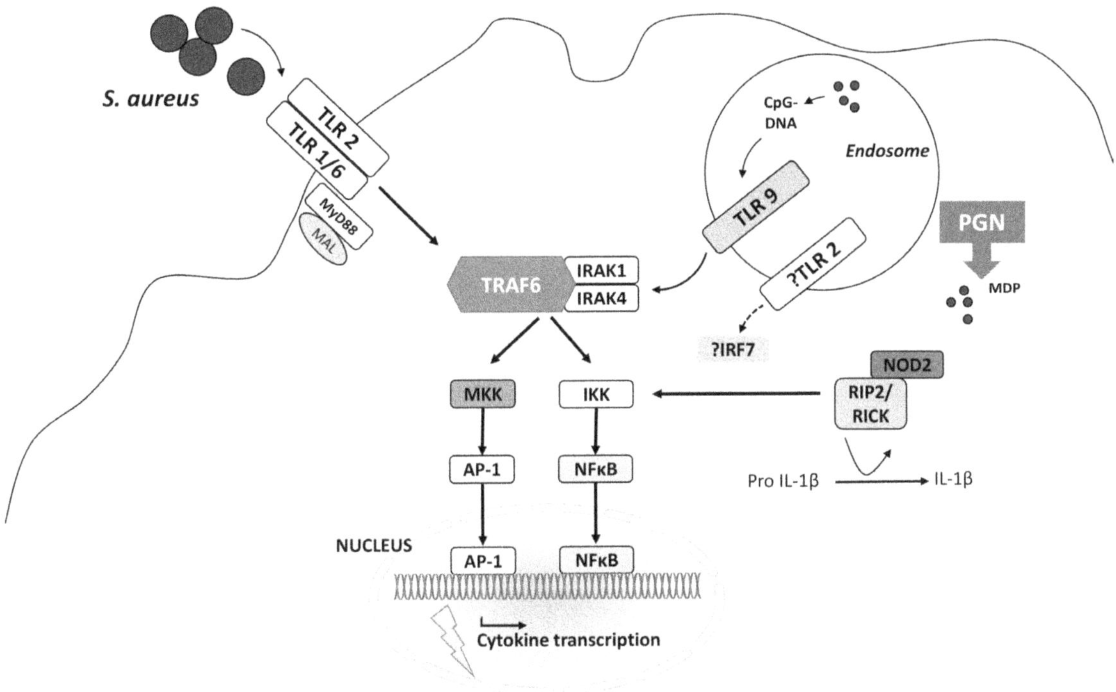

Figure 13.2 Proposed innate immune recognition of *S. aureus*. TLR2 (+1/6) heterodimers recognize extracellular *S. aureus* and undergo conformational changes to initiate a cascade of signalling events signal via MyD88 and MAL. This ultimately results in the activation of NF-κB, AP-1 and perhaps IRF7. These are transcription factors mediating the expression of a number of inflammatory genes. Components of engulfed bacteria are released in the cytosol, possibly activating other PRRs, e.g. TLR9, NOD2, endosomal TLR2, TLR9. The specific role of these PRRs in staphylococcal infections is not well established. Abbreviations: TLR, toll-like receptor; MyD88, myeloid differentiation factor-88; MAL, MyD88 adaptor-like; TRAF6, TNF receptor-associated factor 6; IRAK, IL-1R-associated kinase; MAPK, mitogen-activated protein kinase; IKK, IκB kinase; AP-1, activator-protein 1; NF-κB, nuclear factor kappa-light-chain-enhancer of activated B cells; PGN, peptidoglycan; MDP, muramyl dipeptide; NOD, nucleotide-binding oligomerization domain; RIP, receptor-interacting protein; IL, interleukin.

MyD88 or IRAK-4 deficiencies, leading to impaired pathogen recognition and initiation of host defence. These patient groups have normal resistance to common fungi, parasites, viruses, and many bacteria, but are susceptible to recurrent life-threatening bacterial infections caused by a limited number of species including *S. aureus*, *Streptococcus pneumoniae* and *Pseudomonas aeruginosa*. Typical findings of systemic inflammation – including fever and bloodstream neutrophilia – were usually weak or delayed in this cohort, and forty per cent of patients had died of these infections by the age of 10 years. Survivors experienced fewer episodes after their teenage years, suggesting the existence or development of some compensatory mechanisms (Picard *et al.*, 2010). In fact, various components of microbes can activate multiple PRRs simultaneously, such that deficiency in a single PRR usually results in only a modest clinical phenotype. It is important to note that these innate signalling pathways may also induce TLR2-mediated anti-inflammatory signalling, which may actually be protective during certain infections (Netea *et al.*, 2004; Stenzel *et al.*, 2008). The type of cell and site of the body at which staphylococcal TLR2 ligands are encountered may determine whether subsequent signalling is predominantly pro- or anti-inflammatory (Mele and Madrenas, 2010). Loss of TLR2-mediated anti-inflammatory signalling has been associated with greater mortality and bacterial burden in a systemic *S. aureus* infection model using TLR2-deficient mice, despite pro-inflammatory phagocyte recruitment and function being unaffected (Yimin *et al.*, 2013). How *S. aureus* may harness anti-inflammatory signalling to its own advantage is further discussed in the 'Immune regulation' section.

In summary, upon staphylococcal invasion, recognition of its PAMPs by host PRRs (e.g. TLR2/1, TLR2/6+/− TLR9, NOD2 and perhaps others) induces signals that promote gene transcription of antimicrobial peptides, pro- and anti-inflammatory cytokines and chemokines. The co-ordinated release of the inflammatory mediators recruits further immune cells to the site and directs the inflammatory milieu in an effort to eliminate or neutralize infection, while simultaneously ensuring host damage is minimized.

Initial inflammatory response

The purpose of pathogen recognition is to trigger host defence via an initial inflammatory response which mobilizes innate immunity to sites of infection. In *S. aureus* infection, the end result of the TLR2 signalling pathway is the activation of the transcription factor NF-κB, whereas the NLR pathway activates caspase-1 to produce IL-1β. Within hours of TLR2 binding by staphylococcal ligands, the activated cell has transcribed multiple host cytokine and chemokine genes. Tissue-resident macrophages and dendritic cells endowed with high numbers of surface PRRs are responsible for much of this cytokine production. Three of the major inflammatory cytokines induced via this signalling – IL-1, IL-6 and TNFα – are largely responsible for triggering the initial inflammatory response.

This response comprises a number of events characterized by the Latin descriptor 'rubor, dolor, calor, tumor' – redness, pain, heat and swelling. Changes are elicited in blood vessels including vasodilation, increased vascular permeability and activation of endothelial cells. PRR ligation and cytokine release induces upregulation of complementary adhesion molecules on the surfaces of leucocytes and of blood vessel endothelial cells to facilitate the movement of cells out of the blood and into infected tissues. Immune effector cells are recruited and directed to the site of infection by chemokines. These are a family of chemoattractant proteins which direct leucocyte migration, guiding neutrophils and other immune cells to a site of infection along a concentration gradient. Neutrophils are the earliest cells recruited. CXCL8 (IL-8) is a strong neutrophil chemoattractant, while CCL2/MCP-1 (macrophage/monocyte chemoattractant protein-1) is chemoattractive mainly for monocytes. This influx of fluid and cells into tissues often results in swelling and discomfort. The complement system is also activated during this initial response. Given that innate and adaptive immunity is intricately interlinked, such initial signalling also generates cytokines that begin to activate the adaptive immune system (Table 13.2). When this inflammatory response is no longer localized to an injured or infected area, its systemic effects are more critical and may progress to severe sepsis or septic shock (Bone *et al.*, 1992).

Cytokines can act in an autocrine and paracrine manner – affecting the behaviour of the cell producing the cytokine, and of adjacent cells. Some may also act in an endocrine manner – affecting the behaviour of distant cells – if they enter the circulation and have sufficient half-life in the blood. The three major initial inflammatory cytokines – IL-1, IL-6 and TNFα – can circulate and act on the hypothalamus as endogenous pyrogens to cause fever. Fever is thought to be an evolutionary response, impairing bacterial replication and enhancing adaptive immune responses. Unsurprisingly, therefore, clinical outcomes have not been improved by treating fever itself (Mackowiak, 1998). These cytokines, IL-6 in particular, also act on hepatocytes to induce or inhibit synthesis of a number of proteins secreted into the blood (Mackowiak, 1998). These are termed 'acute phase proteins'. A number of these upregulated proteins – notably C-reactive protein – can bind to microbial cell wall elements, opsonizing the pathogens and activating the complement cascade as an early bacterial clearance strategy. Several additional biological effects are associated with endogenous pyrogen release. The bone marrow is stimulated to mobilize neutrophils, fat and protein stores are released and neuroendocrine effects of appetite suppression and glucocorticoid release are observed. Fever and an acute phase response are typical of more serious systemic *S. aureus* infections. In *S. aureus* BSI patients, persistent fever (del Rio *et al.*, 2009) or elevated IL-6 (McNicholas *et al.*, 2014) predict a more complicated clinical course. However, whether this represents a fundamental host-specific difference in inflammatory response, or is simply indicative of a higher inoculum infection is unclear.

Thus, the initial inflammatory response is triggered by pathogen recognition and – via cytokine/chemokine signalling – attracts innate effector proteins and cells to the site of infection.

Table 13.2 Major cytokines of the initial inflammatory response

Family	Cytokine	Major cellular source	Response
Interleukins	IL-1α and IL-1β	Macrophages, epithelial cells	Fever; induces acute phase protein production; macrophage activation; T cell activation and expansion
	IL-6	Macrophages, endothelial cells	Fever; induces acute phase protein production; T and B cell differentiation
	IL-10	Macrophages/monocytes	Inhibits pro-inflammatory cytokine synthesis
	IL-12	Macrophages, dendritic cells	Activates NK cells; Th1 cell differentiation
	IL-18	Macrophages	Induces IFNγ production by T cells
	IL-23	Macrophages, dendritic cells	Differentiation of Th17 cells
	IL-25	Epithelial cells, mast cells	Promotes Th2 responses
	IL-27	Macrophages, monocytes, dendritic cells	Suppresses Th17 and Th2 cell differentiation; induces anti-inflammatory IL-10
Chemokines	CXCL8 (IL-8)	Macrophages, T cells	Granulocyte chemotaxis and phagocytosis
	CCL2 (MCP-1)	Monocytes, macrophages, dendritic cells	Monocyte chemotaxis
TNF family	TNFα	Macrophages, NK cells, T cells	Fever; induces acute phase protein production; increases vascular permeability; triggers endothelial coagulation
Interferons	IFNα, IFNβ		Antiviral proteins; macrophage activation
Colony-stimulating factors	GM-CSF	Macrophages, T cells, endothelial cells, fibroblasts	Stimulates stem cells to produce granulocytes and monocytes
Growth factors	TGFβ	Macrophages/monocytes, chondrocytes, T cells	Anti-inflammatory; inhibits cell growth; blocks activation of lymphocytes

IL, interleukin; CXCL, C-X-C motif ligand; CCL, C-C motif ligand; MCP, monocyte chemotactic protein; TNF, tumour necrosis factor; NK, natural killer; IFN, interferon; GM-CSF, granulocyte monocyte colony stimulating factor; TGF, transforming growth factor.

Complement

The complement system is a collection of soluble 'inactive' proteins in blood and other body fluids which forms a major humoral component of the innate immune system. These proteins are PRRs and become activated via several pathways – directly via the presence of a pathogen (alternative pathway), or indirectly as a result of proteins that bind to bacterial surfaces (lectin and classical pathways) (Ram *et al.*, 2010). Initial activation initiates a cascade of self-amplifying proteases on the pathogen surface, which ultimately results in pathogen destruction using the complement system's three main effector mechanisms (Fig. 13.3). The first of these is opsonization – coating (encapsulated) bacteria with complement component C3b. Complement receptors on the surface of phagocytes bind C3b attached to the bacterial surface to facilitate pathogen phagocytosis and clearance from the circulation. The second effector mechanism is direct bactericidal action – which involves the assembly of a 'membrane attack complex' by a subset of complement proteins in order to form fatal pores in pathogen membranes. This latter mechanism has been shown to be clinically important only for a limited number of infections, e.g. *Neisseria* species, and not for *S. aureus* infection (Ram *et al.*, 2010). The third complement effector mechanism is to further enhance the inflammatory response by promoting phagocyte recruitment and releasing peptide mediators of inflammation. These stimulate histamine release from mast cells, promote vascular dilatation, increase endothelial permeability, and stimulate neutrophil responses. In addition to these proinflammatory activities, complement activity is modulated by a system

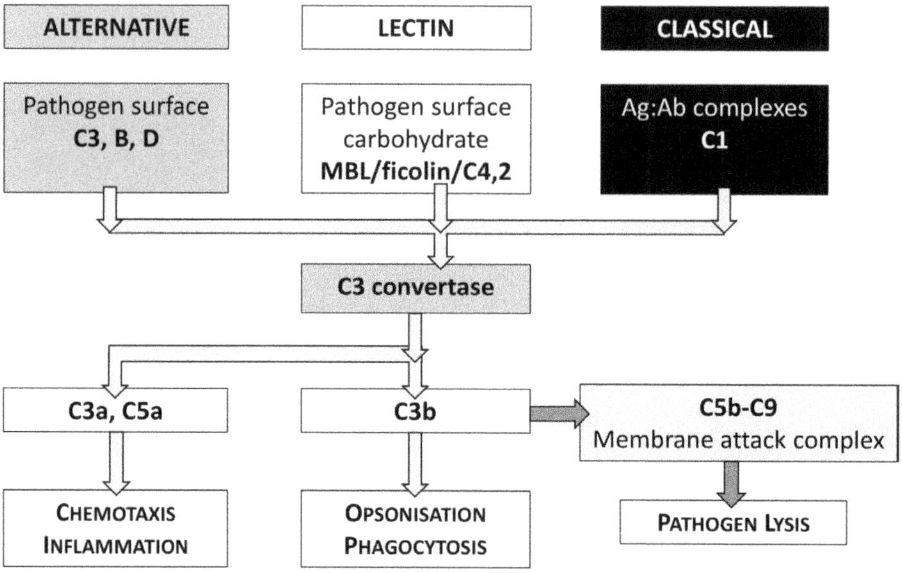

Figure 13.3 Complement activation. Complement proteins may be activated via several pathways – alternative, lectin or classical. The presence of a pathogen directly activates the alternative pathway. Complement proteins binding to carbohydrates on the bacterial surface, or to immune complexes trigger the lectin or classical pathways, respectively. This initiates a cascade of self-amplifying proteases on the pathogen surface, ultimately resulting in its destruction. Opsonization coats bacteria with C3b, which binds to phagocyte receptors to facilitate phagocytosis and clearance from the circulation. The bactericidal 'membrane attack complex' forms fatal pores in bacterial membranes, but is not involved in response to *S. aureus*. Complement also enhances inflammatory response by recruiting phagocytes and releasing peptide mediators of inflammation.

of regulatory proteins to prevent untoward tissue damage.

While all three complement pathways play a role in *S. aureus* recognition, the lectin pathway – where circulating lectin proteins attach to bacterial surface carbohydrates – has been the source of considerable interest. One of these proteins, mannose-binding lectin (MBL), directly binds *S. aureus in vitro*, although C3 deposition on the bacterial surface occurs normally even in the absence of MBL (Brouwer *et al.*, 2008). MBL deficiency has long been putatively linked to predisposition to, or severity of *S. aureus* infection in humans, and impaired MBL production has been associated with delayed clearance of *S. aureus* bloodstream infection in adults (Chong *et al.*, 2014). However, given that the normal range of MBL in healthy individuals may differ 1000-fold, it is difficult to determine the veracity of this association (Ram *et al.*, 2010). Regardless, *S. aureus* is strikingly equipped with multiple mechanisms (discussed in *Immune Evasion* below) to evade clearance by complement, confirming the importance of the complement cascade in immunity against infection.

Phagocytes

Cells of the myeloid lineage (Fig. 13.4) form the cellular arm of the innate immune system. The professional phagocytes – neutrophils, macrophages and dendritic cells – are the most important effector cells of the innate immune system, and are actively recruited to sites of infection. Phagocytosis is the uptake of bacteria, parasites, dead host cells and debris by specialized cells. Professional phagocytes are richly endowed with PRRs. When activated by cell damage or PAMPs, phagocytosis is initiated in addition to signal transduction leading to the initial inflammatory response. Additional phagocyte receptors recognize bacteria opsonized by complement (C3b receptors), antibody (Fc receptors) or extracellular matrix elements (C-type lectins). They then engulf pathogens and utilize chemical cocktails to kill captured micro-organisms or infected cells.

Some phagocytes also function as professional antigen-presenting cells (APCs), which process and display fragments of microbial antigens to T lymphocytes to activate the adaptive immune system. Other phagocytes migrate to lymph nodes

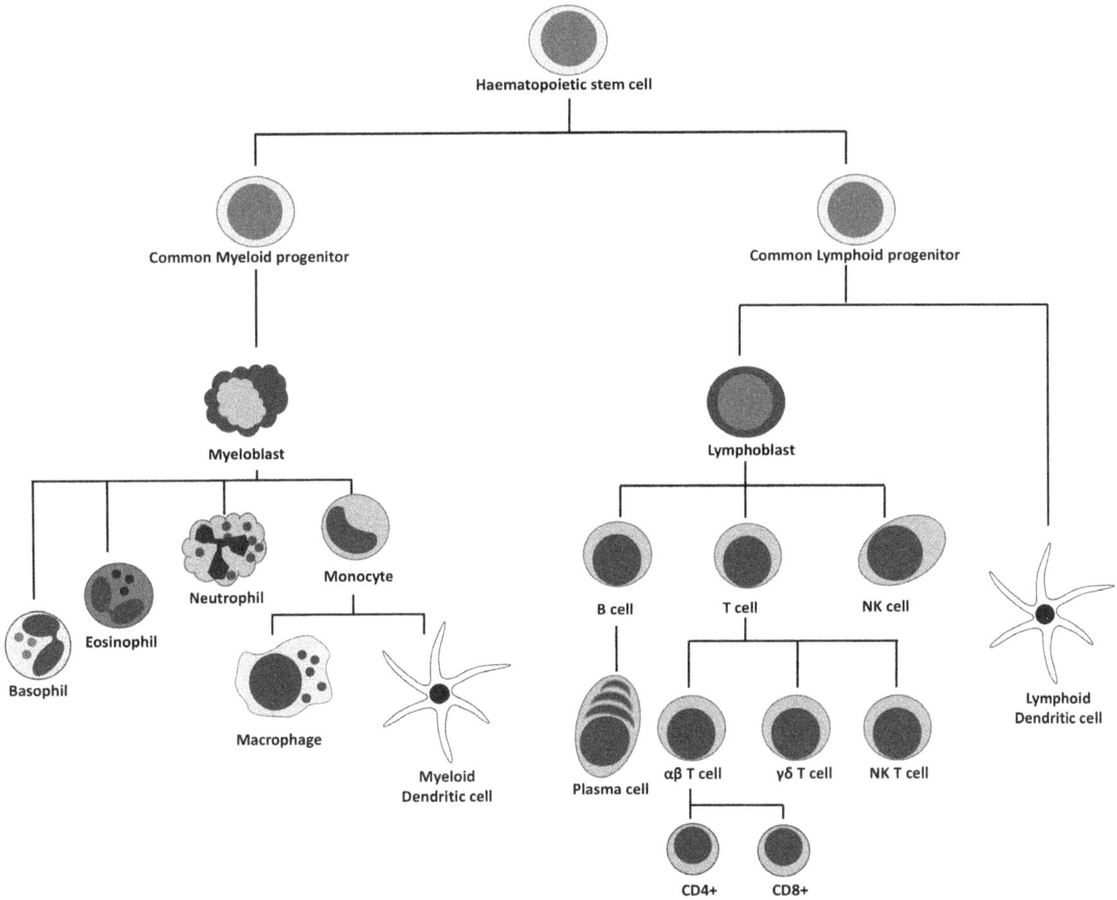

Figure 13.4 Immune cell lineage. Immune cells are derived from haematopoietic stem cells. Myeloid cells comprise the cellular component of the innate immune system. They contain microbicidal substances to directly kill microbes, and some act as phagocytes and antigen presenting cells. Lymphoid cells comprise the cellular component of the adaptive immune system. B cells mature in the bone marrow, T cells in the thymus and NK cells in the bone marrow, lymph nodes, spleen or thymus. Once activated, B cells differentiate into antibody-producing effector cells known as plasma cells. T cells can differentiate into αβ, γδ or NK T cells, which carry out a diverse range of effector functions, often by secreting cytokines to direct the immune response.

and present these antigen fragments – complexed with MHC/HLA (major histocompatibility complex/human leucocyte antigen) – to naive T cells, along with costimulatory signals, in order to trigger antigen-specific adaptive immunity.

Neutrophils

Neutrophils are circulating short-lived cells, and the most numerous leucocytes in human blood. They are identified by their multi-lobed nucleus, cytoplasmic granules and surface markers CD16 (human) and Ly6G (mouse). Their purpose is to identify, engulf and kill invading bacteria. In order to achieve this purpose, they are mobilized from bone marrow, adhere to vascular endothelial cells before squeezing their way out of the bloodstream, and are directed along chemokine concentration gradients in the surrounding infected tissue. Neutrophils are the earliest cells recruited to sites of infection or injury. Once present, they phagocytose bacteria. Opsonized bacteria, particularly IgG-opsonized bacteria, are more efficiently phagocytosed.

A single neutrophil may engulf many bacteria. Neutrophils are equipped with two major methods of killing internalized bacteria – firstly, production of reactive oxygen species (ROS) and secondly, release of packaged enzymes via degranulation (Fig. 13.5) (Rigby and DeLeo, 2012). Phagocytosis stimulates nicotinamide adenine dinucleotide phosphate

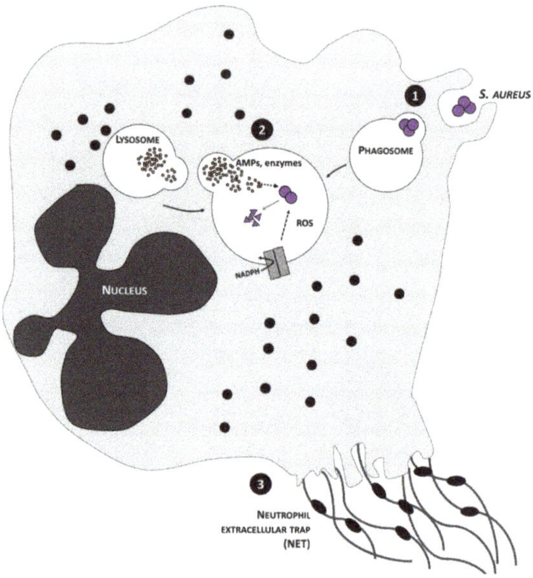

Figure 13.5 Phagocytosis and killing of *S. aureus*. Staphylococci are surrounded by plasma membrane and internalized in a phagosome which is then acidified by the formation of reactive oxygen species (1). Enzyme-containing lysosomes are fused with the phagosome, releasing the lysosomal contents to destroy the bacteria and cytoplasmic granules fuse with the phagosome to release additional enzymes and antimicrobial peptides (2). Dying neutrophils release a mesh of DNA fibres containing histones and antimicrobial proteins that trap, immobilize and kill bacteria (3).

(NADPH)-dependent oxidase to assemble at the phagosomal membrane and transfer electrons to oxygen to form superoxide anions. These dismutate to hydrogen peroxide and other reactive oxygen products. Collectively, these are called reactive oxygen species (ROS) and achieve pathogen killing. Neutrophils are also equipped with cytoplasmic granules. Primary (azurophilic) granules are most important for microbial killing and digestion, and, to that end, largely contain myeloperoxidase, proteolytic enzymes, and antimicrobial peptides. They fuse with the phagosome to release high concentrations of these toxic substances and acidify its contents. Secondary (specific) granules are chiefly concerned with amplifying the neutrophil's potential to respond to stimulation, as opposed to direct microbial killing (Nauseef and Clark, 2010). Optimal killing of ingested bacteria occurs when both reactive species and degranulation are activated. More recently, a novel non-phagocytic microbicidal mechanism has been described – neutrophil extracellular traps (NETs) (Fuchs *et al.*, 2007). In this process, neutrophils rupture their cell membrane in an ROS-dependent cell death process known as NETosis, releasing tangles of chromatin and granular antimicrobial proteins. This extracellular tangle binds microbes, including *S. aureus*, and kills by direct AMP activity.

Host tissues may suffer damage if the release of neutrophil cytotoxic substances is not controlled, hence these cells are tightly regulated. Neutrophils undergo spontaneous apoptosis – a form of programmed cell death following initiation of intracellular apoptotic signalling – as they age. This non-inflammatory process involves organized degradation of cellular components before they are phagocytosed and cleared by other cells. Apoptosis is accelerated following microbial phagocytosis, whereby apoptotic neutrophils full of bacteria are removed by macrophages. Apoptotic neutrophils display 'find me' and 'eat me' signals to attract macrophages (Bratton and Henson, 2011). Such elimination by macrophages augments clearance of infection and prevents inadvertent tissue damage (Grumann *et al.*, 2011). It also triggers macrophage production of anti-inflammatory mediators (e.g. TGF-β, IL-10) and reduces pro-inflammatory signalling, which furthers resolution of inflammation (Fadok *et al.*, 1998). The role of macrophages is further discussed below.

Neutrophil defence in *S. aureus* infection

An effective neutrophil response is considered essential in defence against *S. aureus*, and dead or dying neutrophils are the major component of the pus so characteristic of *S. aureus* infection (Chambers, 2013). However, a balanced, tightly regulated neutrophil response appears crucial, and too few or ironically too many of these cells may, in fact, hinder host defence. Experimental neutrophil depletion in mice increases mortality from *S. aureus* pneumonia (Robertson *et al.*, 2008), while conversely, exogenous G-CSF administration does not improve outcome in systemic sepsis (Alp *et al.*, 2004). Clinical data regarding the influence of neutrophil counts on patient mortality in *S. aureus* bloodstream infection is similarly equivocal, and bacteraemia is more common among non-neutropenic patients (Velasco *et al.*, 2006; Venditti *et al.*, 2003). Functional defects may, however, be more significant in human infection than absolute numbers.

The ideal scenario occurs when neutrophils recognize, home to, ingest and kill invading *S. aureus*, before apoptosing and being efficiently cleared by macrophages. However, a number of diseases are characterized by deficits in these steps. *S. aureus* normally induces numerous chemotactic factors *in vitro* and *in vivo*. Staphylococcal elements such as peptidoglycan, lipoteichoic acid (LTA), capsular polysaccharide and secreted toxins induce C5a, and incite IL-1, IL-6 and CXCL-8 (IL-8) production from monocytes and vascular endothelial cells to recruit neutrophils. Mice deficient in IL-1β have reduced neutrophil recruitment and consequently poorer bacterial clearance (Miller *et al.*, 2007). Endothelial cells infected by *S. aureus* normally express adhesion molecules to direct neutrophils into infected tissue. Inherited leucocyte adhesion deficiencies can result in severe systemic *S. aureus* infections. In these cases, adherence is markedly reduced such that neutrophils are trapped in the circulation, but cannot migrate out of the blood vessels to the site of infection (Etzioni, 2010). Post-phagocytosis, deficiencies in the ability to generate reactive oxygen species markedly impairs ability to eliminate *S. aureus*. Chronic granulomatous disease (CGD) is an immunodeficiency caused by mutations in genes encoding components of the NADPH oxidase complex. Neutrophils from these patients successfully phagocytose bacteria, but are unable to achieve killing of catalase-positive organisms. The rare Chédiak–Higashi syndrome is a multisystem disorder in which staphylococcal infections are also common. Neutrophils of these patients show both impaired chemotaxis and delayed release of lysosomal contents (Nauseef and Clark, 2010).

Macrophages

Monocytes leave the circulation to mature into macrophages, under the direction of CCL2. Macrophages are the major tissue-resident phagocyte populations, and are highly specialized depending on their location – microglia in the central nervous system, or Kupffer cells of the liver. Some are also resident in lymphoid tissue – and therefore they encounter invading bacteria early. Macrophages are richly endowed with PRRs and when they detect PAMPs, they initiate signal transduction of the initial inflammatory response and phagocytosis with lysosomal killing. As outlined above, macrophages remove apoptotic neutrophils full of bacteria, and are largely responsible for recruitment of neutrophils and other inflammatory cells via pro-inflammatory signalling, although they can also produce anti-inflammatory mediators. This scavenger function – whereby pathogens and dying cells are ingested – augments clearance of infection. On the other hand, macrophage anti-inflammatory activity promotes immune regulation and minimizes collateral tissue damage.

Classically activated (type 1/M1) macrophages are primed by IFNγ secreted by T helper cells and activated by PRR ligation. Activated macrophages are potent effector cells with enhanced ability to kill and degrade phagocytosed pathogens, and to present antigen. This killing is predominantly accomplished by the production of reactive nitrogen species (RNS) and superoxide. Although neutrophils are conventionally thought of as the primary protective phagocyte in *S. aureus* infection, human macrophages have been shown to be comparable in terms of their direct intracellular killing of *S. aureus* (Kubica *et al.*, 2008). These cells may also act as professional APCs, thus enhancing the adaptive immune response. Macrophages specialize in managing primarily intracellular pathogens, such as *Listeria* spp. and mycobacteria, but may also deal with extracellular pathogens to a lesser extent. They digest and process phagocytosed bacteria, and re-present fragments of these antigens – known as epitopes – together with MHC-II and co-stimulatory molecules on their cell surface. These are recognized by previously primed T lymphocytes, and cytokine signals from the activated macrophage also polarizes the phenotype of these antigen-specific T cells. Given that macrophages remain tissue-resident once activated, rather than migrating to lymph nodes, they do not prime naive T cells. Instead, their cytokine secretion – which directs local effector T lymphocyte and macrophage response – is thought to be more important overall than their antigen-presenting potential. Alternatively activated (type 2/M2) macrophages are primed by IL-4, and are inefficient at both microbial killing and antigen presentation. They fail to make RNS, inhibit T lymphocyte proliferation and produce anti-inflammatory cytokines (e.g. IL-10, IL-1 receptor antagonist). Alternatively activated macrophages are particularly important in immune regulation, healing and fibrinogenesis (Andrews and Sullivan, 2003). Some pathogens (e.g. *Neisseria gonorrhoeae*) polarize macrophages towards an alternatively activated phenotype which has been proposed to facilitate pathogen persistence (Ortiz *et al.*, 2015). Both types co-exist during infection and exhibit plasticity, such that polarized cells can convert between classic and alternative phenotypes, depending on their microenvironment.

Experimental depletion of macrophages in a murine *S. aureus* pneumonia model did not affect mortality (Robertson *et al.*, 2008). In contrast, depletion prior to systemic challenge resulted in 26–50% mortality, while all wild-type mice survived (Brown *et al.*, 2015). Inherited macrophage disorders in humans are just beginning to be fully characterized, and involve various defective intracellular killing mechanisms. They are overwhelmingly typified by susceptibility to mycobacterial infection, although *S. aureus* infections are also observed (Andrews and Sullivan, 2003).

Dendritic cells

Dendritic cells (DCs) are the primary antigen-presenting cells (APCs) of the immune system. They are the major driver of T cell activation – a new antigenic challenge cannot induce a T cell response without the intervention of these cells. Importantly, they also induce immune tolerance by disabling T cells reactive to self-antigens, thus minimizing autoimmune reactions (Shortman and Liu, 2002).

There are two major classes of DCs – conventional and plasmacytoid. Plasmacytoid DCs (pDCs) are primarily concerned with recognition of and response to viruses. These cells express surface viral entry receptors and intracellular PRRs such as TLR7 and TLR9 which sense viral RNA and DNA in the cytoplasm, and secrete large amounts of antiviral type I interferons (e.g. IFNα, IFNβ). These cells also display Fc receptors to bind immunoglobulin complexed with bacteria, viruses, nucleic acids or allergens. Immune-complex-mediated activation may result in both pro- and anti-inflammatory responses. Plasmacytoid DCs may also play a role in dampening the immune response by priming antigen-specific immunosuppressive T lymphocytes. Conventional dendritic cells (cDCs) phagocytose and process pathogens, but their purpose is not primarily to kill – rather to present antigen to and stimulate naive T lymphocytes in order to activate and direct adaptive immunity. They are laden with PRRs and highly phagocytic. Not only do they phagocytose using PRRs, complement, lectin and immunoglobulin receptors in the same way as neutrophils and macrophages, they also actively sample their external environment via macropinocytosis – a non-specific engulfing of surrounding fluid. In this way, they can detect antigens from virtually all types of pathogens – bacteria, fungi, viruses and parasites – even those which have evolved immune evasion mechanisms to avoid receptor-mediated phagocytosis. Immature human cDCs are abundant in tissues and have many long processes which sample their environment. They are equipped with endosomal compartments loaded with MHC class II molecules, where antigen fragments can be readily complexed with MHC for surface display. Once they are activated by recognition of a pathogen, they leave peripheral tissues to migrate through the lymphatics to regional draining lymph nodes. PRR ligation upregulates the chemokine receptor CCR7 on the DC surface. This is a 'homing' marker for lymphoid tissue, where it binds to its target chemokine CCL21 in the T cell zones. In the lymphoid organs they fully mature and present peptide antigens in association with MHC molecules to naive T cells (or intermediary DCs), along with co-stimulatory signals (e.g. CD40, CD86/80). Some dendritic cells present peptides of self–antigens complexed with MHC molecules in lymphoid tissue. However, because they have not undergone maturation in the same way as cells activated by infection, and do not present co-stimulatory signals, they induce unresponsiveness in naive T cells to self-antigens. Diverse mechanisms for antigen capture and processing in response to a variety of stimuli, along with a capacity for migration to lymphoid organs to programme T lymphocytes, enable DCs to serve as the crucial link between innate and adaptive immunity.

Role of DCs in *S. aureus* infection

DCs contribute to host defence during systemic *S. aureus* infection. An experimental murine model following intravenous inoculation of live bacteria showed that these cells are recruited to and activated in infected organs, secreting IL-6 and TNFα, as well as being the major producers of IL-12 (Schindler *et al.*, 2012). Depletion of conventional DCs exacerbated the severity of infection, with increased organ bacterial burden and accelerated mortality; whereas adoptive transfer of DCs or reconstitution of IL-12 improved control of infection. While DCs efficiently ingested and compartmentalized bacteria within their acidic phagosomes *in vitro*, they did not kill them – indeed, *S. aureus* could replicate within the cells (Schindler *et al.*, 2012). More recently, such killing of *S. aureus* by dendritic cells has been shown to be strain-dependent and to correlate with expression of the *agr* virulence determinant (O'Keeffe *et al.*, 2015). If not by direct killing, dendritic cells may contribute to clearance of bacteria via their indirect effects on neutrophils. Increased numbers of neutrophils were recruited to sites of infection after DC depletion but, while they phagocytosed efficiently, neutrophils were less effective at killing ingested bacteria. This suggests that DCs may contribute to neutrophil regulation in *S. aureus* infection – controlling influx and enhancing microbicidal activity. Plasmacytoid DCs are also activated by *S. aureus*, although their role in host defence is far from established. IgG–*S. aureus* immune complexes bind surface FcγRIIA receptors and initiate bacterial uptake into the cell. This antigen-specific response allows internalized, staphylococcal nucleic acid to trigger IFNα production via TLR9 signalling. Staphylococcal protein A (SpA) may also induce this response in a non-antigen-specific manner, without the requirement for pre-formed anti-staphylococcal IgG. The role of type I interferons in defence against *S. aureus* however, remains to be fully established (Bekeredjian-Ding *et al.*, 2014).

Adaptive immune response to bacterial infection

The adaptive immune system is activated by the innate system in order to effectively eliminate pathogens that have not been cleared by the more primitive defences. It also comprises humoral (e.g. antibodies) and cellular (e.g. lymphocytes) components. The lymphocytes of the adaptive immune system are derived from a common lymphoid progenitor in the bone marrow, and develop into natural killer (NKT), B or T lymphocytes. Because adaptive immunity can rearrange genes, it can construct receptors specific to individual pathogen elements, allowing them to be targeted with much greater precision and, critically, allowing the generation of immunological memory. The antigen-specific memory cells formed during primary infection are capable of persisting as memory cells, even in the absence of antigenic stimulation. They react more readily and effectively on re-infection to rapidly eliminate invading bacteria.

Adaptive immune recognition

Antigen specificity is at the core of the adaptive cellular immune response. In contrast to the limited number of germline-encoded PRRs of the innate system, the adaptive immune system can utilize a much larger repertoire of highly variable antigen-specific receptors. Membrane-bound immunoglobulin (IgM) serves as the B cell receptor (BCR). The variable (V) regions of the receptor are exposed on the cell surface (Fab fragments) and bind to the cognate antigen. The BCR recognizes soluble extracellular complete antigen, and triggers differentiation into immunoglobulin-secreting plasma cells which secrete antibody of the same specificity as the BCR. B cells readily recognize multiple secreted or surface *S. aureus* components – toxins (α-toxin, haemolysins, PVL and enterotoxins), cell wall proteins (clumping factors ClfA/B, fibronectin binding protein), non-protein antigens (capsular polysaccharides, lipoteichoic acid, peptidoglycan) – against which they generate antibodies.

In contrast, conventional (αβ) T cells only recognize peptide fragments of 'digested' antigen presented along with MHC molecules by other cells. This complex is presented to the T cell receptor (TCR). The TCR is structurally similar to the Fab fragment of an antibody molecule. The majority of TCRs comprise an α and β chain, although γ and δ chains can also form receptors. Unlike in the case of B cells, there are few well-established conventional *S. aureus* T cell antigens. Whole heat-inactivated *S. aureus* cells – comprising multiple potential antigens – can drive human Th17 cell responses *in vitro* (Zielinski *et al.*, 2012). A limited number of T cell epitopes have been defined as antigens which activate human or bovine CD4+ T lymphocytes in an MHC II-dependent manner. These include peptides derived from ClfA, staphylokinase, toxic shock syndrome toxin-1 (TSST-1), lipase, phosphodiesterase, IsdB, and putative lipoproteins (Brown *et al.*, 2015b; Lawrence *et al.*, 2012; Ramesh *et al.*, 1992; Warmerdam *et al.*, 2002). In one recent study it was shown that over 3% of the T cell population of healthy adults reacted to secreted or surface-bound staphylococcal proteins (Kolata *et al.*, 2015). Adult patients recovering from bloodstream infection also exhibited CD4+ T cell responses to the pooled antigens contained in heat-killed whole *S. aureus* (Brown *et al.*, 2015b). Recently, *Candida albicans*-derived agglutinin-like sequence 3 protein (Als3p) – a protein structurally homologous to the staphylococcal surface component clumping factor A – has been shown to induce T helper cell responses in mice and humans (Lin *et al.*, 2009b; Schmidt *et al.*, 2012). Non-peptide staphylococcal antigens may also be presented to T cells in conjunction with MHC II. Zwitterionic polysaccharides derived from *S. aureus* can activate CD4+ helper cells in this way thanks to their unique charge structure (Tzianabos *et al.*, 2001; Weidenmaier *et al.*, 2010).

Superantigens

A particular type of recognition by T cells deserves special mention, and can play a significant role in *S. aureus* disease. It has long been established that certain secreted staphylococcal toxins may act as 'superantigens'. These activate T lymphocytes in a pathological way, independent of antigen processing and MHC-dependent presentation. They can active 10–30% of all T lymphocytes in a non-antigen-specific manner by binding directly to the TCR and MHC II, linking them together to precipitate a massive release of pro-inflammatory cytokines that leads to 'toxic shock syndrome' (Ferens and Bohach, 2000) (see Chapter 5). However, the role of superantigens in human disease is not clear. While most *S. aureus* strains produce some superantigens *in vivo*, the incidence of toxic shock syndrome is rare in comparison to invasive bloodstream infection, and superantigen production does not appear to significantly influence clinical outcome (Desachy *et al.*, 2007; DeVries *et al.*, 2011).

Activation of the adaptive immune system

Lymphocytes require activation by innate cells in order to become fully functional, and B and T lymphocytes are activated in complimentary but different ways. Antigen recognition by B and T lymphocytes is also fundamentally different. While B lymphocytes can be activated directly by complete antigens, T lymphocytes only react to digested fragments of antigen – known as epitopes – re-presented on the surface of other cells.

B lymphocyte activation

B cells typically require two signals to become efficient antibody-producing cells. The first signal comes from antigen, and the second signal from T cells. Naive B cells leaving the bone marrow express IgD and IgM on their surface. When they encounter their specific antigen, it cross-links surface IgM, and signalling is carried out by two proteins – Igα and Igβ – associated with the BCR. Antigen-binding ultimately activates transcription factors such as NFκB, nuclear factor of activated T cells (NFAT), and activator protein 1 (AP-1). Antigen-mediated signalling can also be enhanced or inhibited by the participation of several BCR co-receptors (e.g. CD21, CD19, CD81, and inhibitory FcγRIIB). B cells that find a suitable antigen need a helper T cell – the second signal – to support their clonal expansion and switch to production of higher affinity immunoglobulin isotypes. To find their partner T cell, an activated B cell begins to express the chemokine receptor CCR7, which homes towards lymph node T cell zones. Activated T cells, in turn, are drawn towards chemokines produced in the B-cell zone. B cells can further increase their T cell contact by acting as APCs and presenting peptides to the very T cells they need. Activated T cells express CD40 ligand (CD40L, or CD154), which binds to CD40 on B cells and activates the B cell. In addition, the T cell secretes cytokines to polarize the Ig response. The differentiation of B cells and immunoglobulin production and function is further discussed in the 'B lymphocytes and *S. aureus*' section. Activated B cells divide to produce numerous progeny daughter cells. These will compete for antigen, or die via apoptosis in its absence. Survivors become plasma and memory cells.

T lymphocyte activation

The T cell response is also orchestrated by the innate immune system – processing and presenting antigens (signal 1), controlling the expression of costimulatory molecules (signal 2), and secreting appropriate cytokines and chemokines that direct the lymphocytes of the adaptive immune system (signal 3) (Fig. 13.6).

Antigen presentation (signal 1) is crucial to T cell activation and may be instigated by either

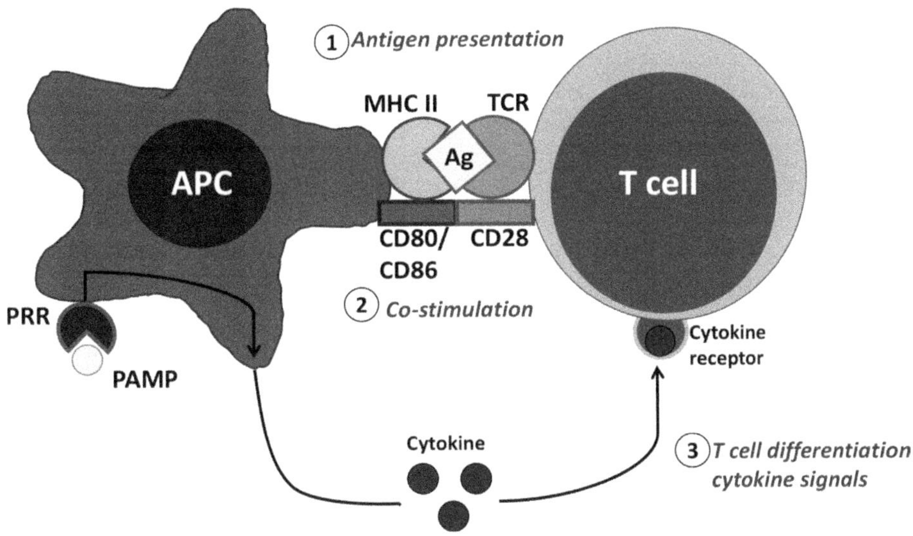

Figure 13.6 T lymphocyte activation. Complete activation of naive T lymphocytes requires 3 signals delivered by the APC. Signal 1 (antigen presentation) is delivered via the TCR when it binds a matching peptide–MHC complex. Signal 2 (co-stimulation) is an accessory signal (typically CD28:CD80/CD86 binding that induces clonal expansion. Signal 3 (T cell differentiation) comprises the cytokine signals delivered from APC to T cell that determine its differentiation into an effector cell subset (e.g. Th1, Th2, Tfh, Th17, Th22, CTL). Abbreviations: APC, antigen-presenting cell; TCR, T cell receptor; MHC, major histocompatibility complex; CD, cluster of differentiation; Th, CD4+ T helper cell; CTL, cytotoxic T lymphocyte.

professional or non-professional APCs. Non-professional APCs utilize MHC class I molecules on the cell surface to present antigen, whereas professional APCs use MHC class II molecules. MHC class I molecules are constitutively expressed by all nucleated cells, and report on intracellular events, such as infection, damage or malignant transformation. The antigen fragments they present, therefore, originate in their own cytosol. Once presented on the surface of infected or injured cells, peptide–MHC I complexes act as alarm 'flags' or 'self-destruct' signals to attract and direct CD8+ cytotoxic T cells (CTLs) to destroy such compromised cells. Expression of MHC II molecules is restricted to professional APCs, including DCs, macrophages and B lymphocytes. This pathway samples the extracellular environment and processes phagocytosed pathogens for presentation to CD4+ helper T cells (Th). The antigen fragments presented by MHC II originate from endocytosed pathogens, and are processed within endosomal compartments, rather than in the cytosol. A few pathways of MHC-linked antigen presentation are more complex. A subset of dendritic cells are also capable of cross presentation – the ability to load peptides derived from exogenous (extracellular) antigens onto MHC I molecules, instead of their usual MHC II (Neefjes *et al.*, 2011). This occurs when DCs take up infected cells or fragments, and their peptides somehow enter the cytosol. It allows DCs to display peptides to CD8+ T cells and initiate a cytotoxic response, instead of their usual T helper cell activation. A more recently described family of non-MHC antigen-presenting molecules are the CD1 family. These resemble MHC class I molecules, are highly expressed on APCs and present antigens to the TCR. In contrast to the peptide antigens presented by the MHC, however, CD1 proteins present lipid and glycolipid antigens.

Presentation of antigenic peptide–MHC complexes (signal 1) alone is not sufficient to fully activate the adaptive immune response. The APC must also deliver a co-stimulatory signal (signal 2) to promote survival and expansion of an antigen-specific T cell. In the absence of such co-stimulation, T cells that recognize antigen either fail to respond and die, or enter a state of unresponsiveness known as anergy. When DCs are stimulated by PRR ligation, synthesis of co-stimulatory molecules is prompted. The best-characterized of these are the B7 molecules – B7.1 (CD80) and B7.2 (CD86). The receptor for B7 molecules (CD28) is constitutively expressed on most T cells. The interaction between B7 (CD80/86) and CD28 induces IL-2 production by the T lymphocyte, generation of high-affinity IL-2 receptors, and promotes cell proliferation (Sharpe and Abbas, 2006). In this way, a single antigen-specific T cell can generate thousands of daughter clones within days of activation.

Following antigen presentation and co-stimulation, APCs – principally DCs – provide signal 3 in the form of cytokines or surface proteins. This determines which type of effector T cell population will be produced (Fig. 13.7) (Yamane and Paul, 2013). The subsets of T helper cells are described below ('Lymphocytes and *S. aureus*'), and are defined by hallmark cytokine production and unique subset-specific transcription factors. Naive T cells activated by antigen are polarized towards Th1 lineage by a number of pathways. IL-12 produced by innate immune cells upon PRR activation, along with NK cell-derived IFNγ activates signal transducer and activator of transcription 4 and 1 (STAT4, STAT1) and T box transcription factor (T-bet). IL-2 is also essential for Th1 activation and proliferation. In contrast, IL-4 activates STAT6 in naive T cells to direct cells towards a Th2 phenotype, which express GATA3 transcription factor. Follicular helper T (Tfh) cells develop under the influence of ICOS, IL-21, IL-23 and perhaps IL-6, which induce expression of B cell lymphoma 6 (Bcl6). Th17 cells differentiate under the influence of IL-6, IL-23 and TGFβ to express the transcription factor RORγt. Th22 maturation requires TNFα and IL-6 to induce expression of aryl hydrocarbon receptor (AHR). Treg polarization occurs in the context of TGFβ and STAT5 activation and results in transcription of Foxp3. More recently, however, it has become accepted that there is a degree of plasticity between T helper cell populations, such that

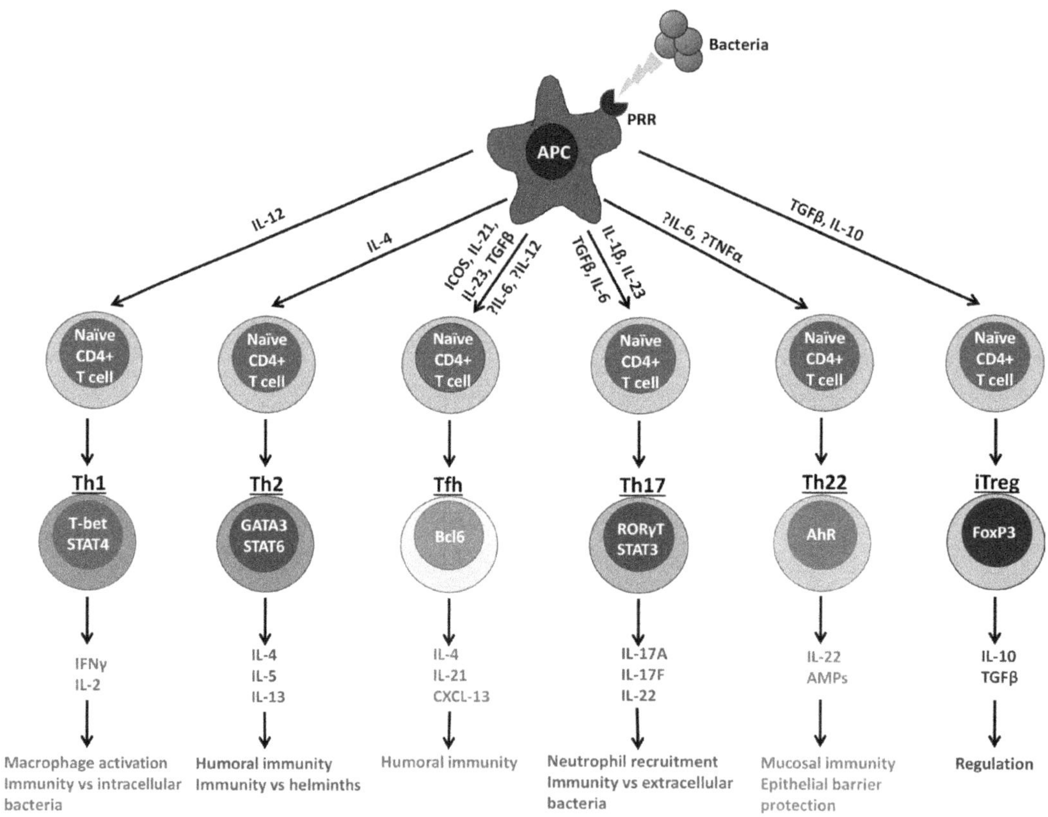

Figure 13.7 CD4+ T cell differentiation depends on the cytokine environment (Signal 3). Following antigen presentation by the APC naive CD4+ T-cell precursor cells can differentiate into subsets of effector T cells: T helper (Th) 1, Th2, Th17, or Th22 cells, T follicular helper cells (Tfh) or inducible T regulatory T cells (Tregs). Subset differentiation is driven by transcription factors and selective cytokines. Each subset carries out specialized immune functions during the inflammatory response to pathogens and/or resolution of inflammation.

once a naive cell has differentiated into a certain subset it is not necessarily committed for its whole lifespan. Under certain environmental conditions, it can convert to an alternative subset, modifying its function to mount the most appropriate response to a pathogen (O'Shea and Paul, 2010). In particular, Th17 cells tend to convert to Th1-like effector phenotypes.

Lymphocytes and *S. aureus*

B lymphocytes

B cells develop in the bone marrow and are identified by the surface markers CD20 and CD19 (non-plasma cells). They recognize complete extracellular soluble antigen via their surface B cell receptor (BCR). This triggers differentiation of B cells into antibody-secreting plasma cells. Antibodies are secreted immunoglobulin molecules formed to recognize microbial tertiary structures in order to neutralize (e.g. toxins) or opsonize them for disposal. They are the humoral constituent of the adaptive response. Antibodies are Y-shaped molecules with variable (V) Fab regions that bind to their cognate antigen, and constant (C) regions. C region structure defines the class of immunoglobulin – IgG, IgA, IgM, IgD or IgE – and is the region of an antibody which recruits other elements of the immune system. IgM has a relatively low affinity for antigen-binding, but aggregates pathogens to facilitate their clearance and activates complement more efficiently than any other isotype. Later, B cells make IgG, IgA, or IgE antibodies with the same antigen specificity by altering just their C region. To switch immunoglobulin isotype, B cells must receive signals from activated helper T cells that are specific for the same antigen. T lymphocytes secrete cytokines to polarize Ig production, (e.g. IFNγ-producing Th1 cells drive antibody class switching to high affinity IgG: IgG1, IgG3 in humans, or IgG2 in mice). IgG is the most abundant and longest-lived immunoglobulin type in serum and extracellular fluid. IgA is involved in mucosal protection and found in secretions. There, it blocks the binding of bacteria, viruses, and toxins to cell receptors, and cross-links pathogens to facilitate their clearance by mucociliary epithelium. IgE is mainly bound to mast cells and mediates hypersensitivity or allergic reactions. Antibodies mediate many activities against pathogens including virus and toxin neutralization, activation of the complement system, and mediation of antibody-dependent cellular cytotoxicity (ADCC). ADCC is carried out by monocyte-macrophages, NK cells, and neutrophils. ADCC facilitates killing of targets, such as virally infected cells or helminths that are too large to be ingested.

Primary IgM antibody responses are evident within 5–7 days of encountering a new antigen, and IgM is the first to appear. Secondary encounters with the same antigen result in a faster response producing greater amounts of primarily IgG antibody, which binds with much greater affinity and can engage Fc receptors on phagocytes to mediate phagocytosis of IgG-opsonized bacteria. Aside from antibody production, B cells can also produce pro-inflammatory cytokines (e.g. IL-1, IL-2, IL-4, IL-6, IFNγ, TNFα). Conversely, a subset of regulatory B cells (Bregs) produce anti-inflammatory IL-10. B cell-derived cytokines play a role in controlling polarization of CD4+ T cells (Fillatreau, 2012). Dysregulated B cells produce self-reactive antibodies that can result in autoimmunity.

S. aureus infection produces a measurable B cell response, but the protective efficacy of this response is far from established. In fact, there is little evidence that B cell deficiency states in humans or in mice result in greater incidence or severity of invasive *S. aureus* disease (Fowler and Proctor, 2014; Laupland *et al.*, 2003; Spellberg *et al.*, 2008). While mice are immunologically naive to the organism, absence of B cells does not worsen outcome in either sepsis or arthritis due to *S. aureus* (Gjertsson *et al.*, 2000). Humans, on the other hand, have been repeatedly exposed to *S. aureus* throughout life. This ensures that healthy children and adults normally have a multitude of circulating anti-*S. aureus* antibodies, although with significant inter-individual variability (Colque-Navarro *et al.*, 2010; Dryla *et al.*, 2005). These antibodies appear to be functional and potentiate opsonophagocytosis *in vitro*, although they may also bind to and thus neutralize each other's activity *in vivo* (Skurnik *et al.*, 2010; Van Strijp *et al.*, 2014). While limited clinical data suggest that higher levels of anti-toxin antibodies may attenuate disease in *S. aureus* BSI, other evidence contests this (Adhikari *et al.*, 2012; Jacobsson *et al.*, 2010; Yi *et al.*, 2012). Overall, it has proved difficult to consistently correlate the presence or titre of total anti-staphylococcal antibodies with improved clinical outcomes (Kolata *et al.*, 2011; Verkaik *et al.*, 2010a). Similarly, neither passive immunization, nor active immunization with vaccines that successfully produced robust humoral responses, have prevented or attenuated the course of infection in clinical trials (DeJonge *et al.*, 2007; Fowler *et al.*, 2013; Shinefield *et al.*, 2002; Weems *et al.*, 2006). Evaluating the role of B cells in defence against *S. aureus* is further complicated by the fact that staphylococcal protein A so effectively disables the humoral immune response during infection (see 'Immune evasion').

T lymphocytes

T cells initially develop in the thymus, and there undergo clonal selection, where self-reactive cells are deleted. T lymphocytes are categorized on the basis of T cell receptor (TCR) as αβ, γδ or NKT (natural killer) T cells. The enormous diversity of the TCR is generated by recombination of variable (V), diversity (D) and joining (J) gene segments, which leads to sequential rearrangement of each receptor in the thymus. Conventional T cells bear a TCR composed of an α and β glycoprotein chain, recognize antigen in association with MHC molecules displayed on the surface of APCs, and comprise the majority of circulating lymphocytes (Chevallier *et al.*, 2013). These αβ T cells are further categorized by co-receptor expression, e.g. CD4 or CD8. CD4+ and CD8+ T cells play distinct roles in defence against infection – CD4+ cells are helper cells which produce cytokines to promote CD8+, B cell and innate immune cell function; whereas CD8+ cells can directly kill compromised or infected cells. A range of T cell subsets can partake in the immune response to *S. aureus*, depending on the site and stage of infection, and these are discussed below.

CD8+ T cells

CD8+ T cells are also known as cytotoxic T lymphocytes (CTLs), and mediate their effector functions through production of cytokines such as IFNγ and TNFα and/or by cytolytic mechanisms such as release of cytotoxic granules. They achieve killing of compromised host cells, such as those infected by viruses, some bacteria or transformed by malignancy. CD8+ T cells interact with MHC I molecules on the surface of cells presenting cytosolic antigens (see 'T lymphocyte activation'). CTLs stimulate target cells to undergo apoptosis. CTL binding to its antigen on an infected cell prompts a controlled release of cytotoxic granules. These contain perforin, granzymes and granulysin. Perforin facilitates entry of granule contents into the target cell; granzymes activate caspases to trigger apoptosis; and human granulysin can directly kill mycobacteria. Each CTL can repeat this process for multiple target cells. Such a method ensures specific killing of compromised cells with minimal collateral damage or pathogen release. CD8+ T cells are also capable of modulating other effectors of the immune response via cytokine (e.g. IFNγ, TNFα) and chemokine secretion (Seder and Ahmed, 2003). Following clearance of antigen, CD8+ effector cells massively contract, with those that survive becoming memory cells.

CTLs are particularly effective in defence against viruses, but also play a role in some bacterial infections – particularly intracellular infections. This has been demonstrated using CD8+ and/or MHC I knockout mice which show increased susceptibility to mycobacteria and *Listeria* spp. (Kaufmann and Ladel, 1994). CD8+ T cells have not been convincingly demonstrated to play a significant role in defence against *S. aureus* infection. Adoptive transfer of vaccine-induced *S. aureus*-specific CD8+ T cells did not demonstrate protection in murine experiments (Joshi *et al.*, 2012), and a CTL response was not generated to any of four *S. aureus* antigens in a human vaccine study (Levy *et al.*, 2015).

CD4+ T cells

The primary function of CD4+ T cells is to provide immunological help to other cells of the adaptive and innate immune systems via the production and release of cytokines. CD4+ T cells can differentiate into several subsets, characterized by their expression of specific transcription factors, signal transduction pathways and cytokine secretion profiles (Fig. 13.7). The main roles of each CD4+ subset are briefly reviewed below, before evaluating their roles in *S. aureus* infection.

Th1 cells

Th1 cells produce IFNγ as their signature cytokine and are particularly important in controlling intracellular infections. IFNγ activates phagocyte killing of intracellular microbes, and antigen presentation to T lymphocytes. IFNγ stimulates macrophage microbicidal activity by enhancing phagocytosis and formation of reactive nitrogen and oxygen species. It also upregulates expression of MHC I and II molecules on a variety of cells, enhancing antigen presentation to T lymphocytes (Spellberg and Edwards, 2001). IFNγ also promotes immunoglobulin isotype switching to high-affinity IgG1 and IgG3 antibodies in humans, or homologous IgG2a antibodies in mice (Finkelman *et al.*, 1988; Spellberg and Edwards, 2001). These IFNγ-induced antibodies bind avidly to the Fc receptors of phagocytes, activating complement for opsonophagocytosis and augmenting cell-mediated killing. Th1 cells can also secrete tumour necrosis factor (TNF), lymphotoxin α, and IL-2, which further contribute to the immune response by a variety of mechanisms – inducing expression of adhesion molecules and chemokines, release of other pro-inflammatory cytokines by non-immune cells, and clonal expansion of lymphocytes (see Table 13.3). Aberrant Th1 responses are implicated in autoimmune disease (Spellberg and Edwards, 2001).

There is accumulating clinical and experimental evidence that Th1 cells play important roles in protection against *S. aureus*, particularly in systemic infection. Underlying conditions such as cancer, HIV infection, end-stage renal disease and diabetes mellitus present the greatest relative risk for *S. aureus* bloodstream infection among patients, and these pathologies are all associated with impaired Th1 cell responses (Chang and Shaio, 1995; Laupland *et al.*, 2008; Litjens *et al.*, 2008; Wiese *et al.*, 2013). Murine studies examining the role of Th1 cells in varied *S. aureus* infection models have generated conflicting results, but the differences seen have largely depended on the site of infection and stage of disease. A number of studies imply a protective role for Th1 cells in systemic *S. aureus* infection. Rapid mortality follows intravenous challenge in T cell knockout mice (Spellberg *et al.*, 2008). Following experimental intravenous *S. aureus* inoculation, Th1-skewed C57BL/6 mice showed greater survival, while administration of IFNγ increased survival,

Table 13.3 Major cytokines of the adaptive immune response

Family	Cytokine	Major cellular source	Response
Interleukins	IL-2	Th0, Th1, CD8+	Expansion and differentiation of activated T and B cells; suppression Th17 development; monocyte-macrophage activation; IFNγ and TNFα induction
	IL-4	Th2	B cell activation; IgG2/4 (human), IgG1 (mice) production; T cell expansion; inhibits macrophage activation
	IL-5	Th2	Eosinophil differentiation;?IgA production
	IL-6	Th17	T and B cell differentiation
	IL-10	Tregs, Th1	Inhibits Th1 differentiation and pro-inflammatory cytokine synthesis
	IL-17	Th17	Neutrophil recruitment; NK cell activation
	IL-22	Th22, Th17	Activates epithelial cell innate response; differentiation of Th17 cells
TNF family	TNFα	Th1, CTL, Th2	Macrophage activation (RNS); endothelial activation
	TNFβ (lymphotoxin α)	Th1, CTLs	Macrophage activation (RNS); neutrophil activation
Interferons	IFNγ	Th1, CTLs	macrophage activation; induces MHC expression; IgG1/3(human), IgG2a (mouse) class switching; inhibits viral replication
Growth factors	TGFβ	Tregs	Anti-inflammatory; inhibits cell growth; blocks activation of lymphocytes and M1 macrophages
	GM-CSF	Th1, CTL, Th2	Phagocyte production in bone marrow

Th0, naive T cell; Th, CD4+ T helper cell; Treg, regulatory T cell; IgG, immunoglobulin; IL, interleukin; RNS, reactive nitrogen species; TNF, tumour necrosis factor; NK, natural killer; IFN, interferon; GM-CSF, granulocyte–monocyte colony-stimulating factor; TGF, transforming growth factor.

phagocytosis and bacterial clearance from organs (von Kockritz-Blickwede *et al.*, 2008; Zhao *et al.*, 1998). Adoptive transfer of *S. aureus*-specific Th1 cells conferred significant protection from systemic infection in another study, leading to significantly accelerated clearance of bacteria and mediated – at least in part – by enhanced macrophage responses (Brown *et al.*, 2015b). Similarly, IFNγ-deficient mice and IFNγ receptor-deficient mice showed greater mortality following intravenous infection as compared to wild type (Lin *et al.*, 2009b; Zhao and Tarkowski, 1995). In this latter study, there was a later trend towards increased mortality in wild type survivors. Additionally, other studies also imply that IFNγ can be detrimental or ineffectual in certain situations. One showed that IFNγ-deficient mice had accelerated bacterial clearance, less tissue damage and improved survival in response to i.v. infection (Sasaki *et al.*, 2000). The role of IFNγ is less evident in localized (non-systemic) infection. In a cutaneous infection model, Th1-derived IFNγ promoted a neutrophil reservoir in which *S. aureus* could persist intracellularly at the wound site and delay bacterial clearance (McLoughlin *et al.*, 2008). In another study, IFNγ-receptor deficiency did not impair murine defence against *S. aureus* skin infection (Cho *et al.*, 2010). The overall impression is that IFNγ may have a differential role in the natural response to infection depending on the site and stage of infection. Th1 response may be most beneficial early in the course of systemic infection, and induction of Th1 immunity may aid clearance of extracellular *S. aureus* by promoting phagocytosis, and of intracellular staphylococci by activating phagocyte killing mechanisms.

Th2 cells

Th2 cells produce IL-4 as their signature cytokine and are important in inducing humoral immunity and controlling large extracellular pathogens, especially helminths and other parasites. IL-4 promotes B lymphocyte proliferation, antibody production and immunoglobulin isotype switching to IgG2, IgG4 and IgE in humans, or IgG1 and IgE in mice (Finkelman *et al.*, 1988; Spellberg and Edwards, 2001). IL-5 and IL-13 are also produced by Th2 cells, and stimulate production of eosinophils, as well as chemotaxis of eosinophils and basophils. Another essential role for Th2 cells is the suppression of Th1 responses and cell-mediated immunity. Th1 and Th2 lymphocytes antagonistically regulate one another. Th1-derived IFNγ directly suppresses IL-4 secretion, thus inhibiting differentiation of naive T cells into Th2 cells. Th2-derived IL-4 suppresses secretion of IL-12 and IFNγ, blocking Th1 development, inhibiting cell-mediated immunity and suppressing inflammatory cytokine production (Spellberg and Edwards, 2001). Importantly, dysregulation of Th2 cells is associated with atopic disease (Spellberg and Edwards, 2001). The adaptive immune system of atopic dermatitis patients is Th2-skewed and they are highly colonized with *S. aureus*. This Th2 skewing may inhibit Th1/Th17 response and increase this cohort's susceptibility to *S. aureus* carriage and infection (Machura *et al.*, 2008). Th2 cells play an important role in driving B cell antibody production in the context of *S. aureus* infection. The clinical importance of this is further discussed in 'B lymphocytes'.

Follicular T helper (Tfh) cells

Tfh cells are specialized providers of help to B cells, and are the real drivers behind humoral immunity. They trigger formation of germinal centres in secondary lymphoid organs. There, they select and support B cells to become plasma cells secreting high affinity antibodies, or long-lived memory B cells (Crotty, 2014). Induction of lasting humoral immune memory cannot occur without Tfh cells, and their response forms the basis of vaccines in current use. Tfh cells produce a number of 'help' signals to sustain B cells. CXCL-13 is chemotactic for B cells, while IL-4 and IL-21 promote B cell proliferation and a switch to high affinity IgG1 antibody production. Dysregulation of Tfh cells may be linked to autoimmune pathology (Craft, 2012).

Patients with AD-HIES are susceptible to recurrent *S. aureus* infections (discussed further in 'Th17 cells', below). They exhibit a severe reduction in memory B cells and titres of antigen-specific antibodies that has been potentially linked with impaired differentiation of Tfh cells (Ma and Deenick, 2014).

Th17 cells

Th17 cells produce IL-17 as their signature cytokine and are important in mucosal protection against extracellular bacterial and fungal infections (Crome *et al.*, 2010). IL-17 recruits neutrophils to sites of infection to promote clearance of these organisms.

It also indirectly enhances neutrophil production in bone marrow and killing activity. Th17 cells also secrete related cytokines including IL-21 and 22. In addition to their role in host defence, dysregulation of Th17 lymphocytes is associated with autoimmune diseases (Crome *et al.*, 2010).

Accumulating evidence suggests that an IL-17 response is particularly important for protection against *S. aureus* infection at mucosal sites, and may prevent haematogenous dissemination from these sites (Yeaman *et al.*, 2014). Gamma-delta T cells (discussed below) are the predominant source of IL-17 during murine mucosal infections (Cheng *et al.*, 2012; Cho *et al.*, 2010; Murphy *et al.*, 2014). However, CD4+ T cells (Th17) appear to be the primary human cellular source of IL-17, although other IL-17-producing lymphocytes have not yet been studied in human *S. aureus* infection. Mice deficient in IL-17A and IL-17F have increased susceptibility to opportunistic mucocutaneous *S. aureus* infection compared to wild type mice, but no difference is observed in systemic infection (Henningsson *et al.*, 2010; Ishigame *et al.*, 2009). Consistent with this, IL-17A$^{-/-}$ mice demonstrated impaired clearance of bacteria following local joint – but not systemic – infection (Henningsson *et al.*, 2010). Adoptive T cell transfer from *S. aureus*-experienced Th1-deficient mice lessened severity of *S. aureus* skin infection – an effect thought to be due to IL-17-producing T cells. Blocking of IL-17 abrogated this protection (Montgomery *et al.*, 2014). This mucocutaneous site-specific protective role for IL-17 during *S. aureus* infection parallels the clinical picture in patients with IL-17 deficiencies (Minegishi *et al.*, 2009). Patients with the rare autosomal dominant hyper-IgE syndrome (AD-HIES) are prone to recurrent staphylococcal skin and lung abscesses (Freeman and Holland, 2009). In these patients, mutations in signal transducer and activator of transcription 3 (STAT3) result in impaired Th17 cell development, but their CD4+ cells retain the ability to differentiate into other subsets (Holland *et al.*, 2007; Milner *et al.*, 2008). Interestingly, these patients are not more prone to *S. aureus* bloodstream infection, strengthening the hypothesis that Th17 response is critically important at skin and respiratory sites only (Minegishi *et al.*, 2009). A heterogeneous group of disorders may cause the syndrome of chronic mucocutaneous candidiasis (CMC), which is also characterized by recurrent and protracted *S. aureus* and *C. albicans* mucocutaneous infections (Kirkpatrick, 2001). A spectrum of inborn errors of IL-17 function – impaired differentiation of Th17 cells, deficient production of IL-17F, or absence of IL-17 receptor – have been described in CMC patients (Cypowyj *et al.*, 2012; Puel *et al.*, 2011).

Th22 cells

Th22 cells are a recently described lineage of T helper cells that home to, and are involved in, skin and mucosal immunity (Rutz *et al.*, 2013). They produce large amounts of IL-22 but little to no IL-17 or IFNγ. IL-22 acts on epithelial cells to induce antimicrobial peptides that may play an important role in protection against cutaneous infections caused by several yeast and bacterial species, including *S. aureus*. In one murine study, IL-22 deficiency did not affect staphylococcal skin lesion infection size (Myles *et al.*, 2013). However, in another study, blocking IL-22 worsened the severity of *S. aureus* skin lesions, resulting in greater abscess volume and necrosis (Yeaman *et al.*, 2014).

Regulatory T cells (Tregs)

Regulatory T (Treg) cells constitute less than 1% of circulating peripheral blood mononuclear cells and 5–10% of all CD4+ T cells. They are identified by the surface markers $CD4^{+}CD25^{+}$ (mouse) or $CD4^{+}CD25^{bright}CD127^{dim/-}$ (human) and the transcription factor Foxp3 (Chevallier *et al.*, 2013). Antigen-specific Tregs suppress immune responses to self-antigens to avoid autoimmunity and, critically, counteract pro-inflammatory responses during infection to prevent collateral damage to host tissues. However, Tregs can also be exploited by some pathogens to suppress protective immune responses, thereby prolonging their own survival. The major Treg effector cytokines are IL-10 and TGFβ. IL-10 suppresses phagocytosis, the oxidative burst, and intracellular killing by phagocytes, and inhibits antigen presentation (Belkaid, 2007). Tregs may be 'natural' – present in the host before pathogen exposure – or 'induced' in naive conventional CD4+ T cells by exposure to signals such as regulatory cytokines, immunosuppressive drugs or APCs conditioned by microbes. Clearance of infection – and subsequent recovery – depends upon the correct balance of effector and regulatory T cells. While an early pro-inflammatory effect is crucial for clearance, failure to control this inflammation damages host tissues and leads to increased mortality. Survivors of sepsis tend to have a robust early pro-inflammatory response, followed by a switch to immunomodulation (de Pablo *et al.*, 2014).

S. aureus can induce anti-inflammatory adaptive responses, either by activating regulatory T cells directly or by controlling the upstream innate cellular response that determines T cell fate. The type of APC presenting *S. aureus* antigens to naive T cells and the signalling pathways used, determines the character of the antigen-specific T cells produced. Following *in vitro* exposure to *S. aureus*, presentation by human dendritic cells drives expansion of pro-inflammatory Th1/Th17 cells, whereas presentation by monocytes results in an anti-inflammatory IL-10 response that inhibits Th1/17 expansion (Frodermann *et al.*, 2011; Peres *et al.*, 2015). Tregs were generated experimentally by exposure of neonatal CD4+ T cells to *S. aureus*, and childhood intestinal colonization with *S. aureus* is associated with increased numbers of circulating immunosuppressive Th2 and Treg cells (Johansson *et al.*, 2012; Rabe *et al.*, 2014). In fact, induction of such permissive immune responses may be a strategy used by *S. aureus* to suppress bacterial clearance and facilitate its persistence during colonization or biofilm formation. A clear consensus, however, has not been reached on the role played by Tregs during *S. aureus* infection.

Gamma-delta (γδ) T cells

A smaller subset of CD4+ T lymphocytes also develop in the thymus, but express TCRs composed of γ and δ chains. They are a minority population in the blood, but abundant at epithelial sites. Gamma-delta cells are at a crossroads between adaptive and innate cells, having some features of both. In addition to their γδ TCR and lymphocyte memory markers (e.g. CD45RO, CD27), they are also equipped with receptors more typical of innate cells – TLRs and NK receptors. They display a range of effector functions. Gamma-delta cells can achieve direct killing by various cytotoxicity mechanisms, and by release of antimicrobial molecules. They may also act as APCs, and secrete cytokines that influence the phenotype of αβ T and innate cells. Human γδ cells, in particular, also have a significant capacity for plasticity. Depending on the infecting organism and cytokine milieu, the same γδ cell subset can readily and rapidly assume Th1-, Th2-, Th17- or Treg-like effector functions (Bonneville *et al.*, 2010). In humans, there are three main γδ subsets – (Vγ9) Vδ1, Vδ2 and Vδ3. Vδ1 cells are primarily distributed in skin and mucosa, Vδ2 account for most of the peripheral blood γδ population, and Vδ3 are a minority ancillary population primarily found in gut and liver. They expand in several bacterial, viral and parasitic infections, and can be activated by PRR ligation, cytokine stimulation and/or specific antigens (Zheng *et al.*, 2013). Vδ1+ cells have been implicated in infection with cytomegalovirus and HIV; Vδ2+cells are associated with defence against influenza, herpes simplex, *Plasmodium falciparum* and *Mycobacterium tuberculosis* infections (Pang *et al.*, 2012; Zheng *et al.*, 2013). Significant Vδ2+ expansion at blood and CNS sites was also seen during bacterial meningitis in children. These antigen-experienced cells produced IL-17 and could recruit and activate neutrophils (Caccamo *et al.*, 2011). Some phospho-antigens can directly activate Vδ2+ cells via the TCR. For instance, the isoprenoid biosynthetic precursor (E)–4-hydroxy-3-methyl-but-2-enyl pyrophosphate (HMB-PP) present in several bacterial species induces IFNγ and TNFα production; and phospho-antigens are also found in some human tumours (Davey *et al.*, 2011; Pang *et al.*, 2012). Synthetic Vδ2 agonists exploit this pathway, and have been evaluated in clinical studies, with some success, as adjunctive treatments for hepatitis C, multiple myeloma and lymphoma (Bonneville *et al.*, 2010). Agonists for Vδ1+ or Vδ3+ subsets are less clear. More recently, evidence suggests that antigen-specific memory cells may persist, just like conventional αβ T cells (Murphy *et al.*, 2014; Sheridan *et al.*, 2013). Murine γδ subsets appear more diverse and are not directly analogous to human subsets; for instance, Vδ2+ cells are only found in primates (Pang *et al.*, 2012), which limits the utility of comparative studies. Dysfunction of γδ lymphocytes has been implicated in the pathogenesis of autoimmunity, allergy and malignancy (Pang *et al.*, 2012).

While *S. aureus* does not produce HMB-PP, it still appears to have the capacity to activate γδ cells, at least in murine systemic and localized infection models. In a *S. aureus* peritonitis model, accelerated bacterial clearance was associated with the expansion of γδ+ (Vγ4+) T cells exclusively producing IL-17, but no detectable IFNγ. Interestingly, recurrent exposure to *S. aureus* induced acquired γδ immunity – a population with memory surface markers and the capacity for IL-1-independent activation (Murphy *et al.*, 2014). Transfer of activated human Vδ2+ cells increased bacterial clearance and improved survival following *S. aureus* intravenous challenge in a humanized SCID mouse model (Wang *et al.*, 2001). This enhanced clearance of infection and improved survival effect was associated with IFNγ production. In mucocutaneous infection models, IL-17-producing γδ T cells are important in early containment and clearance of *S. aureus* (Cheng *et al.*, 2012; Molne *et al.*, 2003). IL-17 may also be differentially induced depending on the infecting *S. aureus* strain (Maher *et al.*, 2013). A role for γδ cells in human *S. aureus* infection is not yet established. Human Vδ2+ cells were not activated by supernatants from neutrophils that had phagocytosed *S. aureus*, and no study has attempted to characterize γδ+ T cells in human *S. aureus* infection (Davey *et al.*, 2011).

Summary of the adaptive immune response in *S. aureus* infection

An arbitrary distinction between innate and adaptive immune responses is somewhat artificial, as they act in concert rather than separately. Activation of the adaptive system is initiated by the innate system – achieving antigen-specificity and polarizing the type of adaptive response required. Adaptive responses have an important part to play in immunity. Extracellular *S. aureus* and its secreted virulence factors may be neutralized by antibody. Once *S. aureus* has entered host cells, either by phagocytosis or invasion, bacteria are sequestered out of the reach of antibodies, and can only be dealt with by the cellular elements of the immune system. Conventional T cells do not directly kill *S. aureus* but, via cytokines, can orchestrate many downstream effects on phagocytes to enhance their microbicidal activity. A balance between the cellular adaptive system's pro- and anti-inflammatory arms is also key, and an early pro- to anti-inflammatory cytokine ratio predicts improved survival. Further clinical studies are vital to advance the knowledge of immune responses during recovery from acute infections in real patients, particularly to correlate the potency of the various T cell subset responses over time with clinical outcomes. It is likely that the most effective immune response will be dynamic over the course of infection – to have the right cells in the right place at the right time for the right duration.

Immune regulation

While the immune response is critical to pathogen clearance and host survival, it must also be under tight control to limit dangerous inflammatory collateral damage. Pro-inflammatory responses efficiently clear infection, but unopposed inflammatory signal amplification may lead to a 'cytokine storm' with multi-organ failure ultimately resulting in host death (Cohen, 2002). The ability of immunity to self-regulate is an essential feature, the loss of which can lead to the development of allergic and autoimmune diseases. A highly complex and often redundant repertoire of immunosuppressive mechanisms involves regulatory cells within both the innate and adaptive arms of the immune system; specifically, tolerogenic DCs, alternatively activated (M2) macrophages, myeloid suppressor dendritic cells (MDSCs), regulatory T (Tregs) and B lymphocytes (Bregs) (Belkaid, 2007; Greten *et al.*, 2011; Iwata *et al.*, 2011). These cells exert their immunosuppressive functions through direct cell-contact mechanisms or through the production of anti-inflammatory cytokines such as IL-10 and TGFβ. Commensal organisms may 'prime' both innate and adaptive cells towards immune tolerance to facilitate their own survival (Surana and Kasper, 2014). However, an unopposed anti-inflammatory response may impair clearance and facilitate damage by pathogenic organisms. The balance between pro- and anti-inflammatory responses is likely to be tailored to the site of infection. This concept has been elegantly explored in *Streptococcus pneumoniae*. The relative proportions of pro- and anti-inflammatory T cells specific to *S. pneumoniae* in tonsillar lymphocytes, the site of colonization, are roughly equal (Zhang *et al.*, 2011). These local Tregs may reduce inflammation-related airway damage during infection, but facilitate the persistence of pneumococcal carriage. In contrast, peripheral blood pneumococcus-specific lymphocytes are overwhelmingly skewed in favour of Th1/Th17 cells (Aslam *et al.*, 2011; Pido-Lopez *et al.*, 2011). This compartment is primed towards a rapid pro-inflammatory response, critical for efficient bacterial clearance in invasive disease. Normal host response may therefore be tailored to a balanced tolerance at sites normally colonized by commensal organisms, and rapid attack at normally sterile sites. It is unknown if this applies to *S. aureus*.

In *S. aureus* infection, mice with increased pro-inflammatory NF-κB signalling and cytokine production have increased mortality from systemic bacterial challenge (Yan *et al*., 2014), suggesting that uncontrolled inflammatory responses may be counter-productive to the host. *S. aureus* can also stimulate the anti-inflammatory arm of the immune system, inhibiting pro-inflammatory cell expansion, and generating a permissive environment which impairs bacterial disposal (Frodermann *et al*., 2011). This anti-inflammatory bias may be exploited by the organism to facilitate persistent infection (e.g. deep-seated or biofilm-related infection). To achieve this, staphylococci induce immunoregulatory innate and adaptive cells. The organism polarizes macrophages towards the poorly-phagocytic M2 phenotype, and expands IL-10-producing MDSCs, both of which inhibit effector T-cell responses (Heim *et al*., 2014; Tebartz *et al*., 2015; Thurlow *et al*., 2011). In the 'Dendritic cells' section, we have seen how TLR2 signalling via different APCs elicits opposing T cell differentiation – *S. aureus*-stimulated DCs produce IL-12 and IL-23 which favour pro-inflammatory Th1/Th17 effector responses, whereas *S. aureus*-stimulated monocytes elicit a robust anti-inflammatory IL-10 response (Frodermann *et al*., 2011; Peres and Madrenas, 2013). Adaptive immunoregulatory cells are also generated by *S. aureus*, as intestinal colonization induces immunosuppressive T cells (Johansson *et al*., 2012).

The hallmark immunosuppressive cytokine IL-10 has been demonstrated to have a role in *S. aureus* infection – limiting local inflammatory tissue destruction. However, it is unclear which cell type(s) are the IL-10 dominant producer. During localized *S. aureus* infections, IL-10-deficient mice were used to demonstrate how IL-10 was beneficial in constraining excessive inflammation and tissue destruction, without affecting bacterial clearance (Gjertsson *et al*., 2002; Stenzel *et al*., 2005). During systemic infection, however, this lack of inflammation may not be so beneficial. One murine study examined how systemic *S. aureus* infection can induce a detrimental immune tolerance. Following intravenous challenge, total T lymphocytes were essential for controlling infection (preventing replication), but did not completely eradicate it. Infecting *S. aureus* switched to a small-colony variant (SCV) phenotype, and T cells underwent *in vivo* anergy over the course of persistent infection to become irreversibly hyporesponsive to staphylococcal antigens (Ziegler *et al*., 2011). The exact mechanism behind this immunosuppression was later demonstrated to be IL-10/TGFβ-independent and to require cell–cell proximity. Tregs contributed to a minor degree, but the predominant inhibitory effect on T cells was due to MDSC activation. Indeed, amplifying this immunosuppressive effect by adoptive transfer of MDSC increased bacterial burden further (Tebartz *et al*., 2015). The effect of IL-10 on bacterial clearance in human *S. aureus* bloodstream infection is unknown, but early elevated serum IL-10 levels are associated with higher mortality, although its cellular source is unknown. This picture may allow *S. aureus* to evade initial pro-inflammatory immune responses and persist, much like in the mouse model (Rose *et al*., 2012). Interestingly, monocytes are the predominant APC in circulation and, as we have seen, *S. aureus*-stimulated monocytes elicit a robust anti-inflammatory IL-10 response. A vigorous immunomodulatory response induced in the bloodstream may delay clearance and explain the tendency for *S. aureus* BSI to persist for much longer periods than other bacteraemias, and to establish deep-seated metastatic infection. Repeated or prolonged *S. aureus* exposure may imprint on the immune system to facilitate bacterial persistence during colonization or infection, and perhaps modulate the outcome later in the course of infection and may have important implications for a subsequent vaccine response (Wertheim *et al*., 2004). There is evidence that pre-existing memory immunosuppressive responses can contribute to impaired killing, increased clinical infection risk and vaccine failure with other organisms, including *Leishmania* spp. and mycobacteria (Carvalho *et al*., 2015; Roberts *et al*., 2005; Zanger *et al*., 2010). This occurs where prior exposure induces innate or adaptive immunomodulatory cells, skewing vaccine-induced immunity in favour of suppression and impeding protective Th1/Th17 responses (Chiacchio *et al*., 2009; Zanger *et al*., 2010). Whether or not a similar effect may occur for putative *S. aureus* vaccines is completely unknown.

Why immunity fails: immune evasion and immunodeficiency

Immune evasion

The bacterial arsenal that contributes to the pathogenesis of *S. aureus* infections is extensive (reviewed in Chapter 5), and includes surface-associated adhesins, capsular polysaccharides, secreted toxins and enzymes. Many of these products are thought of as virulence factors, a term derived from the Latin '*virulentus*' meaning 'full of poison'. While there is extensive knowledge about the activity of individual virulence factors *in vitro* and in animal models, their relative importance in the pathogenesis of clinical infection is, in fact, not well established. An array of secreted virulence enzymes facilitates tissue destruction and furthers the spread of infection, while toxins directly damage host cells. In contrast, several staphylococcal products are not directly 'poisonous' or toxic to host tissues, but instead, impair host pathways designed to clear infection. By utilizing such immune evasion mechanisms, staphylococci may resist opsonophagocytosis, invade and lyse host cells, persist in biofilm or small colony variant forms, and trigger the production of a cascade of host immunomodulating molecules. It may be that the primary danger of staphylococcal virulence factors lies in such 'anti-immunology' – collectively, their ability to inactivate front line host defences.

Evasion of innate immunity by *S. aureus*

Microbial invasion initially encounters innate defences, but *S. aureus* has evolved to overcome many of these, and often has multiple mechanisms to subvert the same host defence mechanism. Immune evasion may occur at various points – thwarting recognition, stunting the initial inflammatory response, inhibiting innate effector mechanisms (complement and phagocytes), resisting intracellular killing, and impeding activation of the adaptive immune response. *S. aureus* is readily equipped to achieve all of these (Foster, 2005; Thammavongsa *et al.*, 2015) (see Tables 13.4 to 13.6).

Table 13.4 Staphylococcal innate immune evasion molecules

Immunomodulatory factor	Human target	Effect
CHIPS	Binds C5a receptor	Blocks neutrophil chemotaxis
FPRLip	Binds FPRL-1	Blocks neutrophil chemotaxis
SSL5, SSL11	Inhibits binding to P-selectin on activated endothelial cells	Blocks neutrophil binding to endothelium
Eap/Map	Binds endothelial ICAM-1	Blocks neutrophil diapedesis
SCIN	Binds C4bC2a and C3bBb to block C5a formation	Blocks complement-mediated neutrophil chemotaxis
Cps 5, 8	Physically blocks phagocyte complement receptors	Inhibits complement-mediated phagocytosis
Efb	Blocks C3 deposition on bacterial surface	Inhibits complement-mediated phagocytosis
ClfA	Binds host fibrinogen to coat bacterial surface	Reduces opsonophagocytosis
Staphylokinase	Activates plasminogen to cleave C3b	Reverses opsonization to block phagocytosis
AdsA	Catalyses adenosine production	Reduces opsonophagocytosis
Hla (α-toxin)	Pore formation	Host cell lysis
Leukocidins	Pore formation	Leucocyte lysis

CHIPS, chemotaxis inhibitory protein of staphylococci; FPRL, formyl peptide receptor-like inhibitory protein; SSL, staphylococcal superantigen-like; Map, MHC class II analogue protein; Eap, extracellular adherence protein; ICAM, intercellular adhesion molecule; SCIN, *Staphylococcus* complement inhibitor; Cps, capsular polysaccharide; Efb, extracellular fibrinogen-binding protein; ClfA, clumping factor A; Sak, staphylokinase; AdsA, adenosine synthase A.

Table 13.5 Staphylococcal leukocidins and human cellular targets

	Leukocidins				
Host target cell	HlgAB	HlgCB	PVL	LukAB	LukED
Innate					
Neutrophils	+++	+++	+++	++	+++
Monocytes/macrophages	++	++	++	++	+++
Dendritic cells	-	-	-	+	++
NK cells	+	-	-	+	+
Adaptive					
CD4+ T cells	-	-	-	+	+++
CD8+ T cells	+	-	-	+	+

Hlg, haemolysin-γ; PVL, Panton–Valentine leucocidin; Luk, leucocidin.

Table 13.6 Staphylococcal factors that directly modulate the host adaptive immune response

Immunomodulatory factor	Prevalence in clinical strains	Reference	Human target	Effect
Protein A (Spa)	91%	Peacock *et al.* (2002)	(i) Fc region free IgG, (ii) B cell-surface IgM V_H3 region	(i) inhibits opsonophagocytosis, (ii) programmed B cell death
Superantigens (staphylococcal enterotoxins and toxic shock syndrome toxins)	73%	Desachy *et al.* (2007)	MHC Class II	Binds MHC Class II to the T cell receptor to cause initial activation followed by anergic unresponsiveness
MHC Class II analogue protein (Map)	94%	Peacock *et al.* (2002)	MHC peptide binding groove	(i) Reduced lymphocyte proliferation, (ii) Th2-predominant response
Leucotoxin ED	30–87%	Gravet *et al.* (1998), Morinaga *et al.* (2003)	CCR5 (T lymphocytes and macrophages)	Cell membrane pore formation causing cytotoxicity

Adapted from Brown *et al.* (2015).

Interference with phagocytosis

Phagocytosis may be blocked at various stages by *S. aureus*. Neutrophil chemotaxis is blocked by the chemotaxis inhibitory protein of staphylococci (CHIPS), which binds to C5a receptors on host leucocytes. Leucocyte adhesion to endothelial cells and subsequent exit from blood vessels is impeded by staphylococcal extracellular adherence protein (Eap), which binds to and blocks intercellular adhesion molecule-1 (ICAM-1) (Foster, 2005). Innate immune recognition can be thwarted by coating recognizable surface components in a capsule or protective matrix. Most clinical strains of *S. aureus* elaborate an amino-sugar capsular polysaccharide. Presence of this capsule or other surface polysaccharides reduces phagocytosis of bacteria by human neutrophils *in vitro*. While complement factors can still assemble on the cell surface beneath the capsule, it physically restricts access of phagocyte complement receptors to these opsonins (Nizet, 2007). Host fibrinogen binds to staphylococcal surface clumping factor A (ClfA), coating the bacteria and inhibiting their phagocytosis by human neutrophils (Higgins *et al.*, 2006). Biofilm phase *S. aureus* bacteria are also coated in positively charged polysaccharides (PIA; polysaccharide intercellular adhesion, also known as PNAG; poly-N-acetylglucosamine) which similarly avert phagocytosis and AMP activity (DeLeo *et al.*, 2009).

Assembly of C3 convertases on the bacterial surface is an essential step in complement-mediated clearance, and staphylococcal impairment of complement function results in effective resistance to phagocytosis. *S. aureus* secretes staphylococcal complement inhibitor (SCIN), which stabilizes complement complexes, preventing C3 cleavage. Similarly, extracellular fibrinogen-binding protein (Efb) blocks C3 deposition on the bacterial surface (Foster, 2005). Once C3b has been formed and has bound immunoglobulin to the bacterial surface, *S. aureus* even has the power to undo this opsonization. Staphylokinase acts on surface-bound plasminogen to activate plasmin, which can cleave bound C3b and IgG from the bacterial surface (Nizet, 2007).

Resistance to intracellular killing and autophagy

Ideally, phagocytes – primarily neutrophils – ingest and kill invading *S. aureus* before undergoing apoptosis and being efficiently cleared by macrophages. However, the organism subverts these processes at multiple steps, so that even when phagocytosis occurs, *S. aureus* may not necessarily succumb to intracellular pathogen killing. Indeed, up to 50% of internalized *S. aureus* can survive, and even divide, in the neutrophils of normal people by avoiding phagosomal killing, apoptosis and macrophage clearance, and later orchestrating cell lysis to ensure bacterial release (Greenlee-Wacker *et al.*, 2014). In fact, neutrophils in particular, may represent a double-edged sword, and be used in one of the organism's key strategies – to facilitate intracellular survival and trafficking of viable *S. aureus*.

S. aureus avoids intracellular killing by inhibiting the fusion of phagosomes with lysosomes, or by withstanding lysosomal degradation and later escaping from the phagosome into the cytoplasm. It is equipped with defences to resist phagosomal antimicrobial peptide and oxidative killing. Amino acid substitutions in staphylococcal surface polysaccharides attenuate the negative charge of the bacterial surface, reducing affinity for phagosomal cationic antimicrobial peptides. Secreted staphylokinase and aureolysin can also inactivate these AMPs. Due to a modification in its peptidoglycan structure, *S. aureus* is also completely resistant to lysozyme, a bactericidal protein that normally cleaves cell wall peptidoglycan to result in cell lysis (Foster, 2005). If these peptides and enzymes are survived, reactive oxygen species are the next hurdle. Adenosine synthase A (AdsA) is an enzyme in the staphylococcal cell wall that catalyses the formation of extracellular adenosine in order to impair intraphagocytic killing (Foster *et al.*, 2014). Adenosine is a potent immunosuppressive signalling molecule that interacts with transmembrane receptors to trigger anti-inflammatory signalling cascades (i.e. a switch from IL-1α to IL-10 production) and, critically, inhibition of neutrophil oxidative bursts. *S. aureus* mutants lacking AdsA are readily killed, whereas wild-type bacteria survive within neutrophils (Foster, 2009). *S. aureus* is also armed with catalase and superoxide dismutases to inactivate free radicals. Interestingly, the carotenoid 'golden' pigment that gives the organism its name scavenges free radicals and confers resistance to reactive species, improving survival within phagocytes. This survival results in increased pathology in an animal model, as experimental subcutaneous wounds were larger and contained more viable bacteria following challenge with a wild-type strain, than wounds challenged with a carotenoid-deficient strain (Liu *et al.*, 2005).

Having circumvented these killing mechanisms, staphylococci then orchestrate their escape into the cytoplasm and later from the cell. Phagosomal escape is dependent on the *agr* virulence determinant, and *agr*-deficient strains are more susceptible to lysosomal degradation (O'Keeffe *et al.*, 2015). A variety of *agr*-encoded toxins, lipases and phenol-soluble modulins have been implicated in escape from phagosomes. Release of viable bacteria from host cells occurs with or without cell death. This takes place by instigating pyronecrosis, by inhibiting apoptosis – upregulating 'do not eat me' (CD47) surface signalling – or by subverting autophagy (Greenlee-Wacker *et al.*, 2014). Staphylococci trapped inside phagosomes secrete toxins which result in phagosomal membranes becoming 'leaky'. Cathepsin and other lethal mediators are released into the cytoplasm, triggering inflammasome activation and caspase-1-mediated IL-1β formation. Pyroptosis (i.e. inflammatory cell death) and the release of bacteria is fuelled by *S. aureus* in this way, although the precise inflammasome stimulus is unclear (Fraunholz and Sinha, 2012). Autophagy is a normal homeostatic process that uses lysosomal machinery to degrade a cell's own used-up or damaged components, collected in double-layered autophagosomes. As bacteria within phagosomes damage and permeabilize these membranes by releasing destructive toxins, autophagy is triggered. Interestingly, *S. aureus* is capable of arresting the maturation of these autophagosomes, replicating within them, subsequently escaping into the cytoplasm and ultimately causing host cell lysis (Fraunholz and Sinha, 2012; Kobayashi *et al.*, 2010). This has previously been demonstrated *in vitro* in epithelial cells and neutrophils, and more recently in macrophages and dendritic cells, and is similarly *agr*-dependent (O'Keeffe *et al.*, 2015). Such intracellular persistence protects the organism from host defence mechanisms and therapeutic antimicrobial agents, and has been proposed as a major contributor to persistent or recrudescent infection. Bacteria may even migrate inside mobile neutrophils to distant sites, before lysing cells, to cause metastatic infection in the 'Trojan horse' hypothesis, although clinical evidence to support this is lacking (Thwaites and Gant, 2011).

While neutrophils are the primary population to phagocytose *S. aureus*, the other professional phagocytes (dendritic cells and monocytes) also readily internalize bacteria (O'Keeffe *et al.*, 2015). Experimental infection of human and murine monocyte-derived macrophages demonstrated that phagocytosed *S. aureus* can subvert autophagy and persist intracellularly in vacuoles for several days before escaping into the cytoplasm and causing host cell lysis. Bacterial survival and escape differs between strains and is dependent on a number of virulence factors, including *agr* and α-haemolysin (Kubica *et al.*, 2008; O'Keeffe *et al.*, 2015). Given that monocytes/macrophages are longer-lived than neutrophils, this may represent an important vehicle for dissemination of infection, and intracellular persistence is associated with higher visceral bacterial burdens (O'Keeffe *et al.*, 2015).

Staphylococci may also invade non-immune cells without the capacity for bacterial killing (Lowy, 1998). Persistence in such non-phagocytic cells challenges the conventional view of *S. aureus* as an extracellular pathogen. Most notably, this uptake occurs in vascular endothelial cells, where staphylococcal fibronectin-binding proteins interact with α5β1 integrins on the host cell surface and result in eventual internalization. IsdB has also been shown to bind to β-integrins to enter a variety of cell types (Foster *et al.*, 2014). This sequestration has been proposed as a major reason why *S. aureus* bacteraemia requires prolonged treatment in order to avoid relapse (Chong *et al.*, 2013). *In vitro*, bacteria may replicate in and lyse cells to release viable bacteria, although the translation of this into clinical infection is as yet unclear. Novel experimental therapy using anti-staphylococcal monoclonal antibodies complexed to a rifamycin that delivers and activates the antibiotic intracellularly results in superior bacterial killing (Lehar *et al.*, 2015).

Leukotoxins

A major weapon used by *S. aureus* to combat host phagocytosis is the production of a variety of potent cytotoxins that can directly target phagocytic cells. As described in Chapter 5, the cytotoxins include haemolysin-γ (Hlg), Panton–Valentine leucocidin (PVL), and other leukocidins (LukAB/GH, LukED, LukM/F-PV-like) that cause lysis by forming pores in host cell membranes. The multiplicity of toxins in various strains and their relative expression *in vivo* are highly variable, although *hlg* and *lukED* genes are present in virtually all clinical strains (Foster, 2005; Spaan *et al.*, 2013; Verkaik *et al.*, 2010b; Yoong and Torres, 2013).

The staphylococcal leukocidins are secreted as monomers which combine on the host cell membrane to form pores which lead to Ca^{2+} influx and osmotic cell lysis. Susceptibility to white blood cell destruction varies dramatically between species, due to differing affinities in binding to species-specific host receptors. This leads to some difficulty in translating the effects of toxins on immune cells from *in vivo* animal models. For instance, PVL and LukAB efficiently lyse human neutrophils, whereas murine neutrophils are relatively resistant. In humans, Hlg binds key host immune receptors highly expressed on neutrophils and monocytes: HlgAB targets human chemokine receptors (CXCR1, CXCR2 and CCR2), while HlgCB and PVL bind complement receptors (C5aR and C5L2); LukAB targets CD11b+ phagocytes (predominantly monocytes); and LukED targets phagocyte CXCR1 and CXCR2 as well as CCR5 (macrophages, DCs and T lymphocytes), which is unique in also targeting adaptive immune responses (Melehani *et al.*, 2015; Reyes-Robles *et al.*, 2013; Spaan *et al.*, 2014). In animal models, increasing toxin-mediated phagocyte lysis generally results in significantly increased lethality (Reyes-Robles *et al.*, 2013; Spaan *et al.*, 2014). Immune cell lysis results in activation of intracellular caspase-1 with associated release of multiple pro-inflammatory cytokines (e.g. IL-1β, IL-18) during pyroptosis/necrosis. The intracellular release of leucocidins can facilitate phagosomal escape and even mediate host cell killing after phagocytosis, independent of NLRP3 inflammasome activation (Melehani *et al.*, 2015). Leucocidins at lower (sublytic) concentrations can also act as immune-activating molecules. For example, PVL and Hlg can act on human neutrophils to promote secretion of pro-inflammatory cytokines, enhance phagocytosis and intracellular killing (Spaan *et al.*, 2014; Yoong and Torres, 2013).

PVL is present in a minority of clinical strains, but became infamous when initially associated with severe skin and soft tissue and necrotizing pneumonia community-acquired MRSA infections in the USA, in which isolates the gene was over-represented. However, the evidence definitively linking this toxin to more severe clinical outcomes is lacking, and *in vitro* experiments using PVL-deficient mutants is also inconclusive in demonstrating differences in pathogenesis or virulence (Watkins *et al.*, 2012). Recently, it has been shown that leucocidins may actually bind, inactivate and antagonize each other, so their relative contributions are complex (Yoong and Torres, 2015). During infection, of course, the host mounts a humoral response to oppose these toxins, such that higher antibody titres against PVL, α-toxin, δ-toxin, and enterotoxin C-1 are associated with a lower risk of sepsis (Adhikari *et al.*, 2012). Given that leucocidins overlap in host receptor targets (see Table 13.5), it is likely that even where a single toxin does contribute to disease, it is only one of a number of significant or redundant virulence factors that do so, and only in the context of a specific host response.

Biofilms and immune evasion

The particular problem of *S. aureus* biofilms deserves special mention. Device-related *S. aureus* biofilm infections pose a significant morbidity risk and usually require device removal to achieve cure. In a biofilm, the organism may be relatively sequestered from the host immune response, may skew this response to facilitate bacterial persistence, and in biofilm form is often impervious to conventional antibiotic therapy. The polymer scaffold of biofilms is relatively resistant to AMPs, and the complex three-dimensional structure means that many organisms are buried in a matrix such that PRR ligands are inaccessible to phagocytes. Activation of the pro-inflammatory IL-1β/MyD88/NFκB pathway also seems attenuated in biofilms. Preliminary data suggests that macrophages are the predominant phagocyte recruited to *S. aureus* biofilms, and that they are skewed towards an anti-inflammatory M2 phenotype, which impedes phagocytic activity, promotes IL-10 secretion and may be involved in fibrotic reactions around the biofilm (Hanke and Kielian, 2012). Intriguingly *mecA*-mediated beta-lactam resistance may markedly influence the phenotype of a biofilm – specifically, MSSA forms primarily carbohydrate (PIA/PNAG) biofilms, whereas PBP2a-expressing MRSA instead forms a proteinaceous (Fnbp) matrix. In a murine infection model, both forms colonized an implanted device equally, but MSSA disseminated systemically from its biofilm much more than the MRSA form, and was associated with significantly increased mortality (Pozzi *et al.*, 2012). Typical antibiotic therapy active against planktonic phase *S. aureus* has little activity against stationary phase biofilm bacteria, and novel therapies that inhibit or degrade the biofilm matrix are showing some promise in pre-clinical models (Kiedrowski and Horswill, 2011).

Evasion of adaptive immunity by *S. aureus*

Most pathogens have evolved mechanisms to evade innate immune defences, which are important for the organism to establish infection. For this reason, adaptive immunity is often employed to clear the pathogen and prevent re-infection. In the case of *S. aureus*, however, this host strategy is not entirely successful, as recurrent infections are commonplace. *S. aureus* is also equipped with mechanisms to thwart the development of an effective adaptive response, or to eliminate these responses once they have formed.

Protein A: subversion of B cell function

Antibodies recognizing staphylococcal cell surface components are present in the sera of most, if not all, individuals, but do not appear sufficient to confer protective immunity (Dryla *et al.*, 2005). Perhaps one of the major reasons behind this is that *S. aureus* is uniquely armed with protein A (SpA) to eliminate antibodies formed against the bacteria during infection. Cell wall-anchored SpA binds to antibodies' Fcγ domains, coating the bacteria in incorrectly orientated antibodies. This prevents opsonophagocytosis by inhibiting classical complement activation and makes Ig Fc portions inaccessible to phagocytes. SpA-deficient *S. aureus* mutants are more efficiently phagocytosed and killed, resulting in a lower bacterial burden, decreased abscess formation and increased survival following experimental infection (Foster, 2005). As well as its anti-opsonic effect, SpA is also released from the cell surface during bacterial growth, and can act as a potent immunomodulatory molecule. It binds the Fab region of B cell-anchored V_H3 IgM antibodies in a non-antigen-specific manner. This 'superantigen' interaction with up to 30% of B cells ultimately causes their clonal expansion, anergy and apoptosis. This significantly depletes the B cell pool, leading to a compromised humoral immune response against other *S. aureus* antigens (Goodyear and Silverman, 2003). These findings have recently been corroborated in a small number of human patients with *S. aureus* infection, where plasmablasts (the B cells responding during infection) were examined. Typically, over half of activated plasmablasts are directed against an infecting pathogen, and these go on to form the longer-lived memory B cell pool. Despite having measurable baseline serological responses to various staphylococcal antigens, however, plasmablasts activated during natural *S. aureus* infection were shown to be completely unresponsive to staphylococcal antigens other than SpA itself (Pauli *et al.*, 2014). This protein A immunodominance drove a V_H3-biased non-specific expansion and activation of B cells (without the ablation seen in mouse models). Finally, SpA may also modulate the immune response by stimulating IFNα secretion to support expansion of IL-10-producing regulatory B cells. This population will also oppose pro-inflammatory T cell responses (Bekeredjian-Ding *et al.*, 2014). All of these are surely valuable mechanisms to prevent the generation of protective anti-staphylococcal humoral and memory responses.

Antibody-mediated neutralization of SpA enables animals to mount effective humoral responses to many staphylococcal antigens, and is considered an essential goal for inclusion in future vaccines. However, wild-type SpA cannot be used as a vaccine antigen due to its B cell superantigen effect. A SpA mutant strain unable to capture immunoglobulins or cross-link B cell receptors to act as superantigens, but able to raise specific antibodies against protein A was recently engineered. This strain is susceptible to (opsono)phagocytosis while not suppressing adaptive immune responses, and is an intriguing vaccine candidate, as infection with it is partially protective against subsequent lethal *S. aureus* challenge (Falugi *et al.*, 2013).

Superantigen subversion of T cell function

S. aureus has up to 23 genes encoding for a family of proteins that can act as T cell superantigens (SAgs), and at least one of these is expressed by most circulating clinical isolates. Some of these are enterotoxins (the causative agents of food poisoning) and toxic shock syndrome toxins (TSST), which are responsible for the syndrome of the same name. A SAg can directly cross-link the Vβ region of a TCR with MHC II molecules without the T cell's specific antigen being involved in the process. These proteins can activate 20–30% of the T cell pool to proliferate in a polyclonal fashion, produce IFNγ and IL-2 and lead to staphylococcal toxic shock syndrome. Instigation of such a 'cytokine storm' leads to further downstream activation of innate cells, and this syndrome carries a mortality of up to 80%. This subverted activation of T cells also prevents normal development – via MHC Class II:TCR presentation of processed antigens – of true microbial antigen-specific effector and memory T cells (Lussow and MacDonald, 1994). This immune evasion occurs because superantigen stimulation induces a combination of clonal anergy (i.e. unresponsiveness to antigenic stimulation) and cytokine-mediated immunosuppression likely produced by regulatory T cells producing IL-10 and TGF-β (Taylor and Llewelyn, 2010). Repeated injection of mice with staphylococcal enterotoxin A (SEA) led to a prolonged state of T cell tolerance in which CD4+ cells showed blunted proliferation and produced little IL-2, IFNγ or IL-4 in response to subsequent *in vitro* stimulation (Miller *et al.*, 1999). While *in vivo* anergy has been shown during persistent *S. aureus* infection, this does not appear to occur in the same SAg-induced fashion (Ziegler *et al.*, 2011).

The primary function of staphylococcal SAgs may be to impair host immune responses in order to increase bacterial survival, but they may also have a role in disease pathogenesis. SAgs appear to induce endothelial inflammation and necrosis. Vascular endothelial cells produced IL-8 in response to SEC, which was associated with increased vegetation size, bacterial load and mortality in a rabbit model of endocarditis, while respiratory SAg challenge produced necrotizing pneumonia (Salgado-Pabón *et al.*, 2013; Strandberg *et al.*, 2010). Passive anti-SAg immunization protected against this. In humans, these superantigens incite a robust and neutralizing immune response. It is perhaps for this reason that despite the prevalence and transcription of multiple superantigens during clinical infections, toxic shock syndrome is much less common than *S. aureus* bacteraemia (Desachy *et al.*, 2007). High titres of anti-TSST-1 seem to be protective against staphylococcal toxic shock syndrome, in that the disease seems to occur in those lacking these antibodies (Stolz *et al.*, 1985). This may be a useful defensive adaptation, protecting against unopposed superantigen-provoked massive pro-inflammatory T cell activation, or even attenuating the severity of sepsis. While limited clinical data suggests that higher levels of anti-toxin antibodies may attenuate disease in invasive *S. aureus* infection without toxic shock, other evidence contests this suggestion (Adhikari *et al.*, 2012; Jacobsson *et al.*, 2010; Yi *et al.*, 2012). Novel therapeutic and vaccination strategies targeting superantigens and their binding sites could reduce *S. aureus*'s ability to evade the immune system, and have shown promise in early animal vaccine work (Lin and Peterson, 2010; Spaulding *et al.*, 2013).

Lymphocyte cytotoxin (LukED)

The majority of staphylococcal leucocidins target innate immune cells, but LukED can also attack adaptive cells and those that bridge innate and adaptive immunity. Its ligands are CXCR1, CXCR2 and CCR5. CXCR1/2 molecules are found on the surfaces of neutrophils, monocytes/macrophages, dendritic cells and cytotoxic CD8+ T lymphocytes, and act as IL-8 receptors. In contrast, CCR5 is best known as a co-receptor used for HIV entry into CD4+ T lymphocytes, but is also displayed on macrophages and dendritic cells.

LukED can kill CCR5+ cells, the majority of which are effector memory Th1/17 CD4+ T lymphocytes. This was elegantly demonstrated by showing the death of cells, including non-immune cells engineered to express this receptor, in the presence of this leucocidin. A *lukED*-deficient *S. aureus* mutant was unable to kill human lymphocytes *in vitro*, and use of the CCR5 antagonist maraviroc or of CCR5-deficient cells blocked LukED-dependent killing. *In vivo* infection showed increased macrophage and lymphocyte death, reduced bacterial clearance and markedly increased overall mortality in CCR5+ mice (Alonzo Iii *et al.*, 2013). This suggests that *S. aureus* can effectively eliminate the adaptive cell populations most implicated in resolution of an infection, as well as possibly preventing the development of useful immunological memory. This could be a significant reason behind the failure of recurrent infections to produce protective immunity (see Tables 13.5 and 13.6).

Blunting of Th1 lymphocyte response

The MHC class II analogue protein (Map) is a secreted staphylococcal component with strong homology to the peptide-binding groove of MHC II. It can thus bind T cell receptors and potentiate bacterial survival by modulating host immunity, more specifically, by attenuating cellular immunity. It reduces T cell proliferation and causes a shift away from Th1 towards Th2 responses. This is associated with delayed bacterial clearance and increased disease severity in a mouse model of septic arthritis and osteomyelitis (Lee *et al.*, 2002).

Summary of *S. aureus* immune evasion

It is clear that *S. aureus* is equipped with multiple strategies to subvert host immune recognition, inhibit chemotaxis and complement, block opsonization, neutralize antibodies, and kill or subvert the activities of phagocytes and lymphocytes. Inhibiting bacterial division with current antimicrobial therapy is not sufficient to reduce pathogenesis, eliminate mortality or prevent recurrent infections. Innovative approaches are now beginning to be explored. These include disabling bacterial virulence mechanisms to render it harmless, rather than eliminating the organism entirely, or targeting the host instead of the pathogen in order to somehow 'boost' immune responses during infection.

Host immunodeficiency and susceptibility to *S. aureus* infection

Immunodeficiency can be defined as a congenital or acquired quantitative or qualitative deficiency of some element(s) of the immune response. Certain immunodeficiencies are associated with increased acquisition or severity of *S. aureus* infection, and may give clues as to which arms of the immune system to target for future immunization and therapeutic strategies. While much research focus to date has been on pathogen attributes, including those that disable the immune system, there is compelling clinical evidence that the outcome of human *S. aureus* infection may, in fact, be more dependent on host rather than microbial factors (Greenberg *et al.*, 2014; Lesens *et al.*, 2003; van Hal *et al.*, 2012).

The impact of various immunodeficiencies can be teased out more readily in animal models, although there are caveats to consider before drawing direct parallels (see 'Challenges in *S. aureus* vaccine research', below). Deficiencies in individual cell types have been discussed in greater detail above (see 'Innate immune response in staphylococcal infection'), but will be briefly recapped here. In terms of innate immune responses, the overall theme is that intact phagocyte responses (i.e. recognition, chemotaxis and oxidative killing) appear essential. Experimental neutrophil depletion in mice increases mortality from *S. aureus* pneumonia, and neutrophil or macrophage dysfunction increases the severity of systemic and pulmonary infections (Köhler *et al.*, 2011; Robertson *et al.*, 2008; Yan *et al.*, 2014). Diminished phagosomal killing in gp91^{phox}-deficient (akin to NADPH oxidative burst reduction in human CGD) mice leads to increased mortality (Köhler *et al.*, 2011). Recently, macrophages have been definitively demonstrated to be crucial to anti-*S. aureus* immunity. Macrophage depletion prior to intraperitoneal *S. aureus* challenge resulted in 26–50% mortality, while all wild-type mice survived (Brown *et al.*, 2015a). While immunodeficiencies can exacerbate *S. aureus* infection, a regulated and balanced innate immune response is also crucial. Overactive NFκB activity is associated with the overproduction of innate inflammatory cytokines and contributes to the susceptibility to *S. aureus* systemic infection seen in A/J mice. The human homologues of several of the genes responsible for overactive inflammatory responses in A/J mice were also significantly expressed in humans with *S. aureus* bloodstream infection as compared with healthy controls (Yan *et al.*, 2014).

Murine studies have also illuminated the protective roles played by adaptive responses, and Th1 and Th17 cell subsets in particular, in *S. aureus* infection. In systemic infection, T cells appear crucial for survival, with rapid mortality following intravenous challenge in T cell knockout mice (Schmaler *et al.*, 2011; Spellberg *et al.*, 2008). Both IFNγ and IL-17 are produced during systemic infection, and deficiency in IFNγ (but not IL-17A) renders mice hypersusceptible to infection. A protective role for Th1 cells is also supported by the observation that Th1-skewed C57Bl/6 mice are naturally more resistant to *S. aureus* infection (Ishigame *et al.*, 2009; Lin *et al.*, 2009b; von Kockritz-Blickwede *et al.*, 2008). Induction of type I immunity may aid clearance of intracellular organisms in addition to promoting phagocytosis of extracellular *S. aureus*. Other *in vivo* studies examining the role of Th1 cells in varied *S. aureus* infection models have generated conflicting results (Brown *et al.*, 2015a). These differences have largely depended on the site of infection and stage of disease (Zhao and Tarkowski, 1995). It is likely that the nature of the infecting stain (Maher *et al.*, 2013) and the site of infection significantly impact upon the nature of the response elicited (McLoughlin *et al.*, 2006; Montgomery *et al.*, 2014).

In humans, genome-wide screening has failed to identify polymorphisms specifically associated with the risk of acquiring, or the severity of, *S. aureus* bloodstream infection (Nelson *et al.*, 2014). Rare but classical inborn errors in the T helper cell–phagocyte axis or in integumental integrity are associated with increased risk of *S. aureus* infection. The most crucial barrier against *S. aureus* infection is the skin, and a multitude of conditions that lead to non-intact skin predispose to infection. The best-characterized is atopic dermatitis (AD), the majority of sufferers from which are also persistent skin carriers of the organism, and prone to opportunistic skin infections during eczematous flares when skin integrity is lost. The skin of a subgroup of AD patients with particular filaggrin mutations exhibits inhibited antimicrobial peptide activity, and *S. aureus* shows increased growth and expression of adherence and immune evasion molecules under these conditions *in vitro* (Miajlovic *et al.*, 2010). Following skin breaches, phagocytes and T helper cells appear to be important for controlling staphylococcal infections, and their role is discussed in more detail in the 'Innate immune response in staphylococcal infection' section above. Briefly, impaired neutrophil chemotaxis (leucocyte adhesion deficiencies) or functional phagocyte deficiencies (e.g. chronic granulomatous disease, Chédiak–Higashi syndrome) lead to increased incidence of *S. aureus* infections, primarily at skin or respiratory sites, although occasionally invasive. Inborn errors of B cell function, or hypogammaglobulinaemia do not appear to increase the incidence or severity of *S. aureus* infection, and perhaps this can be explained by the action of protein A during staphylococcal infection, which induces a B cell paralysis in everyone (Fowler and Proctor, 2014). There is convincing evidence that CD4+ T lymphocytes are important in protection against *S. aureus*. Underlying conditions such as cancer, HIV infection, end-stage renal disease and diabetes mellitus present the greatest relative risk for *S. aureus* bloodstream infection among patients, and these pathologies are all associated with impaired Th1 cell responses (Chang and Shaio, 1995; Laupland *et al.*, 2008; Litjens *et al.*, 2008; Wiese *et al.*, 2013). Congenital errors in Th17 cells – as seen in the rare autosomal dominant hyper-IgE syndrome (AD-HIES) – have been implicated in cutaneous and mucosal immunity to *S. aureus* (discussed in 'Innate immune response in staphylococcal infection'). Despite a global systemic Th17 deficiency, these patients are not more prone to *S. aureus* bacteraemia, perhaps because human skin and respiratory epithelial cells require the synergistic stimuli of both Th17-derived and classical pro-inflammatory cytokines for enhanced production of antimicrobial peptides and neutrophil chemotoxins (Minegishi *et al.*, 2009). In these patients, $CD4^+$ T cells are documented as the cellular source of IL-17, whereas $\gamma\delta^+$ T cells have been identified as the predominant source of IL-17 during murine mucosal and systemic infections (Cheng *et al.*, 2012; Cho *et al.*, 2010).

While congenital immunodeficiencies are often more clearly defined, acquired immunodeficiencies overwhelmingly predominate in clinical practice, and are becoming more prevalent with the continuing expansion of advanced transplantation and widespread use of immunomodulatory therapies. Risk factors for invasive *S. aureus* infection bring up three recurrent themes – namely, patients with recurrent skin breaks, implanted devices that facilitate biofilm persistence and frequent healthcare contact (Laupland *et al.*, 2003). Male patients and those at the extremes of age experience most infections, while advanced age and cumulative co-morbidities are associated with the worst outcomes (Kaasch *et al.*, 2014; van Hal *et al.*, 2012). Haemo- and peritoneal dialysis populations have a risk 150–250 times greater than that of the general population, due to having indwelling tunnelled devices with persistent skin breaks, or repeated needling of arteriovenous fistulas. They are also likely to have multiple medical co-morbidities and may have depressed cell-mediated immune responses (Litjens *et al.*, 2008). HIV-positive patients are also commonly affected with recurrent staphylococcal infections. As well as a tendency towards repeated healthcare contact and problems with skin integrity, HIV is characterized by continuous loss of CD4+ cell numbers and function (Cummins and Badley, 2010). Other high risk patient groups (e.g. transplant, cancer, alcohol dependence, diabetes mellitus) have similar themes of biofilm-friendly devices and/or impaired cell-mediated immunity in common (Laupland *et al.*, 2003).

The biggest host risk factor for *S. aureus* infection is undoubtedly persistent carriage of the bacterium, and the majority of *S. aureus* infections are caused by invasion of the usually harmless endogenous colonizing strain (Kluytmans *et al.*, 1997; von Eiff *et al.*, 2001; Wertheim *et al.*, 2004). Certain patient populations tend to have higher rates of colonization than healthy adults. Almost all adult patients with atopic dermatitis (AD) are *S. aureus* nares and/or skin carriers (Breuer *et al.*, 2002). Other cohorts with recurrent skin breaches also have higher carriage rates, including insulin-dependent diabetics, renal replacement therapy patients and intravenous drug users (Kluytmans *et al.*, 1997). The host factors that determine bacterial carriage are unknown, although certain host immunity genes have been associated with carriage. Polymorphisms in genes encoding for IL-4, mannose-binding lectin, TLR2, C-reactive protein and some defensin genes have all been linked to nasal carriage (Brown *et al.*, 2014).

Immunological memory and protective immunity

Upon surviving an initial infection, the adaptive immune system can acquire the ability to respond more rapidly, specifically and effectively to that pathogen when it is next encountered. This capacity forms the basis of immunological memory.

Upon encountering an antigen for the first time, naive lymphocytes become activated, differentiate into effector cells and clonally expand. These effector cells mediate a range of protective functions (e.g. antibody formation, cytokine helper responses or cytotoxicity) directed against that pathogen, culminating in efficient microbial clearance. Crucially, after this primary immune response, when antigen is absent, the population of antigen-specific adaptive immune cells contracts, with a small number (<10%) persisting in both T cell and B cell pools as memory lymphocytes. During reinfection, these pathogen-specific memory lymphocytes mount a secondary response, with enhanced and immediate protective effects. Memory responses are characterized by more rapid kinetics and greatly enhanced magnitude for a number of reasons. Firstly, there is a greater frequency of antigen-specific cells present in the memory pool and they are present in the circulation, as opposed to solely in lymphoid organs, ready to clonally expand and traffic directly to sites of inflammation. Secondly, memory cells show increased responsiveness to antigen stimulation and are considerably less dependent on co-stimulation by accessory cells and cytokines to become activated and expand (Croft *et al.*, 1994). Thirdly, they are equipped to immediately perform their final effector functions – specifically, antibody production, cytotoxicity or cytokine production.

The capacity of the adaptive immune system to form memory is exploited in immunization, where an antigen-specific immune response leading to the development of memory is 'artificially' induced in the host, priming the immune system for subsequent challenge. In the event of later exposure to the pathogen, a vaccine-induced secondary immune response can be rapidly mounted to attenuate or prevent establishment of infection. More recently, the quality of the memory response generated and its ability to confer protection has garnered some attention, particularly in infections not easily prevented by the production of neutralizing antibodies. Some of the T lymphocytes generated after initial antigen exposure will become central memory cells (T_{CM}) identifiable by the surface memory marker CD45RO and the lymphoid homing marker CCR7. These cells have both long-term memory potential and polyfunctional effector potential in that they can secrete several different cytokines. In terms of generating a durable and effective T cell response, these are the most productive target cells in which to create an antigen-specific response (Seder *et al.*, 2008). Differing routes of antigen exposure can also determine the character of T cell response generated (Pepper *et al.*, 2010). Finally, a growing body of evidence suggests that cells of the innate immune system – primarily monocytes – can undergo epigenetic reprogramming and become 'trained' for enhanced effector responses (Netea, 2013). Innate immune responses are also critical in resistance to infection, and – although not currently targeted in immunization strategies – should not be overlooked in considering protective immunity in future.

While the accelerated and heightened actions of memory immune responses often correlate with avoidance of repeated infections, their presence alone does not always guarantee protection against re-infection. This is well-illustrated in the case of *S. aureus*. Despite the presence of detectable memory T and B cell responses in most of the population – both in health and during infection – recurrent *S. aureus* infections are common (Brown *et al.*, 2015; Dryla *et al.*, 2005; Kolata *et al.*, 2015; Wiese *et al.*, 2013). Notwithstanding suggestions that certain immune elements are vital in animal protection, the absolute or relative contribution of human memory responses to protective immunity remains unclear. The critical immune response required to ablate pathogenesis in *S. aureus* infection, and to result in sterile protection on subsequent challenge is unknown, and these challenges are further discussed in the 'Anti-*S. aureus* vaccines' section.

Anti-*S. aureus* vaccines

In addition to a huge burden of skin and soft tissue infections, *S. aureus* continues to commonly cause devastating and life-threatening diseases. Even with advanced medical care, the 30-day mortality associated with *S. aureus* bloodstream infection (~30%) exceeds that of most other acute medical presentations including myocardial infarction, pulmonary embolism and stroke (Melzer and Welch, 2013). Unlike the sharp declines the incidences of vaccine-preventable diseases have seen, the incidence of invasive *S. aureus* disease is, in fact, increasing, and an effective preventative vaccine remains elusive (Gagliotti *et al.*, 2011). A vaccine of even relatively limited efficacy (≤ 10%) would significantly decrease the incidence of invasive disease and be extremely cost-effective in high risk populations (Lee *et al.*, 2010a,b; Song *et al.*, 2012). There is an urgent additional factor driving vaccine development. *S. aureus* has developed resistance to every licensed anti-staphylococcal antibiotic to date, and the outlook for future therapy is discouraging given the lack of new classes of antimicrobial agents in the drug discovery pipeline (Alm *et al.*, 2014; Stryjewski and Corey, 2014). Pursuing the development of novel antimicrobials is a lengthy and costly process resulting in therapies with a finite period of utility, limited by the rapid rate of developing resistance. This approach leaves us in a race we can never win. As we enter the post-antibiotic era, it is evident that a strategy to prevent disease is clearly preferable to chasing diminishing curative options.

The scientific challenges to *S. aureus* vaccine development are daunting. The strategy of exposing healthy individuals to a modified pathogen or parts thereof to induce long-lived and protective immunological memory has been successful in reducing the incidence of, or even eliminating, several life-threatening infectious diseases. Replication of this strategy has failed so far in the case of *S. aureus*. Despite extensive knowledge about attributes and *in vitro* behaviour of the bacteria, our knowledge of *in vivo* pathogenic attributes in humans, and of the successful immune response to prevent or overcome infection is scant. Therefore, deciding which antigen to target or which type of immunity to induce in vaccination strategies is incredibly challenging. The apparent solution is that a vaccine with multiple antigens, targeting both cellular and humoral immunity and neutralizing toxins is required; although it is unknown if any of these things would be truly protective in humans. Few candidate vaccines have reached the stage of clinical trials, and thus far all aimed to produce protective antibodies. Despite showing promise in animal challenge models, none have shown efficacy in humans. Particular challenges in vaccine development and *S. aureus* immunization studies are discussed further below.

Challenges in *S. aureus* vaccine research

Two critical and extraordinary issues deserve further discussion with regards to the difficulties encountered in trying to identify correlates of immunity in a human population – the difficulty of translational models and the colonization effect.

Translational difficulties

S. aureus is a species-restricted pathogen, causing natural infections in a limited number of mammals – including humans, cattle, rabbits, pigs, horses, dogs, cats and poultry (see Chapters 2 and 3). Having evolved as one of the most common human bacterial pathogens for millennia, the organism unsurprisingly exhibits some human-adapted machinery. *S. aureus* has a higher affinity for human haemoglobin (Hb) than that of other species, and bacterial growth is enhanced in its presence over that of murine Hb. Staphylococcal iron surface determinant B (IsdB) scavenges iron from human haemoglobin, and mice expressing human Hb are more susceptible to systemic infection (Pishchany *et al.*, 2010). Many staphylococcal immune evasion factors also specifically target human immunity (Lowy, 2011). Despite the fact that mice have been used as an immunological model for infection, they are quite different to species naturally infected or colonized with *S. aureus*. The predominant murine leucocytes are lymphocytes, which account for ~60% of white cells, while neutrophils make up ~30% (Nemzek *et al.*, 2001). In humans, this ratio is reversed and the most numerous leucocytes are neutrophils, which may give a different flavour to the principal immune responses seen during infection. A variety of mouse genetic backgrounds have also been used in animal studies – some with Th1-skewed and relative Th2/antibody deficiency, and some with the opposite – which further muddy the interpretation of what their findings would mean for a balanced human immune system (von Kockritz-Blickwede *et al.*, 2008). Immune cell receptors and cytokine dominance during pathogen exposures may also differ significantly in mice (Mestas and Hughes, 2004). Several receptors or antibody subtypes have no human analogue, and co-stimulation for activation may also differ.

Differences in MHC expression further challenge translational immunology work. T cell antigenic determinants (epitopes) are presented to lymphocytes on the surface of APCs bound to MHC molecules. The diversity of MHC (or HLA) types in the human population is a potential obstacle for the study of epitope-based vaccines, as hundreds of different alleles exist. Inbred mice tend to have fewer MHC II polymorphisms than humans and other mammals naturally infected by *S. aureus* (Lawrence *et al.*, 2012). In recent years, HLA transgenic mice have been used experimentally in an effort to overcome this difficulty (Akkina, 2013). It is clear that some of these interspecies immunological differences influence the outcome of infection. One study compared the outcome of systemic *S. aureus* infection in 'humanized' (engrafted with human haematopoietic stem cells) and wild type mice. Humanized mice exhibited greater mortality, morbidity and lymphocyte death than wild type mice infected with the same bacteria (Knop *et al.*, 2015). The potential relevance of inter-species immunological differences was dramatically seen in a clinical trial of a CD28 agonist monoclonal antibody, and is a cautionary tale (Suntharalingam *et al.*, 2006). Despite demonstrating safety in *in vivo* rodent, primate and *in vitro* human cell models, all human volunteers experienced a life-threatening pro-inflammatory 'cytokine storm'. These effects were due to the highly experienced immune system of adult humans (in contrast to the more naive immune system of animals raised in relatively pathogen-free environments). The large number of differentiated effector memory T lymphocytes distributed in adult human tissues, and poised for cytokine release on antigen encounter with minimal or no co-stimulation were overwhelmed by CD28 stimulation in this study. Differences in antigen exposures, memory lymphocyte character and expression of CD28 between species meant this was not predicted by pre-clinical work (Hunig, 2012). These dissimilarities might offer some explanation as to why many new therapeutic strategies with promising results in animal studies have failed in clinical trials.

The efficacy of protection seen in animal models may not translate well into clinical medicine either. Unlike humans, mice require large bacterial inocula to establish *S. aureus* infection, and many animal vaccine studies define efficacy of their interventions as a reduction in quantitative bacteraemia or organ bacterial load, rather than sterile protection against infection. In humans, however, a threshold of 'tolerable' or 'safe' *S. aureus* bacteraemia has not been found. In fact, the number of microorganisms present in the blood is typically less than 10 CFU/ml, and even high-burden endovascular infections tend to have counts less than 500 CFU/ml (Beeson *et al.*, 1945; Reimer *et al.*, 1997). The reduction in organ bacterial loads reported as success in vaccinated animal studies may thus be completely irrelevant for human infections (Beeson *et al.*, 1945; Haimi-Cohen *et al.*, 2002; Khatib *et al.*, 2005). Nonetheless, it is tempting to speculate that the significant reductions seen in mice may equate to undetectable or subclinical infection, as delayed clearance of bloodstream infection is known to lead to inferior clinical outcomes (Fowler *et al.*, 2003; Kim *et al.*, 2010).

The colonisation issue

An additional major problem with using mice as a model for *S. aureus* infections is that mice are not natural carriers of *S. aureus*. This status of *S. aureus* as a commensal organism of a significant proportion of the human population complicates vaccine design and testing. Regardless of their colonization status at the time of enrolment in a clinical study, unlike their laboratory animal counterparts, human participants are certainly not immunologically naive to *S. aureus*. The primary *S. aureus* reservoir in humans is the anterior nares, although colonization at other anatomical sites is common. Approximately 20% of individuals are persistently colonized with *S. aureus*, and the remainder are either never or only intermittently colonized (Brown *et al.*, 2014). Defining the differences between carrier and non-carrier groups has proved difficult. They are most likely multifactorial; with differences in host genetics and epithelial cell molecules, differences in co-existing microbiome, and perhaps differences in *S. aureus* that make some strains superior colonisers. While nasal carriage of *S. aureus* is strongly associated with infection, the vast majority of carriers will never develop invasive disease. The influence of colonization on subsequent infection or on vaccination has had limited examination, despite virtually the entire population having evidence of *S. aureus*-specific immune memory (Dryla *et al.*, 2005; Kolata *et al.*, 2015). Initial pathogen exposure may prime the immune system and alter its response to subsequent bacterial encounters. Repeated or prolonged staphylococcal exposures have been proposed to modulate T cell responses and improve outcome in systemic *S. aureus* infection in both mice and humans (Brown *et al.*, 2015; Murphy *et al.*, 2014; Wertheim *et al.*, 2004). Repeat antigen exposure may somehow polarize and drive different cellular responses in humans than those we may hope for or expect from naive animal models. No candidate vaccines or immunomodulatory therapies to date have been assessed in challenge models in species that can be naturally colonized with *S. aureus*; hence the impact of an 'experienced' immune system is untested.

S. aureus vaccines

Most successful anti-bacterial vaccines induce memory B or plasma cells, which produce antigen-specific antibodies to neutralize toxins, block adhesion to host cells or opsonize bacteria for uptake and killing by phagocytes. For most of these pathogens, natural infection would induce a similar antibody response and result in long-lived protective immunity. Other nasopharyngeal pathobionts have well-characterized disease-causing serotypes (e.g. pneumococcus), or mediate their disease by locally secreting toxins (e.g. pertussis) rather than by systemic invasion. Protective immunity against these organisms is characterized by the production of toxin-neutralizing or opsonizing antibodies which facilitate phagocytosis. It is uncertain how much of this knowledge can really be useful in guiding *S. aureus* vaccine development, a prospect significantly more challenging for numerous reasons. Natural *S. aureus* infection often does not result in protection against subsequent infections, and recurrent infection is common (Fowler *et al.*, 1999). Immunization efforts against such an organism are thus attempting to make a vaccine that does something no current vaccine really does. Attempts to identify particularly 'virulent' strains associated with invasive infections have largely failed (Aamot *et al.*, 2012; Lalani *et al.*, 2008). In contrast to many other bacteria, the elements essential to pathogenesis in *S. aureus* infection are unknown – there is no single toxin or adhesion or immune evasion factor that causes disease when present, and fails to cause disease when absent. The key host responses determining effective immunity are equally obscure: anti-staphylococcal antibodies are common but not protective, and phagocytosis does not necessarily result in death of the organism (Dryla *et al.*, 2005; Fraunholz and Sinha, 2012). It is now widely accepted that anti-*S. aureus* vaccine strategies will likely require multivalent antigens, given that there is significant variation in surface protein expression (McCarthy and Lindsay, 2010), may need to drive both humoral and cellular immune responses to confer efficient protection, and be tailored to induce immune responses appropriate to multiple body sites, e.g. Th17 cells appear relevant in skin and mucosal disease but are not convincingly important once systemic invasion has occurred (Cho *et al.*, 2010). It seems that gaining a true understanding of host–pathogen interactions both in health and disease should be the immediate focus of research, in order to advance the potential for designing vaccines with specific mechanistic goals and true potential efficacy.

Passive and active immunization strategies have been evaluated for preventing the occurrence of or improving the outcome of *S. aureus* infections. A variety of staphylococcal antigens have been used, including surface components, toxins and secreted proteins. Passive immunization trials directly infused mono- or polyclonal antibodies but have to date proved disappointing. This is perhaps not surprising in light of mounting evidence that *S. aureus* has a significant propensity to 'hide' intracellularly, and antibodies primarily target extracellular pathogens. Additionally, the presence of staphylococcal Protein A on the cell wall is a potent anti-humoral defence mechanism, enabling the organism to immobilize even exogenous antibodies on its surface and thus render them ineffective (see 'Immune evasion' above). Most trials of passive immunization have assessed its role either as adjunctive therapy during bloodstream infection, or as prophylaxis against infection in high-risk neonates.

Only two active anti-*S. aureus* vaccines have progressed to ultimately disappointing phase III clinical trials, and those that have completed phase II have not suggested signs of efficacy to date. All have succeeded in producing an appropriate antibody response, and few have used adjuvants. These mechanisms generally showed promising protection in animal challenge models in preclinical studies, before failing to translate efficacy to clinical studies. Candidate vaccines are discussed further

Table 13.7 Clinical trials of *S. aureus* vaccines

Vaccine	Antigens	Adjuvant	Phase	Population	Result	Reference
Passive immunization						
Altastaph	Anti-CP5 and anti-CP8 IgG	–	II	Adult BSI, adjunctive therapy; neonates preventative	No efficacy, or difference in adverse events	Benjamin *et al*. (2006), Rupp *et al*. (2007)
Tefibazumab	Humanized anti-ClfA mAb	–	II	Adult BSI adjunctive therapy	No efficacy	Weems *et al*. (2006)
Veronate	Pooled anti-adhesin IgG (anti-ClfA; FnbP; SdrG)	–	II	Neonates preventative	No efficacy	DeJonge *et al*. (2007)
Aurograb	Anti-GrfA (transporter component)	–	III	Unpublished, but reported as adjunctive therapy for deep-seated MRSA infection	No efficacy	Schaffer and Lee (2009)
Pagimaximab	Anti-LTA mAb	–	II	Neonates preventative	No efficacy	Weisman *et al*. (2011)
Active immunization						
StaphVax	CP5, CP8	–	III	Adult HD, preventative	No efficacy	Fattom *et al*. (2015), Shinefield *et al*. (2002)
V710	IsdB	–	III	Adult cardiothoracic surgery patients, preventative	No efficacy, increased mortality among vaccines who developed *S. aureus* infection	Fowler *et al*. (2013)
SA4Ag	CP5, CP8, ClfA, MntC	–	II	Healthy adults; pre-spinal fusion procedures	–	Unpublished
Toxoids	rLuks-PV, rAT	–	II	Healthy adults	–	Unpublished
NDV-3	rAls3p (ClfA analogue)	Alum	I	–	Immunogenic (humoral and cellular)	Schmidt *et al*. (2012)
GSK2392102A	CP5, CP8, ClfA, AT	AS03B	I	Healthy adults	Immunogenic (humoral and cellular)	Levy *et al*. (2015)
SA75	Chloroform-inactivated whole *S. aureus* cells	–	I	Healthy adults	Immunogenic (humoral)	Vaccine Research International (2004)

CP, capsular polysaccharide; BSI, bloodstream infection; ClfA, clumping factor A; mAb, monoclonal antibody; FnbP, fibrinogen-binding protein; LTA, lipoteichoic acid; IsdB, iron-regulated surface determinant; MntC, manganese transport protein C; rLukS, recombinant LukS subunit of Panton–Valentine leucocidin; (r)AT, (recombinant) alpha toxoid; AS03, adjuvant system 03 (α-tocopherol, squalene and polysorbate emulsion). It should be noted that no Phase II studies were designed with enough statistical power to truly evaluate efficacy due to small participant numbers, so the outcomes quoted are based on the results and statistical analysis presented in each study with that known caveat.

below, and have been comprehensively reviewed elsewhere (Fowler and Proctor, 2014; Jansen *et al.*, 2013; Proctor, 2015; Schaffer and Lee, 2008). Clinical studies are summarized in Table 13.7.

Capsular polysaccharides

Most clinical isolates carry capsular polysaccharides (CP) which increase pathogenesis in animal models and inhibit opsonophagocytosis. The most prevalent of these are CP 5 and 8. The role of anti-CP antibodies is somewhat controversial, as absence of CP does appear to significantly alter the pathogenicity of *S. aureus*. Passive immunization of BALB/c mice with IgG resulted in significantly higher survival rates in systemic infection. Active immunization similarly improved outcome and was associated with lower levels of organ bacterial burden (Fattom *et al.*, 1996). Given this promise, passive immunization via administration of polyclonal anti-CP 5 and 8 IgG (derived from actively immunized volunteers) was evaluated both as preventative and adjunctive therapy in human *S. aureus* bloodstream infection. Anti-CP IgG adjunctive therapy in bacteraemic adult patients failed to show a difference in outcome despite promise in animal models in early studies (Rupp *et al.*, 2007). It also failed to show any protection in preventing bloodstream infection in very low-birthweight neonates in phase II studies (Benjamin *et al.*, 2006). Subsequent work has found that the presence of antibody to another *S. aureus* surface polysaccharide (PNAG) interferes with opsonic killing by anti-CP antibodies. Anti-PNAG and was found to be naturally present in all healthy volunteer serum samples in a small study, perhaps due to prior exposure to *S. aureus* or coagulase-negative staphylococci (Skurnik *et al.*, 2012). This interference was also observed in the serum of *S. aureus* bacteraemic patients when both antibodies were present, although it is unclear if this contributed to failure of immunization. Interestingly, capsular polysaccharides do not express capsule during exponential phase growth *in vitro*, so this may not be a relevant target during active *in vivo* replication (Schaffer and Lee, 2008). Active immunization with CP 5 and 8 conjugated to the same exoprotein carrier was tested in a haemodialysis cohort of 1800 patients (Shinefield *et al.*, 2002). Despite high antibody titres in vaccinees, a reduction in *S. aureus* bacteraemia was not observed throughout the follow-up period, even though 80% of clinical isolates expressed vaccine-type CP (Shinefield *et al.*, 2002). More recently, CPs 5 and 8 have been included in a multivalent candidate vaccine comprising these polysaccharides conjugated to tetanus toxoid, inactivated α-toxin and ClfA with or without alum adjuvant. In a phase I study, this formulation induced humoral immunity (although pre-vaccination baseline titres were already high) and low-level Th0/Th1 responses to ClfA and α-toxin (Levy *et al.*, 2015).

IsdB immunization

In a murine model, immunization with the highly conserved iron-scavenging protein IsdB showed decreased mortality in a subsequent intravenous challenge with live *S. aureus* and seemed to correlate with anti-IsdB antibody titres in this model (Kuklin *et al.*, 2006). It was also shown to confer T cell-mediated protection by expanding a Th17 IsdB-specific cell population, adoptive transfer of which was protective in intravenous challenge (Joshi *et al.*, 2012). This dual activation of humoral and cellular immunity in animal models, as well as its specificity for and conservation across *S. aureus* strains, made IsdB a highly attractive molecule for vaccination. Healthy humans have baseline anti-IsdB antibodies, presumably due to prior exposure and colonization, and titres have been shown to increase significantly between acute and convalescent samples in patients with *S. aureus* bloodstream infection as compared to patients with alternative infections, suggesting IsdB is expressed *in vitro* (Zorman *et al.*, 2013). This candidate antigen was evaluated in a phase IIa study in chronic haemodialysis patients, and whether or not administered with adjuvant, produced increased antibody responses in vaccinees as well as increasing opsonophagocytic functional activity in a subset (Moustafa *et al.*, 2012). In the phase III trial, ~8000 pre-cardiothoracic surgery patients were vaccinated with IsdB (V710) without adjuvant. It was immunogenic insofar as increased anti-IsdB antibody titres were seen, but it failed to significantly reduce the incidence of *S. aureus* bloodstream, deep sternal wound infections, or all-cause mortality (Fowler *et al.*, 2013). However, the trial was terminated early due to increased mortality observed among vaccine recipients who did develop *S. aureus* infection versus placebo recipients who did (23 vs. 4 per 100 person-years), especially among those with MRSA disease. This raised concerns about 'immune priming' being detrimental to the outcome of subsequent infection, although a causal relationship was not established. Cellular immune responses to the vaccine were not measured. Subsequent work on a humanized monoclonal anti-IsdB antibody has shown that its mechanism of protection in the murine model is not clearly defined – the non-classical complement pathway, phagocytes and lymphocytes are essential – and despite aggregating bacteria it does not decrease the *in vitro* survival or replication of *S. aureus* (Pancari *et al.*, 2012).

Clumping factor A and other adhesins

The staphylococcal cell wall-anchored protein clumping factor A has been proposed as a promising target for inclusion in multivalent vaccines given its importance as a virulence and anti-phagocytic factor that is expressed *in vivo* during infection and is ubiquitously expressed among clinical isolates (Anderson *et al.*, 2012). When used as a vaccine with an alum-based adjuvant, it has also shown protection in murine models of *S. aureus* arthritis and systemic infection, which may be IL-17 dependent (Josefsson *et al.*, 2001; Narita *et al.*, 2010). Intranasal ClfA loaded on nanoparticles also protected against systemic infection independent of antibodies (Misstear *et al.*, 2014). In humans, however, blocking ClfA activity during infection using tefibazumab – a monoclonal anti-ClfA antibody – failed to show a difference in outcome when used as an adjunctive therapy in adult bacteraemic patients (Weems *et al.*, 2006). However, some preliminary human work suggests that the level and function of naturally acquired anti-ClfA antibodies is inferior to that produced post-vaccination (Weems *et al.*, 2006). Passive immunization with human antibodies (Veronate) against ClfA and a *S. epidermidis* surface adhesion SdrG, was tested as a prophylactic strategy against development of *S. aureus* sepsis in neonates, but no differences were found in the incidence of sepsis or mortality (DeJonge *et al.*, 2007). Interestingly, ClfA may differ between strains, such that anti-ClfA antibodies that neutralize one strain may not bind to others, and this may have significant implications for its use as a vaccine antigen (Brady *et al.*, 2013).

A vaccine (NDV-3) based on an antifungal molecule structurally similar to ClfA (rAls3p-N) – with alum adjuvant – has been shown to confer Th1/Th17 cell-mediated protection against invasive infection caused by multiple strains of *S. aureus*, as well as against *Candida albicans* in mice (Spellberg *et al.*, 2008). While it also induces functional anti-rAls3p-N antibodies, these seemed a surrogate marker of vaccine protection, as the mechanism of protection was independent of humoral immunity (Lin *et al.*, 2009a). It has recently been demonstrated to drive anti-*S. aureus* Th17 (IL-17 and IL-22) immunity and effectively contain cutaneous infection, reducing the risk of systemic dissemination (Yeaman *et al.*, 2014). Preliminary phase I studies in humans showed this formulation to effectively produce humoral and cellular (Th1/17) responses in the majority of subjects, and further studies are awaited (Schmidt *et al.*, 2012).

Other antigens

A chimeric monoclonal antibody raised against lipoteichoic acid (pagibaximab) was tested in neonates as prophylaxis against bloodstream infection, but was not protective (Weisman *et al.*, 2011). One theory puts forward that the lack of efficacy and occasionally increased mortality seen in surface-component vaccination studies is a result of IgG-mediated aggregation of staphylococci, which enhances virulence (Spaulding *et al.*, 2013). *S. aureus* ATP-binding cassette (ABC) transporter was noted to elicit antibodies in survivors of staphylococcal bloodstream and wound infection patients, the level of which correlated with survival (Burnie *et al.*, 2000). Recombinant antibodies against this transporter improved outcomes in systemic MRSA infection in mice (Schaffer and Lee, 2009). Novel strategies targeted at neutralizing superantigens and lytic toxins – as opposed to surface components – may reduce *S. aureus* pathogenesis and immune evasion. Active multivalent immunization of rabbits with superantigens (TSST-1, SEB, SEC, enterotoxin-like X) and toxins (α-, β- and γ-toxins) has shown promise in intrapulmonary challenge, although it has not been tested in systemic infection (Spaulding *et al.*, 2013). More recently, 4-C Staph vaccine, comprising two surface proteins (FhuD2 and Csa1A) and secreted virulence factors (α-haemolysin, EsxA, EsxB) adjuvanted with either alum or MF59 has been tested in mice. It induced antibodies and T helper cell responses (Th2, Th1, Th17) and reduced mortality in murine systemic infection (Monaci *et al.*, 2015). Phase I study results have not yet been published.

Whole killed organism

Despite the ongoing search for the most immunogenic purified staphylococcal antigens, protection afforded by using the multitude of antigens present in whole killed *S. aureus* cells has also been evaluated. Repeated intradermal vaccination with formalin-killed bacteria did not significantly influence antibody production or disease in a very small sample of chronic furunculosis patients vaccinated with their own strain (Holtfreter *et al.*, 2011). Most of these patients (75%) were nasally colonized, and indeed had robust pre-vaccination antibody responses to many extracellular proteins of their infecting strain, but these did not appear to be effective in preventing recurrence of disease. Similar ineffective protection was seen in a murine model using irradiated whole cell *S. aureus* (van Diemen *et al.*, 2013). Perhaps these findings suggest that secreted elements, only present when live bacteria are present, are crucial to pathogenesis and should be the preferred targets for immunization. As mentioned above, some rabbit studies support this suggestion, as vaccination against cell surface components worsened pathogenicity during subsequent systemic infection; whereas vaccination against secreted superantigens and toxins was protective in a pulmonary challenge (Spaulding *et al.*, 2013). Given that colonization is the predominant risk factor for developing invasive infection, two studies have examined vaccines as a method to reduce carriage. Despite humoral immunogenicity, nasal colonization rates were not significantly affected (Creech *et al.*, 2009; Weems *et al.*, 2006).

Future candidates

An anti-SEB monoclonal antibody has shown promise in attenuating mortality and disease in systemic and deep tissue infection murine models in the absence of toxic shock syndrome (Varshney *et al.*, 2013). A vaccine comprising surface proteins and secreted products (4C-Staph) showing humoral and cellular immunogenicity and protection in animal infection models has yet to register for human study (Bagnoli *et al.*, 2015). At the time of writing, only one additional anti-*S. aureus* vaccine was registered at clinicaltrials.gov for further clinical study in humans– a recombinant staphylococcal enterotoxin B vaccine recently completed phase I recruitment (NCT00974935), but is only anticipated to be useful in prevention or treatment of the rare staphylococcal toxic shock syndrome.

Discussion

The traditional pathogen-centred view of infectious diseases, where microbes are seen as the sole perpetrators of disease, has changed. Increasingly, the concept of a healthy host incorporates the presence of a diverse microbiome, including *S. aureus* along with other pathobionts (Human Microbiome Project Consortium, 2012). The presence of such potentially pathogenic bacteria does not automatically herald clinical infection, largely due to a successful and balanced immune response. In fact, disease – as characterized by damage to the host – is just one of a number of possible outcomes of host–microbe interactions. The ideal outcome of infection is the elimination of the microbe with minimal host damage and lasting protective immunity. Conversely, detrimental outcomes include persistence of a microbe that directly damages host tissues, or host damage induced by an excessive, uncontrolled, or misplaced host immune response. Thus the immune system largely dictates the outcome of infectious diseases, and the pathogenicity of *S. aureus* may be influenced by directly targeting both the bacteria and the host.

Directing efforts against the bacteria has revolutionized the management of *S. aureus* infection, and the absolute reduction in mortality afforded by antibiotic treatment is unparalleled in the history of medicine. Prior to the introduction of penicillin, mortality from *S. aureus* bloodstream infection was over 80% (Skinner and Keefer, 1941). Clinical use of anti-staphylococcal antibiotics have skewed outcome in the host's favour and substantially reduced mortality. Increasing antimicrobial resistance, however, is now shifting the balance back in favour of the bacteria. The collapse of the antimicrobial research and development pipeline in recent decades led the World Economic Forum, in its report on significant imminent global risks, to argue that 'the greatest risk… to human health comes in the form of antibiotic-resistant bacteria' (Howell, 2013). Antibiotic resistance pre-dates the emergence of human life and, given the wealth of resistance genes present and easily transferable between bacteria, further resistance is likely inevitable, even to drugs yet to be invented (Bhullar *et al.*, 2012; Spellberg, 2014). While novel methods of delivering existing antibiotics (e.g. intracellularly) may improve efficacy, innovation in developing therapies that do not further drive resistance is crucial. However, the true revolution in managing infections in the future will reduce the need to use antibiotics, and must come about by targeting the host.

The host may be targeted both by preventing, or modulating the immune response during, infection. *S. aureus* infection may be substantially reduced if future technological advances enable safe delivery of intensive healthcare without our current dependence on implanted foreign materials (e.g. intravenous catheters, polyethylene vascular grafts or metallic joint implants) which currently act as fertile biofilm reservoirs on which *S. aureus* can hide from antimicrobial therapy and immune response. The use of tailored probiotics to compete with pathogenic species, and sequestration of host nutrients (e.g. iron) to limit the resources available for bacterial reproduction may become additional strategies. As discussed previously, the complexity of vaccine design is significantly compounded by the fact that a *S. aureus*-specific immunological memory is already present in the target population. The search for an efficacious vaccine resulting in sterile protection against *S. aureus* infection remains a challenge.

Manipulating the host immune response during infection is a more revolutionary concept in therapeutics, but there is convincing evidence that the course of the host immune response is key in determining clinical outcome, including in *S. aureus* infection (van Hal *et al.*, 2012). Inflammation is an essential part of the immune response to microbes, without which successful resolution of infection and repair of cellular damage would not be possible. However, disordered immunity influences the risk of, and outcome from, infection. Inborn immunodeficiencies mean certain groups are more vulnerable to developing particular infections in the first place. Outside of congenital defects, a large body of evidence also argues that systemic inflammation is dysregulated in septic shock, resulting in 'immunopathology' that further drives host damage (Cohen, 2002). Survivors of sepsis exhibit an early increased pro-inflammatory bias in immune response (Monneret *et al.*, 2008; Tsalik *et al.*, 2014). This is followed by 'immunoparalysis' – lymphocyte depletion, shift from Th1 to Th2 phenotype, and 'exhaustion' of adaptive effector functions – and later, a balanced lymphocyte recovery (de Pablo *et al.*, 2014; Tsalik *et al.*, 2014). Non-survivors have a prolonged immunoparalysis with failure of this recovery. Understanding the immune changes seen during infection and their impact on prognosis is vital before attempting to manipulate this response. While non-specifically augmenting the immune response with pro-inflammatory innate cytokines (e.g. IL-β, Il-6, TNFα) may seem an attractive option, this appears to recapitulate the pathophysiology associated with severe sepsis and may be detrimental to the host (D'Elia *et al.*, 2013). On the other hand, suppressing an overactive immune response may be a more appropriate immunomodulatory strategy, but only provided it does not hinder bacterial clearance, nor prolong immunoparalysis. Observational data support this concept that relative immunosuppression may be of clinical benefit. For instance, solid organ transplant recipients on immunosuppressive therapy demonstrate fewer staphylococcal bacterial bloodstream infections, attenuated immune responses and significantly reduced infection-attributable mortality (Kalil *et al.*, 2015). Non-specific immunosuppressive treatments (e.g. low-dose corticosteroids, statins, pooled immunoglobulin) have been more broadly tested in septic shock, but remain controversial, and none have become part of routine sepsis care (Angus and van der Poll, 2013). Intriguingly, the considerable heterogeneity in human immunological responses is almost completely determined – not by heritable genetics – but by the acquired and cumulative influence of environmental exposures (Brodin *et al.*, 2015). Human inter-individual immune responses in *S. aureus* infection are extremely variable and consistently host-specific, unaltered by differences in bacterial strains (Ahn *et al.*, 2013; Banchereau *et al.*, 2012; Brown *et al.*, 2015). Therefore, the ability to measure a patient's immune phenotype during infection in order to allow an individualized, nuanced tweaking of immune response will be required to guide therapy in the future, rather than a universal non-specific 'one-size-fits-all' intervention. The ideal immune response is likely to be a fine balance between pro- and anti-inflammatory effects, which is site-specific and dynamic over the course of infection. We are currently only in the very early stages of understanding this process for all infections, but particularly in the case of *S. aureus*.

Future novel anti-*S. aureus* strategies may manipulate the pathogen (block virulence mechanisms), the commensal bacteria (avert colonization) or the host ('boost' host immunity or suppress immunopathology). Unlike current antimicrobial therapy, each of these approaches relies on detailed mechanistic knowledge of the targets. Despite considerable advances in understanding the organism, there is extremely limited knowledge of the *in vivo* host immune response to acute infection, and observations of immune signatures during infection have yet to be properly correlated with prognosis. While an effective vaccine is the ultimate goal in combating the twin epidemics of *S. aureus* infection and antimicrobial resistance, bespoke therapeutic immunomodulation remains a tantalizing option. Future development of effective therapies for *S. aureus* requires a thorough understanding of the immune response.

References

Aamot, H.V., Blomfeldt, A., and Eskesen, A.N. (2012). Genotyping of 353 *Staphylococcus aureus* bloodstream isolates collected between 2004 and 2009 at a Norwegian university hospital and potential associations with clinical parameters. J. Clin. Microbiol. *50*, 3111–3114.

Adhikari, R.P., Ajao, A.O., Aman, M.J., Karauzum, H., Sarwar, J., Lydecker, A.D., Johnson, J.K., Nguyen, C., Chen, W.H., and Roghmann, M.C. (2012). Lower antibody levels to *Staphylococcus aureus* exotoxins are associated with sepsis in hospitalized adults with invasive S. aureus infections. J. Infect. Dis. *206*, 915–923.

Ahn, S.H., Tsalik, E.L., Cyr, D.D., Zhang, Y., van Velkinburgh, J.C., Langley, R.J., Glickman, S.W., Cairns, C.B., Zaas, A.K., Rivers, E.P., *et al.* (2013). Gene expression-based classifiers identify *Staphylococcus aureus* infection in mice and humans. PLOS ONE *8*, e48979.

Akira, S., Uematsu, S., and Takeuchi, O. (2006). Pathogen recognition and innate immunity. Cell *124*, 783–801.

Akkina, R. (2013). Human immune responses and potential for vaccine assessment in humanized mice. Curr. Opin. Immunol. *25*, 403–409.

Alm, R.A., McLaughlin, R.E., Kos, V.N., Sader, H.S., Iaconis, J.P., and Lahiri, S.D. (2014). Analysis of *Staphylococcus aureus* clinical isolates with reduced susceptibility to ceftaroline: an epidemiological and structural perspective. J. Antimicro. Chemoth. *69*, 2065–2075.

Alonzo, F., Kozhaya, L., Rawlings, S.A., Reyes-Robles, T., DuMont, A.L., Myszka, D.G., Landau, N.R., Unutmaz, D., and Torres, V.J. (2013). CCR5 is a receptor for *Staphylococcus aureus* leukotoxin ED. Nature *493*, 51–55.

Alp, E., Gozukucuk, S., Canoz, O., Kirmaci, B., and Doganay, M. (2004). Effect of granulocyte colony-stimulating factor in experimental methicillin resistant *Staphylococcus aureus* sepsis. BMC infectious diseases *4*, 43.

Anderson, A.S., Miller, A.A., Donald, R.G., Scully, I.L., Nanra, J.S., Cooper, D., and Jansen, K.U. (2012). Development of a multicomponent *Staphylococcus aureus* vaccine designed to counter multiple bacterial virulence factors. Human Vaccin. Immunother. *8*, 1585–1594.

Andrews, T., and Sullivan, K.E. (2003). Infections in patients with inherited defects in phagocytic function. Clin. Microbiol. Rev. *16*, 597–621.

Angus, D.C., and van der Poll, T. (2013). Severe sepsis and septic shock. N. Engl. J. Med. *369*, 840–851.

Arikawa, J., Ishibashi, M., Kawashima, M., Takagi, Y., Ichikawa, Y., and Imokawa, G. (2002). Decreased levels of sphingosine, a natural antimicrobial agent, may be associated with vulnerability of the stratum corneum from patients with atopic dermatitis to colonization by *Staphylococcus aureus*. J. Invest. Dermatol. *119*, 433–439.

Aslam, A., Chapel, H., and Ogg, G. (2011). Direct ex-vivo evaluation of pneumococcal specific T-cells in healthy adults. PLOS ONE *6*, e25367.

Bagnoli, F., Fontana, M.R., Soldaini, E., Mishra, R.P.N., Fiaschi, L., Cartocci, E., Nardi-Dei, V., Ruggiero, P., Nosari, S., De Falco, M.G., *et al.* (2015). Vaccine composition formulated with a novel TLR7-dependent adjuvant induces high and broad protection against *Staphylococcus aureus*. Proc. Natl. Acad. Sci. U.S.A.*112*, 3680–3685.

Banchereau, R., Jordan-Villegas, A., Ardura, M., Mejias, A., Baldwin, N., Xu, H., Saye, E., Rossello-Urgell, J., Nguyen, P., Blankenship, D., *et al.* (2012). Host immune transcriptional profiles reflect the variability in clinical disease manifestations in patients with *Staphylococcus aureus* infections. PLOS ONE *7*, e34390.

Beeson, P.B., Brannon, E.S., and Warren, J.V. (1945). Observations on the sites of removal of bacteria from the blood in patients with bacterial endocarditis. J. Exp. Med. *81*, 9–23.

Begier, E.M., Frenette, K., Barrett, N.L., Mshar, P., Petit, S., Boxrud, D.J., Watkins-Colwell, K., Wheeler, S., Cebelinski, E.A., Glennen, A., *et al.* (2004). A High-Morbidity Outbreak of Methicillin-Resistant *Staphylococcus aureus* among Players on a College Football Team, Facilitated by Cosmetic Body Shaving and Turf Burns. Clin. Infect. Dis. *39*, 1446–1453.

Bekeredjian-Ding, I., Greil, J., Ammann, S., and Parcina, M. (2014). Plasmacytoid Dendritic Cells: Neglected Regulators of the Immune Response to *Staphylococcus aureus*. Front. Immunol. *5*, 238.

Belkaid, Y. (2007). Regulatory T cells and infection: a dangerous necessity. Nat. Rev. Immunol. *7*, 875–888.

Benjamin, D.K., Schelonka, R., White, R., Holley, H.P., Bifano, E., Cummings, J., Adcock, K., Kaufman, D., Puppala, B., Riedel, P., *et al.* (2006). A blinded, randomized, multicenter study of an intravenous *Staphylococcus aureus* immune globulin. J. Perinatol. *26*, 290–295.

Bhullar, K., Waglechner, N., Pawlowski, A., Koteva, K., Banks, E.D., Johnston, M.D., Barton, H.A., and Wright, G.D. (2012). Antibiotic resistance is prevalent in an isolated cave microbiome. PLOS ONE *7*, e34953.

Bone, R.C., Balk, R.A., Cerra, F.B., Dellinger, R.P., Fein, A.M., Knaus, W.A., Schein, R.M., and Sibbald, W.J. (1992). Definitions for sepsis and organ failure and guidelines for the use of innovative therapies in sepsis. The ACCP/SCCM Consensus Conference Committee. American College of Chest Physicians/Society of Critical Care Medicine. Chest *101*, 1644–1655.

Bonneville, M., O'Brien, R.L., and Born, W.K. (2010). γδ T cell effector functions: a blend of innate programming and acquired plasticity. Nat. Rev. Immunol. *10*, 467–478.

Brady, R.A., Mocca, C.P., and Burns, D.L. (2013). Immunogenicity analysis of *Staphylococcus aureus* clumping factor A genetic variants. Clin. Vaccine Immunol. *20*, 1338–1340.

Bratton, D.L., and Henson, P.M. (2011). Neutrophil clearance: when the party is over, clean-up begins. Trends Immunol. *32*, 350–357.

Breuer, K., Haussler, S., Kapp, A., and Werfel, T. (2002). *Staphylococcus aureus*: colonizing features and influence of an antibacterial treatment in adults with atopic dermatitis. Br. J. Dermatol. *147*, 55–61.

Brodin, P., Jojic, V., Gao, T., Bhattacharya, S., Angel, C.J., Furman, D., Shen-Orr, S., Dekker, C.L., Swan, G.E., Butte, A.J., *et al.* (2015). Variation in the human immune system is largely driven by non-heritable influences. Cell *160*, 37–47.

Brouwer, N., Dolman, K.M., van Houdt, M., Sta, M., Roos, D., and Kuijpers, T.W. (2008). Mannose-binding lectin (MBL) facilitates opsonophagocytosis of yeasts but not of bacteria despite MBL binding. J. Immunol. *180*, 4124–4132.

Brown, A.F., Leech, J.M., Rogers, T.R., and McLoughlin, R.M. (2014). *Staphylococcus aureus* Colonization: Modulation of Host Immune Response and Impact on Human Vaccine Design. Front. Immunol. *4*, 507.

Brown, A.F., Murphy, A.G., Lalor, S.J., Leech, J.M., O'Keeffe, K.M., Mac Aogáin, M., O'Halloran, D.P., Lacey, K.A., Tavakol, M., Hearnden, C.H., *et al.* (2015). Memory Th1 Cells Are Protective in Invasive *Staphylococcus aureus* Infection. PLOS Pathog. *11*, e1005226.

Burnie, J.P., Matthews, R.C., Carter, T., Beaulieu, E., Donohoe, M., Chapman, C., Williamson, P., and Hodgetts, S.J. (2000). Identification of an Immunodominant ABC Transporter in Methicillin-Resistant *Staphylococcus aureus* Infections. Infect. Immun. *68*, 3200–3209.

Caccamo, N., La Mendola, C., Orlando, V., Meraviglia, S., Todaro, M., Stassi, G., Sireci, G., Fournié, J.J., and Dieli, F. (2011). Differentiation, phenotype, and function of interleukin-17-producing human Vγ9Vδ2 T cells. Blood *118*, 129–138.

Carvalho, A.M., Cristal, J.R., Muniz, A.C., Carvalho, L.P., Gomes, R., Miranda, J.C., Barral, A., Carvalho, E.M., and de Oliveira, C.I. (2015). Interleukin 10–Dominant Immune Response and Increased Risk of Cutaneous Leishmaniasis After Natural Exposure to Lutzomyia intermedia Sand Flies. J. Infect. Dis. *212*, 157–165.

Casadevall, A., and Pirofski, L.A. (2003). The damage-response framework of microbial pathogenesis. Nat. Rev. Microbiol. *1*, 17–24.

Chambers, H.F. (2013). Cellulitis, by any other name. Clin. Infect. Dis. *56*, 1763–1764.

Chang, F.Y., and Shaio, M.F. (1995). Decreased cell-mediated immunity in patients with non-insulin-dependent diabetes mellitus. Diabetes Res. Clin. Pract. *28*, 137–146.

Cheng, P., Liu, T., Zhou, W.Y., Zhuang, Y., Peng, L.S., Zhang, J.Y., Yin, Z.N., Mao, X.H., Guo, G., Shi, Y., *et al.* (2012). Role of gamma-delta T cells in host response against *Staphylococcus aureus*-induced pneumonia. BMC Immunol. *13*, 38.

Chevallier, P., Robillard, N., Illiaquer, M., Esbelin, J., Mohty, M., Bodin-Bressollette, C., Guillaume, T., Stocco, V., Auffray, F., Derenne, S., *et al.* (2013). Characterization of various blood and graft sources: a prospective series. Transfusion *53*, 2020–2026.

Chiacchio, T., Casetti, R., Butera, O., Vanini, V., Carrara, S., Girardi, E., Di Mitri, D., Battistini, L., Martini, F., Borsellino, G., *et al.* (2009). Characterization of regulatory T cells identified as CD4(+)CD25(high) CD39(+) in patients with active tuberculosis. Clin. Exp. Immunol. *156*, 463–470.

Cho, J.S., Pietras, E.M., Garcia, N.C., Ramos, R.I., Farzam, D.M., Monroe, H.R., Magorien, J.E., Blauvelt, A., Kolls, J.K., Cheung, A.L., *et al.* (2010). IL-17 is essential for host defense against cutaneous *Staphylococcus aureus* infection in mice. J. Clin. Invest. *120*, 1762–1773.

Chong, Y.P., Moon, S.M., Bang, K.-M., Park, H.J., Park, S.-Y., Kim, M.-N., Park, K.-H., Kim, S.-H., Lee, S.-O., Choi, S.-H., *et al.* (2013). Treatment Duration for Uncomplicated *Staphylococcus aureus* Bacteremia To Prevent Relapse: Analysis of a Prospective Observational Cohort Study. Antimicrob. Agents Ch. *57*, 1150–1156.

Chong, Y.P., Park, K.H., Kim, E.S., Kim, M.N., Kim, S.H., Lee, S.O., Choi, S.H., Jeong, J.Y., Woo, J.H., and Kim, Y.S. (2014). Association of mannose-binding lectin 2 gene polymorphisms with persistent *Staphylococcus aureus* bacteremia. PLOS ONE *9*, e89139.

Cohen, J. (2002). The immunopathogenesis of sepsis. Nature *420*, 885–891.

Colque-Navarro, P., Jacobsson, G., Andersson, R., Flock, J.I., and Möllby, R. (2010). Levels of antibody against 11 *Staphylococcus aureus* antigens in a healthy population. Clin. Vaccine. Immunol. *17*, 1117–1123.

Craft, J.E. (2012). Follicular helper T cells in immunity and systemic autoimmunity. Nat. Rev. Rheumatol. *8*, 337–347.

Creech, C.B., Johnson, B.G., Alsentzer, A.R., Hohenboken, M., Edwards, K.M., and Talbot, T.R. (2009). Vaccination as infection control: a pilot study to determine the impact of *Staphylococcus aureus* vaccination on nasal carriage. Vaccine *28*, 256–260.

Croft, M., Bradley, L.M., and Swain, S.L. (1994). Naive versus memory CD4 T cell response to antigen. Memory cells are less dependent on accessory cell costimulation and can respond to many antigen-presenting cell types including resting B cells. J. Immunol. *152*, 2675–2685.

Crome, S.Q., Wang, A.Y., and Levings, M.K. (2010). Translational mini-review series on Th17 cells: function and regulation of human T helper 17 cells in health and disease. Clin. Exp. Immunol. *159*, 109–119.

Crotty, S. (2014). T follicular helper cell differentiation, function, and roles in disease. Immunity *41*, 529–542.

Cummins, N.W., and Badley, A.D. (2010). Mechanisms of HIV-associated lymphocyte apoptosis: 2010. Cell Death Dis. *1*, e99.

Cypowyj, S., Picard, C., Maródi, L., Casanova, J.L., and Puel, A. (2012). Immunity to infection in IL-17-deficient mice and humans. Eur. J. Immunol. *42*, 2246–2254.

D'Elia, R.V., Harrison, K., Oyston, P.C., Lukaszewski, R.A., and Clark, G.C. (2013). Targeting the "cytokine storm" for therapeutic benefit. Clin. Vaccine Immunol. *20*, 319–327.

Davey, M.S., Lin, C.Y., Roberts, G.W., Heuston, S., Brown, A.C., Chess, J.A., Toleman, M.A., Gahan, C.G., Hill, C., Parish, T., *et al.* (2011). Human neutrophil clearance of bacterial pathogens triggers anti-microbial $\gamma\delta$ T cell responses in early infection. PLOS Pathog. *7*, e1002040.

de Kraker, M.E., Wolkewitz, M., Davey, P.G., Koller, W., Berger, J., Nagler, J., Icket, C., Kalenic, S., Horvatic, J., Seifert, H., *et al.* (2011). Clinical impact of antimicrobial resistance in European hospitals: excess mortality and length of hospital stay related to methicillin-resistant *Staphylococcus aureus* bloodstream infections. Antimicrob. Agents Ch. *55*, 1598–1605.

de Pablo, R., Monserrat, J., Prieto, A., and Alvarez-Mon, M. (2014). Role of circulating lymphocytes in patients with sepsis. Biomed Res. Int. *2014*, 671087.

DeJonge, M., Burchfield, D., Bloom, B., Duenas, M., Walker, W., Polak, M., Jung, E., Millard, D., Schelonka, R., Eyal, F., *et al.* (2007). Clinical trial of safety and efficacy of INH-A21 for the prevention of nosocomial staphylococcal bloodstream infection in premature infants. J. Pediatr. *151*, 260–265.

del Rio, A., Cervera, C., Moreno, A., Moreillon, P., and Miro, J.M. (2009). Patients at risk of complications of *Staphylococcus aureus* bloodstream infection. Clin. Infect. Dis. *48 Suppl 4*, S246–253.

DeLeo, F.R., Diep, B.A., and Otto, M. (2009). Host defense and pathogenesis in *Staphylococcus aureus* infections. Infect. Dis. Clin. North Am. *23*, 17–34.

Desachy, A., Lina, G., Vignon, P., Hashemzadeh, A., Denis, F., Etienne, J., Francois, B., and Ploy, M.C. (2007). Role of superantigenic strains in the prognosis of community-acquired methicillin-susceptible *Staphylococcus aureus* bacteraemia. Clin. Microbiol. Infect. *13*, 1131–1133.

Deshmukh, H.S., Hamburger, J.B., Ahn, S.H., McCafferty, D.G., Yang, S.R., and Fowler, V.G. (2009). Critical role of NOD2 in regulating the immune response to *Staphylococcus aureus*. Infect. Immun. *77*, 1376–1382.

DeVries, A.S., Lesher, L., Schlievert, P.M., Rogers, T., Villaume, L.G., Danila, R., and Lynfield, R. (2011). Staphylococcal toxic shock syndrome 2000–2006: epidemiology, clinical features, and molecular characteristics. PLOS ONE *6*, e22997.

Dryla, A., Prustomersky, S., Gelbmann, D., Hanner, M., Bettinger, E., Kocsis, B., Kustos, T., Henics, T., Meinke, A., and Nagy, E. (2005). Comparison of antibody repertoires against *Staphylococcus aureus* in healthy individuals and in acutely infected patients. Clin. Diag. Lab. Immun. *12*, 387–398.

Etzioni, A. (2010). Defects in the leukocyte adhesion cascade. Clin. Rev. Allergy Immunol. *38*, 54–60.

Fadok, V.A., Bratton, D.L., Konowal, A., Freed, P.W., Westcott, J.Y., and Henson, P.M. (1998). Macrophages that have ingested apoptotic cells in vitro inhibit proinflammatory cytokine production through autocrine/paracrine mechanisms involving TGF-beta, PGE2, and PAF. J. Clin. Invest. *101*, 890–898.

Falugi, F., Kim, H.K., Missiakas, D.M., and Schneewind, O. (2013). Role of protein A in the evasion of host adaptive immune responses by *Staphylococcus aureus*. MBio *4*, e00575–13.

Fattom, A., Matalon, A., Buerkert, J., Taylor, K., Damaso, S., and Boutriau, D. (2015). Efficacy profile of a bivalent *Staphylococcus aureus* glycoconjugated vaccine in adults on hemodialysis: Phase III randomized study. Human Vaccin. Immunother. *11*, 632–641.

Fattom, A.I., Sarwar, J., Ortiz, A., and Naso, R. (1996). A *Staphylococcus aureus* capsular polysaccharide (CP) vaccine and CP-specific antibodies protect mice against bacterial challenge. Infect. Immun. *64*, 1659–1665.

Feil, E.J., Cooper, J.E., Grundmann, H., Robinson, D.A., Enright, M.C., Berendt, T., Peacock, S.J., Smith, J.M., Murphy, M., Spratt, B.G., *et al.* (2003). How clonal is *Staphylococcus aureus*? J. Bacteriol. *185*, 3307–3316.

Ferens, W.A., and Bohach, G.A. (2000). Persistence of *Staphylococcus aureus* on mucosal membranes: superantigens and internalization by host cells. J. Lab. Clin. Med. *135*, 225–230.

Fillatreau, S. (2013). Cytokine-producing B cells as regulators of pathogenic and protective immune responses. Ann. Rheum. Dis. *72* (Suppl. 2), ii80–ii84.

Finkelman, F.D., Katona, I.M., Mosmann, T.R., and Coffman, R.L. (1988). IFN-gamma regulates the isotypes of Ig secreted during in vivo humoral immune responses. J. Immunol. *140*, 1022–1027.

Flo, T.H., Halaas, O., Torp, S., Ryan, L., Lien, E., Dybdahl, B., Sundan, A., and Espevik, T. (2001). Differential expression of Toll-like receptor 2 in human cells. J. Leukoc. Biol. *69*, 474–481.

Foster, T.J. (2005). Immune evasion by staphylococci. Nat. Rev. Microbiol. *3*, 948–958.

Foster, T.J. (2009). Colonization and infection of the human host by staphylococci: adhesion, survival and immune evasion. Vet. Dermatol. *20*, 456–470.

Foster, T.J., Geoghegan, J.A., Ganesh, V.K., and Höök, M. (2014). Adhesion, invasion and evasion: the many functions of the surface proteins of *Staphylococcus aureus*. Nat. Rev. Microbiol. *12*, 49–62.

Fournier, B. (2012). The function of TLR2 during staphylococcal diseases. Front. Cell. Infect. Microbiol. *2*, 167.

Fowler, V.G., Jr., Kong, L.K., Corey, G.R., Gottlieb, G.S., McClelland, R.S., Sexton, D.J., Gesty-Palmer, D., and Harrell, L.J. (1999). Recurrent *Staphylococcus aureus* bacteremia: pulsed-field gel electrophoresis findings in 29 patients. J. Infect. Dis. *179*, 1157–1161.

Fowler, V.G., Olsen, M.K., Corey, G.R., Woods, C.W., Cabell, C.H., Reller, L.B., Cheng, A.C., Dudley, T., and Oddone, E.Z. (2003). Clinical identifiers of complicated *Staphylococcus aureus* bacteremia. Arch. Intern. Med. *163*, 2066–2072.

Fowler, V.G., Allen, K.B., Moreira, E.D., Moustafa, M., Isgro, F., Boucher, H.W., Corey, G.R., Carmeli, Y., Betts, R., Hartzel, J.S., *et al.* (2013). Effect of an investigational vaccine for preventing *Staphylococcus aureus* infections after cardiothoracic surgery: a randomized trial. JAMA *309*, 1368–1378.

Fowler, V.G., and Proctor, R.A. (2014). Where does a *Staphylococcus aureus* vaccine stand? Clin. Microbiol. Infect. *20* (Suppl. 5), 66–75.

Fraunholz, M., and Sinha, B. (2012). Intracellular *Staphylococcus aureus*: live-in and let die. Front. Cell. Infect. Microbiol. *2*, 43.

Freeman, A.F., and Holland, S.M. (2009). Clinical manifestations, etiology, and pathogenesis of the hyper-IgE syndromes. Pediatr. Res. *65*, 32R–37R.

Frodermann, V., Chau, T.A., Sayedyahossein, S., Toth, J.M., Heinrichs, D.E., and Madrenas, J. (2011). A modulatory interleukin-10 response to staphylococcal peptidoglycan prevents Th1/Th17 adaptive immunity to *Staphylococcus aureus*. J. Infect. Dis. *204*, 253–262.

Fuchs, T.A., Abed, U., Goosmann, C., Hurwitz, R., Schulze, I., Wahn, V., Weinrauch, Y., Brinkmann, V., and Zychlinsky, A. (2007). Novel cell death program leads to neutrophil extracellular traps. J. Cell. Biol. *176*, 231–241.

Gagliotti, C., Balode, A., Baquero, F., Degener, J., Grundmann, H., Gur, D., Jarlier, V., Kahlmeter, G., Monen, J., Monnet, D.L., *et al.* (2011). *Escherichia coli* and *Staphylococcus aureus*: bad news and good news from the European Antimicrobial Resistance Surveillance Network (EARS-Net, formerly EARSS), 2002 to 2009. Euro Surveill. *16*, pii:19819.

Gjertsson, I., Hultgren, O.H., Stenson, M., Holmdahl, R., and Tarkowski, A. (2000). Are B lymphocytes of importance in severe *Staphylococcus aureus* infections? Infect. Immun. *68*, 2431–2434.

Gjertsson, I., Hultgren, O.H., and Tarkowski, A. (2002). Interleukin-10 ameliorates the outcome of *Staphylococcus aureus* arthritis by promoting bacterial clearance. Clin. Exp. Immunol. *130*, 409–414.

Goodyear, C.S., and Silverman, G.J. (2003). Death by a B cell superantigen: In vivo VH-targeted apoptotic supraclonal B cell deletion by a Staphylococcal Toxin. J. Exp. Med. *197*, 1125–1139.

Gravet, A., Colin, D.A., Keller, D., Girardot, R., Monteil, H., Prévost, G., and Giradot, R. (1998). Characterization of a novel structural member, LukE-LukD, of the bi-component staphylococcal leucotoxins family. FEBS Lett. *436*, 202–208.

Greenberg, J.A., David, M.Z., Hall, J.B., and Kress, J.P. (2014). Immune dysfunction prior to *Staphylococcus aureus* bacteremia is a determinant of long-term mortality. PLOS ONE *9*, e88197.

Greenlee-Wacker, M.C., Rigby, K.M., Kobayashi, S.D., Porter, A.R., DeLeo, F.R., and Nauseef, W.M. (2014). Phagocytosis of *Staphylococcus aureus* by human neutrophils prevents macrophage efferocytosis and induces programmed necrosis. J. Immunol. *192*, 4709–4717.

Greten, T.F., Manns, M.P., and Korangy, F. (2011). Myeloid derived suppressor cells in human diseases. Int. Immunopharmacol. *11*, 802–807.

Grumann, D., Ruotsalainen, E., Kolata, J., Kuusela, P., Jarvinen, A., Kontinen, V.P., Broker, B.M., and Holtfreter, S. (2011). Characterization of infecting strains and superantigen-neutralizing antibodies in *Staphylococcus aureus* bacteremia. Clin. Vaccine Immunol. *18*, 487–493.

Haimi-Cohen, Y., Vellozzi, E.M., and Rubin, L.G. (2002). Initial concentration of Staphylococcus epidermidis in simulated pediatric blood cultures correlates with time to positive results with the automated, continuously monitored BACTEC blood culture system. J. Clin. Microbiol. *40*, 898–901.

Hanke, M.L., and Kielian, T. (2012). Deciphering mechanisms of staphylococcal biofilm evasion of host immunity. Front. Cell. Infect. Microbiol. *2*, 62.

Heim, C.E., Vidlak, D., Scherr, T.D., Kozel, J.A., Holzapfel, M., Muirhead, D.E., and Kielian, T. (2014). Myeloid-derived suppressor cells contribute to *Staphylococcus aureus* orthopedic biofilm infection. J. Immunol. *192*, 3778–3792.

Henningsson, L., Jirholt, P., Lindholm, C., Eneljung, T., Silverpil, E., Iwakura, Y., Linden, A., and Gjertsson, I. (2010). Interleukin-17A during local and systemic *Staphylococcus aureus*-induced arthritis in mice. Infect. Immun. *78*, 3783–3790.

Higgins, J., Loughman, A., van Kessel, K.P., van Strijp, J.A., and Foster, T.J. (2006). Clumping factor A of *Staphylococcus aureus* inhibits phagocytosis by human polymorphonuclear leucocytes. FEMS Microbiol. Lett. *258*, 290–296.

Holland, S.M., DeLeo, F.R., Elloumi, H.Z., Hsu, A.P., Uzel, G., Brodsky, N., Freeman, A.F., Demidowich, A., Davis, J., Turner, M.L., *et al.* (2007). STAT3 mutations in the hyper-IgE syndrome. N. Engl. J. Med. *357*, 1608–1619.

Holtfreter, S., Jursa-Kulesza, J., Masiuk, H., Verkaik, N.J., de Vogel, C., Kolata, J., Nowosiad, M., Steil, L., van Wamel, W., van Belkum, A., *et al.* (2011). Antibody responses in furunculosis patients vaccinated with autologous formalin-killed *Staphylococcus aureus*. Eur J. Clin. Microbiol. Infect. Dis. *30*, 707–717.

Howell, L. (2013). Global risks 2013: an initiative of the Risk Response Network. (World Economic Forum). Available online: http://reports.weforum.org/global-risks-2013/. Accessed March 2016.

Huang, H., Cohen, S.H., King, J.H., Monchaud, C., Nguyen, H., and Flynn, N.M. (2008). Injecting drug use and community-associated methicillin-resistant *Staphylococcus aureus* infection. Diagn. Microbiol. Infect. Dis. *60*, 347–350.

Human Microbiome Project Consortium. (2012). Structure, function and diversity of the healthy human microbiome. Nature *486*, 207–214.

Hünig, T. (2012). The storm has cleared: lessons from the CD28 superagonist TGN1412 trial. Nat. Rev. Immunol. *12*, 317–318.

Ishigame, H., Kakuta, S., Nagai, T., Kadoki, M., Nambu, A., Komiyama, Y., Fujikado, N., Tanahashi, Y., Akitsu, A., Kotaki, H., *et al.* (2009). Differential roles of interleukin-17A and -17F in host defense against mucoepithelial bacterial infection and allergic responses. Immunity *30*, 108–119.

Iwata, Y., Matsushita, T., Horikawa, M., Dilillo, D.J., Yanaba, K., Venturi, G.M., Szabolcs, P.M., Bernstein, S.H., Magro, C.M., Williams, A.D., *et al.* (2011). Characterization of a rare IL-10-competent B-cell subset in humans that parallels mouse regulatory B10 cells. Blood *117*, 530–541.

Jacobsson, G., Colque-Navarro, P., Gustafsson, E., Andersson, R., and Möllby, R. (2010). Antibody responses in patients with invasive *Staphylococcus aureus* infections. Eur. J. Clin. Microbiol. Infect. Dis. *29*, 715–725.

Jansen, K.U., Girgenti, D.Q., Scully, I.L., and Anderson, A.S. (2013). Vaccine review: "Staphyloccocus aureus vaccines: problems and prospects". Vaccine *31*, 2723–2730.

Johansson, M.A., Saghafian-Hedengren, S., Haileselassie, Y., Roos, S., Troye-Blomberg, M., Nilsson, C., and Sverremark-Ekström, E. (2012). Early-life gut bacteria associate with IL-4-, IL-10- and IFN-γ production at two years of age. PLOS ONE 7, e49315.

Josefsson, E., Hartford, O., O'Brien, L., Patti, J.M., and Foster, T. (2001). Protection against experimental *Staphylococcus aureus* arthritis by vaccination with clumping factor A, a novel virulence determinant. J. Infect. Dis. *184*, 1572–1580.

Joshi, A., Pancari, G., Cope, L., Bowman, E.P., Cua, D., Proctor, R.A., and McNeely, T. (2012). Immunization with *Staphylococcus aureus* iron regulated surface determinant B (IsdB) confers protection via Th17/IL17 pathway in a murine sepsis model. Human Vaccin. Immunother. *8*, 336–346.

Kaasch, A.J., Barlow, G., Edgeworth, J.D., Fowler, V.G., Hellmich, M., Hopkins, S., Kern, W.V., Llewelyn, M.J., Rieg, S., Rodriguez-Baño, J., *et al.* (2014). *Staphylococcus aureus* bloodstream infection: a pooled analysis of five prospective, observational studies. J. Infect. *68*, 242–251.

Kalil, A.C., Syed, A., Rupp, M.E., Chambers, H., Vargas, L., Maskin, A., Miles, C.D., Langnas, A., and Florescu, D.F. (2015). Is bacteremic sepsis associated with higher mortality in transplant recipients than in nontransplant patients? A matched case-control propensity-adjusted study. Clin. Infect. Dis. *60*, 216–222.

Kaufmann, S.H., and Ladel, C.H. (1994). Role of T cell subsets in immunity against intracellular bacteria: experimental infections of knock-out mice with Listeria monocytogenes and Mycobacterium bovis BCG. Immunobiology *191*, 509–519.

Kazakova, S.V., Hageman, J.C., Matava, M., Srinivasan, A., Phelan, L., Garfinkel, B., Boo, T., McAllister, S., Anderson, J., Jensen, B., *et al.* (2005). A clone of methicillin-resistant *Staphylococcus aureus* among professional football players. N. Engl. J. Med. *352*, 468–475.

Khatib, R., Riederer, K., Saeed, S., Johnson, L.B., Fakih, M.G., Sharma, M., Tabriz, M.S., and Khosrovaneh, A. (2005). Time to positivity in *Staphylococcus aureus* bacteremia: possible correlation with the source and outcome of infection. Clin. Infect. Dis. *41*, 594–598.

Kiedrowski, M.R., and Horswill, A.R. (2011). New approaches for treating staphylococcal biofilm infections. Ann. N.Y. Acad. Sci. *1241*, 104–121.

Kim, J., Gregson, D.B., Ross, T., and Laupland, K.B. (2010). Time to blood culture positivity in *Staphylococcus aureus* bacteremia: association with 30-day mortality. J. Infect. *61*, 197–204.

Kirkpatrick, C.H. (2001). Chronic mucocutaneous candidiasis. Pediatr. Infect. Dis. J. *20*, 197–206.

Kluytmans, J., van Belkum, A., and Verbrugh, H. (1997). Nasal carriage of *Staphylococcus aureus*: epidemiology, underlying mechanisms, and associated risks. Clin. Microbiol. Rev. *10*, 505–520.

Knop, J., Hanses, F., Leist, T., Archin, N.M., Buchholz, S., Gläsner, J., Gessner, A., and Wege, A.K. (2015). *Staphylococcus aureus* Infection in Humanized Mice: A New Model to Study Pathogenicity Associated With Human Immune Response. J. Infect. Dis. *212*, 435–444.

Kobayashi, S.D., Braughton, K.R., Palazzolo-Ballance, A.M., Kennedy, A.D., Sampaio, E., Kristosturyan, E., Whitney, A.R., Sturdevant, D.E., Dorward, D.W., Holland, S.M., *et al.* (2010). Rapid neutrophil destruction following phagocytosis of *Staphylococcus aureus*. J. Innate Immun. *2*, 560–575.

Köhler, J., Breitbach, K., Renner, C., Heitsch, A.-K., Bast, A., van Rooijen, N., Vogelgesang, S., and Steinmetz, I. (2011). NADPH-oxidase but not inducible nitric oxide synthase contributes to resistance in a murine *Staphylococcus aureus* Newman pneumonia model. Microbes Infect. *13*, 914–922.

Kolata, J., Bode, L.G., Holtfreter, S., Steil, L., Kusch, H., Holtfreter, B., Albrecht, D., Hecker, M., Engelmann, S., van Belkum, A., *et al.* (2011). Distinctive patterns in the human antibody response to *Staphylococcus aureus* bacteremia in carriers and non-carriers. Proteomics *11*, 3914–3927.

Kolata, J.B., Kühbandner, I., Link, C., Normann, N., Vu, C.H., Steil, L., Weidenmaier, C., and Bröker, B.M. (2015). The Fall of a Dogma? Unexpected High T-Cell Memory Response to *Staphylococcus aureus* in Humans. J. Infect. Dis. *212*, 830–838.

Kubica, M., Guzik, K., Koziel, J., Zarebski, M., Richter, W., Gajkowska, B., Golda, A., Maciag-Gudowska, A., Brix, K., Shaw, L., *et al.* (2008). A potential new pathway for *Staphylococcus aureus* dissemination: the silent survival of S. aureus phagocytosed by human monocyte-derived macrophages. PLOS ONE 3, e1409.

Kuklin, N.A., Clark, D.J., Secore, S., Cook, J., Cope, L.D., McNeely, T., Noble, L., Brown, M.J., Zorman, J.K., Wang, X.M., *et al.* (2006). A novel *Staphylococcus aureus* vaccine: iron surface determinant B induces rapid antibody responses in rhesus macaques and specific increased survival in a murine *S. aureus* sepsis model. Infect. Immun. *74*, 2215–2223.

Lalani, T., Federspiel, J.J., Boucher, H.W., Rude, T.H., Bae, I.G., Rybak, M.J., Tonthat, G.T., Corey, G.R., Stryjewski, M.E., Sakoulas, G., *et al.* (2008). Associations between the genotypes of *Staphylococcus aureus* bloodstream isolates and clinical characteristics and outcomes of bacteremic patients. J. Clin. Microbiol. *46*, 2890–2896.

Laupland, K.B., Church, D.L., Mucenski, M., Sutherland, L.R., and Davies, H.D. (2003). Population-based study of the epidemiology of and the risk factors for invasive *Staphylococcus aureus* infections. J. Infect. Dis. *187*, 1452–1459.

Laupland, K.B., Ross, T., and Gregson, D.B. (2008). *Staphylococcus aureus* bloodstream infections: risk factors, outcomes, and the influence of methicillin resistance in Calgary, Canada, 2000–2006. J. Infect. Dis. *198*, 336–343.

Lawrence, P.K., Rokbi, B., Arnaud-Barbe, N., Sutten, E.L., Norimine, J., Lahmers, K.K., and Brown, W.C. (2012). CD4 T cell antigens from *Staphylococcus aureus* Newman strain identified following immunization with heat-killed bacteria. Clin. Vaccine Immunol. *19*, 477–489.

Lee, B.Y., Ufberg, P.J., Bailey, R.R., Wiringa, A.E., Smith, K.J., Nowalk, A.J., Higgins, C., Wateska, A.R., and Muder, R.R. (2010a). The potential economic value of a *Staphylococcus aureus* vaccine for neonates. Vaccine *28*, 4653–4660.

Lee, B.Y., Wiringa, A.E., Bailey, R.R., Lewis, G.J., Feura, J., and Muder, R.R. (2010b). *Staphylococcus aureus* vaccine for orthopedic patients: an economic model and analysis. Vaccine *28*, 2465–2471.

Lee, L.Y., Miyamoto, Y.J., McIntyre, B.W., Höök, M., McCrea, K.W., McDevitt, D., and Brown, E.L. (2002). The *Staphylococcus aureus* Map protein is an immunomodulator that interferes with T cell-mediated responses. J. Clin. Invest. *110*, 1461–1471.

Lehar, S.M., Pillow, T., Xu, M., Staben, L., Kajihara, K.K., Vandlen, R., DePalatis, L., Raab, H., Hazenbos, W.L., Morisaki, J.H., *et al.* (2015). Novel antibody-antibiotic conjugate eliminates intracellular S. aureus. Nature *527*, 323–328.

Lesens, O.M.D., Methlin, C.M.D., Hansmann, Y.M.D., Remy, V.M.D., Martinot, M.M.D., Bergin, C.M.D., Meyer, P., and Christmann, D.M.D. (2003). Role of Comorbidity in Mortality Related to *Staphylococcus aureus* Bacteremia: A Prospective Study Using the Charlson Weighted Index of Comorbidity. Infect. Cont. Hosp. Ep. *24*, 890–896.

Levy, J., Licini, L., Haelterman, E., Moris, P., Lestrate, P., Damaso, S., Van Belle, P., and Boutriau, D. (2015). Safety and immunogenicity of an investigational 4-component *Staphylococcus aureus* vaccine with or without AS03B adjuvant: Results of a randomized phase I trial. Human Vaccin. Immunother. *11*, 620–631.

Lin, L., Ibrahim, A.S., Baquir, B., Avanesian, V., Fu, Y., and Spellberg, B. (2009a). Immunological surrogate marker of rAls3p-N vaccine-induced protection against *Staphylococcus aureus*. FEMS Immunol. Med. Microbiol. *55*, 293–295.

Lin, L., Ibrahim, A.S., Xu, X., Farber, J.M., Avanesian, V., Baquir, B., Fu, Y., French, S.W., Edwards, J.E., Jr., and Spellberg, B. (2009b). Th1-Th17 cells mediate protective adaptive immunity against *Staphylococcus aureus* and Candida albicans infection in mice. PLOS Pathog. *5*, e1000703.

Lin, Y.C., and Peterson, M.L. (2010). New insights into the prevention of staphylococcal infections and toxic shock syndrome. Expert Rev. Clin. Pharmacol. *3*, 753–767.

Litjens, N.H., Huisman, M., van den Dorpel, M., and Betjes, M.G. (2008). Impaired immune responses and antigen-specific memory CD4+ T cells in hemodialysis patients. J. Am. Soc. Nephrol. *19*, 1483–1490.

Liu, G.Y., Essex, A., Buchanan, J.T., Datta, V., Hoffman, H.M., Bastian, J.F., Fierer, J., and Nizet, V. (2005). *Staphylococcus aureus* golden pigment impairs neutrophil killing and promotes virulence through its antioxidant activity. J. Exp. Med. *202*, 209–215.

Lowy, F.D. (1998). *Staphylococcus aureus* infections. N. Engl. J. Med. *339*, 520–532.

Lowy, F.D. (2011). How *Staphylococcus aureus* adapts to its host. N. Engl. J. Med. *364*, 1987–1990.

Lussow, A.R., and MacDonald, H.R. (1994). Differential effects of superantigen-induced "anergy" on priming and effector stages of a T cell-dependent antibody response. Eur. J. Immunol. *24*, 445–449.

Ma, C.S., and Deenick, E.K. (2014). Human T follicular helper (Tfh) cells and disease. Immunol. Cell. Biol. *92*, 64–71.

Machura, E., Mazur, B., Golemiec, E., Pindel, M., and Halkiewicz, F. (2008). *Staphylococcus aureus* skin colonization in atopic dermatitis children is associated with decreased IFN-gamma production by peripheral blood CD4+ and CD8+ T cells. Pediatr. Allergy Immunol. *19*, 37–45.

Mackowiak, P.A. (1998). Concepts of fever. Arch. Intern. Med. *158*, 1870–1881.

Maher, B.M., Mulcahy, M.E., Murphy, A.G., Wilk, M., O'Keeffe, K.M., Geoghegan, J.A., Lavelle, E.C., and McLoughlin, R.M. (2013). Nlrp-3-driven interleukin 17 production by gammadeltaT cells controls infection outcomes during *Staphylococcus aureus* surgical site infection. Infect. Immun. *81*, 4478–4489.

McCarthy, A.J., and Lindsay, J.A. (2010). Genetic variation in *Staphylococcus aureus* surface and immune evasion genes is lineage associated: implications for vaccine design and host-pathogen interactions. BMC Microbiol. *10*, 173.

McLoughlin, R.M., Lee, J.C., Kasper, D.L., and Tzianabos, A.O. (2008). IFN-gamma regulated chemokine production determines the outcome of *Staphylococcus aureus* infection. J. Immunol. *181*, 1323–1332.

McLoughlin, R.M., Solinga, R.M., Rich, J., Zaleski, K.J., Cocchiaro, J.L., Risley, A., Tzianabos, A.O., and Lee, J.C. (2006). CD4+ T cells and CXC chemokines modulate the pathogenesis of *Staphylococcus aureus* wound infections. Proc. Natl. Acad. Sci. U.S.A. *103*, 10408–10413.

McNicholas, S., Talento, A., J., O.G., Hannan, M.M., Lynch, M., Greene, C.M., Humphreys, H., and Fitzgerald-Hughes, D. (2014). Cytokine responses to *Staphylococcus aureus* bloodstream infection differ between patient cohorts that have different clinical courses of infection. BMC Infect. Dis. *14*, 580.

Mele, T., and Madrenas, J. (2010). TLR2 signalling: At the crossroads of commensalism, invasive infections and toxic shock syndrome by *Staphylococcus aureus*. Int. J. Biochem. Cell Biol. *42*, 1066–1071.

Melehani, J.H., James, D.B., DuMont, A.L., Torres, V.J., and Duncan, J.A. (2015). *Staphylococcus aureus* Leukocidin A/B (LukAB) Kills Human Monocytes via Host NLRP3 and ASC when Extracellular, but Not Intracellular. PLOS Pathog. *11*, e1004970.

Melzer, M., and Welch, C. (2013). Thirty-day mortality in UK patients with community-onset and hospital-acquired meticillin-susceptible *Staphylococcus aureus* bacteraemia. J. Hosp. Infect. *84*, 143–150.

Menzies, B.E., and Kenoyer, A. (2005). *Staphylococcus aureus* Infection of Epidermal Keratinocytes Promotes Expression of Innate Antimicrobial Peptides. Infect. Immun. 73, 5241–5244.

Mestas, J., and Hughes, C.C. (2004). Of mice and not men: differences between mouse and human immunology. J. Immunol. *172*, 2731–2738.

Miajlovic, H., Fallon, P.G., Irvine, A.D., and Foster, T.J. (2010). Effect of filaggrin breakdown products on growth of and protein expression by *Staphylococcus aureus*. J. Allergy. Clin. Immunol. *126*, 1184–1190.

Midorikawa, K., Ouhara, K., Komatsuzawa, H., Kawai, T., Yamada, S., Fujiwara, T., Yamazaki, K., Sayama, K., Taubman, M.A., Kurihara, H., *et al.* (2003). *Staphylococcus aureus* susceptibility to innate antimicrobial peptides, beta-defensins and CAP18, expressed by human keratinocytes. Infect. Immun. *71*, 3730–3739.

Miller, C., Ragheb, J.A., and Schwartz, R.H. (1999). Anergy and cytokine-mediated suppression as distinct superantigen-induced tolerance mechanisms in vivo. J. Exp. Med. *190*, 53–64.

Miller, L.G., and Kaplan, S.L. (2009). *Staphylococcus aureus*: a community pathogen. Infect. Dis. Clin. North Am. *23*, 35–52.

Miller, L.S., Pietras, E.M., Uricchio, L.H., Hirano, K., Rao, S., Lin, H., O'Connell, R.M., Iwakura, Y., Cheung, A.L., Cheng, G., *et al.* (2007). Inflammasome-mediated production of IL-1beta is required for neutrophil recruitment against *Staphylococcus aureus* in vivo. J. Immunol. *179*, 6933–6942.

Milner, J.D., Brenchley, J.M., Laurence, A., Freeman, A.F., Hill, B.J., Elias, K.M., Kanno, Y., Spalding, C., Elloumi, H.Z., Paulson, M.L., *et al.* (2008). Impaired T(H)17 cell differentiation in subjects with autosomal dominant hyper-IgE syndrome. Nature *452*, 773–776.

Minegishi, Y., Saito, M., Nagasawa, M., Takada, H., Hara, T., Tsuchiya, S., Agematsu, K., Yamada, M., Kawamura, N., Ariga, T., *et al.* (2009). Molecular explanation for the contradiction between systemic Th17 defect and localized bacterial infection in hyper-IgE syndrome. J. Exp. Med. *206*, 1291–1301.

Misstear, K., McNeela, E.A., Murphy, A.G., Geoghegan, J.A., O'Keeffe, K.M., Fox, J., Chan, K., Heuking, S., Collin, N., Foster, T.J., *et al.* (2014). Targeted nasal vaccination provides antibody-independent protection against *Staphylococcus aureus*. J. Infect. Dis. *209*, 1479–1484.

Molne, L., Corthay, A., Holmdahl, R., and Tarkowski, A. (2003). Role of gamma/delta T cell receptor-expressing lymphocytes in cutaneous infection caused by *Staphylococcus aureus*. Clin. Exp. Immunol *132*, 209–215.

Monaci, E., Mancini, F., Lofano, G., Bacconi, M., Tavarini, S., Sammicheli, C., Arcidiacono, L., Giraldi, M., Galletti, B., Rossi Paccani, S., *et al.* (2015). MF59- and Al(OH)3-adjuvanted *Staphylococcus aureus* (4C-Staph) vaccines induce sustained protective humoral and cellular immune responses, with a critical role for effector CD4 T cells at low antibody titers. Front. Immunol. *6*, 439.

Monneret, G., Venet, F., Pachot, A., and Lepape, A. (2008). Monitoring immune dysfunctions in the septic patient: a new skin for the old ceremony. Mol. Med. *14*, 64–78.

Montgomery, C.P., Daniels, M., Zhao, F., Alegre, M.L., Chong, A.S., and Daum, R.S. (2014). Protective immunity against recurrent *Staphylococcus aureus* skin infection requires antibody and interleukin-17A. Infect. Immun. *82*, 2125–2134.

Morinaga, N., Kaihou, Y., and Noda, M. (2003). Purification, cloning and characterization of variant LukE-LukD with strong leukocidal activity of staphylococcal bi-component leukotoxin family. Microbiol. Immunol. *47*, 81–90.

Moustafa, M., Aronoff, G.R., Chandran, C., Hartzel, J.S., Smugar, S.S., Galphin, C.M., Mailloux, L.U., Brown, E., Dinubile, M.J., Kartsonis, N.A., *et al.* (2012). Phase IIa study of the immunogenicity and safety of the novel *Staphylococcus aureus* vaccine V710 in adults with end-stage renal disease receiving hemodialysis. Clin. Vaccine Immunol. *19*, 1509–1516.

Murphy, A.G., O'Keeffe, K.M., Lalor, S.J., Maher, B.M., Mills, K.H., and McLoughlin, R.M. (2014). *Staphylococcus aureus* infection of mice expands a population of memory γδ T cells that are protective against subsequent infection. J. Immunol. *192*, 3697–3708.

Myles, I.A., Fontecilla, N.M., Valdez, P.A., Vithayathil, P.J., Naik, S., Belkaid, Y., Ouyang, W., and Datta, S.K. (2013). Signaling via the IL-20 receptor inhibits cutaneous production of IL-1beta and IL-17A to promote infection with methicillin-resistant *Staphylococcus aureus*. Nat. Immunol. *14*, 804–811.

Narita, K., Hu, D.L., Mori, F., Wakabayashi, K., Iwakura, Y., and Nakane, A. (2010). Role of interleukin-17A in cell-mediated protection against *Staphylococcus aureus* infection in mice immunized with the fibrinogen-binding domain of clumping factor A. Infect. Immun. *78*, 4234–4242.

Nauseef, W.M., and Clark, R.A. (2010). Granulocytic Phagocytes. In Mandell, Douglas and Bennett's Principles and Practices of Infectious Diseases, Seventh edn, Mandell, G.L., Bennett, J.E., and Dolin, R., eds. (Elsevier, Philadelphia, PA), vol. 1, pp. 99–127.

Neefjes, J., Jongsma, M.L., Paul, P., and Bakke, O. (2011). Towards a systems understanding of MHC class I and MHC class II antigen presentation. Nat. Rev. Immunol. *11*, 823–836.

Nelson, C.L., Pelak, K., Podgoreanu, M.V., Ahn, S.H., Scott, W.K., Allen, A.S., Cowell, L.G., Rude, T.H., Zhang, Y., Tong, A., *et al.* (2014). A genome-wide association study of variants associated with acquisition of *Staphylococcus aureus* bacteremia in a healthcare setting. BMC Infect. Dis. *14*, 83.

Nemzek, J.A., Bolgos, G.L., Williams, B.A., and Remick, D.G. (2001). Differences in normal values for murine white blood cell counts and other hematological parameters based on sampling site. Inflamm. Res. *50*, 523–527.

Netea, M.G. (2013). Training innate immunity: the changing concept of immunological memory in innate host defence. Eur. J. Clin. Invest. *43*, 881–884.

Netea, M.G., Sutmuller, R., Hermann, C., Van der Graaf, C.A., Van der Meer, J.W., van Krieken, J.H., Hartung, T., Adema, G., and Kullberg, B.J. (2004). Toll-like receptor 2 suppresses immunity against Candida albicans through induction of IL-10 and regulatory T cells. J. Immunol. *172*, 3712–3718.

Nizet, V. (2007). Understanding how leading bacterial pathogens subvert innate immunity to reveal novel therapeutic targets. J. Allergy Clin. Immunol. *120*, 13–22.

Noore, J., Noore, A., and Li, B. (2013). Cationic antimicrobial peptide LL-37 is effective against both extra- and intracellular *Staphylococcus aureus*. Antimicrob. Agents Ch. *57*, 1283–1290.

Nurjadi, D., Herrmann, E., Hinderberger, I., and Zanger, P. (2013). Impaired beta-defensin expression in human skin links DEFB1 promoter polymorphisms with persistent *Staphylococcus aureus* nasal carriage. J. Infect. Dis. *207*, 666–674.

O'Keeffe, K.M., Wilk, M.M., Leech, J.M., Murphy, A.G., Laabei, M., Monk, I.R., Massey, R.C., Lindsay, J.A., Foster, T.J., Geoghegan, J.A., *et al.* (2015). Manipulation of Autophagy in Phagocytes Facilitates *Staphylococcus aureus* Bloodstream Infection. Infect. Immun. *83*, 3445–3457.

O'Shea, J.J., and Paul, W.E. (2010). Mechanisms underlying lineage commitment and plasticity of helper CD4+ T cells. Science *327*, 1098–1102.

Öncül, O., Öksüz, S., Acar, A., Ülkür, E., Turhan, V., Uygur, F., Ulçay, A., Erdem, H., Özyurt, M., and Görenek, L. (2014). Nosocomial infection characteristics in a burn intensive care unit: analysis of an eleven-year active surveillance. Burns *40*, 835–841.

Ortiz, M.C., Lefimil, C., Rodas, P.I., Vernal, R., Lopez, M., Acuña-Castillo, C., Imarai, M., and Escobar, A. (2015). Neisseria gonorrhoeae Modulates Immunity by Polarizing Human Macrophages to a M2 Profile. PLOS ONE *10*, e0130713.

Pancari, G., Fan, H., Smith, S., Joshi, A., Haimbach, R., Clark, D., Li, Y., Hua, J., McKelvey, T., Ou, Y., *et al.* (2012). Characterization of the mechanism of protection mediated by CS-D7, a monoclonal antibody to *Staphylococcus aureus* iron regulated surface determinant B (IsdB). Front. Cell Infect. Microbiol. *2*, 36.

Pang, D.J., Neves, J.F., Sumaria, N., and Pennington, D.J. (2012). Understanding the complexity of γδ T-cell subsets in mouse and human. Immunology *136*, 283–290.

Parker, D., and Prince, A. (2012). *Staphylococcus aureus* induces type I IFN signaling in dendritic cells via TLR9. J. Immunol. *189*, 4040–4046.

Pauli, N.T., Kim, H.K., Falugi, F., Huang, M., Dulac, J., Henry Dunand, C., Zheng, N.-Y., Kaur, K., Andrews, S.F., Huang, Y., *et al.* (2014). *Staphylococcus aureus* infection induces protein A-mediated immune evasion in humans. J. Exp. Med. *211*, 2331–2339.

Peacock, S.J., Moore, C.E., Justice, A., Kantzanou, M., Story, L., Mackie, K., O'Neill, G., and Day, N.P. (2002). Virulent combinations of adhesin and toxin genes in natural populations of *Staphylococcus aureus*. Infect. Immun. *70*, 4987–4996.

Pepper, M., Linehan, J.L., Pagán, A.J., Zell, T., Dileepan, T., Cleary, P.P., and Jenkins, M.K. (2010). Different routes of bacterial infection induce long-lived TH1 memory cells and short-lived TH17 cells. Nat. Immunol. *11*, 83–89.

Peres, A.G., and Madrenas, J. (2013). The broad landscape of immune interactions with *Staphylococcus aureus*: from commensalism to lethal infections. Burns *39*, 380–388.

Peres, A.G., Stegen, C., Li, J., Xu, A.Q., Levast, B., Surette, M.G., Cousineau, B., Desrosiers, M., and Madrenas, J. (2015). Uncoupling of pro- and anti-inflammatory properties of *Staphylococcus aureus*. Infect. Immun. *83*, 1587–1597.

Picard, C., von Bernuth, H., Ghandil, P., Chrabieh, M., Levy, O., Arkwright, P.D., McDonald, D., Geha, R.S., Takada, H., Krause, J.C., *et al.* (2010). Clinical features and outcome of patients with IRAK-4 and MyD88 deficiency. Medicine *89*, 403–425.

Pido-Lopez, J., Kwok, W.W., Mitchell, T.J., Heyderman, R.S., and Williams, N.A. (2011). Acquisition of pneumococci specific effector and regulatory Cd4+ T cells localising within human upper respiratory-tract mucosal lymphoid tissue. PLOS Pathog. *7*, e1002396.

Pishchany, G., McCoy, A.L., Torres, V.J., Krause, J.C., Crowe, J.E., Fabry, M.E., and Skaar, E.P. (2010). Specificity for human hemoglobin enhances *Staphylococcus aureus* infection. Cell Host Microbe *8*, 544–550.

Pozzi, C., Waters, E.M., Rudkin, J.K., Schaeffer, C.R., Lohan, A.J., Tong, P., Loftus, B.J., Pier, G.B., Fey, P.D., Massey, R.C., *et al.* (2012). Methicillin resistance alters the biofilm phenotype and attenuates virulence in *Staphylococcus aureus* device-associated infections. PLOS Pathog. *8*, e1002626.

Proctor, R.A. (2015). Recent developments for *Staphylococcus aureus* vaccines: clinical and basic science challenges. Eur. Cell. Mater. *30*, 315–326.

Puel, A., Cypowyj, S., Bustamante, J., Wright, J.F., Liu, L., Lim, H.K., Migaud, M., Israel, L., Chrabieh, M., Audry, M., *et al.* (2011). Chronic mucocutaneous candidiasis in humans with inborn errors of interleukin-17 immunity. Science *332*, 65–68.

Rabe, H., Nordström, I., Andersson, K., Lundell, A.C., and Rudin, A. (2014). *Staphylococcus aureus* convert neonatal conventional CD4(+) T cells into FOXP3(+) CD25(+) CD127(low) T cells via the PD-1/PD-L1 axis. Immunology *141*, 467–481.

Ram, S., Lewis, L.A., and Rice, P.A. (2010). Infections of people with complement deficiencies and patients who have undergone splenectomy. Clin. Microbiol. Rev. *23*, 740–780.

Ramesh, N., Spertini, F., Scholl, P., and Geha, R. (1992). A toxic shock syndrome toxin-1 peptide that shows homology to mycobacterial heat shock protein 18 is presented as conventional antigen to T cells by multiple HLA-DR alleles. J. Immunol. *148*, 1025–1030.

Reimer, L.G., Wilson, M.L., and Weinstein, M.P. (1997). Update on detection of bacteremia and fungemia. Clin. Microbiol. Rev. *10*, 444–465.

Reyes-Robles, T., Alonzo, F., Kozhaya, L., Lacy, D.B., Unutmaz, D., and Torres, V.J. (2013). *Staphylococcus aureus* leukotoxin ED targets the chemokine receptors CXCR1 and CXCR2 to kill leukocytes and promote infection. Cell Host Microbe *14*, 453–459.

Rieg, S., Kaasch, A.J., Wehrle, J., Hofmann, S.C., Szymaniak-Vits, M., Saborowski, V., Jonas, D., Kalbacher, H., Seifert, H., and Kern, W.V. (2011). Susceptibility of clinical *Staphylococcus aureus* isolates to innate defense antimicrobial peptides. Microbes Infect. *13*, 761–765.

Rigby, K.M., and DeLeo, F.R. (2012). Neutrophils in innate host defense against *Staphylococcus aureus* infections. Semin. Immunopathol. *34*, 237–259.

Roberts, M.T., Stober, C.B., McKenzie, A.N., and Blackwell, J.M. (2005). Interleukin-4 (IL-4) and IL-10 collude in vaccine failure for novel exacerbatory antigens in murine Leishmania major infection. Infect. Immun. *73*, 7620–7628.

Robertson, C.M., Perrone, E.E., McConnell, K.W., Dunne, W.M., Boody, B., Brahmbhatt, T., Diacovo, M.J., Van Rooijen, N., Hogue, L.A., Cannon, C.L., *et al.* (2008). Neutrophil depletion causes a fatal defect in murine pulmonary *Staphylococcus aureus* clearance. J. Surg. Res. *150*, 278–285.

Rose, W.E., Eickhoff, J.C., Shukla, S.K., Pantrangi, M., Rooijakkers, S., Cosgrove, S.E., Nizet, V., and Sakoulas, G. (2012). Elevated serum interleukin-10 at time of hospital admission is predictive of mortality in patients with *Staphylococcus aureus* bacteremia. J. Infect. Dis. *206*, 1604–1611.

Rupp, M.E., Holley, H.P., Jr., Lutz, J., Dicpinigaitis, P.V., Woods, C.W., Levine, D.P., Veney, N., and Fowler, V.G., Jr. (2007). Phase II, randomized, multicenter, double-blind, placebo-controlled trial of a polyclonal anti-*Staphylococcus aureus* capsular polysaccharide immune globulin in treatment of *Staphylococcus aureus* bacteremia. Antimicrob. Agents Ch. *51*, 4249–4254.

Rutz, S., Eidenschenk, C., and Ouyang, W. (2013). IL-22, not simply a Th17 cytokine. Immunol. Rev. *252*, 116–132.

Salgado-Pabón, W., Breshears, L., Spaulding, A.R., Merriman, J.A., Stach, C.S., Horswill, A.R., Peterson, M.L., and Schlievert, P.M. (2013). Superantigens are critical for *Staphylococcus aureus* Infective endocarditis, sepsis, and acute kidney injury. MBio *4*, e00494-13.

Sasaki, S., Nishikawa, S., Miura, T., Mizuki, M., Yamada, K., Madarame, H., Tagawa, Y.I., Iwakura, Y., and Nakane, A. (2000). Interleukin-4 and interleukin-10 are involved in host resistance to *Staphylococcus aureus* infection through regulation of gamma interferon. Infect. Immun. *68*, 2424–2430.

Schaffer, A.C., and Lee, J.C. (2008). Vaccination and passive immunisation against *Staphylococcus aureus*. Int. J. Antimicrob. Agents 32 *Suppl 1*, S71–78.

Schaffer, A.C., and Lee, J.C. (2009). Staphylococcal vaccines and immunotherapies. Infect. Dis. Clin. North Am. *23*, 153–171.

Schindler, D., Gutierrez, M.G., Beineke, A., Rauter, Y., Rohde, M., Foster, S., Goldmann, O., and Medina, E. (2012). Dendritic cells are central coordinators of the host immune response to *Staphylococcus aureus* bloodstream infection. Am. J. Pathol. *181*, 1327–1337.

Schmaler, M., Jann, N.J., Ferracin, F., and Landmann, R. (2011). T and B cells are not required for clearing *Staphylococcus aureus* in systemic infection despite a strong TLR2-MyD88-dependent T cell activation. J. Immunol. *186*, 443–452.

Schmidt, C.S., White, C.J., Ibrahim, A.S., Filler, S.G., Fu, Y., Yeaman, M.R., Edwards, J.E., and Hennessey, J.P. (2012). NDV-3, a recombinant alum-adjuvanted vaccine for Candida and *Staphylococcus aureus*, is safe and immunogenic in healthy adults. Vaccine *30*, 7594–7600.

Seder, R.A., and Ahmed, R. (2003). Similarities and differences in CD4+ and CD8+ effector and memory T cell generation. Nat. Immunol. *4*, 835–842.

Seder, R.A., Darrah, P.A., and Roederer, M. (2008). T-cell quality in memory and protection: implications for vaccine design. Nat. Rev. Immunol. *8*, 247–258.

Sharpe, A.H., and Abbas, A.K. (2006). T-cell costimulation–biology, therapeutic potential, and challenges. N. Engl. J. Med. *355*, 973–975.

Sheridan, B.S., Romagnoli, P.A., Pham, Q.M., Fu, H.H., Alonzo, F., Schubert, W.D., Freitag, N.E., and Lefrançois, L. (2013). γδ T cells exhibit multifunctional and protective memory in intestinal tissues. Immunity *39*, 184–195.

Shimada, T., Park, B.G., Wolf, A.J., Brikos, C., Goodridge, H.S., Becker, C.A., Reyes, C.N., Miao, E.A., Aderem, A., Gotz, F., *et al.* (2010). *Staphylococcus aureus* evades lysozyme-based peptidoglycan digestion that links phagocytosis, inflammasome activation, and IL-1beta secretion. Cell Host Microbe 7, 38–49.

Shinefield, H., Black, S., Fattom, A., Horwith, G., Rasgon, S., Ordonez, J., Yeoh, H., Law, D., Robbins, J.B., Schneerson, R., *et al.* (2002). Use of a *Staphylococcus aureus* conjugate vaccine in patients receiving hemodialysis. N. Engl. J. Med. *346*, 491–496.

Shortman, K., and Liu, Y.J. (2002). Mouse and human dendritic cell subtypes. Nat. Rev. Immunol. *2*, 151–161.

Skinner, D., and Keefer, C.S. (1941). Significance of bacteremia caused by *staphylococcus aureus*: A study of one hundred and twenty-two cases and a review of the literature concerned with experimental infection in animals. Arch. Intern. Med. *68*, 851–875.

Skurnik, D., Kropec, A., Roux, D., Theilacker, C., Huebner, J., and Pier, G.B. (2012). Natural antibodies in normal human serum inhibit *Staphylococcus aureus* capsular polysaccharide vaccine efficacy. Clin. Infect. Dis. *55*, 1188–1197.

Skurnik, D., Merighi, M., Grout, M., Gadjeva, M., Maira-Litran, T., Ericsson, M., Goldmann, D.A., Huang, S.S., Datta, R., Lee, J.C., *et al.* (2010). Animal and human antibodies to distinct *Staphylococcus aureus* antigens mutually neutralize opsonic killing and protection in mice. J. Clin. Invest. *120*, 3220–3233.

Song, Y., Tai, J.H., Bartsch, S.M., Zimmerman, R.K., Muder, R.R., and Lee, B.Y. (2012). The potential economic value of a *Staphylococcus aureus* vaccine among hemodialysis patients. Vaccine *30*, 3675–3682.

Spaan, A.N., Surewaard, B.G., Nijland, R., and van Strijp, J.A. (2013). Neutrophils versus *Staphylococcus aureus*: a biological tug of war. Annu. Rev. Microbiol. *67*, 629–650.

Spaan, A.N., Vrieling, M., Wallet, P., Badiou, C., Reyes-Robles, T., Ohneck, E.A., Benito, Y., de Haas, C.J.C., Day, C.J., Jennings, M.P., *et al.* (2014). The staphylococcal toxins γ-haemolysin AB and CB differentially target phagocytes by employing specific chemokine receptors. Nat Commun 5, 5438.

Spaulding, A.R., Salgado-Pabón, W., Merriman, J.A., Stach, C.S., Ji, Y., Gillman, A.N., Peterson, M.L., and Schlievert, P.M. (2014). Vaccination against *Staphylococcus aureus* pneumonia. J. Infect. Dis. *209*, 1955–1962.

Spellberg, B. (2014). The future of antibiotics. Crit. Care. *18*, 228.

Spellberg, B., and Edwards, J.E. (2001). Type 1/Type 2 immunity in infectious diseases. Clin. Infect. Dis. *32*, 76–102.

Spellberg, B., Ibrahim, A.S., Yeaman, M.R., Lin, L., Fu, Y., Avanesian, V., Bayer, A.S., Filler, S.G., Lipke, P., Otoo, H., *et al.* (2008). The antifungal vaccine derived from the recombinant N terminus of Als3p protects mice against the bacterium *Staphylococcus aureus*. Infect. Immun. *76*, 4574–4580.

Stack, J., Doyle, S.L., Connolly, D.J., Reinert, L.S., O'Keeffe, K.M., McLoughlin, R.M., Paludan, S.R., and Bowie, A.G. (2014). TRAM is required for TLR2 endosomal signaling to type I IFN induction. J. Immunol. *193*, 6090–6102.

Stenzel, W., Dahm, J., Sanchez-Ruiz, M., Miletic, H., Hermann, M., Courts, C., Schwindt, H., Utermohlen, O., Schluter, D., and Deckert, M. (2005). Regulation of the inflammatory response to *Staphylococcus aureus*-induced brain abscess by interleukin-10. J. Neuropath. Exp. Neur. *64*, 1046–1057.

Stenzel, W., Soltek, S., Sanchez-Ruiz, M., Akira, S., Miletic, H., Schluter, D., and Deckert, M. (2008). Both TLR2 and TLR4 are required for the effective immune response in *Staphylococcus aureus*-induced experimental murine brain abscess. Am. J. Pathol. *172*, 132–145.

Stolz, S.J., Davis, J.P., Vergeront, J.M., Crass, B.A., Chesney, P.J., Wand, P.J., and Bergdoll, M.S. (1985). Development of serum antibody to toxic shock toxin among individuals with toxic shock syndrome in Wisconsin. J. Infect. Dis. *151*, 883–889.

Strandberg, K.L., Rotschafer, J.H., Vetter, S.M., Buonpane, R.A., Kranz, D.M., and Schlievert, P.M. (2010). Staphylococcal superantigens cause lethal pulmonary disease in rabbits. J. Infect. Dis. *202*, 1690–1697.

Stryjewski, M.E., and Corey, G.R. (2014). Methicillin-resistant *Staphylococcus aureus*: an evolving pathogen. Clin. Infect. Dis. *58* (Suppl. 1), S10–S9.

Suntharalingam, G., Perry, M.R., Ward, S., Brett, S.J., Castello-Cortes, A., Brunner, M.D., and Panoskaltsis, N. (2006). Cytokine storm in a phase 1 trial of the anti-CD28 monoclonal antibody TGN1412. N. Engl. J. Med. *355*, 1018–1028.

Surana, N.K., and Kasper, D.L. (2014). Deciphering the tête-à-tête between the microbiota and the immune system. J. Clin. Invest. *124*, 4197–4203.

Takeuchi, O., Hoshino, K., and Akira, S. (2000). Cutting edge: TLR2-deficient and MyD88-deficient mice are highly susceptible to *Staphylococcus aureus* infection. J. Immunol. *165*, 5392–5396.

Takigawa, H., Nakagawa, H., Kuzukawa, M., Mori, H., and Imokawa, G. (2005). Deficient production of hexadecenoic acid in the skin is associated in part with the vulnerability of atopic dermatitis patients to colonization by *Staphylococcus aureus*. Dermatology *211*, 240–248.

Taylor, A.L., and Llewelyn, M.J. (2010). Superantigen-induced proliferation of human CD4+CD25- T cells is followed by a switch to a functional regulatory phenotype. J. Immunol. *185*, 6591–6598.

Tebartz, C., Horst, S.A., Sparwasser, T., Huehn, J., Beineke, A., Peters, G., and Medina, E. (2015). A major role for myeloid-derived suppressor cells and a minor role for regulatory T cells in immunosuppression during *Staphylococcus aureus* infection. J. Immunol. *194*, 1100–1111.

Thammavongsa, V., Kim, H.K., Missiakas, D., and Schneewind, O. (2015). Staphylococcal manipulation of host immune responses. Nat. Rev. Microbiol. *13*, 529–543.

Thurlow, L.R., Hanke, M.L., Fritz, T., Angle, A., Aldrich, A., Williams, S.H., Engebretsen, I.L., Bayles, K.W., Horswill, A.R., and Kielian, T. (2011). *Staphylococcus aureus* biofilms prevent macrophage phagocytosis and attenuate inflammation in vivo. J. Immunol. *186*, 6585–6596.

Thwaites, G.E., and Gant, V. (2011). Are bloodstream leukocytes Trojan Horses for the metastasis of *Staphylococcus aureus*? Nat. Rev. Microbiol. *9*, 215–222.

Tsalik, E.L., Langley, R.J., Dinwiddie, D.L., Miller, N.A., Yoo, B., van Velkinburgh, J.C., Smith, L.D., Thiffault, I., Jaehne, A.K., Valente, A.M., *et al.* (2014). An integrated transcriptome and expressed variant analysis of sepsis survival and death. Genome Med. *6*, 111.

Tzianabos, A.O., Wang, J.Y., and Lee, J.C. (2001). Structural rationale for the modulation of abscess formation by *Staphylococcus aureus* capsular polysaccharides. Proc. Natl. Acad. Sci. U.S.A. *98*, 9365–9370.

Vaccine Research International (2004). Phase I Clinical Trial of Vaccine SA75 against Staphylococcal Infections. Synopsis of Trial (www.vri.org.uk/vriannualreport2004.pdf).

van Belkum, A., Verkaik, N.J., de Vogel, C.P., Boelens, H.A., Verveer, J., Nouwen, J.L., Verbrugh, H.A., and Wertheim, H.F. (2009). Reclassification of *Staphylococcus aureus* nasal carriage types. J. Infect. Dis. *199*, 1820–1826.

van Diemen, P.M., Yamaguchi, Y., Paterson, G.K., Rollier, C.S., Hill, A.V.S., and Wyllie, D.H. (2013). Irradiated wild-type and Spa mutant *Staphylococcus aureus* induce anti-S. aureus immune responses in mice which do not protect against subsequent intravenous challenge. Pathog. Dis. *68*, 20–26.

van Hal, S.J., Jensen, S.O., Vaska, V.L., Espedido, B.A., Paterson, D.L., and Gosbell, I.B. (2012). Predictors of mortality in *Staphylococcus aureus* Bacteremia. Clin. Microbiol. Rev. *25*, 362–386.

van Kessel, K.P., Bestebroer, J., and van Strijp, J.A. (2014). Neutrophil-mediated phagocytosis of *Staphylococcus aureus*. Front. Immunol. *5*, 467.

Varshney, A.K., Wang, X., Scharff, M.D., MacIntyre, J., Zollner, R.S., Kovalenko, O.V., Martinez, L.R., Byrne, F.R., and Fries, B.C. (2013). Staphylococcal Enterotoxin B–Specific Monoclonal Antibody 20B1 Successfully Treats Diverse *Staphylococcus aureus* Infections. J. Infect. Dis. *208*, 2058–2066.

Velasco, E., Byington, R., Martins, C.A., Schirmer, M., Dias, L.M., and Gonçalves, V.M. (2006). Comparative study of clinical characteristics of neutropenic and non-neutropenic adult cancer patients with bloodstream infections. Eur. J. Clin. Microbiol. Infect. Dis. *25*, 1–7.

Venditti, M., Falcone, M., Micozzi, A., Carfagna, P., Taglietti, F., Serra, P.F., and Martino, P. (2003). *Staphylococcus aureus* bacteremia in patients with hematologic malignancies: a retrospective case-control study. Haematologica *88*, 923–930.

Verkaik, N.J., Boelens, H.A., de Vogel, C.P., Tavakol, M., Bode, L.G., Verbrugh, H.A., van Belkum, A., and van Wamel, W.J. (2010a). Heterogeneity of the humoral immune response following *Staphylococcus aureus* bacteremia. Eur. J. Clin. Microbiol. Infect. Dis. *29*, 509–518.

Verkaik, N.J., Dauwalder, O., Antri, K., Boubekri, I., de Vogel, C.P., Badiou, C., Bes, M., Vandenesch, F., Tazir,

M., Hooijkaas, H., *et al.* (2010b). Immunogenicity of toxins during *Staphylococcus aureus* infection. Clin. Infect. Dis. *50*, 61–68.

von Eiff, C., Becker, K., Machka, K., Stammer, H., and Peters, G. (2001). Nasal carriage as a source of *Staphylococcus aureus* bacteremia. Study Group. N. Engl. J. Med. *344*, 11–16.

von Kockritz-Blickwede, M., Rohde, M., Oehmcke, S., Miller, L.S., Cheung, A.L., Herwald, H., Foster, S., and Medina, E. (2008). Immunological mechanisms underlying the genetic predisposition to severe *Staphylococcus aureus* infection in the mouse model. Am. J. Pathol. *173*, 1657–1668.

Wang, L., Kamath, A., Das, H., Li, L., and Bukowski, J.F. (2001). Antibacterial effect of human V gamma 2V delta 2 T cells in vivo. J. Clin. Invest. *108*, 1349–1357.

Warmerdam, P.A., Vanderlick, K., Vandervoort, P., De Smedt, H., Plaisance, S., De Maeyer, M., and Collen, D. (2002). Staphylokinase-specific cell-mediated immunity in humans. J. Immunol. *168*, 155–161.

Watkins, R.R., David, M.Z., and Salata, R.A. (2012). Current concepts on the virulence mechanisms of meticillin-resistant *Staphylococcus aureus*. J. Med. Microbiol. *61*, 1179–1193.

Weems, J.J., Jr., Steinberg, J.P., Filler, S., Baddley, J.W., Corey, G.R., Sampathkumar, P., Winston, L., John, J.F., Kubin, C.J., Talwani, R., *et al.* (2006). Phase II, randomized, double-blind, multicenter study comparing the safety and pharmacokinetics of tefibazumab to placebo for treatment of *Staphylococcus aureus* bacteremia. Antimicrob. Agents Ch. *50*, 2751–2755.

Weidenmaier, C., McLoughlin, R.M., and Lee, J.C. (2010). The zwitterionic cell wall teichoic acid of *Staphylococcus aureus* provokes skin abscesses in mice by a novel CD4+ T-cell-dependent mechanism. PLOS ONE *5*, e13227.

Weisman, L.E., Thackray, H.M., Steinhorn, R.H., Walsh, W.F., Lassiter, H.A., Dhanireddy, R., Brozanski, B.S., Palmer, K.G., Trautman, M.S., Escobedo, M., *et al.* (2011). A randomized study of a monoclonal antibody (pagibaximab) to prevent staphylococcal sepsis. Pediatrics *128*, 271–279.

Wertheim, H.F., Vos, M.C., Ott, A., van Belkum, A., Voss, A., Kluytmans, J.A., van Keulen, P.H., Vandenbroucke-Grauls, C.M., Meester, M.H., and Verbrugh, H.A. (2004). Risk and outcome of nosocomial *Staphylococcus aureus* bacteraemia in nasal carriers versus non-carriers. Lancet *364*, 703–705.

Wiese, L., Mejer, N., Schønheyder, H.C., Westh, H., Jensen, A.G., Larsen, A.R., Skov, R., Benfield, T., Danish Staphylococcal Bacteraemia Study Group. (2013). A nationwide study of comorbidity and risk of reinfection after *Staphylococcus aureus* bacteraemia. J. Infect. *67*, 199–205.

Wilhelmsen, K., Mesa, K.R., Prakash, A., Xu, F., and Hellman, J. (2012). Activation of endothelial TLR2 by bacterial lipoprotein upregulates proteins specific for the neutrophil response. Innate. Immun. *18*, 602–616.

Yamane, H., and Paul, W.E. (2013). Early signaling events that underlie fate decisions of naive CD4(+) T cells toward distinct T-helper cell subsets. Immunol. Rev. *252*, 12–23.

Yan, Q., Sharma-Kuinkel, B.K., Deshmukh, H., Tsalik, E.L., Cyr, D.D., Lucas, J., Woods, C.W., Scott, W.K., Sempowski, G.D., Thaden, J.T., *et al.* (2014). Dusp3 and Psme3 are associated with murine susceptibility to *Staphylococcus aureus* infection and human sepsis. PLOS Pathog. *10*, e1004149.

Yeaman, M.R., Filler, S.G., Chaili, S., Barr, K., Wang, H., Kupferwasser, D., Hennessey, J.P., Fu, Y., Schmidt, C.S., Edwards, J.E., *et al.* (2014). Mechanisms of NDV-3 vaccine efficacy in MRSA skin versus invasive infection. Proc. Natl. Acad. Sci. U.S.A. *111*, E5555–E5563.

Yi, J., Park, J.S., Hong, K.H., Lee, S.H., and Kim, E.C. (2012). Serum capacity to neutralize superantigens does not affect the outcome of *Staphylococcus aureus* bacteremia. Eur. J. Clin. Microbiol. Infect. Dis. *31*, 2061–2068.

(2013). Contribution of toll-like receptor 2 to the innate response against *Staphylococcus aureus* infection in mice. PLOS ONE *8*, e74287.

Yoong, P., and Torres, V.J. (2013). The effects of *Staphylococcus aureus* leukotoxins on the host: cell lysis and beyond. Curr. Opin. Microbiol. *16*, 63–69.

Yoong, P., and Torres, V.J. (2015). Counter inhibition between leukotoxins attenuates *Staphylococcus aureus* virulence. Nat Commun *6*, 8125.

Zanger, P., Holzer, J., Schleucher, R., Scherbaum, H., Schittek, B., and Gabrysch, S. (2010). Severity of *Staphylococcus aureus* infection of the skin is associated with inducibility of human beta-defensin 3 but not human beta-defensin 2. Infect. Immun. *78*, 3112–3117.

Zhang, Q., Leong, S.C., McNamara, P.S., Mubarak, A., Malley, R., and Finn, A. (2011). Characterisation of regulatory T cells in nasal associated lymphoid tissue in children: relationships with pneumococcal colonization. PLOS Pathog. *7*, e1002175.

Zhao, Y.X., Nilsson, I.M., and Tarkowski, A. (1998). The dual role of interferon-gamma in experimental *Staphylococcus aureus* septicaemia versus arthritis. Immunology *93*, 80–85.

Zhao, Y.X., and Tarkowski, A. (1995). Impact of interferon-gamma receptor deficiency on experimental *Staphylococcus aureus* septicemia and arthritis. J. Immunol. *155*, 5736–5742.

Zheng, J., Liu, Y., Lau, Y.L., and Tu, W. (2013). $\gamma\delta$-T cells: an unpolished sword in human anti-infection immunity. Cell. Mol. Immunol. *10*, 50–57.

Ziegler, C., Goldmann, O., Hobeika, E., Geffers, R., Peters, G., and Medina, E. (2011). The dynamics of T cells during persistent *Staphylococcus aureus* infection: from antigen-reactivity to in vivo anergy. EMBO Mol. Med. *3*, 652–666.

Zielinski, C.E., Mele, F., Aschenbrenner, D., Jarrossay, D., Ronchi, F., Gattorno, M., Monticelli, S., Lanzavecchia, A., and Sallusto, F. (2012). Pathogen-induced human TH17 cells produce IFN-γ or IL-10 and are regulated by IL-1β. Nature *484*, 514–518.

Zorman, J.K., Esser, M., Raedler, M., Kreiswirth, B.N., Ala'Aldeen, D.A., Kartsonis, N., Smugar, S.S., Anderson, A.S., McNeely, T., and Arduino, J.M. (2013). Naturally occurring IgG antibody levels to the *Staphylococcus aureus* protein IsdB in humans. Hum. Vaccin. Immunother. *9*, 1857–1864.

Index

www.ingramcontent.com/pod-product-compliance
Ingram Content Group UK Ltd.
Pitfield, Milton Keynes, MK11 3LW, UK
UKHW062000041026
15591UKWH00016B/1101

9 781910 190494